Krishan L. Duggal
Bayram Sahin

Differential

Geometry of

Lightlike

Submanifolds

Birkhäuser
Basel · Boston · Berlin

Authors:

Krishan L. Duggal
Department of Mathematics and Statistics
University of Windsor
401 Sunset Avenue
Windsor, Ontario, N9B 3P4
Canada
e-mail: yq8@uwindsor.ca

Bayram Sahin
Department of Mathematics
Inonu University
44280 Malatya
Turkey
e-mail: bsahin@inonu.edu.tr

2000 Mathematics Subject Classification: 53B25, 53C50, 53B50, 53C42, 53C15

Library of Congress Control Number: 2009942369

Bibliographic information published by Die Deutsche Bibliothek
Die Deutsche Bibliothek lists this publication in the Deutsche Nationalbibliografie;
detailed bibliographic data is available in the Internet at <http://dnb.ddb.de>.

1006102775
ISBN 978-3-0346-0250-1 Birkhäuser Verlag AG, Basel · Boston · Berlin

© 2010 Birkhäuser Verlag AG
Basel · Boston · Berlin
P.O. Box 133, CH-4010 Basel, Switzerland
Part of Springer Science+Business Media
Cover design: Birgit Blohmann, Zürich, Switzerland
Printed on acid-free paper produced from chlorine-free pulp. TCF ∞
Printed in Germany

ISBN 978-3-0346-0250-1   e-ISBN 978-3-0346-0251-8

9 8 7 6 5 4 3 2 1   www.birkhauser.ch

Contents

Preface

Since the second half of the 20th century, the Riemannian and semi-Riemannian geometries have been active areas of research in differential geometry and its applications to a variety of subjects in mathematics and physics. A recent survey in Marcel Berger's book [60] includes the major developments of Riemannian geometry since 1950, citing the works of differential geometers of that time. During the mid 1970s, the interest shifted towards Lorentzian geometry, the mathematical theory used in general relativity. Since then there has been an amazing leap in the depth of the connection between modern differential geometry and mathematical relativity, both from the local and the global point of view. Most of the work on global Lorentzian geometry has been described in a standard book by Beem and Ehrlich [34] and in their second edition in 1996, with Easley.

As for any semi-Riemannian manifold there is a natural existence of null (lightlike) subspaces, in 1996, Duggal-Bejancu published a book [149] on the lightlike (degenerate) geometry of submanifolds needed to fill an important missing part in the general theory of submanifolds. Since then the large number of papers published on lightlike hypersurfaces and general theory of submanifolds of semi-Riemannian manifolds has created a demand for publication of this volume as an update on the study of lightlike geometry.

The objective is to focus on all new geometric results (in particular, those available only after publication of the Duggal–Bejancu book) on lightlike geometry with proofs and their physical applications in mathematical physics.

Chapter 1 covers preliminaries, followed by up-to-date mathematical results in Chapters 2, 4 and 5 on lightlike hypersurfaces, half-lightlike, coistropic and r-lightlike submanifolds of semi-Riemannian manifolds, respectively. Due to degenerate induced metric of a lightlike submanifold M, we use a non-degenerate screen distribution $S(TM)$ to project induced objects on M. Unfortunately, $S(TM)$ is not in general unique. Since 1996 considerable work has been done in the search for canonical or unique screens. We highlight that each of these three chapters contain theorems on the existence of unique screen distributions subject to some reasonable geometric conditions. Chapter 3 is focused on applications of lightlike hypersurfaces in two active ongoing research areas in mathematical physics. First, we deal with *black hole horizons*. We prove a *Global Null Splitting Theorem* and relate it with physically significant works of Galloway [197], Ashtekar and

Krishnan's works [16] on *dynamical horizons* and Sultana-Dyer's work [378, 379] on *conformal Killing horizons*, with references to a host of related researchers. Secondly, we present the latest work on *Osserman lightlike hypersurfaces* [20].

Motivation of Chapters 6–9 comes from the historical development of the general theory of Cauchy-Riemann (CR) submanifolds [45] and their use in mathematical physics, as follows:

In the early 1930s, Riemannian geometry and the theory of complex variables were synthesized by Kähler [250] whose work developed (during 1950) into complex manifold theory [169, 302]. A Riemann surface, C^n and its projective space CP^{n-1} are simple examples of complex manifolds. This interrelation between the above two main branches of mathematics developed into what is now known as Kählerian and Sasakian [367] geometry. Almost complex [407], almost contact [66, 68] and quaternion Kähler manifolds [239, 366] and their complex, totally real, CR and slant submanifolds [45, 99, 102, 314, 410] are some of the most interesting topics of Riemannian geometry. By a CR submanifold we mean a real submanifold M of an almost Hermitian manifold $(\bar{M}, J, \bar{g})$, carrying a J-invariant distribution D (i.e., $JD = D$) and whose $\bar{g}$-orthogonal complement is J-anti-invariant (i.e., $JD^{\perp} \subseteq T(M)^{\perp}$), where $T(M)^{\perp} \to M$ is the normal bundle of M in $\bar{M}$. The CR submanifolds were introduced as an umbrella of a variety (such as invariant, anti-invariant, semi-invariant, generic) of submanifolds. Details on these may be seen in [45, 102, 412, 413]. On the other hand, a CR manifold (independent of its landing space) is a C^{∞} differentiable manifold M with a holomorphic subbundle H of its complexified tangent bundle $T(M) \otimes C$, such that $H \cap \bar{H} = \{0\}$ and H is involutive (i.e., $[X, Y] \in H$ for every $X, Y \in H$). For an update on CR manifolds (which is out of the scope of this book) we refer two recent books [132] by Dragomir and Tomassini and [26] by Barletta, Dragomir and Duggal. Here we highlight that Blair and Chen [69] were the first to interrelate these two concepts by proving that *proper CR submanifolds, of a Hermitian manifold, are CR manifolds*. Since then there has been active interrelation between the geometries of real and complex manifolds, with physical applications. Complex manifolds have two interesting classes of Kähler manifolds, namely, (i) Calabi-Yau manifolds with application to super string theory (see Candelas et al. [90]) and (ii) Teichmuller spaces applicable to relativity (see Tromba [396]).

The study of the above mentioned variety of geometric structures was primarily confined to Riemannian manifolds and their submanifolds, which carry a positive definite metric tensor, until in early 1980, when Beem-Ehrlich [34] published a book on *Global Lorentzian Geometry*, a paper by Barros and Romero [28] on *Indefinite Kähler Manifolds* and a book by O'Neill [317] on *Semi-Riemannian Geometry with Applications to Relativity*. Since then a considerable amount of work has been done on the study of semi-Riemannian geometry and its submanifolds, in particular, see a recent book by García-Río et al. [201], Sharma [373] and Duggal [133, 135, 136], including the use of indefinite Kählerian, Sasakian and quaternion structures. As a result, now we know that there are similarities and differences between the Riemannian and the semi-Riemannian geometries, in particular, reference to the Lorentzian case used in relativity. In the case of light-like submanifolds, its geometry is quite different than the counter part of non-degenerate submanifolds. To highlight this, in Duggal-Bejancu's book [149] there is a discussion on CR lightlike submanifolds of an indefinite Kähler manifold. Unfortunately, contrary to the non-degenerate case, CR lightlike submanifolds are non-trivial (i.e., they do not include invariant (complex) and real parts). To fill in this needed information, in Chapter 6 we present the latest work of the authors [159, 160] leading to a new class called *Generalized CR lightlike submanifolds* which represents an umbrella of invariant, screen real and CR lightlike submanifolds. We also present lightlike versions of slant submanifolds and totally real submanifolds.

Motivated by a significant use of contact geometry in differential equations, optics and phase space of a dynamical system (see details in Arnold [9], Maclane [292] and Nazaikinskii [304] and many more references therein), in Chapter 7, we present the first-ever collection of the authors' recent work [161, 162] on *lightlike submanifolds of indefinite Sasakian manifolds*, leading up to another umbrella of invariant, screen real and contact CR-lightlike submanifolds. We highlight that Chapters 6 and 7 fulfill the purpose (see Bejancu [45]) of having an umbrella of all possible lightlike submanifolds of indefinite Kählerian and Sasakian manifolds.

In Chapter 8, we study lightlike submanifolds of indefinite quaternion Kähler manifolds, using the concept of QR-lightlike, screen QR-lightlike and quaternion CR-lightlike submanifolds and give many examples.

In Chapter 9, we present applications of lightlike geometry to null 2-surfaces in spacetimes, lightlike version of harmonic maps and morphisms, CR-structures in general relativity and lightlike contact geometry in physics. Results included in this volume should stimulate future research on lightlike geometry and its applications. To the best of our knowledge, there does not exist any other book covering the material in this volume, other than a fresh and improved version of a small part from [149]. Equations are numbered within each chapter and its section. To illustrate this, a triplet (a, b, c) stands for each equation such that a, b and c are labeled for the chapter, the section and the number of equation in that section accordingly. There is an extensive list of bibliography and subject index. Our approach, in this book, has the following special features:

- Extensive list of cited references on semi-Riemannian geometry and its non-degenerate submanifolds is provided for the readers to easily understand the main focus on lightlike geometry.

- Each chapter starts with basic material on the no-degenerate submanifolds needed for that chapter. We expect that this approach will help readers to understand each chapter independently without knowing all the prerequisites in the beginning.

- The sequence of chapters is arranged so that the understanding of a chapter stimulates interest in reading the next one and so on.

- Physical applications are discussed separately (see Chapters 3 and 9) from the mathematical theory.

- Overall the presentation is self contained, fairly accessible and in some special cases supported by references.

This book is intended for graduate students and researchers who have good knowledge of semi-Riemannian geometry and its submanifolds and interest in the theory and applications of lightlike submanifolds. The book is also suitable for a senior level graduate course in differential geometry.

This work was supported through an operating grant of the first author (Duggal), awarded by the Natural Sciences and Engineering Research Council (NSERC) of Canada. Most of the work presented in this book was conceived and developed during a visit of the second author (Sahin) to the university of Windsor in 2003, a follow up collaboration and Dr. Sahin's second visit in 2007. Dr. Sahin is grateful to the Scientific and Technological Research Council of Turkey (TUBITAK) for a PDF scholarship and the university of Windsor for appointing him visiting scholar and its department of mathematics and statistics for hospitality and kind support.

Both authors are thankful to all authors of books and articles whose work they have used in preparing this book. Last, but not the least, we are grateful to the publisher for their effective cooperation and excellent care in publishing this volume. Constructive suggestions (towards the improvement of the book) by reviewers is appreciated with thanks. Any further comments and suggestions by the readers will be gratefully received.

Krishan L. Duggal
Bayram Sahin

Notation

R	Real numbers.
R^n	$n-$ tuples of real numbers.
$\|\ \|$;	Norm.
$\wedge$	Null cone.
R$_q^n$	Semi-Euclidean space of constant index q.
Rad W	Radical subspace of W.
SW	Screen subspace of W
$[g]$	Matrix of g.
$W^\perp$	Perpendicular space of W.
M	Manifold.
$T_x M$	Tangent space at x.
TM	Tangent bundle.
R	Riemann curvature tensor.
Ric	Ricci tensor.
Q	Ricci operator.
r	Scalar curvature.
$C(X,Y)Z$	Weyl conformal curvature tensor.
$K(\pi)$	Sectional curvature of plane π.
$S^n(r)$	A sphere of radius r.
H^n	$n-$ dimensional hyperbolic space.
$S_q^n(r)$	Pseudo-sphere.
$H_q^n(r)$	Pseudo-hyperbolic space.
grad f	Gradient of function f.
H^f	Hessian of function f.
L_V	Lie derivative with respect to V.
$[,]$	Lie operator.
D	Distribution.
f_*	derivative map of f.
$M \times N$	Product manifold of M and N.
ϕ^*	Pullback map.
Rad $T_x M$	Radical subspace of $T_x M$.
$d\omega$	Exterior derivation.

$S^-(M)$	Unit timelike bundle of (M, g).
$S^+(M)$	Unit spacelike bundle of (M, g).
$S(TM)$	Unit non-null bundle of (M, g).
$R(., Z)Z$	Jacobi operator with respect to Z.
$\exp_x$	Exponential map of M at x.
M	Lightlike hypersurface.
$\mathrm{Rad}(TM)$	Radical distribution.
$S(TM)$	Screen distribution.
$\mathrm{ltr}(TM)$	Lightlike transversal bundle of M.
$\oplus_{\mathrm{orth}}$	Orthogonal direct sum.
B	Local second fundamental form of a hypersurface.
h^ℓ	Lightlike second fundamental form of a lightlike submanifold.
h^s	Screen transversal second fundamental form of a lightlike submf.
A_V	Shape operator of M
h^*	Screen second fundamental form.
A^*	Screen shape operator of $S(TM)$.
$C(X, PY)$	Local screen fundamental form.
$\mathrm{Ric}(X, Y)$	Ricci tensor of M.
$\tilde{g}$	Pseudo-inverse metric.
$R_Z X$	Jacobi operator to Z.
$\mathcal{R}.R$	Curvature operator corresponding to R.
$\mathcal{S}$	First derivative of a screen distribution.
$\mathbf{V^C}$	Complexification of a real vector space.
J	Complex structure on a vector space.
$(V, \mathbf{R}, J)$	Real vector space with complex structure.
$(V, \mathbf{C}, J)$	Complex vector space.
N_J	Nijenhuis tensor field of J.
$\mathbf{K}$	Kähler scalar.
$\mathbf{C}^n_q$	Indefinite complex space form as $c = 0$.
$\mathbf{CP}^n_q(c)$	Indefinite complex space form as $c > 0$.
$\mathbf{CH}^n_q(c)$	Indefinite complex space form as $c < 0$.
θ	Volume expansion.
$\sigma_{\mathbf{ab}}$	shear tensor.
$\omega_{\mathbf{ab}}$	vorticity tensor.
$\nabla F_\star$	second fundamental form.
τ	tension field.

Chapter 1

Preliminaries

1.1 Semi-Euclidean spaces

Denote by $\mathbf{R}$ the set of real numbers and $\mathbf{R}^n$ their n-fold Cartesian product $\mathbf{R} \times \ldots \times \mathbf{R}$, the set of all ordered n-tuples $(x^1, \ldots, x^n)$. Define a function

$$d : \mathbf{R}^n \times \mathbf{R}^n, \quad \text{where} \quad d(x, y) = ||x - y||$$

for every pair (x, y) of the points $x, y \in \mathbf{R}^n$. This function d is known as the Euclidean metric in $\mathbf{R}^n$. Then, we call $\mathbf{R}^n$ with the metric d the n-dimensional Euclidean space. Consider $\mathbf{V}$ a real n-dimensional vector space with a symmetric bilinear mapping $g : \mathbf{V} \times \mathbf{V} \to \mathbf{R}$. We say that g is positive (negative) definite on $\mathbf{V}$ if $g(v, v) > 0\,(g(v, v) < 0)$ for any non-zero $v \in \mathbf{V}$. On the other hand, if $g(v, v) \geq 0\,(g(v, v) \leq 0)$ for any $v \in \mathbf{V}$ and there exists a non-zero $u \in \mathbf{V}$ with $g(u, u) = 0$, we say that g is positive (negative) semi-definite on $\mathbf{V}$.

Let $B = \{u_1, \ldots, u_n\}$ be an arbitrary basis of V. Then, g can be expressed by an $n \times n$ symmetric matrix $G = (g_{ij})$, where

$$g_{ij} = g(u_i, u_j), \quad (1 \leq i, j \leq n).$$

G is called the *associated matrix* of g with respect to the basis B. In another way $\text{rank}\, G = n \iff g$ is non-degenerate on $\mathbf{V}$. The non-degenerate g on $\mathbf{V}$ is called a *semi-Euclidean metric (scalar product)*. Then $(\mathbf{V}, g)$ is a *semi-Euclidean vector space*, for which $g(u, v) = u \cdot v$ where $\cdot$ is the usual dot product. For a semi-Euclidean $\mathbf{V} \neq 0$, there exists an orthonormal basis $E = \{e_1, \ldots, e_n\}$ with

$$g(v, v) = \langle v, v \rangle_q = -\sum_{i=1}^{q} (v^i)^2 + \sum_{a=q+1}^{q+p} (v^a)^2, \tag{1.1.1}$$

where $q + p = n$ and (v^i) are the coordinate components of v with respect to E. Thus, with respect to (1.1.1), G is a diagonal matrix of canonical form:

$$\text{diag}(- \ldots - + \ldots +). \tag{1.1.2}$$

The sum of these diagonal elements (also called the *trace*) of the canonical form is called the *signature* of g and the number of negative signs in (1.1.2) is called the *index of* **V**. Three special cases are important both for geometry and physics. First, g is positive (negative) definite for which **V** is Euclidean with zero index. Second, if the index of g is 1, then g is called a *Minkowski metric* and **V** is called a *Minkowski space*, used in special relativity. Third, if there exists on **V** a degenerate g, then we say that V is a *lightlike vector space* with respect to g. The third case is the focus of study in this book when we consider differential geometry of lightlike submanifolds in Chapters 2 through 9.

Define a mapping (called *norm*), of a semi-Euclidean **V**, by

$$\| \cdot \| \colon V \to \mathbf{R}\,; \qquad \| v \| = |\, g(v,v)\,|^{\frac{1}{2}}, \qquad \forall v \in \mathbf{V}.$$

$\| v \|$ is called the length of v. A vector v is said to be

spacelike,	if $g(v,v) > 0$ or $v = 0$,
timelike,	if $g(v,v) < 0$,
lightlike (null, isotropic),	if $g(v,v) = 0$ and $v \neq 0$.

The set of all null vectors in **V**, denoted by Λ, is called the *null cone of* **V**, i.e.,

$$\Lambda = \{v \in (\mathbf{V} - \{0\}), g(v,v) = 0\}.$$

For a semi-Euclidean **V**, a unit vector u is defined by $g(u,u) = \pm 1$ and, as in the case of Euclidean spaces, we say that u is of length 1. In case g is semi-definite, orthogonal vectors (i.e., $u \perp v$ if $g(u,v) = 0$) are not necessarily at right angles to each other. For example, a null vector is a non-zero vector that is orthogonal to itself. Since, in general, a semi-Euclidean **V** has three types of vectors (spacelike, timelike, lightlike), it is some times desirable to transform a given orthonormal basis $E = \{e_1, \ldots, e_n\}$ into another basis which contains some null vectors. To construct such a basis we let $\{e_1, \ldots, e_p\}$ and $\{e_{p+1}, \ldots, e_{p+q}\}$, $p + q = n$, be unit timelike and spacelike vectors, respectively. In general, the following three cases arise:

Case 1 $(p < q)$. Construct the vectors

$$N_i = \frac{1}{\sqrt{2}}(e_{p+i} + e_i); \qquad N_i^{\star} = \frac{1}{\sqrt{2}}(e_{p+i} - e_i). \tag{1.1.3}$$

Here each N_i and $N_i^{\star}$ are null vectors which satisfy

$$g(N_i, N_j) = g(N_i^{\star}, N_j^{\star}) = 0, \qquad g(N_i, N_j^{\star}) = \delta_{ij}, \tag{1.1.4}$$

$i, j \in \{1, \ldots, p\}$. Thus $\{N_1, \ldots, N_p, N_1^{\star}, \ldots, N_p^{\star}, e_{2p+1}, \ldots, e_{2p+q} = e_n\}$ is a basis of **V** which contains $2p$ null vectors and $q - p$ spacelike vectors.

Case 2 $(q < p)$. For this case we set

$$N_a = \frac{1}{\sqrt{2}}(e_{p+a} + e_a), \qquad N_a^\star = \frac{1}{\sqrt{2}}(e_{p+a} - e_a), \qquad (1.1.5)$$

to obtain the relations (1.1.4) but with i, j replaced by $a, b \in \{1, \ldots, q\}$. Then, the basis of $\mathbf{V}$ contains $2q$ null vectors $\{N_1, \ldots, N_q, N_1^\star, \ldots, N_q^\star\}$ and $p - q$ timelike vectors $\{e_{q+1}, \ldots, e_p\}$.

Case 3 $p = q$. This is a special case for which $n = 2p = 2q$ is even and the null transformed basis of $\mathbf{V}$ is $\{N_1, \ldots, N_p, N_1^\star, \ldots, N_p^\star\}$.

$\mathbf{V}$ is a *proper semi-Euclidean space* if $p \cdot q \neq 0$. In general, there exists a basis

$$B = \{N_1, \ldots, N_r, N_1^\star, \ldots, N_r^\star, u_1, \ldots, u_s\} \qquad (1.1.6)$$

for a proper semi-Euclidean space $(\mathbf{V}, g)$, called a *quasi-orthonormal basis* which satisfies the following conditions:

$$g(N_i, N_j) = g(N_i^\star, N_j^\star) = 0; \qquad g(N_i, N_j^\star) = \delta_{ij},$$
$$g(u_a, N_i) = g(u_a, N_i^\star) = 0; \qquad g(u_a, u_b) = \epsilon_a \delta_{ab}, \qquad (1.1.7)$$

for any $i, j \in \{1, \ldots, r\}$, $a, b \in \{1, \ldots, s\}$, $2r + s = n$ and $\epsilon_a = \pm 1$.

Throughout this book, we set the form of the signature of g as given by (1.1.2), unless otherwise stated. Also, we identify an n-dimensional vector space $\mathbf{V}$ over $\mathbf{R}$ with $\mathbf{V}_q^n$ a semi-Euclidean space of constant index $q > 0$ if $\mathbf{V}$ has an indefinite inner product and $\mathbf{R}^n$ is a Euclidean space.

Electromagnetism in a Minkowski space $\mathbf{R}_1^4$. Here is an example of semi-Euclidean spaces representing some physical objects. Let $\mathbf{R}_1^4$ be the 4-dimensional Minkowski spacetime with the *Galilean coordinates* $(ct = x^o, x^1, x^2, x^3)$, where c is the velocity of light. Thus, the semi-Euclidean metric is given by

$$g(x, y) = - x^o y^o + x^1 y^1 + x^2 y^2 + x^3 y^3, \quad \forall x, y \in \mathbf{R}_1^4.$$

The associated matrix G, of g, with respect to an orthonormal basis is the diagonal matrix given by

$$G = [g_{ab}] = \begin{bmatrix} -1 & 0 & 0 & 0 \\ 0 & 1 & 0 & 0 \\ 0 & 0 & 1 & 0 \\ 0 & 0 & 0 & 1 \end{bmatrix},$$

where $a, b, c, \ldots \in \{0, 1, 2, 3\}$. In the sequel, we set $\frac{\partial}{\partial x^a} = \partial_a$. Suppose $X = X^a \partial_a$ (Einstein sum in a) is a spacelike vector field, where X^a has partial derivatives with respect to the spacetime coordinates. Then, the *Curl* and the *divergence* of X are given by

$$\text{Curl}\, X = (\partial_2 X^3 - \partial_3 X^2)\partial_1 + (\partial_3 X^1 - \partial_1 X^3)\partial_2 + (\partial_1 X^2 - \partial_2 X^1)\partial_3,$$
$$\text{div}\, X = \partial_1 X^1 + \partial_2 X^2 + \partial_3 X^3,$$

respectively. Also, $\partial_a X = \partial_a X^b \partial_b$. The *Maxwell's equations* are:

$$\text{Curl } H \; - \; \frac{1}{c}\frac{\partial E}{\partial t} = 4\,\pi\,\rho\,\frac{v}{c}\,, \qquad\qquad \text{div } E = 4\,\pi\,\rho,$$

$$\text{Curl } E \; + \; \frac{1}{c}\frac{\partial H}{\partial t} = 0\,, \qquad\qquad \text{div } H = 0\,,$$

where H and E are the *magnetic* and *electric fields* respectively, ρ is the *charged density* and v is the local velocity of the charge. The *current density* $\dot{J} = \rho v$ and ρ are the sources of H and E. The Maxwell equations are consistent with the *continuity equation*

$$\frac{\partial \rho}{\partial t} + \text{div}(\rho\,v) = 0\,.$$

Note that in vacuum, $\rho = 0$ and, therefore, the Maxwell equations reduce to

$$\text{Curl } H \; - \; \frac{1}{c}\frac{\partial E}{\partial t} = 0\,, \quad \text{div } E = 0\,,$$

$$\text{Curl } E \; + \; \frac{1}{c}\frac{\partial H}{\partial t} = 0\,, \quad \text{div } H = 0\,.$$

Furthermore, there exists a 3-dimensional spacelike vector field A and a *scalar potential* ϕ such that

$$H = \text{Curl } A\,,$$

$$E = -\,\text{grad}\,\phi \; - \; \frac{1}{c}\frac{\partial A}{\partial t}\,.$$

Taking into account that our metric is of signature $(-\,+\,+)$, we introduce a 4-vector Φ with components

$$\Phi = (\Phi^0,\ \Phi^1,\ \Phi^2,\ \Phi^3\,) = (\phi,\ A^1,\ A^2,\ A^3)$$

whose covariant components are $\Phi_o = -\,\phi$, $\Phi_i = A^i$ $(i = 1, 2, 3)$. Thus, we obtain the following expression of the components of H and E:

$$H^1 = \partial_2\Phi_3 - \partial_3\Phi_2\,, \qquad\qquad E^1 = \partial_1\Phi_o - \partial_o\Phi_1\,,$$

$$H^2 = \partial_3\Phi_1 - \partial_1\Phi_3\,, \qquad\qquad E^2 = \partial_2\Phi_o - \partial_o\Phi_2\,,$$

$$H^3 = \partial_1\Phi_2 - \partial_2\Phi_1\,, \qquad\qquad E^3 = \partial_3\Phi_o - \partial_o\Phi_3\,,$$

since $\phi = ct$. The above six components of H and E can be presented in a unified form as the covariant components of a *second-order skew-symmetric tensor field* $F = (F_{ab})$, in the following way:

$$F_{a\,b} = \partial_a\,\Phi_b - \partial_b\,\Phi_a\,, \quad a,\,b \in \{0, 1, 2, 3\}\,.$$

The above equation is also applicable to a *general curvilinear coordinate system* used in general relativity. However, here we call F, in the Galilean coordinates, the

Minkowski electromagnetic tensor field. Thus, the 4×4 matrices of the covariant and the contravariant components of F are:

$$[F_{ab}] = \begin{bmatrix} 0 & -E^1 & -E^2 & -E^3 \\ E^1 & 0 & H^3 & -H^2 \\ E^2 & -H^3 & 0 & H^1 \\ E^3 & H^2 & -H^1 & 0 \end{bmatrix},$$

$$[F^{ab}] = \begin{bmatrix} 0 & E^1 & E^2 & E^3 \\ -E^1 & 0 & H^3 & -H^2 \\ -E^2 & -H^3 & 0 & H^1 \\ -E^3 & H^2 & -H^1 & 0 \end{bmatrix}.$$

To express the Maxwell equations in tensor form, we introduce the *electric current vector* $\mathbf{J} = (J^o, J^1, J^2, J^3) = (\rho c, \rho v^1, \rho v^2, \rho v^3)$. Then, by using the above Maxwell equations assume the following form:

$$F^{ab}_{;b} = \frac{4\pi}{c} J^a,$$

$$F_{bc;a} + F_{ca;b} + F_{ab;c} = 0,$$

respectively. Therefore, the Maxwell equations have a tensorial form which is valid for any system of coordinates.

Subspaces. Let (W, g) be a real n-dimensional lightlike vector space. The *Radical* of W, with respect to g, is a subspace $\operatorname{Rad} W$ of W defined by

$$\operatorname{Rad} W = \{\xi \in W; \quad g(\xi, v) = 0, \quad v \in W\}.$$

Then, a subspace of W may not be degenerate. To support this assertion we prove the following general result.

Proposition 1.1.1. [149] *Let (W, g) be a real n-dimensional lightlike vector space such that $\dim(\operatorname{Rad} W) = r < n$. Then any complementary subspace to $\operatorname{Rad} W$ is non-degenerate.*

Proof. Let SW be a complementary subspace to $\operatorname{Rad} W$ in W i.e., we have the decomposition

$$W = \operatorname{Rad} W \oplus_{\operatorname{orth}} SW \tag{1.1.8}$$

where $\oplus_{\operatorname{orth}}$ denotes the *orthogonal direct sum.* Suppose there exists a non-zero $u \in SW$ such that $g(u, v) = 0$ for any $v \in SW$. (1.1.8) implies $g(u, \xi) = 0$, $\forall \xi \in \operatorname{Rad} W$. Therefore, $u \in \operatorname{Rad} W$. But $\operatorname{Rad} W$ and SW being complementary subspaces, SW is non-degenerate, which completes the proof. $\qquad\square$

A complementary subspace SW to $\operatorname{Rad} W$ in W is called a *screen subspace* of W. As SW is non-degenerate with respect to g, it is a semi-Euclidean space. Then, there exists an orthonormal basis $\{u_{r+1}, \ldots, u_n\}$ of SW. Thus a basis of

W adapted to (1.1.8) is $B = \{f_1, \ldots, f_r, u_{r+1}, \ldots, u_n\}$ where $f_i \in \operatorname{Rad} W$, $i \in \{1, \ldots, r\}$. Since $\operatorname{Rad} W$ is orthogonal to W, we conclude that the matrix of g with respect to B is of the form:

$$[g] = \begin{bmatrix} O_{r,r} & O_{r,n-r} \\ O_{n-r,r} & \epsilon_a \delta_{ab} \end{bmatrix}, \quad a, b \in \{r+1, \ldots, n\}, \quad \epsilon_a = g(u_a, u_a).$$

Let (V, g) be an n-dimensional semi-Euclidean space and W be a subspace of V. In case $g_{|W}$ is degenerate we say that W is a *lightlike (degenerate) subspace*. Otherwise, we call W a *non-degenerate subspace*. Next, consider the subspace

$$W^\perp = \{v \in V \,;\, g(v, w) = 0, \quad \forall\, w \in W\}.$$

Then, following O'Neill [317, p.49], $W^\perp$ is called W perp. In general, $W \cap W^\perp \neq \{0\}$. For example, consider $W = \{(x, y, x, y) \in \mathbf{V}_1^4 \,;\, x, y \in \mathbf{R}\}$ and obtain $W \cap W^\perp = \{(x, o, x, o) \,;\, x \in R\} \neq \{0\}$. However, the following properties of W are preserved for any general $\mathbf{V}_q^n$.

Proposition 1.1.2. [149] *Let $(\mathbf{V}_q^n, g)$ be an n-dimensional semi-Euclidean space and W its subspace. Then we have*

$$\dim W + \dim W^\perp = n, \tag{1.1.9}$$

$$\left(W^\perp\right)^\perp = W, \tag{1.1.10}$$

$$\operatorname{Rad} W = \operatorname{Rad} W^\perp = W \cap W^\perp. \tag{1.1.11}$$

Proof. By Lemma 22 of Chapter 2 in O'Neill [317], we have (1.1.9) and (1.1.10). Next, by taking $v \in W \cap W^\perp \subset W^\perp$, we obtain $g(v, w) = 0$ for any $w \in W$, that is, $v \in \operatorname{Rad} W$. Conversely, for any $v \in \operatorname{Rad} W \subset W$, we have $g(v, w) = 0$ for any $w \in W$, which implies $v \in W \cap W^\perp$. Hence $\operatorname{Rad} W = W \cap W^\perp$. The last equality in (1.1.11) follows by using (1.1.10). $\qquad\square$

Proposition 1.1.3. [149] *Let $(\mathbf{V}_q^n, g)$ be a proper semi-Euclidean space of index q. Then, there exists a subspace $\bar{W}$ of $\mathbf{V}_q^n$ of dimension $\min\{q, n - q\}$ and no larger, such that $g|_{\bar{W}} = 0$.*

Proof. Let $E = \{e_1, \ldots, e_n\}$ be an orthonormal basis of $(\mathbf{V}_q^n, g)$ with

$$g(x, y) = -\sum_{i=1}^{q} x^i y^i + \sum_{a=p+1}^{n} x^a y^a, \quad \forall\, x, y \in \mathbf{V}_q^n,$$

where (x^i) and (y^i) are the coordinates of x and y. Suppose $2q < n$. Now define a q-dimensional subspace

$$\bar{W} = \operatorname{Span}\{u_1 = e_1 + e_{q+1}, \ldots, u_q = e_q + e_{2q}\}.$$

It follows that $g|_{\bar{W}} = 0$. Choose a null vector $N = \sum_{i=1}^{n} N^i e_i$ such that $g(N, u_a) = 0$, $\forall\, a \in \{1, \ldots, q\}$. Thus, $N^1 = N^{q+1}, \ldots, N^q = N^{2q}$. Since $\| N \| = 0$ and

$\{e_1, \ldots, e_{2q}\}$ and $\{e_{q+1}, \ldots, e_n\}$ are timelike and spacelike respectively, we conclude that $N^{2q+1} = \ldots = N^n = 0$. Hence, $N = \sum_{a=1}^{q} N^a u_a$. Thus, there is no subspace larger than $\bar{W}$ on which g vanishes. Similarly, for $2q \geq n$. $\square$

Let W be an m-dimensional lightlike subspace of $\mathbf{V}_q^n$. A quasi-orthonormal basis

$$B = \{f_1, \ldots, f_r, f_1^*, \ldots, f_r^*, u_1, \ldots, u_t\}$$

such that $W = \mathrm{Span}\{f_1, \ldots, f_r, u_1, \ldots, u_s\}$, if $m = r + s, 1 \leq s \leq t$, or

$$W = \mathrm{Span}\,\{f_1, \ldots, f_m\}, \quad \text{if} \quad m \leq r,$$

is called a *quasi-orthonormal basis* of $\mathbf{V}_q^n$ along W.

Proposition 1.1.4. [149] *There exists a quasi-orthonormal basis of $\mathbf{V}_q^n$ along its lightlike subspace W.*

Proof. Suppose $\mathrm{null}\, W = r < \min\{m, n - m\}$. Then we have

$$W = \mathrm{Rad}\, W \oplus_{\mathrm{orth}} W', \quad W^\perp = \mathrm{Rad}\, W \oplus_{\mathrm{orth}} W'',$$

where W' and W'' are some screen subspaces. We decompose $\mathbf{V}_q^n$ as follows:

$$\mathbf{V}_q^n = W' \perp (W')^\perp. \tag{1.1.12}$$

As W'' is a non-degenerate subspace of $(W')^\perp$ we obtain

$$(W')^\perp = W'' \perp (W'')^\perp, \tag{1.1.13}$$

where $(W'')^\perp$ is the complementary orthogonal subspace to W'' in $(W')^\perp$. It is easy to see that $\mathrm{Rad}\, W$ is a subspace of $(W'')^\perp$. Denote by U a complementary subspace to $\mathrm{Rad}\, W$ in $(W'')^\perp$. As $(W'')^\perp$ is of dimension $2r$ we may consider the basis $\{f_1, \ldots, f_r\}$ and $\{v_1, \ldots, v_r\}$ of $\mathrm{Rad}\, W$ and U, respectively. Now, we look for $\{f_1^*, \ldots, f_r^*\}$ given by

$$f_i^* = A_i^j f_j + B_i^j u_j, \tag{1.1.14}$$

and satisfying the relations in the first line of (1.1.7). By direct calculations, one obtains that $g(f_i, f_k^*) = \delta_{ik}$ if and only if

$$B_k^j g(f_i, v_j) = \delta_{ik}. \tag{1.1.15}$$

As $\det[g(f_i, v_j)] \neq 0$, (otherwise $(W'')^\perp$ would be degenerate), the system (1.1.15) has a unique solution (B_k^j). Next, by using (1.1.14) and (1.1.15) one obtains $g(f_i^*, f_j^*) = 0$ if and only if

$$A_j^i + A_i^j + B_i^h B_j^k g(v_h, v_k) = 0$$

which proves the existence of A_i^j from (1.1.7). Finally, From (1.1.7) and (1.1.8) and the above construction, we obtain the decomposition

$$\mathbf{V}_q^n = W' \perp W'' \perp (\operatorname{Rad} W \oplus \operatorname{Span}\{f_1^*, \dots f_r^*\}).$$

Hence we have a quasi-orthonormal basis of $\mathbf{V}_q^n$ along W given by

$$\{f_1, \dots, f_r, f_1^*, \dots, f_r^*, u_1, \dots, u_{m-r}, w_1, \dots, w_{n-m-r}\},$$

where $\{u_1, \dots, u_{m-r}\}$ and $\{w_1, \dots, w_{n-m-r}\}$ are two orthonormal bases of W' and W'', respectively. In this case

$$W = \operatorname{Span}\{f_1, \dots, f_r, u_1, \dots, u_{m-r}\}.$$

In case $r = m < n - m$, it follows that $\operatorname{Rad} W = W \subset W^\perp$. We put

$$W^\perp = W \oplus_{\operatorname{orth}} W'',$$

where W'' is an arbitrary screen subspace of $W^\perp$. One obtains the orthogonal decomposition

$$\mathbf{V}_q^n = W'' \perp (W'')^\perp,$$

where $(W'')^\perp$ is the complementary orthogonal subspace to W'' in $\mathbf{V}_q^n$. Moreover $(W'')^\perp$ is of dimension $2n$ and contains W. In a similar way as in the first case we find the quasi-orthonormal basis of $\mathbf{V}_q^n$ along W

$$\{f_1, \dots, f_m, f_1^*, \dots, f_m^*, w_1, \dots, w_{n-2m}\},$$

where $\{w_1, \dots, w_{n-2m}\}$ is an orthonormal basis of W'' and $W = \operatorname{Span}\{f_1, \dots, f_m\}$.

In case $r = n - m < m$, it follows that $\operatorname{Rad} W = W^\perp \subset W$. We set

$$W = W^\perp \perp W',$$

where W' is a screen subspace of W. Thus,

$$\mathbf{V}_q^n = W' \perp (W')^\perp,$$

where $(W')^\perp$ is the complementary orthogonal subspace to W' in $\mathbf{V}_q^n$. Moreover, $(W')^\perp$ is of dimension $2(n-m)$ and it contains $W^\perp$. Then the quasi-orthonormal basis of $\mathbf{V}_q^n$ along W is given by

$$\{f_1, \dots, f_{n-m}, f_1^*, \dots, f_{n-m}^*, u_1, \dots, u_{2m-n+2}\},$$

where $\{u_1, \dots, u_{2m-n}\}$ is an orthonormal basis of W'. In this case

$$W = \operatorname{Span}\{f_1, \dots, f_{n-m}, u_1, \dots, u_{2m-n+2}\}.$$

Finally, if $r = m = \frac{n+2}{2}$, we get $\operatorname{Rad} W = W = W^\perp$ and

$$\mathbf{V}_q^n = W \oplus \operatorname{Span}\{f_1^*, \dots, f_m^*\}.$$

Then the quasi-orthonormal basis of $\mathbf{V}_q^n$ along W is given by

$$\{f_1, \dots, f_m, f_1^*, \dots, f_m^*\},$$

where $\{f_1, \dots, f_m\}$ is a basis of W, which completes the proof. $\square$

1.2 Semi-Riemannian manifolds

Recall that the measurement of distances in a Euclidean space $\mathbf{R}^3$ is represented by the distance element

$$ds^2 = dx^2 + dy^2 + dz^2$$

with respect to a rectangular coordinate system (x, y, z). Back in 1854, Riemann generalized this idea for n-dimensional spaces and he defined an element of length by means of a quadratic differential form $ds^2 = g_{ij}dx^i dx^j$ of a differentiable manifold M, where the coefficients g_{ij} are functions of the coordinates system$(x^1, \ldots, x^n)$, which represent a symmetric covariant tensor field g of type $(0, 2)$. Since then much of the subsequent differential geometry was developed on a real smooth manifold (M, g), called a Riemannian manifold, where g is assumed to be positive definite. We review the preliminary information on differential manifolds as follows:

Differentiable manifolds. A *topology* on a set M is a family $\mathcal{T}$ of open subsets of M such that

1. the empty set $\emptyset$ and M are in $\mathcal{T}$,

2. the intersection of any two members of F is in $\mathcal{T}$,

3. the union of an arbitrary collection of members of $\mathcal{T}$ is in $\mathcal{T}$.

In the above case, $(M, \mathcal{T})$ is called a *topological space* whose elements are the open sets of $\mathcal{T}$. As M depends on the choice of $\mathcal{T}$, M can have many topologies. In the sequel, we assume that M is a topological space with a given $\mathcal{T}$. M is a *Hausdorff topological space* if for every p, $q \in M$, $p \neq q$, there exist non-intersecting neighborhoods $\mathcal{U}_1$ and $\mathcal{U}_2$ respectively. A neighborhood of p in M is an open set that contains p. A system of open sets of $\mathcal{T}$ is called a *basis* if every one of its open sets is a union of sets of the system. A topological space is related with the concept of manifold as follows:

Definition 1.2.1. An n-dimensional manifold M, with a countable basis of open sets, is a topological Hausdorff space if each of its points has a neighborhood homeomorphic to an open set in $\mathbf{R}^n$.

A simple way of understanding a manifold is a set M with the property that each point of M can serve as the origin of local coordinates valid in an open neighborhood which is homeomorphic to an open set in $\mathbf{R}^n$. A trivial example is $M = \mathbf{R}^n$. The Hausdorff condition is not necessary, although is assumed most often. The open neighborhood of each point admits a *coordinate system* which determines the position of points and the topology of that neighborhood. For example, any surface (as a manifold) in $\mathbf{R}^3$ is topologically equivalent to a spherical or a hyperbolic or a planar surface.

In order to have a smooth transformation of two such coordinate systems and also taking care of the intersecting neighborhoods, topology has an interplay with the concept of *differentiable manifolds* as follows.

Let $f : \mathcal{U} \to \mathbf{R}$ be a real-valued function on a non-empty open subset $\mathcal{U}$ of $\mathbf{R}^n$. The function f is said to be of class C^k if and only if f has continuous partial derivatives of all orders $r \leq k$. Moreover, f is said to be smooth or analytic if and only if f is of class C^∞ or of class C^ω respectively. A $1-1$ bicontinuous mapping φ of an open set $\mathcal{U}$ of M onto an open set $\varphi(\mathcal{U})$ of $\mathbf{R}^n$ is called a *homeomorphism* of $\mathcal{U}$ onto $\varphi(\mathcal{U})$. Moreover, the inverse mapping of φ is also a homeomorphism. Intuitively one thinks of a homeomorphism as a mapping in which neighboring points remain neighboring points. A homeomorphism $\varphi : M \to \mathbf{R}^n$, mapping from an open set $\mathcal{U}$ of M onto an open set $\varphi(\mathcal{U})$ of $\mathbf{R}^n$, is called a chart. By assigning to each point x in $\mathcal{U}$ the n local coordinates $x^1, \ldots, x^n$, we call $\mathcal{U}$ a local coordinate neighborhood. Let x be the point of the intersection $\mathcal{U}_1 \cap \mathcal{U}_2$ of two coordinate neighborhoods $\mathcal{U}_1$ and $\mathcal{U}_2$ with respect to two charts φ_1 and φ_2. Then, φ_1 and φ_2 are C^k-compatible if $\mathcal{U}_1 \cap \mathcal{U}_2$ is non-empty and $\varphi_2 \circ \varphi_1^{-1} : \varphi_1(\mathcal{U}_1 \cap \mathcal{U}_2) \to \varphi_2(\mathcal{U}_1 \cap \mathcal{U}_2)$ and its inverse are C^k.

Definition 1.2.2. A differentiable or C^∞ (smooth) structure on a topological manifold M is a family $\mathcal{A} = \{\mathcal{U}_\alpha, \phi_\alpha\}$ of coordinate neighborhoods such that

1. the $\mathcal{U}_\alpha$ covers M,

2. for any α, β the neighborhoods $(\mathcal{U}_\alpha, \phi_\alpha)$ and $(\mathcal{U}_\beta, \phi_\beta)$ are (C^∞) compatible,

3. any coordinate neighborhood $(\mathcal{V}, \psi)$ that is compatible with every $(\mathcal{U}_\alpha, \phi_\alpha) \in \mathcal{A}$ is itself in $\mathcal{A}$.

A C^∞-manifold is a topological manifold with a C^∞-differentiable structure.

The family $\mathcal{A}$ is called a maximal atlas on M. M is called a differentiable manifold or a *smooth manifold* if M is of class C^k or C^∞, respectively. An atlas $\mathcal{A} = \{\mathcal{U}_\alpha, \phi_\alpha\}$ of M is said to be locally finite if for each p in M, there is a local coordinate neighborhood $\mathcal{U}$ which intersects with only finitely many $\mathcal{U}_\alpha$'s. Another atlas $\mathcal{B} = \{\mathcal{V}_\beta, \psi_\beta\}$ of M is called a refinement of the atlas $\mathcal{A}$, if each $\mathcal{V}_\beta$ is contained in some $\mathcal{U}_\alpha$. M is *paracompact* if for every atlas $\mathcal{A}$ there is a locally finite refined atlas $\mathcal{B}$ of $\mathcal{A}$. In this book we assume (unless otherwise stated) that all manifolds are smooth and paracompact.

Examples are the 2-sphere $\mathbf{S}^2$, a cylinder, a torus, and Minkowski spacetime. To illustrate this, consider $\mathbf{S}^2$ in $\mathbf{R}^3$, with coordinates (y^i), centered at $(0, 0, 0)$ having radius a. We need two charts, with respect to the coordinates (x, y), as follows:

$$y^1 = \frac{2a^2 x}{x^2 + y^2 + a^2}, \qquad y^2 = \frac{2a^2 y}{x^2 + y^2 + a^2},$$

$$y^3 = \epsilon a \frac{x^2 + y^2 - a^2}{x^2 + y^2 + a^2}, \qquad (\epsilon = \pm 1).$$

Similarly, one can show that $\mathbf{S}^n$ in $\mathbf{R}^{n+1}$ is a smooth manifold. A manifold M is *orientable* if there exists an atlas $\{\mathcal{U}_\alpha, \varphi_\alpha\}$ of M such that in every non-empty intersection $\mathcal{U}_\alpha \cap \mathcal{U}_\beta$, the Jacobian $|\partial x^i/\partial x'^j|$ is positive, where $(x^1, \ldots, x^n)$ and $(x'^1, \ldots, x'^n)$ are coordinates in $\mathcal{U}_\alpha$ and $\mathcal{U}_\beta$ respectively. The Möbius strip is a non-orientable manifold.

Marcel Berger's book [60] includes the major developments of Riemannian geometry since 1950, citing the works of differential geometers of that time. During the mid 1970s, the interest shifted towards Lorentzian geometry, the mathematical theory used in relativity for which g is not definite. Most of the work on the global Lorentzian geometry has been described in a standard book by Beem and Ehrlich [34] and in their second edition in 1996, with Easley. Later on O'Neill [317] published a book on Semi-Riemannian geometry, which carries an indefinite quadratic differential form g of an arbitrary signature. Since then a considerable number of works have appeared on the study of semi-Riemannian geometry. We review some results on semi-Riemannian geometry that are needed in this book.

Let M be a real n-dimensional smooth manifold and g a symmetric tensor of type $(0,2)$ on M. This means g assigns, to each point $p \in M$, a symmetric bilinear form g_x on the tangent space $T_x(M)$. Suppose g_x is non-degenerate and of constant index for all $x \in M$. This condition implies that each $T_x(M)$ is an n-dimensional semi-Euclidean space. Let $X_x = X^i\partial_i$ and $g_{ij} = g(\partial_i, \partial_j)$, $i, j \in (1, \ldots, n)$ and $\{\partial_i\}$ be the natural basis of $T_x(M)$. Then X_x is called

$$
\begin{array}{lll}
\textit{spacelike} & \text{if} \quad g_{ij}X^iX^j > 0 \ \text{ or } \ X_x = 0, \\
\textit{timelike} & \text{if} \quad g_{ij}X^iX^j < 0, \\
\textit{lightlike} & \text{if} \quad g_{ij}X^iX^j = 0 \quad \text{ and } X_x \neq 0.
\end{array}
$$

The set of all null vectors of $T_x(M)$ is called the *null cone* at x, defined by

$$\Lambda_x = \{X_x \in (T_x(M) - \{0\}), \qquad g_{ij}X^iX^j = 0\}.$$

Based on the above, g is called a *semi-Riemannian metric (metric tensor field)* and (M, g) is called a *semi-Riemannian manifold* (see O'Neill [317]). For example, (M, g) is a *Riemannian* or *Lorentzian manifold* according as g is of index 0 or 1 respectively. In case $0 < $ index $ < n$, then we say that M is a *proper semi-Riemannian manifold*. The metric g splits each tangent space at each point $x \in M$ into three categories; namely (i) spacelike (ii) timelike (iii) lightlike(null) vectors. The category into which a tangent vector falls is called its *causal character*. A curve C in M also belongs to one of the three categories. For a vector field X on M we say that X is spacelike (resp. timelike or null) according as $g(X, X) > (resp. < or 0)$. It is known that Riemannian metrics always exist on a paracompact manifold. In fact, suppose $\{\mathcal{U}_\alpha, \phi_\alpha\}_{\alpha \in I}$ is a smooth atlas of M such that $\{\mathcal{U}_\alpha\}_{\alpha \in I}$ is a local finite open cover of M. Consider g_α as a Riemannian metric on $\mathcal{U}_\alpha$ given by

$$g_\alpha(x)\,(U, V) = \sum_{i=1}^{n} U^i V^i,$$

where $\{U^i\}$ and $\{V^i\}$ are the local components of U and V respectively on $\mathcal{U}_\alpha$ with respect to the natural frames field $\{\frac{\partial}{\partial x^i}\}$. Then the desired Riemannian metric is defined by $g = \sum_{\alpha \in I} f_\alpha \, g_\alpha$, where $\{f_\alpha\}_{\alpha \in I}$ is the partition of unity subordinated to the covering $\{\mathcal{U}_\alpha, \phi_\alpha\}_{\alpha \in I}$. The proof of this result is mainly based on the positive definiteness of g_α and thus does not hold, in general, for a non-degenerate metric. However, the existence of both, a Riemannian metric g on M and a unit vector field E_o on M, i.e., $g(E_o, E_o) = 1$, enables one to construct a Lorentz metric on M (see O'Neill [317, page 148]). Indeed, consider the associate 1-form ω_o to E_o with respect to g, that is,

$$\omega_o(X) = g(X, E_o), \qquad \forall X \in \Gamma(TM),$$

and define $\bar{g}$ by

$$\bar{g}(X, Y) = g(X, Y) - 2\omega_o(X)\,\omega_o(Y), \quad \forall X, Y \in \Gamma(TM).$$

Then it is easy to check that $\bar{g}$ is a Lorentz metric on M. Moreover, E_0 is a timelike vector field with respect to $\bar{g}$. In general, a Lorentz manifold (M, g) may not have a globally defined timelike vector field. If (M, g) admits a global timelike vector field, then, it is called a *time orientable Lorentz manifold*, physically known as a *spacetime manifold*. For a detailed discussion of similarities and differences between Riemannian and Lorentzian geometries due to the causal structure of a spacetime manifold, we refer to [34].

Let (M, g) be a proper n-dimensional semi-Riemannian manifold of constant index $q \in \{1, \ldots, n\}$. We recall the following results. For details see any standard book on semi-Riemannian manifolds such as [317].

A *linear connection* on M is a map $\nabla : \Gamma(M) \times \Gamma(M) \to \Gamma(M)$ such that

$$\nabla_{fX+hY} Z = f(\nabla_X Z) + h(\nabla_Y Z), \qquad \nabla_X f = Xf,$$
$$\nabla_X(fY + hZ) = f\nabla_X Y + h\nabla_X Z + (Xf)Y + (Xh)Z,$$

for arbitrary vector fields X, Y, Z and smooth functions f, h on M. ∇_X is called the *covariant derivative operator* and $\nabla_X Y$ is called the *covariant derivative* of Y with respect to X. Define a tensor field ∇Y, of type $(1, 1)$, and given by $(\nabla Y)(X) = \nabla_X Y$, for any Y. Also, $\nabla_X f = Xf$ is the covariant derivative of f along X. The covariant derivative of a 1-form ω is given by

$$(\nabla_X \omega)(Y) = X(\omega(Y)) - \omega(\nabla_X Y). \tag{1.2.1}$$

For local expressions, we consider the natural basis $\{\partial_i\}$, $i \in \{1, \ldots, n\}$, on a coordinate neighborhood $\mathcal{U}$ and set $\nabla_{\partial_j}\partial_i = \Gamma_{ji}^k \partial_k$, where Γ_{ji}^k are n^3 local components of ∇ on M. For $X = X^i \partial_i$, $Y = Y^j \partial_j$ and $\omega = \omega_i dx^i$ we have

$$\nabla_X f = X^i \partial_i f, \qquad \nabla_X Y = Y_{;k}^i X^k \partial_i,$$
$$Y_{;k}^i = \partial_k Y^i + \Gamma_{kj}^i Y^j, \qquad \omega_{i;j} = \partial_j \omega_i - \Gamma_{ji}^k \omega_k, \tag{1.2.2}$$

where ; is a symbol for the covariant derivative. A simple way of understanding the role of covariant derivative (instead of the ordinary derivative) is as follows: Consider two tangent spaces T_xM, T_yM at points x and y of an n-dimensional differentiable manifold M. We know that $\dim(T_xM) = \dim(T_yM) = n$, but, to these tangent spaces at different points on M we need an additional structure on M called a *connection* so that the moving vectors or tensors always belong to M. This is why we see in (1.2.2) that the covariant derivative ∇_XY has an additional term involving connection coefficients Γ^k_{ji}. In particular, if $M = \mathbf{R}^n_q$ then all connection coefficients vanish and ∇ is the ordinary differential operator.

The covariant derivative of a tensor T of type (r,s) along a vector field X is a tensor field ∇_XT, of type (r,s), given by

$$(\nabla_XT)(\omega^1,\ldots,\omega^r,Y_1,\ldots,Y_s) = X(T(\omega^1,\ldots,\omega^r,Y_1,\ldots,Y_r))$$
$$-\sum_{\alpha=1}^{r} T(\omega^1,\ldots,\nabla_X\omega^\alpha,\ldots,\omega^r,Y_1,\ldots,Y_s)$$
$$-\sum_{t=1}^{s} T(\omega^1,\ldots,Y_1,\ldots,\nabla_XY_t,\ldots,Y_s)$$

for any vector field X, r covariant vectors $\omega^1,\ldots,\omega^r$ and s contravariant vectors $Y_1,\ldots,Y_s$. Note that ∇T of T is a tensor of type $(r,s+1)$. Locally, we have

$$T^{i_1\ldots i_r}_{j_1\ldots j_s;k} = \partial_k T^{i_1\ldots i_r}_{j_1\ldots j_s} + \sum_{h=1}^{r} T^{i_1\ldots i_{h-1}qi_{h+1}\ldots i_r}_{j_1\ldots j_s} \Gamma^{i_h}_{qk}$$
$$-\sum_{t=1}^{s} T^{i_1\ldots i_r}_{j_1\ldots j_{t-1}qj_{t+1}\ldots j_s} \Gamma^{q}_{j_tk}.$$

In particular, for a tensor of type $(1,2)$, we have

$$T^h_{ij;k} = \partial_k T^h_{ij} + \Gamma^h_{kt}T^t_{ij} - \Gamma^t_{NJ}T^h_{ti} - \Gamma^t_{ki}T^h_{jt},$$

where T^h_{ij} are the components of T. A vector field Y on M is said to be parallel with respect to a linear connection ∇ if for any vector field X on M it is covariant constant, i.e., $\nabla_XY = 0$. It follows from the third equation of (1.2.2) that Y is parallel on M if and only if its local components Y^i, with respect to a natural basis $\{\partial_i\}$, satisfy the differential equation

$$\partial_j Y^i + \Gamma^i_{kj}Y^k = 0. \tag{1.2.3}$$

In general a tensor field T on M is parallel with respect to ∇ if it is covariant constant with respect to any vector field X on M. Let C be a smooth curve on M given by the equations

$$x^i = x^i(t), \qquad t \in I \subset R, \qquad i = 1,\ldots,n.$$

Then a tangent vector field V to C is given by

$$V = \frac{dx^i}{dt} \partial_i.$$

Thus, a vector field Y is said to be parallel along C if $\nabla_V Y = 0$. Using this and the third equation of (1.2.2) we conclude that Y is parallel along C if and only if

$$\frac{dY^k}{dt} + \Gamma^k_{ij} Y^i \frac{dx^j}{dt} = 0. \tag{1.2.4}$$

The curve C is called a *geodesic* if V is parallel along C, i.e., if $\nabla_V V = fV$ for some smooth function f along C. It is possible to find a new parameter s along C such that f is zero along C and then the geodesic equation $\nabla_V V = 0$ can be expressed, in local coordinate system (x^i), as

$$\frac{d^2 x^k}{ds^2} + \Gamma^k_{ji} \frac{dx^j}{ds} \frac{dx^i}{ds} = 0. \tag{1.2.5}$$

The parameter s is called an *affine parameter*. Two affine parameters s_1 and s_2 are related by $s_2 = as_1 + b$, where a and b are constants. For a smooth or C^r ∇, the theory of differential equations certifies that, given a point p of M and a tangent vector X_p, there is a *maximal geodesic* $C(s)$ such that $C(0) = p$ and $\frac{dx^i}{ds}|_{s=0} = X^i_p$. If C is defined for all values of s, then it is said to be *complete*, otherwise incomplete.

A linear connection ∇ on (M, g) is called a *metric (Levi-Civita)* connection if g is parallel with respect to ∇, i.e.,

$$(\nabla_X g)(Y, Z) = X(g(Y, Z)) - g(\nabla_X Y, Z) - g(Y, \nabla_X Z) = 0, \tag{1.2.6}$$

for any $X, Y, Z \in \Gamma(TM)$. In terms of local coordinates system, we have

$$g_{ij;\,k} = \partial_k \bar{g}_{ij} - g_{ih} \Gamma^h_{jk} - g_{jh} \Gamma^h_{ik} = 0,$$

where

$$\Gamma^h_{ij} = \frac{1}{2} g^{hk} \left\{ \partial_j g_{ki} + \partial_i g_{kj} - \partial_k g_{ij} \right\}, \quad \Gamma^h_{ij} = \Gamma^h_{ji}.$$

Furthermore, if we set $\Gamma_{k|ij} = g_{kh} \Gamma^h_{ij}$, then, the above equation becomes

$$g_{ij;\,k} = \partial_k g_{ij} - \Gamma_{i|jk} - \Gamma_{j|ik} = 0.$$

The connection coefficients $\Gamma_{k|ij}$ and Γ^h_{ij} are called the *Christoffel symbols of first* and *second type* respectively. A result in semi-Riemannian geometry states (see O'Neill [317]) that there exists a metric connection ∇ which satisfies the following identity, the so-called *Koszul formula*

$$2g(\nabla_X Y, Z) = X(g(Y, Z)) + Y(g(X, Z)) - Z(g(X, Y))$$
$$+ g([X, Y], Z) + g([Z, X], Y) - g([Y, Z], X), \tag{1.2.7}$$

for any $X, Y, Z \in \Gamma(TM)$. The *semi-Riemannian curvature tensor*, denoted by R, of M is a $(1, 3)$ tensor field defined by

$$R(X, Y)Z = \nabla_X \nabla_Y Z - \nabla_Y \nabla_X Z - \nabla_{[X, Y]} Z, \text{ i.e.,} \qquad (1.2.8)$$

$$R^t_{jhk} = \partial_h \Gamma^t_{jk} - \partial_k \Gamma^t_{jh} + \Gamma^m_{jk} \Gamma^t_{mh} - \Gamma^m_{jh} \Gamma^t_{mk},$$

for any $X, Y, Z \in \Gamma(TM)$. The *torsion tensor*, denoted by T, of ∇ is a $(1, 2)$ tensor defined by

$$T(X, Y) = \nabla_X Y - \nabla_Y X - [X, Y].$$

R is skew-symmetric in the first two slots. In case T vanishes on M we say that ∇ is *torsion-free* or *symmetric metric connection* on M, which we assume in this book. The two *Bianchi's identities* are

$$R(X, Y)Z + R(Y, Z)X + R(Z, X)Y = 0, \qquad (1.2.9)$$

$$(\nabla_X R)(Y, Z, W) + (\bar{\nabla}_Y R)(Z, X, W) + (\nabla_Z R)(X, Y, W) = 0, \text{ i.e.,}$$

$$R^i_{jkl} + R^i_{klj} + R^i_{ljk} = 0,$$

$$R^i_{jkl; m} + R^i_{jlm; k} + R^i_{jmk; l} = 0.$$

The semi-Riemannian curvature tensor of type $(0, 4)$ is defined by

$$R(X, Y, Z, U) = g(R(X, Y)Z, U), \quad \forall X, Y, Z, U \text{ on } M, \text{ i.e.,}$$

$$R_{ijhk} = R(\partial_h, \partial_k, \partial_j, \partial_i) = g_{it} R^t_{jhk}.$$

Then by direct calculations we get

$$R(X, Y, Z, U) + R(Y, X, Z, U) = 0,$$

$$R(X, Y, Z, U) + R(X, Y, U, Z) = 0,$$

$$R(X, Y, Z, U) - R(Z, U, X, Y) = 0, \text{ i.e.,}$$

$$R_{ijkh} + R_{jikh} = 0, \quad R_{ijkh} + R_{ijhk} = 0, \quad R_{ijhk} - R_{hkij} = 0.$$

Let $\{E_1, \ldots, E_n\}_x$ be a local orthonormal basis of $T_x M$. Then,

$$g(E_i, E_j) = \epsilon_i \, \delta_{ij} \text{ (no summation in } i), \quad X = \sum_{i=1}^{n} \epsilon_i \, g(X, E_i) \, E_i,$$

where $\{\epsilon_i\}$ is the signature of $\{E_i\}$. Thus, we obtain

$$g(X, Y) = \sum_{i=1}^{n} \epsilon_i \, g(X, E_i) \, g(Y, E_i).$$

Set $n = m + 2$. The *Ricci tensor*, denoted by Ric, is defined by

$$\mathrm{Ric}(X, Y) = \mathrm{tr}\{Z \to R(X, Z)Y\}, \qquad (1.2.10)$$

for any $X, Y \in \Gamma(TM)$. Locally, Ric and its *Ricci operator* Q are given by

$$\text{Ric}(X, Y) = \sum_{i=1}^{m+2} \epsilon_i \, g(R(E_i, X)Y, E_i), \quad g(\bar{Q}X, Y) = \text{Ric}(X, Y), \text{ i.e.,} \qquad (1.2.11)$$

$$R_{ij} = R^t{}_{itj}, \qquad Q^i_j = R_{kj} g^{ki}.$$

M is *Ricci flat* if its Ricci tensor vanishes on M. If $\dim(M) > 2$ and

$$\text{Ric} = kg, \quad k \text{ is a constant,} \qquad (1.2.12)$$

then M is an *Einstein manifold*. For $\dim(M) = 2$, any M is Einstein but k in (1.2.12) is not necessarily constant. The *scalar curvature* r is defined by

$$r = \sum_{i=1}^{m+2} \epsilon_i \, \text{Ric}(E_i, E_i) = g^{ij} R_{ij}. \qquad (1.2.13)$$

(1.2.12) in (1.2.13) implies that M is Einstein if and only if r is constant and

$$\text{Ric} = \frac{r}{m+2} g.$$

The *Weyl conformal curvature tensor* C of type $(1,3)$ is defined by

$$C(X, Y)Z = R(X, Y)Z + \frac{1}{m}\{\text{Ric}(X, Z)Y - \text{Ric}(Y, Z)X + g(X, Z)QY$$
$$- g(Y, Z)QX\} - r\{m(m+1)\}^{-1}\{g(X, Z)Y - g(Y, Z)X\}, \text{ i.e.,}$$
$$C^h_{kij} = R^h_{kij} + \frac{1}{m}\left\{\delta^h_j R_{ki} - \delta^h_i R_{kj} + g_{ki} R^h_j - g_{kj} R^h_i\right\}$$
$$+ r\{m(m+1)\}^{-1}\left\{\delta^h_i g_{kj} - \delta^h_i g_{ki}\right\}. \qquad (1.2.14)$$

The tensor C vanishes for $\dim(M) = 3$. Let $g' = \Omega^2 g$ be a conformal transformation of g where Ω is a smooth positive real function on M. In particular, the conformal transformation is called *homothetic* if Ω is a non-zero constant. It is known that C is invariant under any such conformal transformation of the metric. If g is conformally related with a semi-Euclidean flat metric g' we say that g is *conformally flat* and M is then called a *conformally flat manifold*. M is conformally flat if and only if $C \equiv 0$ for $\dim(M) > 3$.

Suppose π is a non-degenerate plane of $T_x M$. Then, according to Section 1, the associated matrix G_x of g_x, with respect to an arbitrary basis $B = \{u, v\}$, is of rank 2 and given by

$$G_p = \begin{pmatrix} g_{uu} & g_{uv} \\ g_{uv} & g_{vv} \end{pmatrix}, \qquad \det(G_p) \neq 0. \qquad (1.2.15)$$

Define a real number $K(\pi) = K_x(u, v) = \frac{R(u, v, v, u)}{\det(G_x)}$, where $R(u, v, v, u)$ is the 4-linear mapping on $T_x M$ by the curvature tensor. The smooth function K, which assigns to each non-degenerate tangent plane π the real number $K(\pi)$ is called the *sectional curvature* of M, which is independent of the basis $B = \{u, v\}$. If K is a constant c at every point of M then M is of constant sectional curvature c, denote by $M(c)$, whose curvature tensor field R is given by [317, page 80]

$$R(X, Y)Z = c\{g(Y, Z)X - g(X, Z)Y\}, \quad \text{i.e.,}$$
$$R^h{}_{kij} = c\{\delta^h_i g_{jk} - \delta^h_j g_{ki}\}. \tag{1.2.16}$$

In particular, if $K = 0$, then M is called a *flat manifold* for which $R = 0$.

Hyperbolic spaces. A 2-dimensional subspace σ of the tangent space T_pM is called a *tangent plane* to M at p. Suppose σ is non-degenerate at p, then, the number

$$K_\sigma(u, v) = \frac{\langle R(u, v)u, v\rangle}{\langle u, u\rangle\langle v, v\rangle - \langle v, v\rangle^2}$$

is called the *sectional curvature* K_σ at the point p of M, which is independent of the choice of basis for σ.

A sphere $\mathbf{S}^n(r)$ of radius r is defined as a hypersurface in a Euclidean space $\mathbf{R}^{n+1}$ given by

$$\mathbf{S}^n(r) = \{v \in \mathbf{R}^{n+1} | \langle v, v\rangle = \sum_i (v^i)^2 = r^2\}$$

whose sectional curvature $K_\sigma = \frac{1}{r^2}$ at every point p and every plane σ.

An n-dimensional *hyperbolic space* $\mathbf{H}^n$ is defined as the component of

$$\{v \in \mathbf{R}^{n+1}_1 | \langle v, v\rangle_1 = -1\}$$

containing the point $(+1, 0, \ldots, 0)$, that is, the upper component of the two-sheeted hyperboloid. The sectional curvature of $\mathbf{H}^n$, defined as above, is $K = -1$. In analogy with the Euclidean sphere of radius r, we have

$$\{v \in \mathbf{R}^{n+1}_1 | \langle v, v\rangle_1 = -r^2\}$$

with the sectional curvature $K = -\frac{1}{r^2}$. Similarly, we have the following hypersurfaces of semi-Euclidean spaces:

$$\mathbf{S}^n_q(r) = \{v \in \mathbf{R}^{n+1}_q | \langle v, v\rangle_q = r^2\}: \quad \textit{pseudo-sphere,}$$
$$\mathbf{H}^n_q(r) = \{v \in \mathbf{R}^{n+1}_{q+1} | \langle v, v\rangle_{q+1} = -r^2\}: \quad \textit{pseudo-hyperbolic space.}$$

The *exterior derivation* is a differential operator, denoted by d, which assigns to each p-form ω, a $(p+1)$-form $d\omega$ defined by

$$(d\omega)(X_1, \ldots, X_{p+1}) = \frac{1}{p+1}\{\sum_{i=1}^{p+1}(-1)^{i+1}X_i\omega(X_1, \ldots, \widehat{X}_i, \ldots, X_{p+1})$$
$$+ \sum_{1\leq i<j\leq 1+p}(-1)^{i+j}\omega([X_i, X_j], X_1, \ldots, \widehat{X}_i, \ldots, \widehat{X}_j, \ldots, X_{p+1})\}$$

where the caret $\widehat{}$ means the term in that particular slot is omitted. In particular, for a 1-form ω and a 2-form Ω ,

$$(d\omega)(X,Y) = \frac{1}{2}\{X(\omega(Y)) - Y(\omega(X)) - \omega([X,Y])\},$$

$$(d\Omega)(X,Y,Z) = \frac{1}{3}\{X(\Omega(Y,Z)) - Y(\Omega(X,Z)) + Z(\Omega(X,Y))$$
$$- \Omega([X,Y],Z) + \Omega([X,Z],Y) - \Omega([Y,Z],X)\}.$$

For example, if $\omega = \omega_i dx^i$ and $\Omega = \frac{1}{2}\Omega_{ij} dx^i \wedge dx^j$, then

$$d\omega = \frac{1}{2!}(d\omega)_{ij} dx^i \wedge dx^j,$$

$$d\Omega = \frac{1}{3!}(d\Omega)_{ijk} dx^i \wedge dx^j \wedge dx^k,$$

$$(d\omega)_{ij} = \partial_i \omega_j - \partial_j \omega_i,$$

$$(d\Omega)_{ijk} = \partial_i \Omega_{jk} + \partial_j \Omega_{ki} + \partial_k \Omega_{ij}.$$

The exterior derivation d has the following properties:

(1) For a smooth function f on M, df is a 1-form (also called the gradient-form of f) such that $(df)X = Xf$, for any vector field X.

(2) For a p-form ω and a q-form θ

$$d(\omega \wedge \theta) = (d\omega) \wedge \theta + (-1)^p \omega \wedge d\theta.$$

(3) $d(d\omega) = 0$ for any p-form ω, (Poincare Lemma).

(4) d is linear with respect to the addition of any two p-forms.

The *gradient* of a smooth function f is defined as a vector field, denoted by grad f, and given by

$$g(\operatorname{grad} f, X) = X(f), \quad \text{i.e.,} \quad \operatorname{grad} f = \sum_{i,j=1}^{n} g^{ij} \partial_i f \partial_j. \tag{1.2.17}$$

The *divergence* and the *Curl* of a vector field X is a smooth function and a 2-form, respectively, denoted by div X and Curl X, and given by

$$\operatorname{div} X = X^m; m = \frac{\partial X^m}{\partial x^m} + \Gamma_{km}^m X^k,$$

$$\operatorname{Curl} X = \frac{1}{2}(\partial_j X_i - \partial_i X_j) dx^i \wedge dx^j. \tag{1.2.18}$$

The *Laplacian* of f, denoted by $\triangle f$, is given by

$$\triangle f = \operatorname{div}(\operatorname{grad} f)$$

$$= \sum_{i,j=1}^{n} g^{ij}\{\frac{\partial^2 f}{\partial x^i \partial x^j} - \Gamma_{ij}^k \frac{\partial f}{\partial x^k}\}. \tag{1.2.19}$$

The *Hessian*, denoted by $H^f = \nabla(\nabla f)$, of a function f, is its second covariant differential given by

$$H^f(X,Y) = XYf - (\nabla_X Y)f = g(\nabla_X(\operatorname{grad} f), Y), \qquad (1.2.20)$$

for all $X, Y \in \Gamma(TM)$.

Lie Derivatives. Let V be a vector field on a real n-dimensional smooth manifold. The *integral curves (orbits)* of V are given by the following system of ordinary differential equations:

$$\frac{dx^i}{dt} = V^i(x(t)), \qquad i \in \{1, \ldots, n\}, \qquad (1.2.21)$$

where (x^i) is a local coordinate system on M and $t \in I \subset R$. It follows from the well-known theorem on the existence and uniqueness of the solution of (1.2.21) that for any given point, with a local coordinate system, there is a unique integral curve defined over a part of the real line.

Consider a mapping ϕ from $[-\delta, \delta] \times \mathcal{U}$ ($\delta > 0$ and $\mathcal{U}$ an open set of M) into M defined by $\phi : (t, x) \to \phi(t, x) = \phi_t(x) \in M$, satisfying:

(1) $\phi_t : x \in \mathcal{U} \to \phi_t(x) \in M$ is a diffeomorphism of $\mathcal{U}$ onto the open set $\phi_t(\mathcal{U})$ of M, for every $t \in [-\delta, \delta]$,

(2) $\phi_{t+s}(x) = \phi_t(\phi_s(x))$, $\forall t, s, t + s \in [-\delta, \delta]$ and $\phi_s(x) \in \mathcal{U}$.

In the above case the family ϕ_t is a *1-parameter group of local transformations* on M. The mapping ϕ is then called a *local flow* on M. Using the equation (1.2.21) it has been proved (see Kobayashi-Nomizu [263, page 13]) that the vector field V generates a local flow on M. If each integral curve of V is defined on the entire real line, we say that V is a *complete vector field* and it generates a *global flow* on M. A set of local (resp. complete) integral curves is called a *local congruence (resp. congruence) of curves* of V. Now we show how the flow ϕ is used to transform any object, say Ω, on M into another one of the same type as Ω, with respect to a point transformation $\phi_t : x^i \to x^i + tV^i$ along an integral curve through x^i. Denote by $\bar{\Omega}(x^i)$ the pullback of $\Omega(x^i + tV^i)$ to the point x^i through the inverse mapping of ϕ_t. This defines a differentiable operator, denoted by $\pounds_V$, which assigns to an arbitrary Ω another object $\pounds_V \Omega$ of the same type as Ω given by

$$(\pounds_V \Omega)(x^i) = \lim_{t \to 0} \frac{1}{t} [\bar{\Omega}(x^i) - \Omega(x^i)]. \qquad (1.2.22)$$

The operator $\pounds_V$ is called the *Lie derivative* with respect to V. It is important to mention that the above definition (1.2.22) holds for local as well as global flows. Following are basic properties of Lie derivatives:

(1) $\pounds_V(aX + bY) = a\pounds_V X + b\pounds_V Y$, for all $a, b \in R$ and $X, Y \in \Gamma(M)$: (linearity)

(2) $\pounds_V(T \otimes S) = (\pounds_V T) \otimes S + T \otimes \pounds_V S$, where $\otimes$ is the tensor product of any two objects T and S: (Leibnitz rule)

(3) $\mathcal{L}_V$ commutes with the contraction operator.

It follows from the above properties that to compute the Lie derivative of any arbitrary Ω, it is sufficient to know the Lie derivatives of a function, a vector field and a 1-form, as the Lie derivatives of all other objects of higher order can be obtained by the use of tensor analysis and the above three properties. An elementary computation provides these three basic Lie derivatives as follows.

Functions. $\mathcal{L}_V f = V(f)$ where f is a function.

Vector Fields. Let X be a vector field on M. Then,

$$\mathcal{L}_V X = [V, X].$$

1-forms. Let $\omega = \omega_i dx^i$ be a 1-form on M. Then,

$$(\mathcal{L}_V \omega)(X) = V(\omega(X)) - \omega[V, X]$$
$$(\mathcal{L}_V \omega)_i = V^j \partial_j(\omega_i) + \omega_j \partial_i(V^j).$$

It follows from the above three Lie derivatives that if V is a vector field of class, say C^k, then the Lie derivative of a function, a vector field and a 1-form is of the same type but of class C^{k-1}. This is also true for higher tensors. Let T be a tensor (or a geometric object) of type (r, s). Then, using the above results and the theory of tensor analysis, we obtain the following general formulae for its Lie derivative with respect to a vector field V:

$$(\mathcal{L}_V T)(\omega^1, \ldots, \omega^r, X_1, \ldots, X_s) = V(T(\omega^1, \ldots, \omega^r, X_1, \ldots, X_s))$$
$$- \sum_{a=1}^{r} T(\omega^1, \ldots, \mathcal{L}_V \omega^a, \ldots, \omega^r, X_1, \ldots, X_s)$$
$$- \sum_{A=1}^{s} T(\omega^1, \ldots, \omega^r, X_1, \ldots, [V, X_A], \ldots, X_s),$$

where $\omega^1, \ldots, \omega^r$ and $X_1, \ldots, X_s$ are r 1-forms and s vector fields respectively. In particular, if T is a tensor of type $(1, 1)$ then

$$(\mathcal{L}_V T)(X) = V(T(X)) - T([V, X]),$$

for an arbitrary vector field X on M. The Lie derivative of a p-form $\omega = a_{i_1 \ldots i_p} dx^{i_1} \wedge \ldots \wedge dx^{i_p}$, with respect to V, is given by

$$\mathcal{L}_V \omega = (\mathcal{L}_V a_{i_1 \ldots i_p}) dx^{i_1} \wedge \ldots \wedge dx^{i_p}$$

where $\wedge$ is the wedge product operator. The following identity holds,

$$\mathcal{L}_V \omega = d i_V \omega + i_V d \omega,$$

where d denotes the exterior derivative operator and i_V is the inner product such that $(i_V\omega)(X_2,\ldots,X_p) = \omega(V,X_2,\ldots,X_p)$. It, therefore, follows that $\mathcal{L}_V$ and d commute, that is, $\mathcal{L}_V(d\omega) = d(\mathcal{L}_V\omega)$.

Lie derivatives of the metric tensor g and its Levi-Civita connection ∇ are

$$(\mathcal{L}_V g)(X,Y) = V(g(X,Y)) - g([V,X],Y) - g(X,[V,Y])$$
$$= g(\nabla_X V, Y) + g(\nabla_Y V, X) \tag{1.2.23}$$

for arbitrary vector fields X and Y on M and Levi-Civita (metric) connection ∇ of g. Locally, we have

$$\mathcal{L}_V g_{ij} = \nabla_i v_j + \nabla_j v_i$$
$$= v_{j\,;\,i} + v_{i\,;\,j} \quad , \quad v_i = g_{ij} V^j.$$

$$\begin{aligned}(\mathcal{L}_V \nabla)(X,Y) &= \mathcal{L}_V \nabla_X Y - \nabla_{[V,X]} Y - \nabla_X [V,Y]\\ &= [V,\nabla_X Y] - \nabla_{[V,X]} Y - \nabla_X (\nabla_V Y - \nabla_Y V)\\ &= \nabla_V \nabla_X Y - \nabla_X \nabla_V Y - \nabla_{[V\,X]} Y + \nabla_X \nabla_Y V - \nabla_{\nabla_X Y} V\\ &= \nabla_X \nabla_Y V - \nabla_{\nabla_X Y} + R(V,X)Y.\end{aligned} \tag{1.2.24}$$

In terms of local coordinates, we have

$$\mathcal{L}_V \Gamma^i_{jk} = \nabla_j \nabla_k V^i + R^i{}_{kmj} V^m$$
$$= \frac{1}{2} g^{im}\left(\nabla_j(L_V g_{km}) + \nabla_k(\mathcal{L}_V g_{jm}) - \nabla_m(\mathcal{L}_V g_{jk})\right),$$

where Γ^i_{jk} are the Christoffel symbols of the second kind, with respect to the metric tensor g_{ij}. Setting $\nabla_X Y = \nabla(X,Y)$, we have the following formulas:

$$\mathcal{L}_V(\nabla_X Y) - \nabla_X(L_V Y) - \nabla_{[V,X]} Y = (\mathcal{L}_V \nabla)(X,Y),$$
$$(\mathcal{L}_V(\nabla_X \omega) - \nabla_X(\mathcal{L}_V \omega) - \nabla_{[V,X]}\omega)Y = -\omega((\mathcal{L}_V \nabla)(X,Y)),$$
$$\{(\mathcal{L}_V(\nabla_X T) - \nabla_X(\mathcal{L}_V T) - \nabla_{[V,X]} T\}Y = (\mathcal{L}_V \nabla)(X,TY)$$
$$- T((\mathcal{L}_V \nabla)(X,Y)),$$
$$\nabla_X(\mathcal{L}_V \nabla)(Y,Z) - \nabla_Y(\mathcal{L}_V \nabla)(X,Z) = (\mathcal{L}_V R)(X,Y,Z)$$

where ω and T are a 1-form and a $(1,1)$ tensor field respectively and X,Y,Z are arbitrary vector fields on M, or, in local coordinates,

$$\mathcal{L}_V(\nabla_i Y^k) - \nabla_i(\mathcal{L}_V Y^k) = (\mathcal{L}_V \Gamma^k{}_{ij})Y^j,$$
$$\mathcal{L}_V(\nabla_i \omega_j) - \nabla_i(\mathcal{L}_V \omega_j) = -(\mathcal{L}_V \Gamma^k{}_{ij})\omega_k,$$
$$\mathcal{L}_V(\nabla_i T^j_k) - \nabla_i(\mathcal{L}_V T^j_k) = (\mathcal{L}_V \Gamma^j{}_{im})T^m_k - T^j_m \mathcal{L}_V(\Gamma^m{}_{ik}),$$
$$\nabla_i \mathcal{L}_V \Gamma^j{}_{km} - \nabla_k \mathcal{L}_V \Gamma^j{}_{im} = \mathcal{L}_V R^j{}_{mik}.$$

For details on the above we refer to a book by Duggal and Sharma [164] on *Symmetries of spacetimes and Riemannian manifolds* which also has a comprehensive account of the works of a large number of researchers.

Finally, we state the well-known Frobenius theorem on integrable distributions. Let (M, g) be a real n-dimensional smooth manifold with a symmetric tensor field g of type $(0, 2)$ on M such that g_x is of constant index q on $T_x M$ for any $x \in M$. We identify each $T_x M$ as a vector space at x. A *distribution* of rank r on M is a mapping D defined on M which assigns to each point x of M an r-dimensional linear subspace D_x of $T_x M$. Let $f : M' \to M$ be an immersion of M' in M. This means that the tangent mapping

$$(f_*)_x : T_x M' \to T_{f(x)} M,$$

is an injective mapping for any $x \in M'$. Suppose D is a distribution on M. Then M' is called an *integral manifold* of D if for any $x \in M'$ we have

$$(f_*)_x (T_x M') = D_{f(x)}.$$

If M' is a connected integral manifold of D and there exists no connected integral manifold $\bar{M}'$, with immersion $\bar{f} : \bar{M}' \to M$, such that $f(M') \subset \bar{f}(\bar{M}')$, we say that M' is a *maximal integral manifold* or a *leaf* of D. The distribution D is said to be *integrable* if for any point $x \in M$ there exists an integral manifold of D containing x. Recall that the distribution D is involutive if for two vector fields X and Y belonging to D, the Lie-bracket $[X, Y]$ also belongs to D. We quote the following well-known theorem:

Theorem (Frobenius). *A distribution D on M is integrable, if and only if, it is involutive. Moreover, through every point $x \in M$ there passes a unique maximal integral manifold of D and every other integral manifold containing x is an open submanifold of the maximal one.*

1.3   Warped product manifolds

Given two manifolds M and N, the set of all product coordinates in $M \times N$ is an atlas on $\bar{M} = M \times N$ which is called the *product manifold* of M and N. The $\dim(M \times N) = \dim(M) + \dim(N)$. This construction can be generalized to the product of any finite number of manifolds. A simple example is the Euclidean space $\mathbf{R}^n = \mathbf{R} \times \ldots \times \mathbf{R}$, the set of all ordered n-tuples $(x^1, \ldots, x^n)$ of real numbers. Later on we shall construct some more types of product manifolds used in differential geometry and its applications. The concept of product manifolds $M \times N$ has the following properties derived from the manifolds M and N:

(a) The projections

$$\pi : M \times N \to M \quad \text{mapping} \quad (p, q) \quad \text{to} \quad p,$$
$$\sigma : M \times N \to N \quad \text{mapping} \quad (p, q) \quad \text{to} \quad q$$

are smooth submersions.

(b) For each $(p, q) \in M \times N$ the subsets

$$M \times q = \{(r, q) \in M \times N : r \in M\},$$
$$p \times N = \{(p, r) \in M \times N : r \in N\}$$

are submanifolds of $M \times N$.

(c) For each (p, q)

$$\pi|M \times q \quad \text{is a diffeomorphism from} \quad M \times q \quad \text{to} \quad M,$$
$$\sigma|p \times N \quad \text{is a diffeomorphism from} \quad p \times N \quad \text{to} \quad N.$$

(d) The tangent spaces

$$T_{(p,q)}M \equiv T_{(p,q)}(M \times q) \quad \text{and} \quad T_{(p,q)}N \equiv T_{(p,q)}(p \times N)$$

are subspaces of $T_{(p,q)}(M \times N)$, which is their respective direct sum.

Suppose P is a submanifold of a Riemannian manifold (M, g). Regarding each tangent space $T_x P$ at each point x of P as a subspace of M, one can obtain a Riemannian metric tensor g_P on P induced by the metric g of M. Call g_P the pullback $i^*(g)$, where $i : P \to M$ is the inclusion map. However, if g is indefinite, then, $i^*(g)$ need not be an induced metric on P. In general, it is a symmetric $(0, 2)$ tensor field. It is a metric tensor if and only if each $T_x(P)$ is non-degenerate in $T_x M$ with same index for all x. Consequently, for the case of a semi-Riemannian manifold (M, g), the pullback $i^*(g)$ is a metric tensor on its submanifold P if $(P, i^*(g))$ is a semi-Riemannian submanifold of (M, g). Observe that the situation is quite different in case P is a lightlike submanifold of M. This case will be discussed in section 4. Using the above and the concept of product manifolds, one can construct the product of two or more semi-Riemannian manifolds as follows:

Consider two semi-Riemannian manifolds (M_1, g_1) and (M_2, g_2), with π and σ the projection maps of $M_1 \times M_2$ onto M_1 and M_2 respectively. Let

$$g = \pi^*(g_1) + \sigma^*(g_2).$$

Then, it is easy to show that g is a metric tensor of a semi-Riemannian manifold $(M = M_1 \times M_2, g)$. Indeed, if $X, Y \in T_{(x,y)}M$, then

$$g(X, Y) = g_1(d\pi(X), d\pi(Y)) + g_2(d\sigma(X), d\sigma(Y))$$

implies that g is symmetric. To show that g is non-degenerate, we suppose that $g(X, Y) = 0$ for all $Y \in T_{(x,y)}M$. This means that, in particular, $g_1(d\pi(X), d\pi(Y)) = 0$ for all $Y \in T_{(x,y)}M_1$ since $d\sigma(Y) = 0$. On the other hand, since $d\pi(Y)$ fills all of $T_x M_1$ we have $d\pi(X) = 0$. Similarly, one can show that $d\sigma(Y) = 0$. Hence

$Y = 0$. Finally, an orthogonal basis for $T_x M_1$ and $T_y M_2$ provide an orthogonal basis for $T_{(x,y)} M$. Therefore, the index of g has constant value $ind M_1 + ind M_2$. Hence, g is a metric tensor on M. One can extend the above scheme to any finite product of semi-Riemannian manifolds. For example, using the Euclidean product scheme (see Section 1), the semi-Euclidean space $\mathbf{R}_q^n$ is a product space given by

$$\overbrace{\mathbf{R}_1^1 \times \ldots \times \mathbf{R}_1^1}^{q \text{ factors}} \times \overbrace{\mathbf{R}^1 \times \ldots \times \mathbf{R}^1}^{(n-q) \text{ factors}}$$

where each $\mathbf{R}_1^1$ is the real line with a timelike metric tensor, that is, the negative of the usual dot product on $\mathbf{R}^1$. In particular, Minkowski spacetime ($q = 1$) is a $(1, n-1)$-product space of its time and the space parts.

However, unfortunately, not every semi-Riemannian (or Riemannian) manifold can be reducible as a product of two or more of its submanifolds. Take a simple example of a surface of revolution $\mathbf{M}$ by rotating a plane curve C about an axis in $\mathbf{R}^3$ and $r : C \to \mathbf{R}^+$ measures the distance to the axis. In spherical coordinates (r, θ, ϕ) the metric of the landing space $\mathbf{R}^3 - 0$ of $\mathbf{M}$ is given by the line element

$$ds^2 = dr^2 + r^2(d\theta^2 + \sin^2 \theta d\phi^2).$$

From this line element it is clear that, in general, $\mathbf{R}^3 - 0$ is not a product space. In particular, setting $r = 1$ gives the line element of a unit sphere $M = \mathbf{S}^2$ and, therefore, $\mathbf{R}^3 - 0$ is diffeomorphic to a product $\mathbf{R}^+ \times \mathbf{S}^2$ under the natural map $(a, b) \Leftrightarrow ab$. Very soon we will present several other examples of reducible and irreducible semi-Riemannian manifolds.

To construct a rich variety of manifolds from a given set of two or more manifolds, in 1969, Bishop and O'Neill [63] introduced a new concept of product manifolds, called *warped product manifolds* as follows:

Let (M_1, g_1) and (M_2, g_2) be two Riemannian manifolds, $h : M_1 \to (0, \infty)$ and $\pi : M_1 \times M_2 \to M_1$, $\sigma : M_1 \times M_2 \to M_2$ the projection maps given by $\pi(x, y) = x$ and $\sigma(x, y) = y$ for every $(x, y) \in M_1 \times M_2$. Denote the warped product manifold $M = (M_1 \times_h M_2, g)$, where

$$g(X, Y) = g_1(\pi_\star X, \pi_\star Y) + h(\pi(x,y)) g_2(\sigma_\star X, \sigma_\star Y) \qquad (1.3.1)$$

for every X and Y of M and $\star$ is the symbol for the tangent map. The manifolds M_1 and M_2 are called the *base* and the *fiber* of M. They proved that M is a complete manifold if and only if both M_1 and M_2 are complete Riemannian manifolds, and they also constructed a large variety of complete Riemannian manifolds of everywhere negative sectional curvature using a warped product.

Later on O'Neill [317] used the above definition and showed that if M_1 and M_2 are semi-Riemannian, then, their warped product is also semi-Riemannian. At this point we need the following definition:

Definition 1.3.1. A diffeomorphism $\phi : (M_1, g_1) \to (M_2, g_2)$ of two semi-Riemannian manifolds such that $\phi^*(g_2) = c g_1$ for some non-zero constant c is called a *homothety of coefficient* c. If $c = 1$ then ϕ is an isometry.

It is easy to verify the following properties of warped products:

(1) The fibers $x \times M_2 = \pi^{-1}(x)$ and the *leaves* $M_1 \times y = \sigma^{-1}(y)$ are semi-Riemannian.

(2) For each $y \in M_2$, the map $\pi|_{(M_1 \times y)}$ is an isometry onto M_1.

(3) For each $x \in M_1$, the map $\sigma|_{(x \times M_2)}$ is a positive homothety onto M_2, with scalar factor $1/h(x)$.

(4) For each $(x, y) \in M$, the leaf $M_1 \times y$ and the fiber $x \times M_2$ are orthogonal at (x, y).

The tangent vectors of leaves and the tangent vectors of fibers are called *horizontal* and *vertical* vectors respectively. See the next section for some more results on semi-Riemannian warped products.

Lorentzian warped products. This section is taken from a book by Beem and Ehrlich [34] which may be consulted for the details which we can not discuss in this book. A spacetime (M, g) is said to be *globally hyperbolic* if there exists a spacelike hypersurface Σ such that every endless causal curve intersects Σ once and only once. Such a hypersurface (if it exists) is called a *Cauchy surface*. If M is globally hyperbolic, then (a) M is homeomorphic to a product manifold $\mathbf{R} \times \Sigma$, where Σ is a hypersurface of M, and for each t, $\{t\} \times \Sigma$ is a Cauchy surface, (b) if Σ' is any compact hypersurface of M without boundary, then Σ' must be a Cauchy surface. It is obvious from the above discussion that Minkowski spacetime is globally hyperbolic. Now we highlight as to why *globally hyperbolic spacetimes* are physically important and also present a mathematical technique to construct an extension of this class to include a large class of time orientable *Lorentzian warped product manifolds*. Recall the following theorem of Hopf-Rinow [231] on compact and complete Riemannian manifolds.

Hopf-Rinow Theorem. *For any connected Riemannian manifold M, the following are equivalent:*

(a) *M is metric complete, i.e., every Cauchy sequence converges.*

(b) *M is geodesic complete, i.e., the exponential map is defined on the entire tangent space $T_x M$ at each $x \in M$.*

(c) *Every closed bounded subset of M is compact.*

Thus the Hopf-Rinow theorem maintains the equivalence of metric and geodesic completeness and, therefore, guarantees the completeness of all Riemannian metrics, for a compact smooth manifold, with the existence of minimal geodesics. Also, if any one of (a) through (c) holds, then the Riemannian function is obviously finite-valued and continuous. In the non-compact case, it is known through the work of Nomizu-Ozeki [311] that every non-compact Riemannian manifold admits a complete metric. Unfortunately, there is no analogue to the Hopf-Rinow theorem for a general Lorentzian manifold. In fact, we know now that the metric completeness and the geodesic completeness are unrelated for arbitrary Lorentz manifolds and their causal structure requires that a complete manifold must independently be spacelike, timelike and null complete. The singularity theorems (see Hawking-Ellis [228]) confirm that not all Lorentz manifolds are metric and/or geodesic complete. Also, the Lorentz distance function fails to be finite and/or continuous for an arbitrary spacetime.

Based on the above, it is natural to ask if there exists a class of spacetimes which shares some of the conditions of the Hopf-Rinow theorem. It has been shown in the works of Beem-Ehrlich [34] that the globally hyperbolic spacetimes turn out to be the most closely related physical model sharing some properties of the Hopf-Rinow theorem. Indeed, timelike Cauchy completeness and finite compactness are equivalent and the Lorentz distance function is finite and continuous for this class. Consequently, the globally hyperbolic spacetimes are physically important. Although the Minkowski spacetime and the Einstein static universe are globally hyperbolic, to include some more physically important models one needs an extended case of the product spaces, called *Lorentzian warped products* which we now explain.

Beem-Ehrlich [34] used the scheme of semi-Riemannian warped products and constructed a large rich class of globally hyperbolic manifolds as follows:

Let (M_1, g_1) and (M_2, g_2) be Lorentz and Riemannian manifolds respectively. Let $h : M_1 \rightarrow (0, \infty)$ be a C^∞ function and $\pi : M_1 \times M_2 \rightarrow M_1$, $\sigma : M_1 \times M_2 \rightarrow M_2$ the projection maps given by $\pi(x, y) = x$ and $\sigma(x, y) = y$ for every $(x, y) \in M_1 \times M_2$. Then, define the metric g given by

$$g(X, Y) = g_1(\pi_\star X, \pi_\star Y) + h(\pi(x, y)) g_2(\sigma_\star X, \sigma_\star Y), \quad \forall X, Y \in \Gamma(TM)$$

where $\pi_\star$ and $\sigma_\star$ are respectively tangent maps. They proved:

Theorem 1.3.2. (Beem-Ehrlich [34]) *Let (M_1, g_1) and (M_2, g_2) be Lorentzian and Riemannian manifolds respectively. Then, the Lorentzian warped product manifold $(M = M_1 \times_h M_2, g = g_1 \oplus_h g_2)$ is globally hyperbolic if and only if both the following conditions hold:*

(1) *(M_1, g_1) is globally hyperbolic.*

(2) *(M_2, g_2) is a complete Riemannian manifold.*

They presented an extensive global study on causal and completeness properties of globally hyperbolic manifolds as well as null cut loci, conjugate and focal points and Morse theory for non-null and null geodesics. Since then extensive research has been done on the geometric and physical use of Lorentzian warped product manifolds. Here we present some physical examples of warped product spacetimes:

I. Robertson-Walker spacetimes. Since the Einstein field equations are a complicated set of non-linear partial differential equations, we often assume certain relevant symmetry conditions for a satisfactory representation of our universe. Through extragalactic observations we know that the universe is approximately spherically symmetric about an observer. In fact, it would be more reasonable to assume that the universe is isotropic, that is, approximately spherical symmetric about each point in spacetime. This means the universe is *spatially homogeneous* [228], that is, admits a 6-parameter group of isometries whose surfaces of transitivity are spacelike hypersurfaces of constant curvature. This implies that any point on one of these hypersurfaces is equivalent to any other point on the same hypersurface. Such a spacetime is called *Robertson-Walker spacetime* with metric

$$ds^2 = -dt^2 + S^2(t)d\Sigma^2, \tag{1.3.2}$$

where $d\Sigma^2$ is the metric of a spacelike hypersurface Σ with spherical symmetry and constant curvature $c = 1, -1$ or 0. With respect to a local spherical coordinate system (r, θ, ϕ), this metric is given by

$$d\Sigma^2 = dr^2 + f^2(r)(d\theta^2 + \sin^2\theta d\phi^2), \tag{1.3.3}$$

where $f(r) = \sin r$, $\sinh r$ or r according as $c = 1, -1$ or 0. The range of the coordinates is restricted from 0 to 2π or from 0 to ∞ for $c = 1$ or -1 respectively. Using the framework of Lorentzian warped products, we now show that all Robertson-Walker spacetimes are globally hyperbolic. We know from (1.3.2) that $d\Sigma^2$ is a Riemannian metric of the spacelike hypersurface Σ. Set $M_1 = (a, b)$ for $(-\infty \leq a, b \leq \infty)$ as 1-dimensional space with negative definite metric $-dt^2$. Define $S^2(t) = h(t)$ where $h : (a, b) \to (0, \infty)$. Then, it follows from the metric (1.3.2) and the discussion on warped product that a Robertson-Walker spacetime (M, g) can be written as a Lorentzian warped product

$$\left(M = M_1 \times_h \Sigma, \ g = -dt^2 \oplus_h d\Sigma^2\right).$$

The map $\pi : M_0 \times_h \Sigma \to R$, given by $\pi(t, x) = t$, is a smooth timelike function on M each of whose level surfaces $\pi^{-1}(t_0) = \{t_0\} \times \Sigma$ is a Cauchy surface. Consequently, it follows from the above stated Beem-Ehrlich theorem that all Robertson-Walker spacetimes are globally hyperbolic. See [228, pages 134–142] for more details on the Robertson-Walker spacetimes.

II. Asymptotically flat spacetimes. One of the important areas of research in general relativity is the study of isolated systems, such as the sun and a host of stars in our universe. It is now well known that such isolated systems can best be understood by examining the local geometry of the spacetimes which are *asymptotically flat*, that is, their metric is flat at a large distance from a centrally located observer. First we define *stationary and static spacetimes*. A spacetime is stationary if it has a 1-parameter group of isometries with timelike orbits. Equivalently, a spacetime is stationary if it has a timelike Killing vector field, say V. A static spacetime is stationary with the additional condition that V is hypersurface orthogonal, that is, there exists a spacelike hypersurface Σ orthogonal to V. The general form of the metric of a static spacetime can be written as

$$ds^2 = -A^2(x^1,\, x^2,\, x^3)dt^2 + B_{\alpha\beta}(x^1,\, x^2,\, x^3)dx^\alpha dx^\beta\,,$$

where $A^2 = -V_a V^a$ and $\alpha,\, \beta = 1,\, 2,\, 3$. Static spacetimes have both the time translation symmetry ($t \to t+$ constant) and the time reflection symmetry ($t \to -t$). A spacetime is said to be *spherically symmetric* if its isometry group has a subgroup isometric to $SO(3)$ and its orbits are 2-spheres.

Case 1: Schwarzschild spacetimes. Let $(M,\, g)$ be a 4-dimensional isolated system with a Lorentz metric g and 3-dimensional spherical symmetry. Choose local coordinates $(t,\, r,\, \theta,\, \phi)$ for which g is given by

$$ds^2 = -e^{2\lambda}dt^2 + e^{2\nu}dr^2 + A\,dr\,dt + B\,r^2\left(d\theta^2 + \sin^2\theta d\phi^2\right),$$

where λ, ν, A and B are functions of t and r only due to the 3-dimensional symmetry. The inherent freedom in choosing some of the coefficients allows us to consider a Lorentz transformation such that $A = 0$ and $B = 1$. Using this we get

$$ds^2 = -e^{2\lambda}dt^2 + e^{2\nu}dr^2 + r^2\left(d\theta^2 + \sin^2\theta d\phi^2\right), \quad \text{where} \tag{1.3.4}$$

$$g_{00} = -e^{2\lambda}, \qquad g_{11} = e^{2\nu}, \qquad g_{22} = r^2, \qquad g_{33} = r^2\sin^2\theta,$$

$$g_{ab} = 0,\ \forall\, a \neq b, \qquad |g| = -r^4\sin^2\theta e^{2(\lambda+\nu)}.$$

Assume that M is Ricci flat, that is, $R_{ab} = 0$. Finding the Christoffel symbols of the second type, we calculate the four non-zero components of the Ricci tensor and then equating them to zero entails the following three independent equations:

$$\partial_r \lambda = \frac{e^{2\lambda} - 1}{2r},$$

$$\partial_r \nu = \frac{1 - e^{2\lambda}}{2r}, \qquad \partial_t \nu = 0.$$

Adding the first and second equations provides $\partial_r(\lambda + \nu) = 0$. Thus, $\lambda + \nu = f(t)$. Now integrating the first equation and then using $\partial_r \lambda = -\partial_r \nu$, we get

$$e^{2\lambda} = e^{-2\nu} = \left(1 - \frac{2m}{r}\right),$$

where m is a positive constant. Thus, (1.3.4) takes the form

$$ds^2 = -\left(1 - \frac{2m}{r}\right)dt^2 + \left(1 - \frac{2m}{r}\right)^{-1}dr^2 + r^2\left(d\theta^2 + \sin^2\theta d\phi^2\right). \tag{1.3.5}$$

This solution is due to Schwarzschild for which M is the *exterior Schwarzschild spacetime* ($r > 2m$) with m and r as the mass and the radius of a spherical body.

If we consider all values of r, then (1.3.5) is singular at $r = 0$ and $r = 2m$. It is well known that $r = 0$ is an *essential singularity* and the singularity $r = 2m$ can be removed by extending (M, g) to another manifold say (M', g') as follows. Let

$$r' = \int \left(1 - \frac{2m}{r}\right)^{-1} dr = r + 2m \log(r - 2m),$$

be a transformation with a new coordinate system (u, r, θ, ϕ), where $u = t + r'$ is an advanced null coordinate. Then, (1.3.5) takes the form:

$$ds^2 = -\left(1 - \frac{2m}{r}\right) du^2 + 2du\, dr + r^2 \left(d\theta^2 + \sin^2 d\phi^2\right), \qquad (1.3.6)$$

which is non-singular for all values of r. Similarly, if we use a retarded null coordinate $v = t - r'$, then (1.3.5) takes the form

$$ds^2 = -\left(1 - \frac{2m}{r}\right) dv^2 - 2dv\, dr + r^2 \left(d\theta^2 + \sin^2 \theta d\phi^2\right). \qquad (1.3.7)$$

The *exterior Schwarzschild spacetime*, with metric (1.3.5) for $r > 2m$, can be regarded as a Lorentzian warped product in the following way. Let $M_1 = \{(t, r) \in \mathbf{R}^2 : r > 2m\}$ be endowed with the Lorentzian metric

$$g_1 = -\left(1 - \frac{2m}{r}\right) dt^2 + \left(1 - \frac{2m}{r}\right) dr^2$$

and let M_2 be the unit 2-sphere S^2 with the usual Riemannian metric g_2 of constant sectional curvature 1 induced by the inclusion mapping $S^2 \to \mathbf{R}^3$. Then, $(M = M_1 \times_h M_2, g = g_1 \oplus_h g_2, r^2 = h)$ is the exterior Schwarzschild spacetime. (M, g) is globally hyperbolic since M_1 is globally hyperbolic and $M_2 = S^2$ can have a complete Riemannian metric.

Case 2: Reissner-Nordström spacetimes. Another solution of asymptotically flat category is due to Reissner-Nordström, which represents the spacetime (M, g) outside a spherically symmetric body having an electric charge e but no spin or magnetic dipole. Similar to the case of the Schwarzschild solution, the metric of this spacetime can be expressed by

$$ds^2 = -\left(1 - \frac{2m}{r} + \frac{e^2}{r^2}\right) dt^2 + \left(1 - \frac{2m}{r} + \frac{e^2}{r^2}\right)^{-1} dr^2$$
$$+ r^2 \left(d\theta^2 + \sin^2 \theta d\phi^2\right) \qquad (1.3.8)$$

for a local coordinate system (t, r, θ, ϕ). This metric is also asymptotically flat as, for $r \to \infty$, it approaches the Minkowski metric and in particular if $e = 0$, then this is a Schwarzschild metric. It is singular at $r = 0$ and $r = m \pm (m^2 - e^2)^{\frac{1}{2}}$ if

$e^2 \leq m^2$. While $r = 0$ is an essential singularity, the other two can be removed as follows. Consider a transformation

$$r' = \int \left(1 - \frac{2m}{r} + \frac{e^2}{r^2} \right)^{-1} dr.$$

Let $(u,\, r,\, \theta,\, \phi)$ be a new coordinate system, with respect to (1.3.8), such that $u = t + r'$, the advanced null coordinate. Then, this metric transforms into

$$ds^2 = - \left(1 - \frac{2m}{r} + \frac{e^2}{r^2} \right) du^2 + 2\, du\, dr + r^2 \left(d\theta^2 + \sin^2 \theta\, d\phi^2 \right)$$

which is regular for values of r and represents an extended spacetime $(M',\, g')$ such that M is embedded in M' and g' is g on the image of M on M'. We leave it as an exercise to show that Reissner-Nordström spacetime is a globally hyperbolic warped product spacetime. Some more examples may be seen in [34].

Finally, we cite the following papers of Beem, Ehrlich and their collaborators (which also include other related papers) on global Lorentzian geometry (in particular, reference to use of warped products) and its applications to relativity.

Beem and others [31, 32, 33, 34, 35, 36, 37, 38, 39].
Ehrlich and others [171, 172, 173, 174, 175, 176].

1.4   Lightlike manifolds

Let (M, g) be a real n-dimensional smooth manifold with a symmetric tensor field g of type $(0,\, 2)$ such that g_x is of constant index q on $T_x M$ for any $x \in M$. In this section, we assume that g_x is degenerate on $T_x M$, that is, there exists a vector $\xi \neq 0$, of $T_x M$, such that $g_x(\xi, v) = 0$, $\forall v \in T_x M$. The *radical* or the *null space* [317, page 53] of $T_x M$, with respect to the symmetric bilinear form g_x, is a subspace $\operatorname{Rad} T_x M$ of $T_x M$ defined by

$$\operatorname{Rad} T_x M = \{ \xi \in T_x M \,;\, g_x(\xi, v) = 0\,, \forall v \in T_x M \}\,.$$

The dimension of $\operatorname{Rad} T_x M$ is called the *nullity degree* of g_x, denoted by $null\, T_x M$. Clearly, g_x is degenerate on $T_x M$ if and only if $null\, T_x M > 0$ which we assume. The *associated quadratic form* of g_x is the mapping $h_x : T_x M \to \mathbf{R}$ given by $h_x(v) = g_x(v, v)$ for any $v \in T_x M$. Then g_x is expressed in terms of h_x as

$$g_x(v, w) = \frac{1}{2} \left\{ h_x(v + w) - h_x(v) - h_x(w) \right\},$$

for all $v, w, \in T_x M$. By virtue of a well-known result of linear algebra, there exists a basis $E_x = \{ e_1, \ldots, e_n \}$ of $T_x M$ such that h_x has a canonical form

$$h_x(v) = \sum_{i=1}^{n} \lambda_i \left(v^i \right)^2,$$

where $\lambda_i \in \mathbf{R}$ and (v^i) are the coordinate components of v with respect to the basis E_x. We say that h_x is of *type* (p, q, r), where $p + q + r = n$, if there exist in the above equation p, q and r coefficients λ_i which are positive, negative and zero, respectively. Note that here q and r are the index and nullity degree respectively of g_x on $T_x M$. Also, the canonical form of h_x is not unique but the type of h_x is independent of the basis of $T_x M$.

Suppose the mapping $\mathrm{Rad}\, TM$ that assigns to each $x \in M$ the radical subspace $\mathrm{Rad}\, T_x M$ of $T_x M$ with respect to g_x defines a smooth distribution of rank $r > 0$ on M. With the above scheme, we say that (M, g) is an *r-lightlike manifold* [149] and g is an *r-degenerate metric* on M. $\mathrm{Rad}\, TM$ is called the *radical distribution* of M. Thus, we have

$$g\,(\xi, X) = 0, \qquad \forall \xi \in \Gamma\,(\mathrm{Rad}\, TM\,), X \in \Gamma\,(TM)\,. \qquad (1.4.1)$$

Moreover, it is easy to see that (M, g) is r-lightlike if and only if g has a constant rank $n - r$ on M. We use the following range of indices:

$$a, b, \ldots \in \{1, \ldots, q\}, \quad A, B, \ldots \in \{q+1, \ldots, q+p\}, \quad \alpha, \beta, \ldots \in \{1, \ldots, r\},$$
$$i, j, \ldots \in \{r+1, \ldots, m\}, \quad I, J, \ldots \in \{1, \ldots, m\}.$$

The associated quadratic form h of g is, locally, given by

$$h = -\sum_{a=1}^{q} (\omega^a)^2 + \sum_{A=q+1}^{q+p} (\omega^A)^2\,,$$

where $\{\omega^1, \ldots, \omega^{p+q}\}$ are $p + q$ linear independent differential 1-forms locally defined on M. Replace in the above equation $\omega^a = \omega_I^a\, dx^I$; $\omega^A = \omega_I^A\, dx^I$, and obtain $h = g_{IJ}\, dx^I\, dx^J$, where

$$g_{IJ} = g\,(\partial_I, \partial_J)$$
$$= -\sum_{a=1}^{q} \omega_I^a\, \omega_J^a + \sum_{A=q+1}^{q+p} \omega_I^A\, \omega_J^A\,,$$

and $\mathrm{rank}[g_{IJ}] = p + q < n$. Consider a complementary distribution $S(TM)$ to $\mathrm{Rad}\, TM$ in TM. As fibers of $S(TM)$ are screen subspaces of $T_x M$, $x \in M$, we call $S(TM)$ a *screen distribution* on M. Since M is supposed to be paracompact, there exists a non-degenerate screen distribution on M.

Example 1. Let S_1^3 be the unit pseudo-sphere of a Minkowski spacetime $\mathbf{R}_1^4$ given by $-t^2 + x^2 + y^2 + z^2 = 1$. Cut S_1^3 by the hypersurface $t - x = 0$ and obtain a lightlike surface M of S_1^3 with $\mathrm{Rad}\, TM$ spanned by a null vector $\xi = \partial_t + \partial_x$. Take a screen distribution $S(TM)$ spanned by a spacelike vector $W = z\, \partial_y - y\, \partial_z$. Thus, M is lightlike with $r = 1$ and $S(TM)$ Riemannian.

Metric (Levi-Civita) connection on lightlike manifolds. So far we have seen that if $\mathrm{Rad}\, TM$ is non-zero, then, the degenerate metric g of a lightlike manifold M is a symmetric tensor field. To further study the geometry of lightlike manifolds one must know the existence of a torsion-free metric (Levi-Civita) connection ∇ on M for which $\nabla g = 0$. To deal with this important problem we need the following from the theory of distributions on manifolds. We use the following range of indices: a, $b, \ldots \in \{1, \ldots, n\}$; $\alpha, \beta, \ldots \in \{1, \ldots, r\}$, $i, j, \ldots \in \{r+1, \ldots, n\}$.

From the Frobenius theorem it follows that leaves of an integrable distribution D determine a *foliation* on M of dimension r, that is, M is a disjoint union of connected subsets $\{L_t\}$ and each point x of M has a coordinate system $(\mathcal{U}; x^1, \ldots, x^n)$ such that $L_t \cap \mathcal{U}$ is locally given by the equations

$$x^i = c^i, \qquad i \in \{r+1, \ldots, n+1\}, \tag{1.4.2}$$

where c^i are real constants, and (x^α), $\alpha \in \{1, \ldots, r\}$, are local coordinates on L_t. We say that the foliation defined by D is *totally geodesic, totally umbilical* or *minimal*, if any leaf of D is totally geodesic, totally umbilical or minimal, respectively. The transformation of coordinates on M endowed with an integrable distribution has a special form. More precisely, considering another coordinate system $(\bar{\mathcal{U}}, \bar{x}^a)$ on M and using (1.4.2) for both systems, we obtain

$$0 = d\bar{x}^i = \frac{\partial \bar{x}^i}{\partial x^j} dx^j + \frac{\partial \bar{x}^i}{\partial x^\alpha} dx^\alpha = \frac{\partial \bar{x}^i}{\partial x^\alpha} dx^\alpha,$$

which imply

$$\frac{\partial \bar{x}^i}{\partial x^\alpha} = 0, \qquad \forall i \in \{r+1, \ldots, n\}, \qquad \alpha \in \{1, \ldots, r\}. \tag{1.4.3}$$

Hence the transformation of coordinates on M is given by

$$\bar{x}^\alpha = \bar{x}^\alpha(x^1, \ldots, x^n), \ \alpha \in \{1, \ldots, r\}; \quad i \in \{r+1, \ldots, m\},$$
$$\bar{x}^i = \bar{x}^i(x^{r+1}, \ldots, x^n). \tag{1.4.4}$$

Thus we get the following transformation of natural frames fields on M:

$$\frac{\partial}{\partial x^\alpha} = B_\alpha^\beta(x) \frac{\partial}{\partial \bar{x}^\beta},$$
$$\frac{\partial}{\partial x^i} = B_i^j(x) \frac{\partial}{\partial \bar{x}^j} + B_i^\alpha(x) \frac{\partial}{\partial \bar{x}^\alpha}, \tag{1.4.5}$$

where we put

$$B_i^j(x) = \frac{\partial \bar{x}^j}{\partial x^i}; \quad B_i^\alpha(x) = \frac{\partial \bar{x}^\alpha}{\partial x^i}; \quad B_\alpha^\beta(x) = \frac{\partial \bar{x}^\beta}{\partial x^\alpha}.$$

Now, suppose $\bar{D}$ is a complementary distribution to the integrable distribution D on M, that is, we have

$$TM = D \oplus \bar{D}. \tag{1.4.6}$$

We call $\bar{D}$ a *transversal distribution* to D in TM. Then we take a local frames field $\{\frac{\partial}{\partial x^\alpha}, X_i\}$ on M adapted to the decomposition (1.4.6), i.e., $X_i \in \Gamma(\bar{D})$ and $\frac{\partial}{\partial x^\alpha} \in \Gamma(D)$. Thus we have

$$\frac{\partial}{\partial x^i} = N_i^\alpha(x) \frac{\partial}{\partial x^\alpha} + A_i^j(x) X_j, \tag{1.4.7}$$

where A_i^j and N_i^α are smooth functions defined on a coordinate neighborhood $\mathcal{U} \subset M$. Hence the transition matrix from the local natural frames field $\{\frac{\partial}{\partial x^\alpha}, \frac{\partial}{\partial x^i}\}$ to $\{\frac{\partial}{\partial x^\alpha}, X_i\}$ is

$$\Lambda = \begin{bmatrix} \delta_\beta^\alpha & N_i^\alpha(x) \\ 0 & A_i^j(x) \end{bmatrix},$$

where

$$\delta_\beta^\alpha = \begin{cases} 1 & \alpha = \beta, \\ 0 & \alpha \neq \beta, \end{cases}$$

are the components of the so-called *Kronecker delta*. As Λ is an invertible matrix, it follows that the $(n - r) \times (n - r)$ matrix whose entries are $A_i^j(x)$ is also an invertible matrix. Thus the set of local vector fields $\{\frac{\delta}{\delta x^{r+1}}, \dots, \frac{\delta}{\delta x^m}\}$ given by

$$\frac{\delta}{\delta x^i} = A_i^j(x) X_j, \qquad i \in \{r+1, \dots, m\},$$

is a local basis of $\Gamma(\bar{D})$. In this way (1.4.7) becomes

$$\frac{\delta}{\delta x^i} = \frac{\partial}{\partial x^i} - N_i^\alpha(x) \frac{\partial}{\partial x^\alpha}. \tag{1.4.8}$$

We next denote by $\bar{N}_j^\beta$ smooth functions in (1.4.8) with respect to another coordinate system $(\bar{\mathcal{U}}, \bar{x}^a)$ on M. Then by using (1.4.5) and (1.4.7) we obtain

$$N_i^\alpha(x) B_\alpha^\beta(x) = \bar{N}_j^\beta(x) B_i^j(x) + B_i^\beta(x). \tag{1.4.9}$$

Conversely, suppose on the domain of each local chart of M there exist $r(n - r)$ real smooth functions N_i^α satisfying (1.4.9) with respect to the transformation (1.4.4). Then we define by (1.4.8) $n - r$ linear independent local vector fields on M. Moreover, by using (1.4.8) and (1.4.9) we obtain

$$\frac{\delta}{\delta x^i} = B_i^j(x) \frac{\delta}{\delta \bar{x}^j}. \tag{1.4.10}$$

Hence there exists on M a globally defined distribution $\bar{D}$ which is locally spanned by $\{\frac{\delta}{\delta x^i}\}$, $i \in \{r+1, \dots, n\}$. Since the dimension of each fiber of $\bar{D}$ is $n - r$ and $\{\frac{\delta}{\delta x^i}\}$ do not belong to $\Gamma(D)$, we conclude that $\bar{D}$ is a complementary distribution to D in TM. Summing up we state the following:

Theorem 1.4.1. *Let D be an integrable distribution of rank r on M. Then there exists a transversal distribution $\bar{D}$ to D in TM, iff on the domain of each local chart on M there exist $r(m - r)$ real smooth functions N_i^α satisfying (1.4.9) with respect to the transformation of coordinates (1.4.4).*

Suppose $\operatorname{Rad} TM$ is integrable. Then, there exists a local coordinates system $(\mathcal{U} ; x^1, \ldots, x^n)$ on M such that (x^α), $\alpha \in \{1, \ldots, r\}$ are the coordinates on a leaf L of $\operatorname{Rad} TM$ with its local equations $x^i = c^i$, $i \in \{r + 1, \ldots, m\}$. As g is degenerate on TM, using (1.1.2) and (1.1.4) we get

$$g_{\alpha\beta} = g_{\alpha i} = g_{i\alpha} = 0; \quad \forall \alpha, \beta \in \{1, \ldots, r\}, i \in \{r + 1, \ldots, n\},$$

and thus the matrix of g with respect to the natural frames field $\{\frac{\partial}{\partial x^I}\}$, $I \in \{1, \ldots, n\}$ becomes

$$[g_{IJ}] = \begin{bmatrix} O_{r,r} & O_{r,n-r} \\ O_{n-r,r} & g_{ij}(x^1, \ldots, x^n) \end{bmatrix}. \tag{1.4.11}$$

Also, suppose that with respect to the above coordinates system we have

$$\frac{\partial g_{ij}}{\partial x^\alpha} = 0, \quad \forall i, j \in \{r + 1, \ldots, n\}, \alpha \in \{1, \ldots, r\}. \tag{1.4.12}$$

The first group of equations in (1.4.5) imply that (1.4.11) holds for any other system of coordinates adapted to the foliation induced by the integrable distribution $\operatorname{Rad} TM$. As in the case of semi-Riemannian manifolds, a vector field X on a lightlike manifold (M, g) is said to be a *Killing vector field* if $\pounds_X g = 0$. A distribution D on M is said to be a *Killing distribution* if each vector field belonging to D is a Killing vector field. According to the terminology used in [149], we say that a lightlike manifold (M, g) is a *Reinhart lightlike manifold* if its $\operatorname{Rad} TM$ is integrable and there exists a local coordinate system such that (1.4.11) holds. See [333] for information on Reinhart Riemannian spaces. Now we state and prove the existence of a metric (Levi-Civita) connection on (M, g).

Theorem 1.4.2. [149] *Let (M, g) be a lightlike manifold. Then the following assertions are equivalent:*

(i) *(M, g) is a Rienhart lightlike manifold.*

(ii) *$\operatorname{Rad} TM$ is a Killing distribution.*

(iii) *There exists a Levi-Civita connection ∇ on M with respect to g.*

Proof. (i) $\Longrightarrow$ (ii). Suppose M is a Reinhart lightlike manifold. As $\operatorname{Rad} TM$ is integrable, consider a coordinate system $(\mathcal{U} : x^1, \ldots, x^n)$ such that any $X \in \Gamma(\operatorname{Rad} TM)$ is locally expressed by $X = X^\alpha \partial_\alpha$. Then by using (1.2.22) and (1.4.1) we see that $\pounds_X g = 0$ becomes

$$X^\alpha \left\{ \frac{\partial (g(Y, Z))}{\partial x^\alpha} - g\left(\left[\frac{\partial}{\partial x^\alpha}, Y \right], Z \right) - g\left(Y, \left[\frac{\partial}{\partial x^\alpha}, Z \right] \right) \right\} = 0, \tag{1.4.13}$$

for any $Y, Z \in \Gamma(TM)$. By using (1.4.1) and $\mathrm{Rad}\,TM$ integrable, it is easy to check that in case at least one of vector fields Y and Z belongs to $\mathrm{Rad}\,TM$, (1.4.13) is identically satisfied. Now Consider $Y = \frac{\partial}{\partial x^i}$ and $Z = \frac{\partial}{\partial x^j}$, i , $j \in \{r+1, \ldots, n\}$ and (1.4.13) follows by using (1.4.12). Hence $\mathrm{Rad}\,TM$ is a Killing distribution.

(ii) $\Longrightarrow$ (i). Suppose $\mathrm{Rad}\,TM$ is a Killing distribution, that is

$$(\pounds_X g)\,(Y,Z) = X(g\,(Y,Z)) - g\,([X,Y],Z) - g\,(Y,[X,Z]) = 0, \qquad (1.4.14)$$

for any $X \in \Gamma(\mathrm{Rad}\,TM)$ and $Y, Z \in \Gamma(TM)$. Consider $Y \in \Gamma(\mathrm{Rad}\,TM)$ in (1.4.14) and by using (1.4.1) obtain $g([X,Y],Z) = 0$, for any $Z \in \Gamma(TM)$. Hence $[X,Y] \in \Gamma(\mathrm{Rad}\,TM)$, that is $\mathrm{Rad}\,TM$ is involutive, and by the Frobenius theorem it is integrable. Finally, take $X = \frac{\partial}{\partial x^\alpha} \in \Gamma(\mathrm{Rad}\,TM)$, $Y = \frac{\partial}{\partial x^i}$ and $Z = \frac{\partial}{\partial x^j}$ in (1.4.14) and obtain (1.4.12). Hence (M, g) is Reinhart.

(iii) $\Longrightarrow$ (ii). Suppose there exists a Levi-Civita connection ∇ on M, that is, g is parallel with respect to ∇. Then, we easily deduce that

$$\begin{aligned}(\pounds_X g)\,(Y,Z) &= \{X(g(Y,Z)) - g(\nabla_X Y, Z) - g(Y, \nabla_X Z)\} \\ &\quad + \{g(\nabla_Y X, Z) + g(Y, \nabla_Z X)\} \\ &= g(\nabla_Y X, Z) + g(Y, \nabla_Z X) \\ &= Y(g(X,Z)) + Z(g\,(X,Y)) - g\,(X, \nabla_Y Z) - g\,(X, \nabla Z Y) \\ &= 0, \quad \forall X \in \Gamma(\mathrm{Rad}\,TM) \quad \text{and} \quad Y, Z \in \Gamma(TM).\end{aligned}$$

Hence $\mathrm{Rad}\,TM$ is a Killing distribution on M.

(ii) $\Longrightarrow$ (iii). As (ii) is satisfied, from the proof of (ii) $\Longrightarrow$ (i) it follows $\mathrm{Rad}\,TM$ is integrable. Consider $\mathrm{Rad}\,TM$ as an $(n+r)$-dimensional manifold with local coordinates $(x^\alpha, x^i, y^\alpha)$, where (x^α, x^i) are local coordinates on M induced by the foliation determined by $\mathrm{Rad}\,TM$ and (y^α) are coordinates on fibers of vector bundle $\mathrm{Rad}\,TM$. Thus the transformation of coordinates on $\mathrm{Rad}\,TM$ is given by (1.4.4) and

$$\bar{y}^\alpha = B_\beta^\alpha\,(x)y^\beta .$$

It follows that

$$\frac{\partial}{\partial y^\alpha} = B_\alpha^\beta\,(x)\,\frac{\partial}{\partial \bar{y}^\beta} , \qquad (1.4.15)$$

which enables one to consider a vector bundle NM over M, locally spanned by $\{\frac{\partial}{\partial y^\alpha}\}$, $\alpha \in \{1, \ldots, r\}$. Moreover, we have

$$T\,(\mathrm{Rad}\,TM)_{|M} = TM \oplus NM. \qquad (1.4.16)$$

Next, since M is paracompact we consider a Riemannian metric g^* on M and a screen distribution $S(TM)$ as the complementary orthogonal distribution to $\mathrm{Rad}\,TM$ in TM with respect to g^*. Then (1.4.16) becomes

$$T(\mathrm{Rad}\,TM)_{|M} = \mathrm{Rad}\,TM \oplus_{\mathrm{orth}} S(TM) \oplus_{\mathrm{orth}} NM. \qquad (1.4.17)$$

Note that NM and $\operatorname{Rad} TM$ are vector bundles of rank r over M such that the transition matrices from $\{\frac{\partial}{\partial y^\alpha}\}$ to $\{\frac{\partial}{\partial \bar{y}^\beta}\}$ and from $\{\frac{\partial}{\partial x^\alpha}\}$ to $\{\frac{\partial}{\partial \bar{x}^\beta}\}$ are the same. Hence any section $N = N^\alpha \frac{\partial}{\partial y^\alpha}$ of NM defines a section $N^* = N^\alpha \frac{\partial}{\partial x^\alpha}$ of $\operatorname{Rad} TM$. Now, denote by p, s and t the projection morphisms of $T(\operatorname{Rad} TM)_{|M}$ on $S(TM)$, $\operatorname{Rad} TM$ and NM respectively, and define

$$\bar{g} : \Gamma\left(T(\operatorname{Rad} TM)_{|M}\right) \times \Gamma\left(T(\operatorname{Rad} TM)_{|M}\right) \longrightarrow \mathcal{F}(\operatorname{Rad} TM) ;$$
$$\bar{g}(\bar{X}, \bar{Y}) = g(p\bar{X}, p\bar{Y}) + g^*(s\bar{X}, (t\bar{Y})^*) + g^*(s\bar{Y}, (t\bar{X})^*) , \qquad (1.4.18)$$

for any $\bar{X}$, $\bar{Y} \in \Gamma(T(\operatorname{Rad} TM)_{|M})$. It is easy to see that $\bar{g}$ is a semi-Riemannian metric on the manifold $\operatorname{Rad} TM$ and the degenerate metric g is the restriction of $\bar{g}$ to $\Gamma(TM)$. Denote by $\bar{\nabla}$ the Levi Civita connection on $(\operatorname{Rad} TM, \bar{g})$ and set

$$\bar{\nabla}_X Y = \nabla_X Y + B^\alpha(X, Y) \frac{\partial}{\partial y^\alpha} , \quad \forall X, Y \in \Gamma(TM) , \qquad (1.4.19)$$

where $\nabla_X Y \in \Gamma(TM)$ and $B^\alpha(X, Y) \in \mathcal{F}(M)$. It follows that ∇ is a torsion-free linear connection on M and B^α are symmetric bilinear forms on $\Gamma(TM)$. Moreover, by using (1.4.15), (1.4.17) and (1.4.19), and $\bar{\nabla}\bar{g} = 0$ we obtain

$$0 = (L_X g)(Y, Z) = -\bar{g}(X, \bar{\nabla}_Y Z + \bar{\nabla}_Z Y) = -2B^\alpha(Y, Z) g^*\left(X, \frac{\partial}{\partial x^\alpha}\right),$$

for any $X \in \Gamma(\operatorname{Rad} TM)$ and Y, $Z \in \Gamma(TM)$. Now $r > 0$ and the g^* Riemannian metric on $\operatorname{Rad} TM$ implies $B^\alpha(Y, Z) = 0$, for any $\alpha \in \{1, \dots, r\}$. Hence $\bar{\nabla}_Y Z = \nabla_Y Z$, and thus $\nabla g = 0$, which completes the proof. $\square$

The following result is a direct consequence of the above theorem.

Corollary 1.4.3. *Let (M, g) be an n-dimensional lightlike manifold with $\operatorname{Rad} TM$ of rank $r = 1$ and a local coordinate system satisfying (1.4.11). Then, $\operatorname{Rad} TM$ is a Killing distribution and there exists a Levi-Civita connection on M with respect to the degenerate metric tensor field g.*

Lightlike warped products. Similar to the case of semi-Riemannian warped products, one can construct a variety of rich classes of *lightlike warped products* (a concept introduced by the first author, Duggal, of this book in a paper [137]) depending on the type of problem in hand. In the following we present two classes of lightlike warped products:

Class A. Let (M_1, g_1) and (M_2, g_2) be a lightlike and a semi-Riemannian manifold of dimensions n_1 and n_2 respectively, where the $\operatorname{Rad} TM_1$ is of rank r. Let $\pi : M_1 \times M_2 \to M_1$ and $\sigma : M_1 \times M_2 \to M_2$ denote the projection maps given by $\pi(x, y) = x$ and $\sigma(x, y) = y$ for $(x, y) \in M_1 \times M_2$ respectively. Observe that the projection π on M_1 will be with respect to the non-degenerate screen distribution $S(TM_1)$ of M_1.

Definition 1.4.4. A product manifold $M = M_1 \times M_2$ is called a lightlike warped product $M_1 \times_h M_2$, with the degenerate metric g given by

$$g(X, Y) = g_1(\pi_\star X, \pi_\star Y) + h(\pi(x, y)) \, g_2(\sigma_\star X, \sigma_\star Y), \qquad (1.4.20)$$

for every X, Y of M, $\star$ is the symbol for the tangent map and $h : M_1 \to (o, \infty)$ is a smooth function.

It follows that $\operatorname{Rad} TM$ of M still has rank r but its screen $S(TM)$ is of dimension $n - r$, where $\dim M = n_1 + n_2 = n$. Consider a simple case of lightlike manifolds with 1-dimensional radical distribution. These are interesting objects extensively used both in mathematics and physics. Geometrically, their radical distribution is obviously integrable, which is a desirable property. Secondly, such spaces represent physical models of singular regions in a spacetime manifold of general relativity (see some such cases in Chapter 3). Among this class there is a special case defined as follows:

Definition 1.4.5. [133] A lightlike manifold (M, g) is called a globally null manifold if it admits a global null vector field and a complete Riemannian hypersurface.

Example 2. Let $(\bar{M}, \bar{g})$ be an $(n + 1)$-dimensional globally hyperbolic spacetime [34], with the line element of the metric $\bar{g}$ given by

$$ds^2 = -dt^2 + dx^1 + \bar{g}_{ab} \, dx^a \, dx^b, \quad (a, b = 2, \ldots, n)$$

with respect to a coordinate system $(t, x^1, \ldots, x^n)$ on $\bar{M}$. We choose the range $0 < x^1 < \infty$ so that the above metric is non-singular. Take two null coordinates u and v such that $u = t + x^1$ and $v = t - x^1$. Thus, the above metric transforms into a non-singular metric: $ds^2 = -du \, dv + \bar{g}_{ab} \, dx^a \, dx^b$.

The absence of du^2 and dv^2 in the above metric implies that $\{v = \text{constant.}\}$ and $\{u = \text{constant.}\}$ are lightlike hypersurfaces of $\bar{M}$. Let $(M, g, r = 1, v = \text{constant.})$ be one of this lightlike pair and let D be the 1-dimensional distribution generated by the null vector $\{\partial_v\}$ in $\bar{M}$. Denote by L the 1-dimensional integral manifold of D. A leaf M' of the $(n - 1)$-dimensional screen distribution of M is Riemannian with metric $d\Omega^2 = \bar{g}_{ab} x^a x^b$ and is the intersection of the two lightlike hypersurfaces. In particular, there will be many global timelike vector fields in globally hyperbolic spacetimes $\bar{M}$. If one is given a fixed global *time function* then its gradient is a global timelike vector field in a given $\bar{M}$. With this choice of a global timelike vector field in $\bar{M}$, we conclude that both its lightlike hypersurfaces admit a global null vector field. Now, using the celebrated Hopf–Rinow theorem one may choose a screen whose leaf M' is a complete Riemannian hypersurface of M. Thus, it is possible to construct a pair of globally null manifolds as hypersurfaces of a globally hyperbolic spacetime. In particular, a Minkowski spacetime can have a pair of hypersurfaces which are globally null manifolds.

In the following we show how a globally null manifold can be expressed as a global product manifold and, then, construct a rich class of lightlike warped products.

Theorem 1.4.6. *Let $(M, g, S(TM))$ be an n-dimensional $(n \geq 3)$ globally null manifold, with a choice of screen distribution $S(TM)$. Then the following assertions are equivalent:*

(a) *$S(TM)$ is a parallel distribution.*

(b) *$M = M' \times C'$ is a global product manifold, where M' is an integral manifold of $S(TM)$ and C' is a 1-dimensional integral manifold of a global null curve C in M.*

Proof. Choose a parallel screen $S(TM)$. Then, by the Frobenius theorem, $S(TM)$ is integrable and its integral manifolds are geodesic in M. Thus $\nabla_X Y \in \Gamma(S(TM))$ for every $X, Y \in S(TM)$. The global null curve C being 1-dimensional, it is integrable. Moreover, it is known [137] that the integral curve of ξ is a null geodesic. Therefore M is a product of two geodesic submanifolds. Hence (a) implies (b). The converse clearly follows.   □

Now we show that there is a large class of globally null product manifolds. Consider a class of globally null manifolds, denoted by $(M, g, S(TM), G)$, such that each of its members carries a smooth 1-parameter group G of isometries whose orbits are global null curves in M. This means that each $\operatorname{Rad} TM$ is a Killing distribution and so, as per Theorem 1.4.2, each M admits a metric (Levi-Civita) connection ∇ with respect to its degenerate metric g.

Proposition 1.4.7. *Let (M, g, G) be an $(n + 1)$-dimensional $(n > 1)$ globally null manifold, with a smooth 1-parameter group G of isometries whose orbits are global null curves in M. Suppose M' is the spacelike n-dimensional orbit space of the action G. Then, $(M, g, S(TM), G)$ is a global product manifold $M = M' \times C'$, where M' and C' are leaves of a chosen integrable screen distribution $S(TM)$ and the $\operatorname{Rad} TM$ of M respectively.*

Proof. Let M' be the orbit space of the action $G \approx C'$, where C' is a 1-dimensional null leaf of $\operatorname{Rad} TM$ in M. Then, M' is a smooth Riemannian hypersurface of M and the projection $\pi : M \to M'$ is a principle C'-bundle, with null fiber G. The global existence of a null vector field, of M, implies that M' is Hausdorff and paracompact. The infinitesimal generator of G is a global null Killing vector field, say ξ, on M. Then, the metric g restricted to the screen distribution $S(TM)$, of M, induces a Riemannian metric, say g', on M'. Since ξ is non-vanishing on M, we can take $\xi = \partial_\theta$ as a global null coordinate vector field for some global function θ on M. Thus, θ induces a diffeomorphism on M such that $(M = M' \times C', g = \pi^\star g')$ is a global product manifold. Finally, the integrability of $S(TM)$ follows from the Theorem 1.4.6, which completes the proof.   □

Finally, using Definition 1.4.5, we have the following characterization theorem for warped product manifolds of Class A (the proof is straightforward):

Theorem 1.4.8. [133] *Let (M_1, g_1) and (M_2, g_2) be lightlike with $\operatorname{Rad} TM$ of rank 1 and Riemannian manifolds respectively. The warped product $(M = M_1 \times_h M_2, g)$ is a globally null manifold if and only if both the following conditions hold:*

(1) (M_1, g_1) *is a globally null manifold.*

(2) (M_2, g_2) *is a complete Riemannian manifold.*

Now we use the above result and generate a class of triple lightlike warped product. Consider a globally null manifold (M_1, g_1) and a complete Riemannian manifold (M_2, g_2) of dimensions n and m respectively. Using the above theorem, construct an $(n + m)$-dimensional globally null warped product manifold $(M = M_1 \times_h M_2, g)$, where h is a smooth function on M_1. Then, we have the following result (proof is similar to the proof of Theorem 1.4.6):

Theorem 1.4.9. *Let* $(M = M_1 \times_h M_2, g, S(TM))$ *be an* $(n+m)$-*dimensional globally null warped product manifold, where* h *is a smooth function on* M_1 *and* $S(TM)$ *a chosen screen distribution. Then, the following assertions are equivalent:*

(a) *The screen distribution* $S(TM)$ *is a parallel distribution.*

(b) $M = L \times M'$ *is a global null product manifold, where* L *is a 1-dimensional integral manifold of the global null curve* C *in* M *and* (M', g') *is a complete Riemannian hypersurface of* M *which is a triple warped product*

$$(M = L \times B \times_h M_2, \ g), \quad M' = (B \times_h M_2, \ g') \tag{1.4.21}$$

where (B, g_B) *is a complete Riemannian hypersurface of* $M_1 = L \times B$.

Class B: Let (M, g) and (N, g') be two m- and n-dimensional lightlike manifolds, with both $\mathrm{Rad}\,TM$ and $\mathrm{Rad}\,TN$ of rank 1. Suppose $S(TM)$ and $S(TN)$ are their screen distributions of index q_1 and q_2 respectively. We construct an $(m + n)$-dimensional semi-Riemannian manifold $(\bar{M}, \bar{g})$, using M and N as follows:

Let $\mathrm{Rad}\,TM$ and $\mathrm{Rad}\,TN$ be locally generated by their respective null vector fields ℓ and k such that $\bar{g}(\ell, \ell) = \bar{g}(k, k) = 0$, $\bar{g}(\ell, k) = -1$. Construct one timelike vector U and one spacelike vector V such that

$$U = \frac{\ell + k}{\sqrt{2}} \quad , \quad V = \frac{\ell - k}{\sqrt{2}} \quad , \quad \bar{g}(U, U) = -1 \quad , \quad \bar{g}(V, V) = 1 \,.$$

Since 1-dimensional $\mathrm{Rad}\,TM$ and $\mathrm{Rad}\,TN$ are obviously integrable, there exists an integrable distribution D generated by $\{U, V\}$, and of index 1. $S(TM)$ and $S(TN)$ being non-degenerate, it follows that $(\bar{M}, \bar{g})$ is a semi-Riemannian manifold of index q_1+q_2+1 such that $T\bar{M} = S(TM) + S(TN) + D$. Thus, one can use the non-degenerate warped product technique to find two semi-Riemannian manifolds, say (M_1, g_1) and (M_2, g_2), such that $\bar{M} = (M_1 \times_f M_2, \bar{g})$ where the warped function f, but $\bar{g}$ is indefinite. If $q_1 = q_2 = 0$, then $\bar{M}$ is Lorentzian. Furthermore, suppose M and N are globally null. This means that they both admit complete Riemannian hypersurfaces as leafs of $S(TM)$ and $S(TN)$ and both the null vectors ℓ and k are globally defined on respective manifolds. Thus, $(\bar{M}, \bar{g})$ admits a globally defined timelike vector field U and it follows from a result in [34] that $(\bar{M}, \bar{g})$ is a globally hyperbolic warped product of a globally hyperbolic spacetime (M_1, g_1) and a complete Riemannian manifold (M_2, g_2). Thus, we have the following theorem:

Theorem 1.4.10. *Let (M, g) and (N, g') be two globally null manifolds of dimensions m and n respectively. Then, there exists an $(m + n)$-dimensional globally hyperbolic Lorentzian warped product manifold $\bar{M} = (\, M_1 \times_f M_2, \bar{g}\,)$, where M_1 and M_2 are a globally hyperbolic spacetime and a complete Riemannian submanifold of $\bar{M}$, constructed by using M and N.*

Using the above theorem one can construct the following specific examples of spacetimes: Schwarzschild and De-Sitter, Robertson–Walker, Reissner–Nordström and Kerr. Also, see [41, 137, 139, 140, 141] for more on warped products, geometric/ physical results on globally null manifolds and examples.

Chapter 2

Lightlike hypersurfaces

Since for any semi-Riemannian manifold $\bar{M}$ there is a natural existence of null (lightlike) subspaces, their study is equally desirable. In particular, from the point of physics lightlike hypersurfaces are of importance as they are models of various types of horizons, such as Killing, dynamical and conformal horizons, studied in general relativity (see some details in Chapter 3). However, due to the degenerate metric of a lightlike submanifold M, one fails to use, in the usual way, the theory of non-degenerate geometry. The primary difference between the lightlike submanifolds and the non-degenerate submanifolds is that in the first case the normal vector bundle intersects the tangent bundle. In other words, a vector of a tangent space $T_x\bar{M}$ cannot be decomposed uniquely into a component tangent to T_xM and a component of normal space $T_xM^{\perp}$. Therefore, the standard definition of the second fundamental form and the Gauss-Wiengarten formulas do not work, in the usual way, for the lightlike case.

To deal with this anomaly, lightlike manifolds have been studied in several ways corresponding to their use in a given problem. Indeed, see Akivis-Goldberg [2, 3, 4], Bonnor [76, 77], Katsuno [252], Leistner [285], Nurowski-Robinson [312], Penrose [329], Perlick [332], Rosca [339] and more referred to in these papers. In 1991, Bejancu-Duggal [51] introduced a general geometric technique of using a non-degenerate screen distribution $S(TM)$ to deal with the above anomaly for lightlike hypersurfaces (also applicable for a general submanifold). Later on, in 1994, Bejancu [48] used the method of non-degenerate screens for null curves. Motivated by the growing use of lightlike geometry in mathematical physics, in 1996, Duggal-Bejancu published a book [149] on "lightlike submanifolds of semi-Riemannian manifolds and applications" (see Kupeli [273] with a different approach). The purpose of this chapter is to present new geometric results on lightlike hypersurfaces available since the publication of this 1996 book. In Chapter 3, we deal with applications to relativity, with focus on a variety of black hole horizons and the latest work on *Osserman lightlike hypersurfaces* [20].

2.1   Basic general results

Semi-Riemannian hypersurfaces Let $(\bar{M}, \bar{g})$ be a proper $(m+2)$-dimensional semi-Riemannian manifold of constant index $q \in \{1, \ldots, m+1\}$. Suppose M is an $(m+1)$-dimensional smooth manifold and $i : M \to \bar{M}$ a smooth mapping such that each point $x \in M$ has an open neighborhood $\mathcal{U}$ for which i restricted to $\mathcal{U}$ is one-to-one and $i^{-1} : i(\mathcal{U}) \to M$ are smooth. Then, we say that $i(M)$ is an *immersed hypersurface* of $\bar{M}$. If this condition globally holds, then $i(M)$ is called an *embedded hypersurface* of $\bar{M}$, which we assume in this book. The embedded hypersurface has a natural manifold structure inherited from the manifold structure on $\bar{M}$ via the embedding mapping. At each point $i(x)$ of $i(M)$, the tangent space is naturally identified with an $(m+1)$-dimensional subspace $T_{i(x)}M$ of the tangent space $T_{i(x)}\bar{M}$. The embedding i induces, in general, a symmetric tensor field, say g, on $i(M)$ such that

$$g(X, Y)|_x = \bar{g}(i_*X, i_*Y)|_{i(x)}, \quad \forall X, Y \in T_x(M).$$

Here i_* is the differential map of i defined by $i_* : T_x \to T_{i(x)}$ and $(i_*X)(f) = X(f \circ i)$ for an arbitrary smooth function f in a neighborhood of $i(x)$ of $i(M)$. Henceforth, we write M and x instead of $i(M)$ and $i(x)$. Due to the causal character of three categories (spacelike, timelike and lightlike) of the vector fields of $\bar{M}$, there are three types of hypersurfaces M, namely, Riemannian, semi-Riemannian and lightlike and g is a non-degenerate or a degenerate symmetric tensor field on M according as M is of the first two types and of the third type respectively.

First we assume that g is non-degenerate so that (M, g) is a semi-Riemannian hypersurface of $(\bar{M}, \bar{g})$. Define the normal bundle subspace

$$TM^\perp = \left\{ V \in \Gamma(T\bar{M}) : \; g(V, W) = 0, \, \forall W \in \Gamma(T\bar{M}) \right\}$$

of M in $\bar{M}$. Since M is a hypersurface, $\dim(T_x M^\perp) = 1$. Following is the orthogonal complementary decomposition:

$$T\bar{M} = TM \perp TM^\perp, \quad TM \cap TM^\perp = \{0\}. \tag{2.1.1}$$

Here, the tangent and the normal bundle subspaces are non-degenerate and any vector field of $T\bar{M}$ splits uniquely into a component tangent to M and a component perpendicular to M. Let $\bar{\nabla}$ and ∇ be the Levi-Civita connections on $\bar{M}$ and M respectively. Then, there exists a uniquely defined unit normal vector field, say $\mathbf{n} \in \Gamma(T\bar{M})$ and the Gauss-Weingarten formulas are

$$\bar{\nabla}_X Y = \nabla_X Y + B(X, Y)\mathbf{n},$$
$$\bar{\nabla}_X \mathbf{n} = -\epsilon A_{\mathbf{n}} X, \tag{2.1.2}$$

for any tangent vectors X and Y of M and $g(\mathbf{n}, \mathbf{n}) = \epsilon = \pm 1$ such that $\mathbf{n}$ belongs to $TM^\perp$ and $\nabla_X Y$, $A_{\mathbf{n}} X$ belong to the tangent space. Here $B(-, -)\mathbf{n}$ is *the second fundamental form tensor* and B is *the second fundamental form*, related with *the shape operator* $A_{\mathbf{n}}$ by

$$B(X, Y) = \bar{g}(A_{\mathbf{n}} X, Y), \quad \forall X, Y \in \Gamma(TM).$$

We say that M is *totally geodesic hypersurface* in $\bar{M}$ if

$$B = 0 \Leftrightarrow A_{\mathbf{n}} = 0.$$

A point p of M is said to be *umbilical* if

$$B(X, Y)_p = k\, g(X, Y)_p, \qquad \forall X, Y \in T_p M,$$

where $k \in \mathbf{R}$ and depends on p. The above definition is independent of any coordinate neighborhood around p. M is *totally umbilical* in $\bar{M}$ if every point of M is umbilical, i.e., if $B = \rho g$ where ρ is a smooth function.

With respect to an orthonormal basis $\{E_1, \ldots, E_n\}$ of $T_p M$, the *mean curvature vector* μ of M is defined by

$$\mu = \frac{\operatorname{tr}(B)}{n} = \frac{1}{n} \sum_{i=1}^{n} \epsilon_i B(E_i, E_i), \quad g(E_i, E_i) = \epsilon_i,$$

where μ is independent of any coordinate neighborhood around p. The characteristic equations of *Gauss* and *Codazzi* are, respectively

$$< \bar{R}(X, Y)Z, W > = < R(X, Y)Z, W > + \epsilon < B(X, Z), B(Y, W) >$$
$$- \epsilon < B(Y, Z), B(X, W) >$$
$$\epsilon < \bar{R}(X, Y)Z, N > = (\nabla_X B)(Y, Z) - (\nabla_Y B)(X, Z)$$

where $\bar{R}$ and R denote the curvature tensors of $\bar{M}$ and M, respectively, $X, Y, Z, W \in \Gamma(TM)$ and $<, >$ is the symbol of inner product. In the following we notice a marked difference between the above structure equations and those of lightlike hypersurfaces due to a degenerate induced metric.

Lightlike hypersurfaces. Now let g be degenerate on M. Then, there exists a vector field $\xi \neq 0$ on M such that

$$g(\xi, X) = 0, \quad \forall X \in \Gamma(TM).$$

The *radical* or the *null space* (O'Neill [317, page 53]) of $T_x M$, at each point $x \in M$, is a subspace $\operatorname{Rad} T_x M$ defined by

$$\operatorname{Rad} T_x M = \{ \xi \in T_x M \ : \ g_x(\xi, X) = 0, \quad \forall X \in T_x M \}, \tag{2.1.3}$$

whose dimension is called the *nullity degree* of g and M is called a *lightlike hypersurface* of $\bar{M}$. Comparing (2.1.1) with (2.1.3), with respect to degenerate g, and any null vector being perpendicular to itself implies that $T_x M^{\perp}$ is also null and

$$\operatorname{Rad} T_x M = T_x M \cap T_x M^{\perp}.$$

For a hypersurface M $\dim(T_xM^\perp) = 1$, implies that $\dim(\operatorname{Rad} T_xM) = 1$ and $\operatorname{Rad} T_xM = T_xM^\perp$. We call $\operatorname{Rad} TM$ a *radical (null) distribution* of M. Thus, for a lightlike hypersurface M, (2.1.1) does not hold because TM and $TM^\perp$ have a non-trivial intersection and their sum is not the whole of tangent bundle space $T\bar{M}$. In other words, a vector of $T_x\bar{M}$ cannot be decomposed uniquely into a component tangent to T_xM and a component of $T_xM^\perp$. Therefore, the standard text-book definition of the second fundamental form and the Gauss-Weingarten formulas do not work, in the usual way, for the lightlike case.

To deal with this problem, in 1991, Bejancu-Duggal [51] introduced a general geometric technique by splitting the tangent bundle $T\bar{M}$ into three non-intersecting complementary (but not orthogonal) vector bundles (two of them null and one non-null). This result on lightlike hypersurfaces was presented by Duggal-Bejancu in a 1996 book [149]. In the same year Kupeli [273] published a book on singular semi-Riemannian geometry. Duggal-Bejancu's approach was basically extrinsic in contrast to the intrinsic one developed by Kupeli. Recently, there has been a considerable amount of new material on lightlike hypersurfaces and their applications to some problems in general relativity and other areas of mathematical physics, published by a large number of researchers. Starting from the next section, we present new results (with proofs) which are so far available on the differential geometry of lightlike hypersurfaces.

Consider a complementary vector bundle $S(TM)$ of $TM^\perp = \operatorname{Rad} TM$ in TM. This means that

$$TM = \operatorname{Rad} TM \oplus_{\text{orth}} S(TM). \tag{2.1.4}$$

$S(TM)$ is called a *screen distribution* on M. It follows from the equation (2.1.4) that $S(TM)$ is a non-degenerate distribution. Moreover, since we assume that M is paracompact, there always exists a screen $S(TM)$. Thus, along M we have the decomposition

$$T\bar{M}_{|M} = S(TM) \perp S(TM)^\perp, \quad S(TM) \cap S(TM)^\perp \neq \{0\}, \tag{2.1.5}$$

that is, $S(TM)^\perp$ is the orthogonal complement to $S(TM)$ in $T\bar{M}_{|M}$. Note that $S(TM)^\perp$ is also a non-degenerate vector bundle of rank 2. However, it includes $TM^\perp = \operatorname{Rad} TM$ as its sub-bundle.

Theorem [149]. *Let $(M, g, S(TM))$ be a lightlike hypersurface of a semi-Riemannian manifold $(\bar{M}, \bar{g})$. Then there exists a unique vector bundle $\operatorname{tr}(TM)$ of rank 1 over M, such that for any non-zero section ξ of $TM^\perp$ on a coordinate neighborhood $\mathcal{U} \subset M$, there exists a unique section N of $\operatorname{tr}(TM)$ on $\mathcal{U}$ satisfying:*

$$\bar{g}(N, \xi) = 1, \quad \bar{g}(N, N) = \bar{g}(N, W) = 0, \qquad \forall W \in \Gamma(ST(M)|_{\mathcal{U}}). \tag{2.1.6}$$

Proof. Note that $S(TM)^\perp$ is a non-degenerate vector bundle of rank 2 and $TM^\perp$ is a vector sub-bundle of $S(TM)^\perp$. Consider a complementary vector bundle F of

$TM^\perp$ in $S(TM)^\perp$ and take $V \in \Gamma(F_{|\mathcal{U}})$, $V \neq 0$. Then $\bar{g}(\xi, V) \neq 0$ on $\mathcal{U}$, otherwise $S(TM)^\perp$ would be degenerate at a point of $\mathcal{U}$. Define on $\mathcal{U}$, a vector field

$$N = \frac{1}{\bar{g}(\xi,\,V)}\{V - \frac{\bar{g}(V,V)}{2\,\bar{g}(\xi,\,V)}\,\xi\,\},\tag{2.1.7}$$

where $V \in \Gamma(F_{|\mathcal{U}})$ such that $\bar{g}(\xi, V) \neq 0$. It is easy to see that N, given by (2.1.7), satisfies (2.1.6). Moreover, by direct calculations, it follows that any N on $\mathcal{U}$, satisfying (2.1.6), is given by (2.1.7). Then we consider another coordinate neighborhood $\mathcal{U}^* \subset M$ such that $\mathcal{U} \cap \mathcal{U}^* \neq \phi$. As both $TM^\perp$ and F are vector bundles over M of rank 1, we have $\xi^* = \alpha \xi$ and $V^* = \beta V$, where α and β are non-zero smooth functions on $\mathcal{U} \cap \mathcal{U}^*$. It follows that N^* is related with N on $\mathcal{U} \cap \mathcal{U}^*$ by $N^* = (1/\alpha)N$. Therefore, the vector bundle F induces a vector bundle $\operatorname{tr}(TM)$ of rank 1 over M such that, locally, (2.1.6) is satisfied. Finally, we consider another complementary vector bundle E to $TM^\perp$ in $S(TM)^\perp$ and by using (2.1.7), for both F and E, we obtain the same $\operatorname{tr}(TM)$, which completes the proof. $\quad\square$

It follows from (2.1.6) that $\operatorname{tr}(TM)$ is a lightlike vector bundle such that $\operatorname{tr}(TM)_u \cap T_u M = \{0\}$ for any $u \in M$. Moreover, from (2.1.4) and (2.1.5) we have the following decompositions:

$$T\bar{M}_{|M} = S(TM) \oplus_{\text{orth}} (TM^\perp \oplus \operatorname{tr}(TM)) = TM \oplus \operatorname{tr}(TM).\tag{2.1.8}$$

Hence for any screen distribution $S(TM)$ we have a unique $\operatorname{tr}(TM)$ which is the complementary vector bundle to TM in $T\bar{M}_{|M}$ and satisfies (2.1.6). This is why we call $\operatorname{tr}(TM)$ the *lightlike transversal vector bundle* of M with respect to $S(TM)$. We denote by $(M, g, S(TM))$ a lightlike hypersurface of an $(m + 2)$-dimensional semi-Riemannian manifold $(\bar{M}, \bar{g})$ and by $\Gamma(\mathcal{E})$ the $\mathcal{F}(M)$-module of smooth sections of a vector bundle $\mathcal{E}$ over M, $\mathcal{F}(M)$ being the algebra of smooth functions on M. Also, all manifolds are supposed to be paracompact. Let $\bar{\nabla}$ be the metric (Levi-Civita) connection on $\bar{M}$ with respect to $\bar{g}$. We show that the existence of a unique null section N of $\operatorname{tr}(TM)$, for a null tangent section ξ of $\operatorname{Rad} TM$, plays a key role in setting up the *Gauss-Weingarten type formulae* (for the lightlike case) from which one can obtain the induced geometric objects such as linear connections, second fundamental forms, curvature tensor and Ricci tensor etc. By using the second form of the decomposition in (2.1.8), we obtain

$$\bar{\nabla}_X Y = \nabla_X Y + h(X, Y),\tag{2.1.9}$$

$$\bar{\nabla}_X V = -A_V X + \nabla^t_X V, \quad \forall X, Y \in \Gamma(TM)\tag{2.1.10}$$

and $V \in \Gamma(\operatorname{tr}(TM))$, where $\nabla_X Y$ and $A_V X$ belong to $\Gamma(TM)$ while $h(X, Y)$ and $\nabla^t_X V$ belong to $\Gamma(\operatorname{tr}(TM))$. It is easy to check that ∇ is a torsion-free induced linear connection on M, h is a $\Gamma(\operatorname{tr}(TM))$-valued symmetric $\mathcal{F}(M)$-bilinear form on $\Gamma(TM)$, and A_V is a $\mathcal{F}(M)$-linear operator on $\Gamma(TM)$. We call ∇^t an induced linear connection on $\operatorname{ltr}(TM)$. Here h and A_V are called the *second fundamental form* and the *shape operator* respectively, of M in $\bar{M}$. Also, (2.1.9) and (2.1.10) are called the *global Gauss* and *Weingarten formulae*, respectively.

Locally, suppose $\{\xi, N\}$ is a pair of sections on $\mathcal{U} \subset M$ satisfying (2.1.6). Define a symmetric $\mathcal{F}(\mathcal{U})$-bilinear form B and a 1-form τ on $\mathcal{U}$ by

$$B(X, Y) = \bar{g}(h(X, Y), \xi), \tag{2.1.11}$$

$$\tau(X) = \bar{g}(\nabla^t_X N, \xi), \tag{2.1.12}$$

for any $X, Y \in \Gamma(TM_{|\mathcal{U}})$. It follows that

$$h(X, Y) = B(X, Y)N, \quad \nabla^t_X N = \tau(X) N. \tag{2.1.13}$$

Hence, on $\mathcal{U}$, (2.1.9) and (2.1.10) become

$$\bar{\nabla}_X Y = \nabla_X Y + B(X, Y)N, \tag{2.1.14}$$

$$\bar{\nabla}_X N = -A_N X + \tau(X)N, \tag{2.1.15}$$

respectively. As B is the only component of h on $\mathcal{U}$ with respect to N, we call B the *local second fundamental form* of M, and the equations (2.1.14), (2.1.15) the *local Gauss* and *Weingarten formulae* .

Since $\bar{\nabla}$ is a metric connection on $\bar{M}$, it is easy to see that

$$B(X, \xi) = 0, \quad \forall X \in \Gamma(TM_{|\mathcal{U}}). \tag{2.1.16}$$

Consequently, the second fundamental form of M is degenerate.

Define a local 1-form η by

$$\eta(X) = \bar{g}(X, N), \quad \forall X \in \Gamma(TM_{|\mathcal{U}}). \tag{2.1.17}$$

Using (2.1.14), a metric connection $\bar{\nabla}$ on $\bar{M}$ and (2.1.17) we get

$$\begin{aligned}
0 &= (\bar{\nabla}_X \bar{g})(Y, Z) \\
&= X(\bar{g}(Y, Z)) - \bar{g}(\bar{\nabla}_X Y, Z) - \bar{g}(Y, \bar{\nabla}_X Z) \\
&= X(g(Y, Z)) - g(\nabla_X Y, Z) - g(Y, \nabla_X Z) \\
&\quad - B(X, Y)\bar{g}(Z, N) - B(X, Z)\bar{g}(Y, N) \\
&= (\nabla_X g)(Y, Z) - B(X, Y)\eta(Z) - B(X, Z)\eta(Y).
\end{aligned}$$

Thus, the connection ∇ on M is not a metric connection and satisfies

$$(\nabla_X g)(Y, Z) = B(X, Y)\eta(Z) + B(X, Z)\eta(Y). \tag{2.1.18}$$

Remark 2.1.1. Comparing the Gauss and Weingarten formulae (2.1.2) of the semi-Riemannian hypersurfaces with the above corresponding formulae of the lightlike hypersurfaces, we observe that the lightlike case involves an extra 1-form τ defined by (2.1.12). Because of this (and several other differences such as degenerate metric and non-uniqueness of a screen etc) the lightlike case is different from the semi-Riemannian case. In this chapter, the reader will see that the 1-form τ plays an important role in the geometry of lightlike hypersurfaces.

In the lightlike case, we also have another second fundamental form and its corresponding shape operator which we now explain as follows:

Let P denote the projection morphism of $\Gamma(TM)$ on $\Gamma(S(TM))$ with respect to the decomposition (2.1.4). We obtain

$$\nabla_X PY = \nabla_X^* PY + h^*(X, PY), \tag{2.1.19}$$

$$\nabla_X U = -A_U^* X + \nabla_X^{*t} U, \quad \forall X, Y \in \Gamma(TM) \tag{2.1.20}$$

and $U \in \Gamma(TM^\perp)$, where $\nabla_X^* Y$ and $A_U^* X$ belong to $\Gamma(S(TM))$, ∇ and ∇^{*t} are linear connections on $\Gamma(S(TM))$ and $TM^\perp$ respectively, h^* is a $\Gamma(TM^\perp)$-valued $\mathcal{F}(M)$-bilinear form on $\Gamma(TM) \times \Gamma(S(TM))$ and A_U^* is $\Gamma(S(TM))$-valued $\mathcal{F}(M)$-linear operator on $\Gamma(TM)$. We call them the *screen second fundamental form* and *screen shape operator* of $S(TM)$, respectively. Define

$$C(X, PY) = \bar{g}(h^*(X, PY), N), \tag{2.1.21}$$

$$\varepsilon(X) = \bar{g}(\nabla_X^{*t} \xi, N), \quad \forall X, Y \in \Gamma(TM). \tag{2.1.22}$$

One can show that $\varepsilon(X) = -\tau(X)$. Thus, locally we obtain

$$\nabla_X PY = \nabla_X^* PY + C(X, PY)\xi, \tag{2.1.23}$$

$$\nabla_X \xi = -A_\xi^* X - \tau(X)\xi, \quad \forall X, Y \in \Gamma(TM). \tag{2.1.24}$$

Here $C(X, PY)$ is called *the local screen fundamental form of $S(TM)$*.

It is well known that the second fundamental form and the shape operator of a non-degenerate hypersurface are related by means of a metric tensor field. Contrary to this, in the lightlike case, we see from (2.1.11) and (2.1.20) that there are interrelations between these geometric objects and those of screen distributions. Precisely, the two local second fundamental forms of M and $S(TM)$ are related to their shape operators by

$$B(X, Y) = g(A_\xi^* X, Y), \quad \bar{g}(A_\xi^* X, N) = 0, \tag{2.1.25}$$

$$C(X, PY) = g(A_N X, PY), \quad \bar{g}(A_N Y, N) = 0. \tag{2.1.26}$$

Proposition 2.1.2. *Let $(M, g, S(TM))$ be a lightlike hypersurface of a semi-Riemannian manifold. Then*

(a) *The shape operator A_N of M has a zero eigenvalue.*

(b) *The screen second fundamental form of $S(TM)$ is also degenerate.*

(c) *The screen shape operator A_ξ^* of $S(TM)$ is symmetric with respect to the second fundamental form of M.*

(d) *The connection ∇^* from (2.1.20) is a metric connection on $S(TM)$.*

(e) *An integral curve of $\xi \in \Gamma(\operatorname{Rad} TM_{|U})$ is a null geodesic of both M and $\bar{M}$ with respect to the connections ∇ and $\bar{\nabla}$, respectively.*

Proof. The second equality in (2.1.26) implies that A_N is $\Gamma(S(TM))$-valued. Hence, rank $A_N \leq m$ which implies the existence of a non-zero $X_0 \in \Gamma(TM_{|\mathcal{U}})$ such that $A_N X_0 = 0$, which proves (a). Then, (b) is immediate from the first relation in (2.1.26). Next, from (2.1.25) we obtain

$$B(X, A^*_\xi Y) = B(A^*_\xi X, Y), \quad \forall X, Y \in \Gamma(TM),$$

so (c) holds. (d) follows from (2.1.14) and (2.1.23). To prove (e) we note that (2.1.16) and (2.1.25) imply $A^*_\xi \xi = 0$, that is, ξ is an eigenvector field for A^*_ξ corresponding to the zero eigenvalue. Thus, by (2.1.14), (2.1.16), (2.1.24) and the above we obtain

$$\bar{\nabla}_\xi \xi = \nabla_\xi \xi = -\tau(\xi)\,\xi.$$

Suppose $\xi = \sum_{\alpha=0}^m \xi^\alpha \frac{\partial}{\partial u^\alpha}$ and consider an integral curve $\mathcal{C} : u^\alpha = u^\alpha(t)$, $\alpha \in \{0,\dots,m\}$, $t \in I \subset \mathbf{R}$, i.e., $\xi^\alpha = \frac{du^\alpha}{dt}$, or equivalently $\xi = \frac{d}{dt}$. In case $\tau(\xi) \neq 0$, choose a new parameter t^* on the null curve C such that

$$\frac{d^2 t^*}{dt^2} + \tau\left(\frac{d}{dt}\right)\frac{dt^*}{dt} = 0\,.$$

The choice of a parameter [149] t^* such that $\nabla_{\frac{d}{dt^*}} \frac{d}{dt^*} = 0$ proves (e).   □

Example 1. Let $(\mathbf{R}^4_1, \bar{g})$ be the Minkowski spacetime with signature $(-, +, +, +)$ of the canonical basis

$$(\partial_t, \partial_1, \partial_2, \partial_3).$$

$(M, g = \bar{g}_{|M}, S(TM))$ is a lightlike hypersurface, given by an open subset of the lightlike cone

$$\left\{ t(1, \cos u \cos v, \cos u \sin v, \sin u) \in \mathbf{R}^4_1 : t > 0, \ u \in (0, \pi/2), \ v \in [0, 2\pi] \right\}.$$

Then $\mathrm{Rad}(TM)$ and $\mathrm{ltr}(TM)$ are given by

$$\mathrm{Rad}(TM) = \mathrm{Span}\{\xi = \partial_t + \cos u \cos v\,\partial_1 + \cos u \sin v\,\partial_2 + \sin u\,\partial_3\},$$

$$\mathrm{ltr}(TM) = \mathrm{Span}\{N = \frac{1}{2}(-\partial_t + \cos u \cos v\,\partial_1 + \cos u \sin v\,\partial_2 + \sin u\,\partial_3)\},$$

respectively and the screen distribution $S(TM)$ is spanned by two orthonormal spacelike vectors

$$\{W_1 = -\sin u \cos v\,\partial_1 - \sin u \sin v\,\partial_2 + \cos u\,\partial_3, \quad W_2 = -\sin v\,\partial_1 + \cos v\,\partial_2\}.$$

Example 2. Let $(\mathbf{R}^4_2, \bar{g})$ be a 4-dimensional semi-Euclidean space with signature $(-, -, +, +)$ of the canonical basis $(\partial_0, \dots, \partial_3)$. Consider a hypersurface M of $\mathbf{R}^4_2$ given by

$$x_0 = x_1 + \sqrt{2}\,\sqrt{x_2^2 + x_3^2}.$$

For simplicity, we set $\mathbf{f} = \sqrt{x_2^2 + x_3^2}$. It is easy to check that M is a lightlike hypersurface whose radical distribution $\operatorname{Rad} TM$ is spanned by

$$\xi = \mathbf{f}\,(\partial_0 - \partial_1) + \sqrt{2}\,(x_2\,\partial_2 + x_3\,\partial_3).$$

Then the lightlike transversal vector bundle is given by

$$\operatorname{ltr}(TM) = \operatorname{Span}\left\{ N = \frac{1}{4\mathbf{f}^2}\left\{\mathbf{f}(-\partial_0 + \partial_1) + \sqrt{2}\,(x_2\,\partial_2 + x_3\,\partial_3)\right\}\right\}.$$

It follows that the corresponding screen distribution $S(TM)$ is spanned by

$$\{W_1 = \partial_0 + \partial_1, \quad W_2 = -x_3\,\partial_2 + x_2\,\partial_3\}.$$

Example 3. Let $(\mathbf{R}_2^4,\, \bar{g})$ be a 4-dimensional semi-Euclidean space with signature $(-, -, +, +)$ of the canonical basis $(\partial_0, \ldots, \partial_3)$ having a lightlike hypersurface M as given in Example 2. Then, by direct calculations we obtain

$$\bar{\nabla}_X W_1 = \bar{\nabla}_{W_1} X = 0,$$
$$\bar{\nabla}_{W_2} W_2 = -x_2\,\partial_2 - x_3\,\partial_3,$$
$$\bar{\nabla}_\xi \xi = \sqrt{2}\,\xi, \quad \bar{\nabla}_{W_2}\xi = \bar{\nabla}_\xi W_2 = \sqrt{2}\,W_2,$$

for any $X \in \Gamma(TM)$. Then, by using the Gauss formula (2.1.14) we get

$$\nabla_X W_1 = 0, \quad \nabla_{W_2} W_2 = -\frac{1}{2\sqrt{2}}\,\xi, \quad \nabla_{W_1} W_1 = 0,$$
$$\nabla_\xi W_2 = \nabla_{W_2}\xi = \sqrt{2}\,W_2.$$

Thus,

$$B(W_1,\, W_1) = 0 = B(W_1,\, W_2), \quad B(W_2,\, W_2) = -\sqrt{2}\,(x_2^2 + x_3^2).$$

For the benefit of readers we reproduce from [149] the local Gauss-Weingarten equations expressed in a coordinate system on M, all coefficients of the induced linear connection ∇ on M and the two local second fundamental forms.

The 1-dimensional $\operatorname{Rad} TM$ on M is obviously integrable. Therefore, there exists an atlas of local charts $\{\mathcal{U}; u^0, \ldots, u^m\}$ such that $\xi = \partial_{u^0} \in \Gamma(\operatorname{Rad} TM_{|\mathcal{U}})$. Let $\{\xi = \partial_{u^0}, \delta_{u^a}, N\}$ be the local field of frames on $\bar{M}$, where $\{\xi = \partial_{u^0}, \delta_{u^a}\}$ is a local field of frames on M. With respect to the Levi-Civita connection $\bar{\nabla}$ on $\bar{M}$ and by using local Gauss-Weingarten equations and $B(X, \xi) = 0$, we obtain

$$\bar{\nabla}_{\delta_{u^b}} \delta_{u^a} = \Gamma^o_{ab}\partial_{u^o} + \Gamma^c_{ab}\delta_{u^c} + B_{ab}N,$$
$$\bar{\nabla}_{\delta_{u^b}} \partial_{u^o} = \Gamma^o_{ob}\partial_{u^o} + \Gamma^k_{oj}\delta_{u^k},$$
$$\bar{\nabla}_{\partial_{u^o}} \delta_{u^a} = \Gamma^o_{ao}\partial_{u^o} + \Gamma^c_{ao}\delta_{u^c},$$
$$\bar{\nabla}_{\partial_{u^o}} \partial_{u^o} = \Gamma^o_{oo}\partial_{u^o},$$
$$\bar{\nabla}_{\delta_{u^a}} N = -A^c_a\delta_{u^c} + \tau_a N,$$
$$\bar{\nabla}_{\partial_{u^o}} N = -A^c_o\delta_{u^c} + \tau_o N,$$

where $\{\Gamma^c_{ab}, \Gamma^o_{ab}, \Gamma^c_{ob}, \Gamma^c_{io}, \Gamma^o_{ao}, \Gamma^o_{ob}, \Gamma^o_{oo}\}$ are the coefficients of the induced linear connection ∇ on M with respect to the frames field $\{\partial_{u^o}, \delta_{u^a}\}$; $\{A^c_a, A^c_o\}$ are the entries of the matrix of $A_N : \Gamma(TM_{|\mathcal{U}}) \to \Gamma(S(TM)_{|\mathcal{U}})$ with respect to the basis $\{\delta_{u^a}, \partial_{u^o}\}$ and $\{\delta_{u^a}\}$ of $\Gamma(TM_{|\mathcal{U}})$ and $\Gamma(S(TM)_{|\mathcal{U}})$ respectively; $B_{ab} = B(\delta_{u^a}, \delta_{u^b}) = B(\partial_{u^a}, \partial_{u^b})$; $\tau_a = \tau(\delta_{u^a})$, $\tau_o = \tau(\partial_{u^o})$.

Taking into account the Lie-bracket $[X, Y] = (X^a \partial_a Y^b - Y^a \partial_a X^b)\partial_b$ with respect to a natural frames field we obtain

$$[\delta_{u^a}, \delta_{u^b}] = S_{ab}\partial_{u^o}, \quad S_{ab} = \frac{\delta S_a}{\delta u^b} - \frac{\delta S_b}{\delta u^a},$$

and

$$[\delta_{u^a}, \partial_{u^o}] = \frac{\partial S_a}{\partial u^o}\partial_{u^o}.$$

As $\bar{\nabla}$ is torsion-free, using the Bianchi identity (1.2.8) and the above two relations, we obtain

$$\Gamma^c_{ab} = \Gamma^c_{ba} \; ; \; \Gamma^o_{ab} = \Gamma^o_{ba} + S_{ba} \; ; \; \Gamma^c_{ob} = \Gamma^c_{bo} \; ; \; \Gamma^o_{ob} = \Gamma^o_{bo} + \frac{\partial S_b}{\partial u^o},$$

and

$$B_{ab} = B_{ba}.$$

Further on, we decompose the following Lie brackets:

$$[N, \delta_{u^c}] = N^o_c \partial_{u^o} + N^d_c \delta_{u^d} + N_c N,$$
$$[N, \partial_{u^o}] = N^o \partial_{u^o} + N^c_o \delta_{u^c} + N_o N.$$

With respect to the non-holonomic frames field $\{\partial_{u^o}, N, \delta_{u^a}\}$ of $\bar{M}$, the semi-Riemannian metric $\bar{g}$ has the matrix

$$[\bar{g}] = \begin{bmatrix} 0 & 1 & 0 \\ 1 & 0 & 0 \\ 0 & 0 & g_{ab}(u^o, \ldots, u^m) \end{bmatrix},$$

where $g_{ab} = \bar{g}(\delta_{u^a}, \delta_{u^b}) = g(\partial_{u^a}, \partial_{u^b})$ are the functions in (1.4.2). As the matrix $[g_{ab}(u)]$ is invertible, consider its inverse matrix $[g^{ab}(u)]$. Then, by using Koszul's identity (1.2.7) and all the above equations, we obtain

$$\Gamma^k_{ab} = \frac{1}{2} g^{cd} \left\{ \frac{\delta g_{ad}}{\delta u^b} + \frac{\delta g_{bd}}{\delta u^a} - \frac{\delta g_{ab}}{\delta u^d} \right\},$$

$$\Gamma^o_{ab} = \frac{1}{2} \left\{ S_{ba} + N^c_b g_{ca} + N^c_a g_{cb} - N(g_{ab}) \right\} = g_{ac} A^c_b,$$

$$\Gamma^o_{ob} = \frac{1}{2} \left\{ \frac{\partial S_b}{\partial u^o} + N_b + N^c_o g_{ab} \right\} = g_{bc} A^c_o + \frac{\partial S_b}{\partial u^o} = -\tau_b,$$

$$\Gamma^c_{bo} = \Gamma^c_{ob} = \frac{1}{2} g^{ca} \frac{\partial g_{ab}}{\partial u^o},$$

$$\Gamma^o_{oo} = N_o = -\tau_o \,,$$

$$B_{ab} = -\frac{1}{2}\frac{\partial g_{ab}}{\partial u^o}\,.$$

By using these, we obtain

$$\Gamma^{*c}_{ab} = \Gamma^c_{ab}\,; \quad \Gamma^{*c}_{bo} = \Gamma^c_{bo}\,,$$
$$C_{ab} = C\left(\delta_{u^b},\delta_{u^a}\right) = \Gamma^o_{ab} = g_{ac}A^c_b,$$
$$C_a = C\left(\partial_{u^o},\delta_{u^a}\right) = \Gamma^o_{ao} = g_{ik}A^k_o\,,$$
$$A^{*a}_{\ b} = g^{ac}B_{bc} = -\Gamma^a_{ob}\,,$$

where $\{\Gamma^{*c}_{ab},\Gamma^{*c}_{ao}\}$ are the coefficients of the metric connection ∇^* on $S(TM)$ with respect to the frames field $\{\partial_{u^o},\delta_{u^c}\}$ and $A^{*a}_{\ b}$ are the entries of $A^*_{\partial_{u^o}}$ with respect to the basis $\{\partial_{u^a}\}$.

Remark 2.1.3. A comprehensive ground work on the geometry and physics of all possible types of geodesic and non-geodesic null curves in semi-Riemannian manifolds, supported by several examples and their fundamental existence theorems, is available in [153, 156, 242, 243, 244, 245, 246], with potential for further research in this direction. Also, more details on the basic general results on lightlike hypersurfaces are available in [18, 149, 156].

2.2   Screen conformal hypersurfaces

It is well known that the second fundamental form and the shape operator of a non-degenerate submanifold are related by the metric tensor field. Contrary to this we see from (2.1.25) and (2.1.26) that in case of lightlike hypersurfaces, there are interrelations between the second fundamental forms of the lightlike hypersurface M and its screen distribution $S(TM)$ to their respective shape operator A_N and A^*_ξ. Thus, the geometry of a lightlike hypersurface depends on a choice of screen distribution which plays an important role in studying differential geometry of lightlike hypersurfaces. Moreover, $S(TM)$ being non-degenerate, its geometry is classical. Thus, one may propose the following problem:

Find a class of lightlike hypersurfaces, whose geometry is essentially the same as that of their chosen screen distribution.

As the shape operator is an information tool in studying geometry of submanifolds, we are led to consider lightlike hypersurfaces whose shape operators are the same as the one of their screen distribution up to a conformal non-vanishing smooth factor in $\mathcal{F}(M)$. More precisely

Definition 2.2.1. [19] A lightlike hypersurface $(M, g, S(TM))$ of a semi-Riemannian manifold is called screen locally conformal if the shape operators A_N and A^*_ξ of M and $S(TM)$, respectively, are related by

$$A_N = \varphi\,A^*_\xi \tag{2.2.1}$$

where φ is a non-vanishing smooth function on a neighborhood $\mathcal{U}$ in M.

In particular, we say that M is *screen homothetic* if φ is non-zero constant. To avoid trivial ambiguities, we take $\mathcal{U}$ connected and maximal in the sense that there is no larger domain $\mathcal{U}' \supset \mathcal{U}$ on which the relation (2.2.1) holds. In case $\mathcal{U} = M$ the screen conformality is global.

Example 4. Let $\mathbf{R}_1^{n+2}$ be the space $\mathbf{R}^{n+2}$ endowed with the semi-Euclidean metric

$$\bar{g}(x, y) = -x^0 y^0 + \sum_{a=1}^{n+1} x^a y^a \qquad (x = \sum_{A=0}^{n+1} x^A \frac{\partial}{\partial x^A}).$$

The light cone Λ_0^{n+1} is given by the equation $-(x^0)^2 + \sum_{a=1}^{n+1}(x^a)^2 = 0$, $x \neq 0$. It is known that Λ_0^{n+1} is a lightlike hypersurface of $\mathbf{R}_1^{n+2}$ and the radical distribution is spanned by a global vector field

$$\xi = \sum_{A=0}^{n+1} x^A \frac{\partial}{\partial x^A} \tag{2.2.2}$$

on Λ_0^{n+1}. The unique section N satisfying (2.1.7) is given by

$$N = \frac{1}{2(x^0)^2} \{ -x^0 \frac{\partial}{\partial x^0} + \sum_{a=1}^{n+1} x^a \frac{\partial}{\partial x^a} \} \tag{2.2.3}$$

and is also globally defined. As ξ is the position vector field we get

$$\bar{\nabla}_X \xi = \nabla_X X = X, \quad \forall X \in \Gamma(TM).$$

Then, $A_\xi^* X + \tau(X)\xi + X = 0$. As A_ξ^* is $\Gamma(S(TM))$-valued we obtain

$$A_\xi^* X = -PX, \quad \forall X \in \Gamma(TM). \tag{2.2.4}$$

Next, any $X \in \Gamma(S(T\Lambda_0^{n+1}))$ is expressed by $X = \sum_{a=1}^{n+1} X^a \frac{\partial}{\partial x^a}$ where $(X^1, \ldots, X^{n+1})$ satisfy

$$\sum_{a=1}^{n+1} x^a X^a = 0 \tag{2.2.5}$$

and then

$$\nabla_\xi X = \bar{\nabla}_\xi X = \sum_{A=0}^{n+1} \sum_{a=1}^{n+1} x^A \frac{\partial X^a}{\partial x^A} \frac{\partial}{\partial x^a},$$

$$\bar{g}(\nabla_\xi X, \xi) = \sum_{A=0}^{n+1} \sum_{a=1}^{n+1} x^a x^A \frac{\partial X^a}{\partial x^A} = -\sum_{a=1}^{n+1} x^a X^a = 0 \tag{2.2.6}$$

where (2.2.5) is differentiated with respect to each x^A. From (2.2.5) and (2.2.6) we obtain $\nabla_\xi X \in \Gamma(S(T\Lambda_0^{n+1}))$, that is, $A_N \xi = 0$. Compute $A_N X$ for $X \in \Gamma(S(T\Lambda_0^{n+1}))$. Let $X, Y \in \Gamma(S(T\Lambda_0^{n+1}))$. Using (2.2.3) and (2.2.5) we obtain

$$C(X,Y) = g(\nabla_X Y, N) = \bar{g}(\bar{\nabla}_X Y, N) = -\frac{1}{2(x^0)^2} g(X,Y),$$

that is,

$$g(A_N X, Y) = -\frac{1}{2(x^0)^2} g(X,Y) \quad X, Y \in \Gamma(S(T\Lambda_0^{n+1})).$$

Therefore, we have

$$A_N X = -\frac{1}{2(x^0)^2} PX \quad X \in \Gamma(T\Lambda_0^{n+1}). \tag{2.2.7}$$

Taking into account (2.2.4) and (2.2.7) we infer the relation

$$A_N X = \frac{1}{2(x^0)^2} A_\xi^* X, \quad X \in \Gamma(T\Lambda_0^{n+1}),$$

that is, Λ_0^{n+1} is a screen globally conformal lightlike hypersurface of $\mathbf{R}_1^{n+2}$ with positive conformal function $\varphi = \frac{1}{2(x^0)^2}$ globally defined on Λ_0^{n+1}.

Proposition 2.2.2. Let $(M, g, S(TM))$ be a lightlike hypersurface of $(\bar{M}, \bar{g})$. Suppose $S(TM)$ is integrable and any leaf of $S(TM)$ is totally umbilical immersed in $\bar{M}$ as a codimension 2 non-degenerate submanifold with nowhere vanishing spacelike mean curvature vector field. If the screen distribution is parallel along integral curves of the radical distribution, then M is screen locally conformal.

Proof. Let us denote by M' a leaf of $S(TM)$. We have

$$\bar{\nabla}_X Y = \nabla_X^* Y + C(X,Y)\xi + B(X,Y)N, \quad \forall X, Y \in \Gamma(TM'). \tag{2.2.8}$$

The mean curvature vector field of M', say $H^\star$, is a vector field of the rank 2 bundle $(TM^\perp \oplus \mathrm{tr}(TM))$, that is the normal bundle of M' in $\bar{M}$. Therefore, there exist two smooth functions α and ρ such that $H^\star = \alpha\xi + \rho N$. Since M' is totally umbilical immersed in $(\bar{M}, \bar{g})$, we have

$$C(X,Y)\xi + B(X,Y)N = g(X,Y)(\alpha\xi + \rho N) \quad \forall X, Y \in \Gamma(TM').$$

Therefore, it follows that

$$B(X,Y) = \rho g(X,Y) \quad \text{and} \quad C(X,Y) = \alpha g(X,Y) \quad \forall X, Y \in \Gamma(TM'). \tag{2.2.9}$$

Also, as $\langle H^\star, H^\star \rangle = 2\alpha\rho$ and $H^\star$ is nowhere vanishing spacelike, we get $\alpha\rho > 0$ on M. From (2.2.9), $C(X,Y) = \frac{\alpha}{\rho} B(X,Y) \quad \forall X, Y \in \Gamma(TM')$ with $\frac{\alpha}{\rho} > 0$ on

M'. This is equivalent to $A_N X = \frac{\alpha}{\rho} A_\xi^* X \ \forall X \in \Gamma(TM')$. Since $S(TM)$ is parallel along integral curves of $\mathrm{Rad}\, TM$, we have $A_N \xi = 0 = \frac{\alpha}{\rho} A_\xi^* \xi$. Thus,

$$A_N X = \frac{\alpha}{\rho} A_\xi^* X \quad \forall X \in \Gamma(TM).$$

Therefore, M is screen conformal with $\varphi = \frac{\alpha}{\rho}$.   □

Remark 2.2.3. If $\bar{M}$ is of constant sectional curvature c, then for $c \neq 0$ the non-vanishing condition on H^* is not necessary, for in this case, we always have $\alpha \neq 0$ at every point.

Theorem 2.2.4. *Let $(M, g, S(TM))$ be a lightlike hypersurface in a Lorentzian manifold $(\bar{M}, \bar{g})$. The following assertions are equivalent:*

(a) *$(M, g, S(TM))$ is screen locally conformal.*

(b) *There is a (maximal) domain $\mathcal{U}$ in M on which M and its screen distribution have commutative shape operators. Moreover, their corresponding principal curvatures are the same up to a nowhere vanishing smooth function φ on $\mathcal{U} \subset M$.*

Proof. If $(M, g, S(TM))$ is screen locally conformal, then on the conformality domain $\mathcal{U}$, there exists a nowhere vanishing smooth function φ such that $A_N X = \varphi A_\xi^* X$ for all X tangent to $\mathcal{U} \subset M$. Then, the commutativity of the shape operators A_N and A_ξ^* is immediate on $\mathcal{U}$. Hence, there is a local frame field with respect to which A_ξ^* and A_N are simultaneously diagonal. Therefore, consider on $\mathcal{U} \subset M$ an eigenframe field $(E_0, \ldots, E_n)$ both for A_N and A_ξ^*. If μ_i and λ_i denote the principal curvatures corresponding to E_i with respect to A_N and A_ξ^* respectively, then by (2.2.1), $\mu_i = \varphi\, \lambda_i$ and the last assertion in (b) follows.

Conversely, assume (b). The commutativity of the shape operators A_N and A_ξ^* implies the existence on $\mathcal{U}$ of a frame field $(E_0, \ldots, E_n)$ with respect to which A_N and A_ξ^* are simultaneously diagonal. Let μ_i and λ_i denote the principal curvatures corresponding to E_i with respect to A_N and A_ξ^*. The last assertion in (b) requires that there exists on $\mathcal{U}$, a nowhere vanishing smooth function φ such that $\mu_i = \varphi\, \lambda_i$, $0 \leq i \leq n$. Decompose a tangent vector $X \in T\mathcal{U}$ on the frame field $(E_0, \ldots, E_n)$ such that $X = X^i E_i$. Then,

$$A_N X = A_N(X^i E_i) = X^i A_N E_i = X^i \mu_i E_i = X^i \varphi \lambda_i E_i$$
$$= \varphi\left(X^i \lambda_i E_i\right) = \varphi X^i A_\xi^* E_i = \varphi A_\xi^*\left(X^i E_i\right) = \varphi A_\xi^* X.$$

Thus, (2.2.1) holds for a non-vanishing smooth function φ on $\mathcal{U} \subset M$.   □

Remark 2.2.5. On the eigenspace, say $\mathcal{N}_0$, of the zero eigenvalue of A_ξ^*, we note the following. Let $\mathcal{S}$ denote a sub-bundle of $S(TM)$ defined by

$$\mathcal{S} = \mathrm{Span}\{A_\xi^* Y, Y \in \Gamma(TM)\}.$$

Then,
$$\mathcal{N}_0 = \operatorname{Span}\{A_\xi^* Y, Y \in \Gamma(TM)\}^{\perp_M} = \mathcal{S}^{\perp_M}$$

where $\perp_M$ is the orthogonality symbol in M. In particular, if $\mathcal{S}$ coincides with $S(TM)$, then any eigenvector field in $\mathcal{N}_0$ is a multiple of ξ.

Now we discuss a related issue of integrability of a screen distribution which is equally important as a geometric property. Although the 1-dimensional $\operatorname{Rad} TM$ of a lightlike hypersurface M is obviously integrable, in general, any screen distribution is not necessarily integrable. Indeed, we have the following example:

Example 5. Let $(\mathbf{R}_2^4, \bar{g})$ be a 4-dimensional semi-Euclidean space of index 2 with signature $(-, -, +, +)$ of the canonical basis $(\partial_0, \ldots, \partial_3)$. Consider a hypersurface M of $\mathbf{R}_2^4$ given by
$$x_3 = x_0 + \sin(x_1 + x_2), \quad x_1 + x_2 \neq n\pi, \ n \in \mathbf{Z}.$$

It is easy to check that M is a lightlike hypersurface whose radical distribution $\operatorname{Rad} TM$ is spanned by
$$\xi = \partial_0 + \cos(x_1 + x_2)\partial_1 - \cos(x_1 + x_2)\partial_2 + \partial_3.$$

Let $V = \partial_0 + \cos(x_1 + x_2)\partial_1$, then, $g(V, V) = g(\xi, V) = -(1 + \cos^2(x_1 + x_2))$. Thus, as per relation (2.1.7), the lightlike transversal vector bundle is given by $\operatorname{ltr}(TM) = \operatorname{Span}\{N\}$, where
$$N = \frac{-1}{2(1 + \cos^2(x_1 + x_2))} \{\partial_0 + \cos(x_1 + x_2)\partial_1 + \cos(x_1 + x_2)\partial_2 - \partial_3\}.$$

The tangent bundle $T\mathbf{R}_2^4$ is spanned by
$$\left\{ \frac{\partial}{\partial u_0} = \partial_0 + \partial_3, \ \frac{\partial}{\partial u_1} = \partial_1 + \cos(x_1 + x_2)\partial_3, \ \frac{\partial}{\partial u_2} = \partial_2 + \cos(x_1 + x_2)\partial_3 \right\}.$$

It follows that the corresponding screen distribution $S(TM)$ is spanned by
$$\{ W_1 = \cos(x_1 + x_2)\partial_0 - \partial_1, \quad W_2 = \partial_2 + \cos(x_1 + x_2)\partial_3 \}.$$

In this case $[W_1, W_2] = \bar{\nabla}_{W_1} W_2 - \bar{\nabla}_{W_2} W_1 = \sin(x_1 + x_2)\{\partial_0 + \partial_3\}$. Thus,
$$\bar{g}([W_1, W_2], N) = \frac{\sin(x_1 + x_2)}{1 + \cos^2(x_1 + x_2)},$$

which implies that $S(TM)$ is not integrable.

The following general result gives equivalent conditions on other induced objects for the integrability of a screen distribution.

Theorem 2.2.6. [51] *Let $(M, g, S(TM))$ be a lightlike hypersurface of a semi-Riemannian manifold $(\bar{M}, \bar{g})$. The following assertions are equivalent:*

(i) $S(TM)$ *is an integrable distribution.*

(ii) $h^*(X, Y) = h^*(Y, X),\ \forall X, Y \in \Gamma(S(TM)).$

(iii) *The shape operator of M is symmetric with respect to g, i.e.,*

$$g(A_V X, Y) = g(X, A_V Y),\ \forall X, Y \in \Gamma(S(TM)), V \in \Gamma(\mathrm{tr}(TM)).$$

Proof. (2.1.17) implies that a vector field X on M belongs to $S(TM)$, if and only if, on each $\mathcal{U} \subset M$ we have $\eta(X) = 0$. By using (2.1.23) and (2.1.17) we get $C(X, Y) - C(Y, X) = \eta([X, Y]),\ \forall X, Y \in \Gamma(TM_{|\mathcal{U}})$, which together with (2.1.21) implies the equivalence of (i) and (ii). The equivalence of (ii) and (iii) follows from (2.1.26), which completes the proof. □

Definition 2.2.7. If any geodesic of a lightlike hypersurface M with respect to an induced connection ∇ is a geodesic of $\bar{M}$ with respect to $\bar{\nabla}$, we say that M is a totally geodesic lightlike hypersurface of $\bar{M}$.

Observe that, as in the non-degenerate case, we say that the vanishing of h on M is equivalent to M *totally geodesic* in $\bar{M}$.

Note. Just as locally product Riemannian or semi-Riemannian manifolds (see [317]), if $S(TM)$ is integrable, then, M is locally a product manifold $C \times M'$ where C is a null curve and M' is a leaf of $S(TM)$. The following holds from the above theorem.

Proposition 2.2.8. [51] *Let $(M, g, S(TM))$ be a lightlike hypersurface of $(\bar{M}, \bar{g})$. Then the following assertions are equivalent:*

(i) $S(TM)$ *is parallel with respect to the induced connection ∇.*

(ii) h^* *vanishes identically on M.*

(iii) A_N *vanishes identically on M.*

For the screen locally conformal lightlike hypersurfaces we have the following.

Theorem 2.2.9. *Let $(M, g, S(TM))$ be a screen locally conformal lightlike hypersurface of a semi-Riemannian manifold $(\bar{M}, \bar{g})$. Then the screen distribution is integrable. Moreover, M is totally geodesic or totally umbilical in $\bar{M}$ if and only if any leaf M' of $S(TM)$ is a codimension 2 non-degenerate submanifold in $\bar{M}$.*

Proof. We know from Theorem 2.2.6 that the screen distribution is integrable if and only if the shape operator of M is symmetric with respect to the induced metric tensor g. The integrability assertion follows relation $A_N = \varphi A_\xi^*$ and the symmetry of A_ξ^* with respect to g. For the last assertion we let X, Y be tangent vector fields of the leaf M' of a screen distribution and h' be its second fundamental form in $\bar{M}$ as a codimension 2 non-degenerate submanifold. Denote by ∇^* the Levi-Civita connection of M'. We have

$$\bar{\nabla}_X Y = \nabla_X^\star Y + C(X, Y)\xi + B(X, Y)N$$

which leads to

$$\bar{\nabla}_X Y = \nabla_X^* Y + g(A_\xi^* X, Y)(\varphi \xi + N) \forall X, Y \in \Gamma(TM'|_{\mathcal{U}})$$

that is

$$h'(X,Y) = B(X,Y)(\varphi \xi + N) = \sqrt{2|\varphi|} B(X,Y) \left(\frac{\varphi}{\sqrt{2|\varphi|}} \xi + \frac{1}{\sqrt{2|\varphi|}} N \right)$$

$\forall X, Y \in \Gamma(TM'|_{\mathcal{U}})$. where $\dfrac{\varphi}{\sqrt{2|\varphi|}} \xi + \dfrac{1}{\sqrt{2|\varphi|}} N$ is a unit normal vector field on M'.

As $\sqrt{2|\varphi|}$ is nowhere zero and $B(X, \xi) = 0$ for X in $\Gamma(TM')$ the last assertion in the above stated theorem follows. $\qquad \square$

2.3 Unique existence of screen distributions

Although the use of a non-degenerate screen distribution $S(TM)$ has been helpful in defining induced objects on the lightlike spaces, because of the degenerate metric, $S(TM)$ is not unique. Therefore, several induced geometric objects depend on the choice of a screen, which creates a problem. For this reason, it is desirable to look for a unique or canonical screen distribution so that the induced objects on M are well-defined. To clarify this point, we first present a brief review of the dependence (or otherwise) on the choice of a screen distribution, followed by up-to-date information on the existence of large classes of lightlike hypersurfaces which can admit a unique screen subject to some reasonable condition(s).

Using (2.1.4), (2.1.6) and Section 1 of Chapter 1, we say that there exists a quasi-orthonormal basis of $\bar{M}$ along M, given by

$$F = \{\xi, N, W_a\}, \quad a \in \{1, \ldots, m\}, \tag{2.3.1}$$

where $\{\xi\}$, $\{N\}$ and $\{W_a\}$ are the null basis of $\Gamma(\operatorname{Rad} TM_{|\mathcal{U}})$, $\Gamma(\operatorname{tr}(TM)_{|\mathcal{U}})$ and the orthonormal basis of $\Gamma(S(TM)_{|\mathcal{U}})$, respectively. Consider two quasi-orthonormal frames fields $F = \{\xi, N, W_a\}$ and $F' = \{\xi, N', W_a'\}$ induced on $\mathcal{U} \subset M$ by $\{S(TM), \operatorname{tr}(TM)\}$ and $\{S'(TM), (\operatorname{tr})'(TM)\}$, respectively for the same ξ. Using (2.1.6) and (2.1.8) we obtain

$$W_a' = \sum_{b=1}^{m} W_a^b (W_b - \epsilon_b \mathbf{f}_b \xi),$$

$$N' = N + \mathbf{f} \xi + \sum_{a=1}^{m} \mathbf{f}_a W_a,$$

where $\{\epsilon_a\}$ are signatures of the orthonormal basis $\{W_a\}$ and W_a^b, $\mathbf{f}$ and $\mathbf{f}_a$ are smooth functions on $\mathcal{U}$ such that $\left[W_a^b \right]$ are $m \times m$ semi-orthogonal matrices.

Computing $\bar{g}(N', N') = 0$ by using (2.1.6) and $\bar{g}(W_a, W_a) = 1$ we get

$$2\mathbf{f} + \sum_{a=1}^{m} \epsilon_a \, (\mathbf{f}_a)^2 = 0.$$

Using this in the second relation of the above two equations, we get

$$W_a' = \sum_{b=1}^{m} W_a^b \, (W_b - \epsilon_b \, \mathbf{f}_b \, \xi), \tag{2.3.2}$$

$$N' = N - \frac{1}{2} \left\{ \sum_{a=1}^{m} \epsilon_a \, (\mathbf{f}_a)^2 \right\} \xi + \sum_{a=1}^{m} \mathbf{f}_a W_a. \tag{2.3.3}$$

The above two relations are used to investigate the transformation of the induced objects when one changes the pair $\{S(TM), \mathrm{tr}(TM)\}$ with respect to a change in the basis. Following are main results on the dependence (or otherwise) of the second fundamental form of M, the function τ on a chosen screen $S(TM)$ and a chosen section $\xi \in \Gamma(\mathrm{Rad}(TM)_{|\mathcal{U}})$ respectively.

Proposition 2.3.1. [149] *Let $(M, g, S(TM))$ be a lightlike hypersurface of a semi-Riemannian manifold $(\bar{M}, \bar{g})$. Then*

(a) *the second fundamental form B of M on $\mathcal{U}$ is independent of $S(TM)$.*

(b) *B and the 1-form τ (in the Weingarten equation) depend on the choice of a section $\xi \in \Gamma(\mathrm{Rad}(TM)_{|\mathcal{U}})$.*

(c) *$d\tau$ is independent of the section ξ.*

Proof. Let $S(TM)$ and $S(TM)'$ be two screens on M with h and h' the second fundamental forms corresponding to $\mathrm{tr}(TM)$ and $\mathrm{tr}(TM)'$, respectively. Using (2.1.9) and (2.1.11) for both screens we have

$$B(X, Y) = \bar{g}(\bar{\nabla}_X Y, \xi) = B'(X, Y), \quad \forall X, Y \in \Gamma(TM_{|\mathcal{U}}). \tag{2.3.4}$$

Thus, $B = B'$ on $\mathcal{U}$, which proves (a). Take $\bar{\xi} = \alpha\xi$, for some function α. Then, it follows that $\bar{N} = (1/\alpha)N$. From (2.1.14) and (2.1.15) we obtain

$$\bar{B} = \alpha B, \quad \tau(X) = \bar{\tau}(X) + X(\log \alpha), \tag{2.3.5}$$

for any $X \in \Gamma(TM_{|\mathcal{U}})$, which proves that B and τ depend on the section ξ on U, proving (b). Finally, taking the exterior derivative d on both sides of the second term of (2.3.5) we get $d\tau = d\bar{\tau}$ on $\mathcal{U}$, which completes the proof. □

To study the dependence of the induced objects $\{\nabla, \tau, A_N, A_\xi^*\}$ on the screen distribution $S(TM)$, let $\{\nabla', \eta', A_{N'}', A_\xi^{*\prime}\}$ be another set of induced objects with respect to another screen distribution $S(TM)'$ and its transversal bundle $\mathrm{tr}(TM)'$. Consider two quasi-orthonormal frames $F = \{\xi, N, W_a\}$

and $F' = \{\xi, N', W'_a\}$ induced on the coordinate neighborhood $\mathcal{U} \subset M$ by $\{S(TM), \mathrm{tr}(TM)\}$ and $\{S'(TM), (tr)'(TM)\}$ respectively. Using the transformation equations (2.3.4) and (2.3.5) we obtain relationships between the geometrical objects induced by the Gauss-Weingarten equations with respect to $S(TM)$ and $S(TM)'$ as follows:

$$\nabla'_X Y = \nabla_X Y + B(X, Y) \left\{ \frac{1}{2} \left(\sum_{a=1}^{m} \epsilon_a (\mathbf{f}_a)^2 \right) \xi - \sum_{a=1}^{m} \mathbf{f}_a W_a \right\}, \qquad (2.3.6)$$

$$\tau'(X) = \tau(X) + B(X, N' - N), \qquad (2.3.7)$$

$$A'_{N'} X = A_N X + \sum_{a=1}^{m} \left\{ \epsilon_a \mathbf{f}_a X(\mathbf{f}_a) - \tau(X) \epsilon_a (\mathbf{f}_a)^2 \right.$$
$$\left. - \frac{1}{2} \epsilon_a (\mathbf{f}_a)^2 B(X, N - N') - \mathbf{f}_a C(X, W_a) \right\} \xi$$
$$+ \sum_{a=1}^{m} \left\{ \mathbf{f}_a \left(\tau(X) + B(X, N' - N) \right) - X(\mathbf{f}_a) \right\} W_a$$
$$- \sum_{a=1}^{m} \mathbf{f}_a \nabla_X^* W_a - \frac{1}{2} \sum_{a=1}^{m} \epsilon_a (\mathbf{f}_a)^2 A_\xi^* X, \qquad (2.3.8)$$

$$A^{*\prime}_\xi X = A_\xi^* X + B(X, N - N') \xi, \quad \forall X, Y \in \Gamma(TM_{|\mathcal{U}}). \qquad (2.3.9)$$

The following result is immediate from (2.3.6), (2.3.7) and (2.3.9).

Proposition 2.3.2. [149] *Let $(M, g, S(TM))$ be a lightlike hypersurface of a semi-Riemannian manifold. Then, the induced connection ∇ on M, the 1-form τ in (2.1.12) and the shape operator A_ξ^* in (2.1.24) all three are independent of $S(TM)$, if and only if, the second fundamental form h of M vanishes identically on M.*

Note. It follows from (2.3.8) that the shape operator A_N depends on the choice of $S(TM)$ even if h vanishes identically on M.

Now, we ask the following question: Is the local screen second fundamental form C independent of its choice? The answer is negative. Indeed, we prove the following result with respect to a change in screen.

Proposition 2.3.3. [19] *The screen second fundamental forms C and C' of the screens $S(TM)$ and $S(TM)'$, respectively, are related as follows:*

$$C'(X, PY) = C(X, PY) - \frac{1}{2} \|W\|^2 B(X, Y) + g(\nabla_X PY, W), \qquad (2.3.10)$$

where $W = \sum_{a=1}^{m} \mathbf{f}_a W_a$ is called the characteristic vector field.

Proof. Using (2.3.3) and (2.3.6) we get

$$C'(X,\, PY) = \bar{g}(\nabla'_X PY,\, N')$$

$$= \bar{g}\left(\nabla_X PY + B(X,\, Y)\{\frac{1}{2}(\sum_{a=1}^{m}\varepsilon_a(\mathbf{f}_a)^2)\xi - \sum_{a=1}^{m}\mathbf{f}_a W_a\},\, N'\right)$$

$$= \bar{g}(\nabla_X PY,\, N) + \bar{g}(\nabla_X PY,\, \sum_{a=1}^{m}\mathbf{f}_a W_a)$$

$$+\, B(X,\, Y)\left\{\frac{1}{2}(\sum_{a=1}^{m}\varepsilon_a(\mathbf{f}_a)^2) - \sum_{b=1}^{m}\sum_{a=1}^{m}g(\mathbf{f}_a W_a,\, \mathbf{f}_b W_b)\right\}$$

$$= C(X,\, PY) + g(\nabla_X PY,\, W) - \frac{1}{2}||W||^2 B(X,\, Y)$$

which is the desired formula.   □

The above problem of interdependence has been a matter of concern in the lightlike geometry. As an alternative approach, one finds a large variety of papers on lightlike spaces where the authors use a specific method suitable for their problem. For example, Akivis-Goldberg [2, 3, 4] ; Bonnor [77]; Leistner [285]; Bolós [74] are some samples of many more in the literature. Also, Kupeli [273] has shown that $S(TM)$ is canonically isometric to the factor vector bundle $TM^* = TM/TM^\perp$ and used canonical projection $\pi : TM \to TM^*$ in studying the geometry of degenerate semi-Riemannian manifolds. Obviously, one is tempted to use Kupeli's method of replacing screen distribution by the factor bundle $TM/\operatorname{Rad} TM$, but, it leads to the intrinsic geometry whereas this book deals with the extrinsic geometry which is in line with the classical theory of submanifolds [45, 97]. Consequently, although specific techniques are suitable for good applicable results, nevertheless, for the fundamental deeper geometric study of lightlike spaces, one must look for a canonical or a unique screen distribution. Fortunately, since 1996, a considerable amount of new work has been done on a canonical or unique screen for a variety of lightlike hypersurfaces. The objective of this section is to present up-to-date information on results leading to the existence of a unique screen.

As the shape operator is an information tool in studying geometry of submanifolds, we use the technique of Atindogbe-Duggal[19] as presented in the previous section and consider a class of screen conformal lightlike hypersurfaces satisfying the relation (2.2.1). It follows from (2.1.25), (2.1.26) and (2.2.1) that the two second fundamental forms B and C of a screen conformal lightlike hypersurface M and its $S(TM)$, respectively, are related by

$$C(X, PY) = \varphi B(X, Y), \quad \forall X, Y \in \Gamma(TM_{|\mathcal{U}}). \tag{2.3.11}$$

Denote by $\mathcal{S}^1$ the first derivative of $S(TM)$ given by

$$\mathcal{S}^1(x) = \operatorname{Span}\{[X, Y]|_x, \quad X_x,\ Y_x \in S(T_x M)\}, \forall x \in M. \tag{2.3.12}$$

Let $S(TM)$ and $S(TM)'$ be two screen distributions on M, h and h' their second fundamental forms with respect to $\operatorname{tr}(TM)$ and $\operatorname{tr}(TM)'$ respectively, for the same $\xi \in \Gamma(TM^\perp|_\mathcal{U})$. Denote by ω the dual one-form of the vector field $W = \sum_{a=1}^{m} \mathbf{f}_a W_a$ (see equation (2.3.10)) with respect to the metric tensor g, that is

$$\omega(X) = g(X, W), \quad \forall X \in \Gamma(TM). \tag{2.3.13}$$

Following is the main result on the existence of a unique screen distribution:

Theorem 2.3.4. [142] *Let* $(M, g, S(TM))$ *be a screen conformal lightlike hypersurface of a semi-Riemannian manifold* $(\bar{M}, \bar{g})$, *with* $\mathcal{S}^1$ *the first derivative of* $S(TM)$ *given by (2.3.12). Then,*

(1) *a choice of the screen distribution* $S(TM)$ *of* M *satisfying (2.2.1) is integrable.*

(2) *The one-form* ω *in (2.3.13) vanishes identically on* $\mathcal{S}^1$.

(3) *If* $\mathcal{S}^1$ *coincides with* $S(TM)$, *then,* M *can admit a unique screen distribution up to an orthogonal transformation and a unique lightlike transversal vector bundle. Moreover, for this class of hypersurfaces, the screen second fundamental form* C *is independent of its choice.*

Proof. It follows from (2.1.25) that, for any lightlike hypersurface, the shape operator A_ξ^* of $S(TM)$ is symmetric with respect to g, i.e.,

$$g(A_\xi^* X, Y) = g(A_\xi^* Y, X), \quad \forall\, X, Y \in \Gamma(S(TM)), \quad \xi \in \Gamma(\operatorname{Rad} TM).$$

This result and the relation (2.3.10), along with Theorem 2.2.6, imply that any choice of screen distribution of a screen conformal lightlike hypersurface M is integrable, which proves (1).

As $S(TM)$ is integrable, $\mathcal{S}^1$ is its sub-bundle. From the relation (2.3.10), since its right-hand side is symmetric in X and Y, we obtain

$$g(\nabla_X PY - \nabla_Y PX, W) = 0, \quad \forall X, Y \in \Gamma(TM).$$

Thus, $g(\nabla_X Y - \nabla_Y X, W) = 0$, $\quad \forall X, Y \in \Gamma(S(TM))$, that is, $\omega([X, Y]) = g([X, Y], W) = 0$, $\quad \forall X, Y \in \Gamma(S(TM))$, which proves (2).

Assume $\mathcal{S}^1 = S(TM)$, that is, ω vanishes on $S(TM)$, which implies from (2.3.13) that $W = 0$. This means that the functions $\mathbf{f}_a$ vanish. Thus, the transformation equations (2.3.3) and (2.3.4) become $W_a' = \sum_{b=1}^{m} W_a^b W_b$, $(1 \leq a \leq m)$ and $N' = N$, where (W_a^b) is an orthogonal matrix of $S(T_x M)$ at any point x of M, which proves the first part of (3). The independence of C follows by putting $W = 0$ in (2.3.10) which completes the proof. $\qquad\square$

Based on the above theorem, one may ask the following converse question. Does the existence of a canonical or a unique distribution $S(TM)$ of a lightlike hypersurface imply that $S(TM)$ is integrable? Unfortunately, the answer, in general, is negative, which we support by an example. First we recall the following known results from pages 114-117 of [149]:

(a) There exists a canonical screen distribution for any lightlike hypersurface of a semi-Euclidean space $\mathbf{R}_q^{m+2}$. In particular, the canonical screen distribution on a lightlike cone of $\mathbf{R}_q^{m+2}$ is integrable.

(b) The canonical screen distribution on any lightlike hypersurface of $\mathbf{R}_1^{m+2}$ is integrable.

Observe that in Example 1 we have shown that there exists a non-integrable screen distribution of a lightlike hypersurface in $\mathbf{R}_2^4$. Connecting this with the above quoted results we conclude that, in this case, there do exist a canonical or unique screen distribution which is not integrable. Consequently, in general, the answer to the above question is negative.

Now we show that, although every screen conformal lightlike hypersurfaces admits an integrable screen distribution, not every such integrable screen coincides with a corresponding canonical screen, that is, there are cases for which $\mathcal{S}^1 \neq S(TM)$. Following is such a class:

Example 6. Consider a smooth function $F : \Omega \rightarrow \mathbf{R}$, where Ω is an open set of $\mathbf{R}^{m+1}$. Then

$$M = \left\{ (x^0, \ldots, x^{m+1}) \in \mathbf{R}_q^{m+2} : x^0 = F(x^1, \ldots, x^{m+1}) \right\}$$

is a *Monge hypersurface* [149]. The natural parametrization on M is

$$x^0 = F(v^0, \ldots, v^m) \quad ; \quad x^{\alpha+1} = v^\alpha, \qquad \alpha \in \{0, \ldots, m\}.$$

Hence, the natural frames field on M is globally defined by

$$\partial_{v^\alpha} = F'_{x^{\alpha+1}} \partial_{x^0} + \partial_{x^{\alpha+1}}, \qquad \alpha \in \{0, \ldots, m\}.$$

Then

$$\xi = \partial_{x^0} - \sum_{s=1}^{q-1} F'_{x^s} \partial_{x^s} + \sum_{a=q}^{m+1} F'_{x^a} \partial_{x^a}$$

spans $TM^\perp$. Therefore, M is lightlike (i.e., $TM^\perp = \operatorname{Rad} TM$), if and only if, the global vector field ξ spanned by $\operatorname{Rad} TM$ which means, if and only if, F is a solution of the partial differential equation

$$1 + \sum_{s=1}^{q-1} (F'_{x^s})^2 = \sum_{a=q}^{m+1} (F'_{x^a})^2.$$

Along M consider the constant timelike section $V^* = \partial_{x^0}$ of $\Gamma(T\mathbf{R}_q^{m+2})$. Then $\bar{g}(V^*, \xi) = -1$ implies that V^* is not tangent to M. Therefore, the vector bundle $H^* = \operatorname{Span}\{V^*, \xi\}$ is non-degenerate on M. The complementary orthogonal vector bundle $S^*(TM)$ to H^* in $T\mathbf{R}_q^{m+2}$ is a non-degenerate distribution on M and is complementary to $\operatorname{Rad} TM$. Thus $S^*(TM)$ is a screen distribution on M. The transversal bundle $tr^*(TM)$ is spanned by $N = -V^* + \frac{1}{2}\xi$ and $\tau(X) = 0$ for any $X \in \Gamma(TM)$. Indeed, $\tau(X) = \bar{g}(\bar{\nabla}_X N, \xi) = \frac{1}{2}\bar{g}(\bar{\nabla}_X \xi, \xi) = 0$. Therefore, the Weingarten equations reduce to $\bar{\nabla}_X N = -A_N X$ and $\nabla_X \xi = -A_\xi^* X$, which implies

$$A_N X = \frac{1}{2} A_\xi^* X, \quad \forall X \in \Gamma(TM).$$

Hence, any lightlike Monge hypersurface of $\mathbf{R}_q^{m+2}$ is screen globally conformal with constant positive conformal function $\varphi(x) = \frac{1}{2}$.

We call $S^*(TM)$ the *natural screen distribution*. Furthermore, it is easy to see from the above construction that in the case $q = 1$ only, the natural and the canonical screen distributions coincide on lightlike Monge hypersurfaces. However, in general, they are different as explained by the following example:

Example 7 [149]. Let $M : x^3 = x^0 + \frac{1}{2}(x^1 + x^2)^2$ be a lightlike hypersurface in $\mathbf{R}_2^4$. It is easy to see that

$$TM^\perp = \mathrm{Span}\left\{ \xi = \frac{\partial}{\partial x^0} + (x^1 + x^2)\frac{\partial}{\partial x^1} - (x^1 + x^2)\frac{\partial}{\partial x^2} + \frac{\partial}{\partial x^3} \right\} = \mathrm{Rad}\,TM.$$

Take as lightlike transversal vector bundle $tr(TM)$ spanned by a vector field

$$N = \frac{1}{2(1 + (x^1 + x^2)^2)}\left(\frac{\partial}{\partial x^0} + (x^1 + x^2)\frac{\partial}{\partial x^1} + (x^1 + x^2)\frac{\partial}{\partial x^2} - \frac{\partial}{\partial x^3} \right).$$

It follows that the canonical screen distribution $S(TM)$ is spanned by

$$\left\{ W_1 = \frac{\partial}{\partial x^1} - (x^1 + x^2)\frac{\partial}{\partial x^0} \;;\; W_2 = \frac{\partial}{\partial x^2} + (x^1 + x^2)\frac{\partial}{\partial x^3} \right\}.$$

On the other hand, if we follow the above construction for M as a Monge hypersurface of $\mathbf{R}_2^4$, then, the natural screen distribution is spanned by

$$\left\{ W_1^* = \frac{\partial}{\partial x^1} + (x^1 + x^2)\frac{\partial}{\partial x^3} \;;\; W_2^* = \frac{\partial}{\partial x^2} + (x^1 + x^2)\frac{\partial}{\partial x^3} \right\},$$

which does not coincide with the above canonical screen distribution.

Now, one may ask whether there exists a general class of semi-Riemannian manifolds which admit screen conformal lightlike hypersurfaces and, therefore, can admit an integrable unique screen distribution. To answer this question in the affirmative, we prove the following result:

Theorem 2.3.5. [143] *Let $(M, g, S(TM))$ be a lightlike hypersurface of a semi-Riemannian manifold $(\bar{M}_q^{m+2}, \bar{g})$, with E a complementary vector bundle of $TM^\perp$ in $S(TM)^\perp$ such that E admits a covariant constant timelike vector field. Then, with respect to a section ξ of $\mathrm{Rad}\,TM$, M is screen conformal. Thus, M can admit an integrable unique screen distribution.*

Proof. By hypothesis, consider (without any loss of generality), along M, a unit timelike covariant constant vector field $V \in \Gamma(E_{|\mathcal{U}})$. To satisfy the condition given in (2.1.7), we choose a section ξ of $\operatorname{Rad} TM$ such that $\bar{g}(V, \xi) \neq 0$. For convenience in calculations, we set $\bar{g}(V, \xi) = \theta^{-1}$. It follows that the vector bundle $\mathcal{B} = \operatorname{Span}\{V, \xi\}$ is non-degenerate on M. Take the complementary orthogonal vector bundle $S(TM)$ to $\mathcal{B}$ in $T\bar{M}$, which is a non-degenerate distribution on M complementary to $\operatorname{Rad} TM$. This means that $S(TM)$ is a screen distribution on M such that $\mathcal{B} = S(TM)^{\perp}$, which we choose as a screen distribution on M. Using this and (2.1.7), the null transversal vector bundle of M takes the form

$$N = \theta \left(V + \frac{\theta}{2} \xi \right). \tag{2.3.14}$$

Using (2.3.14) in (2.1.14) and (2.1.24) we get

$$\tau(X) = \bar{g}(\bar{\nabla}_X N, \xi) = X(\theta)\, \bar{g}(V, \xi) + \frac{1}{2} (\theta)^2 \, \bar{g}(\bar{\nabla}_X \xi, \xi)$$
$$= X(\theta)\,(\theta)^{-1} = X(\ln \theta). \tag{2.3.15}$$

Using this value of τ, (2.3.14) and (2.1.12) we get

$$\bar{\nabla}_X N = X(\theta)V + \theta X(\theta)\xi + \frac{1}{2} \theta^2 \, \bar{\nabla}_X \xi$$
$$= X(\theta)V + \frac{1}{2}\theta X(\theta)\xi - \frac{1}{2} \theta^2 \, A^*_\xi X. \tag{2.3.16}$$

On the other hand, substituting the value of τ in (2.1.15), we get

$$\bar{\nabla}_X N = -A_N X + X(\theta)V + \frac{1}{2}\theta\, X(\theta)\, \xi. \tag{2.3.17}$$

Equating (2.3.16) and (2.3.17), we obtain $A_N = (\theta^2/2)\, A^*_\xi$. Thus, by definition, M is a screen conformal lightlike hypersurface of $\bar{M}$ with conformal function $\varphi = \theta^2/2$. Finally, the existence of an integrable and possibly a unique screen distribution $S(TM)$ (if there is an $\mathcal{S}^1 = S(TM)$) follows from Theorem 2.3.4, which completes the proof. □

Remark 2.3.6. (a) In [4], Akivis-Goldberg pointed out that a canonical construction of a lightlike hypersurface M in a semi-Euclidean space $\mathbf{R}^{m+2}_q$ was neither invariant nor intrinsically connected with the geometry of M. Therefore, in the same paper [4], they constructed invariant normalizations intrinsically connected with the geometry of M and investigated induced linear connections by these normalizations, using relative and absolute invariants defined by the first and the second fundamental forms of M. They further showed that there are three invariant normalizations and linear connections intrinsically connected with the geometry of lightlike hypersurface M in 4-dimensional semi-Euclidean spaces.

(b) As in the case of semi-Euclidean spaces, following the procedure in Akivis-Goldberg's paper [4], there is a need to construct invariant normalizations intrinsically connected with the geometry of a lightlike hypersurface M of a semi-Riemannian manifold.

(c) In another direction, recently in [150] an attempt was made to study a set of distinguished structures, denoted by $(S(TM), \xi)$, on a Lorentzian manifold M which are useful for a variety of geometric and or physical problems, where ξ is a global null section of $\operatorname{Rad} TM$. It has been shown that there is a large class of lightlike hypersurfaces of Lorentzian manifolds with distinguished structures, including a unique structure (see their Theorem 5), having good properties. In the same paper, they have given a lightlike version of Raychaudhuri's equation and found integrability conditions for some well-chosen distinguished structures. Since the result of Theorem 2.3.5 of this book and the paper [134] was not available at the time of writing the paper [150], its results can be further strengthened.

2.4 Induced scalar curvature

In this section we deal with a new concept of the *"Induced scalar curvature"* of a lightlike hypersurface $(M, g, S(TM))$ of a semi-Riemannian manifold $(\bar{M}, \bar{g})$. Denote by $\bar{R}$, R and R^* the curvature tensors of the Levi-Civita connection $\bar{\nabla}$ on $\bar{M}$ and the induced connections ∇ and ∇^* on M and $S(TM)$, respectively. Since the second fundamental form $h(X, Y)$ belongs to $\operatorname{tr}(TM)$ we set

$$(\nabla_X h)(Y, Z) = \nabla^t_X(h(Y, Z)) - h(\nabla_X Y, Z) - h(Y, \nabla_X Z).$$

Then, using (1.2.7), (2.1.9) and (2.1.10) we obtain

$$\bar{R}(X, Y)Z = R(X, Y)Z + A_{h(X, Z)}Y - A_{h(Y, Z)}X \\ + (\nabla_X h)(Y, Z) - (\nabla_Y h)(X, Z),$$

$\forall X, Y, Z \in \Gamma(TM)$. Consider curvature tensor $\bar{R}$ of type $(0, 4)$. From the above equation and the Gauss-Weingarten equations for M and $S(TM)$ we obtain:

$$\bar{g}(\bar{R}(X, Y)Z, PW) = g(R(X, Y)Z, PW) + \bar{g}(h(X, Z), h^*(Y, PW)) \\ - \bar{g}(h(Y, Z), h^*(X, PW)), \tag{2.4.1}$$

$$\bar{g}(\bar{R}(X, Y)Z, U) = \bar{g}((\nabla_X h)(Y, Z) - (\nabla_Y h)(X, Z), U), \tag{2.4.2}$$

$$\bar{g}(\bar{R}(X, Y)Z, V) = \bar{g}(R(X, Y)Z, V), \tag{2.4.3}$$

for any $X, Y, Z, W \in \Gamma(TM)$, $U \in \Gamma(\operatorname{Rad} TM)$, $V \in \Gamma(\operatorname{tr}(TM))$. We call (2.4.1)–(2.4.3) the global *Gauss-Codazzi type* equations for the lightlike hypersurface M. Now using the local Gauss-Weingarten equations with respect to a null pair $\{\xi, N\}$ on $\mathcal{U} \subset M$ we obtain

$$\bar{g}(\bar{R}(X,Y)Z, PW) = g(R(X,Y)Z, PW) + B(X,Z)C(Y, PW)$$
$$- B(Y,Z)C(X, PW), \tag{2.4.4}$$
$$\bar{g}(\bar{R}(X,Y)Z, \xi) = (\nabla_X B)(Y,Z) - (\nabla_Y B)(X,Z)$$
$$+ B(Y,Z)\tau(X) - B(X,Z)\tau(Y), \tag{2.4.5}$$
$$\bar{g}(\bar{R}(X,Y)Z, N) = \bar{g}(R(X,Y)Z, N), \tag{2.4.6}$$
$$g(R(X,Y)PZ, PW) = g(R^*(X,Y)PZ, PW) + C(X, PZ)B(Y, PW)$$
$$- C(Y, PZ)B(X, PW), \tag{2.4.7}$$

for any $X, Y, Z, W \in \Gamma(TM_{|\mathcal{U}})$, where

$$(\nabla_X B)(Y, Z) = X(B(Y,Z)) - B(\nabla_X Y, Z) - B(Y, \nabla_X Z).$$

Using (1.2.8) and (1.2.9) in the right-hand side of (2.4.6) we obtain

$$\bar{g}(\bar{R}(X,Y)PZ, N) = (\nabla_X C)(Y, PZ) - (\nabla_Y C)(X, PZ)$$
$$+ \tau(Y)C(X, PZ) - \tau(X)C(Y, PZ), \tag{2.4.8}$$
$$\bar{g}(\bar{R}(X,Y)\xi, N) = C(Y, A_\xi^* X) - C(X, A_\xi^* Y) - 2d\tau(X,Y), \tag{2.4.9}$$

for any $X, Y, Z \in \Gamma(TM_{|\mathcal{U}})$, where we set

$$(\nabla_X C)(Y, PZ) = X(C(Y, PZ)) - C(\nabla_X Y, PZ) - C(Y, \nabla_X^* PZ)$$

and in (2.4.9) we use the following well-known exterior derivative formula:

$$d\tau(X, Y) = \frac{1}{2}\{X(\tau(Y)) - Y(\tau(X)) - \tau([X, Y])\}.$$

Note. Observe that the Gauss-Codazzi equations do depend on the choice of a screen. Therefore, the new information (as presented in the previous section) on the availability of a canonical or unique screen is valuable for curvature properties in lightlike geometry.

For the benefit of readers we reproduce from [149] the local expressions of Gauss-Codazzi equations in a coordinate system on M. Using a frames field $\{\partial_{u^0}, \delta_u^a, N\}$ on $\bar{M}$ and $\{X_i\} = \{\partial_{u^0}, \delta_u^a\}$ frames field on M for $i, j, k, h \in \{0, \dots, m\}$, we have the following local components of curvature tensors $\bar{R}$ and R:

$$\bar{R}_{ijkh} = \bar{g}(\bar{R}(X_h, X_k)X_j, X_i), \quad R_{ijkh} = g(R(X_h, X_k)X_j, X_i),$$
$$\bar{R}_{ijk}^o = \bar{g}(\bar{R}(X_k, X_j)X_i, N), \quad R_{ijk}^o = g(R(X_k, X_j)X_i, N).$$

Thus, in terms of local coordinates, the local Gauss-Codazzi equations are:

$$\bar{R}_{abcd} = R_{abcd} + C_{ac}B_{bd} - C_{ad}B_{bc},$$
$$\bar{R}_{abod} = R_{abod} + C_a B_{bd},$$
$$\bar{R}_{aocd} = R_{aocd}; \quad \bar{R}_{aood} = R_{aood},$$
$$\bar{R}_{obcd} = B_{bc;d} - B_{bd;c} + B_{bc}\tau_h - B_{bd}\tau_c = -R_{bocd},$$
$$\bar{R}_{ojoh} = -\partial_{u^0}(B_{bd}) + \Gamma_{bo}^c B_{cd} - \tau_o B_{bd} = -R_{bood},$$

where we put

$$B_{bc\,;\,d} = \delta_{u^d}(B_{bc}) - B_{ac}\Gamma^a_{bd} - B_{ba}\Gamma^a_{cd}.$$

$$\bar{R}^o_{abc} = R^o_{abc} = C_{ab;\,c} - C_{ac;\,b} + \tau_b C_{ac} - \tau_c C_{ab}$$
$$\bar{R}^o_{abo} = R^o_{abo} = C_{ab;\,o} - C_{a;\,b} + \tau_b C_a - \tau_o C_{ab},$$
$$\bar{R}^o_{obc} = R^o_{obc} = C_{db}(A^*)^d_c - C_{dc}(A^*)^d_b + 2d\tau\,(\delta_{u^b},\,\delta_{u^c}),$$
$$\bar{R}^o_{ooc} = R^o_{ooc} = C_d(A^*)^d_c + 2d\tau\,(\partial_{u^o},\,\delta_{u^c}),$$

where we set

$$C_{ab;\,c} = \delta_{u^c}(C_{ab}) - \Gamma^h_{bc}C_{ad} - \Gamma^o_{bc}C_a - \Gamma^d_{ac}C_{db},$$
$$C_{ab;\,o} = \partial_{u^o}(C_{ab}) - \Gamma^d_{bo}C_{ad} - \Gamma^o_{bo}C_a - \Gamma^d_{ao}C_{db},$$
$$C_{a;\,b} = \delta_{u^b}(C_a) - \Gamma^d_{ob}C_{ad} - \Gamma^o_{ob}C_a - \Gamma^d_{ab}C_d\,.$$

Using the local Gauss-Weingarten equations (see Section 1) in the above equations, we have the following Gauss-Codazzi equations expressed locally by local components of h, A_N and τ as follows:

$$\bar{R}_{abcd} = R_{abcd} + g_{at}(A^t_c B_{bd} - A^t_d B_{bc}),$$
$$\bar{R}_{ijoh} = R_{ijoh} + g_{it}A^t_o B_{jh},$$
$$\bar{R}^o_{abc} = R^o_{abc} = g_{ad}\left\{A^d_{b;\,c} - A^d_{c;\,b} + \tau_b A^d_c - \tau_c A^d_b\right\},$$
$$\bar{R}^o_{abo} = R^o_{abo} = g_{ad}\left\{A^d_{b;\,o} - A^d_{o;\,b} + \tau_b A^d_o - \tau_o A^d_b\right\},$$

where we set

$$A^d_{b;\,c} = \frac{\delta A^d_b}{\delta u^c} - \Gamma^i_{bc}A^d_a - \Gamma^o_{bc}A^d_o + \Gamma^d_{ac}A^a_b,$$

$$A^d_{b;\,o} = \frac{\partial A^d_b}{\partial u^o} - \Gamma^i_{bo}A^d_a - \Gamma^o_{bo}A^d_o + \Gamma^d_{ao}A^a_b,$$

$$A^d_{o;\,b} = \frac{\delta A^d_o}{\delta u^b} - \Gamma^i_{ob}A^d_a - \Gamma^o_{ob}A^d_o + \Gamma^d_{ab}A^a_o\,.$$

Finally, we have

$$\bar{R}^o_{obc} = R^o_{obc} = A^a_b B_{ac} - A^a_c B_{ab} + 2d\tau\,(\delta_{u^b},\,\delta_{u^c}),$$
$$\bar{R}^o_{ooc} = R^o_{ooc} = A^a_o B_{ac} + 2d\tau\,(\partial_{u^o},\,\delta_{u^c})\,.$$

Induced Ricci curvature. Using the equation (1.2.10) of the induced Ricci tensor $\bar{\mathrm{Ric}}$ of $\bar{M}$, let $R^{(0,\,2)}$ denote a $(0,\,2)$ tensor on M given by

$$R^{(0,\,2)}(X,\,Y) = \mathrm{tr}\{Z \to R(X,\,Z)Y\}, \quad \forall\, X, Y \in \Gamma(TM). \tag{2.4.10}$$

Using the above local results, we obtain

$$\mathrm{Ric}(X,Y) = g^{ij} g\big(R(X, \tfrac{\delta}{\delta u^i})Y, \tfrac{\delta}{\delta u^j}\big) + g\big(R(X, \tfrac{\partial}{\partial u^o})Y, N\big).$$

From this we obtain

$$\mathrm{Ric}(X,Y) - \mathrm{Ric}(Y,X) = g^{ij}\left\{ C\big(X, \tfrac{\delta}{\delta u^j}\big) B\big(Y, \tfrac{\delta}{\delta u^i}\big) - C\big(Y, \tfrac{\delta}{\delta u^j}\big) B\big(X, \tfrac{\delta}{\delta u^i}\big) \right\}$$

$$+ \bar{g}\big(\bar{R}(X,Y)\tfrac{\partial}{\partial u^o}, N\big).$$

Replacing X and Y by $\tfrac{\delta}{\delta u^h}$ and $\tfrac{\delta}{\delta u^k}$ respectively we get

$$R_{kh} - R_{hk} = A^i_h B_{ik} - A^i_k B_{ih} + \bar{R}^o_{okh} = 2d\tau\left(\tfrac{\delta}{\delta u^k}, \tfrac{\delta}{\delta u^h}\right),$$

where we put $R_{kh} = \mathrm{Ric}\left(\tfrac{\delta}{\delta u^h}, \tfrac{\delta}{\delta u^k}\right)$. Similarly, replacing X and Y by $\tfrac{\delta}{\delta u^h}$ and $\tfrac{\partial}{\partial u^o}$ respectively, we obtain

$$R_{oh} - R_{ho} = -A^i_o B_{ih} + \bar{R}^o_{ooh} = 2d\tau\left(\tfrac{\partial}{\partial u^o}, \tfrac{\delta}{\delta u^h}\right),$$

where $R_{oh} = \mathrm{Ric}\left(\tfrac{\delta}{\delta u^h}, \tfrac{\partial}{\partial u^o}\right)$ and $R_{ho} = \mathrm{Ric}\left(\tfrac{\partial}{\partial u^o}, \tfrac{\delta}{\delta u^h}\right)$. Therefore, we obtain the following important result.

Theorem 2.4.1. [149] *Let $(M, g, S(TM))$ be a lightlike hypersurface of a semi-Riemannian manifold $(\bar{M}, \bar{g})$. Then the Ricci tensor of the induced connection ∇ is symmetric, if and only if, each 1-form τ induced by $S(TM)$ is closed, i.e., $d\tau = 0$, on any $\mathcal{U} \subset M$.*

Suppose the Ricci tensor is symmetric. By the above theorem and Poincaré's lemma we obtain $\tau(X) = X(f)$, where f is a smooth function on $\mathcal{U}$. Let $\alpha = \exp(f)$ in the second equation of (2.3.5), then, $\bar{\tau}(X) = 0$, $\forall X \in \Gamma(TM_{|\mathcal{U}})$. Thus, we have

Proposition 2.4.2. *Let (M, g) be a lightlike hypersurface of $(\bar{M}, \bar{g})$. If the Ricci tensor of ∇ is symmetric, then there exists a pair $\{\xi, N\}$ on $\mathcal{U}$ such that the corresponding 1-form τ from the Weingarten equation vanishes.*

It is clear from the above that $R^{(0,2)}$ is not symmetric. Therefore, in general, it is just a tensor quantity. This can also be verified in the following alternative way. Consider the quasi-orthonormal frame $\{\xi; W_a\}$ on M, where $\mathrm{Rad}\,TM = \mathrm{Span}\{\xi\}$ and $S(TM) = \mathrm{Span}\{W_a\}$ and let $E = \{\xi, N, W_a\}$ be the corresponding frames field on $\bar{M}$. Then, we obtain

$$R^{(0,2)}(X, Y) = \sum_{a=1}^{m} \epsilon_a \, g(R(X, W_a)Y, W_a) + \bar{g}(R(X, \xi)Y, N) \qquad (2.4.11)$$

where ϵ_a denotes the causal character (± 1) of respective vector field W_a. Using Gauss-Codazzi equations, we obtain

$$g(R(X, W_a)Y, W_a) = \bar{g}(\bar{R}(X, W_a)Y, W_a)$$
$$+ B(X, Y)C(W_a, W_a) - B(W_a, Y)C(X, W_a).$$

Substituting this in (2.4.11), using (2.1.25) and (2.1.26) we obtain

$$R^{(0,\,2)}(X, Y) = \bar{\text{Ric}}(X, Y) + B(X, Y)\,\text{tr}\,A_N$$
$$- g(A_N X, A_\xi^* Y) - \bar{g}(R(\xi, Y)X, N), \qquad (2.4.12)$$

for all $X, Y \in \Gamma(TM)$ and $\bar{\text{Ric}}$ is the Ricci tensor of $\bar{M}$. Thus, $R^{(0,\,2)}$ is not symmetric. Therefore, in general, $R^{(0,\,2)}$ has no geometric or physical meaning similar to the symmetric Ricci tensor of $\bar{M}$. In the 1996 book [149], $R^{(0,\,2)}$, given by (2.4.10), was called an induced Ricci tensor which is not correct. For this correction, the following definition was introduced in [142]:

Definition 2.4.3. Let $(M, g, S(TM))$ be a lightlike hypersurface of a semi-Riemannian manifold $(\bar{M}, \bar{g})$. A tensor field $R^{(0,\,2)}$ of M, given by (2.4.10), is called its induced Ricci tensor if it is symmetric.

For historical reasons, we still call $R^{(0,\,2)}$ an induced Ricci tensor, but, we denote it by Ric only if it is symmetric. Now we ask the following question: *Are there any lightlike hypersurfaces with symmetric Ricci tensor?* The answer is affirmative for screen conformal lightlike hypersurfaces of a semi-Riemannian manifold of constant curvature for which we prove the following result:

Theorem 2.4.4. [19] *Let $(M, g, S(TM))$ be a screen conformal lightlike hypersurface of a semi-Riemannian manifold $(\bar{M}^{n+2}(c), \bar{g})$ of constant sectional curvature c. Then, M admits an induced Ricci tensor.*

Proof. Using the equation (2.4.13) one can write the general expression of the Ricci tensor of M as

$$R^{0,2}(X, Y) = \bar{\text{Ric}}(X, Y) - \bar{g}(\bar{R}(\xi, Y)X, N)$$
$$+ B(X, Y)\,\text{tr}\,A_N - g(A_N X, A_\xi^* Y). \qquad (2.4.13)$$

As $\bar{M}^{n+2}(c)$ is non-degenerate we have $\bar{\text{Ric}}(\xi, Y)X = \pm c\bar{g}(X, Y)\xi$, where one can take either the sign $+$ or $-$, depending on the chosen definition of the curvature tensor. Thus, taking the $-$ sign, (2.4.13) reduces to

$$\text{Ric}(X, Y) = \bar{\text{Ric}}(X, Y) + c\bar{g}(X, Y) + B(X, Y)\,\text{tr}\,A_N - g(A_N X, A_\xi^* Y).$$

Since g and B are symmetric, it follows that

$$R^{0,2}(X, Y) - R^{0,2}(Y, X) = g(A_N Y, A_\xi^* X) - g(A_N X, A_\xi^* Y).$$

Thus, $R^{0,2}(X, Y) - R^{0,2}(Y, X) = \varphi g([A_\xi^*, A_\xi^*]Y, X) = 0$, completes the proof. $\square$

Remark 2.4.5. Observe that, as per Theorem 2.4.1, the existence of a symmetric Ricci tensor on M is equivalent to $d\tau = 0$, on any $\mathcal{U} \subset M$ and τ need not vanish. Therefore, only vanishing of $d\tau$ is needed to get a symmetric Ricci tensor for M.

Proposition 2.4.6. *A Ricci tensor of a lightlike hypersurface (M, g), of a semi-Rie-mannian manifold $\bar{M}$, is independent of the choice of the null section of* $\operatorname{Rad} TM$.

Proof. Let $\{\xi, N, W_1, \ldots, W_m\}$ be a quasi-orthonormal basis on $\bar{M}$, along M, adapted to the decomposition (2.1.8). Consider another basis $\{\xi', N', W_1', \ldots, W_m'\}$ and its corresponding Ricci tensor Ric$'$ with respect to a pair $\{\xi', N'\}$. Then, using the relation $\xi' = \alpha\xi$ and $N' = (1/\alpha)N$, for some smooth function $\alpha > 0$, writing the corresponding expression of Ric$'$ in (2.4.11) and comparing them we deduce that Ric$'(X, Y) = \operatorname{Ric}(X, Y) \; \forall X, Y \in \Gamma(TM)$, which completes the proof. $\qquad \square$

Induced scalar curvature. To introduce a concept of induced scalar curvature for a lightlike hypersurface M we observe that, in general, the non-uniqueness of screen distribution $S(TM)$ and its non-degenerate causal structure rules out the possibility of a definition for an arbitrary M of a semi-Riemannian manifold. Although, as we know from the previous section, there are many cases of a canonical or unique screen and canonical transversal vector bundle, but, the problem of scalar curvature must be classified subject to the causal structure of a screen. For this reason, we start with a hypersurface $(M, g, S(TM))$ of a Lorentzian manifold $(\bar{M}, \bar{g})$ for which we know that any choice of $S(TM)$ is Riemannian. This case is also physically important (see [2, 3, 4, 74, 77, 139, 150, 285] and many more cited therein). Let $\{\xi; W_a\}$ be the quasi-orthonormal frame for TM induced from a frame $\{\xi; W_a, N\}$ for $T\bar{M}$ such that $S(TM) = \operatorname{Span}\{W_1, \ldots, W_m\}$ and $\operatorname{Rad} TM = \operatorname{Span}\{\xi\}$. Set $X = Y$ in (2.4.10). Then,

$$R^{(0,\,2)}(X,\,X) = \sum_{b=1}^{m} g(R(X,\,W_b)X,\,W_b) + \bar{g}(R(X,\,\xi)X,\,N). \qquad (2.4.14)$$

Replacing X by ξ and using (2.4.4), (2.4.5) and (2.4.9) we obtain

$$R^{(0,\,2)}(\xi,\,\xi) = \sum_{a=1}^{m} g(R(\xi,\,W_a)\xi,\,W_a) - \bar{g}(R(\xi,\,\xi)\xi,\,N)$$

$$= \sum_{a=1}^{m} g(R(\xi,\,W_a)\xi,\,W_a), \qquad (2.4.15)$$

where the second term vanishes due to (2.4.9). Replacing each X successively by each base vector W_a of $S(TM)$ and taking the sum we get

$$\sum_{a=1}^{m} R^{(0,\,2)}(W_a,\,W_a) = \sum_{a=1}^{m}\left\{\sum_{b=1}^{m} g(R(W_a,\,W_b)W_a,\,W_b)\right\}$$

$$+ \sum_{a=1}^{m} \bar{g}(R(W_a,\,\xi)W_a,\,N). \qquad (2.4.16)$$

Finally, adding (2.4.15) and (2.4.16) we obtain a scalar r given by

$$r = R^{(0,\,2)}(\xi, \xi) + \sum_{a=1}^{m} R^{(0,\,2)}(W_a, W_a)$$

$$= \sum_{a=1}^{m} \left\{ \sum_{b=1}^{m} g(R(W_a, W_b)W_a, W_b) \right\}$$

$$+ \sum_{a=1}^{m} \{ g(R(\xi, W_a)\xi, W_a) + \bar{g}(R(W_a, \xi)W_a, N) \}. \tag{2.4.17}$$

In general, r given by (2.4.17) can not be called a scalar curvature of M since it has been calculated from a tensor quantity $R^{(0,\,2)}$. It can only have a geometric meaning if $R^{(0,\,2)}$ is symmetric and its value is independent of the screen, its transversal vector bundle and the null section ξ. Therefore, we need reasonable geometric conditions on M to recover its scalar curvature. For this purpose we introduce the following general concept.

We say that a lightlike hypersurface $(M, g, S(TM))$ of a semi-Riemannian manifold $(\bar{M}, \bar{g})$ and any of its induced objects, denoted by $\Omega^{\mathbf{s}}$, are of *genus* $\mathbf{s}$ if the induced metric tensor $g_{|S(TM)}$ is of constant signature $\mathbf{s}$.

Although the above general concept is for the study of a scalar curvature of lightlike hypersurfaces of a semi-Riemannian manifold, in this book, we restrict M to a Lorentzian manifold $\bar{M}$. Then, we say that M (labeled by $M^{\mathbf{0}}$) is a lightlike hypersurface of genus zero with screen $S(TM)^{\mathbf{0}}$. Denote by $\mathcal{C}[M]^{\mathbf{0}} = [(M, g, S(TM))]^{\mathbf{0}}$ a class of lightlike hypersurfaces of genus zero such that

(a) M admits a canonical or unique screen distribution $S(TM)$ that induces a canonical or unique lightlike transversal vector bundle N,

(b) M admits an induced symmetric Ricci tensor Ric.

For geometric and physical reasons, the above two conditions are necessary to assign a well-defined scalar curvature to each member of $\mathcal{C}[M^{\mathbf{0}}]$.

Definition 2.4.7. Let (M, g, ξ, N) belong to $\mathcal{C}[M]^{\mathbf{0}}$. Then, the scalar r, given by (2.4.17), is called the induced *scalar curvature of genus zero* of M.

Since $S(TM)$ and N are chosen, we must assure the stability of r with respect to a choice of the second fundamental form B and the 1-form τ. Choose another null section $\xi' = \alpha\xi$. Then, with respect to ξ' there is another $N' = (1/\alpha)N$ satisfying (2.1.6). It follows from the transformation equation (2.3.5) that for a canonical or unique vector bundle N, the function α in that relation will be non-zero constant which implies that, for this case, both B and τ are independent of the choice of ξ, except for a non-zero constant factor. Finally, we know that the Ricci tensor does not depend on the choice of ξ. Consequently, r is a well-defined induced scalar curvature of a class of lightlike hypersurfaces of genus zero. To show that there exists a variety of Lorentzian manifolds which admit lightlike hypersurfaces of class $\mathcal{C}[M]^{\mathbf{0}}$ we prove the following two theorems:

Theorem 2.4.8. [142] *Let* $(M, g, S(TM))$ *be a screen conformal lightlike hypersurface of a Lorentzian manifold* $(\bar{M}(c), \bar{g})$, *with* $\mathcal{S}^1$ *the first derivative of* $S(TM)$ *given by* (2.3.12). *If* $\mathcal{S}^1$ *coincides with* $S(TM)$, *then,* M *belongs to* $\mathcal{C}[M]^0$. *Consequently, this class of lightlike hypersurfaces admit induced scalar curvature of genus zero.*

Proof. Since $\mathcal{S}^1 = S(TM)$, it follows from Theorem 2.3.4 that M admits a unique screen distribution $S(TM)$ that induces a unique lightlike transversal vector bundle, which satisfies the condition (a). The condition (b) also holds as Theorem 2.4.4 says that there exists such a class of screen conformal M of $\bar{M}(c)$ which admits a symmetric Ricci tensor. $\square$

Consider a class of Lorentzian manifolds $(\bar{M}, \bar{g})$ which admit at least one covariant constant timelike vector field. Then, we know from section 3 that this class admits screen conformal lightlike hypersurfaces $(M, g, S(TM))$. Consequently, M can admit a unique screen distribution which induces a unique lightlike transversal vector field. This satisfies the condition (a). Thus, using Theorems 2.3.5 and 2.4.4, we state the following:

Theorem 2.4.9. *Consider a lightlike hypersurface* $(M, g, S(TM))$ *of a Lorentzian manifold* $(\bar{M}, \bar{g})$. *Let* E *be a complementary vector bundle of* $TM^{\perp}$ *in* $S(TM)^{\perp}$ *such that* E *admits a covariant constant timelike vector field. Suppose that* M *admits a symmetric Ricci tensor. Then,* M *is a member of* $\mathcal{C}[M]^0$. *Consequently, this class of lightlike hypersurfaces admit induced scalar curvatures of genus zero.*

Physical models. Let $(\bar{M}, \bar{g})$ be an $(m+2)$-dimensional globally hyperbolic spacetime manifold [34] with the metric $\bar{g}$ given by

$$ds^2 = -dt^2 + e^{\mu}(dx^1)^2 \oplus \bar{g}_{ab}dx^a\,dx^b, \quad 2 \le a, b \le m+1,$$

with respect to local coordinates $(t, x^1, \ldots, x^{m+1})$, where μ is a function of t and x^1 alone. It is well known that the above prescribed model of spacetimes admits a timelike covariant constant vector field. Thus, the first hypothesis of Theorem 2.3.5 is satisfied. Now we construct a set of two lightlike hypersurfaces of $\bar{M}$. Let $\mathbf{E} = \{e_0, e_1, \ldots, e_{m+1}\}$ be an orthonormal basis, such that e_0 is timelike and all others are spacelike unit vectors. Transform $\mathbf{E}$ into another basis $\mathcal{E} = \{\partial_u, \partial_v, e_2, \ldots, e_{m+1}\}$ such that ∂_u and ∂_v are real null vectors satisfying $\bar{g}(\partial_u, \partial_v) = 1$ with respect to new coordinates $\{u, v, x^2, \ldots, x^{m+1}\}$. Then, the line element of $\bar{g}$ transforms into

$$ds^2 = -A^2(u, v)du\,dv + \bar{g}_{ab}dx^a\,dx^b, \quad 2 \le a, b \le m+1,$$

for some function $A(u, v)$ on M. The absence of du^2 and dv^2 in the above line element implies that $\{v = \text{constant}\}$ and $\{u = \text{constant}\}$ are two lightlike hypersurfaces and their intersection provides a leaf of their common screen distribution whose Riemannian metric is given by $d\Omega^2 = \bar{g}_{ab}dx^a\,dx^b$. It follows from Theorem 2.4.4 that each of these hypersurfaces is screen conformal, and, therefore, each admits an integrable canonical screen distribution.

It is also possible to consider another physical model of globally hyperbolic warped product spacetimes (including asymptotically flat spacetimes, such as Schwarzschild and Reissner-Nordström and many more) which also admit a pair of screen conformal lightlike hypersurfaces as explained above. Following are two sub-cases.

Case 1. This model $(\bar{M}, \bar{g})$ belongs to all spacetimes of constant curvature. Examples are: Minkowski, de-Sitter and anti-de-Sitter. Then, as per Theorem 2.4.4, the induced Ricci tensor of each such hypersurface is symmetric. Consequently, each such hypersurface is a member of $\mathcal{C}[M]^0$, so each admits an induced scalar curvature of genus zero.

Case 2. This model $(\bar{M}, \bar{g})$ belongs to all spacetimes of non-constant scalar curvature. For example, a spacetime with the metric

$$ds^2 = -dt^2 + S^2(t)\, d\Sigma^2$$

which is called a *Robertson-Walker spacetime* where $d\Sigma^2$ is the metric of a spacelike hypersurface Σ with spherical symmetry and constant curvature $k = 1, -1$ or 0. For this case, Theorem 2.4.8 says that any such hypersurface admits induced scalar curvature if it also admits a symmetric Ricci tensor. Schwarzschild and Reissner-Nordström spacetimes belong to this model.

Special class. Let $(M,\, g,\, S(TM),\, \xi,\, N)$ be a class of lightlike hypersurfaces of a time orientable spacetime manifold $\bar{M}(c)$ of constant curvature c and satisfying Theorem 2.4.8. The curvature tensor $\bar{R}$ of $\bar{M}(c)$ is given by

$$\bar{R}(X,\, Y)Z = c\,\{\bar{g}(Y,\, Z)X - \bar{g}(X,\, Z)Y\}. \tag{2.4.18}$$

Using (2.4.18) in (2.4.6) we have $\bar{g}(R(X,\, \xi)Y,\, N) = -c\bar{g}(X,\, Y)$. Using this result and (2.4.18) in (2.4.4) reduces the equation (2.4.11) to

$$\begin{aligned}
\mathrm{Ric}(X, X) &= \sum_{b=1}^{m} g(R(X,\, W_b)X,\, W_b) - c\bar{g}(X,\, X) \\
&= \sum_{b=1}^{m} \{B(W_b,\, X)C(X,\, W_b) - B(X,\, X)C(W_b,\, W_b)\} \\
&\quad - c\,m\,\bar{g}(X,\, X).
\end{aligned} \tag{2.4.19}$$

Thus, $\mathrm{Ric}(\xi,\, \xi) = 0$ and the equations (2.4.15) and (2.4.16) reduce to

$$\begin{aligned}
\sum_{a=1}^{m} \mathrm{Ric}(W_a,\, W_a) &= \sum_{a=1}^{m} \Big\{ \sum_{b=1}^{m} [B(W_b,\, W_a)C(W_a,\, W_b) \\
&\quad - B(W_a,\, W_a)C(W_b,\, W_b)]\Big\} - c\,m^2 \\
&= r.
\end{aligned} \tag{2.4.20}$$

It is immediate from the above that, for this class $\mathcal{C}[M]^0$, the induced Ricci tensor and the corresponding scalar curvature of M can be determined if one knows $\mathrm{Ric}_{|S(TM)}$ and $r_{|S(TM)}$. More precisely,

$$\mathrm{Ric} = \mathrm{Ric}_{|S(TM)}, \quad r = r_{|S(TM)} - c\, m^2. \tag{2.4.21}$$

Recall from the proof of Theorem 2.3.5 and (2.3.11) that

$$C(X, PW) = \frac{1}{2}\theta^2 B(X, W), \quad \theta^{-1} = \bar{g}(V, \xi), \quad \forall X, W \in \Gamma(TM).$$

Using this relation we obtain the following value of r from (2.4.20):

$$r = \frac{1}{2}\theta^2 \sum_{a=1}^{m} \left\{ \sum_{b=1}^{m} [(B)^2(W_a, W_b) - B(W_a, W_a)B(W_b, W_b)] \right\}$$
$$- c\, m^2. \tag{2.4.22}$$

For 3- and 4-dimensional $\bar{M}(c)$ the values of r, respectively, are:

$$r = -c \text{ and } r = \theta^2 [(B)^2(W_1, W_2) - B(W_1, W_1)B(W_2, W_2)] - 4c.$$

Example 8. Let $(\mathbf{R}_1^4, \bar{g})$ be a Minkowski space with canonical basis $(\partial_t, \partial_1, \partial_2, \partial_3)$. It follows from [149, pages 115-117] that there exists a lightlike hypersurface of $(\mathbf{R}_1^4, \bar{g})$ with a canonical integrable screen distribution and a canonical lightlike transversal vector bundle $\mathrm{tr}(TM)$. Choose a timelike unit vector field $V = \partial_t$ of $T\mathbf{R}_1^4$. We make this choice for a hypersurface $(M, g, S(TM), \xi, N)$ given by an open subset of the lightlike cone

$$\left\{ t(1, \cos u \cos v, \cos u \sin v, \sin u) \in \mathbf{R}_1^4 : t > 0, \ u \in (0, \pi/2), \ v \in [0, 2\pi] \right\}.$$

$$\mathrm{Rad}(TM) = \mathrm{Span}\{\xi = \partial_t + \cos u \cos v\, \partial_1 + \cos u \sin v\, \partial_2 + \sin u\, \partial_3\},$$
$$\mathrm{ltr}(TM) = \mathrm{Span}\{N = \frac{1}{2}(-\partial_t + \cos u \cos v\, \partial_1 + \cos u \sin v\, \partial_2 + \sin u\, \partial_3)\},$$

respectively and the screen distribution $S(TM)$ is spanned by

$$\{W_1 = -\sin u \cos v\, \partial_1 - \sin u \sin v\, \partial_2 + \cos u\, \partial_3, \quad W_2 = -\sin v\, \partial_1 + \cos v\, \partial_2\}.$$

With respect to a local frames field of $\bar{M}$, the metric $\bar{g}$ has the matrix

$$[\bar{g}] = \begin{bmatrix} 0 & 1 & 0 & 0 \\ 1 & 0 & 0 & 0 \\ 0 & 0 & t^2 & 0 \\ 0 & 0 & 0 & t^2 \cos^2 u \end{bmatrix}.$$

We refer to the local expressions and the notation presented earlier. The Christoffel symbols Γ_{ij}^k and the coefficients B_{ij} and A_j^i of the second fundamental form and its shape operator respectively, i, j, $k \in \{0, 1, 2\}$, are given by

$$\Gamma_{22}^1 = \sin u \cos u, \quad \Gamma_{12}^2 = -\tan u, \quad \Gamma_{1o}^1 = \frac{1}{t}, \quad \Gamma_{2o}^2 = \frac{1}{t};$$

$$\Gamma_{11}^o = -\frac{t}{2}, \quad \Gamma_{22}^o = -\frac{1}{2}t\cos^2 u, \quad B_{11} = -t, \quad B_{22} = -t\cos^2 u;$$

$$A_1^1 = -\frac{1}{2t}, \quad A_2^2 = -\frac{1}{2t}, \quad \text{and all others vanish.}$$

Using the local Gauss-Codazzi equations and $\bar{R}_{abcd} = 0$, we have

$$R_{ac} = \frac{2}{t}\Gamma_{ac}^0 + \frac{1}{t^2}\Gamma_{a1}^0 B_{1c} + \frac{1}{t^2\cos^2 u}\Gamma_{a2}^0 B_{2c}$$

$$= \frac{1}{t}B_{ac} + \frac{1}{t^2}B_{a1}\Gamma_{1c}^0 + \frac{1}{t^2\cos^2 u}B_{a2}\Gamma_{2c}^0.$$

Thus $R_{11} = -\frac{1}{2}$, $R_{12} = R_{21} = 0$ and $R_{22} = -\frac{1}{2}\cos^2 u$. Similarly, we have

$$R_{ao} = \frac{2}{t}\Gamma_{a0}^0 = 0, \quad R_{oa} = R_{oo} = 0.$$

Therefore, the Ricci tensor of this lightlike hypersurface is symmetric. Thus, this class of lightlike hypersurfaces belongs to $\mathcal{C}[M]$. The components of the second fundamental form are

$$B(W_1, W_1) = -1, \quad B(W_2, W_2) = -\frac{1}{\cos u}, \quad B(W_1, W_2) = 0.$$

Moreover, in this example, $\bar{g}(V = \partial_t, \xi) = -1$ implies that in the relation (2.4.21) we have $\theta = 1$. Using all this data in (2.4.21) we conclude that the induced scalar curvature of M in $(\mathbf{R}_1^4, \bar{g})$ is given by $r = -(\cos u)^{-1}$.

2.5 Lightlike Einstein hypersurfaces

Recall from the equation (1.2.12) that an $(m + 2)$-dimensional semi-Riemannian manifold $(\bar{M}, \bar{g})$ is Einstein if $\bar{Ric} = \bar{k}\bar{g}$, $\bar{k}$ is a constant. Moreover, $\bar{M}$ is Einstein if and only if $\bar{k} = \bar{r}/(m + 2)$, where $\bar{r}$ is the scalar curvature of $\bar{M}$. Obviously, a geometric concept of a lightlike Einstein hypersurface $(M, g, S(TM))$ must involve its scalar curvature. Therefore, for a well-defined concept of a lightlike Einstein hypersurface M one should assure that M admits a symmetric Ricci tensor from which an induced scalar curvature can be calculated. Based on what we know from two previous sub-sections, this condition is satisfied if M is a member of $\mathcal{C}[M]^0$ for which $\bar{M}$ is Lorentzian. Therefore, we assume that $\bar{M}$ is Lorentzian and restrict M to be a member of $\mathcal{C}[M]^0$. First we need the following related results:

As in the non-degenerate case, a point p of M is said to be *umbilical* if

$$B(X,Y)_p = \mathbf{a}\, g(X,Y)_p, \quad \forall X, Y \in T_p M,$$

where $\mathbf{a} \in \mathbf{R}$ and depends on p. The above definition is independent of any coordinate neighborhood around p. M is *totally umbilical* in $\bar{M}$ if every point of M is umbilical, i.e., if $B = \rho\, g$ where ρ is a smooth function. As per Proposition 2.3.1(a), the above definition does not depend on the choice of a screen distribution of M, but, it is easy to see from (2.1.25) and (2.1.16) that M is totally umbilical, if and only if, on each $\mathcal{U}$ there exists ρ such that

$$A_\xi^* (PX) = \rho PX, \quad \text{for all} \quad X \in \Gamma(TM_{|\mathcal{U}}). \tag{2.5.1}$$

In particular, M is *totally geodesic* in $\bar{M}$ if

$$B = 0 \Leftrightarrow \rho = 0.$$

We have the following general result on totally geodesic lightlike hypersurfaces:

Theorem 2.5.1. [149] *Let* $(M, g, S(TM))$ *be a lightlike hypersurface of a semi-Riemannian manifold* $(\bar{M}, \bar{g})$. *The following are equivalent:*

(i) *M is totally geodesic.*

(ii) *h vanishes identically on M.*

(iii) *$A_{\mathcal{U}}^*$ vanishes identically on M, for any $\mathcal{U} \in \Gamma(\operatorname{Rad} TM)$.*

(iv) *There exists a unique metric connection ∇ on M.*

(v) *$\operatorname{Rad} TM$ is a parallel distribution with respect to ∇.*

(vi) *$\operatorname{Rad} TM$ is a Killing distribution on M.*

Proof. The equivalence of (i) and (ii) follows from (2.1.9) as in the case of non-degenerate submanifolds. Next, by using (2.1.13) and (2.1.16), we obtain the equivalence of (ii) and (iii). (2.1.20) implies the equivalence of (iii) and (v). By the second result of (2.3.5) we have the equivalence of (ii) and (iv). Finally, from (2.1.18) we obtain $(\nabla_X g)(\xi, Z) = B(X, Z)$ which implies $g(\nabla_X \xi, Z) = -B(X, Z)$. Then by using (1.2.23) of Chapter 1 we obtain

$$(\pounds_\xi g)(X, Z) = g(\nabla_X \xi, Z) + g(\nabla_Z \xi, X) = -2B(X, Z)$$

which proves the equivalence of (ii) and (vi). □

Consequently, totally geodesic lightlike hypersurfaces are important both for geometry and physics since they admit a unique metric connection.

In particular, suppose $(M, g, S(TM))$ is a totally geodesic lightlike hypersurface of a semi-Riemannian manifold $(\bar{M}(c), \bar{g})$ of constant curvature c. Then, using (1.2.16) we have $\bar{R}(\xi, Y)X = \bar{g}(X, Y)\xi$. Since it follows from (2.4.6) that

$\bar{g}(\bar{R}(\xi, Y)X, N) = \bar{g}(R(\xi, Y)X, N)$, using both these results in (2.4.12) and (ii) and (iii) of Theorem 2.5.1, we obtain

$$\mathrm{Ric}(X, Y) = \bar{\mathrm{Ric}}(X, Y) - c\,\bar{g}(X, Y),$$

which is symmetric since $\bar{g}$ is symmetric. Thus, we have

Proposition 2.5.2. *Any totally geodesic lightlike hypersurface of $\bar{M}(c)$ admits an induced symmetric Ricci tensor.*

Some results on totally umbilical lightlike hypersurfaces are available in [149, Section 4.5, pages 106-114]. We highlight the following:

- *There exist no lightlike hypersurfaces in $\bar{M}(c), c \neq 0$, with a totally geodesic screen distribution.*

- *Any lightlike hypersurface of $\bar{M}(c)$ with a proper totally umbilical screen distribution is either totally umbilical or totally geodesic immersed in $\bar{M}(c)$.*

- *Any lightlike surface of a 3-dimensional Lorentz manifold is either totally umbilical or totally geodesic.*

Now we deal with the concept of *lightlike Einstein hypersurfaces*. We give a general definition for hypersurfaces in semi-Riemannian manifolds $\bar{M}$, but, for the above stated reasons we focus our study on when $\bar{M}$ is Lorentzian.

Definition 2.5.3. Let $(M, g, S(TM))$ be a lightlike hypersurface of an $(m + 2)$-dimensional semi-Riemannian manifold $(\bar{M}, \bar{g})$ such that M admits an induced symmetric Ricci tensor Ric. Then, M is called an Einstein hypersurface if

$$\mathrm{Ric}(X, Y) = kg(X, Y), \quad \forall X, Y \in \Gamma(TM). \tag{2.5.2}$$

where k is a constant if $m > 1$.

In general, the above definition can not have a geometric meaning unless we relate the constant k with the induced scalar curvature of M, which is not possible for a hypersurface of an arbitrary $\bar{M}$. Therefore, based on what we know at the time of writing this book, we start with the case of a screen conformal (see Definition 2.2.1 of this chapter) lightlike hypersurface $(M, g, S(TM))$ of an $(m + 2)$-dimensional Lorentzian manifold $(\bar{M}(c), \bar{g})$ of constant curvature c and then discuss the geometric meaning of the above definition. For this class of M we know that the screen distribution $S(TM)$ is Riemannian and integrable. First we prove the following:

Proposition 2.5.4. *Let $(M, g, S(TM))$ be a screen homothetic (φ is non-zero constant) lightlike hypersurface of $(\bar{M}(c), \bar{g})$. Then*

$$2\,\varphi\tau(\xi)B(X, PZ) = -cg(X, PZ).$$

Proof. Since $\bar{M}$ is of constant curvature c, by (2.4.18), we have

$$(\nabla_X B)(Y, Z) - (\nabla_Y B)(X, Z) = B(X, Z)\tau(Y) - B(Y, Z)\tau(X).$$

On the other hand, using (2.4.5), (2.4.8) and the last equation, we have

$$2\varphi\{B(X, PZ)\tau(Y) - B(Y, PZ)\tau(X)\} = c\{g(Y, PZ)\eta(X) - g(X, PZ)\eta(Y)\}.$$

Replacing Y by ξ in this equation, we obtain $2\varphi\tau(\xi)B(X, PZ) = -cg(X, PZ)$.
$\square$

Remark 2.5.5. Under the hypothesis of the above proposition, one can show that
if $\tau = 0$, then, $c = 0$ and if $\tau(\xi) \neq 0$, then, M is totally umbilical in $\bar{M}$. Then,
from Proposition 2.4.2 and this remark, we have $\tau = c = 0$.

The scalar curvature $\bar{r}$ of $\bar{M}$, defined by (1.2.13), and the scalar quantity r
of M, obtained from $R^{(0,\,2)}$ (by the method in the previous section) are given by

$$\bar{r} = \bar{\text{Ric}}(\xi,\, \xi) + \bar{\text{Ric}}(N,\, N) + \sum_{a=1}^{m} \epsilon_a\, \bar{\text{Ric}}(W_a,\, W_a),$$

$$r = R^{(0,\,2)}(\xi,\, \xi) + \sum_{a=1}^{m} \epsilon_a\, R^{(0,\,2)}(W_a,\, W_a),$$

respectively. Using these relations and (2.4.12), we obtain

$$R^{(0,\,2)}(\xi,\, \xi) = \bar{\text{Ric}}(\xi,\, \xi),$$

$$R^{(0,\,2)}(W_a,\, W_a) = \bar{\text{Ric}}(W_a,\, W_a) + g(A_\xi^* W_a,\, W_a)\,\text{tr}(A_N)$$
$$- g(A_N W_a,\, A_\xi^* W_a) - \bar{g}(R(\xi,\, W_a)W_a,\, N).$$

Thus we have

$$r = \bar{r} + \text{tr}\, A_\xi^*\, \text{tr}\, A_N - \text{tr}(A_\xi^* A_N)$$
$$- \sum_{a=1}^{m} \epsilon_a\{\bar{g}(R(\xi,\, W_a)W_a,\, N) + \bar{g}(\bar{R}(N,\, W_a)W_a,\, N)\}. \tag{2.5.3}$$

Since $\bar{M}$ is a Lorentzian space form, we have $\bar{R}(\xi,\, Y)X = c\bar{g}(X,\, Y)\xi$, $\bar{\text{Ric}}(X,\, Y) = (m+1)c\,\bar{g}(X,\, Y)$ and $\bar{r} = c\,m(m+1)$, $\bar{g}(\bar{R}(N,\, W_a)W_a,\, N) = 0$. Thus

$$R^{(0,\,2)}(X, Y) = mc\, g(X, Y) + B(X, Y)\,\text{tr}\, A_N - g(A_N X,\, A_\xi^* Y); \tag{2.5.4}$$

$$r = m^2 c + \text{tr}\, A_\xi^*\, \text{tr}\, A_N - \text{tr}(A_\xi^* A_N). \tag{2.5.5}$$

Using (2.5.4) and (2.5.5) and M is Einstein, we obtain

$$r = \text{Ric}(\xi,\, \xi) + \sum_{a=1}^{m} \epsilon_a\, \text{Ric}(W_a,\, W_a)$$

$$= kg(\xi,\, \xi) + k \sum_{a=1}^{m} \epsilon_a\, g(W_a,\, W_a) = k\, m. \tag{2.5.6}$$

Thus we have
$$\text{Ric}(X,Y) = (r/m)g(X,Y) \tag{2.5.7}$$
which provides a geometric interpretation of lightlike Einstein hypersurfaces as we have shown that the constant $k = r/m$. Since ξ is an eigenvector field of A_ξ^* corresponding to the eigenvalue 0 and A_ξ^* is $\Gamma(S(TM))$-valued real symmetric, A_ξ^* has m real orthonormal eigenvector fields in $S(TM)$ and is diagonalizable by an orthogonal operator. Consider a frame field of eigenvectors $\{\xi, E_1, \ldots, E_m\}$ of A_ξ^* such that $\{E_1, \ldots, E_m\}$ is an orthonormal frame field of $S(TM)$. Then

$$A_\xi^* E_i = \lambda_i E_i, \quad 1 \le i \le m.$$

Since M is screen conformal and $\text{Ric} = kg$, the equation (2.5.4) reduces to

$$g(A_\xi^* X, A_\xi^* Y) - sg(A_\xi^* X, Y) + \varphi^{-1}(k - mc)g(X, Y) = 0, \tag{2.5.8}$$

where $s = \text{tr}\, A_\xi^*$. Put $X = Y = E_i$ in (2.5.8), λ_i is a solution of equation

$$x^2 - sx + \varphi^{-1}(k - mc) = 0. \tag{2.5.9}$$

The equation (2.5.9) has at most two distinct solutions which are real-valued functions on $\mathcal{U}$. Assume there exists $p \in \{0, 1, \ldots, m\}$ such that $\lambda_1 = \ldots = \lambda_p = \alpha$ and $\lambda_{p+1} = \ldots = \lambda_m = \beta$, by renumbering if needed. From (2.5.9), we have

$$s = \alpha + \beta = p\alpha + (m - p)\beta; \qquad \alpha\beta = \varphi^{-1}(k - mc). \tag{2.5.10}$$

The first equation of the last relationships reduces to

$$(p - 1)\alpha + (m - p - 1)\beta = 0.$$

Consider the following four distributions D_α, D_β, D_α^s and D_β^s on M;

$$D_\alpha = \{X \in \Gamma(TM) \mid A_\xi^* X = \alpha PX\}, \quad D_\alpha^s = D_\alpha \cap S(TM),$$
$$D_\beta = \{U \in \Gamma(TM) \mid A_\xi^* U = \beta PU\}, \quad D_\beta^s = D_\beta \cap S(TM).$$

Observe that $E_1, \ldots, E_p \in \Gamma(D_\alpha^s)$ and $E_{p+1}, \ldots, E_m \in \Gamma(D_\beta^s)$. The equation (2.5.5) has only one solution $\iff \alpha = \beta \iff D_\alpha = D_\beta\,(= TM)$. If $0 < p < m$, then $D_\alpha \ne D_\beta$ and $D_\alpha \cap D_\beta = \text{Rad}(TM)$. In case $m \ge 2$ and $D_\alpha \ne D_\beta$: If $p = 0$, then α is not an eigenvalue of A_ξ^* but a root of (2.5.9) and $D_\alpha = \text{Rad}(TM)$; $D_\beta = TM$. If $p = m$, then β is not an eigenvalue of A_ξ^* but a root of (2.5.9) and $D_\alpha = TM$; $D_\beta = \text{Rad}(TM)$.

Lemma 2.5.6. *If $D_\alpha \ne D_\beta$, then, $D_\alpha \perp_g D_\beta$ and $D_\alpha \perp_B D_\beta$.*

Proof. If $0 < p < m$, we have $A_\xi^* PX = A_\xi^* X = \alpha PX$ for any $X \in \Gamma(D_\alpha)$ and $A_\xi^* PU = A_\xi^* U = \alpha PU$ for any $U \in \Gamma(D_\beta)$. Thus the projection P maps D_α onto D_α^s and D_β onto D_β^s. Since PX and PU are eigenvector fields of the real symmetric operator A_ξ^*, they have different eigenvalues α and β respectively. $PX \perp PU$ and $g(X, U) = g(PX, PU) = 0$, that is, $D_\alpha \perp_g D_\beta$. Also, since $B(X, U) = g(A_\xi^* X, U) = \alpha g(PX, PU) = 0$, we have $B(D_\alpha, D_\beta) = 0$, that is, $D_\alpha \perp_B D_\beta$. If $p = 0$ or $p = m$, then $D_\alpha = \mathrm{Rad}(TM)\,; D_\beta = TM$ or $D_\alpha = TM\,; D_\beta = \mathrm{Rad}(TM)$ respectively. Thus, $D_\alpha \perp_g D_\beta$ and $D_\alpha \perp_D D_\beta$.   □

Lemma 2.5.7. *If $D_\alpha \neq D_\beta$, then $TM = \mathrm{Rad}(TM) \oplus_{\mathrm{orth}} D_\alpha^s \oplus_{\mathrm{orth}} D_\beta^s$. If $D_\alpha = D_\beta$, then $TM = \mathrm{Rad}(TM) \oplus_{\mathrm{orth}} D_\alpha^s \oplus_{\mathrm{orth}} \{0\}$.*

Proof. If $0 < p < m$, since $\{E_i\}_{1 \le i \le p}$ and $\{E_a\}_{p+1 \le a \le m}$ are vector fields of D_α^s and D_β^s respectively and D_α^s and D_β^s are mutually orthogonal vector subbundles of $S(TM)$, we show that D_α^s and D_β^s are non-degenerate distributions of rank p and rank $(m - p)$ respectively and $D_\alpha^s \cap D_\beta^s = \{0\}$. Thus we have $S(TM) = D_\alpha^s \oplus_{\mathrm{orth}} D_\beta^s$. If $D_\alpha \neq D_\beta$ and $p = 0$, then $D_\alpha^s = \{0\}$ and $D_\beta^s = S(TM)$. If $D_\alpha \neq D_\beta$ and $p = m$, then $D_\alpha^s = S(TM)$ and $D_\beta^s = \{0\}$. Also we have $S(TM) = D_\alpha^s \oplus_{\mathrm{orth}} D_\beta^s$. Next, if $D_\alpha = D_\beta$, then $D_\alpha^s = D_\beta^s = S(TM)$. Thus, from (2.1.4) we have this lemma.   □

Lemma 2.5.8. $\mathrm{Im}(A_\xi^* - \alpha P) \subset \Gamma(D_\beta^s)\,;$ $\mathrm{Im}(A_\xi^* - \beta P) \subset \Gamma(D_\alpha^s)$.

Proof. From (2.5.8), we get $(A_\xi^*)^2 - (\alpha + \beta)A_\xi^* + \alpha\beta P = 0$. If $0 < p < m$. Let $Y \in \mathrm{Im}(A_\xi^* - \alpha P)$. Then, there exists $X \in \Gamma(TM)$ such that $Y = (A_\xi^* - \alpha P)X$ and $(A_\xi^* - \beta P)Y = 0$ and $Y \in \Gamma(D_\beta)$. Thus $\mathrm{Im}(A_\xi^* - \alpha P) \subset \Gamma(D_\beta)$. Since the morphism $A_\xi^* - \alpha P$ maps $\Gamma(TM)$ onto $\Gamma(S(TM))$, we have $\mathrm{Im}(A_\xi^* - \alpha P) \subset \Gamma(D_\beta^s)$. By duality, we also have $\mathrm{Im}(A_\xi^* - \beta P) \subset \Gamma(D_\alpha^s)$. If $p = 0$, then, since $D_\alpha^s = \{0\}\,;\ D_\beta^s = S(TM)$ and $D_\beta = TM$, we have $\mathrm{Im}(A_\xi^* - \alpha P) \subset \Gamma(S(TM))\,;\ A_\xi^* X = \beta P X, \forall X \in \Gamma(TM)$, that is, $\mathrm{Im}(A_\xi^* - \beta P) = \{0\}$ or if $p = m$, then, since $D_\alpha^s = S(TM)\,;\ D_\beta^s = \{0\}$ and $D_\alpha = TM$, we have $A_\xi^* X = \alpha P X, \forall X \in \Gamma(TM)$, that is, $\mathrm{Im}(A_\xi^* - \alpha P) = \{0\}\,;\ \mathrm{Im}(A_\xi^* - \beta P) \subset \Gamma(S(TM))$.   □

Lemma 2.5.9. *The distributions D_α^s and D_β^s are always integrable. In particular, if $D_\alpha \neq D_\beta$, then D_α and D_β are also integrable.*

Proof. If $D_\alpha \neq D_\beta$, then, for $X, Y \in \Gamma(D_\alpha)$ and $Z \in \Gamma(TM)$, we have

$$(\nabla_X B)(Y, Z) = -g\left((A_\xi^* - \alpha P)\nabla_X Y, U\right) + \alpha B(X, Y)\eta(U)$$
$$+ (X\alpha)\, g(PY, Z) + \alpha^2 \eta(Y)\, g(PX, Z).$$

Using this and the fact that

$$(\nabla_X B)(Y, Z) - (\nabla_Y B)(X, Z) = B(X, Z)\tau(Y) - B(Y, Z)\tau(X),$$

we have

$$g\left((A_\xi^* - \alpha P)[X, Y], Z\right) = \{X\alpha + \alpha\tau(X) - \alpha^2\eta(X)\}g(PY, Z)$$
$$- \{Y\alpha + \alpha\tau(Y) - \alpha^2\eta(Y)\}g(PX, Z). \qquad (2.5.11)$$

If we take $Z = U \in \Gamma(D_\beta^s)$, then we have

$$g\left((A_\xi^* - \alpha P)[X, Y], U\right) = 0.$$

Since D_β^s is non-degenerate and $\mathrm{Im}(A_\xi^* - \alpha P) \subset \Gamma(D_\beta^s)$, we have $(A_\xi^* - \alpha P)[X, Y] = 0$. Thus $[X, Y] \in \Gamma(D_\alpha)$ and D_α is integrable. By duality, we know that D_β is also integrable. Since $S(TM)$ is integrable, for any $X, Y \in \Gamma(D_\alpha^s)$, we have $[X, Y] \in \Gamma(D_\alpha)$ and $[X, Y] \in \Gamma(S(TM))$. Thus $[X, Y] \in \Gamma(D_\alpha^s)$ and D_α^s is integrable. So is D_β^s. While, if $D_\alpha = D_\beta$, then $D_\alpha^s = D_\beta^s = S(TM)$ is integrable. □

Lemma 2.5.10. *If $0 < p < m$, then we have*

$$(d\alpha + \alpha\tau - \alpha^2\eta)|_{D_\alpha} = 0 ; \quad (d\beta + \beta\tau - \beta^2\eta)|_{D_\beta} = 0.$$

Proof. From (2.5.10), for $X, Y \in \Gamma(D_\alpha)$ and $Z \in \Gamma(TM)$, we get

$$\{X\alpha + \alpha\tau(X) - \alpha^2\eta(X)\}\, g(PY, Z) = \{Y\alpha + \alpha\tau(Y) - \alpha^2\eta(Y)\}\, g(PX, Z).$$

Since $S(TM)$ is non-degenerate, we have

$$\{X\alpha + \alpha\tau(X) - \alpha^2\eta(X)\}\, PY = \{Y\alpha + \alpha\tau(Y) - \alpha^2\eta(Y)\}\, PX.$$

Suppose there exists a vector field $X_o \in \Gamma(D_\alpha)$ such that $\{d\alpha + \alpha\tau - \alpha^2\eta\}(X_o) \neq 0$ at each point $x \in M$, then $PY = fPX_o$ for any $Y \in \Gamma(D_\alpha)$, where f is a smooth function. It follows that all vectors from the fiber $(D_\alpha)_x$ are collinear with $(PX_o)_x$. It is a contradiction as $\dim((D_\alpha)_x) = p + 1 > 1$. Thus we have $(d\alpha + \alpha\tau - \alpha^2\eta)|_{D_\alpha} = 0$. By duality, we also have $(d\beta + \beta\tau - \beta^2\eta)|_{D_\beta} = 0$. □

Lemma 2.5.11. *Let M be an Einstein screen homothetic lightlike hypersurface of a Lorentz manifold $(\bar{M}(c), \bar{g})$ of constant curvature c. If $0 < p < m$, then α and β are constants along $S(TM)$ if and only if $\tau = 0$ on $S(TM)$.*

Proof. By Lemma 2.5.10, we know that $(d\alpha + \alpha\tau)|_{D_\alpha^s} = 0$ and $(d\beta + \beta\tau)|_{D_\beta^s} = 0$. Thus $\tau = 0$ on $S(TM)$ if and only if $d\alpha = 0$ on D_α^s and $d\beta = 0$ on D_β^s. Since M is screen homothetic , if $\tau = 0$, then $c = 0$ by Remark 2.5.5. Also, since $\alpha\beta = \varphi^{-1}\gamma$, which is the second equation of (2.5.10), is a constant, we have this lemma. □

From now on we let $(M, g, S(TM))$ be an Einstein screen homothetic lightlike hypersurface with the canonical null pair $\{\xi, N\}$).

Remark 2.5.12. If $0 < p < m$, then α and β are not constants along TM but constants along $S(TM)$. Indeed, as explained before $\tau = 0$. From Lemma 2.5.11, α and β are constants along $S(TM)$. Next, if α and β are constants along TM, from Lemma 2.5.10, we have $\eta(X) = 0$ for all $X \in D_\alpha$ and $\eta(U) = 0$ for all $U \in D_\beta$. Observe that a vector field X on M belongs to $S(TM)$ if and only if, locally on each $\mathcal{U} \subset M$, we have $\eta(X) = 0$. This implies that the distributions D_α and D_β are vector sub-bundles of $S(TM)$. Consequently we have $D_\alpha = D_\alpha^s$ and $D_\beta = D_\beta^s$. It is a contradiction to $\mathrm{Rad}(TM) \subset D_\alpha$ and $\mathrm{Rad}(TM) \subset D_\beta$. Thus α and β are not constants along TM.

Lemma 2.5.13. *If $0 < p < m$, then, for $X \in \Gamma(D_\alpha)$ and $U \in \Gamma(D_\beta)$, we have*

$$\nabla_X U \in \Gamma(D_\beta); \qquad \nabla_U X \in \Gamma(D_\alpha). \tag{2.5.12}$$

Proof. From (2.4.5) equipped with $\tau = 0$, we get

$$(\nabla_X B)(U, Z) = (\nabla_U B)(X, Z), \qquad \text{i.e.,}$$
$$g\left(\{(A_\xi^* - \beta P)\nabla_X U - (A_\xi^* - \alpha P)\nabla_U X\}, Z\right) = 0,$$

for any $Z \in \Gamma(TM)$. Since $S(TM)$ is non-degenerate, we have

$$(A_\xi^* - \beta P)\nabla_X U = (A_\xi^* - \alpha P)\nabla_U X.$$

Since the left term of the last equation is in $\Gamma(D_\alpha^s)$ and the right term is in $\Gamma(D_\beta^s)$ and $D_\alpha^s \cap D_\beta^s = \{0\}$, we have

$$(A_\xi^* - \beta P)\nabla_X U = 0, \qquad (A_\xi^* - \alpha P)\nabla_U X = 0.$$

This implies that $\nabla_X U \in \Gamma(D_\beta)$ and $\nabla_U X \in \Gamma(D_\alpha)$. $\qquad\square$

Lemma 2.5.14. *If $0 < p < m$, then, for $X, Y \in \Gamma(D_\alpha)$ and $U, V \in \Gamma(D_\beta)$, we have*

$$g(\nabla_Y X, U) = 0; \qquad g(X, \nabla_V U) = 0. \tag{2.5.13}$$

Proof. Since $g(X, PU) = 0$, we have

$$\nabla_Y(g(X, PU)) - g(\nabla_Y X, PU) - g(X, \nabla_Y PU)$$
$$= B(X, Y)\eta(PU) + B(Y, PU)\eta(X) = 0,$$
$$\nabla_V(g(U, PX)) - g(\nabla_V U, PX) + g(U, \nabla_V PX)$$
$$= B(V, U)\eta(PX) + B(V, PX)\eta(U) = 0.$$

Since $D_\alpha \perp_g D_\beta$ and $B(D_\alpha, D_\beta) = 0$, we have $g(\nabla_Y X, U) = g(\nabla_Y X, PU) = 0$ and $g(X, \nabla_V U) = g(PX, \nabla_V U) = 0$. $\qquad\square$

Since the leaf M' of $S(TM)$ is Riemannian and $S(TM) = D_\alpha^s \perp D_\beta^s$, where D_α^s and D_β^s are parallel distributions with respect to the induced connection ∇^* of M' due to (2.5.13), by the decomposition theorem of de Rham [334] we have $M' = M_\alpha \times M_\beta$, where M_α and M_β are some leaves of D_α^s and D_β^s respectively. Thus we have the following theorem:

Theorem 2.5.15. *Let $(M, g, S(TM))$ be an Einstein screen conformal lightlike hypersurface of a Lorentz manifold $(\bar{M}(c), \bar{g})$ of constant curvature c. Then M is locally a lightlike triple product manifold $C \times (M' = M_\alpha \times M_\beta)$, where C is a null curve, M' is an integral manifold of $S(TM)$, M_α and M_β are leaves of some distributions of M respectively.*

Lemma 2.5.16. *If $0 < p < m$, then $\gamma = \varphi\alpha\beta = 0$. In particular, $\alpha\beta = 0$.*

Proof. For $X \in \Gamma(D_\alpha)$ and $U \in \Gamma(D_\beta)$, we have

$$g(R(X, U)U, X) = g(\nabla_X \nabla_U U, X).$$

From the second equation of (2.5.12), we know that $\nabla_U U$ has no component of D_α. Since the projection morphism P maps $\Gamma(D_\alpha)$ onto $\Gamma(D_\alpha^s)$ and $\Gamma(D_\beta)$ onto $\Gamma(D_\beta^s)$, and $S(TM) = D_\alpha^s \perp D_\beta^s$, we have

$$\nabla_U U = P(\nabla_U U) + \eta(\nabla_U U)\xi; \quad P(\nabla_U U) \in \Gamma(D_\beta^s).$$

It follows that

$$\begin{aligned}
g(\nabla_X \nabla_U U, X) &= g(\nabla_X P(\nabla_U U), X) + \nabla_X(\eta(\nabla_U U)) g(\xi, X) \\
&\quad + \eta(\nabla_U U)g(\nabla_X \xi, X) = -\alpha \eta(\nabla_U U)g(X, X).
\end{aligned}$$

Since $\eta(\nabla_U U) = \varphi\beta\, g(U, U)$, we have

$$g(R(X, U)U, X) = -\varphi\alpha\beta g(X, X)g(U, U).$$

While, since M is screen homothetic and $\tau = 0$, by Remark 2.5.5, we show that $c = 0$. Thus, from the Gauss-Codazzi equations we have

$$g(R(X, U)U, X) = \varphi\alpha\beta g(X, X)g(U, U).$$

From the last two equations, we get $\gamma = \varphi\alpha\beta = 0$. Since φ is a non-zero constant, we have $\alpha\beta = 0$. $\qquad\square$

Using the above results, we have the following main theorem:

Theorem 2.5.17. [157] *Let $(M, g, S(TM))$ be an Einstein screen homothetic light-like hypersurface of an $(m + 2)$-dimensional Lorentz manifold $(\bar{M}(c), \bar{g})$ of constant curvature c. Then $c = 0$ and M is a locally lightlike triple product manifold $C \times (M' = M_\alpha \times M_\beta)$, where C is a null curve, M' is an integral manifold of $S(TM)$, M_α and M_β are leafs of some distributions of M such that*

(1) *If $k \neq 0$, either M_α or M_β is a totally umbilical Riemannian manifold of constant curvature $\varphi\alpha^2$ or $\varphi\beta^2$ which is isometric to an m-(pseudo)sphere and the other is a point.*

(2) *If $k = 0$, M_α is an $(m - 1)$ or an m-dimensional totally geodesic Euclidean space and M_β is a non-null curve or a point in $\bar{M}$.*

Proof. (1) Let $k \neq 0$: In case $(\text{tr } A_\xi^*)^2 \neq 4\varphi^{-1}\gamma$. The equation (2.5.9) has two non-vanishing distinct solutions α and β. If $0 < p < m$, then, by Lemma 2.5.14, we have $\gamma = 0$. Thus $p = 0$ or $p = m$. If $p = 0$, then α is not an eigenvalue of the shape operator A_ξ^* but a solution of the equation (2.5.7) and the equations (2.5.10) reduce to $s = \alpha + \beta = m\beta$; $\alpha\beta = \varphi^{-1}k$. Also if $p = m$, then β is not an eigenvalue of A_ξ^* but a solution of (2.5.9) and the equations (2.5.10) reduce to

$s = \alpha + \beta = m\alpha$; $\alpha\beta = \varphi^{-1}k$. Consequently, if $p = 0$ or $p = m$, then α and β are constants and $D^s_\alpha = \{0\}$; $S(TM) = D^s_\beta$ or $S(TM) = D^s_\alpha$; $D^s_\beta = \{0\}$ respectively. From (2.4.4) and (2.4.7), we have

$$R^*(X,Y)Z = \varphi\alpha^2\{g(Y,Z)X - g(X,Z)Y\}, \quad \forall X, Y, Z \in \Gamma(D_\alpha);$$
$$R^*(U,V)W = \varphi\beta^2\{g(V,W)U - g(U,W)V\}, \quad \forall U, V, W \in \Gamma(D_\beta).$$

Thus either M_α or M_β, which are leaves of D_α or D_β respectively, is a Riemannian manifold M^* of constant curvature either $\varphi\alpha^2$ or $\varphi\beta^2$ and the other leaf is a point $\{x\}$. If $M^* = M_\alpha$, since $B(X,Y) = \alpha g(X,Y)$ for all $X, Y \in \Gamma(S(TM))$, we have $C(X,Y) = \varphi\alpha g(X,Y)$ for all $X, Y \in \Gamma(S(TM))$. If $M^* = M_\beta$, since $B(U,V) = \beta g(U,V)$ for all $U, V \in \Gamma(S(TM))$, we have $C(U,V) = \varphi\beta g(U,V)$ for all $U, V \in \Gamma(S(TM))$. Thus the leaf M^* is a totally umbilical which is not totally geodesic. Consequently M is a locally product manifold $C \times M^* \times \{x\}$ or $C \times \{x\} \times M^*$, where M^* is a totally umbilical Riemannian manifold of constant curvature $\varphi\beta^2$ or $\varphi\alpha^2$ which is isometric to an m-(pseudo)sphere, and $\{x\}$ is a point.

In case $(\operatorname{tr} A^*_\xi)^2 = 4\varphi^{-1}\gamma$. The equation (2.5.9) has only one non-zero constant solution, named by α and α is only one eigenvalue of the shape operator A^*_ξ. In this case, the equations (2.5.10) reduce to $s = 2\alpha = m\alpha$; $\alpha^2 = \varphi^{-1}\gamma$. Thus we have $m = 2$. From (2.4.4) and (2.4.9), we have

$$R^*(X,Y)Z = k\{g(Y,Z)X - g(X,Z)Y\}, \quad \forall X, Y, Z \in \Gamma(S(TM)).$$

Thus the leaf M^* is a Riemannian 2-surface of constant curvature k. Since $B(X,Y) = \alpha g(X,Y)$ for all $X, Y \in \Gamma(TM)$, we have $C(X,Y) = \varphi\alpha g(X,Y)$ for all $X, Y \in \Gamma(S(TM))$. Thus the leaf M^* is also a totally umbilical which is not totally geodesic. Consequently M is a locally product $C \times M^* \times \{x\}$ where C is a null curve in $\bar{M}$ and M^* is a Riemannian 2-surface of constant curvature which is isometric to a 2-(pseudo)sphere.

(2) Let $k = 0$. The equation (2.5.9) reduces to

$$\lambda_i(\lambda_i - s) = 0, \quad 1 \le i \le m.$$

In case $\operatorname{tr} A^*_\xi \ne 0$. Let $\alpha = 0$ and $\beta = s$. Then we have $s = \beta = (m-p)\beta$, i.e., $(m - p - 1)\beta = 0$. So $p = m - 1$, i.e.,

$$A^*_\xi = \begin{pmatrix} 0 & & & \\ & \ddots & & \\ & & 0 & \\ & & & \beta \end{pmatrix}.$$

Consider the frame field of eigenvectors $\{\xi, E_1, \ldots, E_m\}$ of A^*_ξ such that $\{E_i\}_i$ is an orthonormal frame field of $S(TM)$, then $B(E_i, E_j) = C(E_i, E_j) = 0$ for $1 \le i < j \le m$. Thus the leaf M_α of D_α is a totally geodesic $(m-1)$-dimensional

Riemannian manifold and the leaf M_β of D_β is a curve. From (2.4.4) and (2.4.7), we have $\bar{g}(\bar{R}(E_i, E_j)E_j, E_i) = g(R^*(E_i, E_j)E_j, E_i) = 0$. Thus the sectional curvature K of the leaf M_α of D_α is given by

$$K(E_i, E_j) = \frac{g(R^*(E_i, E_j)E_j, E_i)}{g(E_i, E_i)g(E_j, E_j) - g^2(E_i, E_j)} = 0.$$

Thus M is a locally product $C \times M_\alpha \times M_\beta$, where M_α is an $(m-1)$-dimensional Euclidean space and M_β is a curve in $\bar{M}$.

In case $\operatorname{tr} A_\xi^* = 0$. Then we have $\alpha = \beta = 0$ and $A_\xi^* = 0$ or equivalently $B = 0$ and $D_\alpha^s = D_\beta^s = S(TM)$. Thus M is totally geodesic in $\bar{M}$. Since M is screen homothetic, we also have $C = A_N = 0$. Thus the leaf M^* of $S(TM)$ is also totally geodesic. Thus we have $\bar{\nabla}_X Y = \nabla_X^* Y$ for any tangent vector fields X and Y to the leaf M^*. This implies that M^* is a Euclidean m-space. Thus M is a locally product $C \times M^* \times \{x\}$ where C is a null curve and $\{x\}$ is a point. □

Remark 2.5.18. The classification of Einstein hypersurfaces M in Euclidean spaces $\mathbf{R}^{n+1}$ was first studied by Fialkow [188] and Thomas [393] in the middle of the 1930s. It was proved that if M is a connected Einstein hypersurface $(n \geq 3)$, that is $\operatorname{Ric} = \gamma g$, for some constant γ, then γ is non-negative. Moreover,

- if $\gamma = 0$ then M is locally isometric to $\mathbf{R}^n$ and

- if $\gamma > 0$ then M is contained in an n-sphere.

Also see in [22, 23] some more results on lightlike Einstein hypersurfaces.

2.6   Semi-symmetric hypersurfaces

In this section, we investigate lightlike hypersurfaces which are semi-symmetric, Ricci semi-symmetric, parallel or semi-parallel in a semi-Euclidean space. The class of semi-Riemannian manifolds, satisfying the condition

$$\nabla R = 0, \tag{2.6.1}$$

is a natural generalization of the class of manifolds of constant curvature where R denotes the corresponding curvature tensor. A semi-Riemannian manifold is called a *semi-symmetric manifold* if

$$\mathcal{R} \cdot R = 0, \tag{2.6.2}$$

where $\mathcal{R}$ is the curvature operator corresponding to R. Semi-symmetric hypersurfaces of Euclidean spaces were classified by Nomizu [310] and a general study of semi-symmetric Riemannian manifolds was made by Szabo [380].

A semi-Riemannian manifold is said to be a *Ricci semi-symmetric manifold* [124], if the following condition is satisfied:

$$\mathcal{R} \cdot \operatorname{Ric} = 0. \tag{2.6.3}$$

It is clear that every semi-symmetric manifold is Ricci semi-symmetric; the converse is not true in general. If a manifold M is immersed into a manifold $\bar{M}$, the immersion is called *parallel* if the second fundamental form is covariantly constant, i.e., $\nabla h = 0$, where ∇ is an affine connection $\bar{M}$ and h is the second fundamental form of the immersion. The general classification of parallel submanifolds of Euclidean space was obtained in [187] by D. Ferus. He showed that such an immersion is an isometric immersion into an n-dimensional symmetric R-space imbedded in R^{n+p} in the standard way. As a generalization of parallel hypersurfaces, the *semi-parallel hypersurfaces* were defined in [126].

Curvature conditions of symmetry type. Let (M, g) be a semi-Riemannian manifold. Recall from (1.2.7) that the curvature operator of M is given by

$$R(X, Y) = \nabla_X \nabla_Y - \nabla_Y \nabla_X - \nabla_{[X,Y]}$$

for $X, Y \in \Gamma(TM)$, where ∇ denotes the Levi-Civita connection on M.

For a $(0, k)$-tensor field T on M, $k \geq 1$, the $(0, k+2)$-tensor field $R \cdot T$ is defined by

$$(\mathcal{R} \cdot T)(X_1, \ldots, X_k, X, Y) = -T(R(X, Y)X_1, X_2, \ldots, X_k)$$
$$\ldots - T(X_1, \ldots, X_{k-1}, R(X, Y)X_k) \qquad (2.6.4)$$

for $X, Y, X_1, \ldots, X_k \in \Gamma(TM)$, where R denotes the semi-Riemannian curvature tensor of M. Curvature conditions, involving the form $\mathcal{R}.T = 0$, are called curvature conditions of *semi-symmetric type* [124].

Let M be a semi-symmetric semi-Riemannian manifold. Then, from (2.6.4) and properties of curvature tensor, we have

$$(\mathcal{R}(X, Y) \cdot R)(U, V)W = R(X, Y)R(U, V)W - R(U, V)R(X, Y)W$$
$$- R(R(X, Y)U, V)W - R(U, R(X, Y)V)W = 0 \quad (2.6.5)$$

for any $X, Y, U, V, W \in \Gamma(TM)$.

Let M be a Ricci semi-symmetric semi-Riemannian manifold. Then, from (2.6.4), we have

$$(\mathcal{R}(X, Y) \cdot \mathrm{Ric})(X_1, X_2) = - \mathrm{Ric}(R(X, Y)X_1, X_2)$$
$$- \mathrm{Ric}(X_1, R(X, Y)X_2)$$
$$= 0, \quad \forall X, Y, X_1, X_2 \in \Gamma(TM). \qquad (2.6.6)$$

In [126], Deprez defined and studied semi-parallel hypersurfaces in Euclidean n space. We recall that a hypersurface M of a semi-Riemannian manifold $\bar{M}$ is said to be semi-parallel if the following condition is satisfied for every point $p \in M$ and every vector field $X, Y, Z, W \in \Gamma(TM)$:

$$(R(X, Y)h)(Z, W) = -h(R(X, Y)Z, W) - h(Z, R(X, Y)W) = 0. \qquad (2.6.7)$$

Although the conditions (2.6.2) and (2.6.3) are not equivalent for manifolds in general, P.J. Ryan [341] raised the following question for hypersurfaces of Euclidean spaces in 1972: "Are the conditions $\mathcal{R} \cdot R = 0$ and $\mathcal{R} \cdot \mathrm{Ric} = 0$ equivalent for hypersurfaces of Euclidean spaces". Although there are many results which contributed to the solution of the above question in the affirmative under some conditions (see: [122], [123], [298], [387]), Abdalla and Dillen [1] gave an explicit example of a hypersurface in Euclidean space E^{n+1} ($n \geq 4$) that is Ricci semi-symmetric but not semi-symmetric (See also [124] for another example.). This result shows that the conditions $\mathcal{R} \cdot R = 0$ and $\mathcal{R} \cdot \mathrm{Ric} = 0$ are not equivalent for hypersurfaces of Euclidean space in general. A recent survey on Ricci semi-symmetric spaces and contributions to solution of the above problem can be found in [124]. We note that, in [402], I. Van de Woestijne and L. Verstraelen used the standard forms of a symmetric operator in a Lorentzian vector space to give an algebraic proof that the shape operator of a semi-symmetric hypersurface at a point with type number greater than 2 is diagonalizable with exactly two eigenvalues, one of which is zero. At this point we set $A_N \equiv A$, unless otherwise stated.

Proposition 2.6.1. *Let M be a lightlike hypersurface of a semi-Euclidean space $R_q^{(n+2)}$. Then the Gauss equation of M is given by*

$$R(X,Y)Z = B(Y,Z)AX - B(X,Z)AY \tag{2.6.8}$$

for any $X,Y,Z \in \Gamma(TM)$ and $N \in \Gamma(\mathrm{tr}(TM))$.

Proof. By assumption, $\bar{M} = R_q^{(n+2)}$ is a semi-Euclidean space, hence, $\bar{R} = 0$. Then, from (2.4.1) we have

$$R(X,Y)Z + A_{h(X,Z)}Y - A_{h(Y,Z)}X + (\nabla_X h)(Y,Z) - (\nabla_Y h)(X,Z) = 0.$$

We note that $(\nabla_X h)(Y,Z)$ is defined by

$$(\nabla_X h)(Y,Z) = \nabla_X^t h(Y,Z) - h(\nabla_X Y, Z) - h(Y, \nabla_X Z). \tag{2.6.9}$$

On the other hand, (2.1.9) and (2.1.13) imply that $h(X,Y) = B(X,Y)N$ for $X,Y \in \Gamma(TM)$ and $N \in \Gamma(\mathrm{tr}(TM))$. Thus, we get

$$R(X,Y)Z + B(X,Z)A_N Y - B(Y,Z)A_N X + (\nabla_X h)(Y,Z) - (\nabla_Y h)(X,Z) = 0.$$

Then, (2.6.8) holds by comparing the tangential and transversal parts. $\square$

Definition 2.6.2. A lightlike hypersurface M of a semi-Euclidean space is semi-symmetric if the following condition is satisfied:

$$(R(X,Y) \cdot R)(X_1, X_2, X_3, X_4) = 0, \forall X, Y, X_1, X_2, X_3, X_4 \in \Gamma(TM). \tag{2.6.10}$$

It is easy to see that

$$(R(X,Y) \cdot R)(X_1, X_2, X_3, \xi) = 0, \quad \xi \in \Gamma(\mathrm{Rad}\, TM).$$

Thus the condition (2.6.10) is equivalent to the condition

$$(R(X,Y) \cdot R)(X_1, X_2, X_3, PX_4) = 0 \tag{2.6.11}$$

for $X, Y, X_1, X_2, X_3, X_4 \in \Gamma(TM)$. We also note that (2.6.10) and (2.6.11) do not imply (2.6.5) due to $g(R(X,Y)Z, W) \neq -g(R(X,Y)W, Z)$ in general, for $X, Y, Z, W \in \Gamma(TM)$. From (2.6.11) and (2.6.8), we obtain

$$\begin{aligned}
(R(X,Y) \cdot R)(X_1, X_2, X_3, PX_4) = & \, B(Y, X_1)[B(AX, X_3)g(AX_2, PX_4) \\
& - B(X_2, X_3)g(A^2X, PX_4)] + B(X, X_1)[B(X_2, X_3)g(A^2Y, PX_4) \\
& - B(AY, X_3)g(AX_2, PX_4)] + g(AX_1, PX_4)[-B(Y, X_2)B(AX, X_3) \\
& + B(X, X_2)B(AY, X_3)] + B(X_1, X_3)[B(Y, X_2)g(A^2X, PX_4) \\
& - B(X, X_2)g(A^2Y, PX_4)] + g(AX_1, PX_4)[-B(X_3, Y)B(X_2, AX) \\
& + B(X, X_3)B(X_2, AY)] + g(AX_2, PX_4)[B(X_3, Y)B(X_1, AX) \\
& - B(X, X_3)B(X_1, AY)] + B(X_2, X_3)[-B(Y, X_4)g(AX_1, AX) \\
& + B(X, PX_4)g(AX_1, AY)] + B(X_1, X_3)[B(Y, PX_4)g(AX_2, AX) \\
& - B(X, PX_4)g(AX_2, AY)] \tag{2.6.12}
\end{aligned}$$

for $X, Y, X_1, X_2, X_3, X_4 \in \Gamma(TM)$.

Proposition 2.6.3. *Every screen conformal lightlike hypersurface of the Minkowski spacetime is a semi-symmetric lightlike hypersurface.*

Proof. First, from (2.6.12), we have

$$\begin{aligned}
(R(X,Y) \cdot R)(\xi, X_2, X_3, PX_4) = & \, B(Y, \xi)[B(AX, X_3)g(AX_2, PX_4) \\
& - B(X_2, X_3)g(A^2X, PX_4)] \\
& + B(X, \xi)[B(X_2, X_3)g(A^2Y, PX_4) - B(AY, X_3)g(AX_2, PX_4)] \\
& + g(A\xi, PX_4)[-B(Y, X_2)B(AX, X_3) + B(X, X_2)B(AY, X_3)] \\
& + B(\xi, X_3)[B(Y, X_2)g(A^2X, PX_4) - B(X, X_2)g(A^2Y, PX_4)] \\
& + g(A\xi, PX_4)[-B(X_3, Y)B(X_2, AX) + B(X, X_3)B(X_2, AY)] \\
& + g(AX_2, PX_4)[B(X_3, Y)B(\xi, AX) - B(X, X_3)B(\xi, AY)] \\
& + B(X_2, X_3)[-B(Y, PX_4)B(A\xi, AX + B(X, PX_4)g(A\xi, AY)] \\
& + B(\xi, X_3)[B(Y, PX_4)g(AX_2, AX) - B(X, PX_4)g(AX_2, AY)]
\end{aligned}$$

for any $X, Y, X_2, X_3, X_4 \in \Gamma(TM)$ and $\xi \in \Gamma(\operatorname{Rad} TM)$. From (2.1.25), we get

$$\begin{aligned}
(R(X,Y) \cdot R)(\xi, X_2, X_3, PX_4) = & \, g(A\xi, PX_4)[-B(Y, X_2)B(AX, X_3) \\
& + B(X, X_2)B(AY, X_3)] \\
& + g(A\xi, PX_4)[-B(X_3, Y)B(X_2, AX) + B(X, X_3)B(X_2, AY)] \\
& + B(X_2, X_3)[-B(Y, PX_4)B(A\xi, AX + B(X, PX_4)g(A\xi, AY)].
\end{aligned}$$

Then, equation (2.2.1) of screen conformal M implies that

$$(R(X,Y) \cdot R)(\xi, X_2, X_3, PX_4) = \varphi g(A^*_\xi \xi, PX_4)[-B(Y, X_2)B(AX, X_3)$$
$$+ B(X, X_2)B(AY, X_3)]$$
$$+ \varphi g(A^*_\xi \xi, PX_4)[-B(X_3, Y)B(X_2, AX) + B(X, X_3)B(X_2, AY)]$$
$$+ \varphi B(X_2, X_3)[-B(Y, PX_4)B(A^*_\xi \xi, AX + B(X, PX_4)g(A^*_\xi \xi, AY)].$$

From (2.1.25) and (2.1.26), we have $A^*_\xi \xi = 0$. Thus, we derive

$$R(X,Y) \cdot R)(\xi, X_2, X_3, PX_4) = 0.$$

In a similar way, we obtain

$$(R(X,Y) \cdot R)(X_1, X_2, \xi, PX_4) = 0, \qquad (R(\xi, Y) \cdot R)(X_1, X_2, X_3, PX_4) = 0,$$
$$(R(X,Y) \cdot R)(X_1, \xi, X_3, PX_4) = 0, \qquad (R(X, \xi) \cdot R)(X_1, X_2, X_3, PX_4) = 0$$

for $X_1, X_2, X_3, X_4 \in \Gamma(TM)$ and $\xi \in \Gamma(TM^\perp)$. Let $\{X_1, X_2, \xi, N\}$ be a quasi-orthonormal basis of R^4_1 such that $S(TM) = \text{Span}\{X_1, X_2\}$ and $\text{tr}(TM) = \text{Span}\{N\}$. From (2.6.12), we have

$$(R(X_1, X_2) \cdot R)(X_1, X_2, X_1, X_2) = B(X_2, X_1)[B(AX_1, X_1)g(AX_2, X_2)$$
$$- B(X_2, X_1)g(A^2 X_1, PX_2)]$$
$$+ B(X_1, X_1)[B(X_2, X_1)g(A^2 X_2, X_2) - B(AX_2, X_31)g(AX_2, PX_2)]$$
$$+ g(AX_1, X_2)[-B(X_2, X_2)B(AX_1, X_1) + B(X_1, X_2)B(AX_2, X_1)]$$
$$+ B(X_1, X_1)[B(X_2, X_2)g(A^2 X_1, X_2) - B(X_1, X_2)g(A^2 X_2, X_2)]$$
$$+ g(AX_1, X_2)[-B(X_1, X_2)B(X_2, AX_1) + B(X_1, X_1)B(X_2, AX_2)]$$
$$+ g(AX_2, X_2)[B(X_1, X_2)B(X_1, AX_1) - B(X_1, X_1)B(X_1, AX_2)]$$
$$+ B(X_2, X_1)[-B(X_2, X_2)B(AX_1, AX_1 + B(X_1, X_2)g(AX_1, AX_2)]$$
$$+ B(X_1, X_1)[B(X_2, X_2)g(AX_2, AX_1) - B(X_1, X_2)g(AX_2, AX_2)].$$

Since $A_N X \in \Gamma(S(TM))$ for any $X \in \Gamma(TM)$ and $N \in \Gamma(\text{tr}(TM))$ and $A = A_N$ is self-adjoint on $S(TM)$ as M is screen conformal, we get

$$(R(X_1, X_2) \cdot R)(X_1, X_2, X_1, X_2) = B(X_2, X_1)[B(AX_1, X_1)g(AX_2, X_2)$$
$$- B(X_2, X_1)g(AX_1, AX_2)]$$
$$+ B(X_1, X_1)[B(X_2, X_1)g(AX_2, AX_2) - B(AX_2, X_1)g(AX_2, X_2)]$$
$$+ g(AX_1, X_2)[-B(X_2, X_2)B(AX_1, X_1) + B(X_1, X_2)B(AX_2, X_1)]$$
$$+ B(X_1, X_1)[B(X_2, X_2)g(AX_1, AX_2) - B(X_1, X_2)g(AX_2, AX_2)]$$
$$+ g(AX_1, X_2)[-B(X_1, X_2)B(X_2, AX_1) + B(X_1, X_1)B(X_2, AX_2)]$$
$$+ g(AX_2, X_2)[B(X_1, X_2)B(X_1, AX_1) - B(X_1, X_1)B(X_1, AX_2)]$$
$$+ B(X_2, X_1)[-B(X_2, X_2)g(AX_1, AX_1 + B(X_1, X_2)g(AX_1, AX_2)]$$
$$+ B(X_1, X_1)[B(X_2, X_2)g(AX_2, AX_1) - B(X_1, X_2)g(AX_2, AX_2)].$$

Then, again using (2.2.1), we arrive at

$$
\begin{aligned}
(R(X_1, X_2) \cdot R)(X_1, X_2, X_1, X_2) = {}& \varphi B(X_2, X_1)[B(AX_1, X_1)g(A_\xi^* X_2, X_2) \\
& - B(X_2, X_1)g(A_\xi^* X_1, AX_2)] \\
& + \varphi B(X_1, X_1)[B(X_2, X_1)g(A_\xi^* X_2, AX_2) - B(AX_2, X_1)g(A_\xi^* X_2, X_2)] \\
& + \varphi g(A_\xi^* X_1, X_2)[-B(X_2, X_2)B(AX_1, X_1) + B(X_1, X_2)B(AX_2, X_1)] \\
& + \varphi B(X_1, X_1)[B(X_2, X_2)g(A^* \xi X_1, AX_2) - B(X_1, X_2)g(A_\xi^* X_2, AX_2)] \\
& + \varphi g(A_\xi^* X_1, X_2)[-B(X_1, X_2)B(X_2, AX_1) + B(X_1, X_1)B(X_2, AX_2)] \\
& + \varphi g(A_\xi^* X_2, X_2)[B(X_1, X_2)B(X_1, AX_1) - B(X_1, X_1)B(X_1, AX_2)] \\
& + \varphi B(X_2, X_1)[-B(X_2, X_2)g(A_\xi^* X_1, AX_1 + B(X_1, X_2)g(A_\xi^* X_1, AX_2)] \\
& + \varphi B(X_1, X_1)[B(X_2, X_2)g(A_\xi^* X_2, AX_1) - B(X_1, X_2)g(A_\xi^* X_2, AX_2)].
\end{aligned}
$$

Thus, using (2.1.25), we obtain

$$
\begin{aligned}
(R(X_1, X_2) \cdot R)(X_1, X_2, X_1, X_2) = {}& \varphi B(X_2, X_1)[B(AX_1, X_1)B(X_2, X_2) \\
& - B(X_2, X_1)B(X_1, AX_2)] \\
& + \varphi B(X_1, X_1)[B(X_2, X_1)B(X_2, AX_2) - B(AX_2, X_1)B(X_2, X_2)] \\
& + \varphi B(X_1, X_2)[-B(X_2, X_2)B(AX_1, X_1) + B(X_1, X_2)B(AX_2, X_1)] \\
& + \varphi B(X_1, X_1)[B(X_2, X_2)B(X_1, AX_2) - B(X_1, X_2)B(X_2, AX_2)] \\
& + \varphi B(X_1, X_2)[-B(X_1, X_2)B(X_2, AX_1) + B(X_1, X_1)B(X_2, AX_2)] \\
& + \varphi B(X_2, X_2)[B(X_1, X_2)B(X_1, AX_1) - B(X_1, X_1)B(X_1, AX_2)] \\
& + \varphi B(X_2, X_1)[-B(X_2, X_2)B(X_1, AX_1 + B(X_1, X_2)B(X_1, AX_2)] \\
& + \varphi B(X_1, X_1)[B(X_2, X_2)B(X_2, AX_1) - B(X_1, X_2)B(X_2, AX_2)].
\end{aligned}
$$

Since B is symmetric, by direct computations, we get

$$
\begin{aligned}
(R(X_1, X_2) \cdot R)(X_1, X_2, X_1, X_2) = {}& \varphi\{(B(X_2, X_1))^2 B(X_1, AX_2) \\
& - (B(X_1, X_2))^2 B(X_2, AX_1) \\
& - B(X_2, X_2)B(X_1, X_1)B(X_1, AX_2) \\
& + B(X_1, X_1)B(X_2, X_2)B(X_2, AX_1)\}.
\end{aligned}
\tag{2.6.13}
$$

On the other hand, from (2.1.25) and (2.2.1), we have

$$
B(AX_2, X_1) = g(A_\xi^* X_1, AX_2) = g(\varphi A_\xi^* X_1, A_\xi^* X_2) = g(AX_1, A_\xi^* X_2).
$$

Thus, again using (2.1.25), we get

$$
B(AX_2, X_1) = B(X_2, AX_1). \tag{2.6.14}
$$

Then, from (2.6.13) and (2.6.14), we obtain

$$
(R(X_1, X_2) \cdot R)(X_1, X_2, X_1, X_2) = 0.
$$

In a similar way, we have

$$(R(X_1, X_2) \cdot R)(X_1, X_1, X_2, X_2) = (R(X_1, X_2) \cdot R)(X_2, X_1, X_1, X_2) = 0,$$
$$(R(X_1, X_2) \cdot R)(X_2, X_1, X_2, X_1) = (R(X_1, X_2) \cdot R)(X_2, X_2, X_1, X_1) = 0,$$

and

$$(R(X_1, X_2).R)(X_1, X_2, X_2, X_1) = 0.$$

Thus the proof is complete.   □

Remark 2.6.4. From Proposition 2.6.3, it follows that a lightlike cone of $\mathbf{R}_1^4$, a lightlike Monge hypersurface of $\mathbf{R}_1^4$ and lightlike surfaces of $\mathbf{R}_1^3$ are examples of semi-symmetric lightlike hypersurfaces. We also note that Proposition 2.6.3 is valid for a semi-Euclidean space $\mathbf{R}^4{}_q$, $1 \le q < 4$.

Let M be a screen conformal lightlike hypersurface of an $(m+2)$-dimensional semi-Euclidean space. Then, its screen distribution $S(TM)$ is integrable. We denote a leaf of $S(TM)$ by M'. Then, we have the following theorem:

Theorem 2.6.5. *Let M be a screen conformal lightlike hypersurface of an $(m+2)$-dimensional semi-Euclidean space, $m \ge 3$. Then M is semi-symmetric if and only if any leaf M' of $S(TM)$ is semi-symmetric in semi-Euclidean space, that is, the curvature tensor of M' satisfies the condition (2.6.5) in semi-Euclidean space.*

Proof. Using (2.6.8) and (2.2.1) we obtain

$$g(R(X,Y)PZ, PW) = \varphi\{B(Y, Z)B(X, PW)$$
$$- B(X, Z)B(Y, PW)\} \qquad (2.6.15)$$

for any $X, Y, Z, W \in \Gamma(TM)$. Then, by straightforward computations, using (2.1.23)- (2.1.26) and (2.2.1), we get

$$g(R(X,Y)PZ, PW) = g(R^*(X,Y)PZ, PW) - \varphi\{B(Y, PZ)B(X, PW)$$
$$+ B(X, PZ)B(Y, PW)\} \qquad (2.6.16)$$

for any $X, Y, Z, W \in \Gamma(TM)$. Thus, from (2.6.15) and (2.6.16), we derive

$$g(R(X,Y)PZ, PW) = \frac{1}{2}g(R^*(X,Y)PZ, PW). \qquad (2.6.17)$$

On the other hand, from (2.1.26) and (2.6.8), we get

$$g(R(X,Y)Z, N) = 0, \forall X, Y, Z \in \Gamma(TM), N \in \Gamma(\mathrm{tr}(TM)). \qquad (2.6.18)$$

Thus, from (2.6.17) and (2.6.18), we conclude that

$$R(X,Y)PZ = \frac{1}{2}R^*(X,Y)PZ.$$

Hence, using algebraic properties of the curvature tensor field, we have

$$(R(X,Y) \cdot R)(U,V,W,Z) = \frac{1}{4}(R^*(X,Y) \cdot R^*)(U,V,W,Z)$$

for any $X, Y, U, V, W \in \Gamma(S(TM))$, which completes the proof.   □

Remark 2.6.6. The above theorem shows that the semi-symmetry of a screen conformal lightlike hypersurface of a semi-Euclidean space is related with the semi-symmetry of a leaf M' of its integrable screen distribution. In the Lorentzian case, since the screen is Riemannian, studying semi-symmetry of a screen conformal lightlike hypersurface is the same with a Riemannian manifold. In fact, it follows from the proof of Theorem 2.6.5 that the curvature conditions of a screen conformal lightlike hypersurface reduces to the curvature conditions of a leaf of its screen distribution.

Lemma 2.6.7. *Let M be a lightlike hypersurface of semi-Euclidean $(m+2)$-space. Then the Ricci tensor* Ric *of M is given by*

$$\mathrm{Ric}(X,Y) = -\sum_{i=1}^{m} \epsilon_i B(X,Y)C(w_i, wi) - g(A_{\xi}^* Y, AX), \quad \epsilon_i = \pm 1, \qquad (2.6.19)$$

for any $X, Y \in \Gamma(TM)$ and $\{w_i\}_{i=1}^{m}$ is an orthonormal basis of $S(TM)$.

Proof. From (2.1.25) and (2.6.8), we have

$$\mathrm{Ric}(X,Y) = -\sum_{i=1}^{m} \epsilon_i \{B(X,Y)C(w_i, wi) - B(Y, w_i)C(X, w_i)\}$$

for any $X, Y \in \Gamma(TM), \xi \in \Gamma(TM^{\perp})$ and $N \in \Gamma(\mathrm{tr}(TM))$, where $\{w_i\}_{i=1}^{m}$ is a basis of $S(TM)$. Using (2.1.25) and (2.1.26), we get

$$\mathrm{Ric}(X,Y) = -\sum_{i=1}^{m} \epsilon_i \{B(X,Y)C(w_i, wi) - g(\sum_{i=1}^{m} \epsilon_i g(A_{\xi}^* Y, w_i)w_i, AX)\}.$$

Hence, we have (2.6.19), which completes the proof.   □

Definition 2.6.8. A lightlike hypersurface M of a semi-Euclidean space is Ricci semi-symmetric if the following condition is satisfied:

$$(R(X,Y) \cdot \mathrm{Ric})(X_1, X_2) = 0 \qquad (2.6.20)$$

for $X, Y, X_1, X_2 \in \Gamma(TM)$.

Next we prove a result which shows the effect of the Ricci semi-symmetric condition on the geometry of lightlike hypersurfaces.

Theorem 2.6.9. *Let M be a Ricci semi-symmetric lightlike hypersurface of an $(m + 2)$-dimensional semi-Euclidean space. Then either M is totally geodesic or $\mathrm{Ric}(\xi, A\xi) = 0$ for $\xi \in \Gamma(TM^\perp)$.*

Proof. From (2.6.8), (2.6.6) and (2.6.20), we obtain

$$
\begin{aligned}
(R(X,Y) \cdot \mathrm{Ric})(X_1, X_2) = \alpha\{ & -B(X, X_1)B(AY, X_2) + B(Y, X_1)B(AX, X_2) \\
& - B(X, X_2)B(X_1, AY) + B(Y, X_2)B(X_1, AX)\} \\
& - B(X, X_1)B(AX_2, AY) + B(Y, X_1)B(AX_2, AX) \\
& - B(X, X_2)B(AY, AX_1) + B(Y, X_2)B(AX, AX_1)
\end{aligned}
$$

for $X, Y, X_1, X_2 \in \Gamma(TM)$, where $\alpha = -\sum_{i=1}^{m} \epsilon_i C(w_i, wi)$. Now, suppose that M is a Ricci semi-symmetric lightlike hypersurface. Taking $X_1 = \xi$ in the above equation and using (2.1.25), we obtain

$$
-B(X, X_2)B(AY, A\xi) + B(Y, X_2)B(AX, A\xi) = 0.
$$

Hence for $Y = \xi$ we derive

$$
B(X, X_2)B(A\xi, A\xi) = 0.
$$

So, if $B(X, X_2) = 0$ for any $X, X_2 \in \Gamma(TM)$, then M is totally geodesic. If M is not totally geodesic, it follows that $B(A\xi, A\xi) = 0$, then from (2.6.19) we obtain $\mathrm{Ric}(\xi, A\xi) = 0$. $\qquad\square$

Theorem 2.6.10. *Let M be a lightlike hypersurface of a semi-Euclidean $(m + 2)$-space such that $\mathrm{Ric}(\xi, X) = 0$, $\forall X \in \Gamma(TM)$, $\xi \in \Gamma(TM^\perp$ and $A\xi$ is a non-null vector field. Then M is semi-symmetric if and only if M is totally geodesic.*

Proof. Suppose that M is a semi-symmetric lightlike hypersurface of a semi-Euclidean $(m + 2)$-space. Taking $X_1 = \xi$ in (2.6.12), we obtain

$$
\begin{aligned}
\{-B(Y, X_2)B(AX, X_3) + B(X, X_2)B(AY, X_3)\}g(A\xi, PX_4) \\
\{-B(X_3, Y)B(X_2, AX) + B(X, X_3)B(X_2, AY)\}g(A\xi, PX_4) \\
\{-B(Y, PX_4)g(A\xi, AX) + B(X, PX_4)g(A\xi, AY)\}B(X_2, X_3) = 0.
\end{aligned}
$$

Then, for $Y = \xi$, we have

$$
\begin{aligned}
B(X, X_2)B(A\xi, X_3)g(A\xi, PX_4) + B(X, X_3)B(X_2, A\xi)g(A\xi, PX_4) \\
+ B(X, PX_4)g(A\xi, A\xi)B(X_2, X_3) = 0.
\end{aligned}
$$

Thus, by assumption, $\mathrm{Ric}(\xi, X) = 0$, we have $B(X, A\xi) = 0$. Hence, we get

$$
B(X, PX_4)g(A\xi, A\xi)B(X_2, X_3) = 0.
$$

Since $A\xi$ is a non-null vector field by hypothesis, for $X = X_3$ and $X_4 = X_2$ we arrive at

$$
B(X_2, X_3) = 0.
$$

Thus, M is totally geodesic. The converse is clear from (2.6.12). $\qquad\square$

For Lorentzian space $\mathbf{R}_1^{(m+2)}$, we have the following:

Corollary 2.6.11. *Let M be a lightlike hypersurface of a Lorentzian space $R_1^{(m+2)}$ such that $\mathrm{Ric}(\xi, X) = 0$, $\forall X \in \Gamma(TM)$, $\xi \in \Gamma(TM^\perp)$. Then M is totally geodesic if and only if M is semi-symmetric.*

Proof. If M is a lightlike hypersurface of $\mathbf{R}_1^{(m+2)}$. Then the screen distribution of M is a Riemannian vector bundle. From (2.1.26), we can see that $AX \in \Gamma(S(TM))$, $\forall X \in \Gamma(TM)$. Then, the proof follows from Theorem 2.6.10. $\qquad\square$

Theorem 2.6.12. *The second fundamental form of a lightlike hypersurface M of a Lorentzian manifold is parallel if and only if M is totally geodesic.*

Proof. Suppose that the second fundamental form h of M is parallel. Then, from (2.6.9) and (2.1.14) we have

$$(\nabla_X h)(Y, Z) = X(B(Y, Z)N) - B(\nabla_X Y, Z)N - B(Y, \nabla_X Z)N = 0. \quad (2.6.21)$$

Thus, from (2.1.25), for $Y = \xi$, we obtain $B(\nabla_X \xi, Z)N = 0$. Using (2.1.20), we have $B(A_\xi^* X, Z)N = 0$. Hence, $B(A_\xi^* X, Z) = 0$. From (2.1.25) we assume that $Z \in \Gamma(S(TM))$. Thus, $g(A_\xi^* X, A_\xi^* Z) = 0$. Then, for $X = Z$ we get $g(A_\xi^* X, A_\xi^* X) = 0$. On the other hand, any screen distribution $S(TM)$ of a lightlike hypersurface of a Lorentzian manifold is Riemannian. Then, we have $A_\xi^* X = 0$ for any $X \in \Gamma(TM)$. Thus, proof follows from this and (2.1.25). The converse is clear. $\qquad\square$

Theorem 2.6.13. *Let M be a semi-parallel lightlike hypersurface of a semi-Euclidean $(m + 2)$-space. Then either M is totally geodesic or $C(\xi, A_\xi^* U) = 0$ for any $U \in \Gamma(S(TM))$ and $\xi \in \Gamma(TM^\perp)$, where C and A_ξ^* are the second fundamental form and shape operator of $S(TM)$, respectively.*

Proof. Since M is a semi-parallel lightlike hypersurface, we have

$$h(R(X, Y)Z, W) + h(Z, R(X, Y)W) = 0.$$

By using (2.6.8), we obtain

$$B(X, Z)B(AY, W) - B(Y, Z)B(AX, W) + B(X, W)B(Z, AY)$$
$$- B(Y, W)B(AX, Z) = 0 \qquad (2.6.22)$$

for any $X, Y, Z, W \in \Gamma(TM)$. Then, from (2.1.25) and (2.6.22), for $X = \xi$, we have

$$B(Y, Z)B(A\xi, W) + B(Y, W)B(A\xi, Z) = 0.$$

Thus, for $Z = W$, we obtain $B(Y, Z)B(A\xi, Z) = 0$. Now, if $B(Y, Z) = 0$, then M is totally geodesic. If $B(Y, Z) \neq 0$, then from (2.1.26), we have $C(\xi, A_\xi^* U) = 0$ for any $U \in \Gamma(S(TM))$. $\qquad\square$

Example 10. Consider a hypersurface M in $\mathbf{R}_2^4$ given by

$$x_1 = x_2 + \sqrt{2}\sqrt{x_3{}^2 + x_4{}^2}.$$

Then, it is easy to check that M is a lightlike hypersurface. Its radical distribution is spanned by

$$\xi = \sqrt{x_3{}^2 + x_4{}^2}\partial x_1 - \sqrt{x_3{}^2 + x_4{}^2}\partial x_2 + \sqrt{2}x_3\partial x_3 + \sqrt{2}x_4\partial x_4.$$

Then the lightlike transversal vector bundle is spanned by

$$\operatorname{tr}(TM) = \operatorname{Span}\{N = \frac{1}{4(x_3{}^2 + x_4{}^2)}(-\sqrt{x_3{}^2 + x_4{}^2}\,\partial x_1 + \sqrt{x_3{}^2 + x_4{}^2}\,\partial x_2$$
$$+ \sqrt{2}\,x_3\,\partial x_3 + \sqrt{2}\,x_4\,\partial x_4)\}.$$

It follows that the corresponding screen distribution $S(TM)$ is spanned by

$$\{Z_1 = \partial x_1 + \partial x_2, \; Z_2 = -x_4\,\partial x_3 + x_3\,\partial x_4\}.$$

By direct computations, we obtain

$$\bar{\nabla}_X Z_1 = \bar{\nabla}_{Z_1} X = 0, \; \bar{\nabla}_\xi \xi = \sqrt{2}\xi, \bar{\nabla}_{Z_2}\xi = \bar{\nabla}_\xi Z_2 = \sqrt{2}Z_2,$$

and

$$\bar{\nabla}_{Z_2} Z_2 = -x_3\,\partial x_3 - x_4\,\partial x_4$$

for any $X \in \Gamma(TM)$. Then, by using the Gauss formula, we obtain

$$\nabla_X Z_1 = 0, \; \nabla_{Z_2} Z_2 = -\frac{1}{2\sqrt{2}}\xi, \nabla_\xi Z_2 = \nabla_{Z_2}\xi = \sqrt{2}Z_2, \; \nabla_{Z_1} Z = 0$$

and

$$B(Z_2, Z_2) = -\sqrt{2}(x_3{}^2 + x_4{}^2), B(Z_1, Z_2) = 0, B(Z_1, Z_1) = 0.$$

On the other hand, we have

$$\bar{\nabla}_\xi N = \frac{1}{2\sqrt{2}\sqrt{x_3{}^2 + x_4{}^2}}\,\partial x_1 - \frac{1}{2\sqrt{2}\sqrt{x_3{}^2 + x_4{}^2}}\,\partial x_2$$
$$- \frac{1}{2}\frac{x_3}{(x_3{}^2 + x_4{}^2)}\,\partial x_3 - \frac{1}{2}\frac{x_4}{(x_3{}^2 + x_4{}^2)}\,\partial x_4,$$

$$\bar{\nabla}_{Z_1} N = 0,$$

$$\bar{\nabla}_{Z_2} N = -\frac{x_4}{2\sqrt{2}(x_3{}^2 + x_4{}^2)}\,\partial x_3 + \frac{x_3}{2\sqrt{2}(x_3{}^2 + x_4{}^2)}\,\partial x_4.$$

Thus, from the Weingarten formula (2.1.15), we have

$$A_N\xi = 0, \; A_N Z_1 = 0, \; A_N Z_2 = \frac{1}{2\sqrt{2}(x_3{}^2 + x_4{}^2)}Z_2.$$

Then, from the above equations the following equations are satisfied:

$$(R(Z_1, Z_2)\, h)(Z_1, Z_1) = 0\,,\, (R(Z_1, Z_2)\, h)(Z_1, Z_2) = 0\,,\, (R(Z_1, Z_2)\, h)(Z_2, Z_2) = 0.$$

Finally, using (2.1.25) and definition of $(R(X, Y).h)$, we have $R(X, Y)h)(U, \xi) = 0$ for any $X, Y, U \in \Gamma(TM)$ and $\xi \in \Gamma(TM^\perp)$. Thus, M is a non-totally geodesic semi-parallel hypersurface of $\mathbf{R}_2^4$.

Chapter 3

Applications of lightlike hypersurfaces

In this chapter we present the latest work on applications of lightlike hypersurfaces in two active ongoing research areas in mathematical physics. First, we deal with *black hole horizons*. We prove a *Global Null Splitting Theorem* and relate it with physically significant works of Galloway [197] on null hypersurfaces in general relativity, Ashtekar and Krishnan's work [16] on *dynamical horizons* and Sultana-Dyer's work [378, 379] on *conformal Killing horizons*, with related references. Secondly, we present the latest work on *Osserman lightlike hypersurfaces* [20].

3.1 Global null splitting theorem

Kinematic quantities of spacetimes. Let (M, g) be a 4-dimensional spacetime manifold of general relativity. This means that M is a smooth (C^∞) connected Hausdorff 4-dimensional manifold and g is a time orientable Lorentz metric of normal hyperbolic signature $(- + ++)$. The continuity of M has been observed experimentally for distances down to 10^{-15} cms and, therefore, should be sufficient for the general theory of relativity unless the density reaches to about $10^{58} gms/cm^3$. In this book we assume that the density is sufficient to maintain the continuity of the spacetime under investigation.

The set of all integral curves given by a unit timelike or spacelike or null vector field u is called the *congruence of timelike or spacelike or null curves*. We first consider timelike curves, also called *flow lines*. The *acceleration of the flow lines* along u is given by $\nabla_u u$ or $u^a{}_{;b} u^b$ where ∇ is the Levi-Civita connection on (M, g) and $(0 \le a, b \le 3)$. The *projective tensor*, defined by

$$h_{ab} = g_{ab} + u_a u_b, \tag{3.1.1}$$

is used to project a tangent vector at a point p in the spacetime into a spacelike vector orthogonal to u at p. The rate of change of the separation of flow lines from a timelike curve, say C, tangent to u is given by the *expansion tensor*

$$\theta_{ab} = h_a^c h_b^d u_{(c\,;\,d)}. \tag{3.1.2}$$

The *volume expansion* θ, the *shear tensor* σ_{ab}, the *vorticity tensor* ω_{ab} and the *vorticity vector* ω^a are defined as follows:

$$\theta = \operatorname{div} u = \theta_{ab} h^{ab}, \tag{3.1.3}$$

$$\sigma_{ab} = \theta_{ab} - \frac{\theta}{3} h_{ab}, \tag{3.1.4}$$

$$\omega_{ab} = h_a^c h_b^d u_{[c;d]}, \tag{3.1.5}$$

$$\omega^a = \frac{1}{2}\eta^{abcd} u_b \omega_{cd}, \tag{3.1.6}$$

$$\eta^{abcd} = g^{ae} g^{bf} g^{cg} g^{dh} \eta_{efgh},$$

$$\eta_{efgh} = (4!)\sqrt{-g}\,\delta_{[e}^0 \delta_f^1 \delta_g^2 \delta_{h]}^3,$$

where η^{abcd} is the Levi-Civita volume-form. The equation (3.1.5) measures the rate at which the timelike curves rotate about an integral curve of u. The covariant derivative of u can be decomposed as

$$u_{a;b} = \omega_{ab} + \sigma_{ab} + \frac{\theta}{3} h_{ab} - u_b(u_{;a}^c u_c). \tag{3.1.7}$$

The rate at which the expansion θ changes along the flow lines of u is given by the following *Raychaudhuri equation*:

$$u\theta = \frac{d\theta}{ds} = -R_{ab}u^a u^b + 2\omega^2 - 2\sigma^2 - \frac{1}{3}\theta^2 + \operatorname{div}(\nabla_u u), \tag{3.1.8}$$

where $\omega^2 = \frac{1}{2}\omega_{ab}\omega^{ab}$ and $\sigma^2 = \frac{1}{2}\sigma_{ab}\sigma^{ab}$ are both non-negative and s is a parameter of an integral curve of u. Since we will not be using congruence of spacelike curves in this book, we refer to Greenberg [216] for details on their kinematic quantities. Now we consider a *congruence of null geodesics* given by a null vector field ℓ. Since the arc-length parameterization is not possible for a null curve C generated by $\{\ell\}$, we use a Frenet frame

$$F = \{\frac{d}{dp} = \ell, n, U, V\}$$

with respect to a *distinguished parameter* p. Recall that

$$g(\ell, \ell) = g(n, n) = 0,\, g(\ell, n) = -1,\, g(U, U) = g(V, V) = 1.$$

Furthermore, for the congruence of the null geodesic

$$\dot{\ell} = \sigma U = 0 \Rightarrow \sigma = 0,$$

where σ is the curvature function of the null curve C. We also use the 2-dimensional screen distribution S which is complementary to the tangent space $T(C)$ in $(T(C))^\perp$ at every point of C. Thus, it is possible to define a projection operator $\hat{h}_{ab}$ by

$$\hat{h}_{ab} = u_a u_b + v_a v_b, \tag{3.1.9}$$

where

$$u_a = g_{ab} U^b \qquad v_a = g_{ab} V^b. \tag{3.1.10}$$

Here $\hat{h}_{ab}$ is a positive definite metric induced by g_{ab} on the screen distribution S. The inverse metric is given by $\hat{h}^{ab} = \hat{h}_{cd} g^{ca} g^{db}$. Then, any tensor (or just geometric) quantity on M can be projected onto its hatted component in S by using the projection operator (3.1.9). Then, the covariant derivative of the projection null vector $\hat{\ell}$ can be decomposed as follows:

$$\hat{\ell}_{a;b} = \frac{\hat{\theta}}{2} \hat{h}_{ab} + \hat{\sigma}_{ab} + \hat{\omega}_{ab}, \tag{3.1.11}$$

where $\hat{\theta}$, $\hat{\sigma}_{ab}$ and $\hat{\omega}_{ab}$ are respectively called *expansion*, *shear* and *twist* of the null congruence and given by

$$\hat{\theta} = \hat{h}^{ab} \hat{\ell}_{a;b},$$

$$\hat{\sigma}_{ab} = \hat{\ell}_{(a;b)} - \frac{\hat{\theta}}{2} \hat{h}_{ab},$$

$$\hat{\omega}_{ab} = \hat{\ell}_{[a;b]}$$

and they satisfy (see Hawking-Ellis [228, page 88])

$$\ell \hat{\theta} = \frac{d\hat{\theta}}{dp} = -\frac{\hat{\theta}^2}{2} + 2\hat{\omega}^2 + 2\hat{\sigma}^2 - R_{ab} \ell^a \ell^b, \tag{3.1.12}$$

where $2\hat{\omega}^2 = \hat{\omega}_{ab} \hat{\omega}^{ab}$ and $2\hat{\sigma}^2 = \hat{\sigma}_{ab} \hat{\sigma}^{ab}$ are non-negative. (3.1.12) is the analogue of the *Raychaudhuri equation* (3.1.8) of timelike geodesics.

Matter Tensor and Einstein's Field Equations. The matter distribution on (M, g) can be expressed by tensorial equations with regard to the Levi-Civita connection of the metric tensor g. The matter field is given by a symmetric tensor field, denoted by T_{ab}, called the energy momentum tensor such that

1. T_{ab} vanishes on an open set $\mathcal{U}$ of M iff the matter fields vanish on $\mathcal{U}$.

2. T_{ab} is divergence-free, that is, $T^a_{b;a} = 0$.

Let V be a Killing vector field (*i.e.*, $V_{a;b} + V_{b;a} = 0$) on M. Then, the above equation can be integrated along V as follows. Denoting the vector field $T^a_b V^b$ by W^a and computing its divergence we get

$$W^a_{;a} = T^a_{b;a} V^b + T^{ab} V_{b;a} = 0$$

since T_{ab} is symmetric and divergence-free and V is Killing. Now integrating over a compact orientable region $\mathcal{D}$ with boundary $\partial\mathcal{D}$ and using the Gauss divergence theorem gives

$$\int_{\partial\mathcal{D}} W^a\, d\sigma_a = \int_{\mathcal{D}} W^a_{;a}\, dV = 0,$$

where $d\sigma_a = N_a ds$ (N_a is a unit vector field normal to $\partial\mathcal{D}$ and ds is the elementary surface element of $\partial\mathcal{D}$) and dV is the volume element of $\mathcal{D}$. The above equation shows that the total flux of the component of T_{ab} along a Killing vector V taken over a closed surface is zero. Consequently, this result provides a conservation law.

Energy Conditions. For finding energy conditions we need to determine how a prescribed T_{ab} determines the geometry of spacetime by considering the following *Einstein-Hilbert action*:

$$I(g) = \int_{\mathcal{D}} (A(r + 2\Lambda) + L) dV$$

and vary g over a closed region $\mathcal{D}$, where A, r, L and Λ are a suitable coupling constant, scalar curvature, the matter Lagrangian and the cosmological constant respectively. Varying g over $\mathcal{D}$ implies that the action $I(g)$ is stationary if the *Einstein field equations* (see Hawking-Ellis [228])

$$R_{ab} - (\frac{r}{2} - \Lambda)g_{ab} = 8\pi\, T_{ab}$$

are satisfied, where 8π is due to the fact that Einstein field equations reduce to the Newtonian Poisson equation $\nabla^2\phi = 4\pi\,\rho$ for a weak field. Contracting Einstein field equations we get

$$r = 4\Lambda - 8\pi\, T \quad , \qquad T = T^a_a.$$

The left- and the right-hand sides of the Einstein field equations describe the geometry and the physics respectively of the spacetime under investigation. Considerable work has been done on exact solutions by prescribing physically meaningful choices of T_{ab} (see Kramer et al. [267]). For a physically reasonable exact solution, T_{ab} must satisfy the following energy conditions.

For any non-spacelike vector X in $T_p(M)$, at each point p of M, a *weak energy condition* implies that

$$T_{ab}X^a X^b \geq 0.$$

In particular reference to the focus of this book, for any null vector ξ in $T_p(M)$, weak energy condition implies that $R_{ab}\xi^a\xi^b \geq 0$.

A *dominant energy condition* implies that besides satisfying a weak energy condition for a timelike vector field X, $T^a_b X^b$ must be a non-spacelike vector. Physically this means that the local energy flow vector is non-spacelike along with the non-negative local energy density. For any orthonormal basis $\{E_a\}$, the dominant energy condition is equivalent to $T_{00} \geq |T_{ab}|$. Thus, the dominant energy condition is the weak energy condition plus the condition that the pressure should not exceed the energy density.

A spacetime M satisfies the *strong energy condition* if for any non-spacelike vector X of M,

$$R_{ab}X^a X^b \geq 0$$

holds. Moreover, if the Einstein equations hold then, for all timelike X, this condition implies that

$$T_{ab}X^a X^b \geq (\frac{T}{2} - \frac{\Lambda}{8\pi})V^a V_a,$$

which reduces to $T_{ab}X^a X^b = 0$ for all null vectors. When $\Lambda = 0$, the strong energy condition, for all non-spacelike vectors, is equivalent to the condition

$$T_{ab}X^a X^b \geq (\frac{T}{2})X^a X_a,$$

as stated in Hawking-Ellis [228, page 95]. Observe that the strong energy condition is also called the timelike (resp. null) congruence condition, at any point p of M, according as V is timelike (resp. null). On the other hand, the weak energy condition implies that the matter always has a converging effect on the congruence of null geodesics. Assuming that the vorticity is zero, we get from (3.1.12) that

$$\frac{d\hat{\theta}}{dP} = -\frac{\hat{\theta}^2}{2} - 2\hat{\sigma}^2 - R_{ab}\ell^a \ell^b.$$

Hence $\hat{\theta}$ would decrease along ℓ if $R_{ab}\ell^a \ell^b \geq 0$, which is the null convergence condition. Consequently, the weak energy condition implies null convergence condition.

Global Null Splitting Theorem. We follow the general notation of submanifold theory and let (M, g) be a smooth lightlike hypersurface of an $(m+2)$-dimensional Lorentzian manifold $(\bar{M}, \bar{g})$. We need the following from [273]. Consider

$$\widetilde{TM} = TM/\operatorname{Rad}(TM), \qquad \Pi : \Gamma(TM) \to \Gamma(\widetilde{TM}) \quad \text{(canonical projection)}$$

Set $\widetilde{X} = \Pi(X)$ and $\tilde{g}(\widetilde{X}, \widetilde{Y}) = g(X, Y)$. It is easy to prove that the operator $\widetilde{A}_U : \Gamma(\widetilde{TM}) \to \Gamma(\widetilde{TM})$ defined by $\widetilde{A}_U(\widetilde{X}) = -(\Pi(\bar{\nabla}_X U))$, where $U \in \Gamma(\operatorname{Rad}(TM))$ and $X \in \Gamma(TM)$ is self-adjoint. It is known that all Riemannian self-adjoint operators are diagonalizable. Let $\{k_1, \ldots, k_n\}$ be the eigenvalues. If $\widetilde{S}_{k_i}$, $1 \leq i \leq n$, is the eigenspace of k_i then

$$\widetilde{TM} = \widetilde{S}_{k_1} \perp \ldots \perp \widetilde{S}_{k_n}.$$

Choose a screen distribution $S(TM)$ and denote by $P_S : \Gamma(TM) \to \Gamma(S(TM))$ the corresponding projection. Let $U \in \Gamma(\operatorname{Rad} TM)$ be a null normal section on M. Denote by $A_U = A_U^*|_{S(TM)} : \Gamma(S(TM)) \to \Gamma(S(TM))$ where A_U^* is the shape operator on the distribution $S(TM)$ defined by the equation (2.1.20), which is a self-adjoint and diagonalizable operator. Take $W \in \Gamma(S(TM)$. Let $\{k_1^*, \ldots, k_n^*\}$ be the different eigenvalues on M and $S_{k_i^*}(TM)$, $1 \leq i \leq n$ the eigenspace of k_i^*, respectively. Then

$$S(TM) = S_{k_1^*}(TM) \perp \ldots \perp S_{k_n^*}(TM)$$

and we can find the following local adapted orthonormal basis of eigenvectors

$$\{E_1^1, \ldots, E_{r_1}^1, E_1^2, \ldots, E_{r_2}^2, \ldots \ldots, E_1^n, \ldots, E_{r_n}^n\},$$

where $A_U(E_j^i) = k_i^* E_j^i$, with $1 \leq i \leq n$, $1 \leq j \leq r_i$ and r_i is the dimension of the eigenspace of k_i^*. Consider a map $\widetilde{P}_S : \Gamma(\widetilde{TM}) \to \Gamma(S(TM))$ defined by $\widetilde{P}_S(\widetilde{X}) = P_S X$. Then, $\widetilde{P}_S$ is a vector bundle isomorphism, and we have

$$\widetilde{P}_S(\widetilde{A}_U \widetilde{X}) = \widetilde{P}_S\left(-\Pi(\bar{\nabla}_X U)\right) = P_S\left(-\bar{\nabla}_X U\right) = P_S\left(-\bar{\nabla}_{P_S X} U\right) = A_U(P_S X).$$

Lemma 3.1.1. *Let $(M, g, S(TM))$ be a lightlike hypersurface of a Lorentzian manifold $(\bar{M}, \bar{g})$. With the above notation, k is an eigenvalue of $\widetilde{A}_U$ iff k is an eigenvalue of A_U. Furthermore, $\widetilde{X}$ is an eigenvector of $\widetilde{A}_U$ associated with k if and only if $P_S X$ is an eigenvector of A_U associated with k.*

Proof. It is an immediate consequence of $\widetilde{P}_S$ being an isomorphism. $\qquad\square$

Therefore, the eigenvalues associated with a null normal section are the same for all choices of screen distributions. Such eigenvalues $\{k_1, \ldots, k_n\}$ of A_U are called *the principal curvatures* associated with the null normal section U.

Lemma 3.1.2. *Let U and $\hat{U}$ be two null normal sections such that $\hat{U} = \alpha U$. If k is an eigenvalue of A_U then αk is an eigenvalue of $A_{\hat{U}}$ with the same multiplicity.*

Proof. Let $S(TM)$ be a screen distribution. Consider the shape operator $A_{\hat{U}} : \Gamma(S(TM)) \to \Gamma(S(TM))$. From Lemma 3.1.1 we know that the eigenvalues respect to $\hat{U}$ are independent of the screen, so if W is an eigenvector of A_U with respect to k then

$$A_{\hat{U}}(W) = P_S(-\bar{\nabla}_W \hat{U}) = P_S(-\bar{\nabla}_W(\alpha U)) = P_S(-W(\alpha)U - \alpha \bar{\nabla}_W U) = \alpha k W.$$

Thus, αk is an eigenvalue associated with $\hat{U}$. $\qquad\square$

Let $\{E_j^i; 1 \leq i \leq n, 1 \leq j \leq r_i\}$ be a local orthonormal basis of eigenvectors of $\Gamma(S(TM)|_\mathcal{U})$. In order to facilitate the notation and depending on the context we will also denote it by $\{E_a; 1 \leq a \leq n\}$ and so $A_U(E_a) = k_a E_a$, where k_a may be repeated. It is well known that the *lightlike mean curvature* $\theta_U : M \to \mathbf{R}$ with respect to a null section U is given by

$$\theta_U = -\sum_{a=1}^{n} B(E_a, E_a) = -\sum_{a=1}^{n} \langle A_U(E_a), E_a \rangle.$$

It is easy to show that θ_U does not depend on both the screen distribution and the orthonormal basis, and so $\theta_U = -\sum_{a=1}^{n} k_a$. One of the good properties of the mean curvature is that it does not depend on the screen distribution chosen, but only on the local null section U.

A local null normal section ξ is called *geodesic* if $\bar{\nabla}_\xi \xi = 0$ for which the integral curves of ξ are called the *null geodesic generators*. This condition has interesting geometric and physical meanings and also helps in simplifying the computations. If U is a null normal section on M, then for all $p \in M$ one can scale U to be geodesic on a suitable neighborhood $\mathcal{U}$ of p. Let us suppose that the local section ξ is geodesic, that is, $\bar{\nabla}_\xi \xi = 0$ on $\mathcal{U}$. We will denote the shape operator A_ξ^* as A_ξ on $\Gamma(TM|_\mathcal{U})$ (note that for simplicity we are making an abuse of notation, but this should not cause confusion). Consider the *tidal force operator* $\bar{R}_\xi : \Gamma(TM|_\mathcal{U}) \to \Gamma(TM|_\mathcal{U})$ (see [317, Page. 219]) defined as

$$\bar{R}_\xi(X) = \bar{R}(X, \xi)\xi = \bar{\nabla}_{[\xi, X]}\xi - \bar{\nabla}_\xi \bar{\nabla}_X \xi.$$

This is a linear and self-adjoint operator, and $\mathrm{tr}(\bar{R}_\xi) = \overline{\mathrm{Ric}}(\xi, \xi)$. Define $R_\xi : \Gamma(TM|_\mathcal{U}) \to \Gamma(TM|_\mathcal{U})$ in the same way but using ∇ instead of $\bar{\nabla}$. From (2.4.1) it is easy to show that $\bar{R}_\xi = R_\xi$. Thus, we can define the tidal force by means of induced objects of a lightlike hypersurface.

Proposition 3.1.3. *Let ξ be a local normal null geodesic section. Then, the tidal force operator R_ξ satisfies the equation*

$$\langle R_\xi(X), Y \rangle = \langle -A_\xi^2(X) + (\bar{\nabla}_\xi A_\xi)(X), Y \rangle, \tag{3.1.13}$$

where $X, Y \in \Gamma(TM|_\mathcal{U})$ and

$$(\bar{\nabla}_\xi A_\xi)(X) = \bar{\nabla}_\xi(A_\xi X) - A_\xi(\bar{\nabla}_\xi X).$$

Proof. This is shown by the following easy computation.

$$\begin{aligned}
\langle R_\xi(X), Y \rangle &= \langle \bar{\nabla}_{[\xi, X]}\xi - \bar{\nabla}_\xi \bar{\nabla}_X \xi, Y \rangle \\
&= \langle -A_\xi([\xi, X]) + \bar{\nabla}_\xi(A_\xi X), Y \rangle \\
&= \langle -A_\xi(\bar{\nabla}_\xi X) + A_\xi(\bar{\nabla}_X \xi) - \bar{\nabla}_\xi(A_\xi X), Y \rangle \\
&= \langle -A_\xi(A_\xi(X)) + (\bar{\nabla} A_\xi)(X, \xi), Y \rangle. \qquad \square
\end{aligned}$$

Proposition 3.1.4. *Let M be a lightlike hypersurface of a Lorentzian manifold $\bar{M}$ and ξ a geodesic normal null section on $\mathcal{U}$ and θ_ξ be the lightlike mean curvature associated with ξ. Then,*

$$\xi(\theta_\xi) = -\mathrm{Ric}(\xi, \xi) - \mathrm{tr}(A_\xi^2) = -\mathrm{Ric}(\xi, \xi) - \sum_a k_a^2. \tag{3.1.14}$$

Proof. Let $\{E_1, \ldots, E_n\}$ be a basis of eigenvectors on $S(TM)|_\mathcal{U}$. We can compute the terms of equation (3.1.13) when $X = Y = E_a$ and obtain

$$\langle (\bar{\nabla}_\xi A_\xi)(E_a), E_a \rangle = \langle \bar{\nabla}_\xi (A_\xi E_a) - A_\xi (\bar{\nabla}_\xi E_a), E_a \rangle$$
$$= \langle \bar{\nabla}_\xi (k_a E_a), E_a \rangle - \langle A_\xi (\bar{\nabla}_\xi E_a), E_a \rangle$$
$$= \xi(k_a) - \langle A_\xi (\bar{\nabla}_\xi E_a), E_a \rangle$$
$$= \xi(k_a) - \langle \bar{\nabla}_\xi E_a, A_\xi (E_a) \rangle$$
$$= \xi(k_a).$$

It is clear that $\langle A_\xi^2(E_a), E_a \rangle = k_a^2$. Now taking the trace in (3.1.13) we obtain (3.1.14) which completes the proof. □

Consider the flux of ξ as a local congruence of null geodesic curves. It is known that the *vorticity tensor* ω is the antisymmetric part of $-A_\xi$ and the *shear tensor* σ is the trace-free of the symmetric part of $-A_\xi$. Since A_ξ is symmetric,

$$\omega = 0 \quad \text{and} \quad \sigma = -A_\xi - \frac{\theta}{n} I,$$

where I is the identity operator. From (3.1.14) we obtain a version of the following vorticity-free Raychaudhuri's equation for lightlike hypersurfaces

$$\xi(\theta_\xi) = -\operatorname{Ric}(\xi, \xi) - \operatorname{tr}(\sigma^2) - \frac{\theta^2}{m}.$$

Let $\gamma(t) \subset \mathcal{U}$ be a null generator of $M \subset \bar{M}$, where $\bar{M}$ is a physical spacetime. Then, the null mean curvature restricted to γ is the expansion $\theta(t) = \theta_\xi(\gamma(t))$ and the Raychaudhuri's equation restricted to each null generator is the Raychaudhuri's equation for a null geodesic (see [164, page 60] and [197]). This equation shows how the Ricci curvature influences the deviation of null geodesics of M. Since, in general, the screen distribution is not unique there is a need to look for a unique structure $(S(TM), \xi)$ with a given null normal section ξ of M. Also, it is desirable to find such globally defined unique structures. For this purpose we recall the following:

Definition 3.1.5. [150] Let M be a lightlike hypersurface of a Lorentzian manifold $\bar{M}$. A pair $(S(TM), \xi)$ is said to be a global structure on M if and only if ξ is a non-vanishing global null normal (GNN) section on M.

Definition 3.1.6. [150] Let M be a lightlike hypersurface of a Lorentzian manifold $\bar{M}$ admitting a GNN section ξ on M. A Riemannian distribution $D(TM)$ of TM is said to be ξ-distinguished if each section W of $D(TM)$ satisfies $\bar{\nabla}_W \xi \in \Gamma(D(TM))$. In particular, if a screen distribution $S(TM)$ is ξ-distinguished the pair $(S(TM), \xi)$ is called a distinguished structure on M.

Let $(S(TM), \xi)$ be a global structure of M, then we can consider a globally defined shape operator A_ξ of $S(TM))$. From Lemma 3.1.1 we conclude that the eigenvalues with respect to ξ do not depend on the screen. Thus, we say that the eigenvalues of A_ξ are *the principal curvatures* associated with the GNN section ξ.

Theorem 3.1.7. [150] *Let M be a lightlike hypersurface in a Lorentzian manifold $\bar{M}$ admitting a GNN section ξ. Suppose $\bar{\nabla}_\xi \xi = -\rho \xi$ and $\{k_1, \ldots, k_m\}$ are the principal curvatures associated with ξ on M. Then there exist a unique ξ-distinguished Riemannian distribution $D_\xi(TM)$ such that:*

(1) *If $\rho \neq k_i$ for all $i \in \{1, \ldots, m\}$, we have the decomposition*

$$TM = \mathrm{Rad}(TM) \oplus_{\mathrm{orth}} D_\xi(TM)$$

and so $(S(TM) = D_\xi(TM), \xi)$ is a unique distinguished structure.

(2) *If $\rho = k_{i_0}$ for some $i_0 \in \{1, \ldots, m\}$, we have the decomposition*

$$TM = \mathrm{Rad}(TM) \oplus_{\mathrm{orth}} S_\rho(TM) \perp D_\xi(TM),$$

where $S_\rho(TM)$ is any eigenspace associated with the eigenvalue k_{i_0}.

Proof. Suppose $S(TM)$ and $\hat{S}(TM)$ are two different screens. Let $S_k(TM) \leq S(TM)$ and $\hat{S}_k(TM) \leq \hat{S}(TM)$ be both vector sub-bundles associated with the same eigenvalue k. Let $\{E_1, \ldots, E_r\}$ be an orthonormal basis of $S_k(TM)$ and construct the set $\{\hat{E}_1, \ldots, \hat{E}_r\}$ where $\hat{E}_j = P_{\hat{S}}(E_j)$, $1 \leq j \leq r$. From Lemma 3.1.1 we have that $\{\hat{E}_1, \ldots, \hat{E}_r\}$ is an orthonormal basis of $\hat{S}_k(TM)$ satisfying the formulas:

$$\bar{\nabla}_{E_j}\xi = -kE_j - \tau(E_j)\xi,$$
$$\bar{\nabla}_{\hat{E}_j}\xi = -k\hat{E}_j - \hat{\tau}(\hat{E}_j)\xi, \qquad 1 \leq j \leq r,$$
$$\hat{E}_j = E_j + \mu_j \xi.$$

We are interested in finding μ_j such that $\hat{\tau}(\hat{E}_j)$ vanishes, so that $\hat{S}_k(TM)$ is a ξ-distinguished Riemannian distribution. Then we have

$$\bar{\nabla}_{\hat{E}_j}\xi = \bar{\nabla}_{(E_j + \mu_j\xi)}\xi = \bar{\nabla}_{E_j}\xi + \mu_j\bar{\nabla}_\xi\xi = -kE_j - \tau(E_j)\xi - \rho\mu_j\xi,$$
$$-k\hat{E}_j = -k(E_j + \mu_j\xi)$$

and, therefore, $\mu_j(\rho - k) = \tau(E_j)$. Accordingly, we obtain that if $\rho \neq k$ then it is enough to take $\mu_j = \tau(E_j)/(\rho - k)$ and trivially $\hat{S}_k(TM)$ is unique with $\hat{\tau} = 0$ on $\hat{S}_k(TM)$. Thus, we have actually proved both statements taking

$$D(TM) = \bigoplus_{k \neq \rho} \hat{S}_k(TM).$$

If $k_i \neq \rho$ for all i, then $(\hat{S}(TM), \xi)$ is a unique distinguished structure. $\square$

We know (see Chapter 2) that M is *totally geodesic* if the shape operator A_ξ vanishes identically for one (and so for all) null normal sections ξ on each neighborhood $\mathcal{U} \subset M$ and the second fundamental form of M also vanishes. As an immediate consequence of the formula (3.1.2) and the use of a unique distinguished global structure $(S(TM), \xi)$ for a class of lightlike hypersurfaces we have the following result (a local version appeared in [150]).

Theorem 3.1.8 (Global Null Splitting Theorem). *Let M be a lightlike hypersurface in a spacetime $\bar{M}$ admitting a unique distinguished global structure $(S(TM), \xi)$ and satisfying the null energy condition $\mathrm{Ric}(\xi, \xi) \geq 0$ for every GNN $\xi \in \mathrm{Rad}(TM)$. Then M is totally geodesic if and only if its lightlike mean curvature vanishes identically for all GNN sections $\xi \in \mathrm{Rad}(TM)$.*

Remark 3.1.9. Theorem 3.1.8 is a *Global Null Splitting Theorem* for smooth hypersurfaces as follows: It is known (see Theorem 2.5.1 in Chapter 2) that a lightlike hypersurface M of a semi-Riemannian manifold is totally geodesic if and only if $\mathrm{Rad}\, TM$ is a Killing distribution on M. Thus, using the hypothesis of Theorem 3.1.8 we consider a class of lightlike hypersurfaces (M, g, G) such that each M carries a smooth 1-parameter group G of isometries each of whose orbits is a global null curve generated by a corresponding global null vector field in M. Let M' be the orbit space of the action $G \approx C'$, where C' is a 1-dimensional null leaf of $\mathrm{Rad}\, TM$ in M. Then, M' is a smooth Riemannian hypersurface of M and the projection $\pi : M \to M'$ is a principle C'-bundle, with null fiber G. The global existence of a null vector field, of M, implies that M' is Hausdorff and paracompact. The infinitesimal generator of G is a global null Killing vector field, say U, on M. The metric g restricted to its screen distribution space $S(TM)$ then induces a Riemannian metric, say g', on M'. Since U is non-vanishing on M, we can take $U = \partial_f$ as a global null coordinate vector field for some global function f on M. Thus, f induces a diffeomorphism on M such that $(M = M' \times C', g = \pi^\star g')$ is a global product manifold.

Example 1. Let $(\bar{M}, \bar{g})$ be an $(m+2)$-dimensional globally hyperbolic spacetime [34], with the line element of the metric $\bar{g}$ given by

$$ds^2 = -dt^2 + dx^1 + \bar{g}_{ab}\, dx^a\, dx^b, \quad (a, b = 2, \dots, m+1) \tag{3.1.15}$$

with respect to a coordinate system $(t, x^1, \dots, x^{m+1})$ on $\bar{M}$. Choose the range $0 < x^1 < \infty$ so that the above metric is non-singular. Take two null coordinates u and v such that $u = t + x^1$ and $v = t - x^1$. Thus, (3.1.15) transforms into a non-singular metric:

$$ds^2 = -du\, dv + \bar{g}_{ab}\, dx^a\, x^b.$$

The absence of du^2 and dv^2 in this transformed metric implies that $\{v = \text{constant.}\}$ and $\{u = \text{constant.}\}$ are lightlike sub-spaces of $\bar{M}$. Let $(M, g, r = 1, v = \text{constant.})$ be one of this lightlike pair and let D be the 1-dimensional distribution generated by the null vector $\{\partial_v\}$, in $\bar{M}$. Denote by L the 1-dimensional integral manifold of D. A leaf M' of the m-dimensional screen distribution of M is Riemannian with the metric $d\Omega^2 = \bar{g}_{ab}\, x^a\, x^b$ and is the intersection of the two lightlike sub-spaces. In particular, a Minkowski space and a De-Sitter spacetime $\bar{M}$ have a pair of lightlike sub-spaces. In particular, there will be many global timelike vector fields in globally hyperbolic spacetimes $\bar{M}$. If one is given a fixed global *time function* then its gradient is a global timelike vector field in a given $\bar{M}$. With this choice of a global timelike vector field in $\bar{M}$, we conclude that its lightlike subspace

M (the same is true for the other pair) admits a global null vector field. Thus, there exists a pair of lightlike hypersurfaces of globally hyperbolic spacetime $\bar{M}$ both of which have a global structure.

Note. By using the Hopf-Rinow theorem one may choose a unique distinguished global structure $(S(TM), \xi)$ of one of such hypersurface so that a leaf M' of $S(TM)$ is a complete Riemannian hypersurface of M. Thus, it is possible to construct a pair of globally null hypersurfaces of a globally hyperbolic spacetime. In particular, a Minkowski and a De-Sitter spacetime can have a pair of subspaces which are globally null. Similarly, using the technique of warped products (see Chapter 1), one can show that Robertson-Walker, Schwarzschild and Reissner-Nordström spacetimes can have a pair of globally null hypersurfaces. Also see some more examples of globally null manifolds in [41].

3.2 Killing horizons

For physical applications, we use the symbols (M, g) for an n-dimensional semi-Riemannian (in particular, Lorentzian) manifold and (Σ, γ) its hypersurface. A vector field ξ on (M, g) is called a *conformal Killing vector field*, briefly denoted by CKV, with conformal function σ if

$$\pounds_\xi \, g_{ij} = 2\,\sigma\,g_{ij}\,, \quad \text{or} \quad \xi_{i;j} + \xi_{j;i} = 2\,\sigma\,g_{ij}\,, \quad 1 \le i, j \ge n, \tag{3.2.1}$$

which reduces to a *homothetic* or *Killing vector field* whenever σ is non-zero constant or zero respectively. ξ is called proper CKV if σ is non-constant. The second set of equations of (3.2.1) are called *conformal Killing equations*. ξ satisfies the following integrability conditions (also valid for any Riemannian manifolds [408]):

$$(\pounds_\xi \nabla)(X, Y) = (X\,\sigma)\,Y + (Y\,\sigma)\,X - g\,(X, Y)D\,\sigma,$$
$$(\pounds_\xi R)(X, Y, Z) = (\nabla\nabla\sigma)(X, Z)Y - (\nabla\nabla\sigma)(Y, Z)X$$
$$+ g\,(X, Z)\,\nabla_Y\,D\,\sigma - g\,(Y, Z)\,\nabla_X\,D\,\sigma,$$

where $D\,\sigma$ is the gradient of σ. In local coordinates, we have

$$\pounds_\xi \Gamma^i_{jk} = \delta^i_j\,\sigma_k + \delta^i_k\,\sigma_j - g_{jk}\,\sigma^i,$$
$$\pounds_\xi R^i{}_{ijm} = -\,\delta^i_k\,\sigma_{j\,;\,m} + \delta^i_m\,\sigma_{j\,;\,k} - \sigma^i{}_{;\,k}\,g_{jm} + \sigma^i{}_{;\,m}\,g_{jk}.$$

Solutions of the highly non-linear Einstein's field equations require the assumption that they admit Killing or homothetic vector fields. This is due to the fact that the Killing symmetries leave invariant the metric connection, all the curvature quantities and the matter tensor of the Einstein field equations of a spacetime. Most explicit solutions (see [267]) have been found by using one or more Killing vector fields. Related to the theme of this section, here we show that the Killing symmetry has an important role in the most active area of research on Killing horizons in general relativity.

Killing horizon. The concept of *Killing horizon* in a general semi-Riemannian manifold (M, g) has its roots in a 1969 paper of Carter [94] as follows: Consider an open region $\mathcal{U}$ on an n-dimensional semi-Riemannian manifold (M, g) such that there exists a continuous r-parameter group of Killing vector fields ξ_a, $(1 \leq a \leq r)$, generating an r-dimensional sub-tangent space V_x at every $x \in \mathcal{U}$, where $(1 \leq r \leq m - 1)$. Then, the group is said to be *orthogonally transitive* if the r-dimensional orbits of the group are orthogonal to a family of $(n-r)$-dimensional surfaces. Let the respective decomposition be

$$T_x\, M = W_x + V_x\,, \qquad \dim(W_x) = n - r.$$

The group is said to be *invertible* at a point $x \in \mathcal{U}$ if there is an *isometry* leaving x fixed , at x, but leaves unaltered the sense of the direction orthogonal to V_x at the same point $x \in \mathcal{U}$. The existence of such an isometry obviously implies that it is an involution and uniquely determined. It follows from the above decomposition that a group is invertible at a point x only if both V_x and W_x are non-singular. Indeed, if any one is lightlike (null), then there exists a self orthogonal null direction in that one, and, therefore, the group can not be invertible. Carter [94] has shown that for an *Abelian group*, the non-singular orthogonal transitive condition is also sufficient for the group to be invertible. A trivial example is the case of a 1-parameter group which is Abelian, and, therefore the orthogonal transitivity and the invertibility are equivalent for a non-null Killing vector. Now, let (Σ, γ) be a lightlike hypersurface of (M, g). Then, Σ is said to be a *local isometry horizon* (LIH) with respect to a group of isometry if

(a) Σ is invariant under the group.

(b) Each null geodesic generator is a *trajectory* of the group.

In particular, a lightlike hypersurface (Σ, γ) which is an LIH with respect to a 1-parameter group (or sub-group) is said to be a *Killing horizon*, denoted by KH. This means that a KH is a lightlike hypersurface M whose generating null vector can be normalized so as to coincide with one of the Killing vectors ξ_a. Physically, an *LIH*, with respect to a 4-dimensional spacetime manifold M, has the following significant role. A particle on an *LIH*, of M, may immediately be traveling at the speed of light along the single null generator but standing still relative to its surroundings. This happens because, by definition of an *LIH*, the variable affine parameter along a null geodesic leaves invariant both the intrinsic structure of M and the position of the lightlike hypersurface as an *LIH*. On the existence of *LIH* and a KH, we have the following:

Theorem 3.2.1. (Carter [94]) *Let $\mathcal{U}$ be an open subregion of an n-dimensional C^2 manifold with a C^1 semi-Riemannian metric, such that there is a continuous group of isometries whose surfaces of transitivity have constant dimension r in $\mathcal{U}$. Let N be the subset (closed in $\mathcal{U}$) where the surfaces of transitivity become null and suppose they are never more than single null (i.e., the rank of the induced*

metric on the surface of transitivity drops from r to r − 1 on N, but never lower).
Then if the group is orthogonally transitive in $\mathcal{U}$*, it follows that N is the union*
of non-intersecting hypersurfaces which are LIH's with respect to the group, and
consequently (since N is closed in $\mathcal{U}$*) the boundaries of N are members of the*
family. Moreover, if the group is Abelian, each of the resulting LIH's is a KH.

Our aim is to relate the Global Null Splitting Theorem of the previous section with Galloway's works [197, 198] on lightlike (null) hypersurfaces as physical models in general relativity. To understand this, we first recall the following physical terminology (details can be seen in [34]). Let (M, g) be a spacetime manifold which, by definition, is a connected time-orientable Lorentzian manifold. We use the standard causal relations ' $<<$ ' and ' $<$ '.

- A timelike curve (resp. causal) curve $C(t) : I = [a, b] \to M$ is said to be a *future directed curve* if each tangent vector of C is future directed. Similar definitions follow for *past directed curve* timelike and causal curves.

- A point $p \in M$ is called the end point of C for $t = b$ if $\lim_{t \to b} C(t) = p$.

- A non-spacelike curve is *future (resp. past) inextendible* if it has no future (resp. past) end points.

- Given any two points p, q of M, q is *chronological future (resp. past)* of p, denoted by $p << q$ (resp. $q << p$), if there is a future (resp. past) directed timelike curve from p to q.

- The chronological future (resp. past) of p are the sets $I^+(p) = \{q \in M : p << q\}$ and $I^-(p) = \{q \in M : q << p\}$.

- Similarly, the *causal future (resp. past)* of p are the sets $J^+(p) = \{q \in M : p \leq q\}$ and $J^-(p) = \{q \in M : q \leq p\}$ for non-spacelike curves. This means that M, with no closed non-spacelike curve, is a causal space. The sets $I^+(p)$ and $I^-(p)$ are always open in any spacetime, but the sets $J^+(p)$ and $J^-(p)$ are neither open nor closed in general.

- M is strongly causal if all its points have arbitrarily small neighborhoods in which no spacelike curve intersects more than once. A strong causal M is globally hyperbolic, if for each p and q of M, $J^+(p) \cap J^-(q)$ is compact.

- A subset $N \subset M$ is called *achronal* if no two of its points can be joined by a timelike curve. In particular, an *achronal boundary* is a set of the form $I^+(N)$ (or $I^-(N)$), for some $N \subset M$, which is generated by null geodesic segments such that it has past end points but no future end points.

- Let Σ_1 and Σ_2 be two lightlike hypersurfaces that meet at a point p. We say that Σ_2 lies to the future (resp. past) side of Σ_1 near p if for some neighborhood $\mathcal{U}$ of p in M in which Σ_1 is closed achronal, $\Sigma_2 \cap \mathcal{U} \subset J^+(\Sigma_1 \cap \mathcal{U}, \mathcal{U})$ (resp. $\Sigma_2 \cap \mathcal{U} \subset J^-(\Sigma_1 \cap \mathcal{U}, \mathcal{U})$).

Event horizon. An *event horizon* is a general term for a boundary in spacetime, de-fined with respect to an observer, beyond which events cannot affect the observer. We highlight that an event horizon is an intrinsically global concept, in the sense that its definition requires the knowledge of the whole spacetime (to determine whether null geodesics can reach null infinity). In particular, an event horizon is called a KH if its null hypersurface admits a Killing vector field.

A line in a Riemannian manifold is an *inextendible geodesic* each segment of which has minimal length. Contrary to this, as explained in [34, page 8] the infimum of timelike arc lengths of all piecewise smooth curves joining any two chronologically related points $p << q$ is zero. Also, if both the two joining points p, q are in a geodesically convex neighborhood $\mathcal{U}$, then, the timelike line is de-fined as an inextendible future directed timelike geodesic whose segments in $\mathcal{U}$ have maximal length among causal curves joining those end points. Thus, the minimal length for Riemannian manifolds corresponds to the maximum length for Lorentzian manifolds. Due to this marked difference, to understand the concept of timelike lines let p, q be two points of a Lorentzian manifold (M, g). Denote by $S_{p,q}$ the path space of all future directed non-spacelike curves $C : [0, 1] \to M$ with p and q their end points. Let L_g be the Lorentzian arc length as defined in [228, Page 105]. Now we recall the following definition:

Definition 3.2.2. Given $p \in M$, if q is not in $J^+(p)$, set $d(p, q) = 0$. If $q \in J^+(p)$, $d(p, q) = \sup \{L_g(C) : C \in S_{p,q}\}$, where d denotes the distance function.

We highlight the fact that although, in general, the Lorentzian distance $d(p, q)$ is not finite for any arbitrary M, it has been established by Beem-Ehrlich [34] that the globally hyperbolic spacetimes do have finite-valued distance func-tion. This is why such spacetimes are physically significant.

The famous Cheeger-Gromoll Riemannian splitting theorem [96] describes the rigidity of Riemannian manifolds with non-negative Ricci curvature which contain a line. We also know that a complete Riemannian manifold with strictly positive Ricci curvature is void of any lines. This classical result was extended by Eschenburg [178] for proving an analogous Lorentzian splitting theorem which describes the rigidity of spacetimes with strong energy condition, $Ric(X, X) \geq 0$ for all timelike vectors X, containing a timelike line. Secondly, as a physical appli-cation of the Global Null Splitting Theorem 3.1.8, we now recall the latest works of Galloway [197, 198] and others on the maximum principle for null hypersurfaces for which smoothness is not required. This is due to the fact that null hypersur-faces of spacetimes which represent event horizons are in general C^0 but not C^1 (see an example later on). Galloway's approach has its roots in the well-known geometric maximum principle of E. Hopf, a powerful analytic tool which is often used in the theory of minimal or constant mean curvature hypersurfaces. This principle implies that two different minimal hypersurfaces in a Riemannian man-ifold cannot touch each other from one side. A published proof of this fact is not available, however, for a special case of Euclidean spaces see [370]. In 1989, Eschen-burg [178] proved the following result on the maximum principle for hypersurfaces with suitable mean curvature bounds (also valid for spacelike hypersurfaces of Lorentzian manifolds). We quote the following result on the interior and boundary maximum principle for smooth hypersurfaces with suitable mean curvature bounds (also valid for spacelike hypersurfaces).

Theorem 3.2.3. (Eschenburg [178]) *Let* Σ_+ *and* Σ_- *be disjoint open domains with spacelike connected* C^2-*boundaries having a point in common. If the mean curvatures* θ_+ *of* $\partial\Sigma_+$ *and* θ_- *of* $\partial\Sigma_-$ *satisfy*

$$\theta_- \leq -a \quad , \quad \theta_+ \leq a$$

for some real number a, *then* $\partial\Sigma_- = \partial\Sigma_+$, *and* $-\theta_- = \theta_+ = a$.

Later on, Galloway [197] proved the following geometric maximum principle for smooth null hypersurfaces.

Theorem 3.2.4. [198] *Let* Σ_1 *and* Σ_2 *be smooth lightlike hypersurfaces in a spacetime manifold* M. *Suppose,*

(1) Σ_1 *and* Σ_2 *meet at* $p \in M$ *and* Σ_2 *lies to the future side of* Σ_1 *near* p

(2) *the null mean curvature scalars* θ_1 *of* Σ_1, *and* θ_2 *of* Σ_2, *satisfy,* $\theta_1 \leq 0 \leq \theta_2$.

Then Σ_1 *and* Σ_2 *coincide near* p *and this common lightlike hypersurface has mean curvature* $\theta = 0$.

Proof. Let Σ_1 and Σ_2 have a common null direction at p and let Q be a timelike hypersurface in M passing through p and transverse to this direction. Take Q so small such that the intersections

$$B_1 = \Sigma_1 \cap Q \quad \text{and} \quad B_2 = \Sigma_2 \cap Q$$

are smooth spacelike hypersurfaces in Q, with B_2 to the future side of B_1 near p. These two hypersurfaces may be expressed as graphs over a fixed spacelike hypersurface B in Q, with respect to normal coordinates around B. Precisely, let $B_1 = \mathrm{graph}(u_1)$, $B_2 = \mathrm{graph}(u_2)$ and suppose

$$\theta(u_i) = \theta_i|_{B_i = \mathrm{graph}(u_i)}, \quad i = 1, 2.$$

By suitably normalizing the null vector fields $\xi_i \in \Gamma(TM_i)$ determining θ_1 and θ_2, respectively, a simple computation shows that

$$\theta(u_i) = H(u_i) + \text{ lower order terms,}$$

where H is the mean curvature operator on a spacelike graph over B in Q. Thus θ is a second-order quasi-linear elliptic operator. In this case we have:

(1) $u_1 \leq u_2$, and $u_1(p) = u_2(p)$.

(2) $\theta(u_2) \leq 0 \leq \theta(u_1)$.

Then the well-known classical Alexandrov's [5] strong maximal principle for second order quasi-linear elliptic PDE's implies that $u_1 = u_2$. Thus B_1 and B_2 agree near p. The null normal geodesics to Σ_1 and Σ_2 in M will then also agree. Consequently, Σ_1 and Σ_2 agree near p, which completes the proof. $\qquad\square$

Although the above maximum principle theorem is for smooth null hypersurfaces, in reality null hypersurfaces occurring in relativity are the null portions of achronal boundaries as the sets $= \partial I^{\pm}(A)$, $A \subset M$ which are always C^0 hypersurfaces and contain non-differentiable points. For example (see [197]), consider one such set $\Sigma = \partial I^-(A)$ where A consists of two disjoint closed disks in the $t = 0$ slice of a Minkowski 3-space. This set can be represented as a merging surface of two truncated cones having a curve of non-differentiable points corresponding to the intersection of the two cones, but, otherwise it is a smooth null hypersurface. The most important feature of these C^0 null hypersurfaces is that they are ruled by null geodesics which are either past or future inextendible and contained in the hypersurface. Precisely, a C^0 future (past is defined in a time-dual manner) null hypersurface is a locally achronal topological hypersurface Σ of M which is ruled by future inextendible null geodesics. These null geodesics (entirely contained in Σ) are the null geodesic generators of Σ. We now understand how the meaning of mean curvature inequalities can be extended to C^0 null hypersurfaces. For this we need the following general concept of support domains, whose boundaries are at least C^2-hypersurfaces, earlier used for non-null hypersurfaces of Riemannian or Lorentzian manifolds M. Let $W \subset M$ be an arbitrary open domain with topological boundary ∂W and let $b \in \mathbf{R}$.

Definition 3.2.5. [179] ∂W has generalized mean curvature $\leq b$ if for any $p \in \partial W$ there are open domains $W_{p,i}, i = 1, 2, \ldots$, called support domains, whose boundaries are C^2-hypersurfaces near p with shape operator $A_{p,i}$ and mean curvature $\theta_{p,i}$ at some point p, with the following properties:

(a) $W_{p,1} \subset W_{p,2} \subset \ldots \subset W$,

(b) $p \in \partial W_{p,i}$,

(c) *there is a locally uniform upper bound for $A_{p,1}$,*

(d) $\theta_{p,i} \leq b + \epsilon_i$ *for some sequence $\epsilon_i \to 0$.*

The above concept is used for smooth null support hypersurfaces as follows:

Definition 3.2.6. (Galloway [197]) Let Σ be a C^0 future null hypersurface in M. We say that Σ has a null mean curvature $\theta \geq 0$ in the sense of support hypersurfaces provided for each $p \in \Sigma$ and for each $\epsilon > 0$ there exists a smooth (at least C^2) null hypersurface $\Sigma_{p,\epsilon}$ such that

(1) $S_{p,\epsilon}$ is a past support hypersurface for Σ at $p \in \Sigma_{p,\epsilon}$, i.e., $\Sigma_{p,\epsilon}$ passes through p and lies to the past side of Σ near p, and

(2) the null mean curvature of $\Sigma_{p,\epsilon}$ at p satisfies $\theta_{p,\epsilon} \geq -\epsilon$.

Example 2. The future null cone $\wedge^+(p)$ of a Minkowski space has null mean curvature $\theta \geq 0$ in the sense of support hypersurfaces.

Let $\{B_{p,\epsilon} : p \in \Sigma, \ \epsilon > 0\}$ be the collection of null second fundamental forms which are locally bounded from below, i.e., for all $p \in \Sigma$ there is a neighborhood $\mathcal{U}$ of p in Σ and a constant $k > 0$ such that

$$B_{p,\epsilon} \geq -kg'_{q,\epsilon} \forall q \in \mathcal{U} \quad \text{and} \quad \epsilon > 0$$

where $g'_{q,\epsilon}$ is the Riemannian metric on the screen distribution of Σ.

Based on the above, we have the following maximum principle for C^0 null hypersurfaces (proof is common with the proof of Theorem 3.2.4):

Theorem 3.2.7. [197] *Let M_1 be a C^0 future null hypersurface and let M_2 be a C^0 past null hypersurface in a spacetime M. Suppose*

(1) *M_1 and M_2 meet at $p \in M$ and M_2 lies in the future side of M_1 near p,*

(2) *M_1 has null mean curvature $\theta_1 \geq 0$ in the sense of support hypersurfaces, with null second fundamental forms $\{B_{p,\epsilon} : p \in M_1, \epsilon > 0\}$ locally bounded from below, and*

(3) *M_2 has null mean curvature $\theta_2 \leq 0$ in the sense of support hypersurfaces.*

Then M_1 and M_2 coincide near p, i.e., there is a neighborhood $\mathcal{U}$ of p such that $M_1 \cap \mathcal{U} = M_2 \cap \mathcal{U}$. Moreover, $M_1 \cap \mathcal{U} = M_2 \cap \mathcal{U}$ is a smooth null hypersurface with null mean curvature $\theta = 0$.

Remark 3.2.8. The important use of Theorem 3.2.4 is that it extends in an appropriate manner (see more details in [197]) to C^0 null hypersurfaces. Based on this we now quote the following physically meaningful splitting theorem for null hypersurfaces arising in spacetime geometry:

Theorem 3.2.9. [197] *Let (M, g) be a null geodesically complete spacetime which obeys the null energy condition, $\mathrm{Ric}(X, X) \geq 0$ for all null vectors X. If M admits a null line η, then, η is contained in a smooth closed achronal totally geodesic null hypersurface Σ.*

Remark 3.2.10. Galloway proved the above theorem by using the extended version of Theorem 3.2.4 for C^0 null hypersurfaces. He used Kupeli's approach [272] of working with the tangent space of M modeled by the null normal vector field ξ, i.e, let $T_pM/\xi = \{\bar{X} : X \in T_pM\}$ and $\bar{X} - X = a\xi$ for some real number a be the equivalence class of X. Since the approach in this book is quite different from Kupeli's approach we skip Galloway's proof. However, a closer examination of our Global Null Splitting Theorem 3.1.8 indicates that Theorem 3.2.9 holds (in an appropriate use of the properties of C^0 null hypersurfaces) as its special case for those classes of null geodesically complete spacetimes which admit a null line. The upshot of this comment is that both the Theorems 3.1.8 and 3.2.9 can be seen as Null Splitting Theorems for smooth and C^0 null hypersurfaces respectively of appropriate classes of spacetimes.

A Model. Consider a 4-dimensional stationary spacetime (M, g) which is chrono-logical, that is, M admits no closed timelike curves. It is known [228] that a sta-tionary M admits a smooth 1-parameter group, say G, of isometries whose orbits are timelike curves in M. A static spacetime is stationary with the condition that its timelike Killing vector field, say T, is hypersurface orthogonal, that is, there exists a spacelike hypersurface orthogonal to T. Our model will be applicable to both these types. Denote by M' the Hausdorff and para compact 3-dimensional Riemannian orbit space of the action G. The projection $\pi : M \to M'$ is a princi-pal **R**-bundle, with the timelike fiber G. Let $T = \partial t$ be the non-vanishing timelike Killing vector field, where t is a global time coordinate on M'. Then, g induces a Riemannian metric g'_M on M' such that

$$M = \mathbf{R} \times M', \quad g = -u^2 (dt + \eta)^2 + \bar{\pi}^\star g'_M,$$

where η is a connection 1-form for the **R**-bundle π and

$$u^2 = -\bar{g}(T, T) > 0.$$

It is known that a stationary spacetime (M, g) uniquely determines the orbit data (M', g'_M, u, η) as described above, and conversely. Suppose the orbit space M' has a non-empty metric boundary $\partial M' \neq \emptyset$. Consider the maximal solution data in the sense that it is not extendible to a larger domain $(\mathcal{M}', g'_{\mathcal{M}'}, u', \eta') \supset (M', g'_M, u, \eta)$ with $u' > 0$ on an extended spacetime $\mathcal{M}'$. Under these conditions, it is known [228] that in any neighborhood of a point $x \in \partial M'$, either the metric g'_M or the connection 1-form η degenerates, or $u \to 0$. The third case implies that the timelike Killing vector T becomes null and, there exists a Killing horizon $KH = \{u \to 0\}$ of M, subject to satisfying the hypothesis of Galloway's Null Splitting Theorem 3.2.9. This KH is related to a class of lightlike hypersurfaces Σ admitting a unique distinguished ξ-structure $(S(T\Sigma), \xi)$ with respect to a GNN section ξ of $\mathrm{Rad}\, T\Sigma$ as follows:

Let (Σ, γ) be a totally geodesic lightlike hypersurface of a stationary space-time M. Then, as per Theorem 2.5.1 of Chapter 2, its $\mathrm{Rad}\, T\Sigma$ is a Killing dis-tribution. Now consider the case when the timelike vector field $T \in \Gamma(TM)$ is becoming null, that is,

$$\mathrm{Lim}(T)_{u \to 0} = \xi,$$

where $\xi \in \Gamma(TM)$ is a null Killing vector field. Then, the spacelike hypersurface M' of M degenerates to a 3-dimensional lightlike hypersurface Σ of M such that a leaf M_s of $S(TM)$ is identified with the metric boundary $\partial M'$, that is,

$$\Sigma \supset M_s = \partial M' \subset M \subset \bar{M}. \tag{3.2.2}$$

Thus, $M_s = \partial M'$ is a common 2-dimensional submanifold of both Σ and M. Since Σ is totally geodesic in M, we conclude that Σ also admits a smooth 1-parameter group G_Σ of isometries (induced from the group G of M) whose orbits are global null curves in Σ. Thus, as per conclusion (1) of Theorem 3.1.7 we can take a class of unique distinguished ξ-structure $(S(T\Sigma), \xi)$ totally geodesic lightlike hypersurfaces Σ evolving from the spacelike hypersurface M' of M.

Physically, one must find those stationary spacetimes M which are geodesically complete, chronological (for details on these concepts see [34]) and their orbit space M' has a non-empty metric boundary $\partial M'$. The last condition is necessary for the existence of null hypersurfaces as event horizons of such a spacetime M. For this purpose, we recall the following result of Anderson:

Theorem 3.2.11. [6] *Let* (M, g) *be a geodesically complete, chronological, stationary vacuum spacetime.* M *is the flat Minkowski space* $\mathbf{R}_1^4$, *or a quotient of Minkowski space by a discrete group* Γ *of isometries of* $\mathbf{R}^3$, *commuting with* G. *In particular,* M *is diffeomorphic to* $\mathbf{R} \times M'$, $d\theta = 0$, *with constant* u.

Remark 3.2.12. (a) Anderson's above result implies that only a non-flat M will have a non-empty metric boundary of its orbit space. It turns out that asymptotically flat spacetimes are best physical systems for the non-flat stationary spacetimes, many of them do have Killing horizons [228].

(b) Mathematically, the existence of geodesically complete and chronological stationary spacetimes puts strong conditions on both the topology and geometry of its orbit space M' outside large compact sets. For purely geometric reasons, it is interesting to find those conditions, both with respect to flat and non-flat stationary spacetimes. However, physically, since there do exist Killing horizons of some spacetimes, the above model is an application of a class of unique distinguished ξ-structure $(S(T\Sigma), \xi)$ totally geodesic lightlike hypersurfaces Σ with KHs of physically significant stationary spacetimes. Finally, we quote the following physical result of a class spacetime admitting a KH.

Theorem 3.2.13. [167] *Let (M,g) be a globally hyperbolic spacetime which (i) satisfies the Einstein equations for a stress tensor T obeying the mixed energy condition and the dominant energy condition; (ii) admits a homothetic Killing vector field ξ of g; and (iii) admits a compact hypersurface of constant mean curvature. Then either (M,g) is an expanding hyperbolic model with T vanishing everywhere, or ξ is a Killing vector(KV) field.*

3.3 Dynamical horizons

A black hole is a region of spacetime which contains a huge amount of mass compacted into an extremely small volume. The gravity inside a certain radius of a black hole is so overwhelmingly strong that not even light traveling at $186,000$ miles an hour can escape. Shortly after Einstein's first version of the theory of gravitation was published in 1915, in 1916, Karl Schwarzschild computed the gravitational fields of stars using Einstein's field equations. He assumed that the star is spherical, gravitationally collapsed and non-rotating. His solution is called a *Schwarzschild solution* which obtained an exact solution of static vacuum fields of the point-mass. Since then there has been very active ongoing research on exact solutions, using the Schwarzschild metric; but, the real formal mathematical formalism of black holes was initiated, in 1960, by Kruskal [269] (and independently by Szekeres [382]) by extending the Schwarzschild solution into the region of the black hole. Kruskal-Szekeres formulation is now well-known as a reliable fundamental theory for the justification of the existence of black holes and there has been a large body of research papers on this subject. Black holes are now "seen" everywhere by astronomers, even though no one has ever found an event horizon. On the mathematical theory of black holes we refer to a book by Chandrasekhar [95]. To discuss the geometric description of the black hole horizon as a lightlike hypersurface of a spacetime manifold, one needs to understand the following physical concepts:

Non-Expanding Horizon (NEH). A lightlike hypersurface (Σ, γ) of a spacetime (M, g) is called a *non-expanding horizon*(NEH) if and only if

(1) Σ has a topology $R \times S^2$,

(2) the expansion θ vanishes on Σ and

(3) stress energy tensor T obeys the null dominant energy condition.

It follows from the Raychaudhury equation (3.1.8) that θ vanishing for an NEH implies the shear σ also vanishes. Moreover, it is important to know that the property of being a NEH is an intrinsic property of Σ i,e. it is independent of the choice of the null normal vector of Σ. A simple example of NEH is the event horizon of a Schwarzschild black hole [228].

Isolated Horizon (IH). A NEH Σ implies that only its degenerate metric γ is time-independent. Thus, the NEH provides a limited set of tools to study the geometry of spacetimes containing a black hole. For this reason, geometrically the surface of a black hole has been traditionally described in terms of a more general concept, called an *isolated horizon*, briefly denoted by IH, which satisfies the three conditions of a NEH plus all the induced objects defining the null geometry of (Σ, γ) are time-independent. In a simple way, IH represents the maximum degree of stationarity for the horizon defined in a quasi-local manner. By quasi-local we mean an analysis restricted to a 3-dimensional hypersurface with compact sections (we will discuss this more later on). Contrary to event horizons, IHs constitute a local concept. Every Killing horizon which is topologically $R \times S^2$ is an isolated horizon, but, in general, spacetimes with isolated horizons need not admit any Killing field even in a neighborhood of M. However, contrary to Killing horizons – which are also local – IHs are well defined even in the absence of any spacetime symmetry. The classical relation of a black hole with IH has its roots in Hawking's area theorem, which states that if matter satisfies the dominant energy condition, then, the area of the black hole IH cannot decrease [228]. It is important to understand that the Hawking's area law, in general, holds for an event horizon for which no Killing horizon is involved. Moreover, the most extensively studied family of black holes are the Kerr-Newman black holes, all of which have IHs but they admit no null Killing vector on an open set. Thus, the concepts of event horizons and isolated horizons are more general than that of Killing horizons. We refer [12, 13, 14] for more information on IH and their use in physics. In particular, we refer to Lewandowski [288] for some examples of spacetimes admitting IHs.

Expanding black holes. There is clear evidence that our universe is expanding in time. Indeed, we have proof because very distant stars appear as redshifts. A redshift is a change in colour when an object that emits light is moving away from the observer. It is called a redshift because it shifts to the red part of the spectrum. Stars are moving away from us in every direction, so we know that at one time the universe was smaller. Going back far enough in time, it is evident from the above explanation that the universe may have been something the size of an atom; what we call the Primordial Atom. Along with the expanding universe, since the black holes are surrounded by a local mass distribution and expand by the inflow of galactic debris as well as electromagnetic and gravitational radiation, their area increases in a given physical situation. Thus black holes are also expanding in time. Therefore, although the classical isolated black holes have been extensively studied, they do not represent a realistic model in the context of an expanding universe.

Consequently, in particular, one needs to know the geometry of the surrounding of an expanding black hole to find an explicit formula for the increase in area. To address this issue of representing expanding black holes, recently, a concept of *dynamical horizons* was introduced by Ashtekar in collaboration with Krishnan [16, 268] which are a special type of spacelike hypersurfaces of a spacetime whose asymptotic states are the IH's. Note that in most physical situations one expects the IH state to reach only asymptotically, i.e., in the infinite future. For some exceptions, see examples in [16, 268]. To understand the concept of dynamical horizons we need the following terminology:

Trapped and marginally trapped surfaces. In 1965 Penrose introduced the fundamental concept of a trapped surface and proved that a spacetime containing such a surface must come to an end. To understand what a trapped surface is, let Σ be a compact spacelike hypersurface of a time-orientable 4-dimensional spacetime manifold (M, g). Since each 2-dimensional normal space of Σ is timelike, it admits two smooth non-vanishing future directed null normal vector fields, say, ℓ (outer pointing) and n (inner pointing). Then Σ is a *trapped surface* [227] if $\theta_+ \theta_-|_\Sigma < 0$, where θ_+ and θ_- are the expansion functions in the future outer-pointing and inner-pointing null directions respectively. For this case, both the ingoing and the outgoing light rays converge so that all signals from the surface are trapped inside a closed region. This happens when the gravitational field is very strong and, therefore, such trapped 2-surfaces are associated with black holes. A *trapping boundary* is defined as an inextendible region for which each point lies on some trapped surface. Moreover, if we just consider the outer pointing null normal ℓ, we say that Σ is a *marginally outer trapped surface* if $\theta_+|_\Sigma = 0$. Relevant to the context of this section is the following notion of *future, outer, trapping horizons (FOTH)*, first proposed by Hayward [227] and then used by Ashtekar-Krishnan [16] and others in the study of laws of black hole dynamics:

Definition 3.3.1. A future, outer, trapped horizon (FOTH) is a three manifold Σ, foliated by a family of closed 2-surfaces such that (i) one of its future directed null normal, say ℓ, has zero expansion, $\theta_{(\ell)} = 0$; (ii) the other future directed null normal, n, has negative expansion $\theta_{(n)} < 0$ and (iii) the directional derivative of $\theta_{(\ell)}$ along n is negative; $\pounds_n \theta_{(\ell)} < 0$.

Σ is either spacelike or null for which $\theta_{(n)} = 0$ and $\pounds_n \theta_{(\ell)} = 0$. Using the above definition, Hayward [227] derived general laws of black hole dynamics through the following two main results:

(a) <u>Zeroth law.</u> *The total trapping gravity of a compact outer marginal surface has an upper bound, attained if and only if the trapping gravity is constant.*

(b) <u>First law.</u> *The variation of the area form along an outer trapping horizon is determined by the trapping gravity and an energy flux.*

The Dynamical horizons are closely related to FOTH where the condition (iii) of Definition 3.3.1 is not required. More precisely, we have

Definition 3.3.2. [16] A smooth, three-dimensional, spacelike submanifold Σ in a spacetime M is said to be a dynamical horizon if it can be foliated by a family of closed 2-surfaces such that, on each leaf L (i) one of its future directed null normal, say ℓ, has zero expansion, $\theta_{(\ell)} = 0$; (ii) the other future directed null normal, n, has negative expansion $\theta_{(n)} < 0$.

Thus, a dynamical horizon, briefly denoted by DH, is basically a spacelike 3-manifold Σ which is foliated by closed marginally trapped 2-surfaces. However, as pointed out in [16], there are some special physical cases which the condition *(iii)* also satisfies and, then, DH is the same as FOTH for spacelike H. Following are special features of DHs:

• Contrary to event horizons, it is clear from the above definition that the concept of a DH is quasi-local for which the knowledge of full spacetime is not required.

• The DHs do not refer to the asymptotic structure, which is the case for event horizons.

• Stationary black holes do not admit DHs simply because such spacetimes are non-dynamic.

In [16] Ashtekar and Krishnan have obtained the following results:

(i) *Expressions of fluxes of energy and angular momentum carried by gravitational waves across these horizons were obtained.*

(ii) *Provided a detailed area balance law relating to the change in the area of* Σ *to the flux of energy across it.*

(iii) *The cross sections* $\mathcal{S}$ *of* Σ *have the topology* S^2 *if the cosmological constant (of Einstein's field equations) is positive and of* S^2 *or a 2-torus. The 2-torus case is degenerate as the matter and the gravitational energy flux vanishes, the intrinsic metric on each* $\mathcal{S}$ *is flat, the shear of the (expansion free) null normal vanishes and the derivatives of the expansion along both normals vanish.*

(iv) *Obtained a generalization of the first and the second laws of mechanics.*

(v) *Established a relation between DH's and IH's.*

Example 3 [16] **The Vaidya solutions.** Consider the 4-metric

$$g_{ab} = -\left(1 - \frac{2GM(v)}{r}\right)\nabla_a v \nabla_b v + 2\nabla_{(a}v\nabla_{b)}r$$
$$+ r^2(\nabla_a\theta\nabla_b\theta + \sin^2\theta\nabla_a\phi\nabla_b\phi)$$

where $M(v)$ is any smooth non-decreasing function of and (v, r, θ, ϕ) are the ingoing Edington-Frinkelstein coordinates, with zero cosmological constant.

Observe that, in general relativity, the above coordinates, named for Arthur Stanley Eddington and David Finkelstein, are a pair of coordinate systems for a Schwarzschild geometry which are adapted to radial null geodesics, i.e., the worldliness of photons moving directly towards or away from the central mass.

The stress energy tensor T_{ab} is given by

$$T_{ab} = \frac{(dM/dv)}{4\pi r^2}\nabla_a v \nabla_b v,$$

which satisfies the dominant energy condition if $dM/dv \geq 0$ and vanishes if and only if $dM/dv = 0$. The metric 2-spheres are given by $v = $ constant, $r = $ constant and their outgoing and ingoing normals can be taken as:

$$\ell^a = (\partial_v)^a + \frac{1}{2}\left(1 - \frac{2GM(v)}{r}\right)(\partial_r)^a \quad \text{and} \quad n^a = -2(\partial_r)^a$$

where $\ell^a n_a = -2$. The expansion of ℓ^a is given by

$$\theta_\ell = \frac{-2GM(v)}{r^2}.$$

It follows that the only spherically symmetric marginal trapped surfaces are the 2-spheres $v = $ constant and $r = 2GM(v)$. To show that these surfaces are cross sections of a DH, we notice that n^a is negative, $\theta_{(n)} = -4/r$ and at the marginal trapped surfaces $\mathcal{L}_n\theta_{(\ell)} = -2/r^2 < 0$. Due to spherical symmetry the shear of both the null normals vanish identically. Finally, $T_{ab}\ell^a\ell^b > 0$ if and only if $dM/dv = 0$.

Hence, it follows from Definition 3.3.2 that the surface given by $r = 2GM(v)$ with $dM/dv > 0$ is a cross section of a DH. When $dM/dv = 0$, this surface becomes null and a non-expanding IH which happens (in most physical situations) at the equilibrium state in the infinite future.

We highlight that so far we have the following relationship:

$$DH \longrightarrow IH$$

to the traditional description of a time-independent black hole related to the asymptotic state of spacetime at null infinity. Recently, there has been an attempt to extend the classical theory to fully dynamical horizons in non-stationary spacetimes (see a review paper [268] by Krishnan and many other cited therein), with a focus on numerical relativity applications.

For those interested in details on the above properties of DH's and their applications we refer to [16, 268] and several citations therein. Observe that there is also a case when Σ is timelike for which no additional conditions on the expansion are required. These horizons are called timelike membranes and are not related to black hole surfaces. However, they do exist and are relevant to numerical simulations of black hole mergers (see some references in [268]). Also, we refer to a recent paper of Gourgoulhon and Jaramillo [210] in which they have discussed the geometric description of the black hole horizons as lightlike hypersurfaces embedded in spacetime, mainly on the numerical relativity applications. To understand how black holes are formed we refer to a recent monograph by Demetrios Christodoulou [125] on "The formation of Black Holes in general Relativity".

Related to the focus of this book, we construct a physical model of spacetimes which admit dynamical horizons. Observe that an expanded black hole (whose area increases) will not be time independent and so for such spacetimes an event horizon cannot be described by a Killing horizon. This is due to the fact that an expanding universe (in which DHs can exist) does not admit a global timelike Killing vector field needed to generate a Killing horizon. Thus, to study the geometry of black holes, among other possible ways, a differential geometric approach is to use a higher symmetry than the Killing symmetry. Since the causal structure is invariant under a conformal transformation, there has been interest in the study of the effect of conformal transformations on properties of expanding black holes (see [91, 241, 378, 379] and more cited therein). This suggests the use of those spacetimes which admit a higher symmetry defined by a timelike conformal Killing vector (CKV) field (see equation (3.2.1)). A physically significant and suitable choice is the use of the (1+3)-splitting ADM model (see, Arnowitt-Deser-Misner [10]) of the spacetime (M, g). This assumes a thin sandwich of M evolved from a spacelike hypersurface Σ_t at a coordinate time t to another spacelike hypersurface Σ_{t+dt} at coordinate time $t + dt$ whose metric g is given by

$$ds^2 = (-\lambda^2 + |V|^2)dt^2 + 2\gamma_{ab}V^a dx^b dt + \gamma_{ab}dx^a dx^b, \qquad (3.3.1)$$

where $x^0 = t$, and $x^a (a = 1, 2, 3)(x^1 = x, x^2 = y, x^3 = z)$ are three spatial coordinates of the spacelike hypersurface Σ_t with γ_{ab} its 3-metric induced from g, $\lambda = \lambda(t, x, y, z)$ is the lapse function and $V^a \partial / \partial x^a$ is a shift vector. For brevity we denote Σ_t by Σ. Choice of ADM assures the existence of a CKV field ξ defined by equation (3.2.1). For example, ADM includes Robertson-Walker spacetimes which admit a G_9 of CKV fields [290] (also see [291]).

We consider this choice of ADM model of a spacetime (M, g) and, under some specified conditions on the extrinsic curvature of the hypersurface Σ and sectional curvature of the spacetime M with respect to the sections containing the normal vector field, we show that Σ is (i) conformally diffeomorphic to a sphere or (ii) a flat space or (iii) a hyperbolic space or (iv) the product of an open real interval and a complete 2-surface of M. Selecting the possibility (iv), we justify the existence of Σ as a DH of a ADM spacetime, subject to its physical requirements as stated in Definition 3.3.2. We now proceed with the details of proving the above claim.

Let $N = -\lambda \bar{\nabla} t$ be the future pointing unit timelike normal vector field where $\bar{\nabla}$ is the spacetime covariant derivative operator. Let X, Y, Z, W be arbitrary vector fields to Σ and K the Weingarten (Shape) operator of Σ defined by $k(X, Y) = g(KX, Y)$ where k denotes the second fundamental form of Σ. Then, the Gauss and Weingarten formulas (which appeared in Chapter 2 with different notation) are

$$\bar{\nabla}_X Y = \nabla_X Y + k(X, Y)N, \quad \bar{\nabla}_X N = KX \tag{3.3.2}$$

where $\gamma = i^* g$ is the 3-metric induced as the pullback of g under the embedding $i : \Sigma \to M$ and ∇ is the metric connection on Σ. The pair (γ, k) is the initial data for the evolution of Σ in spacetime over an infinitesimal time dt. Let L be the traceless part of K, i.e., $L = K - \frac{\tau}{3} I$, where $\tau = \mathrm{tr}(K)$, and I is the identity tensor. The Gauss-Codazzi equations are

$$g(\bar{R}(X, Y)Z, W) = g(R(X, Y)Z, W) + k(Y, Z)k(X, W)$$
$$- k(X, Z)k(Y, W), \tag{3.3.3}$$
$$g(\bar{R}(X, Y)N, Z) = (\nabla_X k)(Y, Z) - (\nabla_Y k)(X, Z), \tag{3.3.4}$$

where $\bar{R}$ and R denote curvature tensors for g and γ respectively. Then, contracting the above two equations with respect to a local orthonormal basis of the tangent space of M at each of its points we obtain

$$\bar{\mathrm{Ric}}(X, Y) + g(\bar{R}(N, X)Y, N) = \mathrm{Ric}(X, Y) + \tau(KX, Y)$$
$$- g(KX, KY), \tag{3.3.5}$$
$$\bar{\mathrm{Ric}}(X, N) = (\mathrm{div}\, K)X - X\tau, \tag{3.3.6}$$

where $\bar{\mathrm{Ric}}$ and Ric are the Ricci tensors g and γ respectively. Using the above and Einstein's field equations

$$\bar{\mathrm{Ric}} - \frac{\bar{r}}{2} \bar{g} = 8\pi T \tag{3.3.7}$$

one obtains the constraint equations

$$r + \frac{2\tau^2}{3} - |L|^2 = 16\pi T_{nn},$$

$$(\text{div}\,.L)X - \frac{2}{3}X\tau = 8\pi T(X, N), \qquad (3.3.8)$$

$$(\text{div}\,L)X = \frac{2}{3}X\tau = 8\pi T(X, N), \qquad (3.3.9)$$

where $\bar{r}$ and r are the scalar curvatures of g, and γ respectively. Also, T is the stress-energy tensor, $T_{nn} = T(N, N)$ and $||$ the norm operator with respect to γ. Assume that the mixed energy condition holds, that is, at any point x on Σ the strong energy condition $T_{nn} + T_m^m|_x \geq 0$ holds and equality implies vanishing of all components of T at x. Assume that (M, g) admits a CKV field ξ satisfying the equation (3.2.1). The $(1 + 3)$-splitting allows us to decompose $\xi = V + \rho N$ where V is the tangential component of ξ. Then, as derived in [373] we need the following two evolution equations:

$$(\pounds_V \gamma)(X, Y) = \sigma\gamma(X, Y) - 2\rho\gamma(LX, Y) - 2\frac{\rho\tau}{3}\gamma(X, Y), \qquad (3.3.10)$$

$$(\pounds_V L)X = - \left(\nabla_X D\rho - \frac{\nabla^2\rho}{3}X\right) - 8\pi\rho\left(TX - \frac{T_m^m}{3}X\right)$$

$$+ \rho\tau - \left(\frac{\sigma}{2}\right) + \rho\left(QX - \frac{r}{3}X\right), \qquad (3.3.11)$$

$$\bar{\nabla}_N V = KV + D\rho - \rho D\ln\lambda + (V\ln\lambda)N, \qquad (3.3.12)$$

where D is the spatial gradient operator, Q is the Ricci operator of γ and $\nabla^2 = \nabla^a\nabla_a$ (a summed over $1, 2, 3$). A normal section of (M, g) at a point p of a spacelike hypersurface Σ is the timelike plane section spanned by the normal N and a tangential vector X at p. The normal sectional curvature $\bar{K}(N, X)$ of M at p along a unit tangent vector X of Σ is defined as $g(\bar{R}(N, X)N, X)$. If $\bar{K}(N, X)$ is independent of the choice of X at each point, then it can be shown [408] that $\bar{R}(N, X) = fN$ for some smooth function f and arbitrary vector field X tangent to Σ. We quote the following result needed in the next result:

Theorem 3.3.3. (Kuehnel [270]) *Let* (Σ, γ) *be a complete connected n-dimensional Riemannian manifold admitting a non-constant solution ρ of $\nabla\nabla\rho = (\nabla^2\rho/n)\gamma$. Then the number of critical points of ρ is a ≤ 2, and Σ is conformally diffeomorphic to*

(i) *a sphere if $a = 2$,*

(ii) *the Euclidean space or hyperbolic space if $a = 1$, and*

(iii) *the product of an $(n-1)$-dimensional Riemannian manifold and an open real line, if $a = 0$.*

Now we state (with proof) a recent similar result as follows:

Theorem 3.3.4. (Sharma [373]) *Let (M, g) be an ADM spacetime solution of Einstein's field equations admitting a CKV field ξ and be evolved out of a complete spacelike hypersurface Σ such that*

(a) Σ *is totally umbilical in M,*

(b) *the normal component ρ of ξ is non-constant on Σ, and*

(c) *the normal sectional curvature of M is independent of the tangential direction of each point of Σ.*

Then, Σ is conformally diffeomorphic to

(i) *a 3-sphere, or*

(ii) *a 3-Euclidean space, or*

(iii) *a 3-hyperbolic space, or*

(iv) *the Riemannian product of a complete 2-surface and an open real interval.*

Proof. By hypothesis (b), we have $\bar{R}(N, X) = fN$ for some smooth function f and arbitrary vector field X tangent to Σ. A contraction of this equation implies that $f = (-1/3) \operatorname{Ric}(N, N)$. This with σ umbilical in (3.3.5) provides

$$\bar{\operatorname{Ric}} = \operatorname{Ric} + \frac{1}{3} \left(\frac{2\tau^2}{3} - \bar{\operatorname{Ric}}(N, N) \right). \tag{3.3.13}$$

A contraction of the above and using (3.3.7) provides

$$\bar{r} + 2\bar{\operatorname{Ric}}(N, N) = r + \frac{2\tau^2}{3} = 16\pi T_{nn}.$$

All this and again using (3.3.7) in (3.3.13) we obtain

$$8\pi T X = Q X + \left(\frac{2\tau^2}{9} - 8\pi T_{nn} + \frac{2\bar{\operatorname{Ric}}(n, N)}{3} \right) X,$$

whose trace free part is

$$8\pi \left(T X - \frac{T_m^m}{3} X \right) = Q X - (r/3) X.$$

By hypothesis L vanishes. Thus, using all this in (3.3.11) we obtain

$$\nabla_X D\rho = \frac{\nabla^2 \rho}{3} X.$$

Since by hypothesis ρ is non-constant and it satisfies the above equation, Kuehnel's result [270] (as stated above) is applicable, which completes the proof. □

Consider the above model of spacetimes such that (iv) holds, that is, Σ is a product of a complete 2-surface H and an open real interval. Take a pseudo-orthonormal basis $\{\ell, n, u, v\}$ at each point $p \in M$, where $\{\ell, n\}$ are future directed null vectors, $\{u, v\}$ are unit spacelike vectors. Let T_pH be spanned by $\{u, v\}$. Thus, we have

$$TM = TH \perp TH^{\perp},$$

where the normal bundle $TH^{\perp}$ is generated by $\{\ell, n\}$. Let Σ be foliated by a family of 2-surfaces H_s (which are 2-dimensional submanifolds of spacetime M) and containing H_s as a codimension-one submanifold, where $s \in (0, \delta)$ and $\delta > 0$. Suppose each slice H_s is a marginally trapped 2-surface and Σ satisfies Definition 3.3.2. Then, according to Ashtekar-Krishnan [16], Σ is a DH. Observe that Σ as a DH is justified since it comes from the $(1+3)$-splitting ADM model also used in Ashtekar-Krishnan's frame work. Furthermore, it is important to note that in addition to the occurrence of a DH, the above model of Sharma [373] also provides a CKV field ξ and Σ is totally umbilical in M. However, this model will include a DH only if its metric symmetry vector field ξ is never Killing. In the next section we use this latter new information to study physically real models of expanding black holes of a large class of time dependent black hole spacetimes.

3.4   Conformal Killing horizons

We know from discussion in previous section that an isolated horizon(IH) is not a realistic model and dynamical horizon(DH) models are needed to understand the properties of black holes of expanding spacetimes. However, since the asymptotic states of DHs are again the IHs, there is need to find realistic models whose asymptotic states are non-isolated and time-dependent black hole spacetimes. Among other approaches to get such a model, this raises the possibility of using a metric symmetry. For this approach, the first candidate is the use of a Killing symmetry, but, we know that stationary black holes do not admit DHs simply because such spacetimes are non-dynamic. Thus, in the case of time-dependent expanding spacetimes, a Killing horizon (KH) can not be used to define a non-isolated horizon. The second choice is to use a higher than the Killing symmetry. For this reason we present latest work of using differential geometric technique for studying those null hypersurfaces of spacetimes whose null geodesic trajectories coincide with conformal Killing trajectories of a null conformal Killing vector(CKV) field (see defining equation (3.2.1)). This happens when a spacetime admits a timelike CKV field which becomes null on a boundary as a null geodesic hypersurface. Such a horizon is called a conformal Killing horizon(CKH) [378, 379]. See an example of CKH later on in this section. We use as our working space a $(1+3)$-splitting ADM spacetime (M, g) with a CKV field ξ, evolved out of a complete spacelike hypersurface Σ which is totally umbilical in M, and assume that ξ is a null vector field. This choice is due to the fact that an ADM spacetime admits both DHs and CKVs. We need the following results on totally umbilical submanifolds:

Proposition 3.4.1. [165] *A null CKV on a spacetime is a geodesic vector field.*

Proof. The conformal Killing equation for a CKV field ξ is given by

$$\bar{g}(\bar{\nabla}_X \xi, Y) + \bar{g}(\bar{\nabla}_Y \xi, X) = \sigma \bar{g}(X, Y).$$

Substituting ξ for Y, and using $\bar{g}(\bar{\nabla}_X \xi, \xi) = 0$ (since ξ is null) we obtain $\bar{\nabla}_\xi \xi = \sigma \xi$, which completes the proof. □

Proposition 3.4.2. *Let (M, g) be a lightlike hypersurface of a Lorentzian manifold $\bar{M}$. Then M is totally umbilical if and only if every section $\xi \in \Gamma(\mathrm{Rad}\, TM)|_\mathcal{U})$ is a conformal killing vector field on $\mathcal{U}$.*

Proof. We obtain the result from a straightforward computation

$$\begin{aligned}\mathcal{L}_\xi g(X, Y) &= \xi g(X, Y) - g(\mathcal{L}_\xi X, Y) - g(X, \mathcal{L}_\xi Y) = -g(A_\xi X, Y) - g(X, A_\xi Y)\\ &= -2g(A_\xi X, Y).\end{aligned}$$

Then $\mathcal{L}_\xi g(X, Y) = 2\sigma g(X, Y)$ if and only if $g(2A_\xi X + 2\sigma X, Y) = 0$ for all $X, Y \in \Gamma(TM)$ if and only if $A_\xi X = -\sigma PX$, so as per [149, page 107, equation (5.3)], M is totally umbilical. □

Proposition 3.4.3. [332] *Let (Σ, γ) be a totally umbilical submanifold of a semi-Riemannian manifold (M, g). Then,*

(a) *a null geodesic vector field of M that starts tangential to Σ remains within Σ (for some parameter interval around the starting point).*

(b) *Σ is totally geodesic if and only if every geodesic vector field of M that starts tangential to Σ remains within Σ (for some parameter interval around the starting point).*

Proof. We know from Chapter 2 that for a totally umbilical M, the second fundamental form B is proportional to the degenerate metric g, that is, $B(X, Y) = \lambda g(X, Y)$ for some function λ of M. Putting this in the local Gauss equation (2.1.14) we read that for any null vector field X on M the equation $\nabla_X X = 0$ is equivalent to $\bar{\nabla}_X X = 0$. In other words, the null geodesics of $\bar{M}$ with lightlike initial vectors tangent to M are null geodesics of M and thus remain within M. This proves (a). In the totally geodesic case B vanishes so that the same argument works and also if and only if holds for any null or non-null initial vectors as well. □

First, we deal with the case of a Killing horizon which is defined as the union $\Sigma_0 = \bigcup \Sigma_0^s$, where Σ_0^s is a connected component of the set of points forming a family of null hypersurfaces Σ_0^s whose null geodesic (see Proposition 3.4.3) generators coincide with the Killing trajectories of nowhere vanishing ξ_0^s. For this case, the event horizon of a static or stationary asymptotically flat black hole is represented by the Killing horizon if M is analytic and the mixed energy condition holds for the stress-energy tensor of the Einstein equations (see [167] at the end of Section 2). In the following we consider a physical class of an ADM spacetime (M, g) such that the evolution vector field is a null CKV field ξ on M.

Theorem 3.4.4. [166] *Let (M, g) be an ADM spacetime evolved through a 1-parameter family of spacelike hypersurfaces Σ_t such that the evolution vector field is a null CKV field ξ on M. Then, ξ reduces to a Killing vector field if and only if the part of ξ tangential to Σ_t is asymptotic everywhere on Σ_t for all t. Moreover, ξ is a geodesic vector field.*

Proof. First we write equation (3.3.10) as

$$(\pounds_V \gamma)(X, Y) = \sigma \gamma(X, Y) - 2\rho \gamma(KX, Y) \tag{3.4.1}$$

for any vector fields tangent to Σ_t. As ξ is null, we have

$$\gamma(V, V) = \rho^2, \tag{3.4.2}$$

which gives $\gamma(\nabla_X V, V) = \rho X \rho$. Substituting V for Y in (3.4.1) we get

$$\nabla_V V + \rho D\rho = \sigma V - 2\rho KV, \tag{3.4.3}$$

where D is the gradient operator of the 3-metric γ. Taking the inner product of the equation (3.3.12) with V gives $g(\bar{\nabla}_N V, V) + \rho V \ln \lambda - \gamma(KV, V) - V\rho = 0$, which in view of (3.4.2) assumes the form

$$\rho N \rho + \rho V \ln \lambda - \gamma(KV, V) - V\rho = 0. \tag{3.4.4}$$

As $\gamma(\nabla_V V, V) = \rho V \rho$, taking the inner product of (3.4.3) with V yields

$$\rho \gamma(KV, V) = \frac{\sigma}{2}|V|^2 - \rho V \rho. \tag{3.4.5}$$

As ξ is the evolution vector field, we have $\lambda = \rho$, and hence $\rho > 0$. The use of (3.4.5) in (3.4.4) gives $\rho N \rho = V\rho + \frac{\sigma}{2\rho}|V|^2$. Using this last equation and (3.4.5) we get $\rho N \rho = \frac{\sigma}{\rho}|V|^2 - \gamma(KV, V)$. Now using (3.4.4) in this last equation we obtain $\gamma(KV, V) = \frac{\sigma}{2}\rho$. This shows that $\sigma = 0$ on M if and only if $\gamma(KV, V) = 0$, i.e., V is asymptotic everywhere on Σ_t for all t. Finally, ξ geodesic follows from Proposition 3.4.1, which completes the proof. □

Next we obtain time-dependent spacetimes (M, G) admitting conformal Killing horizons related to a stationary, asymptotically flat black hole spacetime (M, g) admitting a Killing horizon Σ_0 generated by the null geodesic Killing field, with the conformal factor $G = \Omega^2 g$. Under this transformation, the Killing vector field is mapped to a conformal Killing field ξ_0 defined by the equation (3.2.1) provided $\xi_0^a \nabla_a \Omega \neq 0$. Since the causal structure and null geodesics are invariant under a conformal transformation, Σ_0 still remains a null hypersurface of (M, G). As per Proposition 3.4.2, the null geodesic vector field of M that starts tangential to Σ_0 will remain within Σ_0. Also, its null geodesic generators coincide with the conformal Killing trajectories. Thus, Σ_0 is a CKH in (M, G). On the existence of a class of ADM spacetimes with a null CKV field, we have the following result:

Theorem 3.4.5. [146] *Let (M, g) be an ADM spacetime solution of Einstein's field equations admitting a null CKV field ξ and be evolved out of a complete spacelike hypersurface Σ such that*

(i) *Σ is totally umbilical in M,*

(ii) *the normal component ρ of ξ is equal to the lapse function λ.*

Then, the conformal function of ξ is

$$\sigma = 2\xi(\ln \lambda) \quad and \quad \pounds_V \gamma = 2V(\ln \lambda)\gamma.$$

Proof. By hypothesis, $\rho = \lambda$, so the orthogonal decomposition of ξ is $\xi = V + \lambda N$. Using this and $L = 0$ in (3.3.10) we obtain

$$\pounds_V \gamma = (\sigma - \frac{2\lambda\tau}{3})\gamma. \tag{3.4.6}$$

Operating the CKV equation on an arbitrary tangent vector X and N gives

$$\bar{\nabla}_N V = \frac{\tau}{3}V + \alpha N$$

where α is a function on M. Finally, operating CKV equation on N, N gives $\alpha = \frac{\sigma}{2} - N\lambda$. Thus, the above equation becomes

$$\bar{\nabla}_N V = \frac{\tau}{3}V + (\frac{\sigma}{2} - N\lambda)N. \tag{3.4.7}$$

For null ξ, it is immediate from $\xi = V + \lambda N$ that

$$\gamma(V, V) = |V|^2 = \lambda^2. \tag{3.4.8}$$

Differentiating (3.4.8) along N, and using (3.4.7) and again (3.4.8) we find

$$N\lambda = \frac{\tau}{3}\lambda \tag{3.4.9}$$

which shows that $\tau = 3N(\ln \lambda)$. We know from [164] that

$$\bar{\nabla}_N N = D \ln \lambda \tag{3.4.10}$$

where D is the spatial gradient operator. Since $N \perp V$, $g(\bar{\nabla}_N N, V) = -g(\bar{\nabla}_N V, N)$. Using (3.4.10) and (3.4.8) in this equation we find

$$V(\ln \lambda) = \frac{\sigma}{2} - \frac{\tau}{3}\lambda. \tag{3.4.11}$$

Using $\xi = V + \lambda N$ in (3.4.11) we obtain $\sigma = 2\xi(\ln \lambda)$. Then, (3.4.8) and (3.4.11) transforms (3.4.6) into $\pounds_V \gamma = 2V(\ln \lambda)\gamma$ which completes the proof. □

Remark 3.4.6. A straightforward calculation shows that $\mathcal{L}_\xi N = -\frac{\sigma}{2} N$, i.e., ξ preserves the unit normal vector N up to a conformal factor. Such conformal vector fields were studied by Coley and Tupper [119] and were termed inheriting conformal vector fields where N can be seen as the 4-velocity.

It follows from (3.4.8) that the coefficient of dt^2 in the metric (3.3.1) vanishes so it takes the form

$$ds^2 = 2\gamma_{ab} V^a dx^b dt + \gamma_{ab} dx^a dx^b. \tag{3.4.12}$$

In Theorem 3.4.5 we assume that the normal component of ξ is equal to the lapse function. Now we also assume that the tangential component of ξ is equal to the shift vector so that ξ is the evolution vector field. Then, the absence of dt^2 in (3.4.12) implies that there exists a null hypersurface (Σ_0, γ^0), of M, defined by taking say $\{x^1 = \text{constant}\}$ whose induced degenerate metric γ^0 is given by

$$ds^2_{\gamma^0} = \gamma^0_{a\alpha} V^a dx^\alpha dt + \gamma^0_{\alpha\beta} dx^\alpha dx^\beta, \quad 2 \le \alpha, \beta \le 3, \tag{3.4.13}$$

where $\gamma^0_{\alpha\beta}$ are the coefficients of a degenerate metric γ^0 induced on Σ_0 by the metric g and ξ is tangent to Σ_0. In this way, one can generate a null hypersurface $(\Sigma^0, \gamma^0)_s$ of M. Contrary to the non-degenerate case, for null hypersurfaces both the tangent space $T_p(\Sigma_0)$ and the normal space $T_p(\Sigma_0)^\perp$, at every point p of Σ_0, are degenerate. Moreover,

$$T_p(\Sigma_0) \cap T_p(\Sigma_0)^\perp = T_p(\Sigma_0)^\perp, \qquad \dim(T_p(\Sigma_0)^\perp) = 1,$$

and they do not span the whole tangent space $T_p M$. In other words, a vector in $T_p M$ cannot be decomposed uniquely into a component tangent to $T_p(\Sigma_0)$ and a component to its perpendicular. Therefore, the standard Gauss-Weingarten equations of semi-Riemannian hypersurfaces do not work for the null geometry. To deal with this anomaly, null hypersurfaces have been studied in several ways corresponding to their use in a given problem. Here we let $S(T\Sigma_0)$ be the 2-dimensional complementary spacelike distribution to $T(\Sigma_0)^\perp$ in $T(\Sigma_0)$, called a screen distribution [149]. Then,

$$T(M) = S(T\Sigma_0) \perp S(T\Sigma_0)^\perp, \quad S(T\Sigma_0) \cap S(T\Sigma_0)^\perp = \{0\}, \tag{3.4.14}$$

where $S(T\Sigma_0)^\perp$ is a 2-dimensional complementary orthogonal distribution of $T(M)$. Let $T(\Sigma_0)^\perp$ be spanned by $\{\ell\}$, where ℓ is a real null vector. Then, we know that (see Section 1 of Chapter 2) with respect to each local coordinate neighborhood of Σ_0, there exists a unique null distribution $E = \cup_{p \in \Sigma_0} E_p$, where E is spanned by a unique null vector field n such that

$$g(\ell, n) = 1, \qquad g(n, n) = g(n, X) = 0, \quad \forall X \in \Gamma(S(T\Sigma_0)). \tag{3.4.15}$$

Using (3.4.14) and (3.4.15) we have the following decompositions:

$$T(M) = S(T\Sigma_0) \perp (T(\Sigma_0)^\perp \oplus E),$$
$$T(\Sigma_0) = S(T\Sigma_0) \oplus_{\text{orth}} T(\Sigma_0)^\perp. \tag{3.4.16}$$

Using the above and (3.4.12) we say that the metric on $S(T\Sigma_0)$ is given by

$$ds^2|_{S(T\Sigma_0)} = \gamma_{\alpha\beta}dx^\alpha dx^\beta. \tag{3.4.17}$$

Remark 3.4.7. The above anomaly in null geometry can also be handled by considering the quotient space $T_pM^* = T_p\Sigma_0/T_p(\Sigma_0)^\perp$ and using the canonical projection $\pi : T_pM \to T_pM^*$ (see details in Kupeli [272]). However, consistent with the technique used in this book, we use a screen distribution $S(\Sigma_0)$ to relate our work with the concept of a DH with that of a CKH. It is true that the screen distribution of Σ_0 is not unique, but, fortunately now we have a large class of Lorentzian manifolds with the choice of a unique or canonical screen [143].

Relating dynamical horizons with conformal Killing horizons. In the following we establish a smooth transition from spacelike DHs to CKHs when each Σ_s degenerates to a null hypersurface Σ_0^s of M. This is done by using the null limit technique (first introduced by Swift [381]) and the concept of CKH introduced by Sultana-Dyer [378].

Let C_s be a differentiable curve of a slice Σ_s such that for each s, the unit normal vector field N_s is given by

$$N_s = \frac{1}{\sqrt{2}}(s^{-1}n - s\ell),$$

and the component of N_s in the ℓ-direction approaches zero as $s \to 0$. This means that N_s approaches a null vector field N_0^s which is entirely in the n-direction as $s \to 0$. From this data we construct a null hypersurface Σ_0^s of M as follows:

Suppose $\Omega(\Sigma_s)$ is an object defined on each hypersurface Σ_s. Then, the concept of null limit (explained above) can be used to define an analogous object $\Omega(\Sigma_0^s)$, for the null hypersurface Σ_0^s, by defining

$$\Omega(\Sigma_0^s) = \mathrm{Lim}_{s\to o}\Omega(\Sigma_s).$$

In this way, we say that

$$\Sigma_0^s = \mathrm{Lim}_{s\to o}(\Sigma_s) \quad \text{such that} \quad \mathrm{Lim}_{s\to o}(\xi_s) = \xi_0^s$$

is the null CKV field of Σ_0^s, where the spacelike hypersurface Σ_s degenerates to the null hypersurface Σ_0^s. Contrary to the non-degenerate case, ξ_0^s can not be uniquely expressed as a sum of its tangential and normal components. Instead, using (3.4.16) we can decompose null ξ_0^s of Σ_0^s in terms of three components as

$$\xi_0^s = V + \frac{\lambda}{\sqrt{2}}(\ell - n).$$

Furthermore, we choose a screen distribution $S_s(\Sigma_0^s)$, of Σ_0^s, so that its leaf L_s coincides with the spacelike 2-surface $H_s \subset \Sigma_s \subset M$ and its metric induced from g (see equation (3.3.1)) is expressed as

$$ds^2|_{H_s} = \gamma_{\alpha\beta}^s dx^\alpha dx^\beta, \quad 2 \le \alpha, \beta \le 3.$$

With this choice, we have the following common 2-surface of Σ_0^s and Σ_s (which is a 2-dimensional submanifold of M:

$$M \supset \Sigma_0^s \supset L_s = H_s \subset \Sigma_s \subset M. \tag{3.4.18}$$

The above null limit technique provides the relationship

$$DH \xrightarrow{CKV} CKH. \tag{3.4.19}$$

Now we introduce the following definition of a time-dependent event horizon:

Definition 3.4.8. (Homothetic Event Horizon) We say that an event horizon is a *homothetic event horizon*, briefly denoted by HEH, if its corresponding null hypersurface admits a null homothetic vector (HV) field and its null geodesic trajectories coincide with the homothetic trajectories of this HV field.

Recently, Sultana and Dyer [378] have studied this problem for those spacetimes which admit a conformal Killing vector field (CKV). They considered a conformal transformation , $G = \Omega^2 g$, to stationary asymptotically flat black hole spacetimes which admits a Killing horizon, Σ_0, generated by the Killing vector field ξ. Under such a conformal transformation, ξ is mapped to a CKV. Although spacetimes are asymptotically conformally flat, nevertheless, there can be non-asymptotically flat spacetimes having conformal Killing horizons. In this particular paper [378] Sultana-Dyer considered spacetimes admitting a timelike CKV which becomes null on a boundary called the conformal stationary limit hypersurface and locally described as the time-dependent event horizon (which we call a homothetic event horizon(HEH)) by using this boundary, provided that it is a null geodesic hypersurface. They have shown that such a hypersurface M is null geodesic if and only if the twist of the conformal Killing trajectories on M vanish. Following is their main result on the extension of Hawking's strong rigidity theorem (see in [228]):

Theorem 3.4.9. [378] *Let (M, G) be a spacetime which is conformally related to an analytic black hole spacetime (M, g), with a Killing horizon Σ_0, such that the conformal factor in $G = \Omega^2 g$ goes to a constant at null infinity. Then the conformal Killing horizon Σ_0 in (M, G) is globally equivalent to the event horizon, provided that the stress energy tensor satisfies the weak energy condition.*

Proof. The proof follows from the global definition of the event horizon and the properties of conformal transformations. Recall that the event horizon is the boundary of the set of a union of the terminal indecomposable past sets (TPIs) which is defined as the chronological past of a future endless (or future inextensible, i. e., having no future end point) causal curve. The TPIs are represented by the points in the set. Each TPI being a past set is unchanged if the metric is altered by the introduction of a finite conformal factor. Hence the global definition of an event is conformally invariant, provided the conformal factor tends to a constant at null infinity. This means that, at the null infinity state, the CKV field reduces to the homothetic vector(HV) field and, then, the event horizon is a HEH. Since, as opposed to the proper conformal symmetry, the Einstein equations are invariant with respect to a homothetic symmetry, the structure of homothetic infinity in (M, G) is preserved. Therefore, the manifold (M, G), with a CKH, reduces to a HEH at null infinity, which completes the proof. □

Following is the picture view of the relationship proved in the above theorem:

$$KH \xrightarrow{CKV} CKH \xrightarrow{HV} HEH, \tag{3.4.20}$$

at null infinity. This Sultana-Dyer paper also contains the case as to what happens when the conformal stationary limit hypersurface does not coincide with the HEH. For this case, they have proved a generalized weak rigidity theorem which establishes the conformal Killing property of the event horizon and the rigidity of its rotating conformal Killing horizons.

In another paper [379], they gave an example of a dynamical cosmological black hole, which is a spacetime that describes an expanding black hole in the asymptotic background of the Einstein-de Sitter universe. For this case, the black hole is primordial in the sense that it forms *ab initio* with the big bang singularity and its expanding event horizon (which we call homothetic event horizon) is represented by a CKH whose conformal factor goes to a constant at null infinity. The metric representing the black hole spacetime is obtained by applying a time-dependent conformal transformation on the Schwarzschild metric, such that the result is an exact solution with the matter content described by a perfect fluid and the other a null fluid. They have also studied properties of several physical quantities related to black holes.

Finally, using (3.4.19) and (3.4.20) we have the relationship

$$DH \xrightarrow{CKV} CKH \xrightarrow{HV} HEH, \tag{3.4.21}$$

at null infinity. However, since a single black hole may admit more than one DHs as it comes from a chosen foliation of closed 2-surfaces, it is important to deal with this anomaly to assure the well-defined concept of a CKH and a HEH. Ashtekar-Galloway [15] have studied this problem and have concluded that the following are two reasons for the lack of uniqueness:

(1) A dynamical horizon Σ comes with an assigned foliation by marginally trapped surfaces (MTSs). This raises the following question: Can a given spacelike Σ admit two distinct foliations by MTSs?

(2) Can a connected trapped region of the spacetime admit distinct spacelike 3-manifolds Σ, each of which is a DH?

In [15], the authors addressed both the above uniqueness problems as follows: For the first question, they have shown that, subject to a maximum principal (see details in [220]), the intrinsic geometric structure of a DH is unique, which provides a full positive answer to this question. For the second question, they have provided partial answers and proved (subject to physically reasonable constraints) some uniqueness results for the location of trapped and marginally trapped surfaces in the vicinity of any DH.

Here we highlight that, using the null limit technique (when Σ degenerates to a null hypersurface Σ_0) we have chosen a screen distribution $S(\Sigma_0)$ (which is also in general not unique) such that one of its leaves, L_s, is the same as a member of the family of closed 2-surfaces H_s (see equation (3.4.18)) which foliates a DH. This identification is valid since Ashtekar-Krishnan's framework [16] and Sharma's model [373] are based on the ADM $(1+3)$-splitting. Thus, the unique existence of the geometric structure of a DH also assures a unique choice of a screen distribution for the model of a family of null hypersurfaces discussed in this section. Furthermore, for the same reason, the use of the null limit technique and all the induced objects of the null hypersurface Σ_0 are also well defined due to the choice of a unique screen.

Finally, we say that to get a model of time-dependent black holes, in particular, via the full relationship (3.4.21), using the link with a conformal Killing horizon, so far there are only two publications by Sultana-Dyer [378, 379] and the material presented in this section. Further research in this direction is needed to have well-defined concepts of CKH as well as HEH in the relation (3.4.21). Active research work on this topic is still going on.

3.5   Differential operators on hypersurfaces

The introduction of a Riemannian metric on a smooth manifold gives rise to a series of differential operators which reveal deep relationships between the geometry and the topology of the manifold. Consider the Laplace-Beltrami operator $\triangle^p$ on p-forms; it is well known that on a compact manifold, its spectrum $\{\lambda_i^p\}$, contains topological and geometric information on the manifold. According to the Hodge decomposition theorem, the dimension of the kernel of $\triangle^p$ equals the pth Betti number, so that the Laplacian determines the Euler characteristic $\chi(M)$ of compact Riemannian manifolds (M, g).

Among usual differential operators on (M, g), the exterior derivative d which takes p-forms to $(p+1)$-forms is defined only in terms of the smooth structure of the manifold M. Recall from Section 2 of Chapter 1, the gradient, the divergence and the Laplace-Beltrami operator which are defined by means of the metric on the dual bundle, which is the inverse of the given metric in the semi-Riemannian case.

Many situations arise in mathematical physics where the metric is degenerated, and it is not possible to define the inverse. A typical case arises in the coupling of Einstein's theory of gravity to both a quantum mechanics particle with spin and an electromagnetic field, that is, the Einstein–Dirac–Maxwell (EDM) system. If, for instance, we consider a Lorentz manifold $\bar{M}$ endowed with the Eddington-Finkelstein metric (cf. Hawking-Ellis [228, page 150])

$$ds^2 = -(1 - \frac{2m}{r})du^2 + 2dudr + r^2(d\theta^2 + \sin^2\theta d\phi^2),$$

given in a coordinate system (u, r, θ, ϕ), where $u = t + \bar{r}$ is an advanced null coordinate with

$$\bar{r} = \int \frac{dr}{1 - \frac{2m}{r}} = r + 2m\ln(r - 2m) \qquad (r > 2m),$$

the hypersurface $M : u = $ constant is a lightlike hypersurface. The Dirac operator can be written outside M and in its interior region. But it is not easy to match the two operators on M, because of the fact that the inverse metric $\overset{*}{g}$ can not be defined on a lightlike hypersurface M.

Secondly, the concept of harmonic maps constitutes a very useful tool for both Global Analysis and Differential Geometry (see the harmonic maps and harmonic morphisms bibliographies [80] and [220], respectively and a book by Baird-Wood [25]). We know that any harmonic function μ is a local solution to the Laplace-Beltrami equation $\triangle_M\mu = 0$, but one cannot use this equation for the lightlike case. Also, the Jacobi and Szabó type operators (in the next section we discuss their role in the study of lightlike Osserman hypersurfaces) need an inverse of a metric.

It is, therefore, reasonable to look for a possible non-degenerate metric, say $\tilde{g}$, associated with the degenerate metric g and work with the inverse of $\tilde{g}$ in defining all those concepts which need the inverse of a metric. Recently, Atindogbe-Ezin-Tossa [21] followed this approach and introduced a pseudo-inversion of degenerate metrics which we explain as follows:

Let $(M, g, S(TM))$ be a lightlike hypersurface of a semi-Riemannian manifold $(\bar{M}, \bar{g})$. Consider on M a pair $\{\xi, N\}$ satisfying (2.1.6) and define the one-form

$$\eta(\bullet) = \bar{g}(N, \bullet).$$

For all $X \in \Gamma(TM)$, $X = PX + \eta(X)\xi$ and $\eta(X) = 0$ if and only if $X \in \Gamma(S(TM))$. Now define $\flat$ by

$$\flat : \Gamma(TM) \longrightarrow \Gamma(T^*M)$$
$$X \longmapsto X^\flat = g(X, \bullet) + \eta(X)\eta(\bullet). \tag{3.5.1}$$

Clearly, such a $\flat$ is an isomorphism of $\Gamma(TM)$ onto $\Gamma(T^*M)$ which generalizes the usual non-degenerate theory. In the latter case, $\Gamma(S(TM))$ coincides with $\Gamma(TM)$, and as a consequence the 1-form η vanishes identically and the projection morphism P becomes the identity map on $\Gamma(TM)$. We let $\sharp$ denote the inverse of the isomorphism $\flat$ given by (3.5.1). For $X \in \Gamma(TM)$ (resp. $\omega \in T^*M$), $X^\flat$ (resp.

$\omega^\sharp$) is called the dual 1-form of X (resp. the dual vector field of ω) with respect to the degenerate metric g. It follows from (3.5.1) that if ω is a 1-form on M then

$$\omega(X) = g(\omega^\sharp, X) + \omega(\xi)\eta(X), \quad \forall X \in \Gamma(TM).$$

Define a $(0,2)$-tensor $\tilde{g}$ by

$$\tilde{g}(X,Y) = X^\flat(Y), \quad \forall X,Y \in \Gamma(TM). \tag{3.5.2}$$

Clearly, $\tilde{g}$ defines a non-degenerate metric on M associated with the degenerate metric g of M. Also, observe that $\tilde{g}$ coincides with g if the latter is non-degenerate. The $(0,2)$-tensor $g^{[\cdot,\cdot]}$, inverse of $\tilde{g}$ is called *the pseudo-inverse of g*. With respect to a quasi-orthonormal local frame field $\{\xi, W_1, \ldots, W_m, N\}$ adapted to the decompositions (2.1.8) we have

$$\tilde{g}(\xi,\xi) = 1, \quad \tilde{g}(\xi,X) = \eta(X), \tag{3.5.3}$$

$$\tilde{g}(X,Y) = g(X,Y), \quad \forall X,Y \in \Gamma(S(TM)).$$

This pseudo-inverse metric $\tilde{g}$ of g has been used (see more details in [21]) in defining the usual differential operators *gradient*, *divergence* and *Laplacian* on lightlike hypersurfaces as follows:

Consider a local coordinate system $(x^0, \ldots, x^{n+1})$ adapted to decomposition (2.1.8), that is,

$$\left(\xi := \frac{\partial}{\partial x^0}, X_i := \frac{\partial}{\partial x^i}, \quad N := \frac{\partial}{\partial x^{n+1}}\right)$$

is a local quasi-orthogonal basis of $T\bar{M}_{|M}$ such that

$$\begin{aligned} \text{Rad}\, TM &= \text{Span}\,\{\xi\} \\ S(TM) &= \text{Span}\,\{X_1, \ldots, X_n\} \\ \text{ltr}(TM) &= \text{Span}\,\{N\}. \end{aligned} \tag{3.5.4}$$

Using (3.5.2) and a standard computation gives

$$\left(\frac{\partial}{\partial x^0}\right)^\flat \left(\frac{\partial}{\partial x^0}\right) = 0 + \eta\left(\frac{\partial}{\partial x^0}\right)\eta\left(\frac{\partial}{\partial x^0}\right) = 1 = dx^0\left(\frac{\partial}{\partial x^0}\right), \tag{3.5.5}$$

$$\left(\frac{\partial}{\partial x^0}\right)^\flat (X_i) = g(P\xi, X_i) + \eta(\xi)\eta(X_i) = 0 = dx^0\left(\frac{\partial}{\partial x^i}\right), \tag{3.5.6}$$

$$\left(\frac{\partial}{\partial x^0}\right)^\flat = dx^0. \tag{3.5.7}$$

The first terms in (3.5.5) and (3.5.6) are $\tilde{g}_{00}$ and $\tilde{g}_{0i}$ respectively. Hence,

$$\tilde{g}_{00} = 1, \quad \tilde{g}_{0i} = \tilde{g}_{i0} = 0, \quad i = 1, \ldots, n. \tag{3.5.8}$$

We also have

$$\left(\frac{\partial}{\partial x^i}\right)^\flat \left(\frac{\partial}{\partial x^k}\right) = g(PX_i, X_k) + \eta(X_i)\eta(X_k) = g_{ik}. \tag{3.5.9}$$

Consequently

$$\tilde{g}_{ij} = g_{ij} \qquad i, j = 1, \dots, n. \tag{3.5.10}$$

Thus, with respect to the quasi-orthonormal basis $(\xi, X_1, \dots X_n, N)$ the matrices of $\tilde{g}$ and $g^{[\cdot,\cdot]}$ are given by

$$\tilde{g} = \begin{pmatrix} 1 & 0 & \dots & 0 \\ 0 & & & \\ \vdots & & g_{ij} & \\ 0 & & & \end{pmatrix}, \tag{3.5.11}$$

$$g^{[\cdot,\cdot]} = \begin{pmatrix} 1 & 0 & \dots & & 0 \\ 0 & & & \\ \vdots & & (g_{ij})^{-1} & \\ 0 & & & \end{pmatrix}, \tag{3.5.12}$$

and

$$g^{[]} \cdot g = g \cdot g^{[]} = \begin{pmatrix} 0 & \dots & & 0 \\ \vdots & 1 & \ddots & \\ 0 & & & 1 \end{pmatrix}. \tag{3.5.13}$$

From (3.5.9) and (3.5.10) we get

$$\left(\frac{\partial}{\partial x^i}\right)^\flat \left(\frac{\partial}{\partial x^k}\right) = g_{ik} = \tilde{g}_{ij} dx^j \left(\frac{\partial}{\partial x^k}\right), \qquad i, j = 1, \dots, n. \tag{3.5.14}$$

Hence

$$\left(\frac{\partial}{\partial x^i}\right)^\flat = \tilde{g}_{ij} dx^j. \tag{3.5.15}$$

Taking into account (3.5.6) and (3.5.14) one gets

$$dx^\beta = g^{[\alpha\beta]} \left(\frac{\partial}{\partial x^\alpha}\right)^\flat \qquad \alpha, \beta = 0, \dots, n. \tag{3.5.16}$$

From (3.5.11), (3.5.12) and (3.5.16) we say that if ρ is an endomorphism of TM (a bilinear form on TM resp) its trace with respect to g is given by

$$\text{tr}_g \rho = \sum_{\alpha=0}^n \tilde{g}(\rho(X_\alpha), X_\alpha), \tag{3.5.17}$$

$$\mathrm{tr}_g\, \rho = g^{[\alpha\beta]} \rho_{\alpha\beta} \qquad (3.5.18)$$

respectively. Now, let $f : M \to \mathbf{R}$ be a smooth function on M. One defines intrinsically the gradient of f by $\mathrm{grad}^g\, f = (df)^{\#}$. But

$$df = \frac{\partial f}{\partial x^\alpha} dx^\alpha \Rightarrow (df)^{\#} = \frac{\partial f}{\partial x^\alpha}(dx^\alpha)^{\#},$$

hence from (3.5.16) we infer

$$\mathrm{grad}^g\, f = \frac{\partial f}{\partial x^\alpha} g^{[\alpha\beta]}\left(\left(\frac{\partial}{\partial x^\beta}\right)^\flat\right)^{\#} = \frac{\partial f}{\partial x^\alpha} g^{[\alpha\beta]}\frac{\partial}{\partial x^\beta},$$

$$\mathrm{grad}^g\, f = g^{[\alpha\beta]} \frac{\partial f}{\partial x^\alpha}\frac{\partial}{\partial x^\beta}. \qquad (3.5.19)$$

Let X be a smooth vector field defined on $\mathcal{U} \subset M$. The divergence $\mathrm{div}^g\, X$ of X w.r.t the degenerate metric g is intrinsically defined by

$$\mathrm{div}^g\, X = \sum_{\alpha=0}^{n} \varepsilon_\alpha X_\alpha^\flat\, (\nabla_{X_\alpha} X).$$

Therefore, from (3.5.2) we have

$$\mathrm{div}^g\, X = \sum_{\alpha=0}^{n} \varepsilon_\alpha \tilde{g}(\nabla_{X_\alpha} X, X_\alpha), \quad \varepsilon_0 = 1. \qquad (3.5.20)$$

To sum up we state the following result on these three operators:

Theorem 3.5.1. *Let $(M, g, S(TM))$ be a lightlike hypersurface of a semi-Riemannian $(n+2)$-dimensional manifold $(\bar{M}, \bar{g})$. There exists an associate metric $\tilde{g}$ and a pseudo-inverse $g^{[\cdot]}$ of g on M such that locally on $\mathcal{U} \subset M$ the following holds:*

(i) *For any smooth function $f : \mathcal{U} \subset M \to \mathbf{R}$ we have*

$$\mathrm{grad}^g\, f = g^{[\alpha\beta]} f_\alpha \partial_\beta \qquad where \quad f_\alpha = \frac{\partial f}{\partial x^\alpha} \quad \partial_\beta = \frac{\partial}{\partial x^\beta} \quad \alpha, \beta = 0, \dots n.$$

(ii) *For any vector field X on $\mathcal{U} \subset M$*

$$\mathrm{div}^g\, X = \sum_{\alpha=0}^{n} \varepsilon_\alpha \tilde{g}(\nabla_{X_\alpha} X, X_\alpha)\ ;\ \varepsilon_0 = 1.$$

(iii) *For a smooth function f defined on $\mathcal{U} \subset M$ we have*

$$\triangle^g f = \sum_{\alpha=0}^{n} \varepsilon_\alpha \tilde{g}(\nabla_{X_\alpha}\, \mathrm{grad}^g\, f, X_\alpha).$$

Here $\triangle^g$ is the D'Alembertian w.r.t g on $\mathcal{U}$.

An application. One of the most important concepts which have been found to be useful in semi-Riemannian geometry (and especially in General Relativity) is the scalar curvature. Physically, the scalar curvature has the following interpretation. Start at a point in a D-dimensional space and move a geodesic distance ε in all directions. In essence you would form the equivalent of a generalized sphere in this space. The area of this sphere can be calculated in flat space. But in curved space the area will deviate from the one we calculated by an amount proportional to the scalar curvature. Precisely,

$$r = \lim_{\varepsilon \to 0} \frac{6D}{\varepsilon^2} \left[1 - \frac{A_{\text{curved}}(\varepsilon)}{A_{\text{flat}}(\varepsilon)} \right].$$

From a geometric point of view, this is just the contraction of the Ricci tensor Ric with (a non-degenerate) g,

$$r = g^{ij} \operatorname{Ric}_{ij}.$$

In Chapter 2, the problem of inducing scalar curvature on lightlike manifolds of a Lorentzian manifold has been presented. Such a problem arises, mainly due to two difficulties: Since the induced connection is not a Levi-Civita connection (unless M be totally geodesic) the $(0, 2)$ induced Ricci tensor is not symmetric in general. Also, as the induced metric is degenerate, its inverse does not exist and it is not possible to proceed in the usual way by contracting the Ricci tensor to get a scalar quantity. To overcome these difficulties, a special class of lightlike hypersurfaces in ambient Lorentzian signature, called lightlike hypersurfaces of genus zero, has been discussed in Chapter 2. Elements of such a class are subject to the following constraints: To admit a canonical screen distribution that induces a canonical transversal vector bundle and to admit an induced symmetric Ricci tensor.

To recover an induced scalar curvature for lightlike hypersurfaces of a large class of semi-Riemannian manifolds, recently Atingdogbe [17] used the concept of pseudo-inversion of a degenerate metric. We present his work as follows:

Consider a lightlike hypersurface $(M, g, S(TM))$ of an $(n + 2)$-dimensional semi-Riemannian manifold $(\bar{M}, \bar{g})$, with induced Ricci tensor Ric. Then we define the *symmetrized induced Ricci tensor* to be the $(0, 2)$-tensor $\operatorname{Ric}^{\text{sym}}$ on M such that for X, Y tangents to M,

$$\operatorname{Ric}^{\text{sym}}(X, Y) = \frac{1}{2}[\operatorname{Ric}(X, Y) + \operatorname{Ric}(Y, X)], \tag{3.5.21}$$

where $\operatorname{Ric}(X, Y) = \operatorname{tr}\{Z \longmapsto R(Z, X)Y\}$. In index notation,

$$\operatorname{Ric}^{\text{sym}}_{\alpha\beta} = \frac{1}{2}[R_{\alpha\beta} + R_{\beta\alpha}], \tag{3.5.22}$$

which we shorten to $R_{\alpha\beta} := \operatorname{Ric}_{\alpha\beta}$. Now, $g^{[]}$ being the pseudo-inverse of g, contracting above equation we obtain the scalar quantity

$$r = g^{[\alpha\beta]} \operatorname{Ric}^{\text{sym}}_{\alpha\beta}. \tag{3.5.23}$$

We know that for a fixed hypersurface $(M, g, S(TM))$, the pair (ξ, N) is not uniquely determined as it is subject to the scaling $\xi \longmapsto \xi' = \alpha\xi$ and $N \longmapsto N' = \frac{1}{\alpha}N$ (α smooth non-vanishing function); the right-hand side of (3.5.23) is independent of the choice of the pair (ξ, N). We define r to be the *extrinsic scalar curvature* of the hypersurface $(M, g, S(TM))$.

Now, we give an expression of symmetrized Ricci and extrinsic scalar curvature r in terms of induced shape operators A_N, A_ξ^* and ambient curvatures. Let $\{E_0 = \xi, E_i\}$ be a quasi-orthonormal frame for TM induced from a frame $\{E_0 = \xi, E_i, E_{n+1} = N\}$ for $T\bar{M}$ such that $S(TM) = \text{Span}\{E_1, \ldots, E_m\}$ and $\text{Rad}(TM) = \text{Span}\{\xi\}$. By use of (3.5.3) we have

$$\text{Ric}(X, Y) = \sum_{\gamma=0}^{n} \tilde{g}_{\gamma\gamma}\tilde{g}(R(E_\gamma, X)Y, E_\gamma)$$

$$= \tilde{g}_{00}\tilde{g}(R(\xi, X)Y, \xi) + \sum_{i=1}^{n} \tilde{g}_{ii}\tilde{g}(R(E_i, X)Y, E_i)$$

$$= \bar{g}(R(\xi, X)Y, N) + \sum_{i=1}^{n} g_{ii}g(R(E_i, X)Y, E_i), \quad \tilde{g}_{00} = 1.$$

Then, using Gauss-Codazzi equations (see Chapter 2) provides

$$\text{Ric}(X, Y) = \bar{\text{Ric}}(X, Y) - \bar{g}(\bar{R}(N, X)Y, \xi) + B(X, Y)\,\text{tr}\,A_N - g(A_N X, A_\xi^* Y)$$
$$= \bar{\text{Ric}}(X, Y) + B(X, Y)\,\text{tr}\,A_N - g(A_N X, A_\xi^* Y) - \eta(\bar{R}(\xi, Y)X).$$

Thus, we obtain the following expression of the symmetrized induced Ricci curvature,

$$\text{Ric}^{\text{sym}}(X, Y) = \bar{\text{Ric}}(X, Y) + B(X, Y)\,\text{tr}\,A_N$$
$$- \frac{1}{2}[\eta(\bar{R}(\xi, Y)X) + \eta(\bar{R}(\xi, X)Y) \qquad\qquad (3.5.24)$$
$$+ g(A_N X, A_\xi^* Y) + g(A_N Y, A_\xi^* X)],$$

where $\bar{\text{Ric}}$ denotes the Ricci curvature of $\bar{M}$. In local coordinates,

$$\text{Ric}^{\text{sym}}_{\alpha\beta} = \bar{\text{Ric}}_{\alpha\beta} + B_{\alpha\beta}\,\text{tr}\,A_N$$
$$- \frac{1}{2}[\eta(\bar{R}(\xi, \partial_\beta)\partial_\alpha) + \eta(\bar{R}(\xi, \partial_\alpha)\partial_\beta) \qquad\qquad (3.5.25)$$
$$+ g(A_\xi^* A_N \partial_\alpha, \partial_\beta) + g(A_\xi^* A_N \partial_\beta, \partial_\alpha)],$$

where we make use of the symmetry of A_ξ^* with respect to g. In the sequel, we let

$$\theta = \frac{1}{\sqrt{2}(n+1)} g^{[\alpha\beta]} B_{\alpha\beta}$$

denote the mean curvature function of M and

$$\bar{e} = \bar{\mathrm{Ric}}(N, N)$$

represents the transverse energy in null direction N. Now applying (3.5.23) by contracting (3.5.25) with respect to $g^{[\alpha\beta]}$ we get the following expression of the extrinsic scalar curvature on the structure $(M, g, S(TM))$,

$$r = \bar{R} + \sqrt{2}(n + 1)\theta \operatorname{tr} A_N - \operatorname{tr}(A_\xi^* A_N)$$
$$- \bar{\theta} - \frac{1}{2}g^{[\alpha\beta]}[\eta(\bar{R}(\xi, \partial_\alpha)\partial_\beta) + \eta(\bar{R}(\xi, \partial_\beta)\partial_\alpha)]. \tag{3.5.26}$$

For an ambient manifold $\bar{M}$ with constant sectional curvature c, we have

Proposition 3.5.2. *Let $(M, g, S(TM))$ be a lightlike hypersurface of an $(n + 2)$-dimensional space form $\bar{M}(c)$. Then*

$$r = n^2 c + \sqrt{2}(n + 1)\theta \operatorname{tr} A_N - \operatorname{tr}(A_\xi^* A_N). \tag{3.5.27}$$

Proof. For $\bar{M}(c)$, we have $\bar{\mathrm{Ric}} = (n + 1)c\bar{g}$, $\eta(\bar{R}(\xi, Y)X) = cg(X, Y)$. Then

$$\mathrm{Ric}^{\mathrm{sym}}(X, Y) = ncg(X, Y) + B(X, Y) \operatorname{tr} A_N$$
$$- \frac{1}{2}[g(A_\xi^* A_N X, Y) + g(A_\xi^* A_N Y, X)].$$

In local coordinates,

$$\mathrm{Ric}^{\mathrm{sym}}_{\alpha\beta} = ncg_{\alpha\beta} + B_{\alpha\beta} \operatorname{tr} A_N$$
$$- \frac{1}{2}[g(A_\xi^* A_N \partial_\alpha, \partial_\beta) + g(A_\xi^* A_N \partial_\beta, \partial_\alpha)],$$

and as given by (3.5.27), which completes the proof. $\square$

Example 4. The null cone $\wedge_0^{n+1}$ in $\mathbb{R}_1^{n+2}$ is given by the equation $-(x^0)^2 + \sum_{a=1}^{n+1} = 0$, $x \neq 0$. For convexity we use either $x^0 < 0$ or $x^0 > 0$. It is known that $\wedge_0^{n+1}$ is a lightlike hypersurface with radical distribution spanned by the global position vector field $\xi = \sum_{A=0}^{n+1} x^A \partial x^A$ on $\wedge_0^{n+1}$. The corresponding null transversal section is given by $N = \frac{1}{2(x^0)^2}\{-x^0\partial_0 + \sum_{a=1}^{n+1} x^a \partial_a\}$ and is globally defined. The associated screen distribution $S(T\wedge_0^{n+1})$ is such that $X = \sum_{a=1}^{n+1} X^a \partial_a$ belongs to $S(T\wedge_0^{n+1})$ if and only if $\sum_{a=1}^{n+1} x^a X^a = 0$. Then direct calculations (using some structure equations in Chapter 2) leads to the following expressions of two shape operators, $A_\xi^* X = -PX$ and $A_N X = -\frac{1}{2(x^0)^2}PX$ where P denotes the projection morphism of the tangent bundle $T\wedge_0^{n+1}$ onto $S(T\wedge_0^{n+1})$. Then we get $\theta = \frac{-n}{\sqrt{2}(n+1)}$, $A_\xi^* A_N = \frac{1}{2(x^0)^2} \circ P$ and $\operatorname{tr} A_N = \frac{-n}{2(x^0)^2}$. Finally, since $\mathbb{R}_1^{n+2}$ is flat, the extrinsic scalar curvature of the lightlike cone $\wedge_0^{n+1}$ is given by

$$r = \frac{n^2 - n}{2(x^0)^2}. \tag{3.5.28}$$

Notes. We highlight the following observations from the above example:

- Induced scalar curvature (3.5.28) is actually independent of the elements defining the screen distribution and depends only upon $\wedge_0^{n+1}$.

- The above screen distribution on $\wedge_0^{n+1}$ is integrable and induces a foliation $\mathcal{F}$. By (3.5.28), the extrinsic scalar curvature r is constant along leaves of $S(T\wedge_0^{n+1})$. Actually, these leaves are n-spheres.

- r vanishes at infinity on $\wedge_0^{n+1}$.

Example 4. Lightlike Monge hypersurfaces of $\mathbb{R}_q^{n+2}$. Consider a smooth function $F : \Omega \to \mathbb{R}$, where Ω is an open set of $\mathbb{R}^{n+1}$, then

$$M = \{(x^0, \ldots, x^{n+1}) \in \mathbb{R}_q^{n+2}, x^0 = F(x^1, \ldots, x^{n+1})\}$$

is a hypersurface which is called a Monge hypersurface [149]. Such a hypersurface is lightlike if and only if F is a solution of the partial differential equation

$$1 + \sum_{i=1}^{q-1} (F'_{x^i})^2 = \sum_{a=q}^{n+1} (F'_{x^a})^2.$$

The radical distribution is spanned by a global vector field

$$\xi = \partial x^0 - \sum_{i=1}^{q-1} F'_{x^i} \partial x^i + \sum_{a=q}^{n+1} F'_{x^a} \partial x^a. \tag{3.5.29}$$

Along M consider the constant timelike section $V^* = \partial x^0$ of $\mathbb{R}_q^{n+2}$. The vector bundle $H^* = \text{Span}\{V^*, \xi\}$ is non-degenerate on M. The complementary orthogonal vector bundle $S^*(TM)$ to H^* in $T\mathbb{R}_q^{n+2}$ is an integrable screen distribution on M called the natural screen distribution on M. The transversal bundle $tr^*(TM)$ is spanned by $N = -V^* + \frac{1}{2}\xi$ and $\tau(X) = 0 \; \forall X \in \Gamma(TM)$. It follows that $S^*(TM)$ is a globally conformal screen on M with constant positive conformal factor $\frac{1}{2}$, that is $A_N = \frac{1}{2}A_\xi^*$. The natural parameterization on M is given by

$$x^0 = F(v^0, \ldots, v^n), \qquad x^{\alpha+1} = v^\alpha, \; \alpha \in \{0, \ldots, n\}.$$

Then the natural frame field on M is globally defined by

$$\partial v^\alpha = F'_{x^{\alpha+1}} \partial x^0 + \partial x^{\alpha+1}, \quad \alpha \in \{0, \ldots, n\}. \tag{3.5.30}$$

Now, direct calculations, using the above two equations, lead to

$$B_{\alpha\beta} = B(\partial v^\alpha, \partial v^\beta) = -\frac{\partial^2 F}{\partial v^\alpha \partial v^\beta}.$$

Let $A_\xi^* \partial v^\alpha = K_\alpha^\mu \partial v^\mu$. Since $A_\xi^* \partial v^\alpha$ belongs to $S(TM)$ we have from (3.5.3),

$$B(\partial v^\alpha, \partial v^\beta) = g(A_\xi^* \partial v^\alpha, \partial v^\beta) = \tilde{g}(\partial v^\alpha, \partial v^\beta).$$

Thus,

$$-\frac{\partial^2 F}{\partial v^\alpha \partial v^\beta} = K^\mu_\alpha \tilde{g}(\partial v^\mu, \partial v^\beta) = K^\mu_\alpha \tilde{g}_{\mu\beta}.$$

That is

$$K^\mu_\alpha = -g^{[\mu\beta]}\frac{\partial^2 F}{\partial v^\alpha \partial v^\beta},$$

and

$$A^*_\xi \partial v^\alpha = 2A_N \partial v^\alpha = -g^{[\mu\beta]}\frac{\partial^2 F}{\partial v^\alpha \partial v^\beta}\frac{\partial}{\partial v^\mu}. \tag{3.5.31}$$

Let $\tilde{F} = (F_{\alpha\beta})$ be the matrix with entries $F_{\alpha\beta} = \frac{\partial^2 F}{\partial v^\alpha \partial v^\beta}$. It follows that

$$\begin{aligned}
\operatorname{tr} A_N = \frac{1}{2}\operatorname{tr} A^*_\xi &= -\frac{1}{2}g^{[\alpha\beta]}g^{[\mu\gamma]}\frac{\partial^2 F}{\partial v^\alpha \partial v^\gamma}\tilde{g}_{\mu\beta}\\
&= -\frac{1}{2}g^{[\alpha\beta]}F^\mu_\alpha\tilde{g}_{\mu\beta} = -\frac{1}{2}\operatorname{tr}_g \tilde{F}. \tag{3.5.32}
\end{aligned}$$

Also,

$$\begin{aligned}
A^*_\xi A_N \partial v^\alpha = \frac{1}{2}A^*_\xi \partial v^\alpha &= -g^{[\mu\beta]}\frac{\partial^2 F}{\partial v^\alpha \partial v^\beta}A^*_\xi\frac{\partial}{\partial v^\mu}\\
&= \frac{1}{2}g^{[\mu\beta]}g^{[\gamma\delta]}\frac{\partial^2 F}{\partial v^\alpha \partial v^\beta}\frac{\partial^2 F}{\partial v^\mu \partial v^\delta}\frac{\partial}{\partial v^\gamma}.
\end{aligned}$$

Then,

$$\begin{aligned}
\operatorname{tr}(A^*_\xi A_N) &= g^{[\alpha\nu]}\tilde{}(A^*_\xi A_N \partial v^\alpha, \partial v^\nu)\\
&= \frac{1}{2}g^{[\alpha\nu]}g^{[\mu\beta]}g^{[\gamma\delta]}\frac{\partial^2 F}{\partial v^\alpha \partial v^\beta}\frac{\partial^2 F}{\partial v^\mu \partial v^\delta}\tilde{g}_{\gamma\nu}.\\
\operatorname{tr}(A^*_\xi A_N) &= \frac{1}{2}g^{[\alpha\nu]}F^\mu_\alpha F^\gamma_\mu \tilde{g}_{\gamma\nu} = \frac{1}{2}\operatorname{tr}_g(\tilde{F}^2). \tag{3.5.33}
\end{aligned}$$

From the last two equations and the definition of θ it follows that

$$r = \frac{1}{2}\Big[[\operatorname{tr}_g(\tilde{F})]^2 - \operatorname{tr}_g(\tilde{F}^2)\Big]. \tag{3.5.34}$$

We now examine how the operators and induced geometric objects involved in (3.5.26) defining the extrinsic scalar curvature r change with a change in screen distribution. First, note that the local fundamental form B of M is independent of the choice of screen distribution. Hence the mean curvature function σ of M is invariant. Now, starting with a $S(TM)$ with local orthonormal basis $\{W_i\}$, consider another screen distribution $S(TM)'$ with orthonormal basis

$$W'_i = \sum_{j=1}^n P^j_i(W_j - \varepsilon_j c_j \xi),$$

where c_i and P_i^j are smooth functions and $\{\varepsilon_j\}$ represents the signature of $\{W_j\}$. Below, we let $W = \sum_{i=1}^n c_i W_i$ and $\rho = \sum_{i=1}^n \varepsilon_i(c_i)^2$ denote characteristic vector and scalar fields in respect of this screen change. Then, the following local transformations are derived:

$$\eta' = \eta + \omega, \qquad A_\xi^{*'} = A_\xi^* - \mu \otimes \xi$$

with $\omega = g(W, \cdot)$ and $\mu = B(W, \cdot)$. Also,

$$A_{N'}X = A_N + \sum_i^n \{\varepsilon_i c_i X(c_c) - \tau(X)\varepsilon_i(c_i)^2 + \frac{1}{2}\varepsilon_i(c_i)^2 \mu(X)\}$$

$$- c_i C(X, W_i)\}\xi + \sum_i^n \{c_i(\tau(X) + \mu(X)) - X(c_i)\}W_i$$

$$- \sum_i c_i \nabla *_X W_i - \frac{1}{2}\rho A_\xi^* X, \forall X \in \Gamma(TM|_{\mathcal{U}})$$

and

$$\bar{e}' = \bar{e} - \frac{1}{2}\rho^2 \,\bar{\mathrm{Ric}}(\xi, \xi) + \bar{\mathrm{Ric}}(W, W) - \rho\bar{\mathrm{Ric}}(N, \xi)$$

$$- \theta \,\bar{\mathrm{Ric}}(\xi, W) + 2\,\bar{\mathrm{Ric}}(W, N).$$

3.6   Osserman lightlike hypersurfaces

Introduction. A primary interest in differential geometry is to determine the curvature and the metric of a given smooth manifold, which distinguishes the geometry of this subject from the others that are analytic , algebraic or topological. It is now well known that the research on the *Osserman condition* (which involves sectional curvature and *Jacobi operator*) has provided substantial information on the curvature and metric tensors of Riemannian [114, 115, 116] and semi-Riemannian [201, 203] manifolds. This introduction is intended to provide some background information for understanding the key role of *Osserman manifolds* in semi-Riemannian geometry and, in particular, Lorentzian geometry. This information will help readers to understand a new class of *lightlike Osserman hypersurfaces* studied in a recent paper [20]. For details on the Osserman conjecture in Riemannian geometry, we refer to [114, 115, 116]. Since an up-to-date account of the semi-Riemannian Osserman geometry is available in a recent book [201] we only state the needed results with appropriate references to the proofs.

Let (M, g) be a semi-Riemannian manifold. We write:

$$\mathcal{S}_p^-(M) = \{x \in T_pM| \ g(x, x) = -1\},$$
$$\mathcal{S}_p^+(M) = \{x \in T_pM| \ g(x, x) = 1\},$$
$$\mathcal{S}_p(M) = \{x \in T_pM| \ |g(x, x)| = 1\} = \mathcal{S}_p^-(M) \cup \mathcal{S}_p^+(M),$$

$$\mathcal{S}^-(M) = \cup_{p \in M} \mathcal{S}_p^-(M) \quad \text{the unit timelike bundle of } (M, g),$$
$$\mathcal{S}^+(M) = \cup_{p \in M} \mathcal{S}_p^+(M) \quad \text{the unit spacelike bundle of } (M, g),$$
$$\mathcal{S}(M) = \cup_{p \in M} \mathcal{S}_p(M) \quad \text{the unit non-null bundle of } (M, g).$$

Let $z \in T_p M$ and ∇ be a metric connection on M. Consider

$$R(\cdot, z)z : T_p M \to T_p M, \quad p \in M$$

a linear map defined by $(R(\cdot, z)z)x = R(x, z)z$ for any $X \in T_p M$, where R denotes the curvature tensor of ∇. Then, using the curvature identities we have

$$R(\cdot, z)z : T_p M \to Z^\perp \tag{3.6.1}$$

where $Z^\perp$ is a non-degenerate orthogonal space to $\text{Span}\{z\}$ in $T_p M$.

Definition 3.6.1. Let (M, g) be a semi-Riemannian manifold and $Z \in \mathcal{S}(M)$. Then, the restriction $R : z^\perp \to z^\perp$ of the linear map (3.6.1) is called the Jacobi operator to z, that is,

$$R_z x = R(x, z)z, \quad \forall x \in z^\perp.$$

It is easy to see that the Jacobi operator is a self-adjoint map. We say that $F \in \otimes^4 T_p^\star M$ is an *algebraic curvature map (tensor)* on $T_p M$ if it satisfies the following symmetries:

$$F(x, y, z, w) = -F(y, x, z, w) = F(z, w, x, y),$$
$$F(x, y, z, w) + F(y, z, x, w) + F(z, x, y, w) = 0 \tag{3.6.2}$$

for all $x, y, z, w \in T_p M$, $p \in M$. The Riemann curvature tensor R is an algebraic curvature tensor on the tangent space $T_p M$, for every $p \in M$. For an algebraic curvature map F on $T_p M$, the associated Jacobi operator F_z is the linear map on $T_p M$ characterized by the identity

$$g(F_z x, y) = F(x, z, z, y), \tag{3.6.3}$$

where $z \in T_p M$ and $x, y \in z^\perp$. F_z is a self-adjoint map and F is a spacelike (respectively timelike) Osserman tensor if $Spec\{F_z\}$ is constant on the pseudo-sphere of unit spacelike (respectively unit timelike) vectors in $T_p M$. These are equivalent notions [203] and such a tensor is called an Osserman tensor. It is easy to calculate that $\text{tr}\, R_z = \text{Ric}(z, z)$. Also, let $\pi = \text{Span}\{x, z\}$ be a non-degenerate plane in $T_p M$. Then, the sectional curvature of π is given by

$$K(\pi) = \frac{g(R(x, z)z, x)}{g(x, x)g(z, z) - g(z, z)^2} = \frac{g(R_z x, x)}{g(x, x)g(z, z)}.$$

The basic problem is to what extent general sectional curvatures can provide information on the curvature and metric tensors.

In general, for a semi-Riemannian case, R_z may not be diagonalizable. Thus, we state the following Osserman conditions in terms of the characteristic polynomial of R_z instead of the eigenvalues as in Riemannian geometry.

Definition 3.6.2. A semi-Riemannian manifold (M, g) is called timelike (respectively spacelike) Osserman at a point $p \in M$ if the characteristic polynomial of R_z is independent of $z \in S_p^-(M)$ (respectively $S_p^+(M)$).

Theorem 3.6.3. [201] *A semi-Riemannian manifold (M, g) is timelike Osserman at $p \in M$ if and only if (M, g) is spacelike Osserman at p.*

Remark 3.6.4. It follows from the above theorem that (M, g) is Osserman at $p \in M$ if (M, g) is both timelike and spacelike Osserman at p. Generalization of the above local result for a *global Osserman condition* was first proved in [205] for Riemannian manifolds and then its semi-Riemannian version was presented as follows (proof is common with the Riemannian case).

Theorem 3.6.5. [201] *Let (M, g) be a connected semi-Riemannian pointwise Osserman manifold such that,*

   (i) *the Jacobi operators have only one eigenvalue and* $\dim M \geq 3$, *or*

   (ii) *the Jacobi operators have exactly two distinct eigenvalues, which are either complex or, real with constant multiplicities, at every $p \in M$ and* $\dim M > 4$.

Then (M, g) is globally Osserman.

Lorentzian Osserman manifolds. The solution to the Osserman condition in Lorentzian geometry was originally studied separately for the timelike and the spacelike cases (see [71, 200]). However, in [201] the authors have given a simple proof by using Theorem 3.6.5. Following are two of their main results.

Theorem 3.6.6. [201] *A Lorentzian manifold (M, g) is Osserman at $p \in M$ if and only if (M, g) is of constant sectional curvature at p.*

Theorem 3.6.7. [201] *If (M, g) is a connected $(n \geq 3)$-dimensional Lorentzian pointwise Osserman manifold, then, (M, g) is a real space form.*

Null Osserman condition. Let $\xi \in T_pM$ be a null vector of an n-dimensional Lorentzian manifold (M, g). Then $\xi^\perp = (\mathrm{Span}\{\xi\})^\perp$ is a degenerate vector space containing $\mathrm{Span}\{\xi\}$. Denote by $\bar{\xi}^\perp = \xi^\perp / \mathrm{Span}\{\xi\}$ the $(n-2)$-dimensional quotient space with the projection $\pi : \xi^\perp \to \bar{\xi}^\perp$.

Definition 3.6.8. Let ξ be a null vector of a Lorentzian manifold (M, g). Then, the linear map $\bar{R}_\xi : \bar{\xi}^\perp \to \bar{\xi}^\perp$ defined by

$$\bar{R}_{\bar{x}} = \pi \left(R(x, \xi)\xi \right), \quad x \in \Gamma(TM), \quad \pi(x) = \bar{x},$$

where R is the curvature tensor of a linear connection ∇ on M, is called the Jacobi operator with respect to ξ.

Remark 3.6.9. Note that, since $R(\xi, \xi)\xi = 0$, $\bar{R}_\xi$ is well defined for any $\xi \in \Gamma(\mathrm{Rad}\, TM)$. Also, $\bar{R}_\xi$ is self-adjoint. Since $g(\xi, \xi) = 0$ for every null vector ξ, there is no canonical way of determining a set of null vectors with respect to which the null Osserman condition can be defined. For this reason we use the property that $g(\xi, z) \neq 0$ for every null ξ and for every timelike z on a Lorentzian manifold (see [317]) and define a canonical set of null vectors as follows:

Definition 3.6.10. Let (M, g) be a Lorentzian manifold with a null vector ξ and a timelike unit vector z. Then, the null congruence $N(z)$, determined by z at p, is defined by

$$N(z) = \{\xi \in T_p(M) : g(\xi, \xi) = 0 \text{ and } g(\xi, z) = -1\}.$$

Definition 3.6.11. Let (M, g) be a Lorentzian manifold of $\dim(M) \geq 3$. M is called null Osserman with respect to a unit timelike vector $z \in T_pM$ if the characteristic polynomial of $\bar{R}_\xi$ is independent of $\xi \in N(z)$.

Theorem 3.6.12. [201] *Let (M, g) be a Lorentzian manifold of* $\dim M \geq 4$ *and* $z, z' \in T_pM$ *be linearly independent timelike unit vectors. A null Osserman manifold (M, g), with respect to z and z', is of constant sectional curvature at $p \in M$.*

Remark 3.6.13. See Chapters 1 through 3 of [201] for details on above introductory notes. Also, for some special cases, such as semi-Riemannian Osserman manifolds of signature $(2, 2)$, new classification results for higher dimensions and the affine Osserman condition, we suggest Chapters 4 through 6 of [201].

Since any semi-Riemannian manifold has lightlike subspaces, one reasonably expects a role for Jacobi and Szabó type operators in the study of lightlike hypersurfaces M. However, since these operators need the use of an inverse of a metric, the definition of these operators is not possible in the usual way for the lightlike case. To deal with this problem, recently Atindogbe-Duggal [20] have used the concept of pseudo-inverse of a degenerate metric g [21] (see details in previous section), defined pseudo-Jacobi operators and studied the Osserman condition in lightlike hypersurfaces. To the best of our knowledge, at the time of publishing this book, reference [20] is the only paper on the Osserman condition in lightlike geometry. The objective of this section is to present the results of this sole paper.

Let $(M, g, S(TM))$ be a lightlike hypersurface of an $(m+2)$-dimensional semi-Riemannian manifold $(\bar{M}, \bar{g})$. Then, besides the problem of its degenerate metric g, in general, induced Riemann curvature tensors on lightlike hypersurfaces are not algebraic curvature tensors, i.e (3.6.2) does not hold. Therefore, as a first step, we find conditions on a lightlike hypersurface to have an induced algebraic Riemann curvature tensor so that (3.6.2) holds, using the pseudo-inverse of the degenerate g.

Pseudo-Jacobi operators. We choose a screen distribution $S(TM)$ of a lightlike manifold M for obtaining an algebraic curvature tensor for M. Later on we choose a canonical screen for results on Osserman hypersurfaces. Consider the non-degenerate metric $\tilde{g}$ associated to g such that (3.5.1) and (3.5.2) hold. Let us start by an intrinsic interpretation of relation (3.6.3) which in a semi-Riemannian setting characterizes the Jacobi operator J associated to an algebraic curvature map $R \in \otimes^4 T_p^\star M$, $(p \in M)$. Indeed, we have equivalently for x in the unit bundle, y, w in $T_p M$,

$$(J_R(x)y)^{\flat_g}(w) = R(y, x, x, w), \quad \text{that is,} \tag{3.6.4}$$

$$J_R(x)y = R(y, x, x, \bullet)^{\sharp_g}, \tag{3.6.5}$$

where $\flat_g$ and $\sharp_g$ are the usual isomorphisms between $T_p M$ and its dual $T_p^\star M$, for a non-degenerate g. As stated above, the metric g and its associated metric $\tilde{g}$ coincide if the former is non-degenerate, and equivalently, relation (3.6.5) can be written in the form

$$\tilde{g}(J_R(x)y, w) = R(y, x, x, w),$$

in which $J_R(x)y$ is well defined. This leads to the following definitions.

Definition 3.6.14. Let $(M, g, S(TM))$ be a lightlike hypersurface of a semi-Riemannian manifold $(\bar{M}, \bar{g})$, $p \in M$, $x \in S_p(M)$ and $R \in \otimes^4 T_p^\star M$ an algebraic curvature map. By a pseudo-Jacobi operator of R with respect to x, we mean the self-adjoint linear map $J_R(x)$ of $x^\perp$ defined by

$$J_R(x)y = R(y, x, x, \bullet)^{\sharp_g},$$

where $\sharp_g$ denotes the isomorphism on the triplet $(M, g, S(TM))$.

Definition 3.6.15. We say that a lightlike hypersurface $(M, g, S(TM))$ of a semi-Riemannian manifold $(\bar{M}, \bar{g})$ is null transversally closed if its transversal lightlike bundle $\text{tr}(TM)$ is parallel along the radical direction, that is

$$\bar{\nabla}_U V \in \text{tr}(TM), \quad \forall\, U \in \text{Rad}\, TM \quad \text{and} \quad V \in \text{tr}(TM).$$

Contrary to the non-null case, the induced curvature tensor of a lightlike hypersurface $(M, g, S(TM))$ may not be an algebraic curvature tensor. For this we prove the following result.

Theorem 3.6.16. *Let $(M, g, S(TM))$ be a lightlike hypersurface of a semi-Riemannian manifold $(\bar{M}, \bar{g})$. If the induced Riemann curvature tensor of M is an algebraic curvature tensor, then at least one of the following holds:*

(a) *M is totally geodesic.*

(b) *M is null transversally closed.*

Proof. Let $\bar{\nabla}$ and ∇ be the Levi-Civita connection on $\bar{M}$ and the induced connection on M, with $\bar{R}$ and R their curvature tensors, respectively. We use the following range of indices: indices $0 \leq \alpha, \beta, \gamma, \ldots \leq m$; $1 \leq i, j, k, \ldots \leq m$ and $0 \leq A, B, C, \ldots \leq m+1$. Consider on $\bar{M}$ a local frame $\{\xi = \frac{\partial}{\partial u^0}, \frac{\partial}{\partial u^i}, N\}$ such that $\{\xi = \frac{\partial}{\partial u^0}, \frac{\partial}{\partial u^i}\}$ is a frame on M. Write ∂_A for $\frac{\partial}{\partial u^A}$. Using the local expressions of $\bar{R}$ and R and Gauss-Weingarten equations, we obtain

$$R_{ijkh} = \bar{R}_{ijkh} + C_{ih} B_{jk} - C_{ik} B_{jh}, \tag{3.6.6}$$

$$R_{ij0h} = \bar{R}_{ij0h} - C_i B_{jh} \tag{3.6.7}$$

where $B_{ij} = B(\partial_i, \partial_j)$, $C_{ij} = C(\partial_i, \partial_j)$, $C_\alpha = C(\partial_\alpha, \xi)$ are the components of second fundamental forms of M and $S(TM)$. Thus, the tensor $R \in \otimes^4 T_p^* M$ does not have the usual symmetries of a semi-Riemannian curvature tensor. Now assume that the induced Riemann curvature tensor R defines an algebraic curvature map. Using (3.6.7) we obtain

$$R_{ij0h} = \bar{R}_{ij0h} - C_i B_{jh} = -\bar{R}_{ji0h} - C_i B_{jh} = -(\bar{R}_{ji0h} + C_i B_{jh}).$$

Thus,

$$R_{ij0h} = -R_{ji0h} \iff C_i B_{jh} = -C_j B_{ih}, \qquad \forall 1 \le i, j, h \le n.$$

Using the symmetry of B and (2.1.6) leads to

$$\begin{aligned} C_i B_{jh} &= -C_j B_{ih} = -C_j B_{hi} = C_h B_{ji} \quad \text{(From (2.1.6))} \\ &= C_h B_{ij} = -C_i B_{hj} = -C_i B_{jh} \quad \forall\, 1 \le i, j, h \le n. \end{aligned}$$

Hence,

$$C_i B_{jh} = 0, \ \forall\, i, j, h, \tag{3.6.8}$$

that is, $C_i = 0 \,\forall\, i$ or $B_{jh} = 0 \,\forall j, h$. Since $B(\xi, \bullet) = 0$, $B_{jh} = 0 \,\forall j, h$ leads to M totally geodesic. Now, assume that in (3.6.8) there exist h_0 and j_0 such that $B_{j_0 h_0} \ne 0$. Then $C_i = 0 \,\forall i$. This leads to the following:

$$\begin{aligned} C_i = C(\xi, \partial_i) &= \bar{g}(\nabla_\xi \partial_i, N) = \bar{g}(\bar{\nabla}_\xi \partial_i, N) \\ &= -\bar{g}(\partial_i, \bar{\nabla}_\xi N), \quad \forall\, 1 \le i \le n. \end{aligned}$$

Hence, the null transversally closed condition (see Definition 3.6.15) is equivalent to $C_i = 0 \,\forall i$ and the proof is complete. $\qquad \square$

We know from the previous chapter that a large number of lightlike hypersurfaces M of semi-Riemannian manifolds do have integrable screen distributions. Moreover, we know that every screen distribution of a screen conformal M (see Section 2 of Chapter 2) is integrable. So it seems reasonable to prove the following characterization result.

Theorem 3.6.17. *Let $(M, g, S(TM))$ be a lightlike hypersurface of a semi-Riemannian manifold $(\bar{M}, \bar{g})$, with non-totally geodesic integrable screen distribution $S(TM)$. Then, the induced Riemann curvature tensor of M defines an algebraic curvature map if and only if M is either totally geodesic or locally screen conformal, with ambient holonomy condition*

$$\bar{R}(X, PY)(\mathrm{Rad}\, TM) \subset \mathrm{Rad}\, TM \quad \forall X, Y \in \Gamma(TM).$$

Proof. Assume the induced Riemann curvature tensor defines an algebraic curvature map and consider relation (3.6.6). We have

$$\begin{aligned} R_{ijkh} &= \bar{R}_{ijkh} + C_{ih} B_{jk} - C_{ik} B_{jh} \\ &= -\bar{R}_{jikh} + C_{ih} B_{jk} - C_{ik} B_{jh} \\ &= -\left(\bar{R}_{jikh} + C_{ik} B_{jh} - C_{ih} B_{jk} \right). \end{aligned}$$

Thus,
$$R_{ijkh} = -R_{jikh} \iff C_{ik}B_{jh} - C_{ih}B_{jk} = C_{jh}B_{ik} - C_{jk}B_{ih}.$$

Also,
$$\begin{aligned} R_{ijkh} &= \bar{R}_{ijkh} + C_{ih}B_{jk} - C_{ik}B_{jh} \\ &= \bar{R}_{khij} + C_{ih}B_{jk} - C_{ik}B_{jh}. \end{aligned}$$

So,
$$R_{ijkh} = R_{khij} \iff C_{ih}B_{jk} - C_{ik}B_{jh} = C_{kj}B_{hi} - C_{ki}B_{hj}.$$

Then, since $S(TM)$ is integrable, C is symmetric and we have
$$C_{jk}B_{ih} = C_{ih}B_{kj}, \quad \forall i, j, k, h.$$

Now, we distinguish two cases: M is totally geodesic or not. If M is totally geodesic then M is not screen conformal since $C \neq 0$. If M is not totally geodesic, there exist j_0 and k_0 such that $B_{j_0 k_0} \neq 0$. Then, we have
$$C_{ih} = \frac{C_{k_0 j_0}}{B_{k_0 j_0}} B_{ih}, \quad \forall\, i, h.$$

Observe that $C_{k_0 j_0} \neq 0$, otherwise C would vanish identically at some $p \in M$. Also, by continuity, $B_{k_0 j_0}$ is non-zero in a neighborhood $\mathcal{U}$ of p in M. Define the function φ on $\mathcal{U}$ by $\varphi(x) = \frac{C_{k_0 j_0}}{B_{k_0 j_0}}(x)$. Then $C(X, Y) = \varphi B(X, Y)$ for all X, Y in $\Gamma(S(TM|_{\mathcal{U}}))$, which is equivalent to $A_N X = \varphi \overset{\star}{A}_\xi X$, for all X, Y in $\Gamma(S(TM|_{\mathcal{U}}))$. Finally note that $\overset{\star}{A}_\xi \xi = 0$. Also, since M is non-totally geodesic, it is null transversally closed, that is $A_N \xi = 0$. Thus $A_N X = \varphi \overset{\star}{A}_\xi X$ for all X, Y in $\Gamma(TM|_{\mathcal{U}})$, that is, M is screen conformal as per (2.2.1). In addition,

$$\begin{aligned} \langle \bar{R}(X, PY)\xi, Z \rangle = -\langle \bar{R}(Z, \xi)X, PY \rangle &\overset{(2.4.4)}{=} -\langle R(Z, \xi)X, PY \rangle \\ &\quad - B(Z, X)C(\xi, PY) + B(\xi, X)C(Z, PY) \\ &= -\langle R(Z, \xi)X, PY \rangle = -\langle R(X, PY)Z, \xi \rangle = 0. \end{aligned}$$

Thus, $\bar{R}(X, PY)\operatorname{Rad}TM \subset \operatorname{Rad}TM$.

Conversely, assume that M is either totally geodesic or screen locally conformal with required ambient holonomy condition. Observe that the first Bianchi identity is straightforward. Also, if M is totally geodesic there is nothing more to prove since $\bar{R}|_{TM} = R$. Now we consider M to be screen conformal with $B \neq 0$ and show that R defines an algebraic curvature map. From (2.4.4) we have for X, Y, $Z \in \Gamma(TM)$ and $W \in S(TM)$,

$$\langle \bar{R}(X, Y)Z, W \rangle = \langle R(X, Y)Z, W \rangle + B(X, Z)C(Y, W) - B(Y, Z)C(X, W).$$

Thus, since $C(X, W) = \varphi B(X, W)$, the above equation becomes

$$\langle \bar{R}(X, Y)Z, W \rangle = \langle R(X, Y)Z, W \rangle + \varphi \left[B(X, Z)B(Y, W) - B(Y, Z)B(X, W) \right].$$

Put

$$\mathcal{B}(X, Y, Z, W) = B(X, Z)B(Y, W) - B(Y, Z)B(X, W).$$

We have

$$\langle R(X, Y)Z, W \rangle = \langle \bar{R}(X, Y)Z, W \rangle - \varphi \mathcal{B}(X, Y, Z, W), \qquad (3.6.9)$$

and it is straightforward that $\mathcal{B}$ has the required symmetries. So the left-hand side of (3.6.9) has the required symmetries. For the right-hand side, first we have $\langle R(X, Y)Z, \xi \rangle = -\langle R(Y, X)Z, \xi \rangle = 0$. Now,

$$\begin{aligned}
\langle R(Z, \xi)X, Y \rangle &= \langle R(Z, \xi)X, PY \rangle \\
&= \langle \bar{R}(Z, \xi)X, PY \rangle - B(Z, X)C(\xi, PY) + B(\xi, X)C(Z, PY), \\
&= \langle \bar{R}(Z, \xi)X, PY \rangle = -\langle \bar{R}(X, PY)\xi, Z \rangle = 0,
\end{aligned}$$

by the ambient holonomy condition and the proof is complete. $\qquad \square$

Example 6. A simple but basic example is the lightlike cone $\wedge_0^{n+1}$ at the origin of $\mathbb{R}_1^{n+2}$ for which the null transversal normalization

$$N = \frac{1}{2(x_0)^2} \left[-x^0 \frac{\partial}{\partial x_0} + \sum_{a=1}^{n+1} x^a \frac{\partial}{\partial x^a} \right]$$

induces the algebraic curvature tensor

$$R(X, Y)Z = \frac{1}{2(x_0)^2} \left[g(Y, Z)PX - g(X, Z)PY \right],$$

where P is the projection morphism on the screen associated to N. Its associated pseudo-Jacobi operator is then given, for $x \in \mathcal{S}_p(\wedge_0^{n+1})$, by

$$J_R(x) = \frac{1}{2(x_0)^2} \langle x, x \rangle \circ P.$$

Remark 3.6.18. (a) If the screen distribution $S(TM)$ is integrable and for any $x \in \mathcal{S}_p(M)$, $(p \in M)$, $J_R(x)$ preserves the radical distribution, then, since g and $\tilde{g}$ coincide on $S(TM)$, relation (3.6.4) (or equivalently (3.6.5) shows that the pseudo-Jacobi operator J_R induces a Jacobi operator $J_{R'}$ on $(M', g' = g|_{M'})$, where M' is a leaf of $S(TM)$ and R' the restriction on $S(TM)$ of R.

(b) Let R be the induced (algebraic) Riemann curvature tensor of M, $(p \in M)$ and $\xi \in \mathrm{Rad}\, T_p M$. Then, we have

$$J_R(x)\xi = 0. \qquad (3.6.10)$$

Indeed, for all $x \in \mathcal{S}_p(M)$, $z \in T_pM$,

$$\tilde{g}(J_R(x)\xi, z) = R(\xi, x, x, z) = g(R(z, x)x, \xi) = 0,$$

and since $\tilde{g}$ is non-degenerate on TM, we have $J_R(x)\xi = 0$.

(c) Note also that the screen subspace is preserved by J_R. For this, it suffices to show that for all $x \in \mathcal{S}_p(M)$, $z \in S(T_pM)$, $\eta(J_R(x)z) = 0$ which is equivalent to $\tilde{g}(J(x)z, \xi) = 0$ using (2.5.2). But

$$\tilde{g}(J_R(x)z, \xi) = R(z, x, x, \xi) \overset{def.}{=} g(R(\xi, x)x, z) \overset{(3.6.2)}{=} -g(R(x, z)x, \xi) = 0.$$

Lightlike Osserman hypersurfaces. By the approach following [149]), the extrinsic geometry of lightlike hypersurfaces $(M, g, S(TM))$ depends on a choice of screen distribution, or equivalently, normalization. Since the screen distribution is not uniquely determined, a well-defined concept of the Osserman condition is not possible for an arbitrary lightlike hypersurface of a semi-Riemannian manifold. Thus, one must look for a class of normalization for which the induced Riemann curvature and associated Jacobi operator has the desired symmetries and properties. Precisely, we introduce the following:

Definition 3.6.19. A screen distribution $S(TM)$ of a lightlike hypersurface M is said to be admissible if the associated induced Riemann curvature of M is an algebraic curvature.

Based on Theorem 3.6.17, we observe that any screen conformal lightlike hypersurface in a semi-Euclidean space admits an admissible screen distribution since its induced curvature tensor defines an algebraic curvature map. In particular, the canonical screens on the lightlike cones, Monge hypersurfaces and totally geodesic lightlike hypersurface all admit admissible screens. Thus, there exists a large class of semi-Riemannian manifolds of an arbitrary signature which admit admissible canonical screen distributions. We make the following definition:

Definition 3.6.20. A lightlike hypersurface (M, g) of a semi-Riemannian manifold $(\bar{M}, \bar{g})$ of constant index is called timelike (resp. spacelike) Osserman at $p \in M$ if, for each admissible screen distribution $S(TM)$ and associated induced algebraic curvature R, the characteristic polynomial of $J_R(x)$ is independent of $x \in \mathcal{S}_p^-(M)$ (resp. $x \in \mathcal{S}_p^+(M)$). Moreover, if this holds at each $p \in M$, then (M, g) is called pointwise Osserman .

Remark 3.6.21. Based on the discussion so far, it is clear that the above definition of Osserman condition is independent of the choice of admissible screen distribution. This conclusion is noteworthy for the entire study of the geometry of Osserman lightlike hypersurfaces.

Example 7. Being totally umbilical is independent of the choice of screen distribution. Now for a given admissible screen on the lightlike cone $\wedge_0^{n+1}$ of $\mathbb{R}_1^{n+2}$, the induced curvature tensor is given by

$$R(X,Y)Z = \frac{1}{2(x^0)^2}[g(Y,Z)PX - g(X,Z)PY]$$

with P the projection morphism of the tangent bundle $T\wedge_0^{n+1}$ onto the screen distribution and the pseudo-Jacobi operator is given for $z \in S_p(\wedge_0^{n+1})$ by

$$J_R(z) = \frac{1}{2(x^0)^2}\langle z, z \rangle \circ P.$$

It follows that the characteristic polynomial is given by

$$f_z(t) = -t[\frac{\varepsilon}{2(x^0)^2} - t]^{n-1}, \qquad \varepsilon = sign(z) = \pm 1,$$

which is independent of both admissible screen distributions and $z \in S_p^-(\wedge_0^{n+1})$ (resp. $z \in S_p^+(\wedge_0^{n+1})$). The lightlike cone is then timelike (resp. spacelike) pointwise Osserman.

An adaptation of the technique in [201, pp. 4-5] to the lightlike case and following the same steps show that (M, g) being timelike Osserman at $p \in M$ is equivalent to (M, g) being spacelike Osserman at p. More precisely, we have

Theorem 3.6.22. *Let (M, g) be a lightlike hypersurface of a semi-Riemannian manifold $(\bar{M}, \bar{g})$. Then, (M, g) is timelike Osserman at p if and only if it is spacelike Osserman at p.*

From now on we refer to Osserman at p as both timelike and spacelike. Recall [149] that the screen distribution $S(TM)$ is totally umbilical if and only if on any coordinate neighborhood $\mathcal{U} \in M$, there exists a smooth function λ such that $C(X, PY) = \lambda g(X, Y), \quad \forall X, Y \in TM|_{\mathcal{U}}$. Then, since C is symmetric in $S(TM)$, it follows that any totally umbilical $S(TM)$ is integrable. In case $\lambda = 0$ there is a totally geodesic screen foliation on M. For this latter case, the following holds:

Theorem 3.6.23. *Let (M, g) be a lightlike hypersurface of a semi-Riemannian manifold $(\bar{M}, \bar{g})$, whose admissible screen distributions are totally geodesic in a neighborhood $\mathcal{U}$ of a $p \in M$. Then, (M, g) is Osserman at p if and only if the semi-Riemannian screen leaves are Osserman at this point. In particular, if $(\bar{M}, \bar{g})$ has constant index $\nu = 2$, then, (M, g) is Osserman at p if and only if semi-Riemannian admissible screen leaves are of constant sectional curvature at p.*

Proof. Consider a totally geodesic admissible screen distribution $S(TM)$ on $\mathcal{U} \subset M$. Let R, R' and R^* denote the algebraic curvature tensors induced on (M, g) by $S(TM)$, the restriction of R on $S(TM)$ and the Riemann curvature tensor given by the Levi-Civita connection ∇ on the screen distribution, respectively. We first show that under the hypothesis, we have $R' = R^*$ at p. Let $x, y, z \in S(T_pM)$. By straightforward calculation using equations of (2.4.1) and (2.4.2), we have

$$R'(x,y)z = R(x,y)z = R^*(x,y)z + \left[C(x,z)A_\xi^* y - C(y,z)A_\xi^* x \right]$$
$$+ \left[(\nabla_x C)(y,z) - (\nabla_y C)(x,z) + \tau(y)C(x,z) \right.$$
$$\left. - \tau(x)C(y,z) \right] \xi.$$

Thus, we get $R'(x,y)z = R^*(x,y)z$ from $C = 0$. Also, $x \in S_p(M)$ if and only if $x^* \in S_p(M^\star)$, with $x^* = Px$ and $M^\star$ the leaf of $S(TM)$ through p. Moreover, $x^\perp = (Px)^\perp$ and $J_R(x) = J_R(Px)$. We infer that $J_{R^*}x^*$ is the restriction of $J_R(x)$ to $x^{*\perp S(TM)}$. On the other hand, observe that

$$x^\perp = x^{*\perp S(TM)} \quad \oplus_{\text{orth}} \quad TM^\perp$$

and from (3.6.10) we have $J_R(x)\xi = 0$ for all $\xi \in \mathrm{Rad}\,TM$. Then, let $f_x(t)$ and h_{x^*} denote the characteristic polynomials of $J_R(x)$ ($x \in S_p^-(M)$) and $J_{R^*}x^*$ ($x^* \in S_p^-(M^\star)$) with $x^* = Px$, respectively. We have

$$f_x(t) = t\, h_{x^*}(t)$$

which shows that the characteristic polynomial of $J_R(x)$ is independent of $x \in S_p^-(M)$ if and only if the characteristic polynomial of $J_{R^*}x^*$ is independent of $x^* \in S_p^-(M^\star)$. Hence, (M,g) is timelike Osserman at p if and only if $M^\star$ (as a semi-Riemannian manifold) is timelike Osserman at p. Similar is the case for $S_p^+(M)$ and $x^* \in S_p^+(M^\star)$. Since $S(TM)$ is an arbitrary admissible screen, the first part of the theorem is proved. Now, assume that $(\bar{M}, \bar{g})$ has constant index $\nu = 2$. Then, T_pM is a degenerate space of signature $(0, -, +, \ldots, +)$. It follows that screen leaves through p are Lorentzian manifolds. but it is well known [201, p.41] that the latter are Osserman at p if and only if they are constant sectional curvature at this point. This completes our proof.   □

Theorem 3.6.24. *Let (M,g) be a totally umbilical lightlike hypersurface of an $(m+2)$-dimensional semi-Riemannian manifold of constant sectional curvature $(\bar{M}(c), \bar{g})$. The set of admissible screen distributions reduces to totally umbilical ones. Also, M is pointwise Osserman and for each admissible $S(TM)$, $\mathrm{Ric}^{S(TM)}$ is symmetric and M is locally Einstein.*

Proof. If $\rho \neq 0$ then M is proper totally umbilical. Using (2.5.1), we get

$$R(X,Y)Z = c\{g(Y,Z)X - g(X,Z)Y\}$$
$$+ \rho\{g(Y,Z)A_N X - g(X,Z)A_N Y\}. \qquad (3.6.11)$$

Now pick an admissible screen $S(TM)$ and let R denote the associate induced curvature tensor. Then, under the hypothesis, $g(R(X,Y)Z,V) = g(R(Z,V)X,Y)$ for all X, Y, Z, V, we have

$$\rho\{g(Y,Z)g(A_N X, V) - g(X,Z)g(A_N Y, V)$$
$$- g(V,X)g(A_N Z, Y) + g(Z,X)g(A_N V, Y)\} = 0$$

$\forall X, Y, Z, V \in \Gamma(TM)$. Since $\rho \neq 0$, choose a $Z \perp X$ and $g(Y, Z) = 1$ to get

$$g\left(A_N X - g(A_N Z, Y)X, V\right) = 0$$

for all $X, V \in TM|_{\mathcal{U}}$. Thus, $A_N X = \lambda PX$ with $\lambda = g(A_N Z, Y)$, that is the screen distribution is totally umbilical.

Conversely, suppose $A_N X = \lambda PX$ for some smooth λ in $C^\infty(M)$. Then, (3.6.11) becomes

$$\begin{aligned}
R(X, Y)Z &= c\{g(Y, Z)X - g(X, Z)Y\} \\
&\quad + \lambda\rho\{g(Y, Z)PX - g(X, Z)PY\},
\end{aligned} \tag{3.6.12}$$

which defines an algebraic curvature map, that is $S(TM)$ is admissible.

Now, let $S(TM)$ be an arbitrary admissible screen distribution on M. We compute the induced Ricci curvature with respect to $S(TM)$ using (3.6.12). Consider a quasi-orthonormal basis $\{\xi, W_1, \ldots, W_n\}$ on $TM|_{\mathcal{U}}$. Then,

$$\begin{aligned}
\mathrm{Ric}(X, Y) &= \sum_{i=1}^{m} g(R(X, W_i)Y, W_i) + \bar{g}(R(X, \xi)Y, N) \\
&= c\left[g(X, Y) - ng(X, Y)\right] + \lambda\rho\left[g(X, Y) - ng(X, Y)\right] - cg(X, Y) \\
&= \left[(1 - n)\lambda\rho - nc\right] g(X, Y).
\end{aligned}$$

Hence, the Ricci tensor is symmetric. Moreover M is locally Einstein. Finally, let $x \in S_p(M)$, $p \in M$, $y \in x^\perp$. Then,

$$\begin{aligned}
J_R(x)y &= R(y, x, x, \cdot)^{\sharp_g} \\
&\stackrel{(3.6.12)}{=} \left[c\{g(x, x)g(\cdot, y) - g(\cdot, x)g(x, y)\} \right. \\
&\qquad \left. \lambda\rho\{g(x, x)g(\cdot, y) - g(\cdot, x)g(x, y)\}\right]^{\sharp_g} \\
&= (c + \lambda\rho)g(x, x)g(\cdot, y)^{\sharp_g} \\
&= (c + \lambda\rho)g(x, x)Py.
\end{aligned}$$

Hence, in an adapted quasi-orthonormal basis and using Remark 3.6.18(b), the matrix of $J_R(x)$ has the form

$$\begin{pmatrix}
0 & & \cdots & & \cdots & 0 \\
\vdots & & & & & \\
\vdots & & (c + \lambda\rho)g(x, x)I_{n-1} & & & \\
0 & & & & &
\end{pmatrix}.$$

Then, the characteristic polynomial f_x of $J_R(x)$ is given by

$$f_x(t) = -t\left[(c + \lambda\rho)g(x, x) - t\right]^{n-1},$$

with $g(x, x) = \pm 1$ and for arbitrary given admissible screen distribution. Thus M is pointwise Osserman, which completes the proof. $\qquad\square$

Corollary 3.6.25. *A lightlike surface M of a 3-dimensional Lorentz manifold $\bar{M}(c)$ is pointwise Osserman if and only if it is null transversally closed.*

Proof. It is well known [149, page 111] that any lightlike surface of a 3-dimensional Lorentz manifold $\bar{M}$ is either proper totally umbilical or totally geodesic. Hence, it remains only to find a necessary and sufficient condition for existence of an umbilical screen line bundle $S(TM)$ on M. As such a $S(TM)$ is non-degenerate, let $\lambda = \frac{C(W,W)}{g(W,W)}$ with $S(TM) = \text{Span}\{W\}$. Then $C(X, PY) = \lambda g(X,Y) \; \forall X, Y \in TM|_\mathcal{U}$ if and only if $C(\xi, W) = 0$, that is M is null transversally closed. $\qquad\square$

    In the semi-Riemannian case, we know [201] that being Osserman at a point simplifies the geometry at that point as the manifold is Einstein at that point. Moreover, if the latter is connected and of at least dimension 3, by Schur's lemma in [62], it is Einstein. For lightlike hypersurface, this is not always the case as is shown in the next theorem using the following lemma.

Lemma 3.6.26. *Let $(M, g, S(TM))$ be a lightlike hypersurface of an $(n+2)$-dimensional semi-Riemannian manifold $(\bar{M}, \bar{g})$, with induced algebraic Riemannian curvature map R. For all $x \in S_pM$, $p \in M$ we have*

$$\text{tr} \, J_R(x) = \text{Ric}(x, x) - \eta(\bar{R}(\xi, x)x).$$

Proof. Let $(e_0 = \xi, e_1 = Px, e_2, \ldots, e_n, N)$ be a $\bar{g}$-quasi orthonormal basis of $T_p\bar{M}$ with $T_pM = \text{Span}\{(e_0, e_1, e_2, \ldots, e_n)\}$ and $S(T_pM) = \text{Span}\{(e_1, e_2, \ldots, e_n)\}$. We have

$$\text{tr} \, J_R(x) = \sum_{\substack{\alpha=0 \\ \alpha\neq 1}}^{n} g^{[\alpha\alpha]} \tilde{g}(J_R(x)e_\alpha, e_\alpha)$$

$$= \sum_{\alpha=2}^{n} g^{[\alpha\alpha]} \tilde{g}(J_R(x)e_\alpha, e_\alpha) + g^{[0\;0]} \tilde{g}(J_R(x)\xi, \xi)$$

$$= \sum_{\alpha=2}^{n} g^{[\alpha\alpha]} R(e_\alpha, x, x, e_\alpha) = \sum_{\alpha=2}^{n} g^{[\alpha\alpha]} g(R(e_\alpha, x)x, e_\alpha)$$

$$= g^{[e_1 e_1]} g(R(x,x)x, x) + \sum_{\alpha=2}^{n} g^{[\alpha\alpha]} g(R(e_\alpha, x)x, e_\alpha)$$

$$+ \bar{g}(R(\xi, x)x, N) - \bar{g}(R(\xi, x)x, N)$$

$$= \text{Ric}(x, x) - \eta(R(\xi, x)x) \overset{(2.4.1)}{=} \text{Ric}(x, x) - \eta(\bar{R}(\xi, x)x). \qquad\square$$

Theorem 3.6.27. *Let (M, g) be a lightlike hypersurface that is Osserman at $p \in M$. If for an admissible screen distribution $S(TM)$, $R^{S(TM)}(\xi, \bullet)\xi$ is zero for a $\xi \in \text{Rad}\,TM$, and $|\eta(\bar{R}(\xi, x)x)| < \mu \in \mathbb{R}$ for every $x \in S_p^-(M)$ (or every $x \in S_p^+(M)$), then $(M, g, S(TM))$ is Einstein at $p \in M$.*

Proof. Let $\bar{M}$ be the ambient semi-Riemannian manifold of M. Denote by R' and g' the restriction on $S(TM)$ of the induced algebraic curvature tensor R and the metric tensor g on M, respectively. The Osserman condition at p implies that the characteristic polynomial of J_R is the same for every $x \in S_p^-(M)$ (or every $x \in S_p^+(M)$). Then $|\operatorname{tr} J_R(x)|$ is bounded on $S_p^-(M)$ (and $S_p^+(M)$). Now, using Lemma 3.6.26, we have for every $x \in S_p^-(M)$ (or every $x \in S_p^+(M)$),

$$|\operatorname{Ric}(x,x)| \leq |\operatorname{tr} J_R(x)| + |\eta(\bar{R}(\xi,x)x)|.$$

It follows that there exist $\alpha \in \mathbb{R}$ such that $|\operatorname{Ric}(x,x)| \leq \alpha$ for every $x \in S_p^-(M)$ (or every $x \in S_p^+(M)$). In particular, we have

$$|\operatorname{Ric}'(x,x)| \leq \alpha$$

for every $x \in S(T_pM) \cap S_p^-(M)$ (or every $x \in S(T_pM) \cap S_p^+(M)$). Therefore, since $(S(T_pM), g')$ is non-degenerate, it follows from a well-known algebraic result (see [121]) that

$$\operatorname{Ric}'(x,y) = \lambda g'(x,y) \quad \forall x, y \in S(T_pM), \quad \text{with } \lambda \in \mathbb{R}. \tag{3.6.13}$$

Consider $(e_0 = \xi, e_1, \ldots, e_n, N)$ a $\bar{g}$-quasi orthonormal basis of $T_p\bar{M}$ with $T_pM = \operatorname{Span}\{(e_0, e_1, \ldots, e_n)\}$ and $S(T_pM) = \operatorname{Span}\{(e_1, \ldots, e_n)\}$. We show that for all $x \in T_pM$, $\operatorname{Ric}(\xi, x) = \operatorname{Ric}(x, \xi) = 0$. Indeed, we have

$$\operatorname{Ric}(\xi, x) = g^{[00]}\tilde{g}(R(\xi,\xi)x,\xi) + \sum_{i=1}^{n} g^{[ii]}\tilde{g}(R(e_i,\xi)x,e_i)$$

$$= \sum_{i=1}^{n} g'^{ii} g(R(e_i,\xi)x,e_i) = \sum_{i=1}^{n} g'^{ii} g(R(x,e_i)e_i,\xi) = 0.$$

Now,

$$\operatorname{Ric}(x,\xi) = g^{[00]}\tilde{g}(R(\xi,x)\xi,\xi) + \sum_{i=1}^{n} g'^{ii} g(R(x,e_i)\xi,e_i)$$

$$= \eta(R(\xi,x)\xi) \stackrel{(2.4.9)}{=} \eta(\bar{R}(\xi,x)\xi) = 0$$

by hypothesis. Hence, since $g(\xi, \bullet) = g(\bullet, \xi) = 0$, the latter together with (3.6.13) leads to $\operatorname{Ric}(x,y) = \lambda g(x,y)$, for all $x, y \in T_pM$, that is, $(M, g, S(TM))$ is Einstein at $p \in M$. $\square$

Corollary 3.6.28. *Let (M, g) be a lightlike hypersurface of a flat semi-Riemannian manifold $\bar{M}$. If (M, g) is Osserman at $p \in M$ then it is Einstein at p.*

Proof. This is an immediate consequence of the previous theorem since the flat condition implies $\bar{R}(\xi, \cdot)\xi = 0$, $\forall \xi \in \operatorname{Rad} TM$ and $\eta(\bar{R}(\xi,x)x) = 0$. $\square$

Chapter 4

Half-lightlike submanifolds

There are two cases of codimension 2 lightlike submanifolds M since for this type the dimension of their radical distribution $\operatorname{Rad} TM$ is either one or two. A codimension 2 lightlike submanifold is called half-lightlike [147] if $\dim(\operatorname{Rad} TM) = 1$. The objective of this chapter is to present up-to-date results of this sub-case.

4.1   Basic general results

Let $(\bar{M}, \bar{g})$ be an $(m+2)$-dimensional semi-Riemannian manifold of index $q \geq 1$ and (M, g) a lightlike submanifold of codimension 2 of $\bar{M}$. Since g is degenerate, there exists locally a vector field $\xi \in \Gamma(TM), \xi \neq 0$, such that $g(\xi, X) = 0$, for any $X \in \Gamma(TM)$. Then, for each tangent space $T_x M$ we consider

$$T_x M^{\perp} = \{u \in T_x \bar{M} : \bar{g}(u, v) = 0, \forall v \in T_x M\}$$

which is a degenerate 2-dimensional subspace of $T_x \bar{M}$. Since M is lightlike, both $T_x M$ and $T_x M^{\perp}$ are degenerate orthogonal subspaces but no longer complementary. In this case the dimension of $\operatorname{Rad} T_x M = T_x M \cap T_x M^{\perp}$ depends on the point $x \in M$. Denote by $\operatorname{Rad} TM$ a radical (null) distribution of the tangent bundle space TM of M. The radical subspace $\operatorname{Rad} T_x M$ is either a 1-dimensional or a 2-dimensional subspace of $T_x M$. There exists a complementary non-degenerate distribution $S(TM)$ to $\operatorname{Rad} TM$ in TM, called a screen distribution of M, with the orthogonal distribution

$$TM = \operatorname{Rad} TM \oplus_{\mathrm{orth}} S(TM).$$

The submanifold $(M, g, S(TM))$ is called a *half-lightlike submanifold* [147, 154] if $\dim(\operatorname{Rad} TM) = 1$. The term half-lightlike has been used since for this class $(TM)^{\perp}$ is half lightlike. On the other hand, if $\dim(\operatorname{Rad} TM) = 2$, then, $\operatorname{Rad} TM = TM^{\perp}$ and $(M, g, S(TM))$ is called a *coisotropic submanifold* [154]. The latter

[147, 154] class is discussed in the next chapter. In this section, we present results on half-lightlike submanifolds (published in [147, 154, 158]) for which there exist ξ, $u \in T_x M^\perp$ such that

$$\bar{g}(\xi, v) = 0, \qquad \bar{g}(u, u) \neq 0, \qquad \forall v \in T_x M^\perp.$$

The above relations imply that $\xi \in T_x M$, so $\xi \in \operatorname{Rad} T_x M$. Therefore, locally there exists a lightlike vector field ξ on M such that

$$\bar{g}(\xi, X) = \bar{g}(\xi, u) = 0, \qquad \forall X \in \Gamma(TM), \quad u \in \Gamma(TM^\perp).$$

Thus, the 1-dimensional $\operatorname{Rad} TM$ of a half-lightlike submanifold M is locally spanned by ξ. In this case there exists a supplementary distribution $S(TM)$ to $\operatorname{Rad} TM$ in TM. Next, consider the orthogonal complementary distribution $S(TM^\perp)$ to $S(TM)$ in $T\bar{M}$. Certainly ξ and u belong to $\Gamma(S(TM)^\perp)$. From now on, we choose u as a unit vector field and put

$$\bar{g}(u, u) = \epsilon,$$

where $\epsilon = \pm 1$. Since $\operatorname{Rad} TM$ is a 1-dimensional vector sub-bundle of $TM^\perp$ we may consider a supplementary distribution D to $\operatorname{Rad} TM$ such that it is locally represented by u. We call D a *screen transversal bundle* of M. Hence we have the orthogonal decomposition

$$S(TM)^\perp = D \perp D^\perp,$$

where $D^\perp$ is the orthogonal complementary distribution to D in $S(TM)^\perp$. Taking into account that $D^\perp$ is non-degenerate and $\xi \in \Gamma(D^\perp)$, there exists a unique locally defined vector field $N \in \Gamma(D^\perp)$, satisfying

$$\bar{g}(N, \xi) \neq 0, \qquad \bar{g}(N, N) = \bar{g}(N, u) = 0 \qquad (4.1.1)$$

if and only if N is given by

$$N = \frac{1}{\bar{g}(\xi, V)}\{V - \frac{\bar{g}(V, V)}{2\,\bar{g}(\xi, V)}\xi\}, \quad V \in \Gamma(F_{|u}) \qquad (4.1.2)$$

such that $\bar{g}(\xi, V) \neq 0$. Here, F is a complementary vector bundle of $\operatorname{Rad} TM$ in $D^\perp$. Hence N is a lightlike vector field which is neither tangent to M nor collinear with u since $\bar{g}(u, \xi) = 0$. If we choose $\xi^* = \alpha \xi$ on another neighborhood of coordinates, then we obtain $N^* = \frac{1}{\alpha} N$. Thus we say that the vector bundle $\operatorname{tr}(TM)$ defined over M by

$$\operatorname{tr}(TM) = D \oplus_{\operatorname{orth}} \operatorname{ltr}(TM),$$

where $\operatorname{ltr}(TM)$ is a 1-dimensional vector bundle locally represented by N, is the *lightlike transversal bundle* of M with respect to the screen distribution $S(TM)$. Therefore,

$$\begin{aligned}
T\bar{M} &= S(TM) \perp (\operatorname{Rad} TM \oplus \operatorname{tr}(TM)) \\
&= S(TM) \perp D \perp (\operatorname{Rad} TM \oplus \operatorname{ltr}(TM)). \qquad (4.1.3)
\end{aligned}$$

As per decomposition (4.1.3), choose the field of frames $\{\xi, F_1, \ldots, F_{m-1}\}$ and $\{\xi, F_1, \ldots, F_{m-1}, u, N\}$ on M and $\bar{M}$ respectively, where $\{F_1, \ldots, F_{m-1}\}$ is an orthonormal basis of $\Gamma(S(TM))$. Denote by P the projection of TM on $S(TM)$ with respect to the decomposition (4.1.3) and obtain

$$X = PX + \eta(X)\xi, \qquad \forall X \in \Gamma(TM),$$

where η is a local differential 1-form on M given by

$$\eta(X) = \bar{g}(X, N). \tag{4.1.4}$$

Suppose $\bar{\nabla}$ is the metric connection on $\bar{M}$. Using (4.1.3), we put

$$
\begin{aligned}
\bar{\nabla}_X Y &= \nabla_X Y + h(X, Y), \\
\bar{\nabla}_X N &= -A_N X + \nabla_X N, \\
\bar{\nabla}_X u &= -A_u X + \nabla_X u,
\end{aligned}
\tag{4.1.5}
$$

for any $X, Y \in \Gamma(TM)$, where $\nabla_X Y$, $A_N X$ and $A_u X$ belong to $\Gamma(TM)$, while $h(X, Y), \nabla_X N$ and $\nabla_X u$ belong to $\Gamma(tr(TM))$. It is easy to check that ∇ is a torsion-free linear connection on M, h is a $\Gamma(tr(TM))$-valued symmetric $F(M)$-bilinear form on $\Gamma(TM)$. Since $\{\xi, N\}$ is locally a pair of lightlike sections on $\mathcal{U} \subset M$, we define symmetric $F(M)$-bilinear forms D_1 and D_2 and 1-forms ρ_1, ρ_2, ε_1 and ε_2 on $\mathcal{U}$ by

$$
\begin{aligned}
D_1(X, Y) &= \bar{g}(h(X, Y), \xi), \\
D_2(X, Y) &= \epsilon\,\bar{g}(h(X, Y), u), \\
\rho_1(X) &= \bar{g}(\nabla_X N, \xi), \quad \rho_2(X) = \epsilon\,\bar{g}(\nabla_X N, u), \\
\varepsilon_1(X) &= \bar{g}(\nabla_X u, \xi), \quad \varepsilon_2(X) = \epsilon\,\bar{g}(\nabla_X u, u),
\end{aligned}
$$

for any $X, Y \in \Gamma(TM)$. It follows that

$$
\begin{aligned}
h(X, Y) &= D_1(X, Y) N + D_2(X, Y) u, \tag{4.1.6} \\
\nabla_X N &= \rho_1(X) N + \rho_2(X) u, \\
\nabla_X u &= \varepsilon_1(X) N + \varepsilon_2(X) u.
\end{aligned}
$$

Hence, on $\mathcal{U}$, equations (4.1.5) become

$$
\begin{aligned}
\bar{\nabla}_X Y &= \nabla_X Y + D_1(X, Y) N + D_2(X, Y) u, \tag{4.1.7} \\
\bar{\nabla}_X N &= -A_N X + \rho_1(X) N + \rho_2(X) u, \tag{4.1.8} \\
\bar{\nabla}_X u &= -A_u X + \varepsilon_1(X) N + \varepsilon_2(X) u, \tag{4.1.9}
\end{aligned}
$$

for any $X, Y \in \Gamma(TM)$. We call h, D_1 and D_2 the *second fundamental form*, the *lightlike second fundamental form* and *the screen second fundamental form* of M with respect to $tr(TM)$ respectively. Both A_N and A_u are linear operators on

$\Gamma\,(TM)$. The first one is $\Gamma\,(S(TM))$-valued, called *the shape operator* of M. Since u is a unit vector field, (4.1.8) implies $\varepsilon_2\,(X)=0$. In a similar way, since ξ and N are lightlike vector fields, from (4.1.6)–(4.1.8) we obtain

$$D_1\,(X\,,\,\xi)=0, \tag{4.1.10}$$
$$\bar{g}\,(\,A_N X\,,\,N)=0, \tag{4.1.11}$$
$$\bar{g}\,(\,A_u X\,,Y)=\epsilon\,D_2\,(X\,,Y)+\varepsilon_1\,(X)\,\eta\,(Y). \tag{4.1.12}$$

Next, by using (4.1.4), (4.1.7) - (4.1.8) and (4.1.12), we obtain

$$\rho_1\,(X)=-\,\eta\,(\nabla_X\xi), \tag{4.1.13}$$
$$\rho_2\,(X)=\epsilon\,\eta\,(A_u X), \tag{4.1.14}$$
$$\varepsilon_1\,(X)=-\,\epsilon\,D_2(X\,,\xi),\qquad \forall X\in\Gamma\,(TM). \tag{4.1.15}$$

Since $\bar{\nabla}$ is a metric connection, using (4.1.7) we obtain

$$(\nabla_X g)\,(Y\,,Z)=D_1(X\,,Y)\,\eta\,(Z)+D_1\,(X\,,Z)\,\eta\,(Y) \tag{4.1.16}$$

for any $X\,,\,Y\,,\,Z\in\Gamma\,(TM)$. Thus, in general, the induced connection ∇ is linear but not a metric (Levi-Civita) connection. From (4.1.7) it follows that D_1 and D_2 are symmetric bilinear forms on $\Gamma\,(TM)$ and they do not depend on the screen distribution. In fact, we have

$$\bar{g}\,(\bar{\nabla}_X Y\,,\xi)=\,D_1\,(X\,,Y),\qquad \bar{g}\,(\bar{\nabla}_X Y\,,u)=\epsilon\,D_2\,(X\,,Y),$$

for any $X\,,Y\in\Gamma\,(TM)$. However, we note that both D_1 and ρ_1 depend on the section $\xi\in\Gamma\,(\mathrm{Rad}\,TM)$. Indeed, in case we take $\xi^*=\alpha\,\xi$, it follows that $N^*=\frac{1}{\alpha}\,N$. Hence we obtain $D_1^*=\alpha\,D_1$ and

$$\rho_1\,(X)=\rho_1^*\,(X)+X\,(log\,\alpha),$$

for any $X\in\Gamma\,(TM)$. Thus, using the differential 2-form

$$d\,\rho_1\,(X\,,\,Y)=\frac{1}{2}\,\{X(\rho_1(Y))-Y(\rho_1(X))-\rho_1([X\,,\,Y])\},$$

we obtain

Theorem 4.1.1. *Let M be a half-lightlike submanifold of a semi-Riemannian manifold $(\bar{M}\,,\,\bar{g})$ of codimension 2. Suppose ρ_1 and ρ_1^* are 1-forms on $\mathcal{U}$ with respect to ξ and ξ^*, respectively. Then, $d\,\rho_1^*=d\,\rho_1$ on $\mathcal{U}$.*

Next, consider the decomposition (4.1.3) and obtain

$$\nabla_X PY=\nabla_X^* PY+h^*\,(X\,,PY),$$
$$\nabla_X\xi=-\,A_\xi^* X+\nabla_X^\perp\xi \tag{4.1.17}$$

for any $X, Y \in \Gamma(TM)$, where $\nabla_X^* PY$ and A_ξ^* belong to $\Gamma(S(TM))$, while $h^*(X, PY)$ and $\nabla_X^\perp \xi$ belong to $\Gamma(\mathrm{Rad}\, TM)$. Also ∇^* and $\nabla^\perp$ are linear connections on the screen and radical distribution respectively, A_ξ^* is a linear operator on $\Gamma(TM)$, h^* is a bilinear form on $\Gamma(TM) \times \Gamma(S(TM))$ and ∇^* is a metric connection on $S(TM)$. Locally, we define on $\mathcal{U}$,

$$E_1(X, PY) = \bar{g}(h^*(X, PY), N), \qquad \forall X, Y \in \Gamma(TM),$$

and

$$u_1(X) = \bar{g}(\nabla_X^\perp \xi, N), \qquad \forall X \in \Gamma(TM).$$

It follows that

$$h^*(X, PY) = E_1(X, PY)\xi, \qquad \forall X, Y \in \Gamma(TM),$$

and

$$\nabla_X^\perp \xi = u_1(X)\xi.$$

Hence, on $\mathcal{U}$, locally, equations (4.1.17) become

$$\nabla_X PY = \nabla_X^* PY + E_1(X, PY)\xi, \tag{4.1.18}$$
$$\nabla_X \xi = -A_\xi^* X + u_1(X)\xi. \tag{4.1.19}$$

We call h^* and E_1 *the second fundamental form* and *the local second fundamental form* of $S(TM)$ with respect to $\mathrm{Rad}(TM)$ and A_ξ^* *the shape operator* of the screen distribution. The geometric objects from Gauss and Weingarten equations (4.1.7)–(4.1.9) on one side and (4.1.18) and (4.1.19) on the other side are related by

$$E_1(X, PY) = g(A_N X, PY), \tag{4.1.20}$$
$$D_1(X, PY) = g(A_\xi^* X, PY), \tag{4.1.21}$$
$$u_1(X) = -\rho_1(X),$$

for any $X, Y \in \Gamma(TM)$. Hence (4.1.19) becomes

$$\nabla_X \xi = -A_\xi^* X - \rho_1(X)\xi.$$

From (4.1.10) and (4.1.21) we derive

$$A_\xi^* \xi = 0. \tag{4.1.22}$$

Thus, ξ is an eigenvector of A_ξ^* corresponding to the zero eigenvalue. Using torsion-free linear connection ∇ and (4.1.18) we obtain

$$[X, Y] = \{\nabla_X^* PY - \nabla_Y^* PX + \eta(X)A_\xi^* Y - \eta(Y)A_\xi^* X\}$$
$$+ \{E_1(X, PY) - E_1(Y, PX) + X(\eta(Y))$$
$$- Y(\eta(X)) + \eta(X)\rho_1(Y) - \eta(Y)\rho_1(X)\}\xi.$$

Taking the scalar product of the last equation with PZ and N respectively and using (4.1.21), we obtain

$$g(\nabla_X^* PY, PZ) - g(\nabla_Y^* PX, PZ) - g([X, Y], PZ)$$
$$= \eta(Y) D_1(X, PZ) - \eta(X) D_1(Y, PZ),$$
$$2d\eta(X, Y) = E_1(Y, PX) - E_1(X, PY)$$
$$+ \rho_1(X)\eta(Y) - \rho_1(Y)\eta(X). \qquad (4.1.23)$$

From the second equation in (4.1.23) and (4.1.4) we obtain

$$\eta([PX, PY]) = E_1(PX, PY) - E_1(PY, PX).$$

From the last equation and (4.1.20) we have:

Theorem 4.1.2. [147] *Let M be a half-lightlike submanifold of $\bar{M}$. Then the following assertions are equivalent:*

(1) *The screen distribution $S(TM)$ is integrable.*

(2) *The second fundamental form of $S(TM)$ is symmetric on $\Gamma(S(TM))$.*

(3) *The shape operator A_N of the immersion of M in $\bar{M}$ is symmetric with respect to g on $\Gamma(S(TM))$.*

Proof. From the second equation in (4.1.23) and (4.1.4) we obtain

$$\eta([PX, PY]) = E_1(PX, PY) - E_1(PY, PX)$$

which is (1) $\Leftrightarrow$(2). Then, using (4.1.20) in the above equation we obtain

$$\eta([PX, PY]) = g(A_N PX, PY) - g(PX, A_N PY)$$

which proves (2)$\Leftrightarrow$ (3). $\square$

Next by using (4.1.10), (4.1.16), (4.1.19) and (4.1.21) we obtain

Theorem 4.1.3. [147] *Let M be a half-lightlike submanifold of a semi-Riemannian manifold $\bar{M}$. Then the following assertions are equivalent:*

(1) *The induced connection ∇ on M is a metric connection.*

(2) *D_1 vanishes identically on M.*

(3) *A_ξ^* vanishes identically on M.*

(4) *ξ is a Killing vector field.*

(5) *$TM^\perp$ is a parallel distribution with respect to ∇.*

Proof. (1)$\Rightarrow$(2): Suppose that ∇ is a metric connection, then for $Y = \xi$ in (4.1.16), we obtain

$$0 = D_1(X,\xi)\eta(Z) + D_1(X,Z)\eta(\xi).$$

Then, $\eta(\xi) = 1$ and (4.1.10) imply that $D_1(X,Z) = 0$ for every $X, Z \in \Gamma(TM)$. (2)$\Rightarrow$(3) follows from (4.1.21). (3)$\Rightarrow$ (4): By direct computation, we get

$$(\mathcal{L}_\xi \bar{g})(X,Y) = \xi\bar{g}(X,Y) - \bar{g}(\mathcal{L}_\xi X, Y) - \bar{g}(X, \mathcal{L}_\xi Y)$$

for $\xi \in \Gamma(\text{Rad}\,TM)$, $X, Y \in \Gamma(TM)$. Since $\bar{g}$ is a semi-Riemannian metric (then $\bar{\nabla}$ is a metric connection), we obtain

$$(\mathcal{L}_\xi \bar{g})(X,Y) = \bar{g}(\bar{\nabla}_X \xi, Y) - \bar{g}(X, \bar{\nabla}_Y \xi).$$

Then, from (4.1.7), (4.1.10) and (4.1.19), we derive

$$(\mathcal{L}_\xi \bar{g})(X,Y) = -g(A_\xi^* X, Y) + g(X, A_\xi^* Y).$$

Thus, (4) $\Rightarrow$(5) and (5)$\Rightarrow$ (1) follows from Theorem 1.4.2 of Chapter 1. $\qquad\square$

Denote by $\bar{R}$ and R the curvature tensors of $\bar{\nabla}$ and ∇ respectively. Then, using (4.1.7), (4.1.8), (4.1.9) and (4.1.10)–(4.1.15), we obtain

$$\begin{aligned}
\bar{R}(X,Y)Z = {}& R(X,Y)Z \\
& + D_1(X,Z)A_N Y - D_1(Y,Z)A_N X \\
& + D_2(X,Z)A_u Y - D_2(Y,Z)A_u X \\
& + \{(\nabla_X D_1)(Y,Z) - (\nabla_Y D_1)(X,Z) \\
& + \rho_1(X)D_1(Y,Z) - \rho_1(Y)D_1(X,Z) \\
& + \varepsilon_1(X)D_2(Y,Z) - \varepsilon_1(Y)D_2(X,Z)\}N \\
& + \{(\nabla_X D_2)(Y,Z) - (\nabla_Y D_2)(X,Z) \\
& + \rho_2(X)D_1(Y,Z) - \rho_2(Y)D_1(X,Z)\}u,
\end{aligned} \qquad (4.1.24)$$

$$\begin{aligned}
\bar{R}(X,Y)N = {}& -\nabla_X(A_N Y) + \nabla_Y(A_N X) + A_N[X,Y] \\
& + \rho_1(X)A_N Y - \rho_1(Y)A_N X \\
& + \rho_2(X)A_u Y - \rho_2(Y)A_u X \\
& + \{D_1(Y, A_N X) - D_1(X, A_N Y) \\
& + 2d\rho_1(X,Y) + \varepsilon_1(X)\rho_2(Y) - \varepsilon_1(Y)\rho_2(X)\}N \\
& + \{D_2(Y, A_N X) - D_2(X, A_N Y) \\
& + 2d\rho_2(X,Y) + \rho_1(Y)\rho_2(X) - \rho_1(X)\rho_2(Y)\}u,
\end{aligned} \qquad (4.1.25)$$

$$\begin{aligned}
\bar{R}(X,Y)u = {}& -\nabla_X(A_u Y) + \nabla_Y(A_u X) + A_u[X,Y] \\
& + \varepsilon_1(X)A_N Y - \varepsilon_1(Y)A_N X \\
& + \{D_1(Y, A_u X) - D_1(X, A_u Y) \\
& + 2d\varepsilon_1(X,Y) + \rho_1(X)\varepsilon_1(Y) - \rho_1(Y)\varepsilon_1(X)\}N
\end{aligned}$$

$$+ \{D_2(Y, A_u X) - D_2(X, A_u Y)$$
$$+ \varepsilon_1(Y)\rho_2(X) - \varepsilon_1(X)\rho_2(Y)\}u \tag{4.1.26}$$

for any $X, Y, Z \in \Gamma(TM)$. Consider the Riemannian curvature of type $(0, 4)$ of $\bar{\nabla}$. Using (4.1.24)–(4.1.26) and the definition of curvature tensors, we derive the following structure equations:

$$\bar{g}(\bar{R}(X,Y)PZ, PW) = g(R(X,Y)PZ, PW) \tag{4.1.27}$$
$$+ D_1(X, PZ)E_1(Y, PW)$$
$$- D_1(Y, PZ)E_1(X, PW)$$
$$+ \epsilon\{D_2(X, PZ)D_2(Y, PW)$$
$$- D_2(Y, PZ)D_2(X, PW)\},$$

$$\bar{g}(\bar{R}(X,Y)PZ, \xi) = g(R(X,Y)PZ, \xi) \tag{4.1.28}$$
$$+ \varepsilon_1(X)D_2(Y, PZ) - \varepsilon_1(Y)D_2(X, PZ)$$
$$= (\nabla_X D_1)(Y, PZ) - (\nabla_Y D_1)(X, PZ)$$
$$+ \rho_1(X)D_1(Y, PZ) - \rho_1(Y)D_1(X, PZ)$$
$$+ \varepsilon_1(X)D_2(Y, PZ) - \varepsilon_1(Y)D_2(X, PZ),$$

$$\bar{g}(\bar{R}(X,Y)PZ, u) = g(\nabla_X(A_u Y) - \nabla_Y(A_u X) - A_u[X,Y], PZ)$$
$$- \varepsilon_1(X)E_1(Y, PZ) + \varepsilon_1(Y)E_1(X, PZ)$$
$$= \epsilon\{(\nabla_X D_2)(Y, PZ) - (\nabla_Y D_2)(X, PZ)$$
$$+ \rho_2(X)D_1(Y, PZ)$$
$$- \rho_2(Y)D_1(X, PZ)\}, \tag{4.1.29}$$

$$\bar{g}(\bar{R}(X,Y)PZ, N) = \bar{g}(R(X,Y)PZ, N)$$
$$+ \epsilon\{\rho_2(Y)D_2(X, PZ) - \rho_2(X)D_2(Y, PZ)\}$$
$$= g(\nabla_X(A_N Y) - \nabla_Y(A_N X)$$
$$- A_N[X,Y], PZ) + \rho_1(Y)E_1(X, PZ)$$
$$- \rho_1(X)E_1(Y, PZ)\epsilon\{\rho_2(Y)D_2(X, PZ)$$
$$- \rho_2(X)D_2(Y, PZ)\}, \tag{4.1.30}$$

$$\bar{g}(\bar{R}(X,Y)\xi, N) = \bar{g}(R(X,Y)\xi, N)$$
$$+ \rho_2(X)\varepsilon_1(Y) - \rho_2(Y)\varepsilon_1(X) \tag{4.1.31}$$
$$= D_1(X, A_N Y) - D_1(Y, A_N X)$$
$$- 2d\rho_1(X,Y) + \rho_2(X)\varepsilon_1(Y) - \rho_2(Y)\varepsilon_1(X),$$

$$\bar{g}(\bar{R}(X,Y)\xi, u) = \epsilon\{(\nabla_X D_2)(Y, \xi) - (\nabla_Y D_2)(X, \xi)\}$$
$$= D_1(X, A_u Y) - D_1(Y, A_u X) \tag{4.1.32}$$
$$- 2d\varepsilon_1(X,Y) + \rho_1(Y)\,\varepsilon_1(X) - \rho_1(X)\,\varepsilon_1(Y),$$

$$\bar{g}(\bar{R}(X,Y)N, u) = \epsilon\{D_2(Y, A_N X) - D_2(X, A_N Y)$$
$$+ 2d\rho_2(X,Y) + \rho_1(Y)\rho_2(X) - \rho_1(X)\rho_2(Y)\}$$

$$= \bar{g}(\nabla_X(A_uY), N) - \bar{g}(\nabla_Y(A_uX), N)$$
$$- \epsilon\rho_2([X,Y]). \tag{4.1.33}$$

Denote by R^* the curvature tensors of ∇^*. Similarly, using (4.1.24)- (4.1.26) and (4.1.27)–(4.1.33), (4.1.18) and (4.1.19), we obtain

$$R(X,Y)PZ = R^*(X,Y)PZ + E_1(X,PZ)A_\xi Y \tag{4.1.34}$$
$$- E_1(Y,PZ)A_\xi X + \{X(E_1(Y,PZ))$$
$$- Y(E_1(X,PZ)) - E_1([X,Y],PZ)$$
$$+ E_1(X, \nabla_Y^* PZ) - E_1(Y, \nabla_X^* PZ)$$
$$- \rho_1(X)E_1(Y,PZ) + \rho_1(Y)E_1(X,PZ)\}\xi,$$

$$R(X,Y)\xi = -\nabla_X^*(A_\xi Y) + \nabla_Y^*(A_\xi X) + A_\xi[X,Y]$$
$$- \rho_1(X)A_\xi Y + \rho_1(Y)A_\xi X \tag{4.1.35}$$
$$+ \{E_1(Y, A_\xi X) - E_1(X, A_\xi Y) - 2d\rho_1(X,Y)\}\xi,$$

$$g(R(X,Y)PZ, PW) = g(R^*(X,Y)PZ, PW)$$
$$+ E_1(X,PZ)D_1(Y,PW)$$
$$- E_1(Y,PZ)D_1(X,PW), \tag{4.1.36}$$

$$\bar{g}(R(X,Y)PZ, N_1) = X(E_1(Y,PZ)) - Y(E_1(X,PZ))$$
$$+ E_1([X,Y],PZ) + E_1(X, \nabla_Y^* PZ)$$
$$- E_1(Y, \nabla_X^* PZ) - \rho_1(X)E_1(Y,PZ)$$
$$+ \rho_1(Y)E_1(X,PZ)$$
$$= g(\nabla_X(A_N Y) - \nabla_Y(A_N X)$$
$$- A_N[X,Y], PZ) - \rho_1(X)E_1(Y,PZ)$$
$$+ \rho_1(Y)E_1(X,PZ), \tag{4.1.37}$$

$$g(R(X,Y)PZ, \xi) = g(\nabla_X^*(A_\xi Y), PZ) - g(\nabla_Y^*(A_\xi X), PZ)$$
$$- D_1([X,Y],PZ) \tag{4.1.38}$$
$$+ \rho_1(X)D_1(Y,PZ) - \rho_1(Y)D_1(X,PZ)$$
$$= (\nabla_X D_1)(Y,PZ) - (\nabla_Y D_1)(X,PZ)$$
$$+ \rho_1(X)D_1(Y,PZ) - \rho_1(Y)D_1(X,PZ),$$

$$\bar{g}(R(X,Y)\xi, N_1) = E_1(Y, A_\xi X) - E_1(X, A_\xi Y)$$
$$- 2d\rho_1(X,Y)$$
$$= D_1(X, A_N Y) - D_1(Y, A_N X)$$
$$- 2d\rho_1(X,Y). \tag{4.1.39}$$

4.2   Unique existence of screen distributions

The general theory of lightlike submanifolds uses a non-degenerate screen distribution which (due to the degenerate induced metric) is not unique. Therefore, the induced objects of the submanifold depend on the choice of a screen that creates a problem. As presented in Chapter 2, now there are large classes of lightlike hypersurfaces of semi-Riemannian manifolds with the choice of a canonical or unique screen distribution. The objective of this section is to show that there exist unique screen distributions for a large variety of half-lightlike submanifolds subject to some reasonable geometric conditions.

Let a screen $S(TM)$ change to another screen $S(TM)'$, where $\{\xi,\, N,\, W_a,\, u\,\}$ and $\{\xi,\, N',\, W'_a,\, u'\,\}$ respectively are two quasi-orthonormal frame fields for the same null section ξ. Following are the transformation equations due to this change:

$$W'_a = \sum_{b=1}^{m-1} A^b_a \left(W_b - \epsilon_b \, \mathbf{f}_b \, \xi \right); \quad u' = u - \epsilon f \xi, \tag{4.2.1}$$

$$N' = N - \frac{1}{2} \left\{ \sum_{a=1}^{m-1} \epsilon_a \mathbf{f}^2_a + \epsilon \mathbf{f}^2 \right\} \xi + \sum_{a=1}^{m+1} \mathbf{f}_a W_a + \mathbf{f} u, \tag{4.2.2}$$

$$\nabla'_X PY = \nabla_X PY + \frac{1}{2} D_1(X, PY) \left(\sum_{a=1}^{m-1} \epsilon_a \mathbf{f}^2_a + \epsilon \mathbf{f}^2 \right) \xi$$

$$+ \epsilon D_2(X, PY) \mathbf{f} \xi - D_1(X, PY) \left(\sum_{a=1}^{m-1} \mathbf{f}_a \, W_a \right), \tag{4.2.3}$$

$$D'_1(X, Y) = D_1(X, Y),\; D'_2(X, Y) = D_2(X, Y) - D_1(X, Y) \mathbf{f}, \tag{4.2.4}$$

$$E'(X,\, PY) = E_1(X,\, PY) - \frac{1}{2} \left(\|W\|^2 - \epsilon \mathbf{f}^2 \right) D_1(X,\, PY)$$

$$+ g(\nabla_X PY, W) + \epsilon D_2(X, PY) \mathbf{f}, \tag{4.2.5}$$

where $W = \sum_{a=1}^{m} \mathbf{f}_a \, W_a$. Let ω be the dual 1-form of W given by

$$\omega(X) = g(X, W), \quad \forall X \in \Gamma(TM). \tag{4.2.6}$$

Denote by $\mathcal{S}$ the first derivative of a screen distribution $S(TM)$ given by

$$\mathcal{S}(x) = \operatorname{Span}\{[X, Y]_x, \quad X_x, Y_x \in S(TM), \quad x \in M\}, \tag{4.2.7}$$

where $[,]$ denotes the Lie-bracket. If $S(TM)$ is integrable, then, $\mathcal{S}$ is a sub-bundle of $S(TM)$. We state and prove the following theorem:

Theorem 4.2.1. [144] *Let $(M, g, S(TM))$ be a half-lightlike submanifold of a semi-Riemannian manifold $\bar{M}^{m+2}$ with $m > 1$. Suppose the sub-bundle F of $D^\perp$ admits a covariant constant non-null vector field. Then, with respect to a section ξ of $\operatorname{Rad} TM$, M admits an integrable screen $S(TM)$. Moreover, if the first derivative $\mathcal{S}$ defined by (4.2.7) coincides with an integrable screen $S(TM)$, then, $S(TM)$ is a unique screen of M, up to an orthogonal transformation with a unique lightlike transversal vector bundle and invariant screen second fundamental form.*

Proof. By hypothesis, consider along M, a unit covariant constant vector field $V \in \Gamma(F_{|\mathcal{U}})$, that is, $\bar{g}(V, V) = e = \pm 1$. To satisfy the condition given in (4.1.2), we choose a section ξ of $\mathrm{Rad}\,TM$ such that $\bar{g}(V, \xi) \neq 0$. Set $\bar{g}(V, \xi) = \theta^{-1}$. Using this in (4.1.2), the null transversal vector bundle of M takes the form

$$N = \theta\left(V - \frac{e\,\theta}{2}\xi\right). \tag{4.2.8}$$

Then using (4.2.8) in (4.1.8) and (4.1.19) we get

$$\rho_1(X) = \bar{g}(\bar{\nabla}_X N, \xi) = X(\theta)\,\bar{g}(V, \xi) - \frac{e}{2}(\theta)^2\,\bar{g}(\bar{\nabla}_X \xi, \xi)$$

$$= X(\theta)\,(\theta)^{-1} = X(\log\theta), \tag{4.2.9}$$

$$\rho_2(X) = \bar{g}(\bar{\nabla}_X N, u) = \bar{g}\left(\bar{\nabla}_X(\theta V) - \frac{e}{2}\bar{\nabla}_X(\theta^2\xi), u\right)$$

$$= -\frac{e\theta^2}{2}\varepsilon D_2(X, \xi). \tag{4.2.10}$$

Using the above value of τ, (4.2.8) and (4.1.19) we obtain

$$\bar{\nabla}_X N = X(\theta)V - \frac{e}{2}\theta^2 D_2(X, \xi)u + \frac{e}{2}\theta^2\,A_\xi^* X - \frac{e}{2}\theta\,X(\theta)\,\xi. \tag{4.2.11}$$

On the other hand, substituting the value of ρ_1 and ρ_2 in (4.1.8), we get

$$\bar{\nabla}_X N = -A_N X + X(\log\theta)N - \frac{\varepsilon\theta^2 e}{2}D_2(X, \xi)u. \tag{4.2.12}$$

Equating (4.2.11) and (4.2.12) we obtain

$$A_N X = -\frac{e\,\theta^2}{2}A_\xi^* X - \frac{e}{2}\theta\,X(\theta)\,\xi + X(\log\theta)\,N - X(\theta)\,V \tag{4.2.13}$$

for $X \in \Gamma(TM_{|\mathcal{U}})$. Thus, for $X, Y \in \Gamma(S(TM))$, we have

$$g(A_N X, Y) = -\frac{e\,\theta^2}{2}g(A_\xi^* X, Y).$$

Since A_ξ^* is symmetric with respect to g on $S(TM)$, it follows that A_N is also self-adjoint on M. Then Theorem 4.1.2 says that $S(TM)$ is integrable. This means that $\mathcal{S}$ is a sub-bundle of $S(TM)$. Using (4.2.13) in (4.1.20) and (4.1.21), we get

$$E_1(X, PY) = -\frac{e\,\theta^2}{2}D_1(X, Y), \quad \forall X, Y \in \Gamma(S(TM)_{|\mathcal{U}}). \tag{4.2.14}$$

Using (4.2.13), (4.2.5) and $D_1' = D_1$ we obtain

$$g(\nabla_X PY, W) = \frac{1}{2}\left(\|W\|^2 - \varepsilon\mathbf{f}^2\right)D_1(X, Y) - \varepsilon D_2(X, Y)\mathbf{f} \tag{4.2.15}$$

$\forall X, Y \in \Gamma(S(TM)_{|u})$. Since the right-hand side of (4.2.15) is symmetric in X and Y, we have $g([X,Y],W) = \omega([X,Y]) = 0$, $\forall X, Y \in \Gamma(S(TM)_{|u})$, that is, ω vanishes on S. By hypothesis, if we take $S = S(TM)$, then, ω vanishes on this choice of $S(TM)$ which implies from (4.2.6) that $W = 0$. Therefore, the functions $\mathbf{f}_a$ vanish. Finally, substituting this data in (4.2.5) it is easy to see that the function $\mathbf{f}$ also vanishes. Thus, the transformation equations (4.2.1), (4.2.2) and (4.2.3) become $W'_a = \sum_{b=1}^{m-1} A^b_a W_b$ $(1 \le a \le m-1)$, $N' = N$ and $E' = E$ where (A^b_a) is an orthogonal matrix of $S(TM)$ at any point $x \in M$. Therefore, $S(TM)$ is a unique screen up to an orthogonal transformation with a unique transversal vector field N and invariant screen fundamental form E. This completes the proof. $\quad\square$

To understand some examples of half-lightlike submanifolds, satisfying the above theorem, we first need the following result.

Proposition 4.2.2. [147] *Let $(M, g, S(TM))$ be a half-lightlike submanifold of a semi-Riemannian manifold $\bar{M}$, with $\bar{g}$ of index $q \in \{1, \ldots, m+1\}$. Then we have the following:*

(i) *If u is spacelike then $S(TM)$ is of index $q-1$. In particular $S(TM)$ is Riemannian for $q = 1$ and Lorentzian for $q = 2$.*

(ii) *If u is timelike then $S(TM)$ is of index $q-2$. In particular $S(TM)$ is Riemannian for $q = 2$ and Lorentzian for $q = 3$.*

Proof. Consider the vector fields

$$e_1 = \frac{1}{\sqrt{2}}(\xi - N), \ e_2 = \frac{1}{\sqrt{2}}(\xi + N).$$

It is clear that $\operatorname{Rad} TM \oplus \operatorname{ltr}(TM) = \operatorname{Span}\{e_1, e_2\}$ and $\operatorname{ind}(\operatorname{Rad} TM \oplus \operatorname{ltr}(TM)) = 1$. On the other hand, we have

$$\operatorname{ind}(\bar{M}) = \operatorname{ind}(S(TM)) + \operatorname{ind}(\operatorname{Rad} TM \oplus \operatorname{ltr}(TM)) + \operatorname{ind}(D).$$

Then (i) and (ii) follow, which completes the proof. $\quad\square$

Remark 4.2.3. It follows from (i) of the above proposition that M can be a half-lightlike submanifold of a Lorentzian manifold for which $\bar{g}(u, u) = \epsilon = 1$. Thus, it is obvious from the structure equations that we choose $\bar{g}(V, V) = e = -1$, a covariant constant timelike unit vector field. There are many examples of n-dimensional product Lorentzian spaces (such as warped product globally hyperbolic space-times [34]) which possess at least one timelike covariant constant vector field and, therefore, can satisfy the hypothesis of Theorem 4.2.1.

Example 1. Consider in $\mathbf{R}^5_1$ the submanifold M given by the equations

$$x_1 = x_3, \ x_5 = \sqrt{1 - \{x_4^2 + x_2^2\}}.$$

Then we have

$$TM = \text{Span}\{\xi = \partial x_1 + \partial x_3, \ Z_1 = x_5\,\partial x_2 - x_2\,\partial x_5, \ Z_2 = x_5\,\partial x_4 - x_4\,\partial x_5\}.$$

It follows that M is 1-lightlike. Take the complementary sub-bundle F spanned by $V = -\partial x_1$ and observe that $g(\xi, V) \neq 0$. We obtain

$$N = \frac{1}{2}(-\partial x_1 + \partial x_3)$$

and

$$u = x_2\,\partial x_2 + x_4\,\partial x_4 + \sqrt{1 - \{x_4^2 + x_2^2\}}\,\partial x_5.$$

It is easy to see that V is covariant constant. Then, we have

$$\bar\nabla_{Z_2}Z_1 = -x_4\,\partial x_2, \ \bar\nabla_{Z_1}Z_2 = -x_2\,\partial x_4.$$

Hence,

$$[Z_1, Z_2] = \frac{x_5 x_4}{x_5^2 + x_2^2}Z_1 - \frac{x_5 x_2}{x_5^2 + x_4^2}Z_2$$

which implies that $S(TM)$ is integrable, so Theorem 4.2.1 is satisfied.

4.3 Totally umbilical submanifolds

The results of this section have been taken from [154]. A half-lightlike submanifold (M, g) of a semi-Riemannian manifold $(\bar M, \bar g)$ is said to be *totally umbilical* in $\bar M$ if there is a normal vector field $\mathcal{Z} \in \Gamma(tr(TM))$ on M, called an *affine normal curvature vector field* of M, such that

$$h(X, Y) = \mathcal{Z}\,\bar g(X, Y), \quad \forall X, Y \in \Gamma(TM).$$

In particular, (M, g) is said to be *totally geodesic* if its second fundamental form $h(X, Y) = 0$ for any $X, Y \in \Gamma(TM)$. By direct calculation it is easy to see that M is totally geodesic if and only if both the lightlike and the screen second fundamental tensors D_1 and D_2 respectively vanish on M, i.e.,

$$D_1(X, Y) = D_2(X, Y) = 0, \quad \forall X, Y \in \Gamma(TM).$$

Moreover, from (4.1.12), (4.1.14), (4.1.15) and (4.1.21) we obtain

$$A_\xi = A_u = \varepsilon_1 = \rho_2 = 0.$$

A straight calculation and using (4.1.6) implies that M is totally umbilical if and only if on each coordinate neighborhood $\mathcal{U}$ there exist smooth functions H_1 and H_2 on $\text{ltr}(TM)$ and D, respectively, such that

$$D_1(X, Y) = H_1\,\bar g(X, Y), \quad D_2(X, Y) = H_2\,\bar g(X, Y), \tag{4.3.1}$$

for any $X, Y \in \Gamma(TM)$. The above definition does not depend on the screen distribution of M. On the other hand, from (4.1.12), (4.1.21) and (4.1.10), (4.1.11) and non-degenerate $S(TM)$ we obtain

Theorem 4.3.1. *Let (M, g) be a half-lightlike submanifold of a semi-Riemannian manifold $(\bar{M}, \bar{g})$. Then M is totally umbilical, if and only if, on each $\mathcal{U}$ there exist H_1, H_2 such that*

$$A_\xi^* X = H_1 \, PX, \qquad \forall X \in \Gamma(TM),$$
$$P(A_u X) = \epsilon \, H_2 \, PX, \qquad \forall X \in \Gamma(TM) \tag{4.3.2}$$

and $\varepsilon_1 = 0$ on $\Gamma(TM)$.

Remark 4.3.2. Note that in case M is totally umbilical, we have

$$D_2(X, \xi) = 0, \qquad \rho_2(\xi) = 0, \qquad A_u\xi = 0, \tag{4.3.3}$$
$$\epsilon \, A_u X = H_2 \, PX + \rho_2(X)\,\xi, \qquad \forall X \in \Gamma(TM). \tag{4.3.4}$$

The curvature tensors of M and $\bar{M}$ are related by the following equations:

$$\begin{aligned}
\bar{g}(\bar{R}(X,Y)PZ,\xi) &= g(R(X,Y)PZ,\xi) \\
&= (\nabla_X D_1)(Y, PZ) - (\nabla_Y D_1)(X, PZ) \tag{4.3.5} \\
&\quad + \rho_1(X)D_1(Y, PZ) - \rho_1(Y)D_1(X, PZ),
\end{aligned}$$

$$\begin{aligned}
\bar{g}(\bar{R}(X,Y)PZ,u) &= g(\nabla_X(A_uY) - \nabla_Y(A_uX) - A_u[X,Y], PZ) \\
&= \epsilon\{(\nabla_X D_2)(Y, PZ) - (\nabla_Y D_2)(X, PZ) \\
&\quad + \rho_2(X)D_1(Y, PZ) \\
&\quad - \rho_2(Y)D_1(X, PZ)\}, \tag{4.3.6}
\end{aligned}$$

$$\begin{aligned}
\bar{g}(\bar{R}(X,Y)\xi, N_1) &= \bar{g}(R(X,Y)\xi, N_1) \\
&= D_1(X, A_{N_1}Y) - D_1(Y, A_{N_1}X) \\
&\quad - 2d\rho_1(X,Y), \tag{4.3.7}
\end{aligned}$$

$$\begin{aligned}
\bar{g}(\bar{R}(X,Y)\xi, u) &= \epsilon\{(\nabla_X D_2)(Y, \xi) - (\nabla_Y D_2)(X, \xi)\} \\
&= D_1(X, A_uY) - D_1(Y, A_uX) \tag{4.3.8}
\end{aligned}$$

and (4.1.27), (4.1.30) and (4.1.33) are unchanged. Using these equations we obtain

Theorem 4.3.3. *Let M be a totally umbilical half-lightlike submanifold of an $(m+2)$-dimensional semi-Riemannian manifold of constant curvature $(\bar{M}(\bar{c}), \bar{g})$. Then the functions H_1, H_2 from (4.3.1) satisfy the differential equations*

$$\xi(H_1) - (H_1)^2 + H_1 \, \rho_1(\xi) = 0, \qquad H_1 = \xi(\log H_2), \tag{4.3.9}$$

and the curvature tensor of M is given by

$$\begin{aligned}
R(X, Y)Z &= \bar{c}\,\{g(Y, Z)X - g(X, Z)Y\} \\
&\quad + H_1\,\{g(Y, Z)A_{N_1}X - g(X, Z)A_N Y\} \\
&\quad + H_2\,\{g(Y, Z)A_u X - g(X, Z)A_u Y\},
\end{aligned}$$

for any $X, Y \in \Gamma(TM)$. Moreover, $PX(H_i) + H_1 \, \rho_i(PX) = 0$.

Proof. Taking account of (4.3.1) in (4.3.5) and (4.3.6), and using the fact that $\bar{M}$ is a space of constant curvature, we obtain

$$\{X(H_i) - H_1 H_i \eta(X) + H_1 \rho_i(X)\} g(Y, PZ)$$
$$= \{Y(H_i) - H_1 H_i \eta(Y) + H_1 \rho_i(Y)\} g(X, PZ) \qquad (4.3.10)$$

for any $X, Y, Z \in \Gamma(TM)$. Take $X = \xi$ and $Z = Y \in \Gamma(S(TM))$ such that $g(Y, Y) \neq 0$ on $\mathcal{U}$ and use (4.3.5) to obtain the first equation in (4.3.9). Next, the second equation in (4.3.9) follows from (4.1.24) taking into account that $\bar{M}$ is a space of constant curvature and using its first equation. Taking $X = PX$ and $Y = PY$ in (4.3.10) and by using (4.1.4) and the fact that $S(TM)$ is non-degenerate, we obtain

$$\{PX(H_i) + H_1 \rho_i(PX)\} PY = \{PY(H_i) + H_1 \rho_i(PY)\} PX.$$

Now suppose there exists a vector field $X_o \in \Gamma(TM)$ such that $PX_o(H_i) + H_1 \rho_i(PX_o) \neq 0$ at each point $u \in M$. Then, from the last equation it follows that all vectors from the fiber $(S(TM))_u$ are collinear with $(PX_o)_u$. This is a contradiction as $\dim((S(TM))_u) = m - 1$. Hence the last equation in the theorem is true at any point of $\mathcal{U}$, which completes the proof. $\qquad \square$

From (4.3.1), (4.3.7) and $\bar{M}$ a space of constant curvature we obtain

$$2 d\rho_1(X, Y) = H_1 \{g(X, A_{N_1} Y) - g(Y, A_{N_1} X)\}.$$

Using (4.3.1) and Theorems 4.1.2 and 4.1.3, we have

Theorem 4.3.4. *Let M be a totally umbilical half-lightlike submanifold of a semi-Riemannian manifold of constant curvature $(\bar{M}(\bar{c}), \bar{g})$. Then each 1-form ρ_1 induced by $S(TM)$ is closed, i.e., $d\rho_1 = 0$ on any $\mathcal{U} \subset M$, if and only if, either (1) the screen distribution $S(TM)$ is integrable, or (2) the induced connection ∇ on M is a metric connection.*

M is *proper totally umbilical* if $H_1 \neq 0$ and $H_2 \neq 0$ on $\mathcal{U}$. From the above theorem and the equation (4.1.33) and (4.3.1) we obtain

Theorem 4.3.5. *Let M be a proper totally umbilical half-lightlike submanifold of a semi-Riemannian manifold of constant curvature $(\bar{M}(\bar{c}), \bar{g})$. Then the following assertions are equivalent:*

(1) *The screen distribution $S(TM)$ is integrable.*

(2) *Each 1-form ρ_1 induced by $S(TM)$ is closed, i.e., $d\rho_1 = 0$ on any $\mathcal{U} \subset M$.*

(3) *Each 1-form ρ_2 induced by $S(TM)$ satisfies*

$$2d\rho_2(X, Y) = \rho_1(X)\rho_2(Y) - \rho_2(X)\rho_1(Y), \forall X, Y \in \Gamma(TM).$$

Proof. Taking account of (4.3.1) in (4.1.33) and using the fact that $\bar{M}$ is a space of constant curvature, we obtain

$$2d\,\rho_2\,(X,\,Y) - \rho_1\,(X)\,\rho_2\,(Y) + \rho_2\,(X)\,\rho_1\,(Y)$$
$$= H_2\,\{\,g\,(X,\,A_N Y\,) - g\,(Y,\,A_N X\,)\}$$

which proves the assertion. $\qquad\qquad\qquad\qquad\qquad\qquad\qquad\qquad\square$

Next, the screen distribution $S(TM)$ is said to be *totally umbilical* in M if there is a smooth vector field $W \in \Gamma\,(\mathrm{Rad}\,TM)$ on M, such that, $h^*\,(X,\,PY) = W g\,(X,\,PY)$ for all $X,\,Y \in \Gamma\,(TM)$. A straight calculation implies that $S(TM)$ is totally umbilical, if and only if, on any coordinate neighborhood $\mathcal{U} \subset M$, there exists a function K such that

$$E_1\,(X,\,PY) = K\,g\,(X,\,PY),\qquad\qquad\qquad (4.3.11)$$

for any $X,\,Y \in \Gamma\,(TM)$. It follows that E_1 is symmetric on $\Gamma\,(S(TM))$ and hence according to Theorem 4.1.2, the screen space $S(TM)$ is integrable. In case $K = 0\,(K \neq 0)$ on $\mathcal{U}$ we say that $S(TM)$ is *totally geodesic (proper totally umbilical)*. From (4.1.11), (4.1.20) and (4.3.11) we obtain

$$A_N X = K\,PX,\qquad E_1\,(\xi,\,PX) = 0,\qquad \forall X \in \Gamma\,(TM).\qquad (4.3.12)$$

From the second equation in (4.1.23) and the fact that the screen distribution $S(TM)$ is totally umbilical we have

$$2\,d\eta\,(X,\,Y) = \rho_1\,(X)\,\eta\,(Y) - \eta\,(X)\,\rho_1\,(Y).$$

Thus we have

Theorem 4.3.6. *Let $(M,\,g)$ be a half-lightlike submanifold of a semi-Riemannian manifold $(\bar{M},\,\bar{g})$ with totally umbilical screen $S(TM)$. Then*

$$d\eta = 0 \quad\Longleftrightarrow\quad \rho_1 = 0.$$

Theorem 4.3.7. *Let $(M,\,g)$ be a half-lightlike submanifold of a semi-Riemannian manifold $(\bar{M}\,(\bar{c}),\,\bar{g})$ of constant curvature $\bar{c}$, with a proper totally umbilical screen distribution $S(TM)$. If M is also totally umbilical, then the mean curvature vector K of $S(TM)$ is a solution of the differential equations*

$$\xi\,(K) - K\,\rho_1\,(\xi) - \bar{c} = K\,H_1,$$
$$PX(K) - K\,\rho_1\,(PX) = \epsilon\,H_2\,\rho_2\,(PX).$$

Proof. Taking account of (4.3.11) and (4.3.12) into (4.1.30), and using (4.1.4), (4.1.18), (4.1.23) and $\bar{M}$ a space of constant curvature, we obtain

$$\{X(K) - K\,\rho_1\,(X) - \bar{c}\,\eta\,(X)\,\}\,g\,(Y,\,PZ)$$
$$- \{Y(K) - K\,\rho_1\,(Y) - \bar{c}\,\eta\,(Y)\,\}\,g(X,\,PZ)$$
$$= K\,\{\eta\,(X)\,D_1\,(Y,\,PZ) - \eta\,(Y)\,D_1\,(X,\,PZ)\,\}$$
$$+ \epsilon\,\{\rho_2\,(X)\,D_2\,(Y,\,PZ) - \rho_2\,(Y)\,D_2\,(X,\,PZ)\}.$$

This implies

$$\{\xi(K) - K\rho_1(\xi) - \bar{c}\}g(Y, PZ)$$
$$= KD_1(Y, PZ) + \epsilon\{\rho_2(\xi)D_2(Y, \ PZ) - \rho_2(Y)D_2(\xi, PZ)\}.$$

From the last two relations and by the method of Theorem 4.3.2, (4.3.1) and (4.3.2), the equations in the theorem hold. □

From the last two equations we have the following corollaries.

Corollary 4.3.8. *Under the hypothesis of Theorem 4.3.7, the induced connection ∇ on M is a metric connection, if and only if, the mean curvature vector K of $S(TM)$ is a solution of the following equations*

$$\xi(K) - K\rho_1(\xi) - \bar{c} = 0.$$

Corollary 4.3.9. *Let (M, g) be a totally umbilical half-lightlike submanifold of a semi-Riemannian manifold $(\bar{M}(\bar{c}), \bar{g})$ of constant curvature $\bar{c}$. If the screen distribution $S(TM)$ is totally geodesic, then, $\bar{c} = 0$, i.e., the ambient semi-Riemannian manifold $\bar{M}$ is semi-Euclidean space.*

For those who prefer to work with local coordinate systems, we find the following local expressions of structure equations. $\mathrm{Rad}\,TM$ being of rank 1, it is integrable and therefore there exists an atlas of local charts $\{\mathcal{U}\,;\,u^0, \ldots, u^{m-1}\}$ such that $\frac{\partial}{\partial u^0} \in \Gamma(\mathrm{Rad}\,TM_{|u})$. Thus, the matrix of the metric g on M with respect to the natural frame field $\{\frac{\partial}{\partial u^0}, \ldots, \frac{\partial}{\partial u^{m-1}}\}$ is

$$[g] = \begin{bmatrix} 0 & 0 \\ 0 & g_{ij}(u^0, \ldots, u^{m-1}) \end{bmatrix},$$

where $g_{ij} = g(\frac{\partial}{\partial u^i}, \frac{\partial}{\partial u^j})$, $i, j \in \{1, 2, \ldots m-1\}$, and $det(g_{ij}) \neq 0$. We use the range of indices: $i, j, k, \ldots \in \{1, \ldots, m-1\}$. According to the general transformations on a foliated manifold (see Chapter 1), we have

$$\begin{aligned} \bar{u}^0 &= \bar{u}^0(u^0, u^1, \ldots, u^{m-1}), \\ \bar{u}^i &= \bar{u}^i(u^1, u^2, \ldots, u^{m-1}). \end{aligned} \tag{4.3.13}$$

It follows that

$$\frac{\partial}{\partial u^0} = B(u)\frac{\partial}{\partial \bar{u}^0},$$

$$\frac{\partial}{\partial u^i} = B_i^j(u)\frac{\partial}{\partial \bar{u}^j} + B_i(u)\frac{\partial}{\partial \bar{u}^0},$$

where we put $B_i^j(u) = \frac{\partial \bar{u}^j}{\partial u^i}$, $B_i = \frac{\partial \bar{u}^0}{\partial u^i}$. As the screen distribution is a transversal distribution to the involutive distribution $\mathrm{Rad}\,TM$, there exist $m-1$ differentiable functions $S_i(u^0, \ldots, u^{m-1})$ satisfying

$$S_i(u)B(u) = \bar{S}_j(u)B_i^j(u) + B_i(u),$$

with respect to the transformations (4.3.13). Then a local basis of $\Gamma\left(S(TM)\right)$ is given by the vector fields

$$\frac{\delta}{\delta\,u^i} \;=\; \frac{\partial}{\partial\,u^i} \;-\; S_i\left(u\right)\frac{\partial}{\partial\,u^o}. \tag{4.3.14}$$

Moreover, we derive

$$\frac{\delta}{\delta\,u^i} \;=\; B_i^j\left(u\right)\frac{\delta}{\delta\,\bar{u}^j},$$

with respect to (4.3.13). Hence, we obtain the local field of frames $\{\frac{\partial}{\partial\,u^o},\frac{\delta}{\delta\,u^i},N_1,u\}$ on $\bar{M}$, where $\{\frac{\partial}{\partial\,u^o},\frac{\delta}{\delta\,u^i}\}$ is a local field of frames on M with $\xi\equiv\frac{\partial}{\partial\,u^o}$.

Next, with respect to the metric connection $\bar{\nabla}$ on $\bar{M}$ and using (4.1.7)–(4.1.10) and (4.1.15), we obtain

$$\bar{\nabla}_{\frac{\delta}{\delta\,u^j}}\frac{\delta}{\delta\,u^i} \;=\; \Gamma_{ij}^o\frac{\partial}{\partial\,u^0} + \Gamma_{ij}^k\frac{\delta}{\delta\,u^k} + D_{ij}^\ell\,N + D_{ij}^s\,u,$$

$$\bar{\nabla}_{\frac{\delta}{\delta\,u^j}}\frac{\partial}{\partial\,u^0} \;=\; \Gamma_{oj}^o\frac{\partial}{\partial\,u^0} + \Gamma_{oj}^k\frac{\delta}{\delta\,u^k} + D_{oj}^s\,u, \tag{4.3.15}$$

$$\bar{\nabla}_{\frac{\partial}{\partial\,u^0}}\frac{\delta}{\delta\,u^i} \;=\; \Gamma_{io}^o\frac{\partial}{\partial\,u^0} + \Gamma_{io}^k\frac{\delta}{\delta\,u^k} + D_{io}^s\,u,$$

$$\bar{\nabla}_{\frac{\partial}{\partial\,u^0}}\frac{\partial}{\partial\,u^0} \;=\; \Gamma_{oo}^o\frac{\partial}{\partial\,u^0} - \epsilon\,\varepsilon_o^\ell\,u,$$

and

$$\bar{\nabla}_{\frac{\delta}{\delta\,u^i}}N \;=\; -A_i^k\frac{\delta}{\delta\,u^k} + \rho_i^\ell\,N + \rho_i^s\,u,$$

$$\bar{\nabla}_{\frac{\partial}{\partial\,u^0}}N \;=\; -A_o^k\frac{\delta}{\delta\,u^k} + \rho_o^\ell\,N + \rho_o^s\,u, \tag{4.3.16}$$

$$\bar{\nabla}_{\frac{\delta}{\delta\,u^i}}u \;=\; -\epsilon\,\rho_i^s\frac{\partial}{\partial\,u^0} - \bar{A}_i^k\frac{\delta}{\delta\,u^k} + \varepsilon_i^\ell\,N,$$

$$\bar{\nabla}_{\frac{\partial}{\partial\,u^0}}u \;=\; -\epsilon\,\rho_o^s\frac{\partial}{\partial\,u^0} - \bar{A}_o^k\frac{\delta}{\delta\,u^k} + \varepsilon_o^\ell\,N,$$

where $\{\Gamma_{ij}^k,\Gamma_{ij}^o,\Gamma_{oj}^k,\Gamma_{io}^k,\Gamma_{oj}^o,\Gamma_{io}^o,\Gamma_{oo}^o\}$ are the coefficients of the induced linear connection ∇ on M with respect to the frames field $\{\frac{\partial}{\partial\,u^o},\frac{\delta}{\delta\,u^i}\}$; $\{A_i^k,A_o^k\}$ and $\{\bar{A}_i^k,\bar{A}_o^k\}$ are the entries of the matrices of A_{N_1}, $A_u:\Gamma\left(TM_{|\mathcal{U}}\right)\longrightarrow\Gamma\left(S(TM)_{|\mathcal{U}}\right)$ with respect to the basis $\{\frac{\partial}{\partial\,u^o},\frac{\delta}{\delta\,u^i}\}$ and $\{\frac{\delta}{\delta\,u^i}\}$ of $\Gamma\left(TM_{|\mathcal{U}}\right)$ and $\Gamma\left(S(TM)_{|\mathcal{U}}\right)$ respectively.

$$D_{ij}^\ell \;=\; D_1\left(\frac{\delta}{\delta\,u^j},\frac{\delta}{\delta\,u^i}\right) \;=\; D_1\left(\frac{\partial}{\partial\,u^j},\frac{\partial}{\partial\,u^i}\right),$$

$$D_{ij}^s \;=\; D_2\left(\frac{\delta}{\delta\,u^j},\frac{\delta}{\delta\,u^i}\right),\qquad D_{oj}^s \;=\; D_2\left(\frac{\delta}{\delta\,u^j},\frac{\partial}{\partial\,u^0}\right),$$

$$D_{io}^s \;=\; D_2\left(\frac{\partial}{\partial\,u^0},\frac{\delta}{\delta\,u^i}\right),\qquad \rho_i^\ell \;=\; \rho_1\left(\frac{\delta}{\delta\,u^i}\right),$$

$$\rho_i^s = \rho_2 \left(\frac{\delta}{\delta\, u^i} \right), \qquad \rho_o^\ell = \rho_1 \left(\frac{\partial}{\partial\, u^0} \right), \qquad \rho_o^s = \rho_2 \left(\frac{\partial}{\partial\, u^0} \right),$$

$$\varepsilon_i^\ell = \varepsilon_1 \left(\frac{\delta}{\delta\, u^i} \right), \qquad \varepsilon_o^\ell = \varepsilon_1 \left(\frac{\partial}{\partial\, u^0} \right).$$

By straightforward calculations from (4.3.14) we have

$$\left[\frac{\delta}{\delta\, u^i}, \frac{\delta}{\delta\, u^j} \right] = S_{ij} \frac{\partial}{\partial\, u^0}, \tag{4.3.17}$$

$$\left[\frac{\delta}{\delta\, u^i}, \frac{\partial}{\partial\, u^0} \right] = \frac{\partial\, S_i}{\partial\, u^0} \frac{\partial}{\partial\, u^0}, \tag{4.3.18}$$

where

$$S_{ij} = \frac{\delta\, S_i}{\delta\, u^j} - \frac{\delta\, S_j}{\delta\, u^i}.$$

As $\bar\nabla$ is torsion-free, by using (4.3.17), we obtain

$$\Gamma_{ij}^k = \Gamma_{ji}^k, \qquad \Gamma_{ij}^o = \Gamma_{ji}^o + S_{ji},$$

$$\Gamma_{oi}^k = \Gamma_{io}^k, \qquad \Gamma_{oi}^o = \Gamma_{io}^o + \frac{\partial\, S_i}{\partial\, u^0}$$

$$D_{ij}^\ell = D_{ji}^\ell, \qquad D_{ij}^s = D_{ji}^s, \qquad D_{io}^s = D_{oi}^s.$$

Further, we decompose the following Lie brackets:

$$\left[N, \frac{\delta}{\delta\, u^k} \right] = N_k^o \frac{\partial}{\partial\, u^0} + N_k^h \frac{\delta}{\delta\, u^h} + N_k^\ell N + N_k^s u,$$

$$\left[N, \frac{\partial}{\partial\, u^0} \right] = N^o \frac{\partial}{\partial\, u^0} + N_o^h \frac{\delta}{\delta\, u^h} + N_o^\ell N + N_o^s u,$$

$$\left[u, \frac{\delta}{\delta\, u^k} \right] = L_k^o \frac{\partial}{\partial\, u^0} + L_k^h \frac{\delta}{\delta\, u^h} + L_k^\ell N + L_k^s u, \tag{4.3.19}$$

$$\left[u, \frac{\partial}{\partial\, u^0} \right] = L^o \frac{\partial}{\partial\, u^0} + L_o^h \frac{\delta}{\delta\, u^h} + L_o^\ell N + L_o^s u,$$

$$[N, u] = M^o \frac{\partial}{\partial\, u^0} + M^h \frac{\delta}{\delta\, u^h} + M^\ell N + M^s u.$$

Using the non-holonomic frames field $\{N, u, \frac{\partial}{\partial\, u^0}, \frac{\delta}{\delta\, u^k}\}$ of $\bar M$, the semi-Riemannian metric $\bar g$ has the matrix

$$[\bar g] = \begin{bmatrix} 0 & 1 & 0 & 0 \\ 1 & 0 & 0 & 0 \\ 0 & 0 & 0 & g' \end{bmatrix},$$

where

$$g_{ij} = \bar g \left(\frac{\delta}{\delta\, u^i}, \frac{\delta}{\delta\, u^j} \right) = g \left(\frac{\partial}{\partial\, u^i}, \frac{\partial}{\partial\, u^j} \right)$$

and $g' = g_{ij}(u^0, \ldots, u^{m-1})$. Consider the inverse matrix $(g^{ij}(u))$ of the invertible matrix $(g_{ij}(u))$. Then, by using the Koszul formula

$$2g(\bar{\nabla}_X Y, Z) = X(g(Y,Z)) + Y(g(X,Z)) - Z(g(X,Y))$$
$$+ g([X,Y],Z) + g([Z,X],Y) - g([Y,Z],X),$$

for any $X, Y, Z \in \Gamma(TM)$ and using (4.3.17) - (4.3.19), and next, by using (4.1.18), (4.1.19), (4.3.17), we obtain

$$\Gamma_{ij}^{k} = \frac{1}{2} g^{kh} \left(\frac{\delta g_{ih}}{\delta u^j} + \frac{\delta g_{jh}}{\delta u^i} - \frac{\delta g_{ij}}{\delta u^h} \right) = \Gamma_{ij}^{*k}, \tag{4.3.20}$$

$$\Gamma_{ij}^{o} = \frac{1}{2} \left\{ S_{ji} + N_j^k g_{ki} + N_i^k g_{kj} - N_1(g_{ij}) \right\} = g_{jk} A_j^k = E_{ij}, \tag{4.3.21}$$

$$\Gamma_{jo}^{o} = \frac{1}{2} \left\{ -\frac{\partial S_j}{\partial u^0} + N_j^\ell + N_o^k g_{kj} \right\} = g_{jk} A_o^k = E_j, \tag{4.3.22}$$

$$\Gamma_{oj}^{o} = \frac{1}{2} \left\{ \frac{\partial S_j}{\partial u^0} + N_j^\ell + N_o^k g_{kj} \right\} = g_{jk} A_o^k + \frac{\partial S_j}{\partial u^0} = -\rho_j^\ell, \tag{4.3.23}$$

$$\Gamma_{jo}^{k} = \Gamma_{jo}^{k} = \frac{1}{2} g^{ki} \frac{\partial g_{ji}}{\partial u^0} = -A_j^{*k}, \tag{4.3.24}$$

$$D_{ij}^{\ell} = -\frac{1}{2} \frac{\partial g_{ij}}{\partial u^0} = g_{ik} A_j^{*k}, \tag{4.3.25}$$

$$D_{ij}^{s} = \frac{1}{2} \epsilon \left\{ L_j^k g_{ki} + L_i^k g_{kj} - u(g_{ij}) \right\} = \epsilon g_{ik} \bar{A}_j^k, \tag{4.3.26}$$

$$D_{oj}^{s} = D_{jo}^{s} = \frac{1}{2} \epsilon \left\{ L_j^\ell + L_o^k g_{kj} \right\} = \epsilon g_{jk} \bar{A}_o^k = -\epsilon \varepsilon_j^\ell, \tag{4.3.27}$$

$$\Gamma_{oo}^{o} = N_o^\ell = -\rho_o^\ell, L_o^\ell = -\varepsilon_o^\ell,$$
$$2\rho_i^s = -\epsilon M^k g_{ki} + \epsilon L_i^o - N_i^s,$$
$$2\rho_o^s = -\epsilon M^\ell + \epsilon L^o - N_o^s,$$

where $\{\Gamma_{ij}^{*k}, \Gamma_{io}^{*k}\}$ are the coefficients of the linear connection ∇^* on $S(TM)$ with respect to the frames field $\{\frac{\partial}{\partial u^0}, \frac{\delta}{\delta u^k}\}$ and A_j^{*i} are the entries of $A_{\frac{\partial}{\partial u^0}}^*$ with respect to the basis $\{\frac{\delta}{\delta u^i}\}$ and $E_{ji} = E_1(\frac{\delta}{\delta u^j}, \frac{\delta}{\delta u^i})$, $E_i = E_1(\frac{\partial}{\partial u^0}, \frac{\delta}{\delta u^i})$.

Replacing Y and Z from (4.1.16) by $\frac{\delta}{\delta u^i}$ and $\frac{\delta}{\delta u^j}$ respectively, and using (4.1.4), we obtain

$$g_{ij;k} = \frac{\delta g_{ij}}{\delta u^k} - \Gamma_{ik}^h g_{hj} - \Gamma_{jk}^h g_{ih} = 0 \tag{4.3.28}$$

and

$$g_{ij;o} = \frac{\partial g_{ij}}{\partial u^0} - \Gamma_{io}^h g_{hj} - \Gamma_{jo}^h g_{ih} = 0. \tag{4.3.29}$$

For the local expression of Gauss–Codazzi equations of a half-lightlike submanifold, consider the frames field $\{\frac{\partial}{\partial u^0}, \frac{\delta}{\delta u^i}, N, U\}$ on $\bar{M}$. In the sequel we use the range of indices: $i, j, k, \dots \in \{1, 2, \dots, m-1\}; A, B, C, D, \dots \in \{0, 1, 2, \dots, m-1\}$. Denote by $\{X_A\}$ the frame fields $\{\frac{\partial}{\partial u^0}, \frac{\delta}{\delta u^i}\}$ on M, i.e., $X_0 = \frac{\partial}{\partial u^0}, X_i = \frac{\delta}{\delta u^i}$. Then consider the local components of curvature tensors $\bar{R}$ and R as follows:

$$\bar{R}_{ABCD} = \bar{g}(\bar{R}(X_D, X_C)X_B, X_A),$$
$$R_{ABCD} = g(R(X_D, X_C)X_B, X_A),$$
$$\bar{R}_{1BCD} = \bar{g}(\bar{R}(X_D, X_C)X_B, N),$$
$$R_{1BCD} = \bar{g}(R(X_D, X_C)X_B, N),$$
$$\bar{R}_{2BCD} = \bar{g}(\bar{R}(X_D, X_C)X_B, U),$$
$$R_{2BCD} = \bar{g}(R(X_D, X_C)X_B, U).$$

We are now concerned with local expression of a Ricci-tensor of a half-lightlike submanifold M of an $(m+2)$-dimensional semi-Riemannian manifold $(\bar{M}, \bar{g})$. By using the frames field $\{\frac{\partial}{\partial u^0}, \frac{\delta}{\delta u^i}\}$ on M we obtain the following local expression for the Ricci tensor,

$$\text{Ric}(X, Y) = g^{ij} g\left(R(X, \frac{\delta}{\delta u^i})Y, \frac{\delta}{\delta u^j}\right) + \bar{g}\left(R(X, \frac{\partial}{\partial u^0})Y, N\right).$$

By using the symmetries of curvature tensor and the first Bianchi identity with respect to $\bar{R}$ and taking into account (4.1.27) we obtain

$$\text{Ric}(X, Y) - \text{Ric}(Y, X)$$
$$= g^{ij}\left\{E_1(X, \frac{\delta}{\delta u^j})D_1(Y, \frac{\delta}{\delta u^i}) - E_1(Y, \frac{\delta}{\delta u^j})D_1(X, \frac{\delta}{\delta u^i})\right\}$$
$$+ \bar{g}(R(X, Y)\frac{\partial}{\partial u^0}, N).$$

Replacing X and Y by $\frac{\delta}{\delta u^h}$ and $\frac{\delta}{\delta u^k}$ respectively and using (4.1.39) and (4.3.21), we infer

$$R_{kh} - R_{hk} = A_h^i D_{ik}^\ell - A_k^i D_{ih}^\ell + R_{1okh} = 2 d\rho_1\left(\frac{\delta}{\delta u^k}, \frac{\delta}{\delta u^h}\right),$$

where we put $R_{kh} = \text{Ric}(\frac{\delta}{\delta u^h}, \frac{\delta}{\delta u^k})$. Similarly, replacing X and Y by $\frac{\delta}{\delta u^h}$ and $\frac{\partial}{\partial u^0}$ respectively and using (4.1.10), (4.1.39) and (4.3.22), we obtain

$$R_{oh} - R_{ho} = -A_o^i D_{ih}^\ell + R_{1ooh} = 2 d\rho_1\left(\frac{\partial}{\partial u^0}, \frac{\delta}{\delta u^h}\right),$$

where $R_{oh} = \text{Ric}(\frac{\delta}{\delta u^h}, \frac{\partial}{\partial u^0})$ and $R_{ho} = \text{Ric}(\frac{\partial}{\partial u^0}, \frac{\delta}{\delta u^k})$. Therefore, from the above last two equations and by using Theorem 4.2.1 we obtain

Theorem 4.3.10. *Let (M, g) be a half-lightlike submanifold of a semi-Riemannian manifold $(\bar{M}, \bar{g})$. Then the Ricci tensor of the induced connection ∇ on M is symmetric, if and only if, each 1-form ρ_1 induced by $S(TM)$ is closed, i.e., $d\rho_1 = 0$ on any $\mathcal{U} \subset M$.*

Combining Theorems 4.3.5 and the above theorem, we have

Theorem 4.3.11. *Let (M, g) be a proper totally umbilical half-lightlike submanifold of a semi-Riemannian manifold of constant curvature $(\bar{M}(\bar{c}), \bar{g})$. Then the following assertions are equivalent:*

(1) *The Ricci tensor of the connection ∇ on M is symmetric.*

(2) *The screen distribution $S(TM)$ is integrable.*

(3) *Each 1-form ρ_1 induced by $S(TM)$ is closed, i.e., $d\rho_1 = 0$ on any $\mathcal{U} \subset M$.*

(4) *Each 1-form ρ_2 induced by $S(TM)$ satisfies*

$$2d\rho_2(X,Y) = \rho_1(X)\rho_2(Y) - \rho_2(X)\rho_1(Y), \ \forall\, X, Y \in \Gamma(TM).$$

Example 2. Consider a surface M in R_2^4 given by the equation

$$x^3 = \frac{1}{\sqrt{2}}(x^1 + x^2); \qquad x^4 = \frac{1}{2}\log(1 + (x^1 - x^2)^2).$$

Then $TM = \mathrm{Span}\{U_1, U_2\}$ and $TM^\perp = \mathrm{Span}\{\xi, u\}$ where

$$U_1 = \sqrt{2}(1 + (x^1 - x^2)^2)\partial_1 + (1 + (x^1 - x^2)^2)\partial_3 + \sqrt{2}(x^1 - x^2)\partial_4,$$
$$U_2 = \sqrt{2}(1 + (x^1 - x^2)^2)\partial_2 + (1 + (x^1 - x^2)^2)\partial_3 - \sqrt{2}(x^1 - x^2)\partial_4,$$
$$\xi = \partial_1 + \partial_2 + \sqrt{2}\,\partial_3,$$
$$u = 2\,(x^2 - x^1)\partial_2 + \sqrt{2}\,(x^2 - x^1)\partial_3 + (1 + (x^1 - x^2))\,\partial_4.$$

By direct calculations we check that $\mathrm{Rad}\,TM$ is a distribution on M of rank 1 spanned by ξ. Hence M is a half-lightlike submanifold of R_2^4. Choose $S(TM)$ and D spanned by U_2 and u which are timelike and spacelike respectively. We obtain the null canonical affine normal bundle

$$\mathrm{ltr}\,(TM) = \mathrm{Span}\{\, N = -\frac{1}{2}\partial_1 + \frac{1}{2}\partial_2 + \frac{1}{\sqrt{2}}\partial_3 \,\},$$

and the canonical affine normal bundle $\mathrm{tr}(TM) = \mathrm{Span}\{\, N, u \,\}$.

Denote by $\bar{\nabla}$ the Levi–Civita connection on R_2^4. Then by straightforward calculations we obtain

$$\bar{\nabla}_{U_2}U_2 = 2(1 + (x^1 - x^2)^2)\{2(x^2 - x^1)\partial_2 + \sqrt{2}(x^2 - x^1)\partial_3 + \partial_4 \},$$

$$\bar{\nabla}_\xi U_2 = 0, \ \bar{\nabla}_X\xi = \bar{\nabla}_X N_1 = 0, \ \forall\, X \in \Gamma(TM).$$

Then using the Gauss and Weingarten formulae we infer

$$D_1 = 0; \quad A_\xi = 0; \quad A_N = 0; \quad \nabla_X \xi = 0; \quad \rho_i(X) = 0;$$

$$D_2(X, \xi) = 0; \quad D_2(U_2, U_2) = 2;$$

$$\nabla_X U_2 = \frac{2\sqrt{2}(x^2 - x^1)^3}{1 + (x^1 - x^2)^2} X^2 U_2;$$

for any $X = X^1 \xi + X^2 U_2$ tangent to M. As $D_1 = 0$, by Theorem 4.1.3 it follows that the induced connection ∇ is a metric connection. Since $\bar{g}(U_2, U_2) = -(1 + (x^1 - x^2)^4)$ we have

$$D_2(U_2, U_2) = H_2\, \bar{g}(U_2, U_2), \qquad H_2 \equiv -\frac{2}{(1 + (x^1 - x^2)^4)}.$$

Therefore, M is a totally umbilical half-lightlike submanifold of R^4_2.

As the Riemannian curvature tensor R of an arbitrary manifold, M can be considered as an $F(M)$-multilinear function on individual vector fields. If $X, Y \in \Gamma(TM)$ the operator

$$R_{XY} : \Gamma(TM) \to \Gamma(TM),$$

sending each Z to $R(X, Y)Z$, is called a *curvature operator*. From the equation (4.1.39) we have

Theorem 4.3.12. *Let (M, g) be a half-lightlike submanifold of $(\bar{M}, \bar{g})$, with its induced connection ∇ which is a metric connection. Then the radical distribution $\mathrm{Rad}\,TM$ of M is an eigenspace for the curvature operator R_{XY}, of M, with respect to the eigenfunction $-2d\rho_1(X, Y), \forall X, Y \in \Gamma(TM)$.*

4.4 Screen conformal submanifolds

It is well known that the second fundamental form and its shape operator of a non-degenerate submanifold are related by means of the metric tensor field. Contrary to this we see from equations (4.1.20) and (4.1.21) in Section 1 that in the case of half-lightlike submanifolds M there are interrelations between these geometric objects and those of its screen distribution. As the shape operator is an information tool in studying the geometry of submanifolds, in this section, we consider a class of half-lightlike submanifolds with conformal screen shape operator defined as follows:

Definition 4.4.1. [158] A half-lightlike submanifold M, of a semi-Riemannian manifold, is called screen locally (resp. globally) conformal if on any coordinate neighborhood $\mathcal{U}$ (resp. $\mathcal{U} = M$) there exists a non-zero smooth function φ such that for any null vector field $\xi \in \Gamma(TM^\perp)$ the relation

$$A_N X = \varphi A_\xi^* X, \quad \forall X \in \Gamma(TM|_\mathcal{U}) \tag{4.4.1}$$

holds between the shape operators A_N and A_ξ^* of M and $S(TM)$ respectively.

Remark 4.4.2. In case of a half-lightlike submanifold, since $A_N X$ and $A_\xi^* X$ belong to the screen distribution for any $X \in \Gamma(TM)$, this definition is well defined. The results of this section have appeared in [158].

Example 3. Consider in R_2^5 a submanifold M given by the equations

$$x_4 = (x_1^2 + x_2^2)^{\frac{1}{2}} , \ x_3 = (1 - x_5^2)^{\frac{1}{2}} , \ x_5 , \ x_1 , \ x_2 > 0.$$

Then we have

$$TM = \mathrm{Span}\{\xi = x_1 \partial x_1 + x_2 \partial x_2 + x_4 \partial x_4, \ U = x_4 \partial x_1 + x_1 \partial x_4,$$
$$V = -x_5 \partial x_3 + x_3 \partial x_5\},$$
$$TM^\perp = \mathrm{Span}\{\xi, \ u = x_3 \partial x_3 + x_5 \partial x_5\}.$$

Thus $\mathrm{Rad}\, TM = \mathrm{Span}\{\xi\}$ is a distribution on M and $S(TM^\perp) = \mathrm{Span}\{u\}$. Hence M is a half-lightlike submanifold of R_2^5 with $S(TM) = \mathrm{Span}\{U, V\}$. Also, the lightlike transversal bundle $n\,\mathrm{tr}(TM)$ is spanned by

$$N = \frac{1}{2x_2^2}\{x_1 \partial x_1 - x_2 \partial x_2 + x_4 \partial x_4\}.$$

By direct calculations, we obtain

$$\bar\nabla_U \xi = U, \quad \bar\nabla_V \xi = 0, \quad \bar\nabla_\xi \xi = \xi,$$
$$\bar\nabla_U N = \frac{1}{2x_2^2} U, \ \bar\nabla_V N = 0, \ \bar\nabla_\xi N = -N.$$

Then, from (4.1.8) and (4.1.19) we obtain

$$A_\xi^* U = -U, \ A_\xi^* V = 0,$$
$$A_N U = -\frac{1}{2x_2^2} U, \quad \rho_1(U) = 0, \quad \rho_2(U) = 0,$$
$$A_N V = 0, \quad \rho_1(V) = 0, \quad \rho_2(V) = 0,$$
$$A_N \xi = 0, \quad \rho_1(\xi) = -1, \quad \rho_2(\xi) = 0.$$

Hence we derive $A_N X = \frac{1}{2x_2^2} A_\xi^* X, \quad \forall X \in \Gamma(TM)$. Thus M is a screen conformal lightlike submanifold with $\varphi = \frac{1}{2x_2^2}$.

Proposition 4.4.3. *Let (M, g) be a half-lightlike submanifold of a semi-Riemannian manifold $\bar M$. Then, M is screen conformal if and only if*

$$E_1(X, PY) = \varphi\, D_1(X, PY), \quad \forall X, Y \in \Gamma(TM). \tag{4.4.2}$$

Proof. Suppose M is a screen conformal half-lightlike submanifold. Then, from (4.1.20), (4.1.21) and (4.4.1), we get

$$E_1(X, PY) = g(A_N X, PY) = \varphi\, g(A_\xi^* X, PY) = \varphi\, D_1(X, PY)$$

$\forall X, Y \in \Gamma(TM)$. Conversely, if $E_1(X, PY) = \varphi D_1(X, PY)$, $\forall X, Y \in \Gamma(TM)$, then (4.1.20) and (4.1.21) imply $g(A_N X, PY) = g(\varphi A_\xi^* X, PY)$. Thus, we get $A_N X = \varphi A_\xi^* X$ which completes the proof. $\qquad \square$

Let M be screen conformal. Then, from (4.1.18) and (4.4.2) we get

$$\nabla_X PY = \nabla_X^* PY + \varphi D_1(X, PY)\xi, \quad \forall X, Y \in \Gamma(TM). \qquad (4.4.3)$$

Now we show that a screen conformal half-lightlike submanifold can admit a unique screen distribution.

Theorem 4.4.4. *Let* $(M, g, S(TM))$ *be a screen conformal half-lightlike subman-ifold of a semi-Riemannian manifold* $(\bar{M}^{m+2}, \bar{g})$. *Then, any screen distribution* $S(TM)$ *of* M *is integrable. Moreover, if the first derivative* $\mathcal{S}$ *defined by (4.2.7) coincides with* $S(TM)$, *then,* $S(TM)$ *is a unique screen of* M, *up to an orthogo-nal transformation with a unique lightlike transversal vector bundle and invariant screen second fundamental form.*

Proof. Using (4.1.7) and (4.4.3), we obtain

$$\bar{g}([X, Y], N) = \bar{g}(\nabla_X Y, N) - \bar{g}(\nabla_Y X, N)$$
$$= \varphi D_1(X, Y)\bar{g}(\xi, N) - \varphi D_1(Y, X)\bar{g}(\xi, N)$$
$$= \varphi\{D_1(X, Y) - D_1(Y, X)\}, \forall X, Y, Z \in \Gamma(TM).$$

Since D_1 is symmetric we get $\bar{g}([X, Y], N) = 0$. Hence $S(TM)$ is integrable. The remainder of the proof follows from the proof of Theorem 4.2.1 of this chapter. $\quad \square$

Definition 4.4.5. *Let* M *be a half-lightlike submanifold of a semi-Riemannian manifold* $\bar{M}$. *Then, we say that* M *is a minimal half-lightlike submanifold if* $\mathrm{tr}\,|_{S(TM)}\, h = 0$ *and* $\varepsilon_1(\xi) = 0$.

From (4.1.7) and (4.1.12), it follows that M is minimal if and only if

$$\sum_{i=1}^{n-1} D_1(e_i, e_i) = 0, \sum_{i=1}^{n-1} D_2(e_i, e_i) = 0, \quad \text{and} \quad \varepsilon_1(\xi) = 0,$$

where $\{e_i\}_{i=1}^{n-1}$ is an orthonormal basis of $S(TM)$.

Theorem 4.4.6. *Let* M *be a screen conformal half-lightlike submanifold of a semi-Riemannian manifold* $\bar{M}$, *with a leaf* M' *of* $S(TM)$. *Then*

1. M *is totally geodesic,*

2. M *is totally umbilical,*

3. M *is minimal,*

if and only if M' *is so immersed as a submanifold of* $\bar{M}$ *and* ε_1 *vanishes on* M.

Proof. Using (4.4.2) we obtain

$$\bar{\nabla}_X Y = \nabla_X^* Y + \varphi \, D_1(X,Y)\xi + D_1(X,Y)N + D_2(X,Y)u \qquad (4.4.4)$$

for any $X, Y \in \Gamma(TM')$. Note that a screen conformal half-lightlike submanifold has an integrable screen distribution. Thus a leaf of $S(TM)$ is a semi-Riemannian submanifold. Therefore, we have

$$\bar{\nabla}_X Y = \nabla'_X Y + h'(X,Y) \qquad (4.4.5)$$

where h' and ∇' are second fundamental form and the Levi-Civita connection of M' in $\bar{M}$. Thus, from (4.4.4) and (4.4.5) we obtain

$$h'(X,Y) = (\varphi \, \xi + N)D_1(X,Y) + D_2(X,Y)u \qquad (4.4.6)$$

for any $X, Y \in \Gamma(TM')$. On the other hand, from (4.1.12) we have

$$\epsilon D_2(\xi, PZ) = g(A_u \xi, PZ) \text{ and } \epsilon D_2(PZ, \xi) = -\varepsilon_1(PZ).$$

Since D_2 is symmetric we obtain $-\varepsilon_1(PZ) = g(A_u \xi, PZ)$. Similarly we get that $\epsilon D_2(\xi, \xi) = \varepsilon_1(\xi)$. Consequently, we obtain

$$D_2(\xi, PZ) = D_2(PZ, \xi) = D_2(\xi, \xi) = 0 \Leftrightarrow \varepsilon_1(Z) = 0, \qquad (4.4.7)$$

for all $Z \in \Gamma(TM)$. Thus the proof follows from equations (4.4.6) and (4.4.7). □

Definition 4.4.7. [273] A lightlike submanifold M is said to be irrotational if $\bar{\nabla}_X \xi \in \Gamma(TM)$ for any $X \in \Gamma(TM)$, where $\xi \in \Gamma(\text{Rad}\, TM)$.

For a half-lightlike M, since $D_1(X, \xi) = 0$, the above definition is equivalent to $D_2(X, \xi) = 0 = \varepsilon_1(X)$, $\forall X \in \Gamma(TM)$. Using this in (4.4.7) we state

Corollary 4.4.8. *Let M be an irrotational screen conformal half-lightlike submanifold of a semi-Riemannian manifold $\bar{M}$. Then*

1. *M is totally geodesic,*

2. *M is totally umbilical,*

3. *M is minimal,*

if and only if a leaf M' of any $S(TM)$ is so immersed as a submanifold of $\bar{M}$.

Theorem 4.4.9. *Let M be a screen conformal half-lightlike submanifold of a semi-Riemannian manifold. The following assertions are equivalent:*

(1) *Any leaf of $S(TM)$ is totally geodesic in M.*

(2) *M is a lightlike product manifold of M' and L, where M', a leaf of $S(TM)$, is a non-degenerate manifold and L is a one-dimensional lightlike manifold.*

(3) *D_1 vanishes identically on M.*

(4) *The induced connection ∇ on M is a metric connection.*

Proof. From (4.1.7) we have $g(\nabla_\xi \xi, X) = \bar{g}(\bar{\nabla}_\xi \xi, X)$ for $X \in \Gamma(S(TM))$ and $\xi \in \Gamma(\text{Rad}(TM))$. Now $\bar{\nabla}$ a metric connection implies $g(\nabla_\xi \xi, X) = \bar{g}(\xi, \bar{\nabla}_\xi X)$. Now using (4.1.7), (4.1.10) and (4.1.11) we obtain $g(\nabla_\xi \xi, X) = -D_1(\xi, X)\bar{g}(\xi, N)$. Thus we get

$$g(\nabla_\xi \xi, X) = 0. \tag{4.4.8}$$

Similarly, from (4.1.8) we derive $\bar{g}(\nabla_X Y, N) = g(A_N X, Y) \ \forall X, Y \in \Gamma(S(TM))$ and $N \in \Gamma(\text{ltr}(TM))$. Then from (4.4.1) we obtain $\bar{g}(\nabla_X Y, N) = \varphi g(A_\xi^* X, Y)$. Thus (4.1.21) implies that

$$\bar{g}(\nabla_X Y, N) = \varphi D_1(X, Y). \tag{4.4.9}$$

Now, from (4.4.8) and (4.4.9), the equivalent of (1) and (2) follows. If M is a lightlike product, then any leaf of $S(TM)$ is parallel. Thus from (4.4.9) $D_1 = 0$, since $D_1(X, \xi) = 0$.

Conversely, if $D_1 = 0$ then from (4.4.9) a leaf of $S(TM)$ is parallel and considering (4.4.8) we obtain (2). Thus (2) $\Leftrightarrow$ (3). Finally, the equivalent of (3) and (4) comes from Theorem 4.1.3, which completes the proof. $\qquad \square$

Let M be a screen conformal half-lightlike submanifold. Consider the Riemannian curvature of type $(0, 4)$ of $\bar{\nabla}$, by using (4.1.24) and the definition of curvature tensors, we derive the following structure equations:

$$\begin{aligned}
\bar{g}(\bar{R}(X, Y)Z, PW) = \ &g(R(X, Y)Z, PW) \\
&+ \varphi\{D_1(X, Z)D_1(Y, PW) \\
&- D_1(Y, Z)D_1(X, PW)\} \\
&+ \epsilon\{D_2(X, Z)D_2(Y, PW) \\
&- D_2(Y, Z)D_2(X, PW)\}, \tag{4.4.10}
\end{aligned}$$

$$\begin{aligned}
\bar{g}(\bar{R}(X, Y)PZ, N) = \ &\bar{g}(R(X, Y)PZ, N) \\
&+ \epsilon\{\rho_2(Y)D_2(X, PZ) - \rho_2(X)D_2(Y, PZ)\} \\
= \ &g(\nabla_X(A_N Y) - \nabla_Y(A_N X) \\
&- A_N[X, Y], PZ) + \varphi\{\rho_1(Y)D_1(X, PZ) \\
&- \rho_1(X)D_1(Y, PZ)\} + \epsilon\{\rho_2(Y)D_2(X, PZ) \\
&- \rho_2(X)D_2(Y, PZ)\}, \tag{4.4.11}
\end{aligned}$$

$$\begin{aligned}
\bar{g}(\bar{R}(X, Y)\xi, PZ) = \ &g(R(X, Y)\xi, PZ) + \epsilon D_2(X, \xi)D_2(Y, PZ) \\
&- \epsilon D_2(Y, \xi)D_2(X, PZ) \tag{4.4.12}
\end{aligned}$$

Let R^* be the curvature tensors of ∇^*. Using (4.1.18) and (4.4.2) we obtain

$$R(X,Y)PZ = R^*(X,Y)PZ - \varphi\{D_1(Y,PZ)A_\xi^* X - D_1(X,PZ)A_\xi^* Y\}$$
$$+ \varphi\{(\nabla_X D_1)(Y,PZ) - (\nabla_Y D_1)(X,PZ)\}\xi$$
$$+ D_1(Y,PZ)\{X(\varphi) - \varphi\rho_1(X)\}\xi$$
$$- D_1(X,PZ)\{Y(\varphi) - \varphi\rho_1(Y)\}\xi. \tag{4.4.13}$$

Theorem 4.4.10. *Let M be a screen conformal half-lightlike submanifold of a semi-Riemannian space form $\bar{M}(c)$. Then, the induced Ricci tensor of M is symmetric if and only if*
$$(D_2 \wedge \rho_2)(\xi, X, Y) = D_2(X,Y)\rho_2(\xi).$$

Proof. The Ricci tensor of a half-lightlike submanifold is given by

$$\mathrm{Ric}(X,Y) = \sum_{i=1}^{m-1} \varepsilon g(R(X,e_i)Y, e_i) + \bar{g}(R(X,\xi)Y, N), \quad \forall X, Y \in \Gamma(TM).$$

For a space form $\bar{M}(c)$, from (4.4.2), (4.4.10) and (4.4.11), we have

$$\mathrm{Ric}(X,Y) = (1-m)cg(X,Y) + \sum_{i=1}^{m-1}\{(-D_1(X,Y)D_1(e_i,e_i)$$
$$+ D_1(e_i,Y)D_1(X,e_i))\varphi$$
$$- \varepsilon(D_2(X,Y)D_2(e_i,e_i) + D_2(e_i,Y)D_2(X,e_i))\}$$
$$- \varepsilon D_2(X,Y)\rho_2(\xi) + \varepsilon D_2(\xi,Y)\rho_2(X).$$

Thus we get

$$\mathrm{Ric}(X,Y) - \mathrm{Ric}(Y,X) = \varepsilon\{D_2(\xi,Y)\rho_2(X) - D_2(\xi,X)\rho_2(Y)\}$$

or

$$\mathrm{Ric}(X,Y) - \mathrm{Ric}(Y,X) = (D_2 \wedge \rho_2)(\xi, X, Y) - D_2(X,Y)\rho_2(\xi) \tag{4.4.14}$$

which proves the theorem.   □

The following result holds from Definition 4.4.7 and (4.4.14).

Corollary 4.4.11. *The Ricci tensor of any irrotational screen conformal half-lightlike submanifold, of $\bar{M}(c)$, is symmetric.*

Let $p \in M$ and ξ be a null vector of T_pM. A plane H of T_pM is called a null plane directed by ξ if it contains ξ, $\bar{g}(\xi,W) = 0$ for any $W \in H$ and there exits $W_0 \in H$ such that $\bar{g}(W_0,W_0) \neq 0$. Then the null sectional curvature of H with respect to ξ and $\bar{\nabla}$ is defined by [34, page 431]

$$K_\xi(H) = \frac{R_p(W,\xi,\xi,W)}{g_p(W,W)}.$$

Theorem 4.4.12. *Let M be a screen conformal half-lightlike submanifold of an $\bar{M}(c)$. Then, the null sectional curvature of M is given by*

$$K_\xi(H) = \epsilon\{D_2(\xi,\xi)D_2(X,X) - D_2(X,\xi)D_2(\xi,X)\}, \tag{4.4.15}$$

for $X \in \Gamma(S(TM))$ and $\xi \in \Gamma(\mathrm{Rad}\,TM)$.

Proof. From (4.4.10) we have

$$K_\xi(H) = \varphi\{D_1(X,\xi)D_1(\xi,X) - D_1(\xi,\xi)D_1(X,X)\}$$
$$+ \epsilon\{D_2(\xi,\xi)D_2(X,X) - D_2(X,\xi)D_2(\xi,X)\}.$$

Using (4.1.10) we obtain (4.4.15) which proves the theorem. □

Moreover, using (4.1.15) in (4.4.15) and Definition 4.4.5, we have

Corollary 4.4.13. *The null sectional curvature of a screen conformal half-lightlike submanifold M, of $\bar{M}(c)$, vanishes identically if and only if*

$$(D_2 \wedge \varepsilon_1)(X,\xi,X) = -\epsilon\varepsilon_1{}^2(X), \ \forall X \in \Gamma(S(TM)), \ \xi \in \Gamma(\mathrm{Rad}\,TM).$$

Consequently, the null sectional curvature of any irrotational conformal half-lightlike submanifold, of $\bar{M}(c)$, vanishes identically.

Theorem 4.4.14. *Let $(M, g, S(TM))$ be a screen conformal half-lightlike submanifold of $\bar{M}(c)$ with $D_2 = 0$. Then, M is flat if and only if a leaf M' of $S(TM)$ is flat and $c = 0$.*

Proof. Suppose M is flat. For $\bar{M}(c)$, from (4.4.11) we derive

$$g(R(X,Y)PZ, N) = -\epsilon D_2(X, PZ)\rho_2(Y) + \epsilon D_2(Y, PZ)\rho_2(X)$$
$$+ c\{g(Y, PZ)\eta(X) - g(X, PZ)\eta(Y)\}$$
$$= 0 \quad \forall X, Y, Z \in \Gamma(TM) \tag{4.4.16}$$

and $N \in \Gamma(\mathrm{ltr}(TM))$. Since $D_2 = 0$ and M is flat, we obtain

$$c\{g(Y, PZ)\eta(X) - g(X, PZ)\eta(Y)\} = 0.$$

Thus, for $X = \xi$ and $Y = PZ$ we derive $cg(PZ, PZ) = 0$ hence $c = 0$. On the other hand, from (4.4.10) we have

$$g(R(X,Y)PZ, PW) = c\{g(Y, PZ)g(X, PW) - g(X, PZ)g(Y, PW)\}$$
$$- \varphi\{D_1(Y, PW)D_1(X, PZ)$$
$$- D_1(Y, PZ)D_1(X, PW)\}$$
$$- \epsilon D_2(X, PZ)D_2(Y, PW)$$
$$+ \epsilon D_2(Y, PZ)D_2(X, PW). \tag{4.4.17}$$

Using (4.4.17) in (4.4.13) we get

$$
\begin{aligned}
2g(R(X,Y)PZ, PW) = {} & g(R^*(X,Y)PZ, PW) + \epsilon(D_2(X, PW)D_2(Y, PZ) \\
& - D_2(X, PZ)D_2(Y, PW) + c\{g(Y, PZ)g(X, PW) \\
& - g(X, PZ)g(Y, PW)\}.
\end{aligned}
\tag{4.4.18}
$$

Thus, from (4.4.17) we have $R^* = 0$ due to $c = 0$ and $D_2 = 0$. Now suppose that M' is flat and $c = 0$. Using (4.4.16) and (4.4.18) we obtain

$$
g(R(X,Y)PZ, PW) = 0, \quad g(R(X,Y)PZ, N) = 0.
\tag{4.4.19}
$$

On the other hand, since $\bar{M}$ is a space form and $D_1(X, \xi) = 0$, we have

$$
\bar{g}(R(X,Y)\xi, N) = 0, \quad \forall X \in \Gamma(TM).
\tag{4.4.20}
$$

Moreover, since $D_2 = 0$, from (4.4.12) we get

$$
g(R(X,Y)\xi, PZ) = 0.
\tag{4.4.21}
$$

Thus, (4.4.19) to (4.4.21) implies $R = 0$ which proves the theorem. $\qquad\square$

Example 4. Consider the screen conformal half-lightlike submanifold M of R_2^5 given in Example 3, and by direct calculations, we obtain

$$
\bar{\nabla}_U V = \bar{\nabla}_V U = \bar{\nabla}_\xi V = \bar{\nabla}_V \xi = 0, \; \bar{\nabla}_U \xi = U,
$$

$$
\bar{\nabla}_U U = \frac{1}{2}\xi + x_2^2\, N, \; \bar{\nabla}_V V = -u, \; \bar{\nabla}_\xi \xi = \xi,
$$

$$
\bar{\nabla}_U u = 0, \; \bar{\nabla}_V u = V, \; \bar{\nabla}_\xi u = 0.
$$

Thus from (4.1.7)–(4.1.9), (4.1.19) and (4.1.20) we derive

$$
\nabla_U U = \frac{1}{2}\xi, \; E_1(U, U) = \frac{1}{2}, \; A_u U = 0, \; A_u V = -V, \; A_u \xi = 0,
$$

$$
D_1(U, U) = x_2^2, \; D_1(V, V) = 0, \; D_2(U, U) = 0, \; D_2(V, V) = -1,
$$

$$
D_2(X, \xi) = 0, \; \epsilon_1(X) = 0 \,\forall X \in \Gamma(TM).
$$

Hence M is irrotational with a symmetric Ricci tensor (Theorem 4.4.10) and vanishing null sectional curvature (Corollary 4.4.13). $D_1 \neq 0$ implies that ∇ is not a metric connection and M is not totally geodesic. Also M' is not totally geodesic in $\bar{M}$ (Theorem 4.4.8). Moreover, $S(TM)$ is not parallel in M due to $E_1(U, U) \neq 0$. Thus M is not a lightlike product (Theorem 4.4.9).

Theorem 4.4.15. *Let M be a half-lightlike submanifold of a semi-Riemannian manifold $\bar{M}$. Suppose $S(TM)$ is integrable and any leaf M' of $S(TM)$ is totally umbilical immersed in $\bar{M}$ as a codimension 3 non-degenerate submanifold with $\alpha\beta > 0$. Then M is screen locally conformal if and only if $E_1(\xi, PX) = 0$ for $\xi \in \Gamma(\operatorname{Rad}TM)$ and $X \in \Gamma(TM)$, where α and β are components of a mean curvature vector field of the leaf, in the direction to ξ and N.*

Proof. Let M' be a leaf of $S(TM)$. Then we have

$$\bar{\nabla}_X Y = \nabla_X^* Y + E_1(X,Y)\xi + D_1(X,Y)N + D_2(X,Y)u$$

for any $X, Y \in \Gamma(TM')$. The mean curvature vector field H^* is $H^* = \alpha\xi + \beta N + \gamma u$. Since M' is totally umbilical in $\bar{M}$ we get

$$E_1(X,Y)\xi + D_1(X,Y)N + D_2(X,Y)u = g(X,Y)\{\alpha\xi + \beta N + \gamma u\}.$$

Thus we have

$$E_1(X,Y) = \alpha g(X,Y) \tag{4.4.22}$$
$$D_1(X,Y) = \beta g(X,Y) \quad D_2(X,Y) = \gamma g(X,Y). \tag{4.4.23}$$

(4.4.22) and (4.4.23) imply $E_1(X,Y) = \frac{\alpha}{\beta}D_1(X,Y)$. Hence, $E_1(X,Y) = \frac{\alpha}{\beta}D_1(X,Y)$ for all $X, Y \in \Gamma(TM')$. Since $A_\xi^* \xi = 0$ and $E_1(\xi,Y) = 0$ we obtain $A_N X = \varphi A_\xi^* X$ for $X \in \Gamma(TM)$. Conversely, if M is screen conformal, then, it can be seen that $E_1(\xi,X) = 0$, which completes the proof. $\square$

For screen conformal M, (4.3.11) and (4.4.2) imply M' is totally umbilical if

$$D_1(X, PY) = \frac{K}{\varphi} g(X, PY), \quad \forall X \in \Gamma(TM). \tag{4.4.24}$$

Theorem 4.4.16. *Let M be a screen conformal half-lightlike submanifold of $\bar{M}$. Then M is totally umbilical if and only if*

$$P(A_u X) = H_2 PX, \; \varepsilon_1(X) = 0, \quad X \in \Gamma(TM)$$

and a leaf M' of any $S(TM)$ is totally umbilical in M.

Proof. From (4.3.1) we obtain that $D_2(X,Y) = g(X,Y)H_2$ if and only if $P(A_u X) = H_2 PX$ and $\varepsilon_1(X) = 0$, $\forall X \in \Gamma(TM)$. Suppose $D_1(X,Y) = g(X,Y)H_1$. Then, M is screen conformal and (4.4.2) implies $E_1(X,Y) = \varphi H_1 g(X,Y)$. Hence M' is totally umbilical with $K = \varphi H_1$. Conversely, if M' is totally umbilical then using (4.4.2), (4.3.11) and (4.1.10) we obtain $D_1(X,Y) = H_1 g(X,Y)$, where $H_1 = \frac{K}{\varphi}$, which completes the proof. $\square$

Theorem 4.4.17. *Let M be a screen conformal totally umbilical half-lightlike submanifold of a semi-Riemannian manifold $\bar{M}$. Then:*

1. *M' is totally umbilical in $\bar{M}$.*

2. *M is totally geodesic if and only if M' is totally geodesic in $\bar{M}$.*

Proof. Totally umbilical M implies $D_2(X,\xi) = 0$ and (4.4.6), (4.3.1) imply $h'(X,Y) = g(X,Y)(H_1 \varphi\xi + H_1 N + H_2 u), \forall X, Y \in \Gamma(TM')$, which completes the proof. $\square$

Let M be a totally umbilical half-lightlike submanifold in $\bar{M}(c)$. Then, by direct calculations, using (4.1.7), (4.1.8), (4.1.9), the definition of totally umbilical submanifold and taking the tangential parts, we obtain

$$
\begin{aligned}
R(X,Y)Z = c\{g(Y,Z)X - g(X,Z)Y\} &- g(X,Z)H_1 A_N Y \\
&+ g(Y,Z)H_1 A_N X + H_2\{ g(Y,Z)A_u X \\
&- g(X,Z)A_u Y\}.
\end{aligned}
\tag{4.4.25}
$$

From (4.1.15), (4.4.25), (4.3.1) , (4.1.21), (4.4.1) and (4.4.24) we get

$$
\begin{aligned}
g(R(X,Y)Z,W) = c\{g(Y,Z)g(X,W) &- g(X,Z)g(Y,W)\} \\
&- g(X,Z)\varphi g(Y,W)(H_1)^2 + g(Y,Z)\varphi g(X,W)(H_1)^2 \\
&- g(X,Z)H_2\epsilon D_2(Y,W) + g(Y,Z)H_2\epsilon D_2(X,W),
\end{aligned}
$$

for all $X,Y,Z \in \Gamma(TM)$ and $W \in \Gamma(S(TM))$. Thus we obtain

$$
g(R(X,Y)Z,W) = [g(Y,Z)g(X,W) - g(X,Z)g(Y,W)][c + \varphi(H_1)^2 + \epsilon(H_2)^2].
\tag{4.4.26}
$$

On the other hand, from (4.4.13) we obtain

$$
\begin{aligned}
g(R(X,Y)Z,W) = R^*(X,Y)Z,W) &- \varphi\, g(Y,Z)H_1 g(A_\xi^* X, W) \\
&+ \varphi\, g(X,Z)H_1 g(A_\xi^* Y, W) \quad \forall X,Y \in \Gamma(TM)
\end{aligned}
$$

and $Z,W \in \Gamma(S(TM))$. Here, using (4.3.2) and (4.4.1) we get

$$
g(R(X,Y)Z,W) = R^*(X,Y)Z,W) - \varphi\,(H_1)^2[g(Y,Z)g(X,W) - g(X,Z)g(Y,W)].
\tag{4.4.27}
$$

Thus from (4.4.26) and (4.4.27) we obtain

$$
\begin{aligned}
g(R^*(X,Y)Z,W) = \{g(Y,Z)g(X,Z) &- g(X,Z)g(Y,W)\}\{c + 2\varphi\,(H_1)^2 \\
&+ \epsilon(H_2)^2\}, \quad \forall X,Y \in \Gamma(TM)
\end{aligned}
\tag{4.4.28}
$$

and $Z,W \in \Gamma(S(TM))$. As a result of (4.4.28) we have the following result.

Theorem 4.4.18. *Let $(M,g,S(TM))$ be a screen conformal totally umbilical half-lightlike submanifold of a semi-Riemannian $\bar{M}(c)$, with a leaf M' of $S(TM)$. If $\dim(M') > 2$, then M' is a semi-Riemannian space form if and only if $\varphi = $ constant.*

From the proofs of Theorems 4.4.17 and above, the following results hold:

(a) *The Ricci tensor of a screen conformal totally umbilical half-lightlike submanifold M of $\bar{M}(c)$ is symmetric.*

(b) *The null sectional curvature of a screen conformal totally umbilical half-lightlike submanifold M of $\bar{M}(c)$ vanishes identically.*

Example 5. Consider in $\mathbf{R}_1^4$ a surface M given by the equations

$$x_1 = x_3 \, , \ x_2 = (1 - x_4^2)^{\frac{1}{2}}.$$

Then we have

$$TM = \mathrm{Span}\{\xi = \partial x_1 + \partial x_3 \, , \ u = -x_4 \partial x_2 + x_2 \partial x_4\},$$
$$TM^\perp = \mathrm{Span}\{\xi = \partial x_1 + \partial x_3 \, , \ v = x_2 \partial x_2 + x_4 \partial x_4\}.$$

Thus, M is a half-lightlike submanifold of $\mathbf{R}_1^4$ with $\mathrm{Rad}\,TM = \mathrm{Span}\{\xi\}$ and

$$S(TM) = \mathrm{Span}\{u\} \, , \quad S(TM^\perp) = \mathrm{Span}\{v\},$$
$$\mathrm{ltr}(TM) = \mathrm{Span}\{N = \frac{1}{2}(-\partial x_1 + \partial x_3)\}.$$

Hence, we obtain $A_\xi^* u = A_N u = 0$. Thus M is a trivial screen conformal half-lightlike submanifold. On the other, by direct calculations, we derive

$$D_1 = 0 \, , \quad D_2(\xi, X) = 0, \quad \forall X \in \Gamma(TM), \quad D_2(u, u) = -\, g(u, u).$$

Thus M is a screen conformal totally umbilical half-lightlike submanifold. More-over, $D_1 = 0$ implies that ∇ is a metric connection.

Remark 4.4.19. Active research on half-lightlike submanifolds is in progress. See some recent papers [247, 248, 249].

Chapter 5

Lightlike submanifolds

The objective of this chapter is to present an up-to-date account of the works published on the general theory of lightlike submanifolds of semi-Riemannian manifolds. This includes unique existence theorems for screen distributions, geometry of totally umbilical, minimal and warped product lightlike submanifolds.

5.1   The induced geometric objects

Let $(\bar{M}, \bar{g})$ be a real $(m+n)$-dimensional semi-Riemannian manifold, where $m > 1$, $n \geq 1$ with $\bar{g}$ a semi-Riemannian metric on $\bar{M}$ of constant index $q \in \{1, \ldots, m + n - 1\}$. Hence $\bar{M}$ is never a Riemannian manifold. Suppose M is an m-dimensional submanifold of $\bar{M}$. The condition $m > 1$ implies that M is not a curve of $\bar{M}$. For $p \in M$, we now consider

$$T_p M^{\perp} = \{V_p \in T_p\bar{M} ; \quad \bar{g}_p(V_p, W_p) = 0, \quad \text{for all} \quad W_p \in T_p M\} .$$

For a lightlike M there exists a smooth distribution such that

$$\operatorname{Rad} T_p M = T_p M \cap T_p M^{\perp} \neq \{0\}, \forall p \in M.$$

If the rank of $\operatorname{Rad} TM$ is $r\,(> 0)$, then, M is called an r-*lightlike submanifold* [149]. Following are four sub-cases with respect to the dimension and codimension of M and rank of $\operatorname{Rad} TM$:

(A) *r-lightlike submanifold, $0 < r < \min\{m, n\}$.*

(B) *Coisotropic submanifold, $1 < r = n < m$.*

(C) *Isotropic submanifold, $1 < r = m < n$.*

(D) *Totally lightlike submanifold, $1 < r = m = n$.*

We now give details for these classes of lightlike submanifolds.

Case (A) $(0 < r < \min\{m, n\})$. Consider a complementary distribution $S(TM)$ of $\operatorname{Rad} TM$ in TM which is called a *screen distribution*. As M is supposed to be paracompact, such a distribution always exists on M. Clearly, $S(TM)$ is orthogonal to $\operatorname{Rad} TM$ and non-degenerate with respect to $\bar{g}$. Besides, we suppose $S(TM)$ is of a constant index on M, i.e., $\bar{g}_p$ has the same index on the fiber $S(TM)_p$, for any $p \in M$. In this way $S(TM)$ has a causal character. Thus we have the direct orthogonal sum

$$TM = \operatorname{Rad} TM \oplus_{\text{orth}} S(TM). \tag{5.1.1}$$

Certainly, $S(TM)$ is not unique, however it is canonically isomorphic to the factor vector bundle $TM^* = TM/\operatorname{Rad} TM$ [273]. Also, see the next section on existence of unique screens. Consider the vector bundle

$$TM^\perp = \cup_{p \in M} T_p M^\perp .$$

Notice that, for the lightlike M, $TM^\perp$ is not complementary to TM in $T\bar{M}_{|M}$ since $\operatorname{Rad} TM = TM \cap TM^\perp$ is now a distribution on M of rank $r > 0$. Consider a complementary vector bundle $S(TM^\perp)$ of $\operatorname{Rad} TM$ in $TM^\perp$. It follows that $S(TM^\perp)$ is also non-degenerate with respect to $\bar{g}$ and $TM^\perp$ has the orthogonal direct decomposition

$$TM^\perp = \operatorname{Rad} TM \oplus_{\text{orth}} S(TM^\perp).$$

The vector sub-bundle $S(TM^\perp)$ is called a *screen transversal vector bundle* of M. As $S(TM)$ is a non-degenerate vector sub-bundle of $T\bar{M}_{|M}$, we have the decomposition

$$T\bar{M}_{|M} = S(TM) \perp S(TM)^\perp,$$

where $S(TM)^\perp$ is the complementary orthogonal vector bundle of $S(TM)$ in $T\bar{M}_{|M}$. Note that $S(TM^\perp)$ is a vector sub-bundle of $S(TM)^\perp$ and since both are non-degenerate we have the orthogonal direct decomposition

$$S(TM)^\perp = S(TM^\perp) \perp S(TM^\perp)^\perp.$$

Since the theory of lightlike submanifold M is mainly based on both $S(TM)$ and $S(TM^\perp)$, a lightlike submanifold is denoted by $\left(M, g, S(TM), S(TM^\perp)\right)$.
    We use the following range for indices used in this section:

$$i, j, k, \ldots \in \{1, \ldots, r\}\,;\, a, b, c, \ldots \in \{r+1, \ldots, m\}\,;\, \alpha, \beta, \gamma, \ldots \in \{r+1, \ldots, n\}.$$

It is known that for non-degenerate submanifolds, the normal bundle is a complementary orthogonal bundle to the tangent bundle of a submanifold. Contrary to this, we have seen from the above that the normal bundle $TM^\perp$ is orthogonal to but not a complement to TM, since it intersects the null tangent bundle $\operatorname{Rad} TM$. This creates a problem as a vector of $T_x M$ cannot be decomposed uniquely into a component tangent to $T_x M$ and a component of $T_x M^\perp$. Therefore, the standard text-book definition of second fundamental forms and the Gauss-Weingarten formulas do not work, in the usual way, for lightlike submanifolds.

To deal with this problem, we use the approach in the 1996 book [149] of a geometric technique by splitting the tangent bundle $T\bar{M}$ into four non-intersecting complementary (but not orthogonal) vector bundles (two null and two non-null). For the benefit of readers we start with proofs of some results taken from [149] which are needed to understand new results of this book.

Theorem 5.1.1. [149] *Let* $\big(M, g, S(TM), S(TM^\perp)\big)$ *be an r-lightlike submanifold of* $(\bar{M}, \bar{g})$ *with* $r > 1$. *Suppose* $\mathcal{U}$ *is a coordinate neighborhood of* M *and* $\{\xi_i\}$, $i \in \{1, \ldots, r\}$ *is a basis of* $\Gamma(\mathrm{Rad}\, TM_{|\mathcal{U}})$. *Then there exist smooth sections* $\{N_i\}$ *of* $S(TM^\perp)^\perp_{|\mathcal{U}}$ *such that*

$$\bar{g}(N_i, \xi_j) = \delta_{ij} \tag{5.1.2}$$

and

$$\bar{g}(N_i, N_j) = 0, \quad \forall i, j \in \{1, \ldots, r\}. \tag{5.1.3}$$

Proof. Consider a complementary vector bundle F of $\mathrm{Rad}\, TM$ in $S(TM^\perp)^\perp$ and choose a basis $\{V_i\}$, $i \in \{1, \ldots, r\}$ of $\Gamma(F_{|\mathcal{U}})$. Thus the sections we are looking for are expressed as

$$N_i = \sum_{k=1}^{r} \{A_{ik}\, \xi_k + B_{ik}\, V_k\},$$

where A_{ik} and B_{ik} are smooth functions on $\mathcal{U}$. Then $\{N_i\}$ satisfy (5.1.2) if and only if

$$\sum_{k=1}^{r} B_{ik}\, \bar{g}_{jk} = \delta_{ij},$$

where $\bar{g}_{jk} = \bar{g}(\xi_j, V_k)$, $j, k \in \{1, \ldots, r\}$. Observe that $G = det\, [\bar{g}_{jk}]$ is everywhere non-zero on $\mathcal{U}$, otherwise $S(TM^\perp)^\perp$ would be degenerate at least at a point of $\mathcal{U}$. It follows that the above system has the unique solution

$$B_{ik} = \frac{(\bar{g}_{ik})'}{G},$$

where $(\bar{g}_{ik})'$ is the cofactor of the element $\bar{g}_{ik}$ in G. Finally, we see that (5.1.3) is equivalent with

$$A_{ij} + A_{ji} + \sum_{k,h=1}^{r} \{B_{ik} B_{jh}\, \bar{g}\,(V_k, V_h)\} = 0,$$

which proves the existence of A_{ij}. $\qquad\square$

Let $\mathrm{tr}(TM)$ and $\mathrm{ltr}(TM)$ be complementary (but not orthogonal) vector bundles to TM in $T\bar{M}_{|M}$ and to $\mathrm{Rad}\, TM$ in $\mathrm{tr}(TM)$ respectively. Then, we obtain

$$\text{tr}(TM) = \text{ltr}(TM) \oplus_{\text{orth}} S(TM^{\perp}), \tag{5.1.4}$$

$$T\bar{M}|_M = TM \oplus \text{tr}(TM)$$

$$= S(TM) \perp S(TM^{\perp}) \perp (\text{Rad}\, TM \oplus \text{ltr}(TM)). \tag{5.1.5}$$

Consider the following local quasi-orthonormal frame of $\bar{M}$ along M:

$$\{\xi_1, \ldots, \xi_r, N_1, \ldots, N_r, X_{r+1}, \ldots, X_m, W_{r+1}, \ldots, W_n\} \tag{5.1.6}$$

where $\{\xi_1, \ldots, \xi_r\}$ is a lightlike basis of $\Gamma\,(\text{Rad}\, TM)$, $\{N_1, \ldots, N_r\}$ a lightlike basis of $\Gamma\,(\text{ltr}(TM))$, $\{X_{r+1}, \ldots, X_m\}$ and $\{W_{r+1}, \ldots, W_n\}$ orthonormal basis of $\Gamma\,(S(TM)|\mathcal{U})$ and $\Gamma\,(S(TM^{\perp})|\mathcal{U})$ respectively.

Example 1. [149] Consider a surface (M, g) in $\mathbf{R}_2^4$ given by the equations

$$x^3 = \frac{1}{\sqrt{2}}(x^1 + x^2); \quad x^4 = \frac{1}{2}\log(1 + (x^1 - x^2)^2\,),$$

where $(x^1, \ldots, x^4)$ is a local coordinate system for $\mathbf{R}_2^4$. Using a simple procedure of linear algebra, we choose a set of vectors $\{U, V, \xi, W\}$ given by

$$U = \sqrt{2}(1 + (x^1 - x^2)^2)\,\partial_1 + (1 + (x^1 - x^2)^2)\,\partial_3 + \sqrt{2}(x^1 - x^2)\,\partial_4,$$

$$V = \sqrt{2}(1 + (x^1 - x^2)^2)\,\partial_2 + (1 + (x^1 - x^2)^2)\,\partial_3 - \sqrt{2}(x^1 - x^2)\,\partial_4,$$

$$\xi = \partial_1 + \partial_2 + \sqrt{2}\,\partial_3,$$

$$W = 2(x^2 - x^1)\,\partial_2 + \sqrt{2}(x^2 - x^1)\,\partial_3 + (1 + (x^1 - x^2)^2)\,\partial_4,$$

so that TM and $TM^{\perp}$ are spanned by $\{U, V\}$ and $\{\xi, W\}$ respectively. By direct calculations it follows that $\text{Rad}\, TM$ is a distribution on M of rank 1 and spanned by the lightlike vector ξ. Choose $S(TM)$ and $S(TM^{\perp})$ spanned by the timelike vector V and the spacelike vector W, respectively. Then,

$$\text{ltr}(TM) = \text{Span}\{N = -\frac{1}{2}\partial_1 + \frac{1}{2}\partial_2 + \frac{1}{\sqrt{2}}\partial_3\},$$

$$\text{tr}(TM) = \text{Span}\{N, W\},$$

where N is a lightlike vector such that $g(N, \xi) = 1$. Thus, M is a 1-lightlike submanifold of Case A, with basis $\{\xi, N, V, W\}$ of $\mathbf{R}_2^4$ along M.

For Case **B**, we have $\text{Rad}\, TM = TM^{\perp}$. Therefore, $S(TM^{\perp}) = \{0\}$ and from (5.1.4) $\text{tr}(TM) = \text{ltr}(TM)$. Thus, (5.1.5) and (5.1.6) reduce to

$$T\bar{M}|_M = TM \oplus \text{tr}(TM) = (TM^{\perp} \oplus \text{ltr}(TM)) \perp S(TM), \tag{5.1.7}$$

$$\{\xi_1, \ldots, \xi_r, N_1, \ldots, N_r, X_{r+1}, \ldots, X_m\}. \tag{5.1.8}$$

Example 2. Consider the unit pseudo-sphere $\mathbf{S}_1^3$ of Minkowski space $\mathbf{R}_1^4$ given by the equation $-t^2 + x^2 + y^2 + z^2 = 1$. Cut $\mathbf{S}_1^3$ by the hypersurface $t - x = 0$

and obtain a lightlike surface (M, g) of S_1^3 with $\operatorname{Rad} TM$ spanned by a lightlike vector $\xi = \partial_t + \partial_x$. Clearly, $\operatorname{Rad} TM = TM^\perp$ and, therefore, this example belongs to Case **B**. Consider a screen distribution $S(TM)$ spanned by a spacelike vector $X = z\,\partial_y - y\,\partial_z$. Then, we obtain a lightlike transversal vector bundle $\operatorname{tr}(TM) = \operatorname{ltr}(TM)$ spanned by $N = -\frac{1}{2}\{(1+t^2)\,\partial_t + (t^2-1)\,\partial_x + 2ty\,\partial_y + 2tz\,\partial_z\}$ such that $g(N, \xi) = 1$, with a basis $\{\xi, N, X\}$ for $\mathbf{S}_1^3$ along M.

For Case **C**, we have $\operatorname{Rad} TM = TM$. Therefore, $S(TM) = \{0\}$. Therefore, (5.1.5) and (5.1.6) reduce to

$$T\bar{M}|_M = TM \oplus \operatorname{tr}(TM) = (TM \oplus \operatorname{ltr}(TM)) \perp S(TM^\perp), \tag{5.1.9}$$

$$\{\xi_1, \ldots, \xi_r, N_1, \ldots, N_r, W_{r+1}, \ldots, W_n\}. \tag{5.1.10}$$

Example 3. Suppose (M, g) is a surface of $\mathbf{R}_2^5$ given by equations

$$x^3 = \cos x^1, \qquad x^4 = \sin x^1, \qquad x^5 = x^2.$$

We choose a set of vectors $\{\xi_1, \xi_2, U_1, U_2\}$ given by

$$\xi_1 = \partial_2 + \partial_5, \quad \xi_2 = \partial_1 - \sin x^1\,\partial_3 + \cos x^1\,\partial_4,$$
$$U_1 = -\sin x^1\,\partial_1 + \partial_3, \quad U_2 = \cos x^1\,\partial_1 + \partial_4,$$

so that $\operatorname{Rad} TM = TM = \operatorname{Span}\{\xi_1, \xi_2\}$, $TM^\perp = \operatorname{Span}\{\xi_1, U_1, U_2\}$. Therefore, M belongs to Case C. Construct two null vectors

$$N_1 = \frac{1}{2}\{-\partial_2 + \partial_5\},$$

$$N_2 = \frac{1}{2}\{-\partial_1 - \sin x^1\,\partial_3 + \cos x^1\,\partial_4\},$$

such that $g(N_i, \xi_j) = \delta_{ij}$ for $i, j \in \{1, 2\}$ and $\operatorname{ltr}(TM) = \operatorname{Span}\{N_1, N_2\}$. Let $W = \cos x^1\,\partial_3 + \sin x^1\,\partial_4$ be a spacelike vector such that $S(TM^\perp) = \operatorname{Span}\{W\}$. Thus, $\{\xi_1, \xi_2, N_1, N_2, W\}$ is a basis of R_2^5 along M.

For Case **D**, $\operatorname{Rad} TM = TM = TM^\perp$, $S(TM) = S(TM^\perp) = \{0\}$. Therefore, (5.1.5) and (5.1.6) reduce to

$$T\bar{M}|_M = TM \oplus \operatorname{ltr}(TM), \tag{5.1.11}$$

$$\{\xi_1, \ldots, \xi_r, N_1, \ldots, N_r\}. \tag{5.1.12}$$

Example 4. Suppose (M, g) is a surface of $\mathbf{R}_2^4$ given by the equations

$$x^3 = \frac{1}{\sqrt{2}}(x^1 + x^2), \qquad x^4 = \frac{1}{\sqrt{2}}(x^1 - x^2).$$

We choose a set of vectors $\{\xi_1, \xi_2, U, V\}$ given by

$$\xi_1 = \partial_1 + \frac{1}{\sqrt{2}}\,\partial_3 + \frac{1}{\sqrt{2}}\,\partial_4, \quad \xi_2 = \partial_2 + \frac{1}{\sqrt{2}}\,\partial_3 - \frac{1}{\sqrt{2}}\,\partial_4,$$

$$U = \partial_1 + \partial_2 + \sqrt{2}\,\partial_3, \quad V = \partial_1 - \partial_2 + \sqrt{2}\,\partial_4,$$

so that TM and $TM^\perp$ are spanned by $\{\xi_1, \xi_2\}$ and $\{U, V\}$ respectively. By direct calculations we check that $\mathrm{Span}\{\xi_1, \xi_2\} = \mathrm{Span}\{U, V\}$, that is, $TM = TM^\perp$. Finally, the two lightlike transversal vector fields are:

$$N_1 = \partial_1 + \sqrt{2}\,\partial_3 + \sqrt{2}\,\partial_4, \quad N_2 = \partial_2 + \sqrt{2}\,\partial_3 - \sqrt{2}\,\partial_4,$$

such that $g(N_i, \xi_j) = \delta_{ij},\ i, j = 1, 2$. Thus, M is of Case D, with a basis $\{\xi_1, \xi_2, N_1, N_2\}$ of R_2^4 along M.

From now on, we denote an m-dimensional lightlike submanifold simply by M instead of $(M, g, S(TM), S(TM^\perp))$ and $(m + n)$-dimensional semi-Riemannian manifold by $\bar{M}$. Let $\bar{\nabla}$ be the Levi-Civita connection on $\bar{M}$. As TM and $\mathrm{tr}(TM)$ are complementary sub-bundles of $T\bar{M}_{|M}$ we set

$$\bar{\nabla}_X Y = \nabla_X Y + h(X, Y), \quad \forall X, Y \in \Gamma(TM), \tag{5.1.13}$$

$$\bar{\nabla}_X V = -A(V, X) + \nabla_X^t V, \quad \forall V \in \Gamma(\mathrm{tr}(TM)), \tag{5.1.14}$$

where $\{\nabla_X Y, A(V, X)\}$, $\{h(X, Y), \nabla_X^t V\}$ belong to $\Gamma(TM)$ and $\Gamma(\mathrm{tr}(TM))$ respectively. It follows that ∇ and ∇^t are linear connections on M and on the vector bundle $\mathrm{tr}(TM)$ respectively. Besides, ∇ is a torsion-free linear connection. Also, h is a $\Gamma(\mathrm{tr}(TM))$-valued symmetric $\mathcal{F}(M)$-bilinear form on $\Gamma(TM)$. Finally, A is a $\Gamma(TM)$-valued $\mathcal{F}(M)$-bilinear form defined on $\Gamma(\mathrm{tr}(TM)) \times \Gamma(TM)$. We call ∇ and ∇^t the *induced linear connection* and the *transversal linear connection* on M respectively. Also h is called the *second fundamental form* of M with respect to $\mathrm{tr}(TM)$. For any $V \in \Gamma(\mathrm{tr}(TM))$ define the $\mathcal{F}(M)$-linear operator

$$A_V : \Gamma(TM) \longrightarrow \Gamma(TM), \quad A_V(X) = A(V, X), \quad \forall X \in \Gamma(TM),$$

and call it the *shape operator* of M with respect to V.

Suppose $S(TM^\perp) \neq \{0\}$, that is, M is either in Case (A) or in Case (C). Using the decomposition (5.1.5), consider the projection morphisms L and S of $\mathrm{tr}(TM)$ on $\mathrm{ltr}(TM)$ and $S(TM^\perp)$ respectively. Then (5.1.13) and (5.1.14) become

$$\bar{\nabla}_X Y = \nabla_X Y + h^l(X, Y) + h^s(X, Y), \tag{5.1.15}$$

$$\bar{\nabla}_X V = -A_V X + D_X^l V + D_X^s V, \tag{5.1.16}$$

where we put

$$h^l(X, Y) = L(h(X, Y)); \qquad h^s(X, Y) = S(h(X, Y)),$$
$$D_X^l V = L(\nabla_X^t V); \qquad D_X^s V = S(\nabla_X^t V).$$

As h^l and h^s are $\Gamma(\mathrm{ltr}(TM))$-valued and $\Gamma(S(TM^\perp))$-valued respectively, we call them the *lightlike second fundamental form* and the *screen second fundamental form* of M. Also note that D^l and D^s do not define linear connections on $\text{tr}(TM)$. In fact, they are Otsuki connections, see [149].

For any $X \in \Gamma(TM)$, by means of the above Otsuki connections define the following differential operators:

$$\nabla^l_X : \Gamma(\text{ltr}(TM)) \to \Gamma(\text{ltr}(TM)) \,;\, \nabla^l_X(LV) = D^l_X(LV), \tag{5.1.17}$$

$$\nabla^s_X : \Gamma(S(TM^\perp)) \to \Gamma(S(TM^\perp)) \,;\, \nabla^s_X(SV) = D^s_X(SV), \tag{5.1.18}$$

for any $V \in \Gamma(\text{tr}(TM))$. It is easy to check that both ∇^l and ∇^s are linear connections on $\text{ltr}(TM)$ and $S(TM^\perp)$ respectively. We call ∇^l and ∇^s the *lightlike connection* and the *screen transversal connection* on M, respectively. Besides, we define the following $\mathcal{F}(M)$-bilinear mappings:

$$D^l : \Gamma(TM) \times \Gamma(S(TM^\perp)) \longrightarrow \Gamma(\text{ltr}(TM)),$$
$$D^l(X, SV) = D^l_X(SV), \tag{5.1.19}$$
$$D^s : \Gamma(TM) \times \Gamma(\text{ltr}(TM)) \longrightarrow \Gamma(S(TM^\perp)),$$
$$D^s(X, LV) = D^s_X(LV), \quad \forall X \in \Gamma(TM) \tag{5.1.20}$$

and $V \in \Gamma(\text{tr}(TM))$. Due to (5.1.17)–(5.1.20), the equation (5.1.16) becomes

$$\bar{\nabla}_X V = -A_V X + \nabla^l_X(LV) + \nabla^s_X(SV) + D^l(X, SV) + D^s(X, LV). \tag{5.1.21}$$

In particular, from (5.1.21) we derive

$$\bar{\nabla}_X N = -A_N X + \nabla^l_X N + D^s(X, N), \tag{5.1.22}$$
$$\bar{\nabla}_X W = -A_W X + D^l(X, W) + \nabla^s_X W, \tag{5.1.23}$$

for any $X \in \Gamma(TM)$, $N \in \Gamma(\text{ltr}(TM))$ and $W \in \Gamma(S(TM^\perp))$.

Next, suppose M is coisotropic or totally lightlike. Since in these cases there is no screen vector bundle, (5.1.15) and (5.1.22) become

$$\bar{\nabla}_X Y = \nabla_X Y + h^l(X, Y), \tag{5.1.24}$$
$$\bar{\nabla}_X N = -A_N X + \nabla^l_X N, \quad \forall X, Y \in \Gamma(TM) \tag{5.1.25}$$

and $N \in \Gamma(\text{ltr}(TM))$. As in the case of non-degenerate submanifolds we call (5.1.13), (5.1.15), (5.1.24) the *Gauss formulae* and (5.1.14), (5.1.16), (5.1.22), (5.1.23), (5.1.25) the *Weingarten formulae* for the lightlike submanifold M.

Suppose M is either r-lightlike with $r < \min\{m, n\}$ or coisotropic. Then according to (5.1.1) and (5.1.9) we set

$$\nabla_X PY = \nabla^*_X PY + h^*(X, PY), \tag{5.1.26}$$
$$\nabla_X \xi = -A^*(\xi, X) + \nabla^{*t}_X \xi, \tag{5.1.27}$$

for any $X, Y \in \Gamma(TM)$ and $\xi \in \Gamma(\text{Rad}\,TM)$, where $\{\nabla^*_X PY, A^*(\xi, X)\}$ and $\{h^*(X, PY), \nabla^{*t}_X \xi\}$ belong to $\Gamma(S(TM))$ and $\Gamma(\text{Rad}(TM))$ respectively. It follows that ∇^* and ∇^{*t} are linear connections on $S(TM)$ and $\operatorname{Rad} TM$ respectively. On the other hand, h^* and A^* are $\Gamma(\operatorname{Rad} TM)$-valued and $\Gamma(S(TM))$-valued $\mathcal{F}(M)$-bilinear forms on $\Gamma(TM) \times \Gamma(S(TM))$ and $\Gamma(\operatorname{Rad} TM) \times \Gamma(TM)$ respectively. Call h^* and A^* the *second fundamental forms* of distributions $S(TM)$ and $\operatorname{Rad}(TM)$ respectively. For any $\xi \in \Gamma(\operatorname{Rad} TM)$ consider the $\mathcal{F}(M)$-linear operator

$$A_\xi^* : \Gamma(TM) \longrightarrow \Gamma(S(TM)); \quad A_\xi^* X = A^*(\xi,\, X), \quad \forall X \in \Gamma(TM),$$

and call it the *shape operator* of $S(TM)$ with respect to ξ. Also, call ∇^* and ∇^{*t} the *induced connections* on $S(TM)$ and $\operatorname{Rad} TM$ respectively. It is important to note that both ∇^* and ∇^{*t} are metric linear connections.

By using the above and $\bar\nabla$ a metric connection we obtain

$$\bar g(h^s(X,Y),W) + \bar g(Y, D^l(X,W)) = g(A_W X, Y), \tag{5.1.28}$$

$$\bar g(h^l(X,Y),\xi) + \bar g(Y, h^l(X,\xi)) + g(Y, \nabla_X \xi) = 0, \tag{5.1.29}$$

$$\bar g(h^*(X,PY),N) = \bar g(A_N X, PY), \tag{5.1.30}$$

$$\bar g(h^l(X,PY),\xi) = g(A_\xi^* X, PY), \tag{5.1.31}$$

$$\bar g(A_N X, PY) = \bar g(N, \bar\nabla_X PY), \tag{5.1.32}$$

$$g(h^l(X,\xi),\xi) = 0, \quad A_\xi^* \xi = 0, \tag{5.1.33}$$

$\forall X, Y \in \Gamma(TM)$, $\xi \in \Gamma(\operatorname{Rad} TM)$, $W \in \Gamma(S(TM^\perp))$, $N, N' \in \Gamma(\operatorname{ltr}(TM))$.

Next, consider a coordinate neighborhood $\mathcal{U}$ of M and let $\{N_i, W_\alpha\}$ be a basis of $\Gamma(\operatorname{tr}(TM)_{\mid M})$ where $N_i \in \Gamma(\operatorname{ltr}(TM)_{\mid M})$, $i \in \{1,\ldots,r\}$ and $W_\alpha \in \Gamma(S(TM^\perp)_{\mid \mathcal{U}})$, $\alpha \in \{r+1,\ldots,n\}$. Then (5.1.15) becomes

$$\bar\nabla_X Y = \nabla_X Y + \sum_{i=1}^{r} h_i^l(X,Y)N_i + \sum_{\alpha=r+1}^{n} h_\alpha^s(X,Y)W_\alpha \tag{5.1.34}$$

and

$$\bar\nabla_X Y = \nabla_X Y + \sum_{i=1}^{m<n} h_i^l(X,Y)N_i + \sum_{\alpha=m+1}^{n} h_\alpha^s(X,Y)W_\alpha \tag{5.1.35}$$

for an r-lightlike submanifold with $r < \min\{m,n\}$ and for an isotropic submanifold respectively. Similarly, (5.1.24) becomes

$$\bar\nabla_X Y = \nabla_X Y + \sum_{i=1}^{n<m} h_i^l(X,Y)N_i \tag{5.1.36}$$

$$\bar\nabla_X Y = \nabla_X Y + \sum_{i=1}^{n=m} h_i^l(X,Y)N_i \tag{5.1.37}$$

for a coisotropic submanifold and a totally lightlike submanifold respectively. We call $\{h_i^l\}$ and $\{h_\alpha^s\}$ the *local lightlike second fundamental forms* and the *local screen second fundamental forms* of M on $\mathcal{U}$.

Since the screen distributions are not unique, the following two results are very important for the entire study of lightlike submanifolds.

Theorem 5.1.2. [149] *The local lightlike second fundamental forms of a lightlike submanifold M do not depend on $S(TM)$, $S(TM^\perp)$ and $\mathrm{ltr}(TM)$.*

Proof. Consider the basis $\{\xi_i\}$, $i \in \{1, \ldots, r\}$ of $\Gamma(\mathrm{Rad}\, TM_{|\mathcal{U}})$ with respect to which we constructed the basis $\{N_i\}$ of $\Gamma(\mathrm{ltr}(TM))$, (see Theorem 5.1.1.) Then from (5.1.34)–(5.1.37) and taking into account (5.1.2), we get

$$h_i^l(X, Y) = \bar{g}(\bar{\nabla}_X Y, \xi_i), \quad \forall X, Y \in \Gamma(TM),$$

which proves our assertion. $\qquad\square$

Proposition 5.1.3. [41] *Let M be a lightlike submanifold of a semi-Riemannian manifold $\bar{M}$. Then, $h^l = 0$ on $\mathrm{Rad}(TM)$.*

Proof. Since $\bar{\nabla}$ is a metric connection, using the Koszul formula we have

$$\bar{g}(\bar{\nabla}_{\xi'}\xi'', K) = \xi'\, \bar{g}(\xi'', K) + \xi''\, \bar{g}(\xi', K) - K\bar{g}(\xi', \xi'') \qquad (5.1.38)$$
$$+ \bar{g}([K, \xi'], \xi'') - \bar{g}([\xi'', K], \xi') + \bar{g}([\xi', \xi''], K),$$

for any $\xi', \xi'' \in \Gamma(\mathrm{Rad}(TM))$ and $K \in \Gamma(T\bar{M}|_M)$. Suppose h^l is not identically zero on $\mathrm{Rad}(TM)$ and let $\xi^1, \xi^2 \in \Gamma(\mathrm{Rad}(TM))$ such that $h^l(\xi^1, \xi^2) \neq 0$. As the direct sum $\mathrm{Rad}(TM) \oplus \mathrm{ltr}(TM)$ is semi-Riemannian and $h^l(\xi^1, \xi^2)$ is a non-zero section of the lightlike vector bundle $\mathrm{ltr}(TM)$, there exists $\xi \in \Gamma(\mathrm{Rad}(TM))$ such that $\bar{g}(h^l(\xi^1, \xi^2), \xi) = 1$. If we substitute $K = \xi$, $\xi' = \xi^1$, $\xi'' = \xi^2$ in (5.1.38), then from (5.1.15) we obtain

$$\bar{g}(h^l(\xi^1, \xi^2), \xi) = \bar{g}(\bar{\nabla}_{\xi^1}\xi^2, \xi) = 0,$$

which is a contradiction. $\qquad\square$

From the geometry of Riemannian submanifolds [97] and non-degenerate submanifolds [317], it is known that the induced connection on a non-degenerate submanifold is a Levi-Civita connection. Unfortunately, in general, this is not true for a lightlike submanifold. Indeed, considering that $\bar{\nabla}$ is a metric connection and by using (5.1.15), (5.1.24) and (5.1.14), we obtain

$$(\nabla_X g)(Y, Z) = \bar{g}(h^l(X, Y), Z) + \bar{g}(h^l(X, Z), Y), \qquad (5.1.39)$$
$$(\nabla_X^t \bar{g})(V, V') = -\{\bar{g}(A_V X, V') + \bar{g}(A_{V'} X, V)\}, \qquad (5.1.40)$$

for any $X, Y, Z \in \Gamma(TM)$ and $V, V' \in \Gamma(\mathrm{tr}(TM))$. Thus, it follows that the induced connection ∇ is not a Levi-Civita connection. From (5.1.39), it also follows that the induced connection on an r-lightlike submanifold with $r < \min\{m, n\}$ or a coisotropic submanifold of $(\bar{M}, \bar{g})$ is a metric connection if and only if h^l vanishes identically on M. However, the induced connection ∇ on an isotropic submanifold and on a totally lightlike submanifold is a Levi-Civita connection. We now quote the following characterization theorem on the existence of an induced Levi-Civita connection on M.

Theorem 5.1.4. [149] *Let M be an r-lightlike submanifold with $r < \min\{m, n\}$ or a coisotropic submanifold of $\bar{M}$. Then the induced linear connection ∇ on M is a metric connection if and only if one of the following conditions is fulfilled:*

(i) *A_ξ^* vanish on $\Gamma(TM)$ for any $\xi \in \Gamma(\operatorname{Rad} TM)$.*

(ii) *$\operatorname{Rad} TM$ is a Killing distribution.*

(iii) *$\operatorname{Rad} TM$ is a parallel distribution with respect to ∇.*

Proof. From Proposition 5.1.3 and (5.1.39), by using (5.1.31) we obtain that ∇ is a metric connection if and only if (i) is satisfied. Next, the equivalence of (i) and (iii) follows from (5.1.27). Finally, by using (1.2.23) of Chapter 1 and (5.1.27) we conclude that $\operatorname{Rad} TM$ is a Killing distribution if and only if

$$g(A_\xi^* X, PY) + g(A_\xi^* Y, PX) = 0, \quad \forall X, Y \in \Gamma(TM). \tag{5.1.41}$$

Thus clearly (i) implies (ii). Conversely, suppose (5.1.41) is satisfied. Then replace X by $\xi' \in \Gamma(\operatorname{Rad} TM)$ and obtain $A_\xi^* \xi' = 0$. Replace X and Y by PX and PY respectively and taking into account that A_ξ^* is a self-adjoint operator, obtain $A_\xi^* PX = 0$. Hence (ii) implies (i), which completes the proof. $\qquad \square$

In general, $S(TM)$ is not necessarily integrable. The following result gives equivalent conditions for the integrability of a $S(TM)$:

Theorem 5.1.5. [149] *Let M be an r-lightlike submanifold with $r < \min\{m, n\}$ or a coisotropic submanifold of $\bar{M}$. Then the following assertions are equivalent:*

(i) *$S(TM)$ is integrable.*

(ii) *h^* is symmetric on $\Gamma(S(TM))$.*

(iii) *A_N is self-adjoint on $\Gamma(S(TM))$ with respect to g.*

Proof. First, note that $S(TM)$ is integrable if and only if, locally on each $U \subset M$ we have $\bar{g}([X, Y], N) = 0$ for $X, Y \in \Gamma(TM)$ and $N \in \Gamma(\operatorname{ltr}(TM))$. By using (5.1.15) and (5.1.26) we obtain

$$\bar{g}([X, Y], N) = g(h^*(X, Y) - h^*(Y, X), N),$$

which implies the equivalence of (i) and (ii). The equivalence of (ii) and (iii) follows from (5.1.30), which completes the proof. $\qquad \square$

Also, from (5.1.26) and (5.1.30) we obtain

Theorem 5.1.6. [149] *Let M be an r-lightlike submanifold with $r < \min\{m, n\}$ or a coisotropic submanifold of $\bar{M}$. Then the following assertions are equivalent:*

(i) *$S(TM)$ is a parallel distribution with respect to ∇.*

(ii) *h^* vanishes identically on M.*

(iii) A_N *is* $\Gamma(\text{Rad}\,TM)$*-valued operator.*

For integrability of $\text{Rad}(TM)$, we have the following.

Theorem 5.1.7. [149] *Let M be an r-lightlike submanifold with $r < \min\{m, n\}$ or a coisotropic submanifold of $\bar{M}$. Then the following assertions are equivalent:*

(i) $\text{Rad}\,TM$ *is integrable.*

(ii) *The lightlike second fundamental form of M satisfies*

$$h^l(PX, \xi) = 0, \quad \forall \xi \in \Gamma(\text{Rad}\,TM), \, X \in \Gamma(TM).$$

(iii) *For any $\xi \in \Gamma(\text{Rad}\,TM)$ the shape operator A_ξ^* of $S(TM)$ vanishes identically on $\Gamma(\text{Rad}\,TM)$.*

Proof. Taking into account that $\bar{\nabla}$ is both metric and torsion-free and by using (5.1.15) we obtain

$$g\left([\xi, \xi'], PX\right) = \bar{g}\left(\xi, h^l(\xi', PX)\right) - \bar{g}\left(\xi', h^l(\xi, PX)\right) \tag{5.1.42}$$

for any $\xi, \xi' \in \Gamma(\text{Rad}\,TM)$ and $X \in \Gamma(TM)$. On the other hand, replace Y by ξ' in (5.1.29) and derive

$$\bar{g}\left(h^l(\xi, PX), \xi'\right) + \bar{g}\left(h^l(\xi', PX), \xi\right) = 0. \tag{5.1.43}$$

Thus the equivalence of (i) and (ii) follows from (5.1.42) and (5.1.43). As a consequence of (5.1.31) we deduce the equivalence of (ii) and (iii), since A_ξ^* is a $\Gamma(S(TM))$-valued linear operator. $\qquad\square$

5.2 Unique screen distributions

Coisotropoic submanifolds. We have seen in the previous two chapters that there do exist large classes of lightlike hypersurfaces and half-lightlike submanifolds which admit unique screen distribution. A next step in this direction is to determine whether the same is true for coisotropic submanifolds. The objective of this section is to present an affirmative answer to this important question.

Let $(M, g, S(TM))$ be an m-dimensional coisotropic submanifold of a semi-Riemannian manifold $(\bar{M}, \bar{g})$ of codimension n. Then, $\text{Rad}\,TM = TM^\perp$ and $S(TM^\perp) = \{0\}$. Therefore, the frame (5.1.6) reduces to

$$\{\xi_1, \ldots, \xi_n, N_1, \ldots, N_n, X_{n+1}, \ldots, X_m\}.$$

We need the following three Gauss and Weingarten type equations:

$$\bar{\nabla}_X Y = \nabla_X Y + \sum_{i=1}^{n} h_i\,(X,\,Y)\,N_i,$$

$$\bar{\nabla}_X N_i = -A_{N_i}X + \sum_{j=1}^{n} \tau_{ij}\,(X)\,N_j, \qquad\qquad (5.2.1)$$

$$\nabla_X \xi_i = -A_{\xi_i}\,X - \sum_{j=1}^{n} \tau_{ij}(X)\,\xi_j, \quad \forall X\,,\,Y \in \Gamma\,(TM),$$

$\forall i = 1, \ldots, n$. Let a screen $S(TM)$ change to another screen $S(TM)'$, where

$$\{\xi_1, \ldots, \xi_n,\, N_1', \ldots, N_n',\, X_1', \ldots, X_{m-n}',\}$$

is another quasi-orthonormal frame for the same set of null sections $\{\xi_1, \ldots, \xi_n\}$. Following are the transformation equations due to this change:

$$X_a' = \sum_{b=1}^{m-n} A_a^b \left(X_b - \epsilon_b \sum_{i=1}^{n} \mathbf{f}_{ib}\,\xi_i \right), \qquad\qquad (5.2.2)$$

$$N_i' = N_i + \sum_{j=1}^{n} N_{ij}\xi_j + \sum_{a=1}^{m-n} \mathbf{f}_{ia} X_a, \qquad\qquad (5.2.3)$$

with the conditions

$$2N_{ii} = -\sum_{a=1}^{m-n} \epsilon_a(\mathbf{f}_{ia})^2, \quad N_{ij} + N_{ji} + \sum_{a=1}^{m-n} \epsilon_a \mathbf{f}_{ia}\mathbf{f}_{ja} = 0,\ \forall i \neq j, \qquad (5.2.4)$$

$$h_i'(X, Y) = h_i(X, Y), \quad \forall X\,,\,Y \in \Gamma\,(TM), \qquad\qquad (5.2.5)$$

$$\nabla_X' PY = \nabla_X PY - \sum_{j=1}^{n} \left(\sum_{i=1}^{n} h_i(X, PY)N_{ij} \right) \xi_j$$

$$- \sum_{a=1}^{m-n} \left(\sum_{i=1}^{n} h_i(X, PY)\mathbf{f}_{ia} \right) X_a. \qquad\qquad (5.2.6)$$

Lemma 5.2.1. *The second fundamental forms h^* and h'^* of the screen distributions $S(TM)$ and $S(TM)'$ respectively are related as follows:*

$$h_i'^*(X,\, PY) = h_i^*(X,\, PY) + \frac{1}{2}||\mathbf{Z}_i||^2 h_i(X, PY) + g(\nabla_X PY,\, \mathbf{Z}_i)$$

$$- \sum_{j \neq i} \{g(\mathbf{Z}_j, \mathbf{Z}_j) - N_{ij}\}\,h_j(X, PY) \qquad\qquad (5.2.7)$$

for a fixed i and j summed from 1 to n and each $\mathbf{Z}_i = \sum_{a=1}^{m-n} \mathbf{f}_{ia} X_a$ are n characteristic vector fields of the screen change.

Proof. Using (5.2.6) and then (5.2.3) we obtain

$$\bar{g}(\nabla'_X PY, N'_i) = \bar{g}(\nabla_X PY, N_i) + \bar{g}(\nabla_X PY, \sum_{a=1}^{m-n} \mathbf{f}_{ia} X_a)$$

$$- \sum_{j=1}^{n} \left(\sum_{i=1}^{n} h_i(X, PY) N_{ij} \right) \bar{g}(\xi_j, N_i)$$

$$- g(\sum_{a=1}^{m-n} \left(\sum_{i=1}^{n} h_i(X, PY) \mathbf{f}_{ia} \right) X_a, \sum_{a=1}^{m-n} \mathbf{f}_{ia} X_a).$$

Hence, we get

$$\bar{g}(\nabla'_X PY, N'_i) = \bar{g}(\nabla_X PY, N_i) + \bar{g}(\nabla_X PY, \mathbf{Z}_i)$$

$$- h^l_i(X, PY)(N_{ii} + \sum_{a=1}^{m-n} \mathbf{f}^2_{ia})$$

$$- \sum_{i \neq j}^{n} h^l_j(X, Y)[N_{ji} + \sum_{a=1}^{m-n} \mathbf{f}_{ia}\mathbf{f}_{ja}].$$

Thus, from (5.2.5) we get

$$\bar{g}(\nabla'_X PY, N'_i) = \bar{g}(\nabla_X PY, N_i) + \bar{g}(\nabla_X PY, \mathbf{Z}_i)$$

$$- h^l_i(X, PY)(-\frac{1}{2} \| \mathbf{Z}i \|^2 + \sum_{a=1}^{m-n} \mathbf{f}^2_{ia})$$

$$- \sum_{i \neq j}^{n} h^l_2(X, Y)N_{ij}.$$

Finally, using (5.1.26) we get (5.2.7), which completes the proof. □

Let ω_i be the respective n dual 1-forms of $\mathbf{Z}_i$ given by

$$\omega_i(X) = g(X, \mathbf{Z}_i), \quad \forall X \in \Gamma(TM), \quad 1 \leq i \leq n. \tag{5.2.8}$$

Denote by $\mathcal{S}$ the first derivative of a screen distribution $S(TM)$ given by

$$\mathcal{S}(x) = \text{Span}\{[X, Y]_x, \quad X_x, Y_x \in S(TM), \quad x \in M\}. \tag{5.2.9}$$

If $S(TM)$ is integrable, then, $\mathcal{S}$ is a sub-bundle of $S(TM)$.

It is known that the second fundamental forms and their respective shape operators of a non-degenerate submanifold are related by means of the metric tensor. Contrary to this we see from equations (5.1.25) and (5.1.26) that there are interrelations between the lightlike and the second fundamental forms of the lightlike M and its screen distribution and their respective shape operators. This interrelation indicates that the lightlike geometry depends on a choice of screen distribution. While we know from equation (5.2.5) that the second fundamental forms of the lightlike M are independent of a screen, the same is not true for the fundamental forms of $S(TM)$ (see equation (5.2.6)), which is the root of non-uniqueness anomaly in the lightlike geometry. Since, in general, it is impossible to remove this anomaly, in two previous cases (i.e., hypersurfaces and half-lightlike submanifolds) we used the conditions (2.2.1) and (4.4.1). However, this condition can not be used for the case of general submanifolds for the following reason.

In a lightlike hypersurface and a half-lightlike submanifold, the equations (2.1.26) and (4.1.11) imply that A_N is $S(TM)$-valued. Thus, it is meaningful to use the conditions (2.2.1) and (4.4.1). However, for a general lightlike submanifold, there is no guarantee that A_N is $S(TM)$-valued. Thus, we can not use shape operators to define the notion 'screen conformal' for the general case.

For the above reason we use a new geometric condition as follows: Consider a class of coisotropic submanifolds M such that the respective fundamental forms of M and the screen distribution $S(TM)$ are related by conformal smooth functions in $\mathcal{F}(M)$. The motivation for this geometric restriction comes from the classical geometry of non-degenerate submanifolds for which there are only one type of fundamental forms with their one type of respective shape operators. Thus, we make the following definition.

Definition 5.2.2. [145]. A coisotropic submanifold $(M, g, S(TM))$ of a semi-Riemannian manifold $(\bar{M}, \bar{g})$ is called a screen locally conformal submanifold if the fundamental forms h_i^* of $S(TM)$ are conformally related to the corresponding lightlike fundamental forms h_i of M by

$$h_i^*(X, PY) = \varphi_i h_i(X, Y), \quad \forall X, Y, \Gamma(TM|_{\mathcal{U}}), \quad i \in \{1, \dots, r\}, \qquad (5.2.10)$$

where $\varphi_i's$ are smooth functions on a neighborhood $\mathcal{U}$ in M.

In order to avoid trivial ambiguities, we will consider $\mathcal{U}$ to be connected and maximal in the sense that there is no larger domain $\mathcal{U}' \supset \mathcal{U}$ on which (5.2.10) holds. In case $\mathcal{U} = M$ the screen conformality is said to be global.

Theorem 5.2.3. [145] *Let* $(M, g, S(TM))$ *be an m-dimensional coisotropic screen conformal submanifold of a semi-Riemannian manifold* $\bar{M}^{m+n}$. *Then:*

(a) *Any choice of a screen distribution of M satisfying (5.2.10) is integrable.*

(b) *All the n-forms ω_i in (5.2.8) vanish identically on the first derivative $\mathcal{S}$ given by (5.2.9).*

(c) *If $\mathcal{S} = S(TM)$, then, there exists a set of n null sections $\{\xi_1, \dots, \xi_n\}$ of $\Gamma(\mathrm{Rad}\,TM)$ with respect to which $S(TM)$ is a unique screen distribution of M, up to an orthogonal transformation with a unique set $\{N_1, \dots, N_n\}$ of lightlike transversal vector bundles and the screen fundamental forms h_i^* are independent of a screen distribution.*

Proof. Substituting (5.2.10) in (5.1.25) and then using (5.1.26) we get

$$g(A_{N_i}X, PY) = \varphi_i g(A_{\xi_i}X, PY) \quad \forall X \in \Gamma(TM_{|\mathcal{U}}). \tag{5.2.11}$$

Since each A_{ξ_i} is symmetric with respect to g, equation (5.2.11) implies that each A_{N_i} is self-adjoint on $\Gamma(S(TM))$ with respect to g, which further follows from Theorem 5.1.5 that any choice of a screen distribution of M, satisfying (5.2.10) is integrable. Thus, (a) holds. Choose an integrable screen $S(TM)$. This means that S is a sub-bundle of $S(TM)$. Using (5.2.10) in (5.2.7) and $h_i' = h_i$ we obtain

$$g(\nabla_X PY, \mathbf{Z}_i) = \frac{1}{2}||\mathbf{Z}_i||^2 h_i(X, PY)$$
$$+ \sum_{j \neq i} \{g(\mathbf{Z}_j, \mathbf{Z}_j) - N_{ji}\} h_j(X, PY) \tag{5.2.12}$$

$\forall X, Y \in \Gamma(TM_{|\mathcal{U}})$ and for each fixed i. Since the right-hand side of (5.2.12) is symmetric in X and Y, we have $g([X, Y], \mathbf{Z}_i) = w_i([X, Y]) = 0, \forall X, Y \in \Gamma(S(TM)_{|\mathcal{U}})$, that is, w_i vanishes on $\mathcal{S}$. Similarly, repeating above steps n-times for each i we claim that each w_i vanishes on $\mathcal{S}$ which proves (b). If we take $S = S(TM)$, then, each w_i vanish on this choice of $S(TM)$ which implies that all the n characteristic vector fields $\mathbf{Z}_i$ vanish. Therefore, all the functions $\mathbf{f}_{ia}$ vanish. Finally, substituting this data in (5.2.4) and (5.2.6) it is easy to see that all the functions N_{ij} also vanish. Thus, the transformation equations (5.2.2), (5.2.3) and (5.2.6) become $X_a' = \sum_{b=1}^{m-n} A_a^b X_b$ $(1 \leq a \leq m - n)$, $N_i' = N_i$ and $h_i'^* = h_i^*$ where (A_a^b) is an orthogonal matrix of $S(TM)$ at any point $x \in M$. Therefore, $S(TM)$ is a unique screen up to an orthogonal transformation with unique transversal vector fields N_i and the screen fundamental forms h_i^* are independent of a screen distribution. This completes the proof. $\qquad\square$

Example 5. Consider the following example of a coisotropic submanifold of codimension two.
$$x_2 = (x_3^2 + x_5^2)^{\frac{1}{2}}, \ x_4 = x_1, \ x_3 > 0, \ x_5 > 0.$$

Then, we have

$$S(TM) = \text{Span}\{X = x_5 \, \partial \, x_2 + x_2 \, \partial \, x_5\},$$
$$\text{Rad}(TM) = \text{Span}\{\xi_1 = \partial \, x_1 + \partial \, x_4, \ \xi_2 = x_2 \, \partial \, x_2 + x_3 \, \partial \, x_3 + x_5 \, \partial \, x_5\},$$
$$\text{ltr}(TM) = \text{Span}\{N_1 = \frac{1}{2}(-\partial \, x_1 + \partial \, x_4),$$
$$N_2 = \frac{1}{2x_3^2}\{-x_2 \, \partial \, x_2 + x_3 \, \partial \, x_3 - x_5 \, \partial \, x_5\}\}.$$

Then, by direct calculations, we get

$$\bar{\nabla}_{\xi_1} X = 0, \bar{\nabla}_{\xi_2} X = X, \bar{\nabla}_{\xi_1} \xi_2 = 0,$$
$$\bar{\nabla}_X X = x_2 \, \partial \, x_2 + x_5 \, \partial \, x_5.$$

Then, using Gauss' formula, we obtain

$$\nabla_X X = \frac{1}{2}\xi_2, \ h_1^*(X, X) = 0, \quad h_1^*(\xi_1, X) = h_2^*(\xi_1, X) = 0,$$

$$h_1^*(\xi_2, X) = h_2^*(\xi_2, X) = h_2(\xi, X) = h_2(\xi_2, X) = 0, \quad h_1 = 0,$$

$$h_2(X, X) = -(x_3^2), \ h_2^*(X, X) = \frac{1}{2}.$$

Thus, M is screen conformal with φ_1 arbitrary and $\varphi_2 = -\frac{1}{2x_3^2}$.

Now one may ask whether there exists another class of coisotropic light-like submanifolds of semi-Riemannian manifolds of an arbitrary signature which admits an integrable unique screen distribution. We answer this question in the affirmative, subject to the following geometric condition (different from the screen conformal condition) on the embedding.

Consider a complementary vector bundle F of $\operatorname{Rad} TM$ in $S(TM)^\perp$ and choose a basis $\{V_i\}$, $i \in \{1, \ldots, n\}$ of $\Gamma(F_{|\mathcal{U}})$. Thus the sections we are looking for are expressed as

$$N_i = \sum_{k=1}^{n} \{A_{ik}\,\xi_k + B_{ik}\,V_k\}\,, \tag{5.2.13}$$

where A_{ik} and B_{ik} are smooth functions on $\mathcal{U}$. Then $\{N_i\}$ satisfy (5.1.2) if and only if $\sum_{k=1}^{n} B_{ik}\,\bar{g}_{jk} = \delta_{ij}$, where $\bar{g}_{jk} = \bar{g}(\xi_j, V_k)$, $j, k \in \{1, \ldots, n\}$. Observe that $G = \det[\bar{g}_{jk}]$ is everywhere non-zero on $\mathcal{U}$, otherwise $S(TM)^\perp$ would be degenerate at least at a point of $\mathcal{U}$. Assume that F is parallel along the tangent direction.

Theorem 5.2.4. [145]. *Let $(M, g, S(TM), F)$ be a coisotropic submanifold of a semi-Riemannian manifold $\bar{M}$ such that the complementary vector bundle F of $\operatorname{Rad} TM$ in $S(TM)^\perp$ is parallel along the tangent direction. Then all the assertions from (a) through (c) of Theorem 5.2.3 will hold.*

Proof. Taking the covariant derivative of N_i (given by (5.2.13)) with respect to $X \in \Gamma(TM)$, we get

$$\bar{\nabla}_X N_i = \sum_{k=1}^{n} \{X(A_{ik})\xi_k + X(A_{ik})\bar{\nabla}_X \xi_k + X(B_{ik})V_k + B_{ik}\bar{\nabla}_X V_k\}.$$

Using the three equations of (5.2.1) we obtain

$$A_{N_i}X = \sum_{k=1}^{n} \left\{ (A_{ik}A_{\xi_k}^* X - A_{ik})\xi_k + \sum_{j=1}^{n} \tau_{kj}(X)\xi_j + X(B_{ik})V_k + B_{ik}\bar{\nabla}_X V_k \right\}$$

$$+ \sum_{j=1}^{n} \tau_{ij}(X)N_j.$$

Since F is parallel, $\bar{\nabla}_X V_k \in \Gamma(F)$. Thus for $Y \in \Gamma(S(TM))$, we get

$$g(A_{N_i}X, Y) = \sum_{k=1}^{n} A_{ik} g(A^*_{\xi_k}X, Y).$$

Then using (5.1.31) we get

$$\bar{g}(h^*(X, Y), N_i) = \sum_{k=1}^{r} A_{ik} \bar{g}(h^l(X, Y), \xi_k). \tag{5.2.14}$$

Since the right side of (5.2.14) is symmetric, it follows that each h^*_i is symmetric on $S(TM)$. This implies from Theorem 5.1.5 that any choice of $S(TM)$ of M, with parallel vector bundle F, is integrable. Thus, (a) holds. Choose an integrable screen $S(TM)$ so that $\mathcal{S}$ is a sub-bundle of $S(TM)$. Now using (5.2.14) in (5.2.7) and $h_i = h'_i$, we obtain

$$g(\nabla_X PY, \mathbf{Z}_i) = \sum_{i \neq j} h_j(X, PY)[N_{ji} + g(\mathbf{Z}_i, \mathbf{Z}_j)]$$

$$+ \frac{1}{2} h_i(X, PY) \| \mathbf{Z}_i \|^2, \tag{5.2.15}$$

Then the rest of the proof is similar to the proof of Theorem 5.2.3. $\qquad\square$

r-lightlike submanifolds. Let (M, g) be an r-lightlike submanifold of $(\bar{M}, \bar{g})$ $(m + n)$-dimensional semi-Riemannian manifold $(\bar{M}, \bar{g})$ of codimension n. Consider two quasi-orthonormal frames $\{\xi_i, N_i, X_a, W_\alpha\}$ and $\{\xi_i, N'_i, X'_a, W'_\alpha\}$ induced on U by $\{S(TM), S(TM^\perp), F\}$ and $\{S'(TM), S'(TM^\perp), F'\}$, respectively. Here F and F' are the complementary vector bundles of $\mathrm{Rad}\, TM$ in $S(TM^\perp)^\perp$ and $S'(TM^\perp)^\perp$, respectively. By direct calculations, using (5.1.5), (5.1.2) we obtain

$$X'_a = \sum_{b=r+1}^{m} \left\{ X^b_a(X_b - \epsilon_b \sum_{i=1}^{r} \mathbf{f}_{ib} \xi_i) \right\}, \tag{5.2.16}$$

$$W'_\alpha = \sum_{\beta=r+1}^{n} \left\{ W^\beta_\alpha(W_\beta - \epsilon_\beta \sum_{i=1}^{r} Q_{i\beta} \xi_i) \right\}, \tag{5.2.17}$$

$$N'_i = N_i + \sum_{j=1}^{r} N_{ij} \xi_j + \sum_{a=r+1}^{m} \mathbf{f}_{ia} X_a + \sum_{\alpha=r+1}^{n} Q_{i\alpha} W_\alpha, \tag{5.2.18}$$

where $\{\epsilon_a\}$ and $\{\epsilon_\alpha\}$ are signatures of basis $\{X_a\}$ and $\{W_\alpha\}$ respectively, X^b_a, W^β_α, N_{ij}, $\mathbf{f}_{ia}$, $Q_{i\alpha}$ are smooth functions on $\mathcal{U}$ such that $[X^b_a]$ and $[W^\beta_\alpha]$ are $(m-r) \times (m-r)$ and $(n-r) \times (n-r)$ semi-orthogonal matrices, and

$$N_{ij} + N_{ji} + \sum_{a=r+1}^{m} \epsilon_a \mathbf{f}_{ia} \mathbf{f}_{ja} + \sum_{\alpha=r+1}^{n} \epsilon_\alpha Q_{i\alpha} Q_{j\alpha} = 0. \tag{5.2.19}$$

Using (5.1.5), we have

$$
\nabla_X Y = \nabla'_X Y + \sum_{j=1}^{r} \left\{ \sum_{i=1}^{r} h_i^l(X,Y) N_{ij} - \sum_{\alpha,\beta=r+1}^{n} \epsilon_\beta h_\alpha'^s(X,Y) W_\alpha^\beta Q_{j\beta} \right\} \xi_j
$$

$$
+ \sum_{a=r+1}^{m} \left\{ \sum_{i=1}^{r} h_i^l(X,Y) \mathbf{f}_{ia} \right\} X_a , \tag{5.2.20}
$$

and

$$
h_\alpha^s(X,Y) = \sum_{i=1}^{r} h_i^l(X,Y) Q_{i\alpha} + \sum_{\beta=r+1}^{n} h_\beta'^s(X,Y) W_\beta^\alpha . \tag{5.2.21}
$$

Also note that, from Theorem 5.1.2, we get

$$
h_i^l(X,Y) = h'^l_i(X,Y), \forall X, Y \in \Gamma(TM). \tag{5.2.22}
$$

Lemma 5.2.5. *The second fundamental forms h^* and h'^* of the screen distributions $S(TM)$ and $S(TM)'$, respectively, in an r-lightlike submanifold M are related as follows:*

$$
h'^*_i(X, PY) = h_i^*(X, PY) + g(\nabla_X PY, \mathbf{Z}_i) - \sum_{i \neq j} h_j^l(X, PY)[N_{ji} + g(\mathbf{Z}_i, \mathbf{Z}_j)]
$$

$$
+ \frac{1}{2} h_i^l(X, PY)[\| \mathbf{W}_i \|^2 - \| \mathbf{Z}_i \|^2]
$$

$$
+ \sum_{\alpha=r+1}^{n} \varepsilon_\alpha h^s{}_\alpha(X, PY) Q_{i\alpha}, \tag{5.2.23}
$$

where $\mathbf{Z}_i = \sum_{a=r+1}^{m} \mathbf{f}_{ia} X_a$ and $\mathbf{W}_i = \sum_{\alpha=r+1}^{n} Q_{i\alpha} W_\alpha$.

Proof. From (5.2.20), we have

$$
\bar{g}(\nabla'_X PY, N'_i) = \bar{g}(\nabla_X PY, N'_i) - \sum_{j=1}^{r} \{ \sum_{i=1}^{r} h_i^l(X, PY) N_{ij}
$$

$$
+ \sum_{\alpha,\beta=r+1}^{n} \epsilon_\beta h_\alpha'^s(X, PY) W_\alpha^\beta Q_{j\beta} \} \bar{g}(\xi_j, N'_i)
$$

$$
- \sum_{a=r+1}^{m} \left\{ \sum_{i=1}^{r} h_i^l(X, PY) \mathbf{f}_{ia} \right\} \bar{g}(X_a, N'_i).
$$

Then, using (5.1.26) and (5.2.18), we obtain

$$
\begin{aligned}
h'^*_i(X, PY) = {} & h^*_i(X, PY) + g(\nabla_X PY, \mathbf{Z}_i) \\
& - h^l_i(X, PY)N_{ii} - \sum_{i \neq j} h^l_i(X, PY)N_{ji} \\
& + \sum_{\alpha, \beta} \varepsilon_\beta h'^s_\alpha(X, PY)W^\beta_\alpha Q_{i\beta} \\
& - \sum_j \sum_a h^l_j(X, PY)g(\mathbf{f}_{ia} X_a, P_{ja} X_a) \\
& - h^l_i(X, PY)g(\mathbf{Z}_i, \mathbf{Z}_i).
\end{aligned}
\tag{5.2.24}
$$

On the other hand, from (5.2.19), we have

$$
N_{ii} = -\frac{1}{2}\Big\{ \sum_{a=r+1}^{m} \varepsilon_a \, (\mathbf{f}_{ia})^2 + \sum_{\alpha=r+1}^{n} \varepsilon_\alpha \, (Q_{i\alpha})^2 \Big\}.
\tag{5.2.25}
$$

Then, using (5.2.25) in (5.2.24), we get

$$
\begin{aligned}
h'^*_i(X, PY) = {} & h^*_i(X, PY) + g(\nabla_X PY, \mathbf{Z}_i) \\
& + \frac{1}{2}h^l_i(X, PY)\{\| \mathbf{Z}_i \|^2 + \| \mathbf{W}_i \|^2\} \\
& - \sum_{i \neq j} h^l_i(X, PY)N_{ji} + \sum_{\alpha, \beta} \varepsilon_\beta h'^s_\alpha(X, PY)W^\beta_\alpha Q_{i\beta} \\
& - \sum_j \sum_a h^l_j(X, PY)g(\mathbf{Z}_i, \mathbf{Z}_j) - h^l_i(X, PY) \| \mathbf{Z}_i \|^2 \ .
\end{aligned}
$$

Finally, using (5.2.21), we obtain (5.2.23). $\qquad\square$

Let ω_i be the respective n dual 1-forms of $\mathbf{Z}_i$ given by

$$
\omega_i(X) = g(X, \mathbf{Z}_i), \quad \forall X \in \Gamma(TM),
\tag{5.2.26}
$$

where $1 \leq i \leq n$. At a point $x \in M$ let $\mathcal{N}^1(x)$ be the space spanned by all vectors $h^s(X, Y)$, $X_x, Y_x \in T_x M$, i.e.,

$$
\mathcal{N}^1(x) = \mathrm{Span}\{h^s(X, Y)_x \ ; \ X_x, Y_x \in T_x M\}.
$$

We call $\mathcal{N}^1$ a *first screen transversal space*.

Theorem 5.2.6. *Let $(M, g, S(TM))$ be an m-dimensional r-lightlike screen conformal submanifold of a semi-Riemannian manifold $\bar{M}^{m+n}$. Then:*

(a) *Any choice of a screen distribution of M, satisfying (5.2.10), is integrable.*

(b) *All the n-forms ω_i in (5.2.26) vanish identically on the first derivative S given by (5.2.9).*

(c) *If $\mathcal{S}$ and $\mathcal{N}^1$ coincide with $S(TM)$ and $S(TM^\perp)$, respectively, then, there exists a set of r null sections $\{\xi_1, \ldots, \xi_r\}$ of $\Gamma(\operatorname{Rad} TM)$ with respect to which $S(TM)$ is a unique screen distribution of M, up to an orthogonal transformation with a unique set $\{N_1, \ldots, N_r\}$ of lightlike transversal vector bundles and the screen fundamental forms h_i^* are independent of a screen distribution.*

Proof. (a) follows as in Theorem 5.2.3(a). To prove (b), take $\mathcal{S} = S(TM)$. Then one can obtain that all $\mathbf{f}_{ia} = 0$. From (5.2.24) we obtain

$$\sum_{i \neq j} h_j^l(X, PY) N_{ji} - \frac{1}{2} h_i^l(X, PY) \parallel \mathbf{W}_i \parallel^2 - \sum_{\alpha, \beta = r+1}^{n} h'^{s}_{\alpha}(X, PY) W_\alpha^\beta Q_{i\beta} = 0.$$

On the other hand, by direct computations, we have

$$\sum_{i \neq j} h_j^l(X, PY) N_{ji} = \sum_{i \neq j} \bar{g}(h'^{l}(X, PY), \xi_j) \bar{g}(N_j', N_i).$$

Adding and subtracting $\bar{g}(h'^{l}(X, PY), \xi_i) \bar{g}(N_i', N_i)$ to the above, we get

$$\sum_{i \neq j} h_j^l(X, PY) N_{ji} = \sum_{k}^{r} \bar{g}(h'^{l}(X, PY), \xi_k) \bar{g}(N_k', N_i) - \bar{g}(h'^{l}(X, PY), \xi_i) \bar{g}(N_i', N_i).$$

Hence we have

$$\sum_{i \neq j} h_j^l(X, PY) N_{ji} = \sum_{k}^{r} \bar{g}(\bar{g}(h'^{l}(X, PY), \xi_k) N_k', N_i) - \bar{g}(h'^{l}(X, PY), \xi_i) \bar{g}(N_i', N_i).$$

Thus we obtain

$$\sum_{i \neq j} h_j^l(X, PY) N_{ji} = \bar{g}(h'^{l}(X, PY), N_i) - \bar{g}(h'^{l}(X, PY), \xi_i) \bar{g}(N_i', N_i).$$

Since h^l is invariant, we arrive at

$$\sum_{i \neq j} h_j^l(X, PY) N_{ji} = -h_i^l(X, PY) N_{ii}.$$

Since $[X_a^b]$, $[W_\alpha^\beta]$ and $[N_{ij}]$ are semi-orthogonal matrices, without any lose of generality, we can assume that these matrices are diagonal. Then the above equation implies that $N_{ii} = 0$. Using this in (5.2.24), we derive

$$\sum_{\alpha = r+1}^{n} \varepsilon_\alpha Q_{i\alpha}^2 = 0$$

which shows that

$$\parallel \mathbf{W}_\alpha \parallel^2 = 0. \tag{5.2.27}$$

Moreover, (5.2.19) and $N_{ij} = 0$ implies that

$$\sum_{\alpha=r+1}^{n} \varepsilon_{\alpha} Q_{i\alpha} Q_{j\alpha} = 0$$

which means that

$$\bar{g}(\mathbf{W}_i, \mathbf{W}_j) = 0. \qquad (5.2.28)$$

Then, using (5.2.23), (5.2.27) and (5.2.28) we obtain

$$\sum_{\alpha=r+1}^{n} \varepsilon_{\alpha} h_{\alpha}^{s}(X, PY) Q_{i\alpha} = \bar{g}(h^{s}(X, PY), \mathbf{W}_i) = 0. \qquad (5.2.29)$$

Then (5.2.29) implies that $\mathbf{W}_i = 0$, on $\mathcal{N}_1$. Similarly, repeating n-times above steps for each i we claim that each $\mathbf{W}_i$ vanishes on $\mathcal{N}^1$ which proves (b). If we take $N^1 = S(TM^{\perp})$, then, each $\mathbf{W}_i$ vanishes on this choice. Therefore, all the functions $Q_{i\alpha}$ vanish. Thus the proof is complete. □

Example 6. Let $\bar{M} = (\mathbf{R}_2^7, \bar{g})$, where $\mathbf{R}_2^7$ is a semi-Euclidean space of signature $(-, -, +, +, +, +, +)$ with respect to the canonical basis

$$\{\partial x_1, \partial x_2, \partial x_3, \partial x_4, \partial x_5, \partial x_6, \partial x_7\}.$$

Let M be a submanifold of $\mathbf{R}_2^7$ given by

$$x^1 = u^1, \ x^2 = u^2, \ x^3 = \frac{u^1}{\sqrt{2}} \sin u^3, \ x^4 = \frac{u^1}{\sqrt{2}} \cos u^3,$$

$$x^5 = \frac{u^1}{\sqrt{2}} \sin u^4, \ x^6 = \frac{u^1}{\sqrt{2}} \cos u^4, \ x^7 = u^2,$$

where $u^3 \in \mathbf{R} - \{k\frac{\pi}{2}\}$ and $u^4 \in \mathbf{R} - \{k\pi, k \in Z\}$. Then TM is spanned by

$$Z_1 = \partial x_1 + \frac{1}{\sqrt{2}} \sin u^3 \, \partial x_3 + \frac{1}{\sqrt{2}} \cos u^3 \, \partial x_4 + \frac{1}{\sqrt{2}} \sin u^4 \, \partial x_5 + \frac{1}{\sqrt{2}} \cos u^4 \, \partial x_6$$

$$Z_2 = \partial x_2 + \partial x_7$$

$$Z_3 = \frac{1}{\sqrt{2}} u^1 \cos u^3 \, \partial x_3 - \frac{1}{\sqrt{2}} u^1 \sin u^3 \partial x_4$$

$$Z_4 = \frac{1}{\sqrt{2}} u^1 \cos u^4 \, \partial x_5 - \frac{1}{\sqrt{2}} u^1 \sin u^4 \, \partial x_6.$$

Thus M is 2-lightlike with $\operatorname{Rad} TM = \operatorname{Span}\{Z_1, Z_2\}$. Choose $S(TM) = \operatorname{Span}\{Z_3, Z_4\}$. Then a screen transversal bundle $S(TM^{\perp})$ is spanned by

$$W = \sin u^3 \, \partial x_3 + \cos u^3 \, \partial x_4 - \sin u^4 \, \partial x_5 - \cos u^4 \, \partial x_6,$$

and a lightlike transversal bundle $\mathrm{ltr}(TM)$ is spanned by

$$N_1 = \frac{1}{2\sqrt{2}}\{-\sqrt{2}\partial x_1 + \sin u^3\,\partial x_3 + \cos u^3\,\partial x_4 + \sin u^4\,\partial x_5 + \cos u^4\,\partial x_6\}$$

$$N_2 = \frac{1}{2}\{-\partial x_2 + \partial x_7\}.$$

Then by direct computations, we have

$$\bar{\nabla}_X Z_2 = \bar{\nabla}_{Z_1} Z_1 = \bar{\nabla}_{Z_2} Z_2 = \bar{\nabla}_{Z_2} Z_3 = \bar{\nabla}_{Z_2} Z_4 = 0,$$

$$\bar{\nabla}_{Z_3} Z_1 = Z_3, \quad \bar{\nabla}_{Z_4} Z_1 = Z_4, \quad \bar{\nabla}_{Z_1} Z_3 = u^1\,Z_3, \ \bar{\nabla}_{Z_1} Z_4 = u^1\,Z_4,$$

$$\bar{\nabla}_{Z_3} Z_3 = -\frac{1}{2}u^1\,\{\sin u^3\,\partial x_3 + \cos u^3\,\partial x_4\},$$

$$\bar{\nabla}_{Z_4} Z_4 = -\frac{1}{2}u^1\,\{\sin u^4\,\partial x_5 + \cos u^4\,\partial x_6\}.$$

Thus, we obtain

$$h^l(Z_1, Z_1) = 0, \quad h^*(Z_1, Z_1) = 0$$
$$h^l(Z_2, Z_2) = 0, \quad h^*(Z_2, Z_2) = 0$$
$$h^l(Z_1, Z_2) = 0, \quad h^*(Z_1, Z_2) = h^*(Z_2, Z_1) = 0$$
$$h^l(Z_3, Z_4) = 0, \quad h^*(Z_3, Z_4) = h^*(Z_4, Z_3) = 0$$
$$h^l(Z_1, Z_3) = 0, \quad h^*(Z_1, Z_3) = h^*(Z_3, Z_1) = 0$$
$$h^l(Z_3, Z_2) = 0, \quad h^*(Z_3, Z_2) = h^*(Z_2, Z_3) = 0$$
$$h^l(Z_1, Z_4) = 0, \quad h^*(Z_1, Z_4) = h^*(Z_4, Z_1) = 0$$
$$h^l(Z_4, Z_2) = 0, \quad h^*(Z_4, Z_2) = h^*(Z_2, Z_2) = 0$$
$$h_2(Z_3, Z_3) = 0, \quad h_2^*(Z_3, Z_3)$$
$$h_2(Z_4, Z_4) = 0, \quad h_2^*(Z_4, Z_4) = 0$$
$$h_1^l(Z_3, Z_3) = -\frac{1}{2\sqrt{2}}u^1, \quad h_1^l(Z_4, Z_4) = -\frac{1}{2\sqrt{2}}u^1$$
$$h_1^*(Z_3, Z_3) = -\frac{1}{4\sqrt{2}}u^1, \quad h_1^*(Z_4, Z_4) = -\frac{1}{4\sqrt{2}}u^1.$$

Hence $\varphi_1 = \frac{1}{2}$ and φ_2 is an arbitrary function on M. Thus M is locally screen conformal and satisfies the hypothesis of the above theorem.

Just as in the coisotropic case, we show that there exists another class of r-lightlike submanifolds of semi-Riemannian manifolds of an arbitrary signature which admit integrable unique screen distributions, subject to a geometric condition (different from the screen conformal condition).

Consider a complementary vector bundle F of $\mathrm{Rad}\,TM$ in $S(TM^{\perp})^{\perp}$ and choose a basis $\{V_i\},\ i \in \{1, \ldots, r\}$ of $\Gamma(F_{|u})$. Thus the sections we are looking for are expressed as

$$N_i = \sum_{k=1}^{r} \{A_{ik}\,\xi_k + B_{ik}\,V_k\}, \tag{5.2.30}$$

where A_{ik} and B_{ik} are smooth functions on $\mathcal{U}$. Then $\{N_i\}$ satisfy (5.1.2) if and only if

$$\sum_{k=1}^{r} B_{ik}\,\bar{g}_{jk} = \delta_{ij},$$

where $\bar{g}_{jk} = \bar{g}(\xi_j, V_k)$, $j, k \in \{1, \ldots, r\}$. Observe that $G = det\,[\bar{g}_{jk}]$ is everywhere non-zero on $\mathcal{U}$, otherwise $S(TM^\perp)^\perp$ would be degenerate at least at a point of $\mathcal{U}$. Assume that F is parallel along the tangent direction.

Theorem 5.2.7. *Let M be an r-lightlike submanifold of a semi-Riemannian manifold $\bar{M}$ such that the complementary vector bundle F of $\operatorname{Rad} TM$ in $S(TM^\perp)^\perp$ is parallel along the tangent direction. Then, all the assertions from (a) through (c) of Theorem 5.2.6 will hold.*

Proof. The covariant derivative of (5.2.30) provides

$$\bar{\nabla}_X N_i = \sum_{k=1}^{r} (X(A_{ik})\xi_k + A_{ik}\bar{\nabla}_X\xi_k + X(B_{ik})V_k + B_{ik}\bar{\nabla}_X V_k).$$

Using (5.1.15), (5.1.22) and (5.1.27), we obtain

$$A_{N_i} X = \nabla_X^l N_i + D^s(X, N_i) - \sum_{k=1}^{r}[X(A_{ik})\xi_k$$
$$+ A_{ik}\{-A_{\xi_k}^* X + \nabla_X^{*t}\xi_k + h^l(X, \xi_k)$$
$$+ h^s(X, \xi_k)\} + X(B_{ik})V_k + B_{ik}\bar{\nabla}_X V_k].$$

Since F is parallel, $\bar{\nabla}_X V_k \in \Gamma(F)$. Thus for $Y \in \Gamma(S(TM))$, we get

$$g(A_{N_i} X, Y) = \sum_{k=1}^{r} A_{ik} g(A_{\xi_k}^* X, Y).$$

Then using (5.1.31) we get

$$\bar{g}(h^*(X, Y), N_i) = \sum_{k=1}^{r} A_{ik}\bar{g}(h^l(X, Y), \xi_k). \tag{5.2.31}$$

Since the right side of (5.2.31) is symmetric, it follows that h^* is symmetric on $S(TM)$. Thus Theorem 5.1.5 implies that $S(TM)$ is integrable. On the other hand, using (5.2.31) in (5.2.23) and $h^l = h'^l$, we obtain

$$g(\bar{\nabla}_X PY, \mathbf{Z}_i) = \sum_{i \neq j} h^l_j(X, PY)[N_{ji} + g(\mathbf{Z}_i, \mathbf{Z}_j)]$$

$$- \frac{1}{2} h^l_i(X, PY)[\| \mathbf{W}_i \|^2 - \| \mathbf{Z}_i \|^2]$$

$$+ \sum_{\alpha=r+1}^{n} \varepsilon_\alpha h^s{}_\alpha(X, PY) Q_{i\alpha}.$$

The rest of the proof is similar to the proofs of Theorems 5.2.4 and 5.2.6.   □

Remark 5.2.8. Observe from Example 6 that the screen conformal condition does not necessarily imply that F is parallel along the tangent direction. Indeed, in this example $\bar{\nabla}_{Z_3} Z_1$ does not belong to F.

5.3   Totally umbilical submanifolds

Let $\{N_i, W_\alpha\}$ be a basis of $\Gamma(tr\,(TM)|_{\mathcal{U}})$ on a coordinate neighborhood $\mathcal{U}$ of M, where $N_i \in \Gamma(ltr(TM)|_{\mathcal{U}})$ and $W_\alpha \in \Gamma(S(TM^\perp)|_{\mathcal{U}})$. Then (5.1.15) becomes

$$\bar{\nabla}_X Y = \nabla_X Y + \sum_{i=1}^{r} h^l_i(X, Y) N_i + \sum_{\alpha=r+1}^{n} h^s_\alpha(X, Y) W_\alpha, \qquad (5.3.1)$$

$$\bar{\nabla}_X Y = \nabla_X Y + \sum_{i=1}^{m<n} h^l_i(X, Y) N_i + \sum_{\alpha=m+1}^{n} h^s_\alpha(X, Y) W_\alpha, \qquad (5.3.2)$$

for an r-lightlike or an isotropic submanifold respectively. (5.1.24) becomes

$$\bar{\nabla}_X Y = \nabla_X Y + \sum_{i=1}^{n<m} h^l_i(X, Y) N_i, \qquad (5.3.3)$$

$$\bar{\nabla}_X Y = \nabla_X Y + \sum_{i=1}^{n=m} h^l_i(X, Y) N_i, \qquad (5.3.4)$$

for a coisotropic and a totally lightlike submanifold respectively. $\{h^l_i\}$ and $\{h^s_\alpha\}$ are the local lightlike second fundamental forms and the local screen second fundamental forms of M. Also (5.1.22) and (5.1.23) become

$$\bar{\nabla}_X N_i = -A_{N_i} X + \sum_{j=1}^{r} \rho_{ij}(X) N_j + \sum_{\alpha=r+1}^{n} \tau_{i\alpha}(X) W_\alpha,$$

$$\bar{\nabla}_X W_\alpha = -A_{W_\alpha} X + \sum_{i=1}^{r} \nu_{\alpha i}(X) N_i + \sum_{\beta=r+1}^{n} \theta_{\alpha\beta}(X) W_\beta,$$

$$\bar{\nabla}_X N_i = -A_{N_i} X + \sum_{j=1}^{m<n} \rho_{ij}(X) N_j + \sum_{\alpha=m+1}^{n} \tau_{i\alpha}(X) W_\alpha,$$

$$\bar{\nabla}_X W_\alpha = -A_{W_\alpha} X + \sum_{i=1}^{m<n} \nu_{\alpha i}(X) N_i + \sum_{\beta=m+1}^{n} \theta_{\alpha\beta}(X) W_\beta,$$

for an r-lightlike and an isotropic submanifold respectively, where

$$\rho_{ij}(X) = \bar{g}(\nabla_X^l N_i, \xi_j), \qquad \epsilon_\alpha \tau_{i\alpha}(X) = \bar{g}(D^s(X, N_i), W_\alpha), \qquad (5.3.5)$$
$$\nu_{\alpha i}(X) = \bar{g}(D^l(X, W_\alpha), \xi_i), \qquad \epsilon_\beta \theta_{\alpha\beta}(X) = \bar{g}(\nabla_X^s W_\alpha, W_\beta),$$

and ϵ_α is the signature of W_α. Similarly, (5.1.26) and (5.1.27) become

$$\nabla_X PY = \nabla_X^* PY + \sum_{i=1}^{r} h_i^*(X, PY)\xi_i,$$

$$\nabla_X \xi_i = -A_{\xi_i}^* X + \sum_{j=1}^{r} \mu_{ij}(X)\xi_j,$$

where $h_i^*(X, PY) = \bar{g}(h^*(X, PY), N_i)$ and $\mu_{ij}(X) = \bar{g}(\nabla_X^{*t}\xi_i, N_j)$. Using the above equations we obtain $\mu_{ij}(X) = -\rho_{ji}(X)$. Thus,

$$\nabla_X \xi_i = -A_{\xi_i}^* X - \sum_{j=1}^{r} \rho_{ji}(X)\xi_j. \qquad (5.3.6)$$

Definition 5.3.1. [155] A lightlike submanifold (M, g) of a semi-Riemannian manifold $(\bar{M}, \bar{g})$ is said to be totally umbilical in $\bar{M}$ if there is a smooth transversal vector field $\mathcal{H} \in \Gamma(\mathrm{tr}(TM))$ on M, called the transversal curvature vector field of M, such that, for all $X, Y \in \Gamma(TM)$,

$$h(X, Y) = \mathcal{H} \bar{g}(X, Y). \qquad (5.3.7)$$

Using (5.1.15) and (5.3.7) it is easy to see that M is totally umbilical, if and only if on each coordinate neighborhood $\mathcal{U}$ there exist smooth vector fields $H^l \in \Gamma(\mathrm{ltr}(TM))$ and $H^s \in \Gamma(S(TM^\perp))$, and smooth functions $H_i^l \in F(\mathrm{ltr}(TM))$ and $H_i^s \in F(S(TM^\perp))$ such that

$$h^l(X, Y) = H^l \bar{g}(X, Y), \qquad h^s(X, Y) = H^s \bar{g}(X, Y),$$
$$h_i^l(X, Y) = H_i^l \bar{g}(X, Y), \qquad h_\alpha^s(X, Y) = H_\alpha^s \bar{g}(X, Y) \qquad (5.3.8)$$

for any $X, Y \in \Gamma(TM)$. The above definition does not depend on the screen distribution and the screen transversal vector bundle of M. On the other hand, from the equation (5.1.29) we obtain

$$g(A_{W_\alpha} X, Y) = \epsilon_\alpha h_\alpha^s(X, Y) + \sum_{i=1}^{r} D_i^l(X, W_\alpha) \eta_i(Y). \qquad (5.3.9)$$

Now replace Y by ξ_j and obtain

$$D_i^l(X, W_\alpha) = -\epsilon_\alpha h_\alpha^s(\xi_i, X). \tag{5.3.10}$$

Using (5.1.28), (5.1.29), (5.3.8) and (5.3.10), we state (the relations (5.3.8) trivially hold in case $S(TM)$ or $S(TM^\perp)$ vanish) the following:

Theorem 5.3.2. [155]. *A lightlike submanifold (M, g), of a semi-Riemannian manifold $(\bar{M}, \bar{g})$, is totally umbilical if and only if on each coordinate neighborhood $\mathcal{U}$ there exist smooth vector fields H^l and H^s such that*

$$D^l(X, W) = 0, \qquad A_\xi^* X = H^l PX, \qquad P(A_W X) = \epsilon\, H^s PX,$$
$$D_i^l(X, W_\alpha) = 0, \qquad A_{\xi_i}^* X = H_i^l PX, \qquad P(A_{W_\alpha} X) = \epsilon_\alpha\, H_\alpha^s PX,$$

for any $X \in \Gamma(TM)$, where ϵ is the signature of $W \in \Gamma(S(TM^\perp))$.

Example 8. Let M be a surface of $\mathbf{R}_2^4$, of Example 1, given by

$$x^3 = \frac{1}{\sqrt{2}}(x^1 + x^2); \quad x^4 = \frac{1}{2}\log(1 + (x^1 - x^2)^2),$$

where $(x^1, \ldots, x^4)$ is a local coordinate system for $\mathbf{R}_2^4$. As explained in Example 1, M is a 1-lightlike surface of Case A, having a local quasi-orthonormal field of frames $\{\xi, N, V, W\}$ along M. Denote by $\bar{\nabla}$ the Levi-Civita connection on $\mathbf{R}_2^4$. Then, by straightforward calculations, we obtain

$$\bar{\nabla}_V V = 2(1 + (x^1 - x^2)^2)\left\{2(x^2 - x^1)\partial_2 + \sqrt{2}(x^2 - x^1)\partial_3 + \partial_4\right\},$$

$$\bar{\nabla}_{\xi_1} V = 0, \; \bar{\nabla}_X \xi_1 = \bar{\nabla}_X N = 0, \; \forall X \in \Gamma(TM).$$

For this example, the equations (5.3.8) reduce to

$$h^1(X, Y) = H^1\, \bar{g}(X, Y)\ ;\ h^2(X, Y) = H^2\, \bar{g}(X, Y)$$

where h^1 and h^2 are $\Gamma(\mathrm{ltr}(TM))$-valued and $\Gamma(S(TM^\perp))$-valued bilinear forms (see equation (5.1.15). Using the Gauss and Weingarten formulae we infer

$$h^1 = 0; \quad A_{\xi_1} = 0; \quad A_N = 0; \quad \nabla_X \xi_1 = 0; \quad \rho_{ij}(X) = 0;$$

where for the symbol ρ_{ij} see the equation (5.3.5). $h^2(X, \xi) = 0;$

$$H^2(V, V) = 2; \quad \nabla_X V = \frac{2\sqrt{2}(x^2 - x^1)^3}{1 + (x^1 - x^2)^2} X^2 V,$$

$\forall X = X^1 \xi_1 + X^2 V \in \Gamma(TM)$. Since $\bar{g}(V, V) = -(1 + (x^1 - x^2)^4)$ we get

$$h^2(V, V) = H^2\, \bar{g}(V, V), \qquad H^2 = -\frac{2}{(1 + (x^1 - x^2)^4)}.$$

Therefore, M is a totally umbilical 1-lightlike submanifold of $\mathbf{R}_2^4$.

In particular, just like the case of hypersurfaces, we say that M is a *totally geodesic* lightlike submanifold of $\bar{M}$ if any geodesic of M with respect to the Levi-Civita connection ∇ is a geodesic of $\bar{M}$. For this we recall the following result:

Theorem 5.3.3. [149]. *Let M be a lightlike submanifold of a semi-Riemannian manifold $\bar{M}$. Then the following assertions are equivalent:*

(i) *M is totally geodesic.*

(ii) *h^l and h^s vanish identically on M.*

(iii) *A_ξ^* vanishes identically on M, for any $\xi \in \Gamma(\operatorname{Rad} TM)$, A_W is $\Gamma(\operatorname{Rad} TM)$-valued for any $W \in \Gamma(S(TM))$ and $D^l(X, SV) = 0$ for any $X \in \Gamma(TM)$ and $V \in \Gamma(\operatorname{tr}(TM))$.*

Proof. (i) $\Longrightarrow$ (ii). Consider $u_o \in M$, $v_o \in T_{u_o}M$ and $\Gamma : u^\alpha = u^\alpha(t)$, $\alpha \in \{1, \ldots, m\}$, the unique geodesic of M such that $u^\alpha(o) = u_o$ and $\frac{du^\alpha}{dt}(o) = v_o$. As Γ is a geodesic of $\bar{M}$ too, from (5.1.15) it follows that $h^l(v, v) = h^s(v, v) = 0$, for any v tangent to Γ. Thus (ii) follows by polarization. (ii) $\Longrightarrow$ (i). It is a consequence of (5.1.15). From Theorem 5.1.4 it follows that $h^l = 0$ if and only if $A_\xi^* = 0$ for any $\xi \in \Gamma(\operatorname{Rad} TM)$. Finally, from (5.1.28) we obtain that $h^s = 0$, if and only if, A_W is $\Gamma(\operatorname{Rad} TM)$ and $D^\ell(X, SV) = 0$. Thus (ii) and (iii) are equivalent too. $\square$

Curvature equations. Denote by $\bar{R}, R$ and R^l the curvature tensors of $\bar{\nabla}, \nabla$ and ∇^l respectively. We obtain

$$
\begin{aligned}
\bar{R}(X, Y)Z = {}& R(X, Y)Z \\
& + A_{h^l(X, Z)}Y - A_{h^l(Y, Z)}X \\
& + A_{h^s(X, Z)}Y - A_{h^s(Y, Z)}X \\
& + (\nabla_X h^l)(Y, Z) - (\nabla_Y h^l)(X, Z) \\
& + D^l(X, h^s(Y, Z)) - D^l(Y, h^s(X, Z)) \\
& + (\nabla_X h^s)(Y, Z) - (\nabla_Y h^s)(X, Z) \\
& + D^s(X, h^l(Y, Z)) - D^s(Y, h^l(X, Z)),
\end{aligned}
\tag{5.3.11}
$$

for any $X, Y, Z \in \Gamma(TM)$. For the curvature tensor $\bar{R}$ of type $(0, 4)$, we have

$$
\begin{aligned}
\bar{R}(X, Y, PZ, PU) = {}& g\left(R(X, Y)PZ, PU\right) + \bar{g}\left(h^*(Y, PU), h^l(X, PZ)\right) \\
& - \bar{g}\left(h^*(X, PU), h^l(Y, PZ)\right) + \bar{g}\left(h^s(Y, PU), h^s(X, PZ)\right) \\
& - \bar{g}\left(h^s(X, PU), h^s(Y, PZ)\right),
\end{aligned}
\tag{5.3.12}
$$

$$
\begin{aligned}
\bar{R}(X, Y, \xi, PU) = {}& g\left(R(X, Y)\xi, PU\right) \\
& + \bar{g}\left(h^*(Y, PU), h^l(X, \xi)\right) - \bar{g}\left(h^*(X, PU), h^l(Y, \xi)\right) \\
& + \bar{g}\left(h^s(Y, PU), h^s(X, \xi)\right) - \bar{g}\left(h^s(X, PU), h^s(Y, \xi)\right)
\end{aligned}
$$

$$= \bar{g}\left((\nabla_Y h^l)(X, PU) - (\nabla_X h^l)(Y, PU), \xi\right)$$
$$+ \bar{g}\left(h^s(Y, PU), h^s(X, \xi)\right) - \bar{g}\left(h^s(X, PU), h^s(Y, \xi)\right),$$
$$\bar{R}(X, Y, N, PU) = -\bar{g}\left(R(X, Y)PU, N\right)$$
$$+ \bar{g}\left(A_N Y, h^l(X, PU)\right) - \bar{g}\left(A_N X, h^l(Y, PU)\right)$$
$$+ \bar{g}\left(h^s(Y, PU), D^s(X, N)\right) - \bar{g}\left(h^s(X, PU), D^s(Y, N)\right)$$
$$= \bar{g}\left((\nabla_Y A)(N, X) - (\nabla_X A)(N, Y), PU\right)$$
$$+ \bar{g}\left(h^s(Y, PU), D^s(X, N)\right) - \bar{g}\left(h^s(X, PU), D^s(Y, N)\right),$$
$$\bar{R}(X, Y, W, PU) = \bar{g}\left((\nabla_Y A)(W, X) - (\nabla_X A)(W, Y), PU\right)$$
$$+ \bar{g}\left(h^*(Y, PU), D^l(X, W)\right) - \bar{g}\left(h^*(X, PU), D^l(Y, W)\right)$$
$$= \bar{g}\left((\nabla_Y h^s)(X, PU) - (\nabla_X h^s)(Y, PU), W\right)$$
$$+ \bar{g}\left(h^l(X, PU), A_W Y\right) - \bar{g}\left(h^l(X, PU), A_W X\right),$$
$$\bar{R}(X, Y, N, \xi) = \bar{g}\left(R^l(X, Y)N, \xi\right) \qquad\qquad (5.3.13)$$
$$+ \bar{g}\left(h^l(Y, A_N X), \xi\right) - \bar{g}\left(h^l(X, A_N Y), \xi\right)$$
$$+ \bar{g}\left(D^s(X, N), h^s(Y, \xi)\right) - \bar{g}\left(D^s(Y, N), h^s(X, \xi)\right)$$
$$= -\bar{g}(R(X, Y)\xi, N)$$
$$+ \bar{g}\left(A_N Y, h^l(X, \xi)\right) - \left(A_N Y, h^l(Y, \xi)\right)$$
$$+ \bar{g}\left(D^s(X, N), h^s(Y, \xi)\right) - \bar{g}\left(D^s(Y, N), h^s(X, \xi)\right),$$

$X, Y, U \in \Gamma(TM)$. Let R^{*t} be the curvature tensor of ∇^{*t}. Then,

$$g(R(X, Y)\xi, PU) = g((\nabla_Y A^*)(\xi, X) - (\nabla_X A^*)(\xi, Y), PU),$$
$$g\left(R(X, Y)\xi, N = \bar{g}(R^{*t}(X, Y)\xi, N)\right. \qquad\qquad (5.3.14)$$
$$+ g(A_N Y, A^*_\xi X) - g(A_N X, A^*_\xi Y),$$
$$\bar{g}\left(R(X, Y)PU, N\right) = \bar{g}\left((\nabla_X A)(N, Y) - (\nabla_Y A)(N, X), PU\right) \qquad (5.3.15)$$
$$+ \bar{g}\left(h^l(X, PU), A_N Y\right) - \bar{g}\left(h^l(Y, PU), A_N X\right)$$
$$= \bar{g}\left((\nabla_X h^*)(Y, PU) - (\nabla_Y h^*)(X, PU), N\right),$$
$$\bar{g}(R(X, Y)\xi, N) + \bar{g}(R^l(X, Y)N, \xi) = g(A^*_\xi X, A_N Y)$$
$$- g(A^*_\xi Y, A_N X). \qquad\qquad (5.3.16)$$

For structure equations of Case B, delete all the components of $S(TM^\perp)$. Similarly, one can find equations for the other two cases.

Induced Ricci Tensor. First note that h^l_i, ρ_{ij} and $\tau_{i\alpha}$ depend on the section $\xi \in \Gamma(\text{Rad } TM)$. Indeed, take $\xi^*_i = \sum_{j=1}^r \alpha_{ij} \xi_j$, where α_{ij} are smooth functions with $\Delta = \det(\alpha_{ij}) \neq 0$ and let A_{ij} be the co-factors of α_{ij} in the determinant of Δ. It follows that $N^*_i = \frac{1}{\Delta} \sum_{j=1}^r A_{ij} N_j$. Hence by straightforward calculation and using (5.3.1)–(5.3.5) we obtain $h^{l*}_i = \sum_{j=1}^r \alpha_{ij} h^l_j$. Denote by ρ^*_{ij} and $\tau^*_{i\alpha}$ the affine combinations of ρ_{ij} and $\tau_{i\alpha}$ with coefficients α_{ij}, A_{ij} and $X(A_{ij})$. Moreover,

$$\text{tr}(\rho_{ij})(X) = \text{tr}(\rho_{ij}^*)(X) + X(\log \Delta), \quad \forall X \in \Gamma(TM).$$

Thus, using the formula $d\rho(X, Y) = \frac{1}{2}\{X(\rho(Y)) - Y(\rho(X)) - \rho([X, Y])\}$ of a differential 2-form, we obtain:

Proposition 5.3.4. *Let (M, g) be a lightlike submanifold of a semi-Riemannian manifold $(\bar{M}, \bar{g})$. Suppose $\text{tr}(\rho_{ij})$ and $\text{tr}(\rho_{ij}^*)$ are 1-forms on $\mathcal{U}$ with respect to ξ_i and ξ_i^*. Then $d(\text{tr}(\rho_{ij}^*)) = d(\text{tr}(\rho_{ij}))$ on $\mathcal{U}$.*

To find a local expression of a Ricci tensor of M, consider the frames field

$$\{\xi_1, , \ldots, \xi_r; N_1, \ldots, N_r; X_{r+1}, \ldots, X_m; W_{r+1}, \ldots, W_n\}$$

on $\bar{M}$, with $\{\mathcal{F}_A\} = \{\xi_1, , \ldots, \xi_r, X_{r+1}, \ldots, X_m\}$ the frame on M. Then,

$$\bar{R}_{ABCD} = \bar{g}(\bar{R}(\mathcal{F}_D, \mathcal{F}_C)\mathcal{F}_B, \mathcal{F}_A), \qquad R_{ABCD} = g(R(\mathcal{F}_D, \mathcal{F}_C)\mathcal{F}_B, \mathcal{F}_A),$$
$$\bar{R}_{iBCD} = \bar{g}(\bar{R}(\mathcal{F}_D, \mathcal{F}_C)\mathcal{F}_B, N_i), \qquad R_{iBCD} = \bar{g}(R(\mathcal{F}_D, \mathcal{F}_C)\mathcal{F}_B, N_i),$$
$$\bar{R}_{\alpha BCD} = \bar{g}(\bar{R}(\mathcal{F}_D, \mathcal{F}_C)\mathcal{F}_B, W_\alpha), \qquad R_{\alpha BCD} = \bar{g}(R(\mathcal{F}_D, \mathcal{F}_C)\mathcal{F}_B, W_\alpha),$$
$$\bar{R}_{i\alpha CD} = \bar{g}(\bar{R}(\mathcal{F}_D, \mathcal{F}_C)W_\alpha, N_i), \qquad R_{i\alpha CD} = \bar{g}(R(\mathcal{F}_D, \mathcal{F}_C)W_\alpha, N_i).$$

Using the above we have the following expression for a $(0, 2)$ tensor:

$$R^{(0,2)}(X, Y) = \sum_{a,b=r+1}^{m} g^{ab}\, g(R(X, X_a)Y, X_b) + \sum_{i=1}^{r} \bar{g}(R(X, \xi_i)Y, N_i).$$

Note. As we discussed in Chapter 2, since the induced connection ∇ on M is not a metric connection, in general, $R^{(0,2)}$ is not symmetric. Indeed, using the symmetries of curvature tensor and the first Bianchi identity and taking into account (5.3.11) and (5.3.12) we obtain

$$R^{(0,2)}(X, Y) - R^{(0,2)}(Y, X)$$

$$= \sum_{a,b=r+1}^{m} g^{ab}\{\bar{g}(h^*(X, X_b), h^l(Y, X_a)) - \bar{g}(h^*(Y, X_b), h^l(X, X_a))\}$$

$$+ \sum_{i=1}^{r}\{g(A_{\xi_i}^* X, A_{N_i}Y) - g(A_{\xi_i}^* Y, A_{N_i}X) + \bar{g}(R^{*t}(X, Y)\xi_i, N_i)\}.$$

Replacing X, Y by X_A, X_B respectively, using (5.1.33), (5.3.6) and

$$\sum_{i=1}^{r} \bar{g}(R^{*t}(X, Y)\xi_i, N_i) = -2 \sum_{i,j=1}^{r} \bar{g}(d(\rho_{ji})(X, Y)\xi_j, N_i)$$

$$= -2\, d(\text{tr}(\rho_{ij})(X, Y),$$

we have

$$R_{AB} - R_{BA} = 2\, d\left(\mathrm{tr}(\rho_{ij})\right)(X_A,\, X_B)$$

where $R_{AB} = \mathrm{Ric}\,(X_B,\, X_A)$. Thus, $R^{(0,2)}$ is not symmetric and has no geometric meaning similar to the symmetric Ricci tensor. Thus, as per Definition 2.4.3 in Chapter 2, a symmetric $R^{(0,2)}$ is called the induced Ricci tensor of M. We conclude from the above that the following holds:

Theorem 5.3.5. *Let $(M,\, g,\, S(TM))$ be an r-lightlike or a coisotropic submanifold of a semi-Riemannian manifold $(\bar{M},\, \bar{g})$. Then the Ricci tensor of the induced connection ∇ on M is symmetric, if and only if, each 1-form $\mathrm{tr}(\rho_{ij})$ induced by $S(TM)$ is closed, i.e., on any $\mathcal{U} \subset M$,*

$$d\left(\mathrm{tr}(\rho_{ij})\right) = 0.$$

Relating the above theorem with Theorem 5.1.5 of integrability conditions for a screen distribution, we quote the following result:

Theorem 5.3.6. [155] *Let $(M,\, g,\, S(TM))$ be a proper totally umbilical r-lightlike or a coisotropic submanifold of a semi-Riemannian manifold $(\bar{M}(\bar{c}),\, \bar{g})$ of a constant curvature $\bar{c}$. Then, the induced Ricci tensor on M is symmetric, if and only if its screen distribution $S(TM)$ is integrable.*

Proof. From (5.3.13), (5.3.14), (5.3.16) and (1.2.16), we get

$$2\, d\left(\mathrm{tr}(\rho_{ij})\right)(X,\, Y) + \sum_{i=1}^{r} H_i^l \left\{ g\,(Y,\, A_{N_i} X) - g\,(X,\, A_{N_i} Y) \right\} = 0.$$

The proof follows from the above equation. $\qquad\qquad\qquad\qquad\qquad\qquad\square$

Remark 5.3.7. We know from Theorems 5.2.4 and 5.2.6 that the geometric condition of a screen conformal or parallel vector field F provides an integrable screen distribution, which is the root requirement for an r-lightlike or coisotropic submanifold to admit a canonical screen. We also know that, in general, the induced Ricci tensor of any lightlike submanifold is not symmetric. Since a symmetric induced Ricci tensor is also a desirable property (along with a canonical or unique screen) the above Theorem 5.3.6 tells us that a large class of totally umbilical r-lightlike or coisotropic lightlike submanifolds of $(\bar{M}(\bar{c}),\, \bar{g})$ are candidates for the existence of an integrable canonical or unique screen distribution and an induced symmetric Ricci tensor.

5.4   Minimal lightlike submanifolds

Let (M, g_M) and (N, g_N) be semi-Riemannian manifolds and suppose that $\varphi :$ $M \longrightarrow N$ is a smooth mapping between them. Then the differential $d\varphi$ of φ can be viewed as a section of the bundle $\mathrm{Hom}\,(TM,\, \varphi^{-1}TN) \longrightarrow M$, where $\varphi^{-1}TN$ is the pullback bundle which has fibers $(\varphi^{-1}TN)_p = T_{\varphi(p)}N, p \in M$. $\mathrm{Hom}(TM, \varphi^{-1}TN)$ has a connection ∇ induced from the Levi-Civita connection ∇^M and the pullback connection. Then the second fundamental form of φ is given by

$$\nabla d\varphi(X, Y) = \nabla^\varphi_X d\varphi(Y) - d\varphi(\nabla^M_X Y), \quad \forall X, Y \in \Gamma(TM).$$

It is known that the second fundamental form is symmetric. A smooth map $\varphi :$ $(M, g_M) \longrightarrow (N, g_N)$ is said to be harmonic if $\mathrm{tr}\,\nabla d\varphi = 0$. On the other hand, the tension field of φ is the section $\tau(\varphi)$ of $\Gamma(\varphi^{-1}TN)$ defined by

$$\tau(\varphi) = \mathrm{div}\, d\varphi = \sum_{i=1}^m \nabla d\varphi(e_i, e_i), \tag{5.4.1}$$

where $\{e_1, \ldots, e_m\}$ is the orthonormal frame on M. Then it follows that φ is harmonic if and only if $\tau(\varphi) = 0$; for more information on harmonic maps and morphisms, see [25]. In the semi-Riemannian context, a minimal isometric immersion is a particular harmonic map. In [323], a harmonic map ϕ between lightlike manifolds is defined with the assumption that ϕ is radical preserving (i.e., ϕ maps the radical of the domain into the radical of the target). This does not apply here to define minimality, since an isometric immersion from M to $\bar{M}$ is not radical preserving. In [141], harmonic maps from a semi-Riemannian manifold into a lightlike manifold are defined only when the target is a Riemannian hypersurface of a globally null manifold. This also does not apply here to define minimality, since our domain M is lightlike. In [149], a minimal lightlike submanifold is defined only in the particular case when M is a hypersurface of the Minkowski space $\bar{M} = \mathbf{R}^4_1$. Recently, a general notion of minimal lightlike submanifolds M of a semi-Riemannian manifold $\bar{M}$ was introduced in [41] as follows:

Definition 5.4.1. [41] We say that a lightlike submanifold $(M, g, S(TM))$ of a semi-Riemannian manifold $(\bar{M}, \bar{g})$ is minimal if:

(i) $h^s = 0$ on $\mathrm{Rad}(TM)$ and

(ii) $\mathrm{tr}\, h = 0$, where trace is written w.r.t. g restricted to $S(TM)$.

In the case **B**, the condition (i) is trivial. It has been shown in [41] that the above definition is independent of $S(TM)$ and $S(TM^\perp)$, but it depends on $\mathrm{tr}(TM)$. As in the semi-Riemannian case, any lightlike totally geodesic M is minimal. Thus, it follows that any totally lightlike M (case **D**) is minimal.

Example 9. Let $(\mathbf{R}^4_1, <, >)$ be the Minkowski space with signature $(+, +, +, -)$ and the canonical basis $\partial_1, \ldots, \partial_4$. Let $S^3_1 = \{p \in \mathbf{R}^4_1 \mid < p, p >= 1\}$ be the 3-dimensional pseudo-sphere of index 1 which is a Lorentzian hypersurface of $(\mathbf{R}^4_1, <, >)$. Consider $(\bar{M} = S^3_1 \times \mathbf{R}^2_1, g)$ the semi-Riemannian cross product, where $\mathbf{R}^2_1$ is semi-Euclidean space with signature $(+, -)$ with respect to the canonical basis $\{\partial_5, \partial_6\}$ and g is the inner product of $\mathbf{R}^6_2 = \mathbf{R}^4_1 \times \mathbf{R}^2_1$ restricted to $\bar{M}$. Then the submanifold $(M, g \mid_M, S(TM), S(TM^\perp))$ is a minimal lightlike submanifold of $\bar{M}$ given by

$$M = S^1 \times \mathcal{H} \times \mathbf{R} = \{(p, t, t) \in S_1^3 \times \mathbf{R}_1^2 \mid t \in \mathbf{R}\}$$

$$p = \frac{\sqrt{2}}{2}(\cos \theta, \sin \theta, \cosh \varphi, \sinh \varphi) \in S_1^3,$$

where $\theta \in [0, 2\pi], \varphi \in \mathbf{R}\}$, $\mathcal{H}$ is the hyperbola and

$$S(TM) = \text{Span}\{e_1 = -\sin \theta \, \partial_1 + \cos \theta \, \partial_2, \ e_2 = -\sinh \varphi \, \partial_3 + \cosh \varphi \, \partial_4\}.$$

Here $\varepsilon_1 = g(e_1, e_1) = 1$, $\varepsilon_2 = g(e_2, e_2) = -1$, $\text{Rad}(TM) = \text{Span}\{\xi = \partial_5 + \partial_6\}$, $\text{ltr}(TM) = \text{Span}\{N = \frac{1}{2}(\partial_5 - \partial_6)\}$ and

$$S(TM^\perp) = \text{Span}\{W = \frac{\sqrt{2}}{2}(\cos \theta \, \partial_1 + \sin \theta \, \partial_2 - \cosh \varphi \, \partial_3 - \sinh \varphi \, \partial_4)\}$$

where $\{e_1, e_2, \partial_5, \partial_6, W\}$ is an orthonormal basis of $\bar{M}$. Let $\bar{p} = \frac{\sqrt{2}}{2}(\cos \theta \, \partial_1 + \sin \theta \, \partial_2 + \cosh \varphi \, \partial_3 + \sinh \varphi \, \partial_4)$ be the position vector of an arbitrary point p of S_1^3 which is normal to S_1^3 in $\mathbf{R}_1^4$. Since the Levi-Civita connection $\bar{M}$ of $\mathbf{R}_1^4$ satisfies $\bar{\nabla}_{e_1} e_1 = -\frac{1}{2}(W + \bar{p})$ and $\bar{\nabla}_{e_2} e_2 = \frac{1}{2}(-W + \bar{p})$, it follows that $h(e_1, e_1) = -\frac{1}{2}W$, $h(e_1, e_2) = 0$ and $h(e_2, e_2) = -\frac{1}{2}W$. Hence, we get

$$\text{tr}_{g \mid_{S(TM)}} h = \varepsilon_1 h(e_1, e_1) + \varepsilon_2 h(e_2, e_2) = h(e_1, e_1) - h(e_2, e_2) = 0.$$

We also have $h(\xi, \xi) = 0$. Therefore, M is a non-totally geodesic minimal lightlike submanifold of $\bar{M}$.

The classical notion of minimality is connected to the geometric interpretation of being an extremal of the volume functional [168]. Here we relate the classical minimality (in the semi-Riemannian case) with the minimality introduced in the lightlike case by Definition 5.4.1, as follows:

Theorem 5.4.2. [41] Let $(M, g, S(TM), S(TM^\perp))$ be a lightlike submanifold of a semi-Riemannian manifold $(\bar{M}, \bar{g})$, with $S(TM)$ integrable. If its leaves are minimal (semi-Riemannian) submanifolds of $(\bar{M}, \bar{g})$ and $h^s = 0$ on $\text{Rad}(TM)$, then, M is a minimal lightlike submanifold of $\bar{M}$. Conversely, if M is a minimal lightlike submanifold of $\bar{M}$, then $\text{Rad}(TM)$ contains the mean curvature vector field of any leaf of $S(TM)$.

Proof. Let $i : \Sigma \to \bar{M}$ denote the inclusion map of any leaf Σ of $S(TM)$. The tension field $\tau(i)$ of Σ can be decomposed from (5.1.15) into:

$$\tau(i) = \tau^\star(i) + \text{tr}_{g \mid_{S(TM)}} h,$$

where $\tau^\star(i) \in \text{Rad}(TM)$. Since Σ is minimal in $\bar{M}$ iff the map i is harmonic, which means $\tau(i) = 0$, the statement follows from Definition 5.4.1. $\qquad \square$

We observe that Example 9 satisfies Theorem 5.4.2 with respect to the leaves $S^1 \times \mathcal{H}$ of $S(TM)$. Let $\Lambda = \{x \in \mathbf{R}_q^{n+1} / <x, x> = 0\}$ be the lightlike cone in the semi-Euclidean space $(\mathbf{R}_q^{n+1}, <>)$. Related to the non-existence result of compact minimal spacelike submanifolds isometrically immersed in semi-Euclidean spaces, we have the following:

Proposition 5.4.3. *There are no minimal lightlike isometric immersions*

$$\phi : (M, <>_{|M}, S(TM), s(TM^\perp)) \rightarrow (\mathbf{R}_q^{n+1}, <>)$$

with $\phi(M) \subset \Lambda$.

Proof. Suppose there exists such a map ϕ. Then, the function p given by

$$p \in M \rightarrow \frac{1}{2} < \phi(p), \phi(p) > \in \mathbf{R}$$

is identically zero and hence,

$$0 = \frac{1}{2} X < \phi(p), \phi(p) > = < X, \phi(p) > \quad \text{and}$$

$$0 = \frac{1}{2} X(X < \phi(p), \phi(p) >) = < \nabla_X^c X, \phi(p) > + < X, X >,$$

$$\forall p \in M, X \in \Gamma(TM),$$

where ∇^c is the canonical Levi-Civita connection of $(\mathbf{R}_q^{n+1}, <>)$. If we replace X consecutively by e_a, where $\{e_a\}$ is the orthonormal basis of $S(TM)$, then from the minimality condition we have:

$$0 = < \mathrm{tr}_{<>_{|S(TM)}} h, \phi(p) > = -\left(\sum_a \epsilon_a < e_a, e_a > \right) < 0,$$

where $\epsilon_a = < e_a, e_a >$, $\forall a$. This contradiction completes the proof. □

In support of Proposition 5.4.3, we give the following example:

Example 10. Let $(\mathbf{R}_1^4, <>)$ be the Minkowski space with signature $(+, +, +, -)$ w.r.t. the canonical basis $(\partial_1, \ldots, \partial_4)$. Then the manifold $(M, <>_{|M}, S(TM))$ is a lightlike hypersurface, given by an open subset of the lightlike cone

$$M = \{t(\cos u \cos v, \cos u \sin v, \sin u, 1) \in \mathbf{R}_1^4 / t > 0, \ u \in (o, \pi/2), \ v \in [0, 2\pi]\},$$

where

$$S(TM) = \mathrm{Span}\{e_1 = -\sin u \cos v \partial_1 - \sin u \sin v \partial_2 + \cos u \partial_3,$$
$$e_2 = -\sin v \partial_1 + \cos v \partial_2\}.$$

We note that e_1 and e_2 are orthonormal,

$$\text{Rad}(TM) = \text{Span}\{\xi = \cos u \cos v \partial_1 + \cos u \sin v \partial_2 + \sin u \partial_3 + \partial_4\},$$

$$\text{ltr}(TM) = \text{Span}\left\{N = \frac{1}{2}(\cos u \cos v \partial_1 + \cos u \sin v \partial_2 + \sin u \partial_3 - \partial_4)\right\},$$

and $< \xi, N >= 1$. We have $h(e_1, e_1) = -(\frac{1}{t \cos u}) N$, $h(e_2, e_2) = -(\frac{1}{t \cos u}) N$ and $h(e_1, e_2) = 0$. It turns out that the open subset of the lightlike cone $(M, <>_{|M})$ is globally null, since ξ is globally defined and $S(TM)$ is a spacelike integrable distribution.

Thus, the above example of the lightlike submanifold M, which is an open subset of the lightlike cone Λ of $(\mathbf{R}_1^4, <>)$, is never minimal.

Since Λ is a proper totally umbilical lightlike submanifold of $(\mathbf{R}_1^4, <>)$, Proposition 5.4.3 can be generalized, by using the definition of totally umbilical lightlike submanifolds, as follows:

Theorem 5.4.4. *There are no minimal lightlike submanifolds contained in a proper totally umbilical lightlike submanifold of a semi-Riemannian manifold.*

Theorem 5.4.5. [160] *Let M be a lightlike submanifold of a semi-Riemannian manifold $\bar{M}$. Then M is minimal if and only if*

$$\text{tr } A^*_{\xi_k} |_{S(TM)} = 0, \ \text{tr } A_{W_j} |_{S(TM)} = 0,$$

$$\bar{g}(Y, D^l(X, W)) = 0, \quad \forall X, Y \in \Gamma(RadTM),$$

where $W \in \Gamma(S(TM^\perp))$, $k \in \{1, \ldots, r\}$ and $j \in \{1, \ldots, (n-r)\}$.

Proof. Since $\text{tr } h |_{S(TM)} = \sum_{i=1}^{m-r} \varepsilon_i (h^l(e_i, e_i) + h^s(e_i, e_i))$ we have

$$\text{tr } h |_{S(TM)} = \sum_{i=1}^{m-r} \varepsilon_i \frac{1}{r} \sum_{k=1}^{r} \bar{g}(h^l(e_i, e_i), \xi_k) N_k + \frac{1}{n-r} \sum_{j=1}^{n-r} \bar{g}(h^s(e_i, e_i), W_j) W_j,$$

where $\{W_1, \ldots, W_{n-r}\}$ is an orthonormal basis of $S(TM^\perp)$. Using (5.1.28) and (5.1.31) we get

$$\text{tr } h |_{S(TM)} = \sum_{i=1}^{m-r} \varepsilon_i \left(\frac{1}{r} \sum_{k=1}^{r} g(A^*_{\xi_k} e_i, e_i) N_k + \frac{1}{n-r} \sum_{j=1}^{n-r} g(A_{W_j} e_i, e_i) W_j\right).$$

On the other hand, from (5.1.14) and (5.1.28) we have

$$\bar{g}(h^s(X, Y), W) = \bar{g}(Y, D^l(X, W)), \ \forall X, Y \in \Gamma(\text{Rad } TM)$$

and $\forall W \in \Gamma(S(TM^\perp))$. Thus our assertion follows from the above two equations and Proposition 5.1.3. $\square$

Theorem 5.4.6. [160] *Let M be a totally umbilical lightlike submanifold of a semi-Riemannian manifold. Then M is minimal if and only if M is totally geodesic.*

Proof. Suppose M is minimal , then $h^s(X, Y) = 0$, $\forall X, Y \in \Gamma(\operatorname{Rad} TM)$ and, from Proposition 5.1.3, $h^l = 0$ on $\operatorname{Rad} TM$. Now choose an orthonormal basis $\{e_1, \ldots, e_{m-r}\}$ of $(\Gamma(S(TM)))$. Then, from (5.3.8) we obtain

$$\operatorname{tr} h(e_i, e_i) = \sum_{i=1}^{m-r} \varepsilon_i g(e_i, e_i) H^l + \varepsilon_i g(e_i, e_i) H^s.$$

Hence we have $\operatorname{tr} h(e_i, e_i) = (m - r)H^l + (m - r)H^s$. Then, since M is minimal and $\operatorname{ltr}(TM) \cap S(TM^{\perp}) = \{0\}$ we get $H^l = 0$ and $H^s = 0$. Hence M is totally geodesic. The converse is clear. $\square$

5.5 Warped product lightlike submanifolds

For basic information on Riemannian (semi-Riemannian), Lorentzian and lightlike warped products see Chapter 1. In this section we first present a new class of lightlike submanifold of semi-Riemannian manifold and investigate the geometry of this class by using warped products.

First of all, we briefly review the notion of lifting which is crucially important for computations on product manifolds; details can be found in [317]. Consider a product manifold $M \times N$. If $f \in C^{\infty}(M, \mathbf{R})$ the lift of f to $M \times N$ is $\tilde{f} = f \circ \pi \in C^{\infty}(M, \mathbf{R})$. If $x \in T_p(M), p \in M$ and $q \in N$ then the lift $\tilde{x}$ to (p, q) is the unique vector in $T_{(p,q)}M$ such that $\pi_*(\tilde{x}) = x$. If $X \in \Gamma(TM)$ the lift of X to $M \times N$ is the vector field $\tilde{X}$ whose value at each (p, q) is the lift of X_p to (p, q). Product coordinate systems show that $\tilde{X}$ is smooth. Let us denote vector fields on M (resp.N), lifted to $M \times N$, by $\Im(M)$ (resp. $\Im(N)$.) Then we have:

Lemma 5.5.1. [317]

(1) *If $\tilde{X}, \tilde{Y} \in \Im(M)$ then $[\tilde{X}, \tilde{Y}] = [\widetilde{X, Y}] \in \Im(M)$, and similarly for $\Im(N)$.*

(2) *If $\tilde{X} \in \Im(M)$ and $\tilde{V} \in \Im(N)$, then $[\tilde{X}, \tilde{V}] = 0$.*

Definition 5.5.2. Let (M_1, g_1) be a totally lightlike submanifold of dimension r and (M_2, g_2) be a semi-Riemannian submanifold of dimension m of a semi-Riemann manifold $(\bar{M}, \bar{g})$. Then the product manifold $M = M_1 \times_f M_2$ is said to be a warped product lightlike submanifold of $\bar{M}$ with the degenerate metric g defined by

$$g(X, Y) = g_1(\pi_* X, \pi_* Y) + (f \circ \pi)^2 g_2(\eta_* X, \eta_* Y) \tag{5.5.1}$$

for every $X, Y \in \Gamma(TM)$ and $*$ is the symbol for the tangent map. Here, $\pi : M_1 \times M_2 \to M_1$ and $\eta : M_1 \times M_2 \to M_2$ denote the projection maps given by $\pi(x, y) = x$ and $\eta(x, y) = y$ for $(x, y) \in M_1 \times M_2$.

It follows that the radical distribution $\operatorname{Rad} TM$ of M has rank r and its screen $S(TM)$ has rank m. Thus M is an r-lightlike submanifold of $\bar{M}$. From now on we consider warped product lightlike submanifolds in the form $M_1 \times_f M_2$, where M_1 is a totally lightlike submanifold and M_2 is a semi-Riemannian submanifold of $\bar{M}$. We say that M is a proper warped product lightlike submanifold if $M_1 \neq \{0\}$, $M_2 \neq \{0\}$ and f is non-constant on M.

Example 11. Let $\bar{M} = (\mathbf{R}^7_2, \bar{g})$, where $\mathbf{R}^7_2$ is a semi-Euclidean space of signature $(-, -, +, +, +, +, +)$ with respect to the canonical basis

$$\{\partial x_1, \partial x_2, \partial x_3, \partial x_4, \partial x_5, \partial x_6, \partial x_7\}.$$

Let M be a submanifold of $\mathbf{R}^7_2$ presented in Example 7 and, keeping the notation, we obtain that $\operatorname{Rad} TM$ and $S(TM)$ are integrable. Now, we denote the leaves of $\operatorname{Rad} TM$ and $S(TM)$ by M_1 and M_2, respectively. Then, the induced metric tensor of M is given by

$$ds^2 = 0(du_2^2 + du_3^2) + \frac{(u^1)^2}{2}(du_3^2 + du_4^2)$$

$$= \frac{(u^1)^2}{2}(du_3^2 + du_4^2).$$

Hence $M = M_1 \times_f M_2$ with $f = \frac{u^1}{\sqrt{2}}$.

Proposition 5.5.3. *There exist no proper isotropic or totally lightlike warped product submanifolds of a semi-Riemannian manifold $\bar{M}$.*

Proof. Let M be an isotropic warped product lightlike submanifold. Then $S(TM) = \{0\}$, so $M_2 = 0$. The proof of the other assertion is similar. □

Proposition 5.5.4. *Let $M = M_1 \times_f M_2$ be a proper warped product lightlike submanifold of a semi-Riemannian manifold $\bar{M}$. Then M_1 is totally geodesic in M as well as in $\bar{M}$.*

Proof. Let ∇ be a linear connection on M induced from $\bar{\nabla}$. We know that ∇ is not a metric connection. From the Kozsul formula we have

$$2\bar{g}(\bar{\nabla}_X Y, Z) = X\bar{g}(Y, Z) + Y\bar{g}(X, Z) - Z\bar{g}(X, Y)$$
$$+ \bar{g}([X, Y], Z) + \bar{g}([Z, X], Y) - \bar{g}([Y, Z], X)$$

for $X, Y \in \Gamma(TM_1)$ and $Z \in \Gamma(S(TM))$. On the other hand, from Lemma 5.5.1, we have $[X, Z] = 0$ for $X \in \Gamma(\operatorname{Rad} TM)$ and $Z \in \Gamma(S(TM))$. Thus we get

$$2\bar{g}(\bar{\nabla}_X Y, Z) = \bar{g}([X, Y], Z).$$

Using again Lemma 5.5.1, we get $[X, Y] \in \Gamma(\operatorname{Rad} TM)$. Hence we derive $2\bar{g}(\bar{\nabla}_X Y, Z) = 0$. Thus, from (5.1.15) we have $g(\nabla_X Y, Z) = 0$. This shows that M_1 is totally geodesic in M. On the other hand, from Proposition 5.1.3 of this chapter we conclude that any totally lightlike submanifold of a semi-Riemannian manifold $\bar{M}$ is totally geodesic in $\bar{M}$. Thus the proof is complete. □

Corollary 5.5.5. *Let* $M = M_1 \times_f M_2$ *be a proper warped product lightlike subman-ifold of a semi-Riemannian manifold* $\bar{M}$. *Then we have*

$$h^l(X, Z) = 0, \ h^*(X, Z) = 0, \forall X \in \Gamma(\operatorname{Rad} TM), Z \in \Gamma(S(TM)). \qquad (5.5.2)$$

Proof. From (5.1.15) we have $\bar{g}(h^l(X, Z), Y) = \bar{g}(\bar{\nabla}_X Z, Y)$ for $X, Y \in \Gamma(\operatorname{Rad} TM)$ and $Z \in \Gamma(S(TM))$. Hence, $\bar{g}(h^l(X, Z), Y) = -\bar{g}(\bar{\nabla}_X Y, Z) = -g(\nabla_X Y, Z)$. From Proposition 5.5.4, we have known that M_1 is totally geodesic in M. Hence we get $\bar{g}(h^l(X, Z), Y) = 0$, thus we obtain the first assertion of (5.5.2). In a similar way, we obtain the second assertion. $\qquad \square$

Proposition 5.5.6. *Let* $M = M_1 \times_f M_2$ *be a proper warped product lightlike sub-manifold of a semi-Riemannian manifold* $\bar{M}$. *Then* M *is totally umbilical in* $\bar{M}$ *if and only if* $h^s(X, Y) = g(X, Y)H^s$ *for* $X, Y \in \Gamma(TM)$, *where* H^s *is a smooth vector field on coordinate neighborhood* $\mathbf{U} \subset M$.

Proof. Let us suppose that $\nabla_X Z \in \Gamma(\operatorname{Rad} TM)$ for $X \in \Gamma(\operatorname{Rad} TM)$ and $Z \in \Gamma(S(TM))$. Then from the Kozsul formula we have

$$2\bar{g}(\bar{\nabla}_X Z, W) = X\bar{g}(Z, W)$$

for $W \in \Gamma(S(TM))$. Since by assumption $\nabla_X Z \in \Gamma(\operatorname{Rad} TM)$ and the definition of a warped metric tensor, using (5.1.15) we get

$$g(\nabla_X Z, W) = 0.$$

Hence we derive

$$X(f \circ \pi)^2 g_2(Z, W) = 0.$$

Since g_2 is constant on M_1 we obtain

$$\frac{X(f)}{f} g(Z, W) = 0;$$

here we have put f for $f \circ \pi$. Thus $X(f) = 0$ or $g_2(Z, W) = 0$. Since g_2 is non-degenerate and f is not constant, we get a contradiction, so $\nabla_X Z$ does not belong to $\operatorname{Rad} TM$. Now, since $\bar{\nabla}$ is a metric connection, we have

$$\bar{g}(\bar{\nabla}_Z W, X) = -\bar{g}(W, \bar{\nabla}_Z X)$$

for $X \in \Gamma(\operatorname{Rad} TM)$ and $Z, W \in \Gamma(S(TM))$. Using (5.1.15) we have

$$\bar{g}(h^l(Z, W), X) = -g(W, \nabla_Z X) = -g(W, \nabla_X Z).$$

Hence

$$\bar{g}(h^l(Z, W), X) = -\frac{X(f)}{f} g(Z, W). \qquad (5.5.3)$$

Thus the proof follows from (5.5.3), Corollary 5.5.5 and the definition of a totally umbilical lightlike submanifold. $\qquad \square$

Theorem 5.5.7. *Let $M = M_1 \times_f M_2$ be a proper warped product lightlike subman-ifold of a semi-Riemannian manifold $\bar{M}$. Then the induced connection ∇ is never a metric connection.*

Proof. Let us suppose that ∇ is a metric connection on M. Then $h^l = 0$. Thus from (5.5.3) we obtain $\frac{X(f)}{f} g(Z, W) = 0$ for $X \in \Gamma(\operatorname{Rad} TM)$ and $Z, W \in \Gamma(S(TM))$. Hence $X(f) f g_2(Z, W) = 0$. Thus, $f \neq 0$ implies that $X(f) = 0$ or $g_2(Z, W) = 0$. Since M is a proper warped product lightlike submanifold and g_2 is non-degenerate, this is a contradiction. □

Corollary 5.5.8. *There exist no totally geodesic proper warped product lightlike submanifolds of a semi-Riemannian manifold $\bar{M}$.*

It follows from Lemma 5.5.1 that the radical distribution and the screen distribution of M are integrable.

Theorem 5.5.9. *Let $M = M_1 \times_f M_2$ be a proper warped product lightlike submanifold of a semi-Riemannian manifold $\bar{M}$. Then M is totally umbilical if any leaf of screen distribution is so immersed as a submanifold of $\bar{M}$ and $\bar{g}(D^l(Z, W), X) = 0$ for$X \in \Gamma(\operatorname{Rad} TM), Z \in \Gamma(S(TM))$ and $W \in \Gamma(S(TM^\perp))$.*

Proof. We note that from (5.5.1), we get $g(X, Y) = (f \circ \pi)^2 g_2(PX, PY)$ for $X, Y \in \Gamma(TM)$ From Proposition 5.5.4 we know that $\operatorname{Rad} TM$ defines a totally geodesic foliation in $\bar{M}$, hence $h^l(X, Y) = h^s(X, Y) = 0$ for $X, Y \in \Gamma(\operatorname{Rad} TM)$. Moreover from Corollary 5.5.5, we have $h^l(X, Z) = 0$ and $h^*(X, Z) = 0$ for $X \in \Gamma(\operatorname{Rad} TM)$ and $Z \in \Gamma(S(TM))$. Now, from (5.1.15) and (5.1.23) we obtain

$$\bar{g}(h^s(X, Z), W) = \bar{g}(X, D^l(Z, W)) \tag{5.5.4}$$

for $X \in \Gamma(\operatorname{Rad} TM), Z \in \Gamma(S(TM))$ and $W \in \Gamma(S(TM^\perp))$. On the other hand, from (5.1.15) we write

$$\bar{\nabla}_Z V = \nabla'_Z V + h'(Z, V)$$

for $Z, V \in \Gamma(S(TM))$, where h' is the second fundamental form of M_2 in $\bar{M}$ and ∇' is the metric connection of M_2 in $\bar{M}$. Hence we have

$$h'(Z, V) = h^*(Z, V) + h^l(Z, V) + h^s(Z, V). \tag{5.5.5}$$

Thus the proof follows (5.5.4) and (5.5.5). □

Theorem 5.5.10. *Let $M = M_1 \times_f M_2$ be a coisotropic warped product lightlike submanifold of a semi-Riemannian manifold $\bar{M}$. Then M is totally umbilical if any leaf of screen distribution is so immersed as a submanifold of $\bar{M}$.*

Proof. If M is coisotropic, then $S(TM^\perp) = \{0\}$. Thus, $h^s = 0$ on M. Then the proof follows from Proposition 5.5.4, Corollary 5.5.5 and (5.5.5). □

Example 12. Let $\bar{M} = (\mathbf{R}_3^6, \bar{g})$ be a semi-Riemannian manifold, where $\mathbf{R}_3^6$ is a semi-Euclidean space of signature $(-, -, -, +, +, +)$ with respect to the canonical basis

$$\{\partial x_1, \partial x_2, \partial x_3, \partial x_4, \partial x_5, \partial x_6\}.$$

Let M be a submanifold of $\mathbf{R}_3^6$ given by

$$x^1 = u^1, \qquad x^2 = u^2, \qquad x^3 = u^1 \cos u^3 \sinh u^4,$$
$$x^4 = u^1 \cos u^3 \cosh u^4, \qquad x^5 = u^2, \qquad x^6 = u^1 \sin u^3,$$

where $u^3 \in \mathbf{R} - \{k \frac{\pi}{2} \, k \in Z\}$. Then TM is spanned by

$$Z_1 = \partial x_1 + \cos u^3 \sinh u^4 \, \partial x_3 + \cos u^3 \cosh u^4 \, \partial x_4 + \sin u^3 \, \partial x_6$$
$$Z_2 = \partial x_2 + \partial x_5$$
$$Z_3 = -u^1 \sin u^3 \sinh u^4 \, \partial x_3 - u^1 \sin u^3 \cosh u^4 \, \partial x_4 + u^1 \cos u^3 \, \partial x_6$$
$$Z_4 = u^1 \cos u^3 \cosh u^4 \, \partial x_3 + u^1 \cos u^3 \sinh u^4 \, \partial x_4.$$

Thus M is 2-lightlike. Choose $S(TM) = \mathrm{Span}\{Z_3, Z_4\}$, then it follows that a lightlike transversal vector bundle $\mathrm{ltr}(TM)$ is spanned by

$$N_1 = \frac{1}{2}\{-\partial x_1 + \cos u^3 \sinh u^4 \, \partial x_3 + \cos u^3 \cosh u^4 \, \partial x_4 + \sin u^3 \, \partial x_6\}$$
$$N_2 = \frac{1}{2}\{-\partial x_2 + \partial x_5\}.$$

Hence M is a coisotropic submanifold. By direct calculations, we conclude that $\mathrm{Rad}\, TM$ and $S(TM)$ are integrable in M. Now denote the leaves of $\mathrm{Rad}\, TM$ and $S(TM)$ by M_1 and M_2. We also obtain that the induced metric tensor is

$$ds^2 = (u^1)^2 (du_3^2 - \cos^2 u^3 \, du_4^2).$$

Thus M is a coisotropic warped product submanifold of $\mathbf{R}_3^6$ with $f = u^1$. By direct calculations, using Gauss formulas (5.1.15) and (5.1.26) we obtain

$$h^l(X, Z_1) = h^l(X, Z_2) = h^l(Z_3, Z_4) = 0, \forall X \in \Gamma(TM),$$
$$h^l(Z_3, Z_3) = -u^1 N_1, \quad h^l(Z_4, Z_4) = u^1 \cos^2 u^3 N_1 \qquad (5.5.6)$$

and

$$h^*(X, Z_1) = h^*(X, Z_2) = h^*(Z_3, Z_4) = h^*(Z_4, Z_3) = 0, \forall X \in \Gamma(TM),$$
$$h^*(Z_3, Z_3) = -\frac{1}{2}u^1 Z_1, \quad h^*(Z_4, Z_4) = \frac{1}{2} u^1 \cos^2 u^3 Z_1.$$

Denote the second fundamental form of M_2 in $\bar{M}$ by h'. Then,

$$h'(Z_3, Z_4) = 0$$

and

$$h'(Z_3, Z_3) = -(u^1)^2(N_1 + \frac{1}{2}Z_1), \; h'(Z_4, Z_4) = (u^1)^2 \cos^2 u^3 (N_1 + \frac{1}{2}Z_1).$$

Hence we have

$$h'(X, Y) = g(X, Y)H',$$

where $H' = -N_1 - \frac{1}{2}Z_1$ for $X, Y \in \Gamma(S(TM))$. Thus, it follows that M_2 is totally umbilical in $\bar{M}$. On the other hand, from (5.5.6) we have

$$h^l(X, Y) = g(X, Y)H^l,$$

where $H^l = -\frac{1}{u_1}N_1$ for $X, Y \in \Gamma(TM)$. Thus, it follows that M is also totally umbilical in $\bar{M}$.

Chapter 6

Submanifolds of indefinite Kähler manifolds

In the 1996 book [149] there is a brief discussion on Cauchy-Riemann(CR) lightlike submanifolds of an indefinite Kähler manifold. Contrary to the non-degenerate case [45, 133, 373], CR-lightlike submanifolds are non-trivial (i.e., they do not include invariant (complex) and real parts). Since then considerable work has been done on new concepts to obtain a variety of classes of lightlike submanifolds. In this chapter we present up-to-date new results on all possible (complex, screen real and totally real) lightlike submanifolds of an indefinite Kähler manifold.

6.1 Introduction

In this section, we review indefinite Kähler manifolds and CR-submanifolds of Kählerian manifolds needed for the lightlike submanifolds.

Definition 6.1.1. Let $\mathbf{V}$ be a real vector space on $\mathbf{R}$. Then its *complexification* is the complex vector space $\mathbf{V}^C$, denoted by $(\mathbf{V}^C, C)$, where

(i) $Z = X + iY \in \mathbf{V}^C \Longleftrightarrow X, Y \in \mathbf{V}$,

(ii) $(X_1 + iY_1) + (X_2 + iY_2) = (X_1 + X_2) + i(Y_1 + Y_2), \forall X_1, X_2, Y_1, Y_2 \in \mathbf{V}$,

(iii) $(\alpha + i\beta)(X + iY) = (\alpha X - \beta Y) + i(\beta X + \alpha Y) \forall X, Y \in \mathbf{V}, \alpha, \beta \in \mathbf{R}$.

On the other hand, *complex conjugation* in $\mathbf{V}^C$ is defined by $\bar{Z} = X - iY$. Also note that $\mathbf{V}^C$ satisfies the axioms for a complex vector space [309].

Definition 6.1.2. Let $\mathbf{V}$ be a real vector space and $J : V \longrightarrow V$ be an endomorphism such that $J^2 = -I$, where I denotes the identity endomorphism. Then, J is called a complex structure on $\mathbf{V}$.

Denote a real vector space with complex structure by $(\mathbf{V}, R, J)$. It is known that $(\mathbf{V}, R, J)$ is even-dimensional. Define complex scalar multiplication as $(a + ib)v = av + bJv, \quad v \in \mathbf{V}$. Then, it is obvious that an $\mathbf{R}$-linear operator commutes with J and is $\mathbf{C}$-linear. Thus, a complex structure on $\mathbf{V}$ gives a complex vector space structure. Note that if $\dim(\mathbf{V}, R) = n = 2m$, then $\dim(\mathbf{V}, C, J) = m$. Thus it follows that $(\mathbf{V}, C, J)$ can not be the same as $(\mathbf{V}^C, C)$. However, we show that there is a natural relation between these two vector spaces. For this, we need the following. Let $(\mathbf{V}, R, J)$ be a real vector space with a complex structure J. Define the following subsets of $\mathbf{V}^C$;

$$W(J) = \{Z \mid Z = X - iJ(X) \quad \text{and} \quad X \in V\},$$
$$\bar{W}(J) = \{Z \mid Z = X + iJ(X) \quad \text{and} \quad X \in V\}.$$

It is easy to verify that $W(J)$ and $\bar{W}(J)$ are subspaces of the complexificiation $\mathbf{V}^C$ of $\mathbf{V}$. Also, $W(J)$ and $\bar{W}(J)$ are complex conjugate to each other.

Theorem 6.1.3. [190] *If $(\mathbf{V}, R, J)$ is a real vector space with a complex structure J, then, $\mathbf{V}^C = W(J) \oplus \bar{W}(J)$. Conversely, given $(\mathbf{V}, R)$, U and $\bar{U}$ such that U and $\bar{U}$ are subspaces of $(\mathbf{V}^C, C)$ as complex conjugate to each other and $\mathbf{V}^C = U \oplus \bar{U}$, then there exists a complex structure J for V such that $W(J) = U$ and $\bar{W}(J) = \bar{U}$.*

It follows from the above theorem that a complex structure J on $\mathbf{V}$ is equivalent to a choice of complex conjugate subspaces $W(J)$ and $\bar{W}(J)$ of the complexification $\mathbf{V}^C$ with the property $\mathbf{V}^C = W(J) \oplus \bar{W}(J)$. We also note that a complex structure J determines an orientation of $\mathbf{V}$.

Let $\mathbf{V}$ be a complex vector space and V be the underlying real vector space of $\mathbf{V}$, that is, V is a real vector space when we consider $\mathbf{V}$ as a real vector space by forgetting about multiplication with i [273]. Then, the induced isomorphism J on V by $\mathbf{V}$ as defined by $J\alpha = i\alpha$ is a complex structure on V, called the induced complex structure on V by $\mathbf{V}$. For example, the underlying real vector space of $\mathbf{C}^n$ is identified with $\mathbf{R}^{2n}$ by $(x_1 + iy_1, \ldots, x_n + iy_n) \longrightarrow (x_1, \ldots, x_n, y_1, \ldots, y_n)$. Thus, the induced complex structure, say J_o, on the underlying real vector space of $\mathbf{C}^n$ is said to be the canonical complex structure of $\mathbf{R}^{2n}$.

Definition 6.1.4. A Hermitian structure on a real vector space V with a complex structure J is a linear map $h : \mathbf{V} \times \mathbf{V} \to \mathbf{C}$ with the properties

(a) $h(JX, Y) = ih(X, Y)$,

(b) $h(X, Y) = \overline{h(Y, X)}, \quad \forall X, Y \in V$.

Let h be a Hermitian scalar product on $\mathbf{V}$. Then, from [309], it is known that there exists an ordered basis $\{v_1, \ldots, v_n\}$ of $\mathbf{V}$ such that

(1) $h(v_i, v_j) = 0$ for $i \neq j$,

(2) $h(v_i, v_i) = 0$, for $1 \leq i \leq r$,

(3) $h(v_i, v_i) = -1$ for $r + 1 \leq i \leq r + q$,

(4) $h(v_i, v_i) = 1$ for $r + q + 1 \le i \le n = r + q + p$,

where $r = \dim(\text{Rad }\mathbf{V})$, q is the index of $\mathbf{V}$ and p is the number of unit spacelike vectors. In this case, $(\mathbf{V}, h)$ is called a Hermitian inner product.

Let (V, h) be an inner product space and let $h = g + i\Omega$, where $g = Re(h)$ and $\Omega = \text{Im}(h)$. If $\mathbf{V}$ is a real vector space then J is a complex structure on $\mathbf{V}$. By using (b), we have

$$g(X, Y) - i\Omega(X, Y) = g(Y, X) + i\Omega(Y, X).$$

Equating real and imaginary parts, we get

$$g(X, Y) = g(Y, X), \quad \Omega(X, Y) = -\Omega(Y, X).$$

It follows that g is symmetric, Ω is a skew form on V such that $g(JX, JY) = g(X, Y)$, $\Omega(JX, JY) = \Omega(X, Y)$ and $\Omega(X, Y) = g(X, JY)$ for all $X, Y \in V$. Conversely, let (V, J) be a complex vector space and g be an inner product on V with $g(JX, JY) = g(X, Y)$ for every $X, Y \in V$. By setting $\Omega(X, Y) = g(X, JY)$, we obtain $h = g + i\Omega$ which is a Hermitian inner product on $\mathbf{V} = (V, J)$. Also it is easy to see that h is non-degenerate on V if and only if g is non-degenerate on V. Note that, on $\mathbf{C^n}$ the dot product is a positive definite Hermitian product $u \cdot v = \sum u_a \bar{v}_a$. Thus, the real part of a Hermitian scalar product h is a real scalar product.

Let M be a C^∞ real $2n$-dimensional manifold, covered by coordinate neighborhoods with coordinates (x^i), where i runs over $1, 2, \ldots, n, \bar{1}, \bar{2}, \ldots, \bar{n}$ and the index a runs over $1, 2, \ldots, n$. M can be considered as a *complex manifold* of dimension n if we define complex coordinates $(z^a = x^a + iy^a)$ on a neighborhood of $z \in M$ such that the intersection of any two such coordinate neighborhoods is regular. During the 1950s it was popular to construct complex manifolds in this sense, as real manifolds with an additional differential geometric structure, since such approach was amenable to the fiber bundle theory. We explain this additional structure as follows:

Let there exist an endomorphism J (a tensor field J of type $(1, 1)$) of the tangent space $T_p(M)$, at each point p of M, such that

$$J(\partial / \partial x^a) = \partial / \partial y^a, \qquad J(\partial / \partial y^a) = -(\partial / \partial x^a).$$

Hence $J^2 = -I$, where I is the identity morphism of $T(M)$. Alternatively, M taken as an n-dimensional complex manifold admits a globally defined tensor field,

$$J = i\partial / \partial z^a \otimes dz^a - i\partial / \partial \overline{z}^a \otimes \overline{dz}^a$$

which retains the property $J^2 = -I$ and remains as a real tensor with respect to any of the z^a-coordinate charts. Moreover, the tensor field J (expressed as above) reminds the real manifold that it has a complex structure.

Now the question is whether the above property is sufficient for the existence of a complex structure on M. Fukami-Ishihara [195], in 1955, answered this question in the negative by proving that a 6-dimensional sphere S^6 has no complex structure but its tangent bundle admits such an endomorphism J. Therefore, we call a real $2n$-dimensional M, endowed with J satisfying $J^2 = -I$, an *almost complex manifold* and J its *almost complex structure* tensor. It is known that an almost complex manifold is even-dimensional. A tensor field N of type $(1, 2)$ defined by

$$N_J(X, Y) = [JX, JY] - J^2[X, Y] - J([X, JY] + [JX, Y]), \qquad (6.1.1)$$

for any $X, Y \in \Gamma(TM)$, is called the *Nijenhuis tensor field* of J. Then, J defines a complex structure [305] on M if and only if N vanishes on M.

From a mathematical standpoint, the purpose of introducing complex manifolds is to study holomorphic functions. In fact, a complex structure for a manifold can be thought as a consistent rule for deciding which complex-valued functions on the manifold are holomorphic. In this perspective, it is important to show the relationship between a holomorphic map and a complex manifold. To see this relationship, we give the following definition.

Definition 6.1.5. Let M and M' be almost complex manifolds with almost complex structures J and J', respectively. A mapping $f : M \longrightarrow M'$ is said to be *almost complex* if $J'f_* = f_*J$.

Proposition 6.1.6. [413] *Let M and M' be complex manifolds. A mapping $f : M \longrightarrow M'$ is holomorphic if and only if f is almost complex with respect to the complex structures of M and M'.*

From the point of differential geometry (which is the focus of study in this book), it is desirable to define a metric on M. Therefore, consider a Riemannian metric g on an almost complex manifold (M, J). We say that the pair (J, g) is an *almost Hermitian structure* on M, and M is an *almost Hermitian manifold* if

$$g(JX, JY) = g(X, Y), \qquad \forall X, Y \in \Gamma(M). \qquad (6.1.2)$$

Moreover, if J defines a complex structure on M, then (J, g) and M are called *Hermitian structure* and *Hermitian manifold*, respectively. The fundamental 2-form Ω of an almost Hermitian manifold is defined by

$$\Omega(X, Y) = g(X, JY), \quad \forall X, Y \in \Gamma(M). \qquad (6.1.3)$$

A Hermitian metric on an almost complex M is called a *Kähler metric* and then M is called a *Kähler manifold* if Ω is closed, i.e.,

$$d\,\Omega(X, Y, Z) = 0, \quad \forall X, Y \in \Gamma(M), \qquad (6.1.4)$$

where ∇ is the Riemannian connection of g. It is known (see Kobayashi-Nomizu [263]) that the Kählerian condition (6.1.4) is equivalent to

$$(\nabla_X J)Y = 0, \forall X, Y \in \Gamma(M). \qquad (6.1.5)$$

For a Kähler manifold, the following identities hold:

$$R(X, Y)\, JZ = J\, R(X, Y)Z, \quad Q\, J = J\, Q,$$

$$\mathrm{Ric}(X,\, Y) = \frac{1}{2}\{\mathrm{tr}(J) \circ R(X,\, JY)\},$$

where Q is the Ricci operator. Now we deal with the question of why we impose a Kählerian structure on M. For this purpose, we prefer using a complex coordinate system and express the 2-form Ω from the Hermitian metric as follows: Consider the existence of a smooth metric of the form

$$ds^2 = g_{a\bar{b}}(z^c, \overline{z^c})dz^a\, \overline{dz^b} + g_{\bar{a}b}(z^c,\, \overline{z^c})\, \overline{dz^a}dz^b$$

such that the Hermitian property $g_{a\bar{b}} = \overline{g_{\bar{a}b}}$ is satisfied to assure that the metric is real-valued with respect to the underlying real coordinates. Then,

$$\Omega = ig_{a\bar{b}}dz^a \wedge \overline{dz^b} - ig_{\bar{a}b}\,\overline{dz^a} \wedge dz^b$$

which is closed if and only if M is Kählerian. The following is another way of characterizing a Kählerian manifold. A Hermitian manifold is Kählerian if and only if its Hermitian metric g is locally derivable from a real scalar potential $\mathbf{K}$ (called, the *Kähler scalar*) according to

$$g_{a\bar{b}} = \frac{\partial^2 \mathbf{K}}{\partial z^a \overline{\partial z^b}}.$$

Then, the Riemann curvature tensor is locally given by

$$R_{abcd} = \frac{\partial^4 \mathbf{K}}{\partial z^a\, \partial z^b\, \overline{\partial z^c}\, \partial z^d} - g^{\bar{s}t} \frac{\partial^3 \mathbf{K}}{\partial z^{\bar{s}}\, \partial z^b \partial z^d} \frac{\partial^3 \mathbf{K}}{\partial z^t \overline{\partial z^a}\, \overline{\partial z^c}}$$

together with those components $(R_{\bar{b}\bar{a}cd}, R_{\bar{a}bd\bar{c}}, R_{\bar{b}\bar{a}d\bar{c}})$ which can be obtained from these by using the symmetries of the curvature tensor. All other components vanish. The Ricci tensor is given by

$$R_{\bar{a}b} = \frac{\partial^2}{\partial z^a\, \partial z^b}\, (\log(\det\, g_{r\bar{s}})) = R_{b\bar{a}}, \quad R_{ab} = 0 = R_{\bar{a}\bar{b}}.$$

Consequently, it is useful to have a Kählerian manifold since for this case the Ricci tensor, like the metric tensor, is derivable from a local scalar potential $\mathbf{K}$. Also, just like the metric tensor, the Ricci tensor has its uniquely associated real closed 2-form, called the Ω' form, given by

$$\Omega' = iR_{a\bar{b}}dz^a \wedge \overline{dz^b} - iR_{\bar{a}b}\,\overline{dz^a} \wedge dz^b, \quad d(\Omega') = 0.$$

Theorem 6.1.7. [413] *Let M be a real $2n$-dimensional Kählerian manifold. If M is of constant curvature, then, M is flat provided $n > 1$.*

The above theorem tells us that the notion of constant curvature for Kählerian manifolds is not essential. Therefore, the notion of constant holomorphic sectional curvature was introduced for Kählerian manifolds. For this purpose, we first state the notion of holomorphic section as follows:

Consider a vector U at a point p of a Kähler manifold M. Then, the pair (U, JU) determines a plane π (since JU is obviously orthogonal to U) element called a *holomorphic section*, whose curvature K is given by

$$K = \frac{g(R(U, JU)JU, U)}{(g(U, U))^2},$$

and is called the *holomorphic sectional curvature* with respect to U. If K is independent of the choice of U at each point, then $K = c$, an absolute constant. A simply connected complete Kähler manifold of constant sectional curvature c is called a *complex space-form*, denoted by $M(c)$, which can be identified with the complex projective space $P_n(c)$, the open ball D_n in C^n or C^n according as $c > 0$, $c < 0$ or $c = 0$. The curvature tensor of $M(c)$ is

$$R(X, Y)Z = \frac{c}{4} [g(Y, Z)X - g(X, Z)Y + g(JY, Z)JX$$
$$- g(JX, Z)JY + 2g(X, JY)JZ]. \tag{6.1.6}$$

A contraction of the above equation shows that $M(c)$ is Einstein, that is

$$\operatorname{Ric}(X, Y) = \frac{n+1}{2} c g(X, Y).$$

The above results also hold for a class of semi-Riemannian manifolds (see Barros-Romero [28]), subject to the following restrictions on the signature of its indefinite metric g. For the endomorphism J, satisfying $J^2 = -I$, the eigenvalues of J are $i = \sqrt{-1}$ and $-i$ each one of multiplicity n. As J is real, and satisfies $J^2 = -I$, the only possible signatures of g are $(2p, 2q)$ with $p + q = n$. For example, g can not be a Lorentz metric and therefore, for real J, a spacetime can not be almost Hermitian, and therefore, can not be Hermitian or Kählerian. Subject to this restriction, according to Barros and Romero [28], (J, g) and M are called *indefinite Hermitian (or Kähler) structure* and *indefinite Kähler*, respectively.

The sectional curvature function K is defined for a non-degenerate plane π of T_pM the same as in Riemannian geometry. The restriction of K to the non-degenerate holomorphic (respectively totally real) planes is called the *holomorphic (respectively totally real) curvature* of M. We quote the following result on holomorphic sectional curvature:

Proposition 6.1.8. [28] *Let (M, g, J) be an indefinite Kähler manifold. Then all the non-degenerate holomorphic planes, for every $p \in M$, have the same sectional curvature $c \in R$ iff $R = cR^o$, where R^o is defined by*

$$R^o(u, v)w = \frac{1}{4}\{g(v, w)u - g(u, w)v + g(Jv, w)Ju$$
$$- g(Ju, w)Jv + 2g(u, Jv)Jw\}, \quad \forall u, v, w \in T_pM.$$

Proof. Assume $H_u = c$ for all $u \in TpM$ with $g(u, u) \neq 0$, where H_u is the restriction of K to the non-degenerate holomorphic planes. Then the tensor $R' = R - cR_0$ satisfies

$$g(R(u, Ju)Ju, u) = 0, \forall u \in T_pM \quad \text{with} \quad g(u, u) \neq 0. \tag{6.1.7}$$

Since $\{u \in T_pM \mid g(u, u) \neq 0\}$ is dense in T_pM, (6.1.7) holds without restriction on u. But R' satisfies the curvature identities defined in chapter 1 and therefore (6.1.7) implies by the well-known argument $R' = 0$. The converse is trivial $\quad\square$

From Proposition 6.1.8, we have the following result:

Corollary 6.1.9. [28] *If the connected complex n-dimensional indefinite Kähler manifold M with $n \geq 2$ admits a function $f : M \to \mathbf{R}$ such that $K(\pi) = f(p)$ for all non-degenerate holomorphic planes of T_pM, then, f must be of constant value $c \in \mathbf{R}$, that is, M is an indefinite complex space form of constant holomorphic sectional curvature c.*

Barros and Romero [28] constructed $\mathbf{C_q^n}$, $\mathbf{CP_q^n}(\mathbf{c})$ and $\mathbf{CH_q^n}(\mathbf{c})$ as examples of simply connected *indefinite complex space-forms* according as $c = 0$, $c > 0$ and $c < 0$, respectively. Here, $2q$ is the index of g. In the following we reproduce the details on these examples:

Example 1. The complex manifold $\mathbf{C^n}$, with the real part of the Hermitian form

$$b_{q,n}(z, w) = -\sum_{k=1}^{q} \bar{z}_k w_k + \sum_{j=q+1}^{n} \bar{z}_j w_j, \quad \forall z, w \in \mathbf{C^n} \tag{6.1.8}$$

defines a flat indefinite complex space form of index $2q$, denoted by C_q^n. In fact, C_q^n can be identified with (R_{2q}^{2n}, J, g) where J and g are given by

$$J(x^1, \bar{x}^1, \ldots, x^n, \bar{x}^n) = (-\bar{x}^1, x^1, \ldots, -\bar{x}^n, x^n),$$

$$g((x^1, \bar{x}^1, \ldots, x^n, \bar{x}^n), (u^1, \bar{u}^1, \ldots, u^n, \bar{u}^n)) = -\sum_{i=1}^{q} (x^i u^i + \bar{x}^i \bar{u}^i)$$
$$+ \sum_{a=q+1}^{q} (x^a u^a + \bar{x}^a \bar{u}^a).$$

Example 2. Define a complex n-dimensional indefinite space form $\mathbf{CP_q^n(c)}$, with index $2q$ and a constant holomorphic sectional curvature c. Its underlying complex manifold is the open submanifold

$$\{z \in \mathbf{C^{n+1}} / b_{q,n+1}(z, z) > 0\} / \mathbf{C^*} \quad \text{of} \quad \mathbf{CP^n} = \{\mathbf{C^{n+1}} - \{0\}\} / \mathbf{C^*},$$

the ordinary complex projective space. Now consider the semi-Riemannian submanifold

$$S_{2q}^{2n+1}(c/4) = \{z \in \mathbf{C}^{n+1}/b_{z,n+1}(z,z) = (4/c)\} \subset \mathbf{C}_q^{n+1},$$

which is of constant curvature $c/4$ [317, 2.4.5a]. Evidently

$$\phi : \mathbf{S}_{2q}^{2n+1}(c/4) \to \mathbf{CP_q^n}(z \mapsto z \cdot \mathbf{C}^*) \tag{6.1.9}$$

is a submersion which is the indefinite *Hopf fibration* of $\mathbf{S}_{2q}^{2n+1}(c/4)$ with spacelike fibers $z \cdot \mathbf{S}^1$ for $z \in \mathbf{S}_{2q}^{2n+1}(c/4)$. Then there exists a unique indefinite Kähler metric of index $2q$ on $\mathbf{CP_q^n(c)}$ such that (3.1.8) becomes its semi-Riemannian submanifold. As in the case for $\mathbf{CP^n}$, it follows analogously that $\mathbf{CP_q^n(c)}$ endowed with this metric becomes an indefinite complex space form of index $2q$ with constant holomorphic sectional curvature c.

Example 3. Similarly, the indefinite hyperbolic space form $\mathbf{CH_q^n(c)}$ of complex dimension n with index $2q$ and of constant holomorphic sectional curvature $c < 0$ can be obtained from $\mathbf{CP_{n-q}^n(-c)}$ by replacing its metric by its negative.

Thus, we state (the proof is same as in the Riemannian case):

Theorem 6.1.10. [28] *Every connected, simply-connected, complete indefinite Kähler manifold of complex dimension n, of index $2q$, and of constant holomorphic sectional curvature c is holomorphically isometric to $\mathbf{C_q^n}$, $\mathbf{CP_q^n(c)}$ or $\mathbf{CH_q^n(c)}$ according as $c = 0$, $c > 0$ and $c < 0$, respectively.*

Finally, we know from a lemma of Bishop-Goldberg [64] that the holomorphic sectional curvature defined on the unit bundle of an almost Hermitian manifold (M, g, J), with positive definite metric g, is bounded at each point of M. This is not true for an indefinite metric as it is known that its holomorphic sectional curvature is bounded (above or below) if and only if it is constant [28].

Cauchy Riemann Submanifolds of Kähler Manifolds. According to the behavior of the tangent bundle of a submanifold with respect to the action of the almost complex structure $\bar{J}$ of the ambient manifold, there are four popular classes of submanifolds, namely, Kähler submanifolds, totally real submanifolds, CR-submanifolds and slant submanifolds. A submanifold of a Kähler manifold is called a complex submanifold if each of its tangent spaces is invariant under the almost complex structure of the ambient manifold. A complex submanifold of a Kähler manifold is a Kähler manifold with induced metric structure, called a Kähler submanifold. It is well known that a Kähler submanifold of a Kähler manifold is always minimal[413] and [313]. A totally real submanifold M is a submanifold such that the almost complex structure $\bar{J}$ of the ambient manifold $\bar{M}$ carries a tangent space of M into the corresponding normal space of M. A totally real submanifold is called Lagrangian if $\dim_R M = \dim_C \bar{M}$. Real curves of Kähler manifolds are examples of totally real submanifolds. The first contribution to the geometry of totally real submanifolds was given in the early 1970s [111]. For details, see [410].

In 1978, generalizing these two types of submanifolds, Bejancu [45] defined Cauchy-Riemann (CR) submanifolds as follows: By a CR submanifold we mean a real submanifold M of an almost Hermitian manifold $(\bar{M}, J, \bar{g})$, carrying a J-invariant distribution D (i.e., $JD = D$) and whose $\bar{g}$-orthogonal complement is J-anti-invariant (i.e., $JD^{\perp} \subseteq T(M)^{\perp}$), where $T(M)^{\perp} \rightarrow M$ is the normal bundle of M in $\bar{M}$. The CR submanifolds serve as an umbrella of a variety (such as invariant (complex), anti-invariant (totally real), semi-invariant, generic) of submanifolds. Basic details on these may be seen in [45, 412, 413]. Here we only briefly present the following needed to understand the sections on lightlike submanifolds.

It is easy to see that a CR-submanifold is holomorphic or totally real according as $D^{\perp} = \{0\}$ or $D = \{0\}$ respectively, otherwise it is called a proper CR-submanifold. Every real hypersurface of an almost Hermitian manifold is an example of a proper CR-submanifold [45].

On the other hand, a CR manifold (independent of its landing space) is a C^{∞} differentiable manifold M with a holomorphic sub-bundle H of its complexified tangent bundle $T(M) \otimes C$, such that $H \cap \bar{H} = \{0\}$ and H is involutive (i.e., $[X, Y] \in H$ for every $X, Y \in H$). For an update on CR manifolds (out of the scope of this book) we refer [73] and two recent books [132] by Dragomir and Tomassini and [26] by Barletta, Dragomir and Duggal. Here we recall that a CR-manifold can be characterized by means of the real invariant distribution D as follows:

Theorem 6.1.11. [45] *A smooth manifold L is a CR manifold, if and only if, it is endowed with an almost complex distribution $(\mathcal{D}, J)$ such that*

$$[JX, JY] - [X, Y] \in \Gamma(\mathcal{D}), \tag{6.1.10}$$

$$N_J(X, Y) = 0, \tag{6.1.11}$$

for all $X, Y \in \Gamma(\mathcal{D})$, where N_J is the operator for the Nijenhuis tensor field of J.

Blair and Chen [69] were the first to interrelate these two concepts by proving that *proper CR submanifolds, of a Hermitian manifold, are CR manifolds.* They obtained the following result for CR-submanifolds of complex space forms:

Theorem 6.1.12. [69] *Let M be a submanifold of a complex space form $\bar{M}(c)$ with $c \neq 0$. Then M is a CR-submanifold of $\bar{M}$ if and only if the maximal holomorphic subspace $\mathcal{D}_p = T_pM \cap JT_pM, p \in M$, defines a non-trivial differentiable distribution $\mathcal{D}$ on M such that*

$$g(\bar{R}(X,Y)Z, W) = 0, \forall X, Y \in \Gamma(\mathcal{D}) \quad \text{and} \quad \forall Z, W \in \Gamma(\mathcal{D}^{\perp})$$

where $\mathcal{D}^{\perp}$ denotes the orthogonal complementary distribution of $\mathcal{D}$ in M.

On the integrability of two distributions we recall the following:

(i) *The totally real distribution $\mathcal{D}^{\perp}$ on M is always integrable[99].*

(ii) *The holomorphic distribution $\mathcal{D}$ is integrable if and only if the second fundamental form of M satisfies [45]*

$$h(X, JY) = h(JX, Y), \forall X, Y \in \Gamma(\mathcal{D}).$$

Thus, from (i), it follows that every proper CR-submanifold of a Kähler manifold is foliated by totally real submanifolds. Note that Chen's integrability theorem was extended to CR-submanifolds of several other Riemannian manifolds. For example, this theorem was used for CR-submanifolds of locally conformal Kähler manifolds [131]. Although, in general, the holomorphic distribution is not integrable, Chen proved the following:

Proposition 6.1.13. [99] *The holomorphic distribution of a CR-submanifold of a Kähler manifold is always minimal.*

On totally umbilical CR-submanifolds we need the following:

Theorem 6.1.14. [45] *Let M be a totally umbilical CR-submanifold of a Kähler manifold $\bar{M}$. Then either*

(a) *M is totally geodesic, or*

(b) *M is totally real, or*

(c) *the totally real distribution is 1-dimensional.*

A CR-submanifold M of a Kähler manifold is called a *mixed geodesic CR-submanifold* if its second fundamental form satisfies $h(X, Y) = 0$ for $X \in \Gamma(\mathcal{D})$ and $Y \in \Gamma(\mathcal{D}^\perp)$. On this class, we quote the following:

Theorem 6.1.15. [45] *A CR-submanifold M of a Kähler manifold is mixed totally geodesic if and only if each of the totally real distributions is totally geodesic in M.*

A CR-submanifold M, of a Kähler manifold $\bar{M}$, is called a *CR-product* [99] if it is locally a Riemannian product of a Kähler submanifold M_T and a totally real submanifold $M_\perp$ of $\bar{M}$. We quote the following three results:

Proposition 6.1.16. [69] *A totally geodesic CR-submanifold, of a Kähler manifold, is a CR-product.*

Theorem 6.1.17. [99] *A proper CR-submanifold, of a Kähler manifold $\bar{M}$, is a CR-product if and only if $\nabla P = 0$, where P is the endomorphism on TM induced from the almost complex structure J of $\bar{M}$.*

Theorem 6.1.18. [99] *A CR-product of a complex hyperbolic space is either a Kähler submanifold or totally real submanifold. A CR-product complex Euclidean space C^m is a product submanifold of a complex linear space C^r and totally real submanifold in a complex subspace C^{m-r} of C^m. If M is a CR-product of complex space form CP^m, then:*

(i) *$m \geq h + p + hp$, where $\dim(\mathcal{D}) = 2h$ and $\dim(\mathcal{D}^\perp) = p$.*

(ii) *The square length $\| h \|^2$ of the second fundamental form satisfies $\| h \|^2 \geq 4hp$.*

A CR-submanifold M is called a *mixed foliate CR-submanifold* if M is mixed geodesic and its holomorphic distribution is integrable. Mixed foliate CR-submanifolds of complex space forms were studied by many authors and are completely determined. We quote the following main results:

(i) [57]. *A complex projective space admits no mixed foliate proper CR-submanifolds.*

(ii) [99]. *A CR-submanifold in C^n is mixed foliate if and only if it is a CR-product.*

(iii) [30], [113]. *A CR-submanifold in a complex hyperbolic space CH^m is mixed foliate if and only if it is either a Kähler submanifold or a totally real submanifold.*

Another generalization of holomorphic and totally real submanifolds, named slant submanifolds, introduced by Chen is as follows:

Definition 6.1.19. [102] Let $\bar{M}$ be a Kähler manifold and M be a real submanifold of $\bar{M}$. For any non-zero vector X tangent to M at a point $p \in M$, the angle $\theta(X)$ between JX and the tangent space T_pM is called the *Wirtinger angle of X*. A submanifold M of $\bar{M}$ is called a *slant submanifold* if the Wirtinger angle $\theta(X)$ is constant, i.e., it is independent of the choice of $p \in M$ and $X \in T_pM$. This Wirtinger angle is called the *slant angle* of the slant submanifold.

By the above definition, it follows that holomorphic (resp. totally real) submanifolds are slant submanifolds with $\theta = 0$ (resp., $\theta = \frac{\pi}{2}$). A slant submanifold is called proper if it is neither totally real nor a holomorphic submanifold of $\bar{M}$. It is known that every proper slant submanifold is even-dimensional. We highlight that there is no inclusion relation between CR-submanifolds and slant submanifolds.

Let M be a submanifold of a Kähler manifold $\bar{M}$. Then, for any vector field X tangent to M, we write

$$JX = TX + FX,$$

where TX and FX denote the tangential and normal components of JX. Then T is an endomorphism of the tangent bundle. Using the endomorphism T, we obtain the following simple characterization for slant submanifolds.

Theorem 6.1.20. [102] *Let M be a real submanifold of a Kähler manifold $\bar{M}$. Then M is a slant submanifold if and only if $T^2X = \lambda X$ where $\lambda \in [-1, 0]$ is a fixed number and X is a vector field tangent to M.*

It is easy to see that any holomorphic or totally real submanifold satisfies the condition $\nabla T = 0$. If any proper slant submanifold satisfies $\nabla T = 0$, it is called *Kählerian slant submanifold* of $\bar{M}$. It follows that the Kählerian slant submanifolds are Kähler manifolds with respect to the induced metric and the almost complex structure defined by $\tilde{J} = (\sec \theta)T$, [102].

Theorem 6.1.21. [102] *Let M be a submanifold of a Kähler manifold $\bar{M}$. Then M satisfies condition $\nabla T = 0$ if and only if M is locally the Riemannian product $M_1 \times \ldots \times M_k$, where each M_i is a Kähler submanifold, a totally real submanifold or a Kählerian slant submanifold of $\bar{M}$.*

It is known [102] that totally umbilical slant surfaces and Kählerian slant submanifolds are totally geodesic. Following is a recent general result:

Theorem 6.1.22. [358] *Every totally umbilical proper slant submanifold of a Kähler manifold is totally geodesic.*

This opens a research problem as to the validity of the above general result for slant lightlike submanifolds. There are a large number of papers (and some books) where detailed information on the above quoted results, related results and other classes (such as semi-slant, generic and skew CR-submanifolds, Lorentzian CR-submanifolds, semi-Riemannian CR-submanifolds etc) is available. For those interested see [11, 29, 30, 42, 43, 45, 52, 57, 69, 75, 81, 97, 99, 100, 101, 102, 104, 105, 106, 107, 113, 109, 128, 133, 147, 226, 256, 262, 278, 299, 303, 306, 312, 324, 356, 349, 368, 373, 383, 384, 385, 391, 392, 410, 412, 413].

6.2   Invariant lightlike submanifolds

Let $(\bar{M}, \bar{J}, \bar{g})$ be a real $2m$-dimensional, $m > 1$, indefinite Kähler manifold, where $\bar{g}$ is a semi-Riemannian metric of index $v = 2q$, $0 < q < m$. Suppose $(M, g, S(TM))$ is a lightlike submanifold of $\bar{M}$, where g is the degenerate induced metric on M. As a part of our objective in presenting all possible lightlike submanifolds, in this section we collect up-to-date results on invariant (complex) lightlike submanifolds of Kähler manifolds. Later on these results will be related with the results on various types of Cauchy-Riemann (CR) lightlike submanifolds discussed in Sections 5 and 6.

Following the classical Riemannian definition of invariant submanifolds [413], we say that M is an *invariant (complex)* lightlike submanifold of $\bar{M}$ if

$$\bar{J} \operatorname{Rad}(TM) = \operatorname{Rad} TM, \quad \bar{J}(S(TM)) = S(TM).$$

It is easy to see that for an invariant M, $\bar{J}(\operatorname{tr}(TM)) = \operatorname{tr}(TM)$ and these distributions are even-dimensional. Thus, the dimension of an invariant M is ≥ 4.

Example 4. Let M be a submanifold of $\mathbf{R}_2^6$ given by

$$x_5 = x_1 \cos \alpha - x_2 \sin \alpha, \ x_6 = x_1 \sin \alpha + x_2 \cos \alpha.$$

Then, the tangent bundle TM of M is spanned by

$$U_1 = \partial x_1 + \cos \alpha \partial x_5 + \sin \alpha \partial x_6,$$
$$U_2 = \partial x_2 - \sin \alpha \partial x_5 + \cos \alpha \partial x_6,$$
$$U_3 = \partial x_3, \ U_4 = \partial x_4.$$

Therefore, M is a 2-lightlike submanifold with $\operatorname{Rad} TM = \operatorname{Span}\{U_1, U_2\}$. Moreover, using the canonical complex structure of $\mathbf{R}_2^6$ we have $\bar{J}(\operatorname{Rad} TM) = \operatorname{Rad} TM$. Choose the complementary vector bundle to TM in $S(TM^\perp)^\perp$ which is spanned by $\{V_1 = \partial x_1, V_2 = \partial x_2\}$. Then, $\operatorname{ltr}(TM)$ is spanned by

$$N_1 = \frac{1}{2}\left(-\partial x_1 + \cos\alpha\frac{\partial}{\partial x_5} + \sin\alpha\partial x_6\right),$$

$$N_1 = \frac{1}{2}\left(-\partial x_2 - \sin\alpha\partial x_5 + \cos\alpha\partial x_6\right).$$

It is easy to check that $\bar{J}(\operatorname{ltr}(TM)) = \operatorname{ltr}(TM)$. Thus M is an invariant lightlike submanifold of $\mathbf{R}_2^6$.

We start by discussing the integrability conditions of the radical distribution $\operatorname{Rad} TM$ and the screen distribution $S(TM)$ of M.

Proposition 6.2.1. *Let M be an invariant lightlike submanifold of an indefinite Kähler manifold $\bar{M}$. Then $\operatorname{Rad} TM$ is integrable if and only if*

$$A_\xi^* \bar{J}\xi' = \bar{J}A_{\xi'}^*\xi, \quad \forall \xi, \xi' \in \Gamma(\operatorname{Rad}(TM)).$$

Proof. $\operatorname{Rad} TM$ is integrable if and only if $g([\xi, \xi'], X) = 0$ for any $\xi, \xi' \in \Gamma(\operatorname{Rad} TM)$ and $X \in \Gamma(S(TM))$. Using the Kähler character of $\bar{M}$, from (5.1.25) and (5.1.27) we obtain

$$g([\xi, \bar{J}\xi'], X) = g(A_\xi^* \bar{J}\xi' - \bar{J}A_{\xi'}^*\xi, X)$$

which proves our assertion. □

Proposition 6.2.2. *Let $(M, g, S(TM))$ be an invariant lightlike submanifold of an indefinite Kähler manifold $\bar{M}$. Then, its screen distribution $S(TM)$ is integrable if and only if*

$$\bar{J}h^*(X, Y) = h^*(\bar{J}Y, X), \quad \forall X, Y \in \Gamma(S(TM)).$$

Proof. By the definition of invariant lightlike submanifolds, a screen distribution $S(TM)$ is integrable if and only if

$$\bar{g}([X, Y], N) = 0, \quad \forall X, Y \in \Gamma(S(TM)), \ N \in \Gamma(\operatorname{ltr}(TM)).$$

Since $\bar{M}$ is indefinite Kählerian, from (6.1.5) and (5.1.25) we obtain

$$\bar{g}([X, \bar{J}Y], N) = \bar{g}(\bar{J}h^*(X, Y) - h^*(\bar{J}Y, X), N). \qquad □$$

Consider a lightlike submanifold $(M, g, S(TM))$ whose screen distribution $S(TM)$ is integrable. Let (M', g') be an integral manifold of $S(TM)$, where g' is the induced non-degenerate semi-Riemannian metric on M'. We say that $(M, g, S(TM))$ is a *screen Kählerian manifold* if (M', g', J') has a Kählerian structure induced by the almost complex operator J' on M'. We need above sub-structure on M to relate the next theorem with the following result of non-degenerate submanifolds:

It is known that invariant (complex) submanifolds of Kähler [314] (resp., of indefinite Kähler) manifolds [336]) are Kähler manifolds (resp, indefinite Kähler manifolds). For the lightlike case, we have:

Theorem 6.2.3. *Let $(M, g, S(TM))$ be an invariant lightlike submanifold of an indefinite Kähler manifold such that M admits an integrable screen distribution $S(TM)$. Then M is a screen Kählerian manifold.*

Proof. For an invariant lightlike submanifold M, we have

$$\bar{J}S(TM) = S(TM), \quad \bar{J}\operatorname{Rad}TM = \operatorname{Rad}TM.$$

Let M' be a integral manifold of a screen distribution $S(TM)$. Denote by J' the induced operator of $\bar{J}$ on M'. Consider an immersion f as follows:

$$f : M' \to \bar{M};$$

$\bar{J}S(TM) = S(TM)$ implies M' is an invariant submanifold of $\bar{M}$. Thus,

$$\bar{J}f_* = f_*J' \tag{6.2.1}$$

where f_* is the differential of immersion f. Since M' is non-degenerate, it has a semi-Riemannian metric. We denote this metric by g' and obtain

$$g' = f^*\bar{g} \quad \text{or}$$
$$g'(X, Y) = f^*\bar{g}(X, Y) = \bar{g}(f_*X, f_*Y), \forall \in \Gamma(TM'). \tag{6.2.2}$$

Since $\bar{g}$ is Hermitian we have $g'(X, Y) = \bar{g}(\bar{J}f_*X, \bar{J}f_*Y)$. Using (6.2.1) we obtain $g'(X, Y) = \bar{g}(f_*J'X, f_*J'Y)$. From (6.2.2) we get $g'(X, Y) = g'(J'X, J'Y)$, which shows that g' is Hermitian. Let the fundamental 2-forms of $\bar{M}$ and M' be Ω and Ω', respectively. Then from (6.2.1) and (6.2.2) we obtain

$$\Omega(f_*X, f_*Y) = \Omega'(X, Y).$$

Ω closed and ∇^* a metric connection on M' imply that Ω' is closed, which proves that M' is Kählerian. Thus, M is screen Kählerian. ☐

Denote by $\bar{\nabla}$ and ∇ semi-Riemann connection on $\bar{M}$ and induced connection on M, respectively. Using the Kähler structure of $\bar{M}$, for an invariant M we have

$$\bar{\nabla}_X\bar{J}Y = \nabla_X\bar{J}Y + h^l(X, \bar{J}Y) + h^s(X, \bar{J}Y),$$
$$\bar{J}\bar{\nabla}_XY = \bar{J}\nabla_XY + \bar{J}h^l(X, Y) + \bar{J}h^s(X, Y).$$

From the above two results the following three relations hold:

Lemma 6.2.4. *Let $\bar{M}$ be an indefinite Kähler manifold and M be a complex r-lightlike submanifold of $\bar{M}$. Then we have*

$$\nabla_X \bar{J}Y = \bar{J}\nabla_X Y, \tag{6.2.3}$$

$$h^l(X, \bar{J}Y) = \bar{J}h^l(X, Y), \tag{6.2.4}$$

$$h^s(X, \bar{J}Y) = \bar{J}h^s(X, Y), \quad \forall X, Y \in \Gamma(TM). \tag{6.2.5}$$

It is well known that every invariant submanifold of a Riemannian Kähler (or non-degenerate submanifold of indefinite Kähler) manifold is minimal [314], [336]. The following proposition shows that this classical result is not true for any invariant lightlike submanifold:

Proposition 6.2.5. *An invariant lightlike submanifold M of $\bar{M}$ is minimal if and only if*

$$D^l(X, W) = 0, \quad \forall X \in \Gamma(\operatorname{Rad} TM), W \in \Gamma(S(TM^\perp)).$$

Proof. Let M be an invariant lightlike submanifold of $\bar{M}$, with a quasi-orthonormal basis

$$\{\xi_1, \dots, \xi_k, \bar{J}\xi_1, \dots, \bar{J}\xi_k, e_1, \dots, e_{(m-r)}, \bar{J}e_1, \dots, \bar{J}e_{m-r}\}$$

of TM such that $\{\xi_1, \dots, \xi_k, \bar{J}\xi_1, \dots, \bar{J}\xi_k\} \in \operatorname{Rad} TM$ and

$$\{e_1, \dots, e_{(m-r)}, \bar{J}e_1, \dots, \bar{J}e_{(m-r)}\} \in S(TM).$$

Since $h^l = 0$ on $\operatorname{Rad} TM$, we have

$$\operatorname{tr} h = \sum_{i=1}^{2(m-r)} h^s(e_i, e_i) = \sum_{i=1}^{(m-r)} h^s(e_i, e_i) + h^s(\bar{J}e_i, \bar{J}e_i).$$

On the other hand, (6.2.5) implies that $h^s(\bar{J}X, \bar{J}Y) = -h^s(X, Y)$. Using this in the above equation, we have $\operatorname{tr} h = 0$ on $S(TM)$. On the other hand, from (5.1.28) we have $\bar{g}(h^s(X, Y), W) = \bar{g}(D^l(X, W), Y)$ for $X, Y \in \Gamma(\operatorname{Rad} TM)$ and $W \in \Gamma(S(TM^\perp))$, which completes the proof. $\qquad \square$

However, there are classes of invariant lightlike submanifolds for which the classical result holds. To show this, we first recall (see Definition 4.4.7 of Chapter 4) that M is irrotational if

$$\bar{\nabla}_X \xi \in \Gamma(\operatorname{Rad} TM), \quad \forall X \in \Gamma(TM), \xi \in \Gamma(\operatorname{Rad} TM).$$

Considering (5.1.15), it follows that M is irrotational if and only if the following is satisfied:

$$h^l(X, \xi) = 0 \quad \text{and} \quad h^s(X, \xi) = 0. \tag{6.2.6}$$

Based on the above definition, we have the following result on minimal M:

Corollary 6.2.6. [160] *Any irrotational or totally umbilical invariant lightlike submanifold of an indefinite Kähler manifold is minimal.*

Complex lightlike hypersurfaces. An invariant lightlike submanifold M of $\bar{M}$ can be seen as a *complex hypersurface* if $TM^{\perp} = \operatorname{Rad}TM = \operatorname{Span}\{\xi, \bar{J}\xi\}$. We know from Chapter 5 that such a complex hypersurface is a coisotropic submanifold. Following is our first result:

Theorem 6.2.7. [159] *Let $(M, g, S(TM))$ be a complex lightlike hypersurface of an indefinite Kähler manifold $\bar{M}$. Then, radical distribution $\operatorname{Rad}TM$ defines a totally geodesic foliation on M. Moreover, $M = \mathcal{M}_1 \times \mathcal{M}_2$ is a product manifold if and only if h^* vanishes, where $\mathcal{M}_1$ is a 1-dimensional complex leaf of $TM^{\perp}$ and $\mathcal{M}_2$ is a leaf of $S(TM)$.*

Proof. Let M be a complex lightlike hypersurface of $\bar{M}$ and $X, Y \in \Gamma(\operatorname{Rad}TM)$. Since $\operatorname{Rad}TM$ is spanned by ξ and $\bar{J}\xi$ we can write

$$X = \alpha_1\xi + \beta_1\bar{J}\xi,$$
$$Y = \alpha_2\xi + \beta_2\bar{J}\xi.$$

Since ∇ is a linear connection we obtain

$$g(\nabla_X Y, PZ) = \alpha_1\alpha_2 g(\nabla_\xi\xi, PZ) + \alpha_1\beta_2 g(\nabla_\xi\bar{J}\xi, PZ)$$
$$+ \beta_1\alpha_2 g(\nabla_{\bar{J}\xi}\xi, PZ) + \beta_1\beta_2 g(\nabla_{\bar{J}\xi}\bar{J}\xi, PZ).$$

From (5.1.33) we get

$$g(\nabla_X Y, PZ) = \alpha_1\beta_2 g(\nabla_\xi\bar{J}\xi, PZ) + \beta_1\alpha_2 g(\nabla_{\bar{J}\xi}\xi, PZ).$$

Thus using (6.1.2), (5.1.14), (6.2.3) and (5.1.33) we have

$$\begin{aligned}
g(\nabla_X Y, PZ) &= \alpha_1\beta_2\bar{g}(\bar{\nabla}_\xi\bar{J}\xi, PZ) + \beta_1\alpha_2\bar{g}(\bar{\nabla}_{\bar{J}\xi}\xi, PZ)\\
&= -\alpha_1\beta_2\bar{g}(\bar{J}\xi, \bar{\nabla}_\xi PZ) - \beta_1\alpha_2\bar{g}(\xi, \bar{\nabla}_{\bar{J}\xi}PZ)\\
&= -\alpha_1\beta_2\bar{g}(\bar{J}\xi, h^l(\xi, PZ)) - \beta_1\alpha_2\bar{g}(\xi, h^l(\bar{J}\xi, PZ))\\
&= \alpha_1\beta_2\bar{g}(\xi, h^l(\xi, \bar{J}PZ)) - \beta_1\alpha_2\bar{g}(\xi, h^l(\xi, \bar{J}PZ))\\
&= 0.
\end{aligned}$$

Thus, $\operatorname{Rad}TM$ is a totally geodesic foliation in M. The proof follows from the fact (see Chapter 5) that a screen distribution $S(TM)$ of a lightlike submanifold of $\bar{M}$ defines a totally geodesic foliation in M if and only if h^* vanishes. Thus, proof is complete. □

Let $D_1, D_2, \theta_1, \theta_2, \phi_1, \phi_2$ be $F(M)$-bilinear symmetric forms on $\mathcal{U} \subset M$, defined by

$$\begin{aligned}
D_1(X, Y) &= \bar{g}(h^l(X, Y), \xi), & D_2(X, Y) &= \bar{g}(h^l(X, Y), \bar{J}\xi),\\
\theta_1(X) &= \bar{g}(\bar{\nabla}_X N, \xi), & \theta_2(X) &= \bar{g}(\bar{\nabla}_X N, \bar{J}\xi),\\
\phi_1(X) &= \bar{g}(\bar{\nabla}_X \bar{J}N, \xi), & \phi_2(X) &= \bar{g}(\bar{\nabla}_X \bar{J}N, \bar{J}\xi),
\end{aligned}$$

for any $X, Y \in \Gamma(TM), \xi \in \Gamma(\mathrm{Rad}\,TM)$ and $N \in \Gamma(\mathrm{ltr}(TM))$. Then from (5.1.14) and (5.1.20) we have

$$\bar{\nabla}_X Y = \nabla_X Y + D_1(X,Y)N + D_2(X,Y)\bar{J}N, \tag{6.2.7}$$
$$\bar{\nabla}_X N = -A_N X + \theta_1(X)N + \theta_2(X)\bar{J}N, \tag{6.2.8}$$
$$\bar{\nabla}_X \bar{J}N = -A_{\bar{J}N}X + \phi_1(X)N + \phi_2(X)\bar{J}N. \tag{6.2.9}$$

On the other hand, using (6.1.2) and (6.1.5) we obtain

$$D_2(X,Y) = -D_1(X,\bar{J}Y). \tag{6.2.10}$$

In a similar way, we get

$$\phi_1(X) = -\theta_2(X), \phi_2(X) = \theta_1(X). \tag{6.2.11}$$

Thus, from (6.2.10) and (6.2.11), on $\mathcal{U}$, (6.2.7)–(6.2.9) become

$$\bar{\nabla}_X Y = \nabla_X Y + D_1(X,Y)N - D_1(X,\bar{J}Y)\bar{J}N, \tag{6.2.12}$$
$$\bar{\nabla}_X N = -A_N X + \theta_1(X)N + \theta_2(X)\bar{J}N, \tag{6.2.13}$$
$$\bar{\nabla}_X \bar{J}N = -A_{\bar{J}N}X - \theta_2(X)N + \theta_1(X)\bar{J}N. \tag{6.2.14}$$

Now locally define

$$E_1(X,PY) = \bar{g}(\nabla_X PY, N), \qquad E_2(X,PY) = \bar{g}(\nabla_X PY, \bar{J}N),$$
$$\psi_1(X) = \bar{g}(\nabla_X \xi, N), \qquad \psi_2(X) = \bar{g}(\nabla_X \xi, \bar{J}N),$$
$$\tau_1(X) = \bar{g}(\nabla_X \bar{J}\xi, N), \qquad \tau_2(X) = \bar{g}(\nabla_X \bar{J}\xi, \bar{J}N).$$

Then, (5.1.15), (5.1.26) and (5.1.27) become

$$\nabla_X PY = \nabla_X^* Y + E_1(X,PY)N + E_2(X,PY)\bar{J}N, \tag{6.2.15}$$
$$\nabla_X \xi = -A_\xi^* X + \psi_1(X)\xi + \psi_2(X)\bar{J}\xi, \tag{6.2.16}$$
$$\nabla_X \bar{J}\xi = -A_{\bar{J}\xi}^* X + \tau_1(X)\xi + \tau_2(X)\bar{J}\xi. \tag{6.2.17}$$

Using (6.1.2) and (6.1.5) we get

$$E_2(X,PY) = -E_1(X,\bar{J}PY), \qquad \tau_1(X) = -\psi_2(X), \tag{6.2.18}$$
$$\tau_2(X) = \psi_1(X). \tag{6.2.19}$$

Now, since $\bar{g}(\xi, \bar{J}N) = 0$ and $\bar{g}(\xi, N) = 1$, from (6.1.5) we have

$$\psi_2(X) = \theta_2(X), \tag{6.2.20}$$
$$\psi_1(X) = -\theta_1(X). \tag{6.2.21}$$

Thus (6.2.15)–(6.2.17) become

$$\nabla_X PY = \nabla_X^* Y + E_1(X,PY)N - E_1(X,\bar{J}PY)\bar{J}N, \tag{6.2.22}$$
$$\nabla_X \xi = -A_\xi^* X - \theta_1(X)\xi + \theta_2(X)\bar{J}\xi, \tag{6.2.23}$$
$$\nabla_X \bar{J}\xi = -A_{\bar{J}\xi}^* X - \theta_2(X)\xi - \theta_1(X)\bar{J}\xi. \tag{6.2.24}$$

Theorem 6.2.8. [159] *Let M be a complex lightlike hypersurface of an indefinite Kähler manifold with an integrable screen $S(TM)$. Suppose (M', g') is a non-degenerate submanifold of $\bar{M}$, with M' a leaf of $S(TM)$. Then M is totally geodesic, with an induced metric connection if M' is an immersed submanifold of $\bar{M}$.*

Proof. We denote the second fundamental form of M' by h'. Since M is a complex lightlike hypersurface, $\dim(\operatorname{ltr}(TM)) = 2$. Thus we can write

$$h^l(X, Y) = \alpha N + \beta \bar{J} N$$

where α and β are functions on M. Thus $h^l(X, \xi) = 0$ if and only if $\bar{g}(h^l(X, \xi), \xi) = \bar{g}(h^l(X, \xi), \bar{J}\xi) = 0$ for $X \in \Gamma(\operatorname{Rad} TM)$. From (5.1.33) we have $\bar{g}(h^l(X, \xi), \xi) = 0$. Using (6.2.4) we get

$$\bar{g}(h^l(X, \xi), \bar{J}\xi) = -\bar{g}(\bar{J}h^l(X, \xi), \xi) = -\bar{g}(h^l(\bar{J}X, \xi), \xi) = 0.$$

In a similar way we have $h^l(X, \bar{J}\xi) = 0$. From (5.1.15) we get

$$\bar{\nabla}_X Y = \nabla'_X Y + h'(X, Y)$$

for $X, Y \in \Gamma(S(TM))$, where ∇' is a metric connection on M'. Thus from (6.2.12) and (6.2.22) we obtain

$$\begin{aligned} h'(X, Y) &= E_1(X, Y)\xi - E_1(X, Y)\bar{J}\xi + D_1(X, Y)N - D_1(X, \bar{J}Y)\bar{J}N \\ &= E_1(X, Y)\xi - E_1(X, Y)\bar{J}\xi + h(X, Y) \end{aligned} \qquad (6.2.25)$$

for any $X, Y \in \Gamma(S(TM))$. On the other hand we have

$$\begin{aligned} g(X, Y) &= g(X, \sum_{i=1}^{m} \epsilon_i \bar{g}(Y, X_i)X_i + \bar{g}(Y, N)\xi + \bar{g}(Y, \bar{J}N)\bar{J}\xi) \\ &= \sum_{i=1}^{m} \epsilon_i \bar{g}(Y, X_i)\bar{g}(X, X_i) \end{aligned}$$

which implies

$$g(X, Y) = g'(X, Y) \qquad (6.2.26)$$

for any $X, Y \in \Gamma(TM)$. Thus from (6.2.25) and (5.1.27) the proof is complete. $\square$

Example 5. Let (M, g) be a 4-dimensional real lightlike submanifold of a 6-dimensional Kähler manifold $\bar{M}$ with $\operatorname{Rad} TM$ of rank 2. Thus $\dim(S(TM)) = 2$ and M is coisotropic. Let S_x be a fiber of $S(TM)$ at $x \in M$. Then, the complexified tangent subspace $T_x^c(S)$ admits a canonical endomorphism J_s, satisfying $J_s^2 = -I_s$, which defines a complex structure on S_x^c, induced from $\bar{J}$, such that

$$T_x^c(S) = T_x(S^c)_+ \oplus T_x(S^c)_-, \quad T_x(S^c)\pm = \{z \in T_x^c(S) : J_s z = \pm i z\}.$$

Let $(K^c_{\pm} \subset T^c_x(M)$ be defined by a canonical map: $(K^c_{\pm})_x \to (S^c_{\pm})_x$. Using the above equation and the onto projection $(K^c_{\pm})_x \to (S^c_{\pm})_x$, one can show that each $(K^c_{\pm})_x$ is a complex 1-dimensional lightlike holomorphic sub-bundle of TM. Choose $K^c_+ = H$ and let $\text{Rad}\,TM = Re(H + \bar{H})$. It is easy to see that $\text{Rad}\,TM$ is involutive and has a complex structure $J_{\text{Rad}\,TM}$ satisfied by $J^2_{\text{Rad}\,TM} = -I_{\text{Rad}\,TM}$, which is induced from the complex structure $\bar{J}$, with $H = \{X - iJ_{\text{Rad}\,TM}X : X \in \text{Rad}\,TM\}$. Thus M is an invariant (complex) coisotropic submanifold of $\bar{M}$.

Null sectional curvature and induced Ricci tensor. Let M be a complex lightlike hypersurface of an indefinite complex space form $\bar{M}(c)$. We denote the null sectional curvature of M by $K_\xi(M)$ given by

$$K_\xi(M) = \frac{g(R(X,\xi)\xi, X)}{g(X, X)}$$

where X is a non-null vector field on M. From (6.1.6) and (5.3.11) we have

$$g(R(X,\xi)\xi, X) = -\bar{g}(h^*(\xi, X), h^l(X,\xi)) + g(h^l(\xi,\xi), h^*(X, X)).$$

Since rank of $\text{Rad}\,TM$ is two we can write

$$h^*(Y, PZ) = h^*{}_1(Y, PZ)\xi + h^*{}_2(Y, PZ)\bar{J}\xi$$

where $h^*{}_1(Y, PZ)$ and $h^*{}_2(Y, PZ)$ are functions on M. Thus we obtain

$$\begin{aligned} g(R(X,\xi)\xi, X) = &-h^*{}_1(\xi, X)\bar{g}(\xi, h^l(X,\xi)) - h^*{}_2(\xi, X)\bar{g}(\bar{J}\xi, h^l(X,\xi)) \\ &+ h^*{}_1(X, X)\bar{g}(\xi, h^l(\xi,\xi)) + h^*{}_2(X, X)\bar{g}(\bar{J}\xi, h^l(\xi,\xi)). \end{aligned}$$

Using (6.2.4) we have

$$\begin{aligned} g(R(X,\xi)\xi, X) = &-h^*{}_1(\xi, X)\bar{g}(\xi, h^l(X,\xi)) - h^*{}_2(\xi, X)\bar{g}(\xi, h^l(\bar{J}X,\xi)) \\ &+ h^*{}_1(X, X)\bar{g}(\xi, h^l(\xi,\xi)) + h^*{}_2(X, X)\bar{g}(\xi, h^l(\xi, \bar{J}\xi)). \end{aligned}$$

Thus from (5.1.33) we obtain $K_\xi(M) = 0$. This proves the following:

Corollary 6.2.9. [159] *There exist no complex lightlike hypersurface with positive or negative null sectional curvature in an indefinite $\bar{M}(c)$.*

For geometric reasons, we assume that the tensor $R^{(0,2)}$ (introduced in Section 2) is symmetric and, therefore, it is the Ricci tensor (denoted by Ric) of a complex lightlike hypersurface of an indefinite complex space form. Then we have

$$\text{Ric}(X, Y) = \sum_{i=1}^{m} \epsilon_i g(R(X, e_i)Y, e_i) + \bar{g}(R(X, \xi)Y, N) + \bar{g}(R(X, \bar{J}\xi)Y, \bar{J}N).$$

Using (6.1.6), (5.1.28) and (5.3.11) we obtain

$$\text{Ric}(X,Y) = -\frac{c}{2}(p+2)g(PX,PY) + \sum_{i=1}^{m}\epsilon_i\{-g(h^*(e_i,e_i),h^l(X,Y))$$
$$+ \bar{g}(h^*(X,e_i),h^l(e_i,Y))\} + \bar{g}(A_N\xi,h^l(X,Y))$$
$$- \bar{g}(A_N\bar{J}\xi,h^l(\bar{J}X,Y)). \tag{6.2.27}$$

Corollary 6.2.10. [159] *Let M be a complex lightlike hypersurface of an indefinite complex space form. Then the Ricci tensor of M is degenerate.*

Proof. Setting $Y = \xi$ in (6.2.27) we obtain

$$\text{Ric}(X,\xi) = +\sum_{i=1}^{m}\epsilon_i\{-g(h^*(e_i,e_i),h^l(X,\xi)) + \bar{g}(h^*(X,e_i),h^l(e_i,\xi))\}$$
$$+ \bar{g}(A_N\xi,h^l(X,\xi)) - \bar{g}(A_N\bar{J}\xi,h^l(\bar{J}X,\xi)). \qquad \square$$

Thus, the same way as in the proof of the previous corollary, from (6.1.2), (5.1.33) and (6.2.3) we obtain $\text{Ric}(X,\xi) = 0$. Hence we have the following corollary.

Corollary 6.2.11. [159] *There exists no complex lightlike hypersurface with positive or negative Ricci tensor in an indefinite complex space form.*

Finally, we note that invariant lightlike submanifolds have been studied in [180] and [272], by using a different method.

6.3 Screen real submanifolds

In this section, we study a class of submanifolds which may be considered as a lightlike version of totally real (non-invariant) submanifolds of Riemannian Kähler manifolds [410]. We look for those lightlike submanifolds for which the non-degenerate screen distribution is totally real with respect to the complex structure operator $\bar{J}$. Such submanifolds are called *screen real submanifolds*, defined as follows:

Let $(M,g,S(TM))$ be a lightlike submanifold of an indefinite Kähler manifold $(\bar{M},\bar{g})$. Then, M is called a *screen real submanifold* of $\bar{M}$ [159] if $\text{Rad}\,TM$ is invariant and $S(TM)$ is anti-invariant with respect to $\bar{J}$, i.e.,

$$\bar{J}(\text{Rad}\,TM) = \text{Rad}\,TM, \tag{6.3.1}$$
$$\bar{J}(S(TM)) \subseteq S(TM^{\perp}). \tag{6.3.2}$$

It follows from the above definition that,

$$\bar{J}(\text{ltr}(TM) = \text{ltr}(TM) \quad \text{and} \quad \bar{J}(\mu) = \mu,$$

μ is the complementary orthogonal vector sub-bundle to $\bar{J}(S(TM))$ in $S(TM^{\perp})$. Moreover, any screen real lightlike 3-dimensional submanifold must be 2-lightlike.

Example 6. Consider in $\mathbf{R}_2^6$ the submanifold M given by the equations

$$x_1 = x_5 \tanh \alpha - x_6 \operatorname{sech} \alpha,$$
$$x_2 = x_5 \operatorname{sech} \alpha - x_6 \tanh \alpha,$$
$$x_3 = x_4.$$

Then tangent bundle TM and normal bundle $TM^{\perp}$ of M are spanned by

$$Z_1 = \tanh \alpha \partial x_1 + \operatorname{sech} \alpha \partial x_2 + \partial x_5,$$
$$Z_2 = -\operatorname{sech} \alpha \partial x_1 - \tanh \alpha \partial x_2 + \partial x_6,$$
$$Z_3 = \partial x_3 + \partial x_4,$$
$$\xi_1 = \tanh \alpha \partial x_1 + \operatorname{sech} \alpha \partial x_2 + \partial x_5,$$
$$\xi_2 = -\operatorname{sech} \alpha \partial x_1 - \tanh \alpha \partial x_2 + \partial x_6,$$
$$W = -\partial x_3 + \partial x_4.$$

Then we can easily see that M is a 2-lightlike submanifold with $\operatorname{Rad} TM = \operatorname{Span}\{\xi_1, \xi_2\}$. Moreover $\operatorname{Rad} TM$ is invariant with respect to canonical complex structure $\bar{J}$ of $\mathbf{R}_2^6$. On the other hand $S(TM)$ is spanned by Z_3 and $\bar{J}Z_3 = W$. We obtain the lightlike transversal bundle spanned by

$$N_1 = -\frac{1}{2}(\tanh \alpha \partial x_1 + \operatorname{sech} \alpha \partial x_2 - \partial x_5),$$
$$N_2 = -\frac{1}{2}(-\operatorname{sech} \alpha \partial x_1 + \tanh \alpha \partial x_2 - \partial x_6).$$

Hence we obtain that $\operatorname{ltr}(TM)$ is also invariant with respect to $\bar{J}$. Thus, M is a screen real lightlike submanifold of $\mathbf{R}_2^6$.

For a screen real lightlike submanifold M, using (5.1.15), (5.1.23), (6.3.1) and (6.3.2) we obtain

$$-A_{\bar{J}Y} X + \nabla_X^s \bar{J}Y + D^l(X, \bar{J}Y) = f\nabla_X Y + \omega\nabla_X Y + Bh^s(X,Y)$$
$$+ Ch^s(X,Y) + \bar{J}h^l(X,Y)$$

for $X, Y \in \Gamma(S(TM))$. Comparing the tangential and transversal parts of both sides of this equation, we find

$$-A_{\bar{J}Y} X = f\nabla_X Y + Bh^s(X,Y), \qquad (6.3.3)$$
$$\nabla_X^s \bar{J}Y = \omega\nabla_X Y + Ch^s(X,Y), \qquad (6.3.4)$$
$$D^l(X, \bar{J}Y) = \bar{J}h^l(X,Y), \quad \forall X, Y \in \Gamma(S(TM)). \qquad (6.3.5)$$

In a similar way, from (5.1.22) we have

$$h^s(\xi_1, \bar{J}\xi_2) = -\bar{J}A^*_{\xi_1}\xi_2, \quad \forall \xi_1, \xi_2 \in \Gamma(\mathrm{Rad}\,TM). \tag{6.3.6}$$

Now we deal with the integrability conditions of two distributions.

Proposition 6.3.1. *Let $(M, g, S(TM))$ be a screen real lightlike submanifold of an indefinite Kähler manifold $\bar{M}$. Then the screen distribution $S(TM)$ of M is integrable if and only if*

$$A_{\bar{J}X}Y - A_{\bar{J}Y}X \in \Gamma(S(TM)), \quad \forall X, Y \in \Gamma(S(TM)). \tag{6.3.7}$$

Proof. Interchanging X and Y in (6.3.3) and subtracting we have

$$-A_{\bar{J}Y}X + A_{\bar{J}X}Y = f[X, Y], \quad \forall X, Y \in \Gamma(S(TM)). \tag{6.3.8}$$

Thus $[X, Y] \in \Gamma(S(TM))$ if and only if (6.3.7) is satisfied. □

Proposition 6.3.2. *Let $(M, g, S(TM))$ be a screen real lightlike submanifold of an indefinite Kähler manifold $\bar{M}$. Then, the radical distribution $\mathrm{Rad}\,TM$ of M is integrable if and only if*

$$A^*_{\xi_1}\xi_2 = A^*_{\xi_2}\xi_1, \quad \forall \xi_1, \xi_2 \in \Gamma(\mathrm{Rad}\,TM). \tag{6.3.9}$$

Proof of above is similar to the case of invariant lightlike submanifolds.

We know from previous sections that the induced connection ∇ on M is not a metric connection in general. Thus it is important to find which class of lightlike submanifolds have induced metric connection. In this respect, the following theorem is important because it shows that the induced connection on irrotational screen real submanifolds of indefinite Kähler manifolds is a metric connection.

Theorem 6.3.3. *There exists a Levi-Civita metric induced connection on an irrotational screen real lightlike submanifold M of a Kähler manifold. Consequently, the radical distribution of M is Killing.*

Proof. From (6.2.6), an irrotational M implies $h^l(\xi, X) = h^s(X, \xi) = 0$ for any $X \in \Gamma(TM)$ and $\xi \in \Gamma(\mathrm{Rad}\,TM)$. From (5.1.15) $\bar{g}(h^l(X, Y), \xi) = \bar{g}(\bar{\nabla}_X Y, \xi)$. Thus, from (6.1.5) we obtain $\bar{g}(h^l(X, Y), \xi) = \bar{g}(\bar{\nabla}_X \bar{J}Y, \bar{J}\xi)$. $\bar{\nabla}$ being a metric connection, we have $\bar{g}(h^l(X, Y), \xi) = -\bar{g}(\bar{J}Y, \bar{\nabla}_X \bar{J}\xi)$. (5.1.15) implies $\bar{g}(h^l(X, Y), \xi) = \bar{g}(\bar{J}Y, h^l(X, \bar{J}\xi))$. Since M is irrotational, we have $\bar{g}(h^l(X, Y), \xi) = 0$. Hence, $h^l(X, Y) = 0$ Then our first assertion follows from (6.2.6). The second assertion is then immediate, which completes the proof. □

Due to geometric and physical importance of the existence of a metric connection, we now focus on the study of irrotational screen real lightlike submanifolds. See [159, 160] on this topic, which also have some more general results.

Proposition 6.3.4. *The radical distribution of an irrotational lightlike submanifold M, of an indefinite Kähler manifold $\bar{M}$, defines a totally geodesic foliation on M.*

Proof. The radical distribution defines a totally geodesic foliation if and only if $g(\nabla_{\xi_1}\xi_2, W) = 0$ for $\xi_1, \xi_2 \in \Gamma(\text{Rad}\,TM)$ and $W \in \Gamma(S(TM))$. From (5.1.15) we have $g(\nabla_{\xi_1}\xi_2, W) = \bar{g}(\bar{\nabla}_{\xi_1}\xi_2, W)$. Since $\bar{\nabla}$ is a metric connection, we have $g(\nabla_{\xi_1}\xi_2, W) = -\bar{g}(\xi_2, \bar{\nabla}_{\xi_1} W)$. Using (5.1.15) here, we obtain $g(\nabla_{\xi_1}\xi_2, W) = -\bar{g}(\xi_2, h^l(\xi_1, W))$. Since M is irrotational, $h^l(\xi_1, W) = 0$. Thus, we conclude that $g(\nabla_{\xi_1}\xi_2, W) = 0$ which proves our assertion. $\square$

Theorem 6.3.5. *Let $(M, g, S(TM))$ be an irrotational screen real lightlike subman-ifold of an indefinite Kähler manifold $\bar{M}$ with an integrable screen distribution $S(TM)$ and M' be a leaf of $S(TM)$. Then,*

- *M' is totally umbilical in $\bar{M}$ if and only if M' and M are totally umbilical in M and $\bar{M}$ respectively.*

- *If M' is a totally geodesic submanifold of $\bar{M}$, then M is also a totally geodesic submanifold of $\bar{M}$.*

Proof. Let $X, Y \in \Gamma(TM)$. Then, we have

$$g(X, Y) = g(\sum_{i=1}^{r} \eta_i(X)\xi_i + PX, Y) = g(PX, PY) = g'(X, Y)$$

where g' is the semi-Riemannian metric of M'. On the other hand for irrotational M we have $h^l = 0$, $h^s(\xi, X) = 0$ for $\xi \in \Gamma(\text{Rad}\,TM)$ and $X \in \Gamma(TM)$. From (5.1.15) we have $\bar{\nabla}_X Y = \nabla'_X Y + h'(X, Y)$, $\forall X, Y \in \Gamma(S(TM))$, where ∇' and h' are the metric connection of M' and second fundamental form of M', respectively. Thus we have $h'(X, Y) = h^*(X, Y) + h^s(X, Y)$, for $X, Y \in \Gamma(S(TM))$, which completes the proof. $\square$

Theorem 6.3.6. *An irrotational screen real lightlike submanifold M of $\bar{M}$ is mini-mal if and only if $\text{tr}\, A_{W_i} = 0$ on $S(TM)$, $W_i \in \Gamma(S(TM^\perp))$.*

Proof. Theorem 6.3.3 tells that an irrotational screen real lightlike submanifold M admits an induced metric connection, so $h^l = 0$. Moreover, M irrotational implies $h^s(X, \xi) = 0$ for $X \in \Gamma(TM)$ and $\xi \in \Gamma(\text{Rad}\,TM)$. Thus, $h^s = 0$ on $\text{Rad}\,TM$. Hence, M is minimal if and only if $\text{tr}\, h^s(e_i, e_i) = 0$, where $\{e_i\}$ is an orthonormal basis of $S(TM)$. It follows from (5.1.28) that

$$\text{tr}\, h^s(e_i, e_i) = \frac{1}{n-r} \sum_{j=1}^{n-r} \sum_{i=1}^{m-r} \bar{g}(A_{W_j} e_i, e_i)W_j,$$

which completes the proof. $\square$

Curvature properties. Suppose that the ambient manifold $\bar{M}$ is of constant holo-morphic sectional curvature c. Then from (6.1.6) and (5.3.11) we have

$$R(X, Y)Z = \frac{c}{4}\{g(Y, Z)X - g(X, Z)Y\} + A_{h^l(Y,Z)}X \qquad (6.3.10)$$
$$- A_{h^l(X,Z)}Y + A_{h^s(Y,Z)}X - A_{h^s(X,Z)}Y$$

for any $X, Y \in \Gamma(TM)$. Using (5.1.22) and (6.3.10) we obtain

$$
\begin{aligned}
\bar{g}(R(X,Y)Z, N) = \frac{c}{4}\{ &g(Y,Z)\bar{g}(X,N) - g(X,Z)\bar{g}(Y,N)\} \\
&+ \bar{g}(A_N X, h^l(Y,Z)) - \bar{g}(A_N Y, h^l(X,Z)) \\
&+ \bar{g}(D^s(X,N), h^s(Y,Z)) \\
&- \bar{g}(D^s(Y,N), h^s(X,Z)), \quad X, Y \in \Gamma(TM)
\end{aligned}
\tag{6.3.11}
$$

and $N \in \Gamma(\mathrm{ltr}(TM))$. From (5.1.28) and (5.1.31) we derive

$$
\begin{aligned}
\bar{g}(R(X,Y)Z, T) = \frac{c}{4}\{ &g(Y,Z)g(X,T) - g(X,Z)g(Y,T)\} \\
&- \bar{g}(h^*(Y,T), h^l(X,Z)) + \bar{g}(h^*(X,T), h^l(Y,Z)) \\
&+ \bar{g}(h^s(X,T), h^s(Y,Z))Y \\
&- \bar{g}(h^s(Y,T), h^s(X,Z)),
\end{aligned}
\tag{6.3.12}
$$

for any $Y, T \in \Gamma(S(TM))$. From (6.3.12) we have the null sectional curvature of a screen real lightlike submanifold as follows:

$$
\begin{aligned}
K_\xi(M) = \frac{1}{\| X \|}\{ &-\bar{g}(h^*(\xi, X), h^l(X, \xi)) + \bar{g}(h^*(X, X), h^l(\xi, \xi)) \\
&- \bar{g}(h^s(\xi, X), h^s(X, \xi)) + \bar{g}(h^s(X, X), h^s(\xi, \xi))\}
\end{aligned}
\tag{6.3.13}
$$

where X is a non-null vector field on $S(TM)$. Thus we have the following:

Theorem 6.3.7. *There exist no irrotational screen real lightlike submanifolds of an indefinite $\bar{M}(c)$ with positive or negative null sectional curvature.*

As we have seen from section 1 of Chapter 5, the induced connection on a lightlike submanifold is not a metric Levi-Civita connection in general. As a result of this, the $(0,2)$ type $R^{(0,2)}$ tensor (deduced from the curvature tensor) is not symmetric. For an irrotational M we prove the following:

Theorem 6.3.8. *An irrotational screen real lightlike submanifold of $\bar{M}(c)$ admits a symmetric Ricci tensor.*

Proof. From the definition of screen real lightlike submanifold, we have

$$
R^{(0,2)}(X,Y) = \sum_{i=1}^{m-r} \epsilon_i g(R(X, X_i)Y, X_i) + \sum_{j=1}^{r} \bar{g}(R(X, \xi_j)Y, N_j)
$$

for any $X, Y \in \Gamma(TM)$. Thus, from (6.3.12) and (6.3.13) we have

$$
R^{(0,2)}(X,Y) = \frac{c}{4}(m-1) - \sum_{j=1}^{r} \bar{g}(A_{N_j}\xi_j, h^l(X,Y))
$$
$$
- \bar{g}(A_{N_j}X, h^l(\xi_j, Y)) + \bar{g}(D^s(\xi_j, N_j), h^s(X,Y))
$$
$$
- \bar{g}(D^s(X, N_j), h^s(\xi_j, Y)) + \sum_{i=1}^{m-r} \epsilon_i\{\bar{g}(h^*(X_i, X_i), h^l(X,Y))
$$
$$
- \bar{g}(h^*(X, X_i), h^l(Y, X_i) + \bar{g}(h^s(X_i, X_i), h^s(X,Y))
$$
$$
- \bar{g}(h^s(X, X_i), h^s(X_i, Y))\}. \tag{6.3.14}
$$

$h^l = 0$ for an irrotational M screen real lightlike submanifold, hence,

$$
R^{(0,2)}(X,Y) = \frac{c}{4}(m-1)g(X,Y) + \sum_{j=1}^{r} \bar{g}(D^s(\xi_j, N_j), h^s(X,Y))
$$
$$
+ \sum_{i=1}^{m-r} \epsilon_i\{\bar{g}(h^s(X_i, X_i), h^s(X,Y)) - \bar{g}(h^s(X, X_i), h^s(X_i, Y))\}.
$$

Thus, the proof is complete from above equation.   □

6.4   Screen CR-lightlike submanifolds

In two previous sections we have seen that, just like the initial development of Riemannian submanifolds [314, 410], we have two main sub-classes, namely, invariant and screen real lightlike submanifolds of Kählerian manifolds. We also know that Bejancu [45] introduced a generalized concept of Riemannian submanifolds, called *Cauchy-Riemann (CR) submanifolds*, which is an umbrella of a variety of submanifolds including invariant and totally real submanifolds. On the other hand, we know from Duggal-Bejancu's book [149] that CR-lightlike submanifolds do not include invariant and screen real submanifolds. For this reason the authors of this book introduced a new class, called *Screen Cauchy-Riemann lightlike submanifolds* (briefly, SCR) of an indefinite Kähler manifold [159], which includes invariant and screen real as its two sub-cases. In this section, the objective is to collect all the results published in [159].

Definition 6.4.1. [159] Let $(M, g, S(TM))$ be a lightlike submanifold of an indefinite Kähler manifold $(\bar{M}, \bar{g})$. We say that M is a *SCR-lightlike submanifold* of $\bar{M}$ if the following conditions are satisfied:

(1) There exist a real non-degenerate distribution $\mathcal{D} \subseteq S(TM)$ such that

$$
S(TM) = \mathcal{D} \perp \mathcal{D}^\perp, \mathcal{D} \cap \mathcal{D}^\perp = \{0\},
$$
$$
\bar{J}(\mathcal{D}) = \mathcal{D}, \bar{J}\mathcal{D}^\perp \subseteq S(TM^\perp), \tag{6.4.1}
$$

where $\mathcal{D}^{\perp}$ is orthogonal complementary to $\mathcal{D}$ in $S(TM)$.

(2) $\operatorname{Rad} TM$ is invariant with respect to $\bar{J}$.

It follows that $\operatorname{ltr}(TM)$ is invariant with respect to $\bar{J}$, that is,

$$\bar{J}\operatorname{ltr}(TM) = \operatorname{ltr}(TM). \tag{6.4.2}$$

Thus, the tangent bundle of M is decomposed as

$$TM = \mathcal{D}' \oplus_{\text{orth}} \mathcal{D}^{\perp}, \tag{6.4.3}$$

where

$$\mathcal{D}' = \mathcal{D} \oplus_{\text{orth}} \operatorname{Rad} TM. \tag{6.4.4}$$

Denote the orthogonal complement to the $\bar{J}\mathcal{D}^{\perp}$ in $S(TM^{\perp})$ by $\mathcal{D}_0$. Then,

$$\operatorname{tr}(TM) = \operatorname{ltr}(TM) \oplus_{\text{orth}} \bar{J}\mathcal{D}^{\perp} \perp \mathcal{D}_0. \tag{6.4.5}$$

We say that M is a *proper SCR-lightlike submanifold* of $\bar{M}$ if $\mathcal{D} \neq \{0\}$ and $\mathcal{D}^{\perp} \neq \{0\}$, that is, M is an r-lightlike submanifold of Case A. Note the following special features:

(a) Condition (2) implies that $\dim(\operatorname{Rad} TM) = 2p \geq 2$.

(b) For a proper M, $\dim(\mathcal{D}) = 2s \geq 2$ and $\dim(\operatorname{ltr}(TM)) = \dim(\operatorname{Rad} TM) = 2p \geq 2$. Thus, $\dim(M) \geq 5$,　$\dim(\bar{M}) = 2k \geq 8$.

(c) There exist no proper SCR-lightlike hypersurfaces.

(d) Any proper SCR-lightlike 5-dimensional submanifold must be 2-lightlike.

Example 7. Let M be a submanifold of $\mathbf{R}_2^8$ given by equations

$$x_1 = u_1 - u_2, \quad x_5 = -u_2 - u_3,$$
$$x_2 = u_1 + u_2, \quad x_6 = x_7 = u_1,$$
$$x_3 = u_4, \ x_4 = u_5, \quad x_8 = u_2 - u_3.$$

Then the tangent bundle of M is spanned by

$$Z_1 = \partial x_1 + \partial x_2 + \partial x_6 + \partial x_7, \qquad Z_2 = -\partial x_1 + \partial x_2 - \partial x_5 + \partial x_8,$$
$$Z_3 = \partial x_5 - \partial x_8, Z_4 = \partial x_3, \qquad Z_5 = \partial x_4.$$

Thus M is a 2-lightlike submanifold with $\operatorname{Rad} TM = \operatorname{Span}\{Z_1, Z_2\}$. By using the canonical complex structure of $\mathbf{R}_2^8$, we see that $\bar{J}Z_1 = Z_2$ and $\bar{J}Z_4 = Z_5$, that is $\operatorname{Rad} TM$ and $\mathcal{D}_0 = \operatorname{Span}\{Z_4, Z_5\}$ are invariant with respect to $\bar{J}$. Moreover, $S(TM^{\perp})$ is spanned by $W = -\partial x_6 + \partial x_7$. Hence, a lightlike transversal vector bundle $\operatorname{ltr}(TM)$ is spanned by

$$N_1 = \frac{1}{4}(-\partial x_1 - \partial x_2 + \partial x_6 + \partial x_7), \ N_2 = \frac{1}{4}(\partial x_1 - \partial x_2 - \partial x_5 + \partial x_8).$$

It is easy to check that $\mathrm{ltr}(TM)$ is invariant with respect to $\bar{J}$. Moreover, $W = \bar{J}Z_3$. Thus, $\mathcal{D} = \mathrm{Span}\{Z_1, Z_2, Z_4, Z_5\}$ and $\mathcal{D}' = \mathrm{Span}\{Z_3\}$, i.e., M is a SCR-lightlike submanifold of $\mathbf{R}_2^8$.

We denote by n_1 the complex dimension of the distribution $\mathcal{D}$ and by n_2 the real dimension of the distribution $\mathcal{D}^\perp$. Then for $n_1 = 0$, we get $\mathcal{D}' = \mathrm{Rad}\,TM$ which is invariant with respect to $\bar{J}$. In this case, a SCR-lightlike submanifold becomes a screen real submanifold. For $n_2 = 0$, we have $TM = \mathcal{D}'$ which is invariant. Thus, in this case, a SCR-lightlike submanifold becomes an invariant lightlike submanifold. Hence, we conclude that SCR-lightlike submanifolds contain invariant and screen real submanifolds as particular subspaces. We say that M is an *anti-holomorphic SCR-lightlike submanifold* if $n_2 = \dim S(TM^\perp)$, i.e, $\mathcal{D}_0 = \{0\}$.

Let M be a coisotropic SCR-lightlike submanifold of an indefinite Kähler manifold $\bar{M}$. Then, $S(TM^\perp) = \{0\}$ which implies that $\mathcal{D}^\perp = \{0\}$. Hence, $TM = \mathcal{D}'$, so M is invariant. Similarly, if M is an isotropic or totally lightlike submanifold, then, M is invariant. Hence, the proper and screen real SCR-lightlike submanifolds must be r-lightlike. Thus we have

Proposition 6.4.2. *There exist no screen real SCR coisotropic or isotropic or totally lightlike submanifolds of a Kähler manifold.*

On existence of a SCR-lightlike submanifold, we prove the following:

Theorem 6.4.3. *Let M be a $2r$-lightlike submanifold of a real $2k$-dimensional indefinite almost Hermitian manifold $\bar{M}$ such that $\dim(S(TM^\perp)) = 1$. If $\mathrm{Rad}\,TM$ is invariant with respect to $\bar{J}$ and $r < \frac{m}{2}$ then M is a proper SCR-lightlike submanifold of $\bar{M}$, where $m = \dim M$*

Proof. Since $\mathrm{Rad}\,TM$ is invariant with respect to $\bar{J}$, $\mathrm{ltr}(TM)$ is also invariant with respect to $\bar{J}$. We take a screen transversal vector bundle $S(TM^\perp) = \mathrm{Span}\{W\}$. Then from (6.1.2) we have

$$\bar{g}(\bar{J}W, \xi) = \bar{g}(W, \bar{J}\xi) = 0, \bar{g}(\bar{J}W, W) = 0.$$

Since $S(TM^\perp)$ is one-dimensional, we deduce that $\bar{J}S(TM^\perp)$ is a distribution on M. Thus we can choose a screen vector bundle containing $\bar{J}(S(TM^\perp)$. Then there exists a vector bundle $\mathcal{D}$ of rank $m - (2r + 1)$ such that

$$S(TM) = \bar{J}(S(TM^\perp)) \oplus \mathcal{D} \qquad (6.4.6)$$

where $\mathcal{D}$ is a non-degenerate distribution; otherwise $S(TM)$ would be degenerate. Moreover, from (6.1.2) and (6.4.6) we obtain

$$\bar{g}(\bar{J}X, \bar{J}W) = \bar{g}(X, W) = 0,$$
$$\bar{g}(\bar{J}X, W) = -\bar{g}(X, \bar{J}W) = 0,$$
$$\bar{g}(\bar{J}X, N) = -\bar{g}(X, \bar{J}N) = 0.$$

Since $\mathrm{ltr}(TM)$ is invariant with respect to $\bar{J}$, we conclude that $\mathcal{D}$ is an almost complex distribution on M, which completes the proof. $\square$

Now we prove a characterization theorem which is the lightlike version of Blair-Chen's characterization Theorem 6.1.12 for the Riemannian case.

Theorem 6.4.4. *Let M be a lightlike submanifold of an indefinite complex space form with $c \neq 0$. Then M is a SCR-lightlike submanifold with $\mathcal{D} \neq \{0\}$ if and only if the following conditions are fulfilled:*

(a) *The maximal complex subspaces of $T_p M$, $p \in M$ define a distribution $\mathcal{D}' = \operatorname{Rad} TM \oplus_{\mathrm{orth}} \mathcal{D}$, where $\mathcal{D}$ is a non-degenerate almost complex distribution.*

(b) *$\bar{g}(\bar{R}(X,Y)Z,W) = 0$, $\forall X,Y \in \Gamma(\mathcal{D}')$ and $Z,W \in \Gamma(\mathcal{D}^\perp)$, where $\mathcal{D}^\perp$ is the orthogonal complementary distribution to $\mathcal{D}$ in $S(TM)$.*

Proof. If M is a SCR-lightlike submanifold, then $\mathcal{D}' = \operatorname{Rad} TM \oplus_{\mathrm{orth}} \mathcal{D}$. If $X \in \Gamma(TM)$ and $Y \in \Gamma(\operatorname{Rad} TM)$, then from (6.1.2) we have (b). If $X,Y \in \Gamma(\mathcal{D})$ then since $\mathcal{D}^\perp$ is orthogonal to $\mathcal{D}$, using (6.1.2) we obtain (b). Conversely, from (a) $\bar{J} \operatorname{Rad} TM$ is a distribution on M and $\mathcal{D}$ holomorphic implies $\bar{J} \operatorname{Rad} TM \cap \mathcal{D} = \{0\}$. From (b), (6.1.2) and (6.1.6) we derive

$$\bar{g}(\bar{R}(\xi,Y)Z,\bar{J}Z) = -2\bar{g}(\bar{J}\xi,Y)(Z,Z) = 0,$$

for $\xi \in \Gamma(\operatorname{Rad} TM)$, $Y \in \Gamma(\mathcal{D}^\perp)$ and $Z \in \Gamma(\mathcal{D})$. Since $\mathcal{D}$ is a non-degenerate almost complex distribution on $S(TM)$ there exists at least a non-null vector field. Hence $\bar{g}(\bar{J}\xi,Y) = 0$. Thus $\operatorname{Rad} TM$ is invariant with respect to $\bar{J}$. Similarly, we have

$$\bar{g}(\bar{R}(\bar{J}X,X)Z,W) = -2\bar{g}(X,X)\bar{g}(\bar{J}Z,W) = 0$$

for $X \in \Gamma(\mathcal{D})$ and $Z,W \in \Gamma(\mathcal{D}^\perp)$. Hence $\bar{J}\mathcal{D}_p^\perp$ is orthogonal to $\mathcal{D}_p^\perp$. Since $\mathcal{D}$ is holomorphic, $\bar{J}\mathcal{D}_p^\perp$ is also orthogonal to $\mathcal{D}$. Moreover, since $\operatorname{ltr}(TM)$ is invariant with respect to $\bar{J}$, we obtain $\bar{J}\mathcal{D}_p^\perp \in \Gamma(S(TM^\perp))|_p$. $\qquad\square$

As studied in previous sections, it is desirable to find conditions for an integrable screen distribution of M. For this let M be a SCR-lightlike submanifold of an indefinite Kähler manifold $\bar{M}$. For each vector field X tangent to M, we put

$$\bar{J}X = PX + \omega X \tag{6.4.7}$$

where PX and ωX are respectively the tangential and the transversal parts of $\bar{J}X$. From (6.4.4)–(6.4.5) we obtain $PX \in \Gamma(\mathcal{D}')$ and $\omega X \in \Gamma(\bar{J}\mathcal{D}^\perp)$. Also, for each vector field V transversal to M, we put

$$\bar{J}V = BV + CV \tag{6.4.8}$$

where BV and CV are respectively the tangential and the transversal parts of $\bar{J}V$. From (6.4.3) and (6.4.5) we have $BV \in \Gamma(\mathcal{D}^\perp)$ and $CV \in \Gamma(\operatorname{ltr}(TM) \perp \mathcal{D}_0)$. We note, if $V \in \Gamma(S(TM^\perp))$ then $CV \in \Gamma(\mathcal{D}_0)$. Now applying $\bar{J}$ to (6.4.7) we obtain $-X = P^2 X$. Hence we have

$$P^3 + P = 0,$$

that is, P is an f-structure ([413]). Using (6.1.5), (5.1.15), (5.1.22), (5.1.23), (6.4.7), (6.4.8), we obtain

$$(\nabla_X P)Y = A_{\omega Y} X + Bh^s(X, PY), \tag{6.4.9}$$

$$h^l(X, PY) = -D^l(X, \omega Y) + \bar{J}h^l(X, Y), \tag{6.4.10}$$

$$(\nabla_X \omega)Y = -h^s(X, PY) + Ch^s(X, Y) \tag{6.4.11}$$

for any $X, Y \in \Gamma(TM)$, where $(\nabla_X P)Y = \nabla_X PY - P\nabla_X Y$ and $(\nabla_X \omega)Y = \nabla_X^s \omega Y - \omega \nabla_X Y$. P is called parallel if $(\nabla_X P)Y = 0$ for any $X, Y \in \Gamma(TM)$.

Theorem 6.4.5. *Let M be a SCR-lightlike submanifold of an indefinite Kähler manifold $\bar{M}$. Then, the screen distribution of M is integrable if and only if the following three conditions are satisfied:*

$$g(A_N Y, \bar{J}X) = g(A_N X, \bar{J}Y), \quad \forall X, Y \in \Gamma(\mathcal{D}), \tag{6.4.12}$$

$$\bar{g}(D^s(X, N)\bar{J}Y) = \bar{g}(D^s(Y, N)\bar{J}X), \quad \forall X, Y \in \Gamma(\mathcal{D}^\perp) \tag{6.4.13}$$

and

$$g(A_N Y, \bar{J}X) = -\bar{g}(D^s(X, N)\bar{J}Y), \quad \forall X \in \Gamma(\mathcal{D}), \forall Y \in \Gamma(\mathcal{D}^\perp). \tag{6.4.14}$$

Proof. Since $\bar{M}$ is a Kähler manifold we have $d\Omega = 0$, where d is the exterior derivative and Ω is the fundamental 2-form of $\bar{M}$. Hence we have $d\Omega(X, Y, N) = 0$ for $X, Y \in \Gamma(S(TM)$ and $N \in \Gamma(\mathrm{ltr}(TM))$. Thus we get

$$0 = N\bar{g}(X, \bar{J}Y) - \bar{g}([X, Y], \bar{J}N) - \bar{g}([Y, N], \bar{J}X) - \bar{g}([N, X], \bar{J}Y). \tag{6.4.15}$$

Using (6.1.2), (6.1.5), (5.1.22) and ∇ and $\bar{\nabla}$ torsion free, we obtain

$$\begin{aligned} \bar{g}([X, Y], \bar{J}N) &= \bar{g}(A_N Y, \bar{J}X) - \bar{g}(D^s(Y, N), \bar{J}X) \\ &\quad - \bar{g}(A_N X, \bar{J}Y) - \bar{g}(D^s(X, N), \bar{J}Y) \end{aligned}$$

for $X, Y \in \Gamma(S(TM)$ and $N \in \Gamma(\mathrm{ltr}(TM))$. Since $S(TM) = \mathcal{D} \perp \mathcal{D}^\perp$, for $X, Y \in \Gamma(\mathcal{D})$ we have (6.4.12), for $X \in \Gamma(\mathcal{D})$ and $Y \in \Gamma(\mathcal{D}^\perp)$ we have (6.4.13) and for $X, Y \in \Gamma(\mathcal{D}^\perp)$ we get (6.4.14), which completes the proof. □

Theorem 6.4.6. *Let M be a SCR-lightlike submanifold of an indefinite Kähler manifold $\bar{M}$. Then, the screen distribution of M defines a totally geodesic foliation on M if and only if:*

1. $A_{\bar{J}Y} X$ *has no components in $\mathrm{Rad}\, TM$.*

2. $A_N X$ *has no components in $\mathcal{D}$,* $\forall X \in \Gamma(TM)$ *and $Y \in \Gamma(\mathcal{D}^\perp)$.*

Proof. From (6.1.2), (6.1.5) and (5.1.23) we get

$$\bar{g}(\nabla_X Y, N) = -g(A_{\bar{J}Y} X, \bar{J}N)$$

for any $X \in \Gamma(TM)$ and $Y \in \Gamma(\mathcal{D}^\perp)$. On the other hand from (5.1.15) and (5.1.31) we obtain $\bar{g}(\nabla_X Y, N) = \bar{g}(h^*(X, Y), N) = g(A_N X, Y)$, for $X \in \Gamma(TM)$ and $Y \in \Gamma(\mathcal{D})$. Since $S(TM) = \mathcal{D} \perp \mathcal{D}^\perp$, the proof is complete. □

Theorem 6.4.7. *Let M be a SCR-lightlike submanifold of an indefinite Kähler manifold $\bar{M}$. The distribution $\mathcal{D}'$ is integrable if and only if*

$$h(X, \bar{J}Y) = h(\bar{J}X, Y), \quad \forall X, Y \in \Gamma(\mathcal{D}_0). \tag{6.4.16}$$

Proof. From (6.1.5),(5.1.15),(6.4.7) and (6.4.8) we obtain

$$h(X, \bar{J}Y) = \omega \nabla_X Y + Ch(X, Y)$$

for any $X, Y \in \Gamma(\mathcal{D}_0)$. Then taking into account that h is symmetric and ∇ is torsion free, we obtain

$$h(X, \bar{J}Y) - h(\bar{J}X, Y) = \omega[X, Y]$$

which completes the proof. □

In [159], we have already obtained necessary and sufficient conditions for integrable distribution $\mathcal{D}^\perp$. Here, we prove a stronger result.

Theorem 6.4.8. *Let M be a SCR-lightlike submanifold of an indefinite Kähler manifold $\bar{M}$. Then the following assertions are equivalent:*

1. $\bar{g}(D^s(W, N), \bar{J}Z) = \bar{g}(D^s(Z, N), \bar{J}W), \forall Z, W \in \Gamma(\mathcal{D}^\perp), \quad N \in \Gamma(\text{ltr}(TM))$.

2. A_N *is self-adjoint on* $\mathcal{D}'$ *with respect to* g.

3. $\mathcal{D}^\perp$ *is integrable.*

Proof. First, we notice that $\mathcal{D}^\perp$ is integrable if and only if $g([Z, W], \bar{J}X) = \bar{g}([Z, W], \bar{J}N) = 0$ for $Z, W \in \Gamma(\mathcal{D}^\perp)$, $X \in \Gamma(\mathcal{D})$ and $N \in \Gamma(\text{ltr}(TM))$. Since $\bar{M}$ is a Kähler manifold, we have $d\Omega = 0$, hence we obtain $d\Omega(X, Z, W) = 0$ for $X \in \Gamma(\mathcal{D})$ and $Z, W \in \Gamma(\mathcal{D}^\perp)$. Then, we get

$$3d\Omega(X, Z, W) = -\Omega([Z, W], X) = g([Z, W], \bar{J}X) = 0. \tag{6.4.17}$$

Thus, it is enough to investigate the condition $\bar{g}([Z, W], \bar{J}N) = 0$.

(1)$\Rightarrow$(2): From (5.1.22), we have

$$\bar{g}(D^s(W, N), \bar{J}Z) = \bar{g}(\bar{\nabla}_W N, \bar{J}Z).$$

Using (6.1.2) and (6.1.5) we get

$$\bar{g}(D^s(W, N), \bar{J}Z)) = -\bar{g}(\bar{\nabla}_W \bar{J}N, Z).$$

Then, from (5.1.22), we obtain, $\bar{g}(D^s(W, N), \bar{J}Z) = g(A_{\bar{J}N}W, Z)$. Hence, we have

$$\bar{g}(D^s(W, N), \bar{J}Z) = \bar{g}(D^s(Z, N), \bar{J}W)$$

which implies $g(A_{\bar{J}N}W, Z) = g(A_{\bar{J}N}Z, W)$.

(2)⇒(3): Since $\bar{\nabla}$ is a metric connection, from (5.1.14), we have

$$\bar{g}([Z,W],\bar{J}N) = g(W, A_{\bar{J}N}Z) - g(A_{\bar{J}N}W, Z).$$

Since $A_{\bar{J}N}$ is self-adjoint on $\mathcal{D}^\perp$ we derive

$$\bar{g}([Z,W],\bar{J}N) = 0. \tag{6.4.18}$$

Then, from (6.4.17) and (6.4.18) we conclude that $\mathcal{D}^\perp$ is integrable.

(3)⇒(1): From $\bar{\nabla}$ a metric connection, integrable $\mathcal{D}^\perp$ and (6.1.2), we get

$$\bar{g}([Z,W],\bar{J}N) = -\bar{g}(\bar{\nabla}_Z N, \bar{J}W) + \bar{g}(\bar{\nabla}_W N, \bar{J}Z).$$

Using (5.1.22) we get

$$\bar{g}(D^s(W,N), \bar{J}Z) - \bar{g}(D^s(Z,N), \bar{J}W) = 0$$

which completes the proof. □

Corollary 6.4.9. *Let M be a SCR-lightlike submanifold of an indefinite Kähler manifold $\bar{M}$. Then Radical distribution is integrable if and only if*

$$\bar{g}(h^l(\xi, Z), \xi') = \bar{g}(h^l(\xi', Z), \xi), \tag{6.4.19}$$
$$\bar{g}(h^s(\xi, \bar{J}\xi'), \bar{J}W) = \bar{g}(h^s(\bar{J}\xi, \xi'), \bar{J}W) \tag{6.4.20}$$

for any $\xi, \xi' \in \Gamma(\operatorname{Rad} TM)$, $Z \in \Gamma(\mathcal{D})$ and $W \in \Gamma(\mathcal{D}^\perp)$.

Proof. By Definition 6.4.1, $\operatorname{Rad} TM$ is integrable if and only if

$$g([\xi, \xi'], Z) = g([\xi, \xi'], W) = 0, \quad \forall \xi, \xi' \in \Gamma(\operatorname{Rad} TM)$$

for any $Z \in \Gamma(\mathcal{D})$ and $W \in \Gamma(\mathcal{D}^\perp)$. Thus from (6.1.5) and (5.1.13) we have

$$g([\xi, \xi'], \bar{J}Z) = \bar{g}(h^l(\xi, Z), \xi') - \bar{g}(h^l(\xi', Z), \xi). \tag{6.4.21}$$

Using (6.1.2) and (5.1.13) we obtain

$$g([\xi, \xi'], W) = \bar{g}(h^s(\xi, \bar{J}\xi'), \bar{J}W) - \bar{g}(h^s(\bar{J}\xi, \xi'), \bar{J}W). \tag{6.4.22}$$

Thus from (6.4.21) and (6.4.22) the proof of the corollary is complete. □

Similarly, we have the following corollary.

Corollary 6.4.10. *Let M be a SCR-lightlike submanifold of an indefinite Kähler manifold $\bar{M}$. Then the distribution $\mathcal{D}$ is integrable if and only if A_N is self adjoint on $\mathcal{D}$ and*

$$g(A_{\bar{J}W}X, \bar{J}Y) = g(A_{\bar{J}W}Y, \bar{J}X) \tag{6.4.23}$$

for any $X, Y \in \Gamma(\mathcal{D})$, $N \in \Gamma(\operatorname{ltr}(TM))$ and $W \in \Gamma(\mathcal{D}^\perp)$.

Theorem 6.4.11. *Let M be a SCR-lightlike submanifold of an indefinite Kähler manifold $\bar{M}$. Then, the following assertions are equivalent:*

(1) $h^s(X, \bar{J}Y)$ *has no components in* $\bar{J}\mathcal{D}^\perp$ *for* $X, Y \in \Gamma(\mathcal{D}')$.

(2) $\bar{g}(\bar{J}\xi, D^l(X, \bar{J}Z)) = 0$ *and* $A_{\bar{J}Z}X$ *has no components in* $\mathcal{D}$ *for* $X \in \Gamma(\mathcal{D}')$, $\xi \in \Gamma(\operatorname{Rad} TM)$ *and* $Z \in \Gamma(\mathcal{D}^\perp)$.

(3) $\mathcal{D}'$ *defines a totally geodesic foliation on* M.

Proof. $(1) \Rightarrow (2)$. From (5.1.28) we have

$$\bar{g}(h^s(X, \bar{J}Y), \bar{J}Z) + \bar{g}(\bar{J}Y, D^l(X, W)) = g(A_{\bar{J}Z}X, \bar{J}Y)$$

for any $X, Y \in \Gamma(\mathcal{D}')$ and $Z \in \Gamma(\mathcal{D}^\perp)$. Taking $Y = \xi \in \Gamma(\operatorname{Rad} TM)$ in this equation we have

$$\bar{g}(h^s(X, \bar{J}Y), \bar{J}Z) = -\bar{g}(\bar{J}Y, D^l(X, W)).$$

Hence we have $\bar{g}(\bar{J}\xi, D^l(X, \bar{J}Z)) = 0$ for $X \in \Gamma(\mathcal{D}')$, $\xi \in \Gamma(\operatorname{Rad} TM)$ and $Z \in \Gamma(\mathcal{D}^\perp)$. If we take $Y \in \Gamma(\mathcal{D})$ we obtain

$$\bar{g}(h^s(X, \bar{J}Y), \bar{J}Z) = g(A_{\bar{J}Z}X, \bar{J}Y).$$

Hence $A_{\bar{J}Z}X$ has no components in $\mathcal{D}$.

$(2) \Rightarrow (3)$. $\bar{\nabla}$ is a metric connection and from (5.1.15), (5.1.23) we have

$$g(\nabla_X Y, Z) = g(\bar{J}Y, A_{\bar{J}Z}X) - g(D^l(X, \bar{J}Z), \bar{J}Y) = 0$$

for any $X, Y \in \Gamma(\mathcal{D}')$ and $Z \in \Gamma(\mathcal{D}^\perp)$. Hence we have $\nabla_X Y \in \Gamma(\mathcal{D}')$.

$(3) \Rightarrow (1)$. From (6.1.2) and (5.1.15) we get $g(\nabla_X Y, Z) = \bar{g}(h^s(X, \bar{J}Y), \bar{J}Z)$, for any $X, Y \in \Gamma(\mathcal{D}')$ and $Z \in \Gamma(\mathcal{D}^\perp)$. Since $\nabla_X Y \in \Gamma(\mathcal{D}')$, (1) holds. □

Theorem 6.4.12. *Let M be a SCR-lightlike submanifold of an indefinite Kähler manifold $\bar{M}$. Then, the following assertions are equivalent:*

(1) $A_{\bar{J}Y}X$ *has no components in* $\mathcal{D}'$ *for* $X, Y \in \Gamma(\mathcal{D}^\perp)$.

(2) $h^s(X, \bar{J}Z)$ *and* $D^s(X, \bar{J}N)$ *have no components in* $\bar{J}\mathcal{D}^\perp$ *for* $X \in \Gamma(\mathcal{D}^\perp)$ *and* $Z \in \Gamma(\mathcal{D})$.

(3) $\mathcal{D}^\perp$ *defines a totally geodesic foliation on* M.

Proof. $(1) \Rightarrow (2)$. Suppose that $A_{\bar{J}Y}X$ has no components in $\mathcal{D}'$. Then from (5.1.28) we obtain

$$0 = g(A_{\bar{J}Y}X, \bar{J}Z) = \bar{g}(h^s(X, \bar{J}Z), \bar{J}Y)$$

for $X, Y \in \Gamma(\mathcal{D}^\perp)$ and $Z \in \Gamma(\mathcal{D})$. Hence $h^s(X, \bar{J}Z)$ has no components in $\bar{J}\mathcal{D}^\perp$. On the other hand using (6.1.2),(5.1.15) and (5.1.22) we obtain

$$0 = g(A_{\bar{J}Y}X, \bar{J}N) = \bar{g}(\bar{J}Y, D^s(X, \bar{J}N)), \quad \forall X, Y \in \Gamma(\mathcal{D}^\perp).$$

Hence, $D^s(X, \bar{J}N)$ has no components in $\bar{J}\mathcal{D}^\perp$.

(2) $\Rightarrow$ (3). By definition, a SCR-lightlike submanifold, $\mathcal{D}^\perp$ is parallel if and only if $g(\nabla_X Y, Z) = \bar{g}(\nabla_X Y, N) = 0$ for $X, Y \in \Gamma(\mathcal{D}^\perp)$, $Z \in \Gamma(\mathcal{D})$ and $N \in \Gamma(\text{ltr}(TM))$. Using (6.1.2), (6.1.5), (5.1.15) and taking account that $\bar{\nabla}$ is a metric connection we have

$$g(\nabla_X Y, Z) = -\bar{g}(\bar{J}Y, h^s(X, \bar{J}Z)) = 0$$

for $X, Y \in \Gamma(\mathcal{D}^\perp)$ and $Z \in \Gamma(\mathcal{D})$. In a similar way we get

$$\bar{g}(\nabla_X Y, N) = \bar{g}(\bar{J}Y, D^s(X, \bar{J}N)) = 0.$$

Thus $\mathcal{D}^\perp$ defines a totally geodesic foliation on M.

(3) $\Rightarrow$ (1) Suppose that $\mathcal{D}^\perp$ is parallel, then from (5.1.15), (6.1.2), (6.1.5) and (5.1.23) we obtain

$$g(\nabla_X Y, Z) = \bar{g}(A_{\bar{J}Y}X, \bar{J}Z) = 0$$

for $X, Y \in \Gamma(\mathcal{D}^\perp)$ and $Z \in \Gamma(\mathcal{D})$. In a similar way we have

$$\bar{g}(\nabla_X Y, N) = -\bar{g}(A_{\bar{J}Y}X, \bar{J}N) = 0.$$

Hence the proof is complete. $\quad\square$

We say that M is a *mixed geodesic SCR-lightlike submanifold* if its second fundamental form h satisfies $h(X, U) = 0$ for $X \in \Gamma(\mathcal{D}')$ and $U \in \Gamma(\mathcal{D}^\perp)$. From (5.1.14), it is easy to see, if M is mixed geodesic then we have

$$h^s(X, U) = 0 \quad \text{and} \quad h^l(X, U) = 0.$$

Corollary 6.4.13. *Let M be a mixed geodesic anti-holomorphic SCR-lightlike submanifold of an indefinite Kähler manifold $\bar{M}$. Then $\mathcal{D}^\perp$ defines a totally geodesic foliation on M if and only if $D^s(Z, \bar{J}N) = 0$ for $Z \in \Gamma(\mathcal{D}^\perp)$ and $N \in \Gamma(\text{ltr}(TM))$.*

Proof. From (5.1.23), (5.1.15) and (6.1.2) we have

$$g(\nabla_Z W, X) = \bar{g}(\bar{\nabla}_Z \bar{J}W, \bar{J}X)$$

for $X \in \Gamma(\mathcal{D})$ and $Z, W \in \Gamma(\mathcal{D}^\perp)$. Then, using again (5.1.15) we get

$$g(\nabla_Z W, X) = -\bar{g}(h^s(\bar{J}X, Z), \bar{J}W).$$

Since M is mixed geodesic anti-holomorphic, we get $g(\nabla_Z W, X) = 0$. In a similar way, from (6.1.2), (6.1.5), (5.1.15) and (5.1.23) we obtain

$$g(\nabla_Z W, N) = -\bar{g}(\bar{J}W, D^s(Z, \bar{J}N))$$

which proves the assertion. $\quad\square$

We say that a SCR-lightlike submanifold M of a Kähler manifold $\bar{M}$ is a *SCR-lightlike product* if both distributions $\mathcal{D}'$ and $\mathcal{D}^\perp$ are integrable and their leaves are totally geodesic, i.e., M is locally a product manifold $(M_1 \times M_2, g)$ with

$$g = g_1 + g_2,$$

where g_1 is the degenerate metric tensor of the leaf M_1 of $\mathcal{D}'$ and g_2 is the non-degenerate metric tensor of the leaf M_2 of $\mathcal{D}^\perp$. For a mixed geodesic SCR-lightlike submanifold, we have the following:

Theorem 6.4.14. *A mixed geodesic SCR-lightlike submanifold M of an indefinite Kähler manifold $\bar{M}$ is a SCR-lightlike product if and only if:*

(1) *$A_{\bar{J}Z} X$ has no components in $\mathcal{D}$ for $X \in \Gamma(\mathcal{D}')$, $Z \in \Gamma(\mathcal{D}^\perp)$,*

(2) *For $X \in \Gamma(\mathcal{D}')$ and $Z \in \Gamma(\mathcal{D}^\perp)$, $D^l(X, \bar{J}Z) = 0$,*

(3) *$A_{\bar{J}Z} W$ has no components in $\operatorname{Rad} TM$ for $W, Z \in \Gamma(\mathcal{D}^\perp)$.*

Proof. From (6.1.2) and (6.1.5), we have $g(\nabla_X Y, Z) = \bar{g}(\bar{\nabla}_X \bar{J}Y, \bar{J}Z)$ for $X \in \Gamma(\mathcal{D}')$, $Y \in \Gamma(\mathcal{D})$ and $Z \in \Gamma(\mathcal{D}^\perp)$. Then using (5.1.15), we get $g(\nabla_X Y, Z) = \bar{g}(h^s(X, \bar{J}Y), \bar{J}Z)$. Thus, (5.1.28) implies that $g(\nabla_X Y, Z) = \bar{g}(A_{\bar{J}Z} X, \bar{J}Y)$ which implies (1). In a similar way, for $X \in \Gamma(\mathcal{D}')$ and $Y \in \Gamma(\operatorname{Rad} TM)$, we have $g(\nabla_X Y, Z) = \bar{g}(\bar{\nabla}_X \bar{J}Y, \bar{J}Z)$. Hence, we get $g(\nabla_X Y, Z) = \bar{g}(h^s(X, \bar{J}Y), \bar{J}Z)$. Then, from (5.1.28), we derive

$$g(\nabla_X Y, Z) = \bar{g}(\bar{J}Y, D^l(X, \bar{J}Z))$$

which implies (2). Considering the definition of a SCR-lightlike product, it is enough to prove that the third condition is satisfied if and only if $\mathcal{D}^\perp$ defines a totally geodesic foliation on M. We note that $\mathcal{D}^\perp$ defines a totally geodesic foliation if and only if $g(\nabla_Z W, X) = \bar{g}(\nabla_Z W, N) = 0$. The first equation can be seen from the proof of Corollary 6.4.13. On the other hand, from (6.1.2), (6.1.5) and (5.1.15), we obtain $g(\nabla_Z W, N) = \bar{g}(\bar{\nabla}_Z \bar{J}W, \bar{J}N)$ for $Z, W \in \Gamma(\mathcal{D}^\perp)$ and $N \in \Gamma(\operatorname{ltr}(TM))$. Hence, using (5.1.23), we have $g(\nabla_Z W, N) = -\bar{g}(A_{\bar{J}Z} W, \bar{J}N)$ which completes the proof. □

Theorem 6.4.15. *The induced connection ∇ of a mixed geodesic SCR-lightlike submanifold M of an indefinite Kähler manifold $\bar{M}$ is a metric connection if and only if $A_\xi^* X$ has no components in $\mathcal{D}$ for $X \in \Gamma(\mathcal{D})$ and $\xi \in \Gamma(\operatorname{Rad} TM)$.*

Proof. Since M is mixed geodesic, it follows that $h^l(X, Z) = 0$ for $X \in \Gamma(\mathcal{D})$ and $Z \in \Gamma(\mathcal{D}')$. From Proposition 5.1.3, we have $h^l(\xi_1, \xi_2) = 0$ for $\xi_1, \xi_2 \in \Gamma(\operatorname{Rad} TM)$. On the other hand, from (5.1.29) we get

$$g(A_\xi^* X, Y) = \bar{g}(h^l(X, Y), \xi)$$

for any $X, Y \in \Gamma(\mathcal{D})$ and $\xi \in \Gamma(\operatorname{Rad} TM)$. Then, it is easy to see from the above equation that $h^l(X, Y) = 0$ if and only if $A_\xi^* X$ has no components in $\mathcal{D}$. Thus, it is enough to show that $h^l(Z, W) = 0$ for $Z, V \in \Gamma(\mathcal{D}^\perp)$. From (6.1.5) and (5.1.15) we obtain $\bar{g}(h^l(Z, W), \xi) = \bar{g}(\bar{\nabla}_Z \bar{J}W, \bar{J}\xi)$. Since $\bar{\nabla}$ is a metric connection and $\bar{J}(\mathcal{D}^\perp) \oplus_{\text{orth}} \text{Rad}\, TM$, we arrive at $\bar{g}(h^l(Z, W), \xi) = -\bar{g}(\bar{J}W, \bar{\nabla}_Z \bar{J}\xi)$. Then, using (5.1.15), we derive

$$\bar{g}(h^l(Z, W), \xi) = -\bar{g}(\bar{J}W, h^s(Z, \bar{J}\xi)).$$

Then, mixed geodesic M implies that $h^s(Z, \bar{J}\xi) = 0$. Hence, we get

$$\bar{g}(h^l(Z, W), \xi) = 0$$

which proves our assertion. □

If M is totally umbilical, then we have the following important result:

Theorem 6.4.16. *The induced connection on a totally umbilical proper SCR-lightlike submanifold is a metric connection.*

Proof. Using (5.1.14),(6.4.7), (6.4.8) and taking account that $\text{ltr}(TM)$ is invariant with respect to $\bar{J}$ we have

$$h^l(X, \bar{J}Y) = \bar{J}h^l(X, Y)$$

for any $X, Y \in \Gamma(TM)$. Since M is umbilical we get

$$g(X, \bar{J}Y)H^l = g(X, Y)\bar{J}H^l.$$

Interchanging X and Y in this equation and subtracting we derive

$$g(X, \bar{J}Y)H^l = 0.$$

Taking $X, Y \in \Gamma(\mathcal{D})$ we have $H^l = 0$. Then our assertion follows (5.1.39). □

Theorem 6.4.17. *Let M be a proper SCR-lightlike submanifold of an indefinite Kähler manifold. Then M is proper totally umbilical only if its anti-invariant distribution $\mathcal{D}^\perp$ is one-dimensional. Moreover, $\bar{J}$ induces an almost contact metric structure on $S(TM)$.*

Proof. Let M be a totally umbilical proper SCR- lightlike submanifold. Then using (5.1.13), (6.4.7) (6.4.8) and taking the screen transversal part we obtain

$$h^s(X, \bar{J}X) + \omega \nabla_X X+ = Ch^s(X, X)$$

for $X \in \Gamma(\mathcal{D})$. Since M is totally umbilical we get

$$\omega \nabla_X X = 0, g(X, X)CH^s = 0.$$

Hence we obtain

$$\nabla_X X \in \Gamma(\mathcal{D}'), \tag{6.4.24}$$
$$H^s \in \Gamma(\bar{J}\mathcal{D}^\perp). \tag{6.4.25}$$

Now using (5.1.15), (5.1.23), (6.4.7), (6.4.8) and taking the tangent part we obtain

$$-A_{\bar{J}W}Z = f\nabla_Z W + Bh^s(Z,W)$$

for any $Z, W \in \Gamma(\mathcal{D}^\perp)$. Hence we have

$$g(A_{\bar{J}W}Z,Z) = -g(h^s(Z,W), \bar{J}Z).$$

Since M is umbilical, using (5.1.28) we obtain

$$g(Z,Z)\bar{g}(H^s, \bar{J}W) = g(Z,W)g(H^s, \bar{J}Z). \qquad (6.4.26)$$

Interchanging Z and W we get

$$g(W,W)\bar{g}(H^s, \bar{J}Z) = g(Z,W)g(H^s, \bar{J}W). \qquad (6.4.27)$$

Thus from (6.4.26) and (6.4.27) we have

$$\bar{g}(H^s, \bar{J}Z) = \frac{g(Z,W)^2}{g(Z,Z)g(W,W)}\bar{g}(H^s, \bar{J}Z). \qquad (6.4.28)$$

Since $\mathcal{D}^\perp$ is non-degenerate we can choose non-null vector fields Z and W. Thus using (6.4.25) in (6.4.28) we obtain either $H^s = 0$ or Z and W is linearly dependent. Now assume that $\mathcal{D}^\perp = \mathrm{Span}\{T\}$. Let $\bar{J}T = W$. For each vector field X tangent to M, from (6.4.7) we obtain $\omega X = \eta(X)W, \forall X \in \Gamma(TM)$, where $\eta(X) = \varepsilon g(X,T)$. Thus, from the above two equations, we have $\bar{J}X = \phi X + \eta(X)W$. Applying $\bar{J}$ to the last equation we get

$$\phi^2 X = -X + \eta(X)T, \ \eta(T) = 1.$$

Thus, ϕ defines an almost contact structure [66] on M. Suppose (M', g') is the integral manifold of $S(TM)$, where g' is the non-degenerate metric on M'. Then, from (6.4.7) we obtain

$$g'(\phi X, \phi Y) = g'(X,Y) - \varepsilon\eta(X)\eta(Y), \forall X, Y \in \Gamma(TM').$$

Thus $(\phi, T, g', \varepsilon)$ defines a Riemannian [66] or Lorentzian [134] almost contact structure on M' according as ε is $+1$ or -1, which proves the theorem. $\qquad \square$

Theorem 6.4.18. *There is no totally umbilical proper SCR-lightlike submanifold in a positively or negatively curved indefinite Kähler manifold.*

Proof. Suppose M is a totally umbilical proper SCR-lightlike submanifold of an indefinite Kähler manifold $\bar{M}$ with $\bar{K}_{\bar{M}} \neq 0$. Taking into account (6.1.2) and (5.3.11), we have

$$
\begin{aligned}
\bar{R}(X,Z,Z,X) &= \bar{R}(X,Z,\bar{J}Z,\bar{J}X) \\
&= -\bar{R}(X,Z,\bar{J}X,\bar{J}Z) \\
&= -\bar{g}((\nabla_X h^s)(Z,\bar{J}X),\bar{J}Z) + \bar{g}((\nabla_Z h^s)(\bar{J}X,Z),\bar{J}Z) \\
&\quad + \bar{g}(D^s(X,h^l(Z,\bar{J}X)),\bar{J}Z) - \bar{g}(D^s(X,h^l(Z,\bar{J}X)),\bar{J}Z)
\end{aligned}
$$

for any $X \in \Gamma(\mathcal{D})$ and $Z \in \Gamma(\mathcal{D}^{\perp})$. Since M is a totally umbilical proper SCR-lightlike, we have $h^l(Z, \bar{J}X) = 0, h^l(X, \bar{J}X) = 0$. Thus we obtain

$$\bar{R}(X, Z, Z, X) = \bar{g}((\nabla_X h^s)(Z, \bar{J}X), \bar{J}Z) + \bar{g}((\nabla_Z h^s)(\bar{J}X, Z), \bar{J}Z).$$

Hence we have

$$K_{\bar{M}}(X, Z) = -\{g(\nabla_X Z, \bar{J}X) + g(\nabla_X \bar{J}X, Z)\}\bar{g}(H^s, \bar{J}Z)$$
$$+ \{g(\nabla_Z X, \bar{J}X) + g(\nabla_Z \bar{J}X, X)\}\bar{g}(H^s, \bar{J}Z).$$

$g(Z, X) = 0$ and $g(\bar{J}X, X) = 0$ imply $g(\nabla_X Z, \bar{J}X) = -g(Z, \nabla_X \bar{J}X)$ and $g(\nabla_Z X, \bar{J}X) = -g(X, \nabla_Z \bar{J}X)$. Thus, $K_{\bar{M}}(X, Z) = 0$ which is a contradiction. $\square$

Theorem 6.4.19. *There exist no totally umbilical proper SCR-lightlike submanifolds in any positively or negatively null sectional curved indefinite Kähler manifold $\bar{M}$.*

Proof. The proof is similar to that of Theorem 6.4.18. $\square$

Corollary 6.4.20. *Let M be a totally umbilical proper SCR-lightlike submanifold of an indefinite Kähler manifold $\bar{M}$. Then the radical distribution defines a totally geodesic foliation on M.*

Proof. By the definition of a SCR-lightlike submanifold $\operatorname{Rad} TM$ defines a totally geodesic foliation if and only if $g(\nabla_X Y, U) = g(\nabla_X Y, Z) = 0$ for $X, Y \in \Gamma(\operatorname{Rad} TM)$, $U \in \Gamma(\mathcal{D})$ and $Z \in \Gamma(\mathcal{D}^{\perp})$ From (5.1.14) we have

$$g(\nabla_X Y, U) = \bar{g}(\bar{\nabla}_X Y, U) = -\bar{g}(Y, \bar{\nabla}_X U)$$
$$= -\bar{g}(Y, h^l(X, U))$$
$$= 0.$$

In a similar way we obtain $g(\nabla_X Y, Z) = 0$ which proves our assertion. $\square$

Theorem 6.4.21. *Let M be a totally umbilical proper SCR-lightlike submanifold of an indefinite Kähler manifold $\bar{M}$ with integrable screen distribution. Let M' be a leaf of $S(TM)$. If M' is a totally geodesic submanifold of $\bar{M}$, then M is also a totally geodesic submanifold of $\bar{M}$.*

Proof. Let $X, Y \in \Gamma(TM)$. Then we have

$$g(X, Y) = g(\sum_{i=1}^{r} \eta_i(X)\xi_i + PX, Y) = g(PX, PY) = g'(X, Y)$$

where g' is the semi-Riemannian metric of M'. On the other hand, since M is umbilical, we have $h^l = 0$, $h^s(\xi, X) = 0$ for $\xi \in \Gamma(\operatorname{Rad}(TM))$ and $X \in \Gamma(TM)$. From (5.1.14) we have

$$\bar{\nabla}_X Y = \nabla'_X Y + h'(X, Y)$$

for $X, Y \in \Gamma(S(TM))$, where ∇' and h' are the metric connection of M' and second fundamental form of M' respectively. Thus we have

$$h'(X,Y) = h^*(X,Y) + h(X,Y)$$

for any $X, Y \in \Gamma(S(TM))$. Hence the proof is complete. $\qquad\square$

Example 8. Consider a submanifold M, in $\mathbf{R}_2^6$ with the equations

$$x_1 = x_6,\ x_2 = -x_5,\ x_3 = \sqrt{1 - x_4^2}.$$

The tangent bundle of M is spanned by

$$\xi_1 = \partial x_1 + \partial x_6,\ \xi_2 = \partial x_2 - \partial x_5,\ Z = -x_4\, \partial x_3 + x_3 \partial x_4.$$

We see that M is a SCR-lightlike submanifold with $\operatorname{Rad} TM = \operatorname{Span}\{\xi_1, \xi_2\}$. $\bar{J} \operatorname{Rad} TM = \operatorname{Rad} TM$. $\mathcal{D}_0 = \{0\}$ and $\mathcal{D}' = \operatorname{Span}\{Z\}$. On the other hand, $S(TM^{\perp})$ is spanned by $W = x_3\, \partial x_3 + x_3\, \partial x_4$ and the lightlike transversal bundle $\operatorname{ltr}(TM)$ is spanned by

$$N_1 = \frac{1}{2}\{\partial x_1 + \partial x_6\},\ N_2 = \frac{1}{2}\{\partial x_2 - \partial x_5\},$$

is invariant. By direct calculations we get

$$\bar{\nabla}_X \xi_1 = \bar{\nabla}_X \xi_2 = \bar{\nabla}_{\xi_2} Z = 0$$

and $\bar{\nabla}_Z Z = -W$ for any $X \in \Gamma(TM)$. Hence, $h^s(Z,Z) = g(Z,Z)H^s$, where $H^s = -W$. Thus, M is a totally umbilical submanifold.

6.5   Generalized CR-lightlike submanifolds

We have seen in the previous section that SCR-lightlike submanifolds include invariant and screen real cases, but, unfortunately, they exclude real lightlike hypersurfaces, which are extensively used in mathematical physics. Also, since the dimension of such submanifolds must be ≥ 5, they also exclude 2-, 3- and 4-dimensional lightlike submanifolds which have uses in mathematics and physics. Moreover, we will soon see that there is another class, called *Cauchy-Riemann (CR) lightlike submanifolds*, introduced in the book [149], which has no intersection with the SCR-lightlike submanifolds.

   In this section we present the latest results on a new class, called *Generalized CR-lightlike submanifolds* which is an umbrella of lightlike submanifolds, including above sub-cases. Consequently, with the material of this last section, we fulfill the main purpose for which CR-submanifolds were designed by Bejancu [45] and extended by Sharma [373] and Duggal [133, 135, 136], for semi-Riemannian and Lorentzian cases respectively. For easy understanding of our presentation, we first need the following definition of CR-lightlike submanifolds.

Definition 6.5.1. [149] A real lightlike submanifold M of an indefinite almost Hermitian manifold $\bar{M}$ is called a *CR-lightlike submanifold* of $\bar{M}$ if the following two conditions hold:

(A) $\bar{J}(\operatorname{Rad} TM)$ is a distribution on M such that

$$\operatorname{Rad} TM \cap \bar{J}(\operatorname{Rad} TM) = \{0\}.$$

(B) There exist vector bundles $S(TM)$, $S(TM^{\perp})$, $\operatorname{ltr}(TM)$, $\mathcal{D}_o$ and $\mathcal{D}'$ over M, such that

$$S(TM) = \left\{\bar{J}(\operatorname{Rad} TM) \oplus \mathcal{D}'\right\} \perp \mathcal{D}_o,$$
$$\bar{J}(\mathcal{D}_o) = \mathcal{D}_o, \quad \text{and} \quad \bar{J}(\mathcal{D}') = \mathcal{L}_1 \perp \mathcal{L}_2,$$

where $\mathcal{D}_o$ is a non-degenerate distribution on M, $\mathcal{L}_2$ is a vector subbundle $S(TM^{\perp})$ and $\mathcal{L}_1 = \operatorname{ltr}(TM)$, respectively.

From the above definition and (5.1.1) of Section 1 of Chapter 5, the tangent bundle of a CR-lightlike submanifold is decomposed as follows:

$$TM = \mathcal{D} \oplus \mathcal{D}', \quad \text{where} \tag{6.5.1}$$
$$\mathcal{D} = \operatorname{Rad} TM \oplus_{\mathrm{orth}} \bar{J}(\operatorname{Rad} TM) \oplus_{\mathrm{orth}} \mathcal{D}_o. \tag{6.5.2}$$

If M is a real hypersurface ($r = 1$), then, $TM^{\perp} = \operatorname{Rad} TM$ and $S(TM^{\perp}) = \{0\}$. Therefore, $\mathcal{L}_2 = \{0\}$ and $\operatorname{tr}(TM) = \operatorname{ltr}(TM)$. Thus, we say that M is a real lightlike hypersurface of $\bar{M}$ if the condition **(B)** and (6.5.1) reduce to

$$S(TM) = \left\{\bar{J}(\operatorname{Rad} TM) \oplus \bar{J}(\operatorname{ltr}(TM))\right\} \perp \mathcal{D}_o,$$
$$TM = \left\{\bar{J}(\operatorname{Rad} TM)) \oplus \bar{J}(\operatorname{ltr}(TM))\right\} \perp \mathcal{D}_o \oplus_{\mathrm{orth}} \operatorname{Rad} TM,$$
$$T\bar{M}_{|M} = \left\{\bar{J}(TM^{\perp}) \oplus \bar{J}(\operatorname{tr}(TM))\right\} \perp \mathcal{D}_o$$
$$\perp \left\{\operatorname{ltr}(TM) \oplus \operatorname{Rad} TM\right\}.$$

Thus, any 3-dimensional CR-lightlike submanifold is a 1-lightlike submanifold. In order not to repeat, we refer to [149, Chapter 6] for details on the above classes. We also refer to [361, 363, 360] for some work after the publication of [149].

Suppose M is an invariant lightlike submanifold of $\bar{M}$. This means that $\bar{J}(\operatorname{Rad} TM) = \operatorname{Rad} TM$ and $\bar{J}(S(TM)) = S(TM)$. Hence M is not a CR-lightlike submanifold since condition **(A)** does not hold. Similarly, the real (non-invariant) case is not possible. Therefore, we conclude that CR-lightlike submanifolds are always non-trivial, that is, $\mathcal{D}$ and $\mathcal{D}'$ from (6.5.1) are at least of rank 1.

It is quite clear from the above analysis that there is no inclusion relation between SCR and CR lightlike submanifolds. Now we are ready to define and study the following new concept:

Definition 6.5.2. [160] Let $(M, g, S(TM))$ be a real lightlike submanifold of an indefinite Kähler manifold $(\bar{M}, \bar{g}, \bar{J})$. We say that M is a *generalized Cauchy-Riemann (GCR)-lightlike submanifold* if the following holds:

(A) There exist two sub-bundles $\mathcal{D}_1$ and $\mathcal{D}_2$ of $\operatorname{Rad} TM$ such that

$$\operatorname{Rad} TM = \mathcal{D}_1 \oplus \mathcal{D}_2, \quad \bar{J}(\mathcal{D}_1) = \mathcal{D}_1, \quad \bar{J}(\mathcal{D}_2) \subset S(TM). \tag{6.5.3}$$

(B) There exist two sub-bundles $\mathcal{D}_0$ and $\mathcal{D}'$ of $S(TM)$ such that

$$S(TM) = \{\bar{J}\mathcal{D}_2 \oplus \mathcal{D}'\} \perp \mathcal{D}_0 \,, \quad \bar{J}(\mathcal{D}_0) = \mathcal{D}_0, \quad \bar{J}(\mathcal{D}') = \mathcal{L}_1 \perp \mathcal{L}_2 \tag{6.5.4}$$

where $\mathcal{D}_0$ is a non-degenerate distribution on M, $\mathcal{L}_1$ and $\mathcal{L}_2$ are vector sub-bundles of $\operatorname{ltr}(TM)$ and $S(TM^{\perp})$, respectively.

The tangent bundle TM of M is decomposed as

$$TM = \mathcal{D} \oplus \mathcal{D}', \qquad \mathcal{D} = \operatorname{Rad} TM \oplus_{\mathrm{orth}} \mathcal{D}_0 \oplus_{\mathrm{orth}} \bar{J}\mathcal{D}_2. \tag{6.5.5}$$

M is called a *proper GCR-lightlike submanifold* if $\mathcal{D}_1 \neq \{0\}, \mathcal{D}_2 \neq \{0\}, \mathcal{D}_0 \neq \{0\}$ and $\mathcal{L}_2 \neq \{0\}$, which has the following features:

1. The condition (A) implies that $\dim(\operatorname{Rad} TM) \geq 3$.

2. The condition (B) implies that $\dim(\mathcal{D}) = 2s \geq 6$, $\dim(\mathcal{D}') \geq 2$ and $\dim(\mathcal{D}_2) = \dim(\mathcal{L}_2)$. Thus $\dim(M) \geq 8$ and $\dim(\bar{M}) \geq 12$.

3. Any proper 8-dimensional GCR-lightlike submanifold is 3-lightlike.

4. $\bar{M}$ Kähler and (1) imply that $\operatorname{ind}(\bar{M}) \geq 4$.

We denote by m_1 the complex dimension of the distribution $\mathcal{D}_1$ and by m_2 the real dimension of the distribution $\mathcal{D}_2$. Then, for $m_1 = 0$ we have $\bar{J}(\mathcal{D}_2) = \bar{J}(\operatorname{Rad} TM) \subset S(TM)$. Thus, a GCR-lightlike submanifold with $m_1 = 0$ becomes a CR-lightlike submanifold. For $m_2 = 0$, we have $\bar{J}(\mathcal{D}_2) = \bar{J}(\operatorname{Rad} TM) = \operatorname{Rad} TM$. Thus, a GCR-lightlike submanifold with $m_2 = 0$ is a SCR-lightlike submanifold. Consequently, we conclude that GCR-lightlike submanifolds serve as an umbrella of real hypersurfaces, invariant, screen real and CR-lightlike submanifolds.

Example 9. Let M be a submanifold of $\mathbf{R}_4^{14}$ given by

$$x_1 = x_{14} \quad x_2 = -x_{13}, \quad x_3 = x_{12}, \quad x_7 = \sqrt{1 - x_8^2}.$$

Then TM is spanned by $Z_1, Z_2, Z_3, Z_4, Z_5, Z_6, Z_7, Z_8, Z_9, Z_{10}$, where

$$Z_1 = \partial_{x_1} + \partial_{x_{14}}, \quad Z_2 = \partial_{x_2} - \partial_{x_{13}}, \quad Z_3 = \partial_{x_3} + \partial_{x_{12}},$$
$$Z_4 = \partial_{x_4}, \quad Z_5 = \partial_{x_5}, \quad Z_6 = \partial_{x_6}, \quad Z_7 = -x_8 \, \partial_{x_7} + x_7 \, \partial_{x_8},$$
$$Z_8 = \partial_{x_9}, \quad Z_9 = \partial_{x_{10}}, \quad Z_{10} = \partial_{x_{11}}.$$

Hence M is 3-lightlike with $\operatorname{Rad} TM = \operatorname{Span}\{Z_1, Z_2, Z_3\}$ and $\bar{J}Z_1 = Z_2$. Thus, $\mathcal{D}_1 = \operatorname{Span}\{Z_1, Z_2\}$. On the other hand, $\bar{J}Z_3 = Z_4 - Z_{10} \in \Gamma(S(TM))$ implies that $\mathcal{D}_2 = \operatorname{Span}\{Z_3\}$. Moreover, $\bar{J}Z_5 = Z_6$ and $\bar{J}Z_8 = Z_9$. Hence $\mathcal{D}_0 = \operatorname{Span}\{Z_5, Z_6, Z_8, Z_9\}$. By direct calculations, we get $S(TM^\perp) = \operatorname{Span}\{W = x_7\,\partial_{x_7} + x_8\,\partial_{x_8}\}$. Thus, $\bar{J}Z_7 = -W$. Hence, $\mathcal{L}_2 = S(TM^\perp)$. On the other hand, the lightlike transversal bundle $\operatorname{ltr}(TM)$ is spanned by

$$\{N_1 = \frac{1}{2}(-\partial_{x_1} + \partial_{x_{14}}),\; N_2 = \frac{1}{2}(-\partial_{x_2} - \partial_{x_{13}}),\; N_3 = \frac{1}{2}(-\partial_{x_3} + \partial_{x_{12}}\}.$$

From this we have $\operatorname{Span}\{N_1, N_2\}$ is invariant with respect to $\bar{J}$ and $\bar{J}N_3 = -\frac{1}{2}Z_4 - \frac{1}{2}Z_{10}$. Hence, $\mathcal{L}_1 = \operatorname{Span}\{N_3\}$ and $\mathcal{D}' = \operatorname{Span}\{\bar{J}N_3, \bar{J}W\}$. Thus, M is a proper GCR-lightlike submanifold of $\mathbf{R}_4^{14}$.

Let M be a GCR-lightlike submanifold of an indefinite Kähler manifold $\bar{M}$. If M is isotropic then $S(TM) = \{0\}$ implies $\mathcal{D}_0 = \{0\}$, $\bar{J}\mathcal{D}_2 = 0$ (so $\mathcal{D}_2 = \{0\}$) and $\bar{J}(\mathcal{L}_2) = 0$ implies $\mathcal{L}_2 = \{0\}$. Thus we get $TM = \mathcal{D}_1$ which is invariant with respect to $\bar{J}$. If M is totally lightlike, since $\operatorname{Rad} TM = TM = TM^\perp$, we have $\mathcal{D}_0 = \mathcal{D}_2 = \bar{J}\mathcal{L}_2 = \{0\}$. Thus $TM = \mathcal{D}_1$ and, therefore, M is invariant. Hence, we conclude that there exist no isotropic or totally lightlike proper GCR-lightlike submanifolds. Now, let M be coisotropic GCR-lightlike. Since $S(TM^\perp) = \{0\}$, $\bar{J}\mathcal{L}_2 = \{0\}$, therefore, $\mathcal{L}_2 = \{0\}$. Moreover $\mathcal{D}_2 = \{0\}$ if and only if M is invariant.

Let Q, P_1 and P_2 be the projection morphisms on $\operatorname{Rad} TM$, $\bar{J}\mathcal{L}_1 = \mathcal{M}_1$ and $\bar{J}\mathcal{L}_2 = \mathcal{M}_2$, respectively . Then we have

$$X = QX + P_1 X + P_2 X \tag{6.5.6}$$

for $X \in \Gamma(TM)$. On the other hand, for $X \in \Gamma(TM)$, we write

$$\bar{J}X = TX + \omega X, \tag{6.5.7}$$

where TX and ωX are its tangential and transversal parts. Then, from (6.5.6) we obtain

$$\bar{J}X = TX + \omega P_1 X + \omega P_2 X \tag{6.5.8}$$

where $TX \in \Gamma(\mathcal{D})$, $\omega P_1 X \in \Gamma(\mathcal{L}_1)$ and $\omega P_2 X \in \Gamma(\mathcal{L}_2)$. Similarly,

$$\bar{J}V = BV + CV, \quad V \in \Gamma(\operatorname{tr}(TM)) \tag{6.5.9}$$

where BV and CV are sections of TM and $\operatorname{tr}(TM)$, respectively. Now, differentiating (6.5.6) and using (5.1.15)–(5.1.23) and (6.5.9), we have:

$$(\nabla_X T)Y = A_{\omega P_1 Y} X + A_{\omega P_2 Y} X + Bh(X, Y), \tag{6.5.10}$$

$$D^s(X, \omega P_1 Y) = -\nabla^s_X \omega P_2 Y + \omega P_2 \nabla_X Y \tag{6.5.11}$$
$$\qquad\qquad - h^s(X, TY) + Ch^s(X, Y),$$

$$D^l(X, \omega P_2 Y) = -\nabla^l_X \omega P_1 Y + \omega P_1 \nabla_X Y - h^l(X, TY)$$
$$\qquad\qquad + Ch^l(X, Y), \quad \forall X, Y \in \Gamma(TM). \tag{6.5.12}$$

We now study integrability conditions of the distributions $\mathcal{D}$ and $\mathcal{D}'$.

Theorem 6.5.3. *Let M be a GCR-lightlike submanifold of an indefinite Kähler manifold $\bar{M}$. Then:*

(i) *The distribution $\mathcal{D}$ is integrable if and only if*

$$h(X, \bar{J}Y) = h(\bar{J}X, Y), \forall X, Y \in \Gamma(\mathcal{D}).$$

(ii) *The distribution $\mathcal{D}'$ is integrable if and only if*

$$A_{\bar{J}Z}V = A_{\bar{J}V}Z, \forall Z, V \in \Gamma(\mathcal{D}').$$

Proof. From (6.5.11) and (6.5.12) we have $\omega P \nabla_X Y = h(X, TY) - Ch(X, Y)$, for $X, Y \in \Gamma(\mathcal{D})$. Hence, $\omega P[X, Y] = h(X, TY) - h(TX, Y)$ which proves (i). From (6.5.10) we get $-T\nabla_Z V = A_{\omega P_1 V} Z + A_{\omega P_2 V} Z + Bh(V, Z)$, for $Z, V \in \Gamma(\mathcal{D}')$. Then we obtain $T[Z, V] = A_{\bar{J}Z}V - A_{\bar{J}V}Z$, which completes the proof. □

Theorem 6.5.4. *Let M be a GCR-lightlike submanifold of $\bar{M}$. The distribution $\mathcal{D}$ defines a totally geodesic foliation in M if and only if*

$$Bh(X, Y) = 0, \quad \forall X, Y \in \Gamma(\mathcal{D}).$$

Proof. From Definition 6.5.2, $\mathcal{D}$ defines a totally geodesic foliation if and only if $g(\nabla_X Y, \bar{J}\xi) = g(\nabla_X Y, \bar{J}W) = 0$ for $X, Y \in \Gamma(\mathcal{D})$, $\xi \in \Gamma(\mathcal{D}_2)$ and $W \in \Gamma(\mathcal{L}_2)$. From (5.1.15) and (6.1.5) we obtain $g(\nabla_X Y, \bar{J}\xi) = -\bar{g}(\bar{\nabla}_X \bar{J}Y, \xi)$. Again using (5.1.15) we get

$$g(\nabla_X Y, \bar{J}\xi) = -\bar{g}(h^l(X, \bar{J}Y), \xi) \quad \forall X, Y \in \Gamma(\mathcal{D}), \ \xi \in \Gamma(\mathcal{D}_2). \qquad (6.5.13)$$

Similarly, using (5.1.15) and (6.1.5) we derive $g(\nabla_X Y, \bar{J}W) = -\bar{g}(\bar{\nabla}_X \bar{J}Y, W)$. Then, we have

$$g(\nabla_X Y, \bar{J}W) = -\bar{g}(h^s(X, \bar{J}Y), W) \qquad (6.5.14)$$

for $X, Y \in \Gamma(\mathcal{D})$ and $W \in \Gamma(\mathcal{L}_2)$. It follows from (6.5.13) and (6.5.14) that $h^s(X, \bar{J}Y)$ has no components in $\mathcal{L}_2$ and $h^l(X, \bar{J}Y)$ has no components in $\mathcal{L}_1$ for $X, Y \in \Gamma(\mathcal{D})$ if and only if $\mathcal{D}$ defines a totally geodesic foliation in M. Thus, our assertion follows by using these last results and (6.5.9). □

Theorem 6.5.5. *Let M be a GCR-lightlike submanifold of $\bar{M}$. The distribution $\mathcal{D}'$ defines a totally geodesic foliation in M if and only if*

$$A_{\omega Y}X \in \Gamma(\mathcal{D}'), \forall X, Y \in \Gamma(\mathcal{D}').$$

Proof. (6.5.11) implies $T\nabla_X Y = A_{\omega Y}X + Bh(X, Y)$ for $X, Y \in \Gamma(\mathcal{D}')$. Then, $\mathcal{D}'$ being a totally geodesic foliation implies $A_{\omega Y}X = -Bh(X, Y)$. Hence, $A_{\omega Y}X \in \Gamma(\mathcal{D}')$, for all $X, Y \in \Gamma(\mathcal{D}')$. Conversely, $A_{\omega Y}X \in \Gamma(\mathcal{D}')$, for $X, Y \in \Gamma(\mathcal{D}')$ implies $T\nabla_X Y = 0$. Hence, $\nabla_X Y \in \Gamma(\mathcal{D}')$ which completes the proof. □

Combining the results of Theorems 6.5.4 and 6.5.5, we say that M is a *GCR-lightlike product* if $\mathcal{D}$ and $\mathcal{D}'$ are its totally geodesic foliations.

As in the Riemannian [45] and the CR-lightlike cases [361], we say that M is a *$\mathcal{D}$-geodesic GCR-lightlike submanifold* if its second fundamental form h satisfies

$$h(X,Y) = 0, \forall X, Y \in \Gamma(\mathcal{D}). \tag{6.5.15}$$

It is easy to see that M is a $\mathcal{D}$-geodesic GCR-lightlike submanifold if

$$h^l(X,Y) = 0, \; h^s(X,Y) = 0, \quad \forall X, Y \in \Gamma(\mathcal{D}). \tag{6.5.16}$$

Lemma 6.5.6. *The distribution $\mathcal{D}$ of a GCR-lightlike submanifold M of $\bar{M}$ is a totally geodesic foliation in $\bar{M}$ if and only if M is $\mathcal{D}$-geodesic*

Proof. Let $\mathcal{D}$ define a totally geodesic foliation in $\bar{M}$. Then, $\bar{\nabla}_X Y \in \Gamma(\mathcal{D})$ for $X, Y \in \Gamma(\mathcal{D})$. Using (5.1.15) we obtain

$$\bar{g}(h^l(X,Y), \xi) = \bar{g}(\bar{\nabla}_X Y, \xi) = 0, \forall \xi \in \Gamma(\mathrm{Rad}\,TM),$$
$$\bar{g}(h^s(X,Y), W) = \bar{g}(\bar{\nabla}_X Y, W) = 0, \forall W \in \Gamma(S(TM^\perp)).$$

Hence, $h^l(X,Y) = h^s(X,Y) = 0$ which means M is $\mathcal{D}$-geodesic. Conversely, assume M is $\mathcal{D}$-geodesic. Then, from (5.1.15) we derive $\bar{\nabla}_X Y \in \Gamma(TM)$, for $X, Y \in \Gamma(\mathcal{D})$. Using (6.1.5) and (5.1.15) we get

$$\bar{g}(\bar{\nabla}_X Y, \bar{J}\xi) = -\bar{g}(\bar{\nabla}_X \bar{J}Y, \xi) = -\bar{g}(h^l(X, \bar{J}Y), \xi) = 0,$$
$$\bar{g}(\bar{\nabla}_X Y, Z) = \bar{g}(\bar{\nabla}_X \bar{J}Y, \bar{J}Z) = \bar{g}(h^s(X, \bar{J}Y), \bar{J}Z) = 0,$$

$Z \in \Gamma(\bar{J}\mathcal{L}_2)$ and $\xi \in \Gamma(\mathcal{D}_2)$. Thus, $\bar{\nabla}_X Y \in \Gamma(\mathcal{D})$, which completes the proof. □

We say that M is a *mixed geodesic GCR-lightlike submanifold* (see [45] [361] for the Riemannian and the CR-lightlike cases respectively), if its second fundamental form h satisfies

$$h(X,Y) = 0, \quad \forall X \in \Gamma(\mathcal{D}), \quad Y \in \Gamma(\mathcal{D}'). \tag{6.5.17}$$

It is easy to see that M is a mixed geodesic GCR-lightlike submanifold if

$$h^l(X,Y) = 0, \; h^s(X,Y) = 0, \quad \forall X \in \Gamma(\mathcal{D}), \; Y \in \Gamma(\mathcal{D}'). \tag{6.5.18}$$

Proposition 6.5.7. *Let M be a GCR-lightlike submanifold of an indefinite Kähler manifold $\bar{M}$. Then, M is mixed geodesic if and only if*

$$A_{\bar{J}Z}X \in \Gamma(\mathcal{D}) \quad and \quad \nabla^t_X \bar{J}Z \in \Gamma(\mathcal{L}_1 \perp \mathcal{L}_2), \quad \forall Z \in \Gamma(\mathcal{D}'), \; X \in \Gamma(\mathcal{D}).$$

Proof. We have $h(X,Z) = \bar{\nabla}_X Z - \nabla_X Z$ for $X \in \Gamma(\mathcal{D})$ and $Z \in \Gamma(\mathcal{D}')$. Then from (6.1.5) we obtain $h(X,Z) = -\bar{J}\bar{\nabla}_X \bar{J}Z - \nabla_X Z$. Using (5.1.13), (6.5.8), (6.5.9) and taking the transversal part we obtain $h(X,Z) = -\omega P_1 A_{\bar{J}Z}X - \omega P_2 A_{\bar{J}Z}X - C\nabla^t_X \bar{J}Z$, which proves the assertion. □

GCR-lightlike submanifolds of $\bar{M}(c)$. We start with the following characterization theorem in terms of the curvature tensor field of $\bar{M}(c)$:

Theorem 6.5.8. *A lightlike submanifold M, of an indefinite complex space form $\bar{M}(c)$ with $c \neq 0$, is GCR-lightlike, with $\mathcal{D}_0 \neq 0$, if and only if:*

(i) *The maximal complex subspaces of T_pM, $p \in M$ define a distribution*

$$\mathcal{D} = \mathcal{D}_1 \oplus \mathcal{D}_2 \oplus \bar{J}\mathcal{D}_2 \oplus_{\mathrm{orth}} \mathcal{D}_0, \quad \text{where} \quad \mathrm{Rad}\,TM = \mathcal{D}_1 \oplus \mathcal{D}_2$$

   and $\mathcal{D}_0$ is a non-degenerate complex distribution.

(ii) *There exists a lightlike transversal vector bundle $\mathrm{ltr}(TM)$ such that*

$$\bar{g}(\bar{\mathbf{R}}(X,Y)N, N') = 0, \quad \forall X, Y \in \Gamma(\mathcal{D}).$$

(iii) *There exists a vector sub-bundle $\mathcal{M}_2$ on M such that*

$$\bar{g}(\bar{\mathbf{R}}(X,Y)W, W) = 0, \quad \forall X, Y \in \Gamma(\mathcal{D}), \; W, W' \in \Gamma(\mathcal{M}_2),$$

   where $\mathcal{M}_2$ is orthogonal to $\mathcal{D}$ and $\bar{\mathbf{R}}$ is the curvature tensor of $\bar{M}(c)$.

Proof. Suppose M is a GCR-lightlike submanifold of $\bar{M}(c)$,$c \neq 0$. Then $\mathcal{D} = \mathcal{D}_1 \oplus \mathcal{D}_2 \oplus \bar{J}\mathcal{D}_2 \oplus_{\mathrm{orth}} \mathcal{D}_0$ is a maximal subspace. Thus (i) is satisfied. From $(6.1.6)$ we have

$$\bar{g}(\bar{R}(X,Y)N, N') = \frac{c}{2}g(X, \bar{J}Y)\bar{g}(\bar{J}N, N'), \quad \forall X, Y \in \Gamma(\mathcal{D})$$

and $N, N' \in \Gamma(\mathrm{ltr}(TM))$. By the definition of a GCR-lightlike submanifold we have $\bar{g}(\bar{J}N, N') = 0$. Hence, (ii) holds. Similarly, from $(6.1.6)$ we obtain

$$\bar{g}(\bar{R}(X, \bar{J}Y)W, W') = \frac{c}{2}\{g(X, \bar{J}Y)\bar{g}(\bar{J}W, W')\} = 0$$

for $X, Y \in \Gamma(\mathcal{D})$ and $W, W' \in \Gamma(\bar{J}\mathcal{L}_2 = \mathcal{M}_2)$, which proves (iii). Conversely, assume (i), (ii) and (iii). Then, from (i) we see that a part, $\mathcal{D}_2$, of $\mathrm{Rad}\,TM$ is a distribution on M such that $\bar{J}\mathcal{D}_2 \cap \mathrm{Rad}\,TM = \{0\}$. It also shows that the other part of $\mathrm{Rad}\,TM$ is invariant. Thus (A) of the definition of GCR-lightlike submanifold is satisfied. Therefore, we can choose a screen distribution containing $\bar{J}\mathcal{D}_2$ and $\mathcal{D}_0$, (since $\mathcal{D}_0$ is non-degenerate). $\mathrm{ltr}(TM)$ orthogonal to $S(TM)$ implies that $\bar{g}(\bar{J}N, \xi) = -\bar{g}(N, \bar{J}\xi) = 0$ for $\xi \in \Gamma(\mathcal{D}_2)$. Hence we conclude that some part of $\bar{J}\,\mathrm{ltr}(TM)$ defines a distribution on M, say $\mathcal{M}_1$. On the other hand, from (ii) we derive $\frac{c}{2}g(X, \bar{J}Y)\bar{g}(\bar{J}N, N') = 0$ for $X, Y \in \Gamma(\mathcal{D}_0)$ and $N, N' \in \Gamma(\mathrm{ltr}(TM))$. Since $c \neq 0$ and $\mathcal{D}_0$ is non-degenerate we conclude that $\bar{g}(\bar{J}N, N') = 0$, that is $\bar{J}\,\mathrm{ltr}(TM) \cap \mathrm{Rad}\,TM = \{0\}$. Moreover, if $\bar{g}(N, \xi) = 1$ for $\xi \in \Gamma(\mathcal{D}_2)$ and $N \in \Gamma(\bar{J}\mathcal{M}_1)$ then we have $\bar{g}(\bar{J}N, \bar{J}\xi) = 1$. This shows that $\mathcal{M}_1$ is not orthogonal to $\mathcal{D}_2$ and, hence, it is not orthogonal to $\mathcal{D}$. Now, consider a distribution $\mathcal{M}_2$ which is orthogonal to $\mathcal{D}$. Then $\mathcal{M}_2 \cap \mathcal{M}_1 = \{0\}$ which is orthogonal to $\mathcal{M}_1$. Furthermore, from (iii) we have

$$\frac{c}{2}\{g(X, \bar{J}Y)\bar{g}(\bar{J}W, W') = 0, \quad \forall X, Y \in \Gamma(\mathcal{D}_0), \ W, W' \in \Gamma(\mathcal{M}_2).$$

Since $c \neq 0$ and $\mathcal{D}_0$ is non-degenerate we have $\bar{g}(\bar{J}W, W') = 0$. This implies $\bar{J}\mathcal{M}_2 \perp \mathcal{M}_2$. On the other hand, since $\mathcal{M}_2$ is orthogonal to $\mathcal{D}$, we obtain

$$g(\bar{J}W, X) = -\bar{g}(W, \bar{J}X) = 0, \quad \forall X \in \Gamma(\mathcal{D}), \ W \in \Gamma(\mathcal{M}_2).$$

Hence $\bar{J}\mathcal{M}_2 \perp \mathcal{D}$. Thus since $\bar{J}\mathcal{M}_2 \perp \mathcal{D}$, $\bar{J}\mathcal{M}_2 \perp \mathcal{M}_1$ and $\bar{J}\mathcal{M}_2 \perp \mathcal{M}_2$, we conclude that $\bar{J}\mathcal{M}_2 \subset S(TM^\perp)$, which completes the proof. $\square$

Theorem 6.5.9. *Let M be a mixed geodesic proper GCR-lightlike submanifold of an indefinite complex space form $\bar{M}(c)$, whose distribution $\mathcal{D}$ is a totally geodesic foliation in $\bar{M}$. Then, M is a complex semi-Euclidean space.*

Proof. From (6.1.6) we obtain

$$\bar{g}(\bar{R}(X, \bar{J}X)Z, \bar{J}Z) = -\frac{c}{2}g(X, X)g(Z, Z), \ \forall X \in \Gamma(\mathcal{D}_0), \tag{6.5.19}$$

for $Z \in \Gamma(\bar{J}\mathcal{L}_2)$. On the other hand, M mixed geodesic and (5.3.11) imply

$$\bar{g}(\bar{R}(X, \bar{J}X)Z, \bar{J}Z) = \bar{g}((\nabla_X h^s)(\bar{J}X, Z) - (\nabla_{\bar{J}X} h^s)(X, Z), \bar{J}Z) \tag{6.5.20}$$

for $X \in \Gamma(\mathcal{D}_0)$ and $Z \in \Gamma(\bar{J}\mathcal{L}_2)$. Thus from (6.5.19) and (6.5.20) we derive

$$-\frac{c}{2}g(X, X)g(Z, Z) = \bar{g}((\nabla_X h^s)(\bar{J}X, Z) - (\nabla_{\bar{J}X} h^s)(X, Z), \bar{J}Z). \tag{6.5.21}$$

M mixed geodesic implies $(\nabla_X h^s)(\bar{J}X, Z) = -h^s(\nabla_X \bar{J}X, Z) - h^s(\bar{J}X, \nabla_X Z)$. Similarly, $(\nabla_{\bar{J}X} h^s)(X, Z) = -h^s(\nabla_{\bar{J}X} X, Z) - h^s(X, \nabla_{\bar{J}X} Z)$. Hence

$$(\nabla_X h^s)(\bar{J}X, Z) - (\nabla_{\bar{J}X} h^s)(X, Z) = h^s([\bar{J}X, X], Z) - h^s(\bar{J}X, \nabla_X Z) + h^s(X, \nabla_{\bar{J}X} Z).$$

Moreover, the totally geodesic foliation in $\bar{M}$ and mixed geodesic M imply

$$(\nabla_X h^s)(\bar{J}X, Z) - (\nabla_{\bar{J}X} h^s)(X, Z) = -h^s(\bar{J}X, \nabla_X Z) + h^s(X, \nabla_{\bar{J}X} Z).$$

Now using (6.5.14) and mixed geodesic M we have

$$(\nabla_X h^s)(\bar{J}X, Z) - (\nabla_{\bar{J}X} h^s)(X, Z) = h^s(\bar{J}X, TA_W X) - h^s(X, TA_W \bar{J}X),$$

where $\bar{J}W = Z$. On the other hand, from the assumption of theorem and Lemma 6.5.6, it follows that M is $\mathcal{D}$-geodesic. Thus we arrive at

$$(\nabla_X h^s)(\bar{J}X, Z) - (\nabla_{\bar{J}X} h^s)(X, Z) = 0.$$

Then (6.5.21) becomes $\frac{c}{2}g(X, X)g(Z, Z) = 0$. Since $\mathcal{D}_0$ and $\bar{J}\mathcal{L}_2$ are non-degenerate, we have $c = 0$, which completes the proof. $\square$

Corollary 6.5.10. *There exist no totally geodesic proper GCR-lightlike submanifolds in $\bar{M}(c)$ with $c \neq 0$.*

Theorem 6.5.11. *Let M be a totally umbilical proper GCR-lightlike submanifold of an indefinite Kähler manifold $\bar{M}$. If $\mathcal{D}_0$ is integrable, then, the induced connection ∇ is a metric connection. Moreover, $h^s = 0$.*

Proof. (6.5.12) implies $h^l(X, \bar{J}Y) = \omega P_1 \nabla_X Y + Ch^l(X, Y)$ for $X, Y \in \Gamma(\mathcal{D}_0)$. From the first equation of (5.3.8) we have $g(X, \bar{J}Y)H^l = \omega P_1 \nabla_X Y + g(X, Y)CH^l$. Hence, $g(X, \bar{J}Y)H^l - g(Y, \bar{J}X)H^l + \omega P_1[X, Y] = 0$. Since $\mathcal{D}_0$ is integrable, for $X = \bar{J}Y$, from (6.1.5) we get $2g(Y, Y)H^l = 0$. Now $\mathcal{D}_0$ non-degenerate implies $H^l = 0$. Then, using (5.3.8) we obtain $h^l = 0$ which implies from (5.1.39) that ∇ is a metric connection. $h^s = 0$ is immediate. $\qquad \square$

Lemma 6.5.12. *Let M be a totally umbilical proper GCR-lightlike submanifold of an indefinite Kähler manifold $\bar{M}$. Then $H^s \in \Gamma(\mathcal{L}_2)$.*

Proof. From (6.5.11) we obtain $h^s(X, \bar{J}Y) = Ch^s(X, Y) + \omega P_2 \nabla_X Y$ for $X, Y \in \Gamma(\mathcal{D}_0)$. From M totally umbilical we get $g(X, X)CH^s + \omega P_2 \nabla_X X = 0$. Hence, non-degenerate $\mathcal{D}_0$ implies $CH^s = 0$, i.e., $H^s \in \Gamma(\mathcal{L}_2)$. $\qquad \square$

Theorem 6.5.13. *Let M be a totally umbilical proper GCR-lightlike submanifold of an indefinite Kähler manifold $\bar{M}$. One of the following holds:*

(a) *M is totally geodesic.*

(b) *$h^s = 0$ or $\dim(\mathcal{L}_2) = 1$ and $\mathcal{D}_0$ is not integrable.*

Proof. If $\mathcal{D}_0 \neq \{0\}$ and integrable, from Theorem 6.5.11 we obtain that $h^l = h^s = 0$ which is case (a). Now suppose that $\mathcal{D}_0$ is not integrable. From (5.1.15), (5.1.23), (6.5.7), (6.5.9) and taking the tangential part we have

$$-A_{\bar{J}W}Z = T\nabla_Z W + Bh(Z, W)$$

for $Z, W \in \Gamma(\bar{J}\mathcal{L}_2)$. Hence, using (6.1.5), (5.1.28) and (6.5.9) we get

$$\bar{g}(h^s(Z, Z), \bar{J}W) = \bar{g}(\nabla_Z W, \bar{J}Z) + g(h^s(Z, W), \bar{J}Z)$$

which implies

$$\bar{g}(h^s(Z, Z), \bar{J}W) = g(h^s(Z, W), \bar{J}Z).$$

Since M is totally umbilical we derive

$$g(Z, Z)\bar{g}(H^s, \bar{J}W) = g(Z, W)\bar{g}(H^s, \bar{J}Z). \qquad (6.5.22)$$

Interchanging the roles of Z and W in this equation we have

$$g(W, W)\bar{g}(H^s, \bar{J}Z) = g(Z, W)\bar{g}(H^s, \bar{J}W). \qquad (6.5.23)$$

Thus from (6.5.22) and (6.5.23) we obtain

$$\bar{g}(H^s, \bar{J}Z) = \frac{g(Z, W)^2}{g(W, W)g(Z, Z)}\bar{g}(H^s, \bar{J}Z). \tag{6.5.24}$$

Since $\bar{J}\mathcal{L}_2$ is non-degenerate, choosing non-null vector fields Z and W and using Lemma 6.5.12 in (6.5.24) we conclude that either $H^s = 0$ or Z and W are linearly dependent. This proves (b), which completes the proof. □

Theorem 6.5.14. *There exists no totally umbilical proper GCR-lightlike submanifold of an indefinite complex space form $\bar{M}(c)$, $c \neq 0$.*

Proof. Suppose M is a totally umbilical proper GCR-lightlike submanifold of $\bar{M}(c)$, $c \neq 0$. Then from (6.1.6) we obtain

$$\bar{R}(X, \bar{J}X)Z = -\frac{c}{2}g(X, X)\bar{J}Z, \quad \forall X \in \Gamma(\mathcal{D}_0), \ Z \in \Gamma(\bar{J}\mathcal{L}_2). \tag{6.5.25}$$

On the other hand, from (5.3.8) and (5.3.11) we get

$$\bar{R}(X, \bar{J}X)Z = (\nabla_X h^s)(\bar{J}X, Z) - (\nabla_{\bar{J}X} h^s)(X, Z). \tag{6.5.26}$$

Thus from (6.5.25) and (6.5.26) we obtain

$$-\frac{c}{2}g(X, X)\bar{J}Z = (\nabla_X h^s)(\bar{J}X, Z) - (\nabla_{\bar{J}X} h^s)(X, Z). \tag{6.5.27}$$

Since M is totally umbilical, from (5.3.7) we have

$$(\nabla_X h^s)(\bar{J}X, Z) = -g(\nabla_X \bar{J}X, Z)H^s - g(\bar{J}X, \nabla_X Z)H^s.$$

Since $\bar{g}(\bar{J}X, Z) = 0$, $\forall X \in \Gamma(\mathcal{D}_0)$ and $Z \in \Gamma(\bar{J}\mathcal{L}_2)$, differentiating this equation w.r.t. X we get $g(\nabla_X \bar{J}X, Z) = -g(\bar{J}X, \nabla_X Z)$. Thus, $(\nabla_X h^s)(\bar{J}X, Z) = 0$, $\forall X \in \Gamma(\mathcal{D}_0)$ and $\forall Z \in \Gamma(\bar{J}\mathcal{L}_2)$. Similarly, $(\nabla_{\bar{J}X} h^s)(X, Z) = 0$. Hence (6.5.27) becomes $-\frac{c}{2}g(X, X)\bar{J}Z = 0$. Since M is proper and $\mathcal{D}_0$ is non-degenerate, we get $c = 0$. This contradiction completes the proof. □

Example 10. Consider a GCR-lightlike submanifold of $\mathbf{R}_4^{14}$ presented in Example 9. By direct calculations we obtain

$$\bar{\nabla}_X Z_1 = \bar{\nabla}_X Z_2 = \bar{\nabla}_X Z_3 = \bar{\nabla}_X Z_4 = \bar{\nabla}_X Z_5 = 0,$$
$$\bar{\nabla}_X Z_6 = \bar{\nabla}_X Z_8 = \bar{\nabla}_X Z_9 = \bar{\nabla}_X Z_{10} = 0,$$
$$\bar{\nabla}_{Z_7} Z_7 = -W, \quad \forall X \in \Gamma(TM).$$

Hence we have $h^l = 0$ which shows that the induced connection ∇ is a metric connection. Also we obtain

$$h^s(X, Z_1) = h^s(X, Z_2) = 0, \qquad h^s(X, Z_3) = h^s(X, Z_4) = h^s(X, Z_5) = 0,$$
$$h^s(X, Z_6) = h^s(X, Z_8) = 0, \qquad h^s(X, Z_9) = h^s(X, Z_{10}) = 0,$$

$$h^s(Z_7, Z_7) = -W, \quad \forall X \in \Gamma(TM).$$

Thus, $h^s(Z_7, Z_7) = g(Z_7, Z_7)H^s$, where $H^s = -W$. Hence, M is a totally umbilical proper GCR-lightlike submanifold. Moreover, we see that M is not totally geodesic and $\dim \mathcal{L}_2 = 1$.

Finally, we prove a characterization theorem for minimal GCR-lightlike submanifolds. First recall that (as in the non-degenerate case) a totally geodesic lightlike submanifold is minimal. Thus, it follows from Theorem 6.5.13 (a) that "*a proper GCR-lightlike submanifold of an indefinite Kähler manifold, with an integrable distribution $\mathcal{D}_0$, is minimal*".

Theorem 6.5.15. *A totally umbilical proper GCR-lightlike submanifold M, of an indefinite Kähler manifold $\bar{M}$, is minimal if and only if*

$$\operatorname{tr} A_{W_i} = 0 \text{ on } \mathcal{D}_0 \perp \bar{J}\mathcal{L}_2, \quad \text{and} \quad \operatorname{tr} A^*_{\xi_k} = 0 \text{ on } \mathcal{D}_0 \perp \bar{J}\mathcal{L}_2$$

for $W_j \in \Gamma(S(TM^\perp))$, where $k \in \{1, \ldots, r\}$ and $j \in \{1, \ldots, (n-r)\}$.

Proof. By Definition 6.5.2, a GCR-lightlike submanifold is minimal if and only if

$$\sum_{i=1}^{a} h(Z_i, Z_i) + \sum_{j=1}^{b} h(\bar{J}\xi_j, \bar{J}\xi_j) + \sum_{j=1}^{b} h(\bar{J}N_j, \bar{J}N_j) + \sum_{l=1}^{c} h(\bar{J}W_l, \bar{J}W_l) = 0$$

and $h^s = 0$ on $\operatorname{Rad}(TM)$, $a = \dim(\mathcal{D}_0)$, $b = \dim(\mathcal{D}_2)$ and $c = \dim(\mathcal{L}_2)$. M is totally umbilical and (5.3.7) implies $h(\bar{J}\xi_j, \bar{J}\xi_j) = h(\bar{J}N_j, \bar{J}N_j) = 0$. Similarly, $h^s = 0$ on $\operatorname{Rad}(TM)$. Then follow the proof of Theorem 5.4.5 in Chapter 5.   □

It is known that the holomorphic distribution of a CR-submanifold of a Kähler manifold is minimal. The next theorem shows that this is not valid for the holomorphic distribution $\mathcal{D}_0$ in a GCR-lightlike submanifold.

Theorem 6.5.16. *Let M be a GCR-lightlike submanifold of an indefinite Kähler manifold $\bar{M}$. Then the distribution $\mathcal{D}_0$ is minimal if and only if*

$$A_{N'}X - \bar{J}A_{N'}\bar{J}X \quad \text{and} \quad A_N\bar{J}X + \bar{J}A_N X \quad \text{have no components in} \quad \mathcal{D}_0,$$

for $X \in \Gamma(\mathcal{D}_0)$, $N \in \Gamma(\mathcal{L}_1)$ and $N' \in \Gamma(\operatorname{ltr}(TM))$.

Proof. $\mathcal{D}_o$ being almost complex, using (5.1.15), (6.1.5) and (5.1.27) we have $g(\nabla_X X, \bar{J}\xi) = -g(\bar{J}X, A^*_\xi X)$ and $g(\nabla_{\bar{J}X}\bar{J}X, \bar{J}\xi) = g(A^*_\xi \bar{J}X, X)$ for $X \in \Gamma(\mathcal{D}_0)$ and $\xi \in \Gamma(\mathcal{D}_2)$. Since the shape operator of $S(TM)$ is self-adjoint, we get

$$g(\nabla_X X + \nabla_{\bar{J}X}\bar{J}X, \bar{J}\xi) = 0, \quad \forall X \in \Gamma(\mathcal{D}_0), \quad \xi \in \Gamma(\mathcal{D}_2). \tag{6.5.28}$$

In a similar way we obtain

$$g(\nabla_X X + \nabla_{\bar{J}X}\bar{J}X, \bar{J}W) = 0, \quad \forall X \in \Gamma(\mathcal{D}_0), \quad W \in \Gamma(\mathcal{L}_2). \tag{6.5.29}$$

Also, from (5.1.15),(6.1.5) and (5.1.22) we have $g(\nabla_X X, \bar{J}N) = -g(\bar{J}X, A_N X)$ and $g(\nabla_{\bar{J}X} \bar{J}X, \bar{J}N) = g(X, A_N \bar{J}X)$, $\forall X \in \Gamma(\mathcal{D}_0)$ and $N \in \Gamma(\mathcal{L}_1)$. Hence,

$$g(\nabla_X X + \nabla_{\bar{J}X} \bar{J}X, \bar{J}N) = g(X, A_N \bar{J}X + \bar{J}A_N X). \tag{6.5.30}$$

In a similar way we derive

$$g(\nabla_X X + \nabla_{\bar{J}X} \bar{J}X, N') = g(X, A_{N'} X - \bar{J}A_{N'} \bar{J}X), \ \forall X \in \Gamma(\mathcal{D}_0) \tag{6.5.31}$$

and $N' \in \Gamma(\text{ltr}(TM))$. Thus, the assertion follows from (6.5.28)–(6.5.31). □

Example 11. Let $\bar{M} = (\mathbf{R}_4^{10} \times \mathbf{R}^4, g)$ be a semi-Riemannian cross product manifold, where $\mathbf{R}^4$ is Euclidean space and $\mathbf{R}_4^{10}$ is a semi-Euclidean space of signature $(-,-,-,-,+,+,+,+,+,+)$ with respect to the canonical basis

$$\{\partial x_1, \partial x_2, \partial x_3, \partial x_4, \partial x_5, \partial x_6, \partial x_7, \partial x_8, \partial x_9, \partial x_{10}\}$$

and g is the inner product of $\mathbf{R}_4^{14}$. Let M be a submanifold of $\mathbf{R}_4^{14}$ given by

$$x_1 = u_1 \cosh \alpha, \quad x_2 = u_2 \cosh \alpha, \quad x_3 = u_3, \quad x_4 = u_4, \quad x_5 = u_3, \quad x_6 = u_5,$$
$$x_7 = \cos u_6 \cosh u_7, \quad x_8 = \sin u_6 \quad \sinh u_7, \quad x_9 = u_1 \sinh \alpha - u_2,$$
$$x_{10} = u_2 \sinh \alpha + u_1, \quad x_{11} = \cos u_8 \cosh u_9, \quad x_{12} = \cos u_8 \sinh u_9,$$
$$x_{13} = \sin u_8 \cosh u_9, \quad x_{14} = \sin u_8 \sinh u_9$$

where $\sin u_6 \neq 0$, $\cos u_6 \neq 0$, $\sin u_8 \neq 0$ and $\cos u_8 \neq 0$. TM is spanned by

$$Z_1 = \cosh \alpha \, \partial x_1 + \sinh \alpha \, \partial x_9 + \partial x_{10}, \quad Z_2 = \cosh \alpha \, \partial x_2 - \partial x_9 + \sinh \alpha \, \partial x_{10},$$
$$Z_3 = \partial x_3 + \partial x_5, \quad Z_4 = \partial x_4, \quad Z_5 = \partial x_6,$$
$$Z_6 = -\sin u_6 \cosh u_7 \, \partial x_7 + \cos u_6 \sinh u_7 \, \partial x_8,$$
$$Z_7 = \cos u_6 \sinh u_7 \, \partial x_7 + \sin u_6 \cosh u_7 \, \partial x_8,$$
$$Z_8 = -\sin u_8 \cosh u_9 \, \partial x_{11} - \sin u_8 \sinh u_9 \, \partial x_{12}$$
$$\quad + \cos u_8 \cosh u_9 \, \partial x_{13} + \cos u_8 \sinh u_9 \, \partial x_{14},$$
$$Z_9 = \cos u_8 \sinh u_9 \, \partial x_{11} + \cos u_8 \cosh u_9 \, \partial x_{12}$$
$$\quad + \sin u_8 \sinh u_9 \, \partial x_{13} + \sin u_8 \cosh u_9 \, \partial x_{14}.$$

Hence M is 3-lightlike with $\text{Rad} \, TM = \text{Span}\{Z_1, Z_2, Z_3\}$ and $\bar{J}Z_1 = Z_2$ and $\mathcal{D}_1 = \{Z_1, Z_2\}$. On the other hand, $\bar{J}Z_3 = Z_4 + Z_5 \in \Gamma(S(TM))$ implies that $\mathcal{D}_2 = \text{Span}\{Z_3\}$. Moreover, $\bar{J}Z_6 = -Z_7$ implies $\mathcal{D}_0 = \text{Span}\{Z_6, Z_7\}$. Also, $\bar{J}Z_8$ and $\bar{J}Z_9$ are orthogonal to TM and $\{\bar{J}Z_8, \bar{J}Z_9\}$ is not lightlike. Thus we can choose $\mathcal{L}_2 = S(TM^\perp) = \{\bar{J}Z_8, \bar{J}Z_9\}$. Hence $\{Z_8, Z_9\}$ is a screen real sub-bundle. As a result we conclude that M is a proper GCR-lightlike submanifold of $\mathbf{R}_4^{14}$, with lightlike transversal bundle $\mathrm{ltr}(TM)$ spanned by

$$N_1 = \frac{1}{2}\{-\cosh\alpha\,\partial x_1 - \sinh\alpha\,\partial x_9 + \partial x_{10}\},$$

$$N_2 = \frac{1}{2}\{-\cosh\alpha\,\partial x_2 - \partial x_9 - \sinh\alpha\,\partial x_{10}\},$$

$$N_3 = \frac{1}{2}\{-\partial x_3 + \partial x_5\}.$$

$\mathrm{Span}\{N_1, N_2\}$ is invariant with respect to $\bar{J}$ and $\bar{J}N_3 = -\frac{1}{2}Z_4 + \frac{1}{2}Z_5$. Hence $\mathcal{L}_1 = \mathrm{Span}\{N_3\}$ and $\bar{J}\mathcal{D}' = \mathrm{Span}\{\bar{J}Z_8, \bar{J}Z_9, N_3\}$. Now by direct calculations, using Gauss and Weingarten formulas, we have

$$h^l = 0,\ h^s(X, Z_1) = h^s(X, Z_2) = 0,\qquad h^s(X, Z_3) = h^s(X, Z_4),$$
$$h^s(X, Z_5) = h^s(X, Z_6) = 0,\qquad h^s(X, Z_7) = 0, \forall X \in \Gamma(TM),$$

$$h^s(Z_8, Z_8) = \left(\frac{1}{1 + 2\sinh^2 u_9}\right)\bar{J}Z_9,\quad h^s(Z_9, Z_9) = -\left(\frac{1}{1 + 2\sinh^2 u_9}\right)\bar{J}Z_9.$$

Hence, the induced connection is a metric connection and M is not totally geodesic, but, it is a proper minimal GCR-lightlike submanifold of $\mathbf{R}_4^{14}$.

6.6   Totally real lightlike submanifolds

It is important to note that $\bar{J}\,\mathrm{Rad}\,TM$ is a distribution on a lightlike submanifold of an indefinite Kähler manifold $\bar{M}$, in all previous sections discussed so far. This means that (contrary to the Riemannian case of CR-submanifolds) so far we do not have any *totally real (also called anti-invariant)* lightlike submanifolds M of an indefinite Kähler manifold $\bar{M}$, since in all those cases $\bar{J}\,\mathrm{Rad}\,TM$ is always tangent to M. For an extensive study of Riemannian totally real submanifolds, see Yano-Kon [410] and many references therein. In particular, so far the collection of lightlike submanifolds excludes the important subcase of real lightlike curves of Hermitian or Kähler manifolds. For an up-to date study of all types of null curves of semi-Riemannian manifolds and their applications in mathematical physics, see a recent book [156]. In order to add the above two subcases in the variety of submanifolds, we present in this section recent work of Sahin [346, 356]. This will complete our objective of discussing all possible lightlike submanifolds of Kähler manifolds.

Definition 6.6.1. [346] Let M be a lightlike submanifold of an indefinite Kähler manifold $\bar{M}$ such that $\mathrm{Rad}\,TM$ is transversal with respect to $\bar{J}$, i.e., $\bar{J}(\mathrm{Rad}\,TM) = \mathrm{ltr}(TM)$. Then we say that M is a totally real lightlike submanifold if $S(TM)$ is also transversal with respect to $\bar{J}$, i.e, $\bar{J}(S(TM)) \subseteq S(TM^\perp)$.

The above definition implies that $\mathrm{ltr}(TM)$ is also transversal with respect to $\bar{J}$. Moreover, we have the following result.

Proposition 6.6.2. *There do not exist 1-lightlike totally real lightlike submanifolds of an indefinite Kähler manifold $\bar{M}$.*

Proof. Let us suppose that M is a 1-lightlike totally real lightlike submanifold of an indefinite Kähler manifold $\bar{M}$. In this case $\mathrm{Rad}\,TM = \mathrm{Span}\{\xi\}$. This implies that $\mathrm{rank}(\mathrm{ltr}(TM)) = 1$, say $\mathrm{ltr}(TM) = \mathrm{Span}\{N\}$. Then, Definition 6.6.1 implies that $\bar{J}\xi = \alpha N$ for some differentiable function α. On the other hand, we have $\bar{g}(\xi, N) = 1$ and $\bar{g}(\bar{J}\xi, \xi) = 0$. Hence, we get

$$0 = \bar{g}(\bar{J}\xi, \xi) = \alpha\bar{g}(\xi, N) = \alpha.$$

Then, we have $\alpha = 0$. Thus, we obtain $\bar{J}\xi = 0$. Since $\bar{J}$ is non-singular, we derive $\xi = 0$, which is a contradiction. Thus, we conclude that M can not be 1-lightlike. □

From Definition 6.6.1 and Proposition 6.6.2, we note the following:

1. $\dim(\mathrm{Rad}\,TM) \geq 2$.

2. There exist no totally real lightlike hypersurfaces and totally real half-lightlike submanifolds of a Kähler manifold.

3. Any 3-dimensional totally real lightlike submanifold must be 2-lightlike.

Example 12. Consider a plane in R_2^4 given by

$$x_1 = x_4, x_2 = x_3.$$

Then TM is spanned by

$$Z_1 = \partial x_1 + \partial x_4, \; Z_2 = \partial x_2 + \partial x_3.$$

Thus M is a totally lightlike submanifold of R_2^4. Moreover, the lightlike transversal bundle $\mathrm{ltr}(TM)$ is spanned by

$$N_1 = \frac{1}{2}\{-\partial x_1 + \partial x_4\}, \; N_1 = -\frac{1}{2}\{\partial x_2 - \partial x_3\}.$$

Then it is easy to see that $\bar{J}Z_1 = -2N_2$ and $\bar{J}Z_2 = 2N_1$. Hence, $\bar{J}(\mathrm{Rad}\,TM) = \mathrm{ltr}(TM)$ which implies that M is a totally real lightlike submanifold.

Suppose M is a totally real lightlike submanifold of an indefinite Kähler manifold $\bar{M}$, with μ a complementary orthogonal distribution to $\bar{J}(S(TM))$ in $S(TM^\perp)$. Then, for $V \in \Gamma(S(TM^\perp))$, we can write

$$\bar{J}V = BV + CV, \quad BV \in \Gamma(S(TM)), \quad CV \in \Gamma(\mu). \tag{6.6.1}$$

Theorem 6.6.3. *Let M be a totally real lightlike submanifold of an indefinite Kähler manifold $\bar{M}$. Then, the induced connection ∇ on M is a metric connection if and only if $D^s(X, \bar{J}Y) \in \Gamma(\mu)$ for $X \in \Gamma(TM)$, $Y \in \Gamma(\mathrm{Rad}\,TM)$.*

Proof. By Definition 6.6.1, $\nabla_X Y \in \Gamma(\operatorname{Rad} TM) \iff g(\nabla_X Y, Z) = 0$ for $X \in \Gamma(TM), Y \in \Gamma(\operatorname{Rad} TM)$ and $Z \in \Gamma(S(TM))$. From (5.1.15) we have $g(\nabla_X Y, Z) = \bar{g}(\bar{\nabla}_X Y, Z)$. Using (6.1.5) we get $g(\nabla_X Y, Z) = \bar{g}(\bar{\nabla}_X \bar{J}Y, \bar{J}Z)$. From (5.1.22) we derive $g(\nabla_X Y, Z) = \bar{g}(D^s(X, \bar{J}Y), \bar{J}Z)$ which proves the assertion. $\qquad\square$

Remark 6.6.4. The study of totally real lightlike submanifolds of an indefinite Kähler manifold is a good topic of further investigation for which nothing is known more than what we have presented in this book. The classical Riemannian study of Yano-Kon [410] may be followed in the study of this new area of research.

Now we discuss another class of lightlike subspaces of indefinite Kähler manifolds. It will be seen that this new class includes real lightlike curves which are missing in previous sections.

Definition 6.6.5. [356] Let M be an r-lightlike submanifold of an indefinite Kähler manifold $\bar{M}$. Then, we say that M is a screen transversal(ST) lightlike submanifold of $\bar{M}$ if there exists a screen transversal bundle $S(TM^{\perp})$ such that

$$\bar{J}(\operatorname{Rad} TM) \subset S(TM^{\perp}). \tag{6.6.2}$$

Lemma 6.6.6. *Let M be an r-lightlike submanifold of an indefinite Kähler manifold $\bar{M}$. Suppose that $\bar{J}\operatorname{Rad} TM$ is a vector sub-bundle of $S(TM^{\perp})$. Then, $\bar{J}\operatorname{ltr}(TM)$ is also a vector sub-bundle of the screen transversal bundle $S(TM^{\perp})$. Moreover, $\bar{J}\operatorname{Rad} TM \cap \bar{J}\operatorname{ltr}(TM) = \{0\}$.*

Proof. Let us assume that $\operatorname{ltr}(TM)$ is invariant with respect to $\bar{J}$, i.e., $\bar{J}(\operatorname{ltr}(TM)) = \operatorname{ltr}(TM)$. By the definition of a lightlike submanifold, there exist vector fields $\xi \in \Gamma(\operatorname{Rad} TM)$ and $N \in \Gamma(\operatorname{ltr}(TM))$ such that $\bar{g}(\xi, N) = 1$. Also from (6.1.5) we get

$$\bar{g}(\xi, N) = \bar{g}(\bar{J}\xi, \bar{J}N) = 1.$$

However, if $\bar{J}N \in \Gamma(\operatorname{ltr}(TM))$ then by hypothesis, we get $\bar{g}(\bar{J}N, \bar{J}\xi) = 0$. Hence, we obtain a contradiction which implies that $\bar{J}N$ does not belong to $\operatorname{ltr}(TM)$. Now, suppose that $\bar{J}N \in \Gamma(S(TM))$. Then, in a similar way, we have

$$1 = \bar{g}(\xi, N) = \bar{g}(\bar{J}N, \bar{J}\xi) = 0$$

since $\bar{J}\xi \in \Gamma(S(TM^{\perp}))$ and $\bar{J}N \in \Gamma(S(TM))$. Thus, $\bar{J}N$ does not belong to $S(TM)$. We can also obtain that $\bar{J}N$ does not belong to $\operatorname{Rad} TM$. Then, from the decomposition of a lightlike submanifold, we conclude that $\bar{J}N \in \Gamma(S(TM^{\perp}))$. Now, suppose that there exists a vector field $X \in \Gamma(\bar{J}\operatorname{Rad} TM \cap \bar{J}\operatorname{ltr}(TM))$. Then, we have $X \in \Gamma(\bar{J}(\operatorname{Rad} TM))$. Hence, $\bar{g}(X, \bar{J}N) = 0$ since $X \in \Gamma(\bar{J}(\operatorname{ltr}(TM))$. However, for an r-lightlike submanifold there exists some vector fields $JX \in \Gamma(\operatorname{Rad} TM)$ such that $\bar{g}(JX, N) \neq 0$. Then, from (6.1.5), we get $0 \neq \bar{g}(JX, N) = -\bar{g}(X, \bar{J}N) = 0$ which is a contradiction. Thus, the proof is complete. $\qquad\square$

From the above definition and Lemma 6.6.6, it follows that $\bar{J}\operatorname{ltr}(TM) \subset S(TM^{\perp})$. Also it is obvious that there is no coisotropic and totally lightlike screen transversal lightlike submanifolds of indefinite Kähler manifolds due to $S(TM^\perp) = \{0\}$. It is important to emphasize that $\bar{J}(\operatorname{Rad} TM)$ and $\bar{J}(\operatorname{ltr}(TM))$ are not orthogonal. If they were orthogonal, then $S(TM^\perp)$ would be degenerate.

Definition 6.6.7. Let M be screen transversal lightlike submanifold of an indefinite Kähler manifold $\bar{M}$. Then We say that M is a screen transversal totally real lightlike submanifold if $S(TM)$ is screen transversal with respect to $\bar{J}$, that is,

$$\bar{J}(S(TM)) \subset S(TM^\perp). \tag{6.6.3}$$

From Definition 6.6.7, if M is a screen transversal totally real lightlike submanifold then we have

$$S(TM^\perp) = \bar{J}(\operatorname{Rad} TM) \oplus \bar{J}(\operatorname{ltr}(TM)) \perp \bar{J}(S(TM)) \perp \mathcal{D}_0 \tag{6.6.4}$$

where $\mathcal{D}_0$ is a non-degenerate orthogonal complementary distribution to $\bar{J}(\operatorname{Rad} TM) \oplus \bar{J}(\operatorname{ltr}(TM)) \perp \bar{J}(S(TM)))$ in $S(TM^\perp)$.

Proposition 6.6.8. *Let M be a screen transversal totally real lightlike submanifold of an indefinite Kähler manifold $\bar{M}$. Then the distribution $\mathcal{D}_0$ is invariant with respect to $\bar{J}$.*

Proof. For $X \in \Gamma(\mathcal{D}_0)$, $\xi \in \Gamma(\operatorname{Rad} TM)$ and $N \in \Gamma(\operatorname{ltr}(TM))$, we have

$$\bar{g}(\bar{J}X, \xi) = -\bar{g}(X, \bar{J}\xi) = 0 \quad \text{and} \quad \bar{g}(\bar{J}X, N) = -\bar{g}(X, \bar{J}N) = 0$$

which imply that $\bar{J}(\mathcal{D}_0) \cap \operatorname{Rad} TM = \{0\}$ and $\bar{J}(\mathcal{D}_0) \cap \operatorname{ltr}(TM) = \{0\}$. From (6.1.5) we get

$$\bar{g}(\bar{J}X, \bar{J}\xi) = \bar{g}(X, \xi) = 0 \quad \text{and} \quad \bar{g}(\bar{J}X, \bar{J}N) = \bar{g}(X, N) = 0$$

which show that $\bar{J}(\mathcal{D}_0) \cap \bar{J}(\operatorname{Rad} TM) = \{0\}$ and $\bar{J}(\mathcal{D}_0) \cap \bar{J}(\operatorname{ltr}(TM)) = \{0\}$. Moreover, for $Z \in \Gamma(S(TM))$, since $\bar{J}Z \in \Gamma(\bar{J}(S(TM)), \bar{J}(S(TM))$ and $\mathcal{D}_0$ are orthogonal, we obtain $\bar{g}(\bar{J}X, Z) = -\bar{g}(X, \bar{J}Z) = 0$ which shows that $\bar{J}(\mathcal{D}_0) \cap S(TM) = \{0\}$. Hence, we also have $\bar{J}(\mathcal{D}_0) \cap \bar{J}(S(TM)) = \{0\}$. Thus, we arrive at

$$\bar{J}(\mathcal{D}_0) \cap TM = \{0\}, \bar{J}(\mathcal{D}_0) \cap \operatorname{ltr}(TM) = \{0\}$$

and

$$\bar{J}(\mathcal{D}_0) \cap \{\bar{J}(S(TM)) \perp \bar{J}(\operatorname{ltr}(TM)) \oplus \bar{J}(\operatorname{Rad} TM)\} = \{0\}$$

which show that $\mathcal{D}_0$ is invariant. $\qquad\square$

If M is an isotropic ST-lightlike submanifold of an indefinite Kähler manifold, from Definition 6.6.5 and Lemma 6.6.6, we have the decomposition

$$TM = \operatorname{Rad} TM$$

and

$$T\bar{M} = \{TM \oplus \operatorname{ltr}(TM)\} \perp \{\bar{J}(\operatorname{Rad} TM) \oplus \bar{J}(\operatorname{ltr}(TM)) \perp \mathcal{D}_0\}. \tag{6.6.5}$$

Proposition 6.6.9. [356] *A real lightlike (null) curve M of an indefinite Kähler manifold $\bar{M}$ is an isotropic screen transversal lightlike submanifold.*

Proof. Since M is a real lightlike curve, we have

$$TM = \operatorname{Rad} TM = \operatorname{Span}\{\xi\}$$

where ξ is the tangent vector field to M. Then, from (6.1.5), we have $\bar{g}(\bar{J}\xi, \xi) = 0$ which implies that $\bar{J}\xi$ does not belong to $\operatorname{ltr}(TM)$. Since $\dim(TM) = 1$, $\bar{J}\xi$ and ξ are linearly independent and $\bar{J}\xi$ does not belong to TM. Hence, we conclude that $\bar{J}\xi \in \Gamma(S(TM^\perp))$. In a similar way, from (6.1.5) we get $\bar{g}(\bar{J}N, N) = 0$ which shows that $\bar{J}N$ does not belong to $\operatorname{Rad} TM$. Also, we have $\bar{g}(\bar{J}N, \xi) = \bar{g}(N, \bar{J}\xi) = 0$ due to $\bar{J}\xi \in \Gamma(S(TM^\perp))$. This implies that $\bar{J}N$ does not belong to $\operatorname{ltr}(TM)$. Hence $\bar{J}N \in \Gamma(S(TM^\perp)$. Moreover, we get

$$\bar{g}(\bar{J}N, \bar{J}\xi) = \bar{g}(\xi, N) = 1.$$

Thus, we conclude that $S(TM^\perp)$ is expressed as

$$S(TM^\perp) = \bar{J}(\operatorname{Rad} TM) \oplus \bar{J}(\operatorname{ltr}(TM)) \perp \mathcal{D}_0$$

where $\mathcal{D}_0$ is a non-degenerate almost complex distribution. Considering the definition of an isotropic **ST**- lightlike submanifold of an indefinite Kähler manifold, the proof is complete. $\qquad\square$

Observe that Proposition 6.6.9 gives us a new viewpoint for the theory of lightlike curves. For instance, lightlike curves of R_2^4 can now be studied.

6.7   Slant lightlike submanifolds

In this section we present a lightlike version of slant submanifolds which have been studied widely in the Riemannian case [102]. We start with the following two lemmas which will be useful later on.

Lemma 6.7.1. *Let M be an r-lightlike submanifold of an indefinite Hermitian manifold $\bar{M}$ of index $2q$. Suppose that $\bar{J} \operatorname{Rad} TM$ is a distribution on M such that $\operatorname{Rad} TM \cap \bar{J} \operatorname{Rad} TM = \{0\}$. Then $\bar{J} \operatorname{ltr}(TM)$ is a sub-bundle of the screen distribution $S(TM)$ and $\bar{J} \operatorname{Rad} TM \cap \bar{J} \operatorname{ltr}(TM) = \{0\}$.*

Proof. Since by hypothesis $\bar{J} \operatorname{Rad} TM$ is a distribution on M such that $\bar{J} \operatorname{Rad} TM \cap \operatorname{Rad} TM = \{0\}$, we have $\bar{J} \operatorname{Rad} TM \subset S(TM)$. Now we claim that $\operatorname{ltr}(TM)$ is not invariant with respect to $\bar{J}$. Let us suppose that $\operatorname{ltr}(TM)$ is invariant with respect to $\bar{J}$. Choose $\xi \in \Gamma(\operatorname{Rad} TM)$ and $N \in \Gamma(\operatorname{ltr}(TM))$ such that $\bar{g}(N, \xi) = 1$. Then from (6.1.2) we have $1 = \bar{g}(\xi, N) = \bar{g}(\bar{J}\xi, \bar{J}N) = 0$ due to $\bar{J}\xi \in \Gamma(S(TM))$ and $\bar{J}N \in \Gamma(\operatorname{ltr}(TM))$. This is a contradiction, so $\operatorname{ltr}(TM)$ is not invariant with respect to $\bar{J}$. Also $\bar{J}N$ does not belong to $S(TM^\perp)$, since $S(TM^\perp)$ is orthogonal to

$S(TM)$, $\bar{g}(\bar{J}N, \bar{J}\xi)$ must be zero, but from (6.1.2) we have $\bar{g}(\bar{J}N, \bar{J}\xi) = \bar{g}(N, \xi) \neq 0$ for some $\xi \in \Gamma(\operatorname{Rad} TM)$, this is again a contradiction. Thus we conclude that $\bar{J}\operatorname{ltr}(TM)$ is a distribution on M. Moreover, $\bar{J}N$ does not belong to $\operatorname{Rad} TM$. Indeed, if $\bar{J}N \in \Gamma(\operatorname{Rad} TM)$, we would have $\bar{J}^2N = -N \in \Gamma(\bar{J}\operatorname{Rad} TM)$, but this is impossible. Similarly, $\bar{J}N$ does not belong to $\bar{J}\operatorname{Rad} TM$. Thus we conclude that $\bar{J}\operatorname{ltr}(TM) \subset S(TM)$ and $\bar{J}\operatorname{Rad} TM \cap \bar{J}\operatorname{ltr}(TM) = \{0\}$. $\qquad\square$

Remark 6.7.2. Lemma 6.7.1 shows that behavior of the lightlike transversal bundle $\operatorname{ltr}(TM)$ is exactly same as the radical distribution $\operatorname{Rad} TM$.

Lemma 6.7.3. *Under the hypothesis of Lemma 6.7.1, if $r = q$, then any complementary distribution to $\bar{J}(\operatorname{Rad} TM) \oplus \bar{J}\operatorname{ltr}(TM)$ in $S(TM)$ is Riemannian.*

Proof. Let $\dim(\bar{M}) = m + n$ and $\dim(M) = m$. Lemma 6.7.1 implies that $\bar{J}\operatorname{ltr}(TM) \oplus \bar{J}\operatorname{Rad} TM \subset S(TM)$. We denote the complementary distribution to $\bar{J}\operatorname{ltr}(TM) \oplus \bar{J}\operatorname{Rad} TM$ in $S(TM)$ by D'. Then we have a local quasi-orthonormal field of frames on $\bar{M}$ along M,

$$\{\xi_i, N_i, \bar{J}\xi_i, \bar{J}N_i, X_\alpha, W_a\}, i \in \{1, \ldots, r\}, \alpha \in \{3r+1, \ldots, m\}, a \in \{r+1, \ldots, n\},$$

where $\{\xi_i\}$ and $\{N_i\}$ are the basis of $\operatorname{Rad} TM$ and $\operatorname{ltr}(TM)$, respectively whereas $\{\bar{J}\xi_i, \bar{J}N_i, X_\alpha\}$ and $\{W_a\}$ are orthonormal bases of $S(TM)$ and $S(TM^\perp)$, respectively. From the basis $\{\xi_1, \ldots, \xi_r, \bar{J}\xi_1, \ldots, \bar{J}\xi_r, \bar{J}N_1, \ldots, \bar{J}N_r, N_1, \ldots, N_r\}$ of $\operatorname{ltr}(TM) \oplus \operatorname{Rad} TM \oplus \bar{J}\operatorname{Rad} TM \oplus \bar{J}\operatorname{ltr}(TM)$, we construct an orthonormal basis $\{U_1, \ldots, U_{2r}, V_1, \ldots, V_{2r}\}$ as follows.

$$U_1 = \tfrac{1}{\sqrt{2}}(\xi_1 + N_1) \qquad\qquad U_2 = \tfrac{1}{\sqrt{2}}(\xi_1 - N_1)$$
$$U_3 = \tfrac{1}{\sqrt{2}}(\xi_2 + N_2) \qquad\qquad U_4 = \tfrac{1}{\sqrt{2}}(\xi_2 - N_2)$$
$$\cdots \qquad\qquad\qquad\qquad \cdots$$
$$\cdots \qquad\qquad\qquad\qquad \cdots$$
$$U_{2r-1} = \tfrac{1}{\sqrt{2}}(\xi_r + N_r) \qquad\qquad U_{2r} = \tfrac{1}{\sqrt{2}}(\xi_r - N_r)$$
$$V_1 = \tfrac{1}{\sqrt{2}}(\bar{J}\xi_1 + \bar{J}N_1) \qquad\qquad V_2 = \tfrac{1}{\sqrt{2}}(\bar{J}\xi_1 - \bar{J}N_1)$$
$$V_3 = \tfrac{1}{\sqrt{2}}(\bar{J}\xi_2 + \bar{J}N_2) \qquad\qquad V_4 = \tfrac{1}{\sqrt{2}}(\bar{J}\xi_2 - \bar{J}N_2)$$
$$\cdots \qquad\qquad\qquad\qquad \cdots$$
$$\cdots \qquad\qquad\qquad\qquad \cdots$$
$$V_{2r-1} = \tfrac{1}{\sqrt{2}}(\bar{J}\xi_r + \bar{J}N_r) \qquad\qquad V_{2r} = \tfrac{1}{\sqrt{2}}(\bar{J}\xi_r - \bar{J}N_r)$$

Hence, $\operatorname{Span}\{\xi_i, N_i, \bar{J}\xi_i, \bar{J}N_i\}$ is a non-degenerate space of constant index $2r$. Thus we conclude that $\operatorname{Rad} TM \oplus \bar{J}\operatorname{Rad} TM \oplus \operatorname{ltr}(TM) \oplus \bar{J}\operatorname{ltr}(TM)$ is non-degenerate and of constant index $2r$ on $\bar{M}$. Since

$$\operatorname{ind}(T\bar{M}) = \operatorname{ind}(\operatorname{Rad} TM \oplus \operatorname{ltr}(TM)) + \operatorname{ind}(\bar{J}\operatorname{Rad} TM \oplus \bar{J}\operatorname{ltr}(TM)$$
$$+ \operatorname{ind}(D' \perp S(TM^\perp))$$

we have $2q = 2r + \mathrm{ind}(D' \perp S(TM^\perp))$. Thus, if $r = q$, then $D' \perp S(TM^\perp)$ is Riemannian, i.e., $\mathrm{ind}(D' \perp (S(TM)^\perp)) = 0$. Hence D' is Riemannian. □

Remark 6.7.4. To define the notion of a slant lightlike submanifold, one needs to consider angle between two vector fields. As we can see from Chapter 5, a lightlike submanifold has two (radical and screen) distributions: The radical distribution is totally lightlike and therefore it is not possible to define the angle between two vector fields of radical distribution. On the other hand, the screen distribution is non-degenerate. Although there are some definitions for the angle between two vector fields in Lorentzian vector space (See: [317], Proposition 30, P:144), that is not appropriate for our goal, because a manifold with a metric of Lorentz signature cannot admit an almost Hermitian structure (See: [190], Theorem VIII.3, P: 184). Thus one way to define slant notion is to choose a Riemannian screen distribution on a lightlike submanifold, for which we use Lemma 6.7.3.

Definition 6.7.5. [355] Let M be a q-lightlike submanifold of an indefinite Hermitian manifold $\bar{M}$ of index $2q$. Then we say that M is a slant lightlike submanifold of $\bar{M}$ if the following conditions are satisfied:

(A) $\mathrm{Rad}\,TM$ is a distribution on M such that

$$\bar{J}\,\mathrm{Rad}\,TM \cap \mathrm{Rad}\,TM = \{0\}. \tag{6.7.1}$$

(B) For each non-zero vector field tangent to D at $x \in \mathbf{U} \subset M$, the angle $\theta(X)$ between $\bar{J}X$ and the vector space D_x is constant, that is, it is independent of the choice of $x \in \mathbf{U} \subset M$ and $X \in D_x$, where D is complementary distribution to $\bar{J}\,\mathrm{Rad}\,TM \oplus \bar{J}\,\mathrm{ltr}(TM)$ in the screen distribution $S(TM)$.

This constant angle $\theta(X)$ is called a slant angle of the distribution D. A slant lightlike submanifold is said to be proper if $D \neq \{0\}$ and $\theta \neq 0, \frac{\pi}{2}$.

From Definition 6.7.5, we have the decomposition

$$TM = \mathrm{Rad}\,TM \oplus_{\mathrm{orth}} S(TM) \tag{6.7.2}$$
$$= \mathrm{Rad}\,TM \oplus_{\mathrm{orth}} (\bar{J}\,\mathrm{Rad}\,TM \oplus \bar{J}\,\mathrm{ltr}(TM)) \perp D. \tag{6.7.3}$$

Proposition 6.7.6. *There exist no proper slant totally lightlike or isotropic submanifolds in indefinite Hermitian manifolds.*

Proof. We suppose that M is a totally lightlike submanifold of $\bar{M}$. Then $TM = RadTM$, hence $D = \{0\}$. The other assertion follows similarly. □

Remark 6.7.7. As per Proposition 6.7.6, Definition 6.7.5 does not depend on $S(TM)$ and $S(TM^\perp)$, but, it depends on the transformation equations (5.2.12)–(5.2.18), with respect to the screen second fundamental forms h^s.

Example 13. Let $\bar{M} = (\mathbf{R}_2^8, \bar{g})$ be a semi-Euclidean space, with $\mathbf{R}_2^8$ of signature $(-, -, +, +, +, +, +, +)$ with respect to the canonical basis

$$\{\partial x_1, \partial x_2, \partial x_3, \partial x_4, \partial x_5, \partial x_6, \partial x_7, \partial x_8\}.$$

Let M be a submanifold of $\mathbf{R}_2^8$ given by

$$X(u, v, \theta, t, s) = (u, v, \sin\theta, \cos\theta, -\theta\sin t, -\theta\cos t, u, s).$$

Then the tangent bundle TM is spanned by

$$Z_1 = \partial x_1 + \partial x_7 \quad Z_2 = \partial x_2,$$
$$Z_3 = \cos\theta\, \partial x_3 - \sin\theta\, \partial x_4 - \sin t\, \partial x_5 - \cos t\, \partial x_6,$$
$$Z_4 = -\theta\cos t\, \partial x_5 + \theta\sin t\, \partial x_6, \quad Z_5 = \partial x_8.$$

It follows that M is a 1-lightlike submanifold of $\mathbf{R}_2^8$ with $\mathrm{Rad}\,TM = \mathrm{Span}\{Z_1\}$. Moreover we obtain $\bar{J}\,\mathrm{Rad}\,TM = \mathrm{Span}\{Z_2 + Z_5\}$ and therefore it is a distribution on M. Choose $D = \mathrm{Span}\{Z_3, Z_4\}$ which is Riemannian. Then M is a slant distribution with slant angle $\frac{\pi}{4}$, with the screen transversal bundle $S(TM^\perp)$ spanned by

$$W_1 = -\csc\theta\, \partial x_4 + \sin t\, \partial x_5 + \cos t\, \partial x_6,$$
$$W_2 = (2\sec\theta - \cos\theta)\, \partial x_3 + \sin\theta\, \partial x_4 + \sin t\, \partial x_5 + \cos t\, \partial x_6,$$

which is also Riemannian. Finally, $\mathrm{ltr}(TM)$ is spanned by

$$N = \frac{1}{2}(-\partial x_1 + \partial x_7).$$

Hence we have $\bar{J}N = -Z_2 + Z_5 \in \Gamma(S(TM))$ and $\bar{g}(\bar{J}N, \bar{J}Z_1) = 1$. Thus we conclude that M is a proper slant lightlike submanifold of $\mathbf{R}_2^8$.

Proposition 6.7.8. *Slant lightlike submanifolds do not include invariant and screen real lightlike submanifolds of an indefinite Hermitian manifold.*

Proof. Let M be an invariant or screen real lightlike submanifold of an indefinite Hermitian manifold $\bar{M}$. Then, since $\bar{J}\,\mathrm{Rad}\,TM = \mathrm{Rad}\,TM$, the first condition of slant lightlike submanifold is not satisfied which proves our assertion. $\qquad\square$

It is known that CR-lightlike submanifolds also do not include invariant and screen real lightlike submanifolds. Thus we may expect some relations between CR-lightlike submanifolds and slant lightlike submanifolds. Indeed we have:

Proposition 6.7.9. *Let M be a q-lightlike submanifold of an indefinite Kähler manifold $\bar{M}$ of index $2q$. Then any coisotropic CR-lightlike submanifold is a slant lightlike submanifold with $\theta = 0$. In particular, a lightlike real hypersurface of an indefinite Hermitian manifold $\bar{M}$ of index 2 is a slant lightlike submanifold with $\theta = 0$. Moreover, any CR-lightlike submanifold of $\bar{M}$ with $D_o = \{0\}$ is a slant lightlike submanifold with $\theta = \frac{\pi}{2}$.*

Proof. Let M be a q-lightlike CR-lightlike submanifold of an indefinite Hermitian manifold $\bar{M}$. By definition of a CR-lightlike submanifold, $\bar{J} \operatorname{Rad} TM$ is a distribution on M such that $\operatorname{Rad} TM \cap \bar{J} \operatorname{Rad} TM = \{0\}$. If M is coisotropic, then $S(TM^\perp) = \{0\}$. Thus $L_2 = 0$. The complementary distribution to $\bar{J} \operatorname{Rad} TM \oplus \bar{J} \operatorname{ltr}(TM)$ is D_o, which by Lemma 6.7.3 is Riemannian. Since D_o is invariant with respect to $\bar{J}$, it follows that $\theta = 0$. The second assertion is clear as a lightlike real hypersurface of $\bar{M}$ is coisotropic. Now, if M is a CR-lightlike submanifold with $D_o = \{0\}$, then the complementary distribution to $\bar{J} \operatorname{Rad} TM \oplus \bar{J} \operatorname{ltr}(TM)$ is D'. D' being anti-invariant with respect to $\bar{J}$, it follows that $\theta = \frac{\pi}{2}$, which completes the proof. $\qquad \square$

From Proposition 6.7.9, coisotropic CR-lightlike submanifolds, lightlike real hypersurfaces and CR-lightlike submanifolds with $D_o = \{0\}$ are some of the many more examples of slant lightlike submanifolds. For any $X \in \Gamma(TM)$ we write

$$\bar{J}X = TX + FX \tag{6.7.4}$$

where TX is the tangential component of $\bar{J}X$ and FX is the transversal component of $\bar{J}X$. Similarly, for $V \in \Gamma(\operatorname{tr}(TM))$ we write

$$\bar{J}V = BV + CV \tag{6.7.5}$$

where BV is the tangential component of $\bar{J}V$ and CV is the transversal component of $\bar{J}V$. Given a slant lightlike submanifold, we denote by P_1, P_2, Q_1 and Q_2 the projections on the distributions $\operatorname{Rad} TM$, $\bar{J} \operatorname{Rad} TM$, $\bar{J} \operatorname{ltr}(TM)$ and D, respectively. Then we can write

$$X = P_1 X + P_2 X + Q_1 X + Q_2 X \tag{6.7.6}$$

for $X \in \Gamma(TM)$. By applying $\bar{J}$ to (6.7.6) we obtain

$$\bar{J}X = \bar{J}P_1 X + \bar{J}P_2 X + TQ_2 X + FQ_1 X + FQ_2 X \tag{6.7.7}$$

for $X \in \Gamma(TM)$. By direct calculations we have

$$\bar{J}P_1 X \in \Gamma(\bar{J} \operatorname{Rad} TM), \quad \bar{J}P_2 X = TP_2 X \in \Gamma(\operatorname{Rad} TM) \tag{6.7.8}$$
$$FP_1 X = 0, \ FP_2 X = 0, \quad TQ_2 X \in \Gamma(D) \tag{6.7.9}$$

and

$$FQ_1 X \in \Gamma(\operatorname{ltr}(TM)).$$

Moreover, (6.7.7)–(6.7.9) imply

$$TX = TP_1 X + TP_2 X + TQ_2 X. \tag{6.7.10}$$

Now we prove two characterization theorems for slant lightlike submanifolds.

Theorem 6.7.10. *Let M be a q-lightlike submanifold of an indefinite Hermitian manifold $\bar{M}$ of index $2q$. Then M is a slant lightlike submanifold if and only if the following conditions are satisfied:*

(1) $\bar{J}\operatorname{ltr}(TM)$ is a distribution on M.

(2) There exists a constant $\lambda \in [-1,0]$ such that

$$(Q_2T)^2 X = \lambda X, \quad \forall X \in \Gamma(TM). \tag{6.7.11}$$

Moreover, in such a case, $\lambda = -\cos^2\theta$.

Proof. Let M be a q-lightlike submanifold of an indefinite Hermitian manifold $\bar{M}$ of index $2q$. If M is a slant lightlike submanifold of $\bar{M}$, then $\bar{J}\operatorname{Rad}TM$ is a distribution on $S(TM)$, thus from Lemma 6.7.1, it follows that $\bar{J}\operatorname{ltr}(TM)$ is also a distribution on M and $\bar{J}\operatorname{ltr}(TM) \subset S(TM)$. Thus (1) is satisfied. Moreover, the angle between $\bar{J}Q_2X$ and D_x is constant. Hence we have

$$\cos\theta(Q_2X) = \frac{\bar{g}(\bar{J}Q_2X, TQ_2X)}{|\bar{J}Q_2X \,\| TQ_2X|} = \frac{-\bar{g}(Q_2X, \bar{J}TQ_2X)}{|Q_2X \,\| TQ_2X|}$$
$$= \frac{-\bar{g}(Q_2X, TQ_2TQ_2X)}{|Q_2X \,\| TQ_2X|}. \tag{6.7.12}$$

On the other hand, we have

$$\cos\theta(Q_2X) = \frac{|TQ_2X|}{|\bar{J}Q_2X|}. \tag{6.7.13}$$

Thus from (6.7.12) and (6.7.13) we get

$$\cos^2\theta(Q_2X) = \frac{-\bar{g}(Q_2X, TQ_2TQ_2X)}{|Q_2X|^2}.$$

Since $\theta(Q_2X)$ is constant on D, we conclude that

$$(Q_2T)^2 X = \lambda Q_2 X, \; \lambda \in [-1,0].$$

Furthermore, in this case $\lambda = -\cos^2\theta(Q_2X)$. Conversely, suppose that (1) and (2) are satisfied. Then (1) implies that $\bar{J}\operatorname{Rad}TM$ is a distribution on M. From Lemma 6.7.2, it follows that the complementary distribution to $\bar{J}\operatorname{Rad}TM \oplus \bar{J}\operatorname{ltr}(TM)$ is a Riemannian distribution. The rest of the proof is clear. $\square$

Corollary 6.7.11. *Let M be a slant lightlike submanifold of an indefinite Hermitian manifold $\bar{M}$. Then we have*

$$g(TQ_2X, TQ_2Y) = \cos^2\theta g(Q_2X, Q_2Y) \tag{6.7.14}$$

and

$$g(FQ_2X, FQ_2Y) = \sin^2\theta g(Q_2X, Q_2Y), \quad \forall X, Y \in \Gamma(TM). \tag{6.7.15}$$

Proof. From (6.1.2) and (6.7.4) we have

$$g(TQ_2X, TQ_2Y) = -g(Q_2X, T^2Q_2Y), \quad \forall X, Y \in \Gamma(TM).$$

From Theorem 6.7.10, we get (6.7.14) and (6.7.15) follows from (6.7.14).   □

Theorem 6.7.12. *Let M be a q-lightlike submanifold of an indefinite Hermitian manifold $\bar{M}$ of index $2q$. Then M is a slant lightlike submanifold if and only if the following conditions are satisfied:*

(1) *$\bar{J}\,\mathrm{ltr}(TM)$ is a distribution on M.*

(2) *There exists a constant $\mu \in [-1, 0]$ such that*

$$BFQ_2X = \mu Q_2X, \quad \forall X \in \Gamma(TM).$$

In this case $\mu = -\sin^2\theta$, where θ is the slant angle of M and Q_2 the projection on D which is complementary to $\bar{J}\,\mathrm{Rad}\,TM \oplus \bar{J}\,\mathrm{ltr}(TM)$.

Proof. From Lemma 6.7.1, we conclude that $\bar{J}\,\mathrm{Rad}\,TM \cap \bar{J}\,\mathrm{ltr}(TM) = \{0\}$ and $\bar{J}\,\mathrm{ltr}(TM)$ is a sub-bundle of $S(TM)$. Moreover, the complementary distribution to $\bar{J}\,\mathrm{ltr}(TM) \oplus \bar{J}\,\mathrm{Rad}\,TM$ in $S(TM)$ is Riemannian. Furthermore, from the proof of Lemma 6.7.2, $S(TM^\perp)$ is also Riemannian. Thus condition (A) in the definition of slant lightlike submanifold is satisfied. On the other hand, applying $\bar{J}$ to (6.7.7) and using (6.7.4) and (6.7.7) we obtain

$$-X = -P_1X - P_2X + T^2Q_2X + FTQ_2X + JFQ_1X + BFQ_2X + CFQ_2X.$$

Since $JFQ_1X = -Q_1X \in \Gamma(S(TM))$, taking the tangential parts we have

$$-X = -P_1X - P_2X + T^2Q_2X - Q_1X + BFQ_2X.$$

Then considering (6.7.6) we get

$$-Q_2X = T^2Q_2X + BFQ_2X. \tag{6.7.16}$$

Now, if M is slant lightlike, then from Theorem 6.7.10 we have $T^2Q_2X = -\cos^2\theta Q_2X$, hence we derive

$$BFQ_2X = -\sin^2\theta\, Q_2X.$$

Conversely, suppose that $BFQ_2X = \mu Q_2X$, $\mu \in [-1, 0]$, then from (6.7.16) we obtain

$$T^2Q_2X = -(1+\mu)Q_2X.$$

Put $-(1+\mu) = \lambda$ so that $\lambda \in [-1, 0]$. Then the proof follows from Theorem 6.7.10.   □

Example 14. Let $\bar{M} = \mathbf{R}_2^8$ be a semi-Euclidean space of signature $(-, -, +, +, +, +, +, +)$ with respect to the canonical basis

$$\{\partial x_1, \partial x_2, \partial x_3, \partial x_4, \partial x_5, \partial x_6, \partial x_7, \partial x_8\}.$$

Consider a complex structure J_1 defined by

$$J_1(x_1, x_2, x_3, x_4, x_5, x_6, x_7, x_8) = (-x_2, x_1, -x_4, x_3, -x_7 \cos\alpha - x_6 \sin\alpha,$$
$$- x_8 \cos\alpha + x_5 \sin\alpha, x_5 \cos\alpha + x_8 \sin\alpha,$$
$$x_6 \cos\alpha - x_7 \sin\alpha).$$

for $\alpha \in (0, \frac{\pi}{2})$. Let M be a submanifold of $(\mathbf{R}_2^8, J_1)$ given by

$$x_1 = u_1 \cosh\theta, \ x_2 = u_2 \cosh\theta, \ x_3 = -u_3 + u_1 \sinh\theta, \ x_4 = u_1 + u_3 \sinh\theta,$$
$$x_5 = \cos u_4 \cosh u_5, \ x_6 = \cos u_4 \sinh u_5, \ x_7 = \sin u_4 \sinh u_5,$$
$$x_8 = \sin u_4 \cosh u_5,$$

$u_1 \in (0, \frac{\pi}{2})$. Then TM is spanned by

$$Z_1 = \cosh\theta\, \partial x_1 + \sinh\theta \partial x_3 + \partial x_4,$$
$$Z_2 = \cosh\theta\, \partial x_2, \ Z_3 = -\partial x_3 + \sinh\theta\, \partial x_4,$$
$$Z_4 = - \sin u_4 \cosh u_5\, \partial x_5 - \sin u_4 \sinh u_5\, \partial x_6 + \cos u_4 \sinh u_5\, \partial x_7$$
$$+ \cos u_4 \cosh u_5\, \partial x_8,$$
$$Z_5 = \cos u_4 \sinh u_5\, \partial x_5 + \cos u_4 \cosh u_5\, \partial x_6 + \sin u_4 \cosh u_5\, \partial x_7$$
$$+ \sin u_4 \sinh u_5\, \partial x_8.$$

Hence M is 1-lightlike with $\operatorname{Rad} TM = \operatorname{Span}\{Z_1\}$ and $J_1(\operatorname{Rad} TM)$ spanned by $J_1 Z_1 = Z_2 + Z_3$. Thus $J_1 \operatorname{Rad} TM$ is a distribution on M. Then it is easy to see that $D = \{Z_4, Z_5\}$ is a slant distribution with respect to J_1 with slant angle α. The screen transversal bundle $S(TM^\perp)$ is spanned by

$$W_1 = - \cosh u_5\, \partial x_5 + \sinh u_5\, \partial x_6 + \tan u_4 \sinh u_5 \partial x_7$$
$$- \tan u_4 \cosh u_5\, \partial x_8,$$
$$W_2 = - \tan u_4 \sinh u_5\, x_5 + \tan u_4 \cosh u_5\, \partial x_6 - \cosh u_5\, \partial x_7$$
$$+ \sinh u_5\, \partial x_8.$$

On the other hand, $\operatorname{ltr}(TM)$ is spanned by

$$N = \tanh\theta \sinh\theta\, \partial x_1 + \sinh\theta\, x_3 + \partial x_4.$$

Hence $J_1 N = \tanh^2 \theta\, Z_2 + Z_3$. Thus we conclude that M is a slant lightlike submanifold of $(\mathbf{R}_2^8, J_1)$. Now by direct computations, using Gauss and Weingarten formulas, we have

$$h^l = 0, h^s(X, Z_1) = h^s(X, J_1 Z_1) = 0, \ h^s(X, J_1 N) = 0, \forall X \in \Gamma(TM),$$

$$h^s(Z_4, Z_4) = \frac{\cos u_4}{\sinh^2 u_5 + \cosh^2 u_5} W_1, \ h^s(Z_5, Z_5) = \frac{-\cos u_4}{\sinh^2 u_5 + \cosh^2 u_5} W_1.$$

Hence, the induced connection is a metric connection and M is not totally geodesic, but, it is a proper minimal slant lightlike submanifold of $(\mathbf{R}^8_2, J_1)$.

Remark 6.7.13. We note that the method established in Example 14 can be generalized. Namely, let M be a 1-lightlike submanifold of R^8_2. If an integral manifold M_θ of the distribution D complementary to the distribution $\bar{J}\,\mathrm{Rad}\,TM \oplus \bar{J}\,\mathrm{ltr}(TM)$ in $S(TM)$ is an invariant submanifold of $\bar{M}$ with respect to J_o defined by

$$J_o(x_1, x_2, x_3, x_4) = (-x_3, -x_4, x_1, x_2),$$

then M is a slant lightlike submanifold with respect to J_1. Thus, there are many examples of minimal slant lightlike submanifolds of $\mathbf{R}^8_2$.

Lemma 6.7.14. *Let M be a proper slant lightlike submanifold of an indefinite Kähler manifold $\bar{M}$ such that* $\dim(D) = \dim(S(TM^\perp))$. *If $\{e_1, \ldots, e_m\}$ is a local orthonormal basis of $\Gamma(D)$, then $\{\csc\theta F e_1, \ldots, \csc\theta\, F e_m\}$ is an orthonormal basis of $S(TM^\perp)$.*

Proof. Since $e_1, \ldots, e_m$ is a local orthonormal basis of D and D is Riemannian, from Corollary 6.7.11, we obtain

$$\bar{g}(\csc\theta F e_i, \csc\theta F e_j) = \csc^2\theta\,\sin^2\theta\,g(e_i, e_j) = \delta_{ij},$$

which proves the assertion. $\square$

Theorem 6.7.15. *Let M be a proper slant lightlike submanifold of an indefinite Kähler manifold $\bar{M}$. Then M is minimal if and only if*

$$\mathrm{tr}\,A^*_{\xi_j}\,|_{S(TM)} = 0, \ \mathrm{tr}\,A_{W_\alpha}\,|_{S(TM)} = 0,$$
$$\bar{g}(D^l(X, W), Y) = 0, \forall X, Y \in \Gamma(RadTM),$$

where $\{\xi_j\}^r_{j=1}$ is a basis of $\mathrm{Rad}\,TM$ and $\{W_\alpha\}^m_{\alpha=1}$ is a basis of $S(TM^\perp)$.

Proof. From Proposition 5.1.3 of Chapter 5, we have $h^l = 0$ on $\mathrm{Rad}\,TM$. Thus M is minimal if and only if

$$\sum_{i=1}^r h(\bar{J}\xi_i, \bar{J}\xi_i) = \sum_{i=1}^r h(\bar{J}N_i, \bar{J}N_i) + \sum_{k=1}^m h(e_k, e_k) = 0.$$

Using (5.1.28) we obtain

$$\sum_{i=1}^r h(\bar{J}\xi_i, \bar{J}\xi_i) = \sum_{i=1}^r \frac{1}{r}\sum_{j=1}^r g(A^*_{\xi_j}\bar{J}\xi_i, \bar{J}\xi_i)N_j$$

$$+ \frac{1}{m}\sum_{\alpha=1}^m g(A_{W_\alpha}\bar{J}\xi_i, \bar{J}\xi_i)W_\alpha. \tag{6.7.17}$$

In a similar way we obtain

$$\sum_{i=1}^{r} h(\bar{J}N_i, \bar{J}N_i) = \sum_{i=1}^{r} \frac{1}{r} \sum_{j=1}^{r} g(A_{\xi_j}^* \bar{J}N_i, \bar{J}N_i)N_j$$

$$+ \frac{1}{m} \sum_{\alpha=1}^{m} g(A_{W_\alpha} \bar{J}N_i, \bar{J}N_i)W_\alpha, \qquad (6.7.18)$$

$$\sum_{k=1}^{m} h(e_k, e_k) = \sum_{k=1}^{m} \frac{1}{r} \sum_{j=1}^{r} g(A_{\xi_j}^* e_k, e_k)N_j$$

$$+ \frac{1}{m} \sum_{\alpha=1}^{m} g(A_{W_\alpha} e_k, e_k)W_\alpha. \qquad (6.7.19)$$

Thus our assertion follows from (6.7.17), (6.7.18) and (6.7.19). □

Theorem 6.7.16. *Let M be a proper slant lightlike submanifold of an indefinite Kähler manifold $\bar{M}$ such that $\dim(D) = \dim(S(TM^\perp))$. Then M is minimal if and only if*

$$\operatorname{tr} A_{\xi_j}^* \mid_{S(TM)} = 0, \ \operatorname{tr} A_{Fe_i} \mid_{S(TM)} = 0,$$
$$\bar{g}(D^l(X, Fe_i), Y) = 0, \forall X, Y \in \Gamma(\operatorname{Rad} TM),$$

where $\{e_1, \ldots, e_m\}$ is a basis of D.

Proof. From Lemma 6.7.14, $\{\csc \theta Fe_1, \ldots, \csc \theta Fe_m\}$ is an orthonormal basis of $S(TM^\perp)$. Thus we can write

$$h^s(X, X) = \sum_{i=1}^{m} A_i \csc \theta Fe_i, \forall X \in \Gamma(TM)$$

for some functions A_i, $i \in \{1, \ldots, m\}$. Hence we obtain

$$h^s(X, X) = \sum_{i=1}^{m} \csc \theta g(A_{Fe_i} X, X) Fe_i, \quad \forall X \in \Gamma(\bar{J} \operatorname{Rad} M \oplus \bar{J} \operatorname{ltr}(TM) \perp D).$$

Then the assertion of the theorem comes from Theorem 6.7.15. □

6.8 Screen slant lightlike submanifolds

From Proposition 6.7.8, it follows that slant lightlike submanifolds do not contain invariant and screen real submanifolds. However, a slant submanifold in Riemannian complex geometry is a generalization of invariant and totally real submanifolds. Thus it is desirable to investigate a new class of lightlike submanifolds which includes invariant lightlike submanifolds as well as screen real lightlike submanifolds. In this section we present such lightlike submanifolds of indefinite Hermitian manifolds.

Lemma 6.8.1. *Let M be a $2q$-lightlike submanifold of an indefinite Kähler manifold $\bar{M}$ with index $2q$. The screen distribution $S(TM)$ of lightlike submanifold M is Riemannian.*

Proof. Let $\bar{M}$ be a real $2k = m+n$-dimensional indefinite Kähler manifold and $\bar{g}$ be a semi-Riemannian metric on $\bar{M}$ of index $2q$. Assume that M is an m-dimensional and $2q(< m)$-lightlike submanifold of $\bar{M}$. Then we have a local quasi orthonormal field of frames on $\bar{M}$ along M

$$\{\xi_i,\ N_i,\ X_\alpha,\ W_a\}, i \in \{1,\dots,2q\},\ \alpha \in \{2q+1,\dots,m\},\ a \in \{(m-2q)+1,\dots,n\},$$

where $\{\xi_i\}$ and $\{N_i\}$ are lightlike bases of $\operatorname{Rad} TM$ and $\operatorname{ltr}(TM)$, respectively and $\{X_\alpha\}$ and $\{W_a\}$ are orthonormal bases of $S(TM)$ and $S(TM^\perp)$, respectively. From the null basis $\{\xi_1,\dots,\xi_{2q}, N_1,\dots,N_{2q}\}$ of $\operatorname{ltr}(TM) \oplus \operatorname{Rad} TM$, we can construct an orthonormal basis $\{U_1,\dots,U_{4q}\}$ as follows.

$$U_1 = \tfrac{1}{\sqrt{2}}(\xi_1 + N_1) \qquad U_2 = \tfrac{1}{\sqrt{2}}(\xi_1 - N_1)$$
$$U_3 = \tfrac{1}{\sqrt{2}}(\xi_2 + N_2) \qquad U_4 = \tfrac{1}{\sqrt{2}}(\xi_2 - N_2)$$
$$\cdots \qquad\qquad \cdots$$
$$\cdots \qquad\qquad \cdots$$
$$U_{4q-1} = \tfrac{1}{\sqrt{2}}(\xi_{2q} + N_{2q}) \quad U_{4q} = \tfrac{1}{\sqrt{2}}(\xi_{2q} - N_{2q})$$

Thus, $\operatorname{Rad} TM \oplus \operatorname{ltr}(TM)$ is non-degenerate and of constant index $2q$ on $\bar{M}$. Since

$$\operatorname{ind}(T\bar{M}) = \operatorname{ind}(\operatorname{Rad} TM \oplus \operatorname{ltr}(TM)) + \operatorname{ind}(S(TM^\perp) \perp S(TM)),$$

we obtain that $S(TM) \perp S(TM^\perp)$ is of constant index zero, that is, $S(TM)$ and $S(TM^\perp)$ are Riemannian vector bundles. Thus the proof is complete.   □

Thus Lemma 6.8.1 enables us to give the following definition.

Definition 6.8.2. [359] Let $(M, g, S(TM))$ be a $2q$-lightlike submanifold of an indefinite Kähler manifold $\bar{M}$ with index $2q$. We say that M is a *screen slant lightlike submanifold* of $\bar{M}$ if the following conditions are satisfied:

(i) $\operatorname{Rad} TM$ is invariant with respect to $\bar{J}$, i.e., $\bar{J}(\operatorname{Rad} TM) = \operatorname{Rad} TM$.

(ii) For each non-zero vector field X tangent to $S(TM)$ at $x \in U \subset M$, the angle $\theta(X)$ between $\bar{J}X$ and $S(TM)$ is constant, i.e., it is independent of the choice of x and $X \in \Gamma(S(TM))$.

We note that $\theta(X)$ is called the *slant angle* and (a) $\operatorname{Rad} TM$ is even-dimensional, (b) screen slant lightlike submanifolds do not include a real hypersurface.

From now on suppose that $(M, g, S(TM))$ is a $2q$-lightlike submanifold of an indefinite Kähler manifold with constant index $2q$ and denote it by M.

Proposition 6.8.3. *Let M be a screen slant lightlike submanifold of $\bar{M}$. Then M is invariant (resp. screen real) if and only if $\theta = 0$ (resp. $\theta = \frac{\pi}{2}$).*

Proof. If M is invariant, then $\bar{J}(\operatorname{Rad} TM) = \operatorname{Rad} TM$ and $\bar{J}(S(TM)) = S(TM)$, thus $\theta = 0$. Conversely, if M is screen slant lightlike with $\theta = 0$ then it is clear that $\bar{J}(S(TM)) = S(TM)$. Since $\operatorname{Rad} TM$ is invariant with respect to $\bar{J}$, the proof is complete. The other assertion is similar. □

Thus it follows that a screen slant lightlike submanifold M is a natural generalization of invariant and screen real lightlike submanifolds. M is said to be proper if it is neither invariant nor a screen real lightlike submanifold.

Proposition 6.8.4. *There exists no proper screen slant totally lightlike or isotropic submanifold.*

Proof. We suppose that M is a totally lightlike submanifold, then $TM = \operatorname{Rad} TM = TM^{\perp}$. Hence M is an invariant lightlike submanifold. If M is an isotropic submanifold, then $S(TM) = \{0\}$, thus $TM = \operatorname{Rad} TM$ which is invariant. □

For any vector field $X \in \Gamma(S(TM))$, we write

$$\bar{J}X = TX + \omega X, \tag{6.8.1}$$

where $TX \in \Gamma(TM)$ and $\omega X \in \Gamma(\operatorname{tr}(TM))$.

Corollary 6.8.5. *Let M be a screen slant lightlike submanifold of an indefinite Kähler manifold $\bar{M}$. Then, for any $X \in \Gamma(TM)$, we have*

(i) *If $X \in \Gamma(S(TM))$, then $\omega X \in \Gamma(S(TM^{\perp}))$.*

(ii) *If $X \in \Gamma(\operatorname{Rad} TM)$, then $\omega X = 0$.*

Proof. It is easy to see that $\operatorname{ltr}(TM)$ is invariant w.r.t. $\bar{J}$ due to $\bar{J}(\operatorname{Rad} TM) = \operatorname{Rad} TM$. (ii) is clear. □

Proposition 6.8.6. *There exists no coisotropic proper screen slant lightlike submanifold in an indefinite Kähler manifold $\bar{M}$.*

Proof. We suppose that M is a coisotropic submanifold, then $S(TM^{\perp}) = \{0\}$. Thus we have $\bar{J}X = TX$, for $X \in \Gamma(S(TM))$. Hence we can see that M is an invariant lightlike submanifold. □

Proposition 6.8.3 implies that invariant and screen real lightlike submanifolds are examples of screen slant lightlike submanifolds. Now, we want to present some examples of proper screen slant lightlike submanifolds.

Example 15. For any $\alpha > 0$, we consider the immersion

$$x(u, v, t, s) = (t, s, u \cos \alpha, -v \cos \alpha, u \sin \alpha, v \sin \alpha, t, s,)$$

in $\mathbf{R}_2^8$. Then $\operatorname{Rad} TM$ is spanned by

$$\xi_1 = \partial x_1 + \partial x_7, \ \xi_2 = \partial x_2 + \partial x_8$$

and $S(TM)$ is spanned by

$$X_1 = \cos\alpha\,\partial\,x_3 + \sin\alpha\,\partial\,x_5, \ X_2 = -\cos\alpha\,\partial\,x_4 + \sin\alpha\,\partial\,x_6.$$

Then we can see that $\operatorname{Rad} TM$ is invariant with respect to $\bar{J}$ and $S(TM)$ is a slant distribution with slant angle 2α. Thus M is a screen slant hyperplane in $\mathbf{R}_2^8$. Moreover, we obtain that the screen transversal vector bundle $S(TM^\perp)$ is spanned by

$$W_1 = \sin\alpha\,\partial\,x_4 + \cos\alpha\,\partial\,x_6, \ W_2 = -\sin\alpha\,\partial\,x_3 + \cos\alpha\,\partial\,x_5$$

and the lightlike transversal bundle $\operatorname{ltr}(TM)$ is spanned by

$$N_1 = \frac{1}{2}\{-\partial\,x_1 + \partial\,x_7\}, \ N_2 = \frac{1}{2}\{-\partial\,x_2 + \partial\,x_8\}.$$

Example 16. Consider in $\mathbf{R}_2^8$ the submanifold M given by

$$x(u,v,t,s) = (u,v,s\,\sin t,\,s\,\cos t,\,\sin s,\,\cos s,\,u\,\cos\alpha - v\,\sin\alpha,\,u\,\sin\alpha + v\,\cos\alpha)$$

for $\alpha,\,t,s \in (0,\frac{\pi}{2})$. Then TM is spanned by

$$\xi_1 = \partial\,x_1 + \cos\alpha\,\partial\,x_7 + \sin\alpha\partial\,x_8,$$
$$\xi_2 = \partial\,x_2 - \sin\alpha\,\partial\,x_7 + \cos\alpha\partial\,x_8,$$
$$X_1 = s\,\cos t\,\partial\,x_3 - s\,\sin t\,\partial\,x_4,$$
$$X_2 = \sin t\,\partial\,x_3 + \cos t\,\partial\,x_4 + \cos s\,\partial\,x_5 - \sin s\,\partial\,x_6.$$

It follows that $\operatorname{Rad} TM = \operatorname{Span}\{\xi_1, \xi_2\}$, hence M is a 2-lightlike submanifold. Since $\bar{J}\operatorname{Rad} TM = \operatorname{Rad} TM$, $\operatorname{Rad} TM$ is invariant. Moreover, we can choose $S(TM) = \operatorname{Span}\{X_1, X_2\}$ which is a Riemannian vector sub-bundle and it can be easily proved that $S(TM)$ is a slant distribution with slant angle $\theta = \frac{\pi}{4}$. Finally, the screen transversal vector bundle $S(TM^\perp)$ is spanned by

$$W_1 = \sin s\partial\,x_5 + \cos s\,\partial\,x_6$$
$$W_2 = \sin t\,\partial\,x_3 + \cos t\,\partial\,x_4 - \cos s\,\partial\,x_5 + \sin s\,\partial\,x_6$$

and the lightlike transversal bundle $\operatorname{ltr}(TM)$ is spanned by

$$N_1 = \frac{1}{2}\{-\partial\,x_1 + \cos\alpha\,\partial\,x_7 + \sin\alpha\partial\,x_8\},$$

$$N_2 = \frac{1}{2}\{-\partial\,x_2 - \sin\alpha\,\partial\,x_7 + \cos\alpha\partial\,x_8\}.$$

Example 17. For any $k,\alpha > 0$, let $M \subset \mathbf{R}_2^{12}$ be given by

$$x(u,v,t,s) = (u\cosh\alpha,\,v\cosh\alpha,\,u,\,v,\,t,\,s,\,k\cos t,\,k\sin t,\,k\cos s,\,k\sin s,\,u\sinh\alpha,$$
$$v\sinh\alpha).$$

Then TM is spanned by

$$\xi_1 = \cosh \alpha \, \partial \, x_1 + \partial x_3 + \sinh \alpha \, \partial \, x_{11},$$
$$\xi_2 = \cosh \alpha \, \partial \, x_2 + \partial \, x_4 + \sinh \alpha \, \partial \, x_{12},$$
$$X_1 = \partial \, x_5 - k \, \sin t \partial \, x_7 + k \, \cos t \, \partial \, x_8,$$
$$X_2 = \partial \, x_6 - k \, \sin s \partial x_9 + k \, \cos s \, \partial x_{10}.$$

Then we see that $\operatorname{Rad} TM = \operatorname{Span}\{\xi_1, \xi_2\}$. Thus M is a 2-lightlike submanifold. It is easy to see $\bar{J} \operatorname{Rad} TM = \operatorname{Rad} TM$, that is, $\operatorname{Rad} TM$ is invariant. Choose $S(TM) = \operatorname{Span}\{X_1, X_2\}$ and obtain that $S(TM)$ is a slant distribution with slant angle $\theta = \cos^{-1}(\frac{1}{1+k^2})$. Thus M is a screen slant lightlike submanifold of $\mathbf{R}_2^{12}$.

Chen [102] studied the following important problem in complex geometry: Given a surface M of a Kähler manifold $\bar{M}$, when is M slant in $\bar{M}$? Based on the above problem, there are several interesting results on the geometry of a slant surface of a Euclidean space R^4 in [102]. For the screen slant lightlike case, we have the following result.

Proposition 6.8.7. *There exists no screen slant lightlike surface of an indefinite Hermitian (or Kähler) manifold with index 2.*

Proof. Let M be a screen slant lightlike surface of an indefinite Hermitian manifold $\bar{M}$ with index 2. Then M is 2-lightlike or 1-lightlike. If M is 2-lightlike then $S(TM) = 0$ and M is totally lightlike. Hence M is invariant. Now, suppose M is 1-lightlike. Then, $\operatorname{Rad} TM = \operatorname{Span}\{\xi\}$. This is not possible, because $\operatorname{Rad} TM$ is invariant with respect to $\bar{J}$. $\qquad \square$

Let M be a screen slant lightlike submanifold of an indefinite Kähler manifold $\bar{M}$. We denote the projection morphisms on the distributions $\operatorname{Rad} TM$ and $S(TM)$ by Q and P, respectively. Then we have

$$X = QX + PX, \quad \forall X \in \Gamma(TM), \tag{6.8.2}$$

where QX denotes the component of X in $\operatorname{Rad} TM$ and PX denotes the component of X in $S(TM)$. Applying $\bar{J}$ on (6.8.2) we obtain

$$\bar{J}X = \bar{J}QX + \bar{J}PX = TQX + TPX + \omega PX. \tag{6.8.3}$$

Thus we derive

$$\bar{J}QX = TQX, \, \omega QX = 0 \tag{6.8.4}$$

and

$$TPX \in \Gamma(S(TM)). \tag{6.8.5}$$

On the other hand, the screen transversal bundle $S(TM^\perp)$ has the decomposition

$$S(TM^\perp) = \omega P(S(TM)) \perp v. \tag{6.8.6}$$

Then, for any $W \in \Gamma(S(TM^\perp))$ we write

$$\bar{J}W = BW + CW \qquad (6.8.7)$$

where $BW \in \Gamma(S(TM)$ and $CW \in \Gamma(v)$.

Next, we give a characterization of screen slant lightlike submanifolds:

Theorem 6.8.8. *Let M be a $2q$-lightlike submanifold of an indefinite Kähler manifold $\bar{M}$ with constant index $2q$. Then M is a screen slant lightlike submanifold if and only if*

(i) *$\mathrm{ltr}(TM)$ is invariant with respect to $\bar{J}$.*

(ii) *There exists a constant $\lambda \in [-1, 0]$ such that*

$$(P \circ T)^2 X = \lambda X \qquad (6.8.8)$$

for any $X \in \Gamma(S(TM))$. Moreover, in this case $\lambda = -\cos\theta\,|_{S(TM)}$.

Proof. Let M be a $2q$-lightlike submanifold of an indefinite Kähler manifold $\bar{M}$ with index $2q$. Then Lemma 6.8.1 guarantees that $S(TM)$ is a Riemannian vector bundle. Let M be a screen slant lightlike submanifold. Then its radical distribution is invariant with respect to $\bar{J}$ and from Corollary 6.8.5, we have $\omega P X \in \Gamma(S(TM))$. Thus, using (6.1.5) and (6.8.3) we have

$$\bar{g}(\bar{J}N, X) = -\bar{g}(N, \bar{J}X) = -\bar{g}(N, TPX) - \bar{g}(N, \omega PX) = 0$$

for $X \in \Gamma(S(TM))$. Hence we conclude that $\bar{J}N$ does not belong to $S(TM)$. On the other hand, from (6.1.5) and (6.8.7) we obtain

$$\bar{g}(\bar{J}N, W) = -\bar{g}(N, \bar{J}W) = -\bar{g}(N, BW) - \bar{g}(N, CW) = 0$$

for $W \in \Gamma(S(TM^\perp))$. Thus $\bar{J}N$ is not belong to $S(TM^\perp)$. Now suppose that $\bar{J}N \in \Gamma(\mathrm{Rad}\,TM)$. Then we obtain $\bar{J}\bar{J}N = -N \in \Gamma(\mathrm{ltr}\,TM)$, since $\mathrm{Rad}\,TM$ is invariant with respect to $\bar{J}$ we get a contradiction which proves (i). With regards to statement (ii), since M is a screen slant lightlike submanifold, there is a constant angle θ which is independent of $X \in \Gamma(STM)$ and $x \in \mathbf{U} \subset M$. Thus we derive

$$\cos\theta(X) = \frac{\bar{g}(\bar{J}X, TPX)}{|\bar{J}X||TPX|} = -\frac{\bar{g}(X, \bar{J}TPX)}{|\bar{J}X||TPX|} = -\frac{\bar{g}(X, (P \circ T)^2 X)}{|X||TPX|}. \qquad (6.8.9)$$

On the other hand, we have

$$\cos\theta(X) = \frac{|TPX|}{|\bar{J}X|}. \qquad (6.8.10)$$

Thus, from (6.8.9) and (6.8.10) we obtain

$$\cos^2\theta(X) = -\frac{g(X, (P \circ T)^2 X)}{|X|^2}.$$

Since $\theta(X)$ is a constant, we conclude that $(P o T)^2 X = \lambda X, \lambda \in [-1, 0]$, which proves (ii). The converse can be obtained in a similar way. $\qquad \square$

From Theorem 6.8.8 we obtain the following corollary:

Corollary 6.8.9. *Let M be a screen slant lightlike submanifold of $\bar{M}$. Then*

$$g(TPX, TPY) = \cos^2 \theta \mid_{S(TM)} g(X, Y), \tag{6.8.11}$$

$$\bar{g}(\omega PX, \omega PY) = \sin^2 \theta \mid_{S(TM)} g(X, Y), \quad \forall X, Y \in \Gamma(TM). \tag{6.8.12}$$

Proof. From (6.1.5) and (6.8.3) we obtain

$$g(TPX, TPY) = -g(X, (PoT)^2 Y)$$

for any $X, Y \in \Gamma(S(TM))$. Then from Theorem 6.8.8 we derive

$$g(TPX, TPY) = \cos^2 \theta g(X, Y),$$

which proves (6.8.11). In a similar way we obtain (6.8.12). □

Differentiating (6.8.3) and comparing the tangent and transversal parts we have

$$(\nabla_X T)Y = A_{\omega PY} X + B h^s(X, Y), \tag{6.8.13}$$

$$\bar{J} h^l(X, Y) = h^l(X, \bar{J}QY) + h^l(X, TPY) + D^l(X, \omega PY), \tag{6.8.14}$$

$$(\nabla_X \omega)Y = -h^s(X, \bar{J}QY) - h^s(X, TPY) + C h^s(X, Y) \tag{6.8.15}$$

for $X, Y \in \Gamma(TM)$, where $(\nabla_X T)Y = \nabla_X \bar{J}QY + \nabla_X TPY - \bar{J}Q\nabla_X Y - TP\nabla_X Y$ and $(\nabla_X \omega)Y = \nabla_X^s \omega PY - \omega P \nabla_X Y$.

Theorem 6.8.10. *Let M be a screen slant lightlike submanifold of an indefinite Kähler manifold $\bar{M}$. Then:*

(i) *The radical distribution* $\operatorname{Rad} TM$ *is integrable if and only if the screen transversal second fundamental form of M satisfies*

$$h^s(X, \bar{J}Y) = h^s(\bar{J}X, Y), \forall X, Y \in \Gamma(\operatorname{Rad} TM).$$

(ii) *The screen distribution* $S(TM)$ *is integrable if and only if*

$$Q(\nabla_X TPY - \nabla_Y TPX) = Q(A_{\omega PY} X - A_{\omega PX} Y), \forall X, Y \in \Gamma(S(TM)).$$

Proof. From (6.8.15) we obtain

$$h^s(X, \bar{J}Y) - C h^s(X, Y) = \omega \nabla_X Y, \forall X, Y \in \Gamma(\operatorname{Rad} TM).$$

Thus we obtain $h^s(X, \bar{J}Y) - h^s(\bar{J}X, Y) = \omega P[X, Y]$ which proves assertion (i). From (6.8.13) we derive

$$\nabla_X TPY - A_{\omega PY} X = \bar{J}Q\nabla_X Y + TP\nabla_X Y + B h^s(X, Y)$$

for any $X, Y \in \Gamma(STM))$. Hence we get

$$\nabla_X TPY - \nabla_Y TPX + A_{\omega PX}Y - A_{\omega PY}X = \bar{J}Q[X,Y] + TP[X,Y].$$

Thus we obtain

$$Q(\nabla_X TPY - \nabla_Y TPX) + Q(A_{\omega PX}Y - A_{\omega PY}X) = \bar{J}Q[X,Y],$$

which proves (ii). $\square$

Theorem 6.8.11. *Let M be a screen slant lightlike submanifold of an indefinite Kähler manifold $\bar{M}$. Then the screen distribution defines a totally geodesic foliation if and only if $\bar{J}A_{\omega PY}X - A_{\omega PTPY}X$ has no components in $\mathrm{Rad}\,TM$ for $X,Y \in \Gamma(S(TM))$.*

Proof. Using (6.1.5) and (5.1.31) we have $\bar{g}(\nabla_X Y, N) = \bar{g}(\bar{\nabla}_X \bar{J}Y, \bar{J}N)$, for $X, Y \in \Gamma(S(TM))$ and $N \in \Gamma(\mathrm{ltr}(TM))$. Thus from (6.8.3) and (5.1.23), we obtain $\bar{g}(\nabla_X Y, N) = \bar{g}(\bar{\nabla}_X TPY, \bar{J}N) - \bar{g}(A_{\omega PY}X, \bar{J}N)$. Using again (6.1.5), (5.1.15), (6.8.3) and (5.1.23) in the first expression in the above equation we derive

$$\bar{g}(\nabla_X Y, N) = g(\nabla_X (P \circ T)^2 Y, N) - \bar{g}(A_{\omega PTPY}X, N) - \bar{g}(A_{\omega PY}X, \bar{J}N).$$

Thus from Theorem 6.8.8 we get

$$\bar{g}(\nabla_X Y, N) = \cos^2 \theta \bar{g}(\nabla_X Y, N) + \bar{g}(A_{\omega PTPY}X, N) - \bar{g}(A_{\omega PY}X, \bar{J}N).$$

Hence we obtain

$$\sin^2 \theta \bar{g}(\nabla_X Y, N) = -\bar{g}(A_{\omega PTPY}X, N) - \bar{g}(A_{\omega PY}X, \bar{J}N).$$

This completes the proof. $\square$

Next we investigate $\nabla_X T = 0$ on a screen slant lightlike submanifold. In the non-degenerate complex geometry, if a slant submanifold satisfies the above property, it is called a Kählerian slant submanifold (see section 6.1).

Theorem 6.8.12. *Let M be a screen slant lightlike submanifold of an indefinite Kähler manifold $\bar{M}$. Then T is parallel if and only if $D^s(X, N) \in \Gamma(v)$ and*

$$\bar{g}(h^s(X,Y), \omega PZ) = g(h^s(X,Z), \omega PY)$$

for $X \in \Gamma(TM)$ and $Y, Z \in \Gamma(S(TM))$.

Proof. From (6.8.13) we obtain $\bar{g}((\nabla_X T)Y, N) = 0$, for $Y \in \Gamma(\mathrm{Rad}\,TM)$ and $X \in \Gamma(TM)$. For $Y \in \Gamma(S(TM))$, we get $\bar{g}((\nabla_X T)Y, N) = \bar{g}(A_{\omega PY}X, N)$. Using (5.1.22) we derive

$$\bar{g}((\nabla_X T)Y, N) = \bar{g}(D^s(X, N), \omega PY) \qquad (6.8.16)$$

for any $X \in \Gamma(S(TM))$. On the other hand, from (5.1.31), (6.7.1) and (6.1.5) we obtain

$$g((\nabla_X T)Y, Z) = g(A_{\omega PY}X, Z) - g(h^s(X,Y), \omega PZ)$$

for any $X, Y \in \Gamma(TM)$ and $Z \in \Gamma(S(TM))$. By using (5.1.28) we get

$$g((\nabla_X T)Y, Z) = g(h^s(X, Z), \omega PY) - g(h^s(X, Y), \omega PZ) \tag{6.8.17}$$

for any $X, Y \in \Gamma(TM)$ and $Z \in \Gamma(S(TM))$. Thus from (6.8.16) and (6.8.17) we obtain our assertion. $\qquad\square$

Theorem 6.8.13. *Let M be a screen slant lightlike submanifold of an indefinite Kähler manifold $\bar{M}$. If $(\nabla_X T)Y = 0$ for $X \in \Gamma(TM)$ and $Y \in \Gamma(\operatorname{Rad} TM)$, then the induced connection ∇ is a metric connection.*

Proof. If $(\nabla_X T)Y = 0$ for $X \in \Gamma(TM)$ and $Y \in \Gamma(\operatorname{Rad} TM)$, then from (6.8.13) we have $Bh^s(X, Y) = 0$, hence $g(Bh^s(X, Y), Z) = 0$ for $X \in \Gamma(TM)$, $Y \in \Gamma(\operatorname{Rad} TM)$ and $Z \in \Gamma(TM)$. Thus we obtain

$$\bar{g}(\bar{J}h^s(X, Y), Z) = 0, \tag{6.8.18}$$
$$\bar{g}(h^s(X, Y), \omega PZ) = 0. \tag{6.8.19}$$

Now, by using (5.1.15) we get

$$\bar{g}(\omega P\nabla_X Y, \bar{J}h^s(X, Y)) = \bar{g}(\omega P\nabla_X Y, \bar{J}\bar{\nabla}_X Y) - \bar{J}\nabla_X Y - \bar{J}h^l(X, Y))$$

for $X \in \Gamma(TM)$, $Y \in \Gamma(\operatorname{Rad} TM)$, since $\operatorname{ltr}(TM)$ is invariant, from (6.1.5) and (6.8.3) we get

$$\bar{g}(\omega P\nabla_X Y, \bar{J}h^s(X, Y)) = \bar{g}(\omega P\nabla_X Y, \bar{\nabla}_X \bar{J}Y) - \bar{g}(\omega P\nabla_X Y, \omega P\nabla_X Y).$$

Then using (5.1.15) we derive

$$\bar{g}(\omega P\nabla_X Y, \bar{J}h^s(X, Y)) = \bar{g}(\omega P\nabla_X Y, h^s(X, \bar{J}Y)) - \bar{g}(\omega P\nabla_X Y, \omega P\nabla_X Y).$$

Thus from (6.8.19) we have

$$\bar{g}(\omega P\nabla_X Y, \bar{J}h^s(X, Y)) = -\bar{g}(\omega P\nabla_X Y, \omega P\nabla_X Y).$$

Then using (6.8.12) we obtain

$$\bar{g}(\omega P\nabla_X Y, \bar{J}h^s(X, Y)) = -\sin^2 \theta g(P\nabla_X Y, P\nabla_X Y) \tag{6.8.20}$$

for any $X \in \Gamma(TM)$ and $Y \in \Gamma(\operatorname{Rad} TM)$. On the other hand, from (6.1.5) and (6.8.3) we have

$$\bar{g}(\omega P\nabla_X Y, \bar{J}h^s(X, Y)) = -\bar{g}(TP\nabla_X Y, \bar{J}h^s(X, Y))$$

for any $X \in \Gamma(TM)$ and $Y \in \Gamma(\operatorname{Rad} TM)$. Then, from (6.8.19) we derive

$$\bar{g}(\omega P\nabla_X Y, \bar{J}h^s(X, Y)) = 0. \tag{6.8.21}$$

Then (6.8.20) and (6.8.21) imply

$$\sin^2 \theta g(P\nabla_X Y, P\nabla_X Y) = 0.$$

Since M is a proper screen slant lightlike submanifold and $S(TM)$ is Riemannian we obtain $P\nabla_X Y = 0$, hence $\nabla_X Y \in \Gamma(\mathrm{Rad}\,TM)$, i.e., the radical distribution $\mathrm{Rad}\,TM$ is parallel. Thus the assertion of the theorem follows from Theorem 5.1.4.

$\square$

Remark 6.8.14. It is clear that radical distribution and screen distribution are orthogonal. However, we note that if $\mathrm{Rad}\,TM$ is parallel, then it doesn't imply that the screen distribution $S(TM)$ is parallel, contrary to the non-degenerate case.

Example 18. Let $\bar{M} = \mathbf{R}_2^8$ be a semi-Euclidean space of signature $(-, -, +, +, +, +, +, +)$ with respect to the canonical basis $\{\partial x_1, \partial x_2, \partial x_3, \partial x_4, \partial x_5, \partial x_6, \partial x_7, \partial x_8\}$. Consider a complex structure J_1 defined by

$$J_1(x_1, x_2, x_3, x_4, x_5, x_6, x_7, x_8) = (-x_2, x_1, -x_4, x_3, -x_7 \cos\alpha, -x_6 \sin\alpha,$$
$$-x_8 \cos\alpha + x_5 \sin\alpha, x_5 \cos\alpha + x_8 \sin\alpha,$$
$$x_6 \cos\alpha, -x_7 \sin\alpha)$$

for $\alpha \in (0, \frac{\pi}{2})$. Let M be a submanifold of $(\mathbf{R}_2^8, J_1)$ given by

$$x_1 = u_1, x_2 = u_2, x_3 = u_1 \cos\theta - u_2 \sin\theta, x_4 = u_1 \sin\theta + u_2 \cos\theta$$
$$x_5 = u_3, x_6 = \sin u_3 \sinh u_4, x_7 = u_4, x_8 = \cos u_3 \cosh u_4.$$

Then TM is spanned by

$$Z_1 = \partial x_1 + \cos\theta \,\partial x_3 + \sin\theta \,\partial x_4$$
$$Z_2 = \partial x_2 - \sin\theta \,\partial x_3 + \cos\theta \,\partial x_4$$
$$Z_3 = \partial x_5 + \cos u_3 \sinh u_4 \,\partial x_6 + \sin u_3 \cosh u_4 \,\partial x_8$$
$$Z_3 = \sin u_3 \cosh u_4 \,\partial x_6 \partial x_6 + \partial x_7 + \cos u_3 \sinh u_4 \,\partial x_8.$$

Then M is a 2-lightlike submanifold and $\mathrm{Rad}\,TM = \{Z_1, Z_2\}$. It follows that $\mathrm{Rad}\,TM$ is J_1-invariant. It is easy to see that $S(TM) = \mathrm{Span}\{Z_3, Z_4\}$ is a slant distribution with respect to J_1 with slant angle α. The screen transversal bundle is spanned by

$$W_1 = -\cosh u_4 \sinh u_4 \,\partial x_5 + \cos u_3 \cosh u_4 \,\partial x_6 - \sin u_3 \cos u_3 \,\partial x_7$$
$$- \sin u_3 \sinh u_4 \,\partial x_8,$$
$$W_2 = \sin u_3 \cos u_3 \,\partial x_5 + \sin u_3 \sinh u_4 \,\partial x_6 - \cosh u_4 \sinh u_4 \,\partial x_7$$
$$+ \cos u_3 \cosh u_4 \,\partial x_8$$

and the lightlike transversal bundle is spanned by

$$N_1 = \frac{1}{2}\{-\partial x_1 + \cos\theta\,\partial x_3 + \sin\theta\,\partial x_4\}$$

$$N_2 = \frac{1}{2}\{-\partial x_2 - \sin\theta\,\partial x_3 + \cos\theta\,\partial x_4\}.$$

Now, by direct calculations, using Gauss formulas we get

$$h^l = 0,\ h^s(X, Z_1) = 0,\ h^s(X, Z_2) = 0, \forall X \in \Gamma(TM)$$

and

$$h^s(Z_3, Z_3) = \frac{-(\sinh^2 u_4 + \cos^2 u_3)}{\cosh^4 u_4 - \sin^4 u_3} W_2,\ h^s(Z_4, Z_4) = \frac{\sinh^2 u_4 + \cos^2 u_3}{\cosh^4 u_4 - \sin^4 u_3} W_2$$

$$h^s(Z_3, Z_4) = \frac{\sinh^2 u_4 + \cos^2 u_3}{\cosh^4 u_4 - \sin^4 u_3} W_1.$$

Hence the induced connection is a metric connection, M is not totally geodesic and it is also not totally umbilical, but it is a minimal proper screen slant lightlike submanifold of $\mathbf{R}_2^8$.

Next, we prove two characterization results for minimal slant lightlike submanifolds. First we give the following lemma which will be useful later.

Lemma 6.8.15. *Let M be a proper screen slant lightlike submanifold of an indefinite Kähler manifold $\bar{M}$ such that $\omega(S(TM)) = S(TM^\perp)$. If $\{e_1, \ldots, e_m\}$ is a local orthonormal basis of $S(TM)$, then $\{\csc\theta\omega e_1, \ldots, \csc\theta\,\omega e_m\}$ is an orthonormal basis of $S(TM^\perp)$.*

Proof. Since $e_1, \ldots, e_m$ is a local orthonormal basis of $S(TM)$ and $S(TM)$ is Riemannian , from Corollary 6.8.9, we obtain

$$\bar{g}(\csc\theta\omega e_i, \csc\theta\omega e_j) = \csc^2\theta\sin^2\theta\, g(e_i, e_j) = \delta_{ij},$$

which proves the assertion. □

Theorem 6.8.16. *Let M be a proper screen slant lightlike submanifold of an indefinite Kähler manifold $\bar{M}$. Then M is minimal if and only if*

$$\operatorname{tr} A^*_{\xi_j}\,|_{S(TM)} = 0,\ \operatorname{tr} A_{W_\alpha}\,|_{S(TM)} = 0$$

and

$$\bar{g}(D^l(X, W), Y) = 0, \forall X, Y \in \Gamma(RaDTM),$$

where $\{\xi_j\}_{j=1}^r$ is a basis of $\operatorname{Rad} TM$ and $\{W_\alpha\}_{\alpha=1}^m$ is a basis of $S(TM^\perp)$.

Proof. In Proposition 5.1.3, we have $h^l = 0$ on $\operatorname{Rad} TM$. Thus M is minimal if and only if

$$\sum_{k=1}^{m} h(e_k, e_k) = 0$$

and $h^s = 0$ on $\operatorname{Rad} TM$. Using (5.1.28) and (5.1.31) we obtain

$$\sum_{k=1}^{m} h(e_k, e_k) = \sum_{k=1}^{m} \frac{1}{r} \sum_{j=1}^{r} g(A^*_{\xi_j} e_k, e_k) N_j$$

$$+ \frac{1}{m} \sum_{\alpha=1}^{m} g(A_{W_\alpha} e_k, e_k) W_\alpha. \tag{6.8.22}$$

From (5.1.28) we obtain $\bar{g}(h^s(X,Y), W) = \bar{g}(D^l(X, W), Y)$ for $X, Y \in \Gamma(\operatorname{Rad} TM)$. Thus our assertion follows from (6.8.22). □

Theorem 6.8.17. *Let M be a proper slant lightlike submanifold of an indefinite Kähler manifold $\bar{M}$ such that $\omega(S(TM)) = S(TM^\perp)$. Then M is minimal if and only if*

$$\operatorname{tr} A^*_{\xi_j} \big|_{S(TM)} = 0, \ \operatorname{tr} A_{Fe_i} \big|_{S(TM)} = 0$$

and

$$\bar{g}(D^l(X, \omega e_i), Y) = 0, \forall X, Y \in \Gamma(\operatorname{Rad} TM)$$

where $\{e_1, \ldots, e_m\}$ is a basis of $S(TM)$.

Proof. From Lemma 6.8.15, $\{\csc \theta \omega e_1, \ldots, \csc \theta \, \omega e_m\}$ is an orthonormal basis of $S(TM^\perp)$. Thus we can write

$$h^s(X, X) = \sum_{i=1}^{m} A_i \csc \theta \omega e_i, \forall X \in \Gamma(TM).$$

for some functions A_i, $i \in \{1, \ldots, m\}$. Hence we obtain

$$h^s(X, X) = \sum_{i=1}^{m} \csc \theta g(A_{\omega e_i} X, X) \omega e_i, \forall X \in \Gamma(S(TM)).$$

Then the assertion of the theorem comes from Proposition 5.1.3 and Theorem 6.8.16. □

Remark 6.8.18. Observe that between slant lightlike and screen slant lightlike submanifolds there exists no inclusion relation because a lightlike real hypersurface is a slant lightlike submanifold and it is not a screen slant lightlike submanifold. Moreover, invariant and screen real lightlike submanifolds are screen slant lightlike submanifolds, but they are not slant lightlike submanifolds. Notice that it follows from Proposition 6.8.7 that the screen slant lightlike geometry is different from its counterpart of Chen's Riemannian case. For example, there does not exist any screen slant lightlike surface of a semi-Euclidean space R^4_2.

Chapter 7

Submanifolds of indefinite Sasakian manifolds

In this chapter, we first give a review of indefinite Sasakian manifolds, contact CR-submanifolds and a variety of other submanifolds of Sasakian manifolds. Then, similar to the case of the previous chapter, we focus on up-to-date published results (with complete proofs) on the geometry of invariant, contact Cauchy-Riemann (CR) and contact screen Cauchy-Riemann (SCR) lightlike submanifolds of indefinite Sasakian manifolds.

7.1 Introduction

Indefinite Sasakian manifolds. A $(2n + 1)$-dimensional differentiable manifold M is said to have an *almost contact structure* (ϕ, ξ, η) if it admits a vector field ξ, a 1-form η and a field ϕ of endomorphism of the tangent vector space satisfying

$$\phi^2 = -I + \eta \otimes \xi, \quad \eta(\xi) = 1. \tag{7.1.1}$$

It follows that $\phi(\xi) = 0$, $\eta \circ \phi = 0$ and $\operatorname{rank}(\phi) = 2n$. Then, the manifold M, with a (ϕ, ξ, η)-structure is called an *almost contact manifold* and the following holds (also see Blair [67] and Gray [211]):

Theorem 7.1.1. (Sasaki [367]) *A $(2n + 1)$-dimensional manifold M is an almost contact manifold if and only if the structure group of its tangent bundle is reducible to $U(n) \times \{I\}$.*

From the point of view of differential geometry it is desirable to define a metric on a paracompact manifold M. We say that a Riemannian metric g is an associated metric of an almost contact structure (ϕ, ξ, η) of M if

$$g(\phi X, \phi Y) = g(X, Y) - \eta(X) \eta(Y), \quad \forall X, Y \in \Gamma(TM). \tag{7.1.2}$$

It is always possible to choose such a Riemannian metric on M, although such a metric is not unique. (M, g), with g satisfying (7.1.2), is called an *almost contact metric manifold* with a (ϕ, ξ, η, g)-structure and g is called its *compatible (or associated) metric*, whose fundamental 2-form Ω is defined by

$$\Omega(X, Y) = g(X, \phi Y), \quad \forall X, Y \in \Gamma(TM). \tag{7.1.3}$$

Since Ω, satisfying (7.1.3), also satisfies $\eta \wedge \Omega^n \neq 0$, this almost contact metric manifold M is orientable. A $(2n + 1)$-dimensional manifold M is called a *contact manifold* if there exists a 1-form η on M such that $\eta \wedge (d\eta)^n \neq 0$ everywhere. When $\Omega = d\eta$ (i.e., Ω is closed), then, (M, ϕ, ξ, η, g) is called a *contact metric manifold*. These metrics can be constructed by the polarization of $\Omega = d\eta$ evaluated on a local orthonormal basis of the tangent space with respect to an arbitrary metric, on the contact sub-bundle D. Thus, a contact metric manifold is an analogue of an almost Kähler manifold, in odd dimensions.

Now we recall the concept of *normal almost contact structures* whose condition is an analogue of the integrability condition in almost complex structures. To understand this we let (M, ϕ, ξ, η) be an almost contact manifold. Construct the product manifold $\bar{M} = M \times \mathbf{R}$ such that $(X, f\partial_t)$ is any vector field on $\bar{M}$, where $X \in \Gamma(TM)$ and f is a smooth function on $\bar{M}$. Define an almost complex structure J on $\bar{M}$ by

$$J(X, f\partial_t) (\phi X - f\xi, \eta(X)\partial_t).$$

Then, the almost contact structure (ϕ, ξ, η) on M is said to be normal if J is integrable. It is known (see Blair [66, page 49] for details) that M has a *normal contact structure* if

$$N_\phi + 2\, d\eta \otimes \xi = 0, \tag{7.1.4}$$

where $N_\phi = [\phi, \phi]$ is the Nijenhuis tensor field of ϕ. A normal contact metric manifold is called a *Sasakian manifold*. The following is well known.

Theorem 7.1.2. (Sasaki [367]) *An almost contact metric manifold (M, ϕ, ξ, η, g) is Sasakian if and only if*

$$(\nabla_X \phi)Y = g(X, Y)\xi - \eta(Y)X$$

where ∇ is the metric connection on M.

Sasakian space forms. A plane section S in $T_x M$ of a Sasakian manifold M is called a ϕ-section if it is spanned by a unit vector X orthogonal to ξ and ϕX, where X is a non-null vector field on M. The sectional curvature $K(X, \phi X)$ of a ϕ-section is called a ϕ-sectional curvature. If M has a ϕ-sectional curvature c which does not depend on the ϕ-section at each point, then, c is constant in M and M is called a *Sasakian space form*, denoted by $M(c)$. Well-known examples are: the unit sphere in $\mathbf{C}^{n+1}$ and R^{2n+1}. See some more examples in Blair [66]. Finally, we state the following result on the curvature tensor R of $M(c)$.

Theorem 7.1.3. (Ogiue [313]) *If the ϕ-sectional curvature c at any point of a Sasakian manifold M $(\dim(M) \geq 5)$ is independent of the choice of ϕ-section at the point, then c is constant on M and its curvature tensor is*

$$
\begin{aligned}
R(X,Y)Z = \frac{(c+3)}{4}\{g(Y,Z)X - g(X,Z)Y\} + \frac{(c-1)}{4}\{\eta(X)\eta(Z)Y \\
- \eta(Y)\eta(Z)X + g(X,Z)\eta(Y)V - g(Y,Z)\eta(X)V \\
+ g(\phi Y, Z)\phi X + g(\phi Z, X)\phi Y - 2g(\phi X, Y)\phi Z\} \qquad (7.1.5)
\end{aligned}
$$

for any X,Y and Z vector fields on M.

In 1990, Duggal [134] introduced a larger class of contact manifolds as follows. Let (M, g) be a semi-Riemannian manifold of a constant index. We use the same notation of geometric objects as in the Riemannian case. A $(2n + 1)$-dimensional smooth semi-Riemannian manifold (M, g) is called an *(ϵ)-almost contact metric manifold* with a (ϕ, ξ, η, g)-structure if

$$
\begin{aligned}
\phi^2 &= -I + \eta \otimes \xi, \quad \eta(\xi) = 1, \\
g(\xi, \xi) &= \epsilon, \quad g(\phi X, \phi Y) = g(X, Y) - \epsilon \eta(X)\eta(Y), \qquad (7.1.6)
\end{aligned}
$$

where $\epsilon = 1$ or -1 according as ξ is spacelike or timelike and $\text{rank}(\phi) = 2n$. It is important to mention that in the above definition ξ is never a null vector field. The index q of g is an odd or even number according as ξ is timelike or spacelike. This holds since on M one can consider an orthonormal field of frame $\{E_1, \ldots, E_n, \phi E_1, \ldots, \phi E_n, \xi\}$ with $E_i \in \Gamma(D)$ and $g(E_i, E_i) = g(\phi E_i, \phi E_i)$.

Unfortunately, contrary to the Riemannian case ($\epsilon = 1, q = 0$) for which there always exists a Riemannian metric satisfying (7.1.6), there is no such guarantee for a non-degenerate metric satisfying (7.1.6) for a proper semi-Riemannian manifold M. At best the following is known.

Theorem 7.1.4. [149] *Let (ϕ, ξ, η) be an almost contact structure and h_0 a metric on a semi-Riemannian manifold M such that ξ is non-null. Then there exists on M a $(1, 2)$ type symmetric tensor field g satisfying (7.1.6).*

Proof. Define two semi-Riemannian metrics $h_1 = \frac{\epsilon}{\alpha}h_0$ where $\alpha = h_0(\xi, \xi)$ and h such that

$$
h(X,Y) = h_1(\phi^2 X, \phi^2 Y) + \epsilon\eta(X)\eta(Y), \quad \forall X, Y \in \Gamma(TM).
$$

To prove that h is semi-Riemannian we first note that $\eta(X) = \epsilon h(X, \xi)$ and $h(\xi, \xi) = \epsilon$. Secondly, let L be the distribution spanned by ξ on M and D the complementary orthogonal distribution to L with respect to h_1. Then, it is easy to show that $h(X, X) = h_1(X, X)$, $\forall X \in \Gamma(D)$, since $h_1(X, \xi) = 0$ and $h_1(\xi, \xi) = -\epsilon$. Thus h is a semi-Riemannian metric on M of the same index as h_1 is on D. Finally, by defining a symmetric tensor field

$$
g(X,Y) = \frac{1}{2}\{h(X,Y) + h(\phi X, \phi Y) + \epsilon\eta(X)\eta(Y)\},
$$

we obtain

$$g(\phi X, \phi Y) = \frac{1}{2}\{h(\phi X, \phi Y) + h(-X + \eta(X)\xi, \eta(Y)\xi)\}$$
$$= g(X, Y) - \epsilon\eta(X)\eta(Y),$$

which proves the theorem.   □

However, on the brighter side, we now prove that there always exists a Lorentzian metric g on a M satisfying (7.1.6).

Corollary 7.1.5. *Under the hypothesis of Theorem 7.1.4, there exists a Lorentzian metric g on M satisfying (7.1.6).*

Proof. Since M is paracompact there exists a Riemannian metric say h_0 on M. Define three metrics h_1, h and g as in Theorem 7.1.4 with $\epsilon = -1$. Then it is easy to see that both h and g are Lorentzian metrics on M, which completes the proof.   □

Moreover, there are restrictions on the signature of g as follows. For an almost contact metric manifold M, its $2n$-dimensional distribution $(D, \bar{g}/D)$ has an indefinite *Hermitian structure*, defined by $J = \phi/D$, i.e., we have

$$g(JX, JY) = g(X, Y), \forall X, Y \in \Gamma(D).$$

For the *endomorphism* J, satisfying $J^2 = -I$ on D, as explained in the previous section, the only possible signatures of g/D are $(2p, 2s)$ with $p + s = n$. In particular, $(D, g/D, J)$ satisfying the above condition, can not carry a Lorentz metric, if J is real. Subject to these restrictions on g, we have the following two classes of (ϵ)-almost contact metric manifolds.

(1) $\epsilon = 1$, $q = 2r$. M is a *spacelike almost contact metric manifold.*

(2) $\epsilon = -1$, $q = 2r + 1$. M is a *timelike almost contact metric manifold.*

A subcase of the second class is the *Lorentz almost contact manifold* for which $\epsilon = -1$ and $q = 1$.

Now consider the fundamental 2-form Ω of M exactly as in the Riemannian case and defined by the equation (7.1.3) and its normality condition as defined by the equation (7.1.4). Then, we say that (ϕ, ξ, η, g) is an (ϵ)-*contact metric structure* if $\Omega = d\eta$ and M is, then, called an (ϵ)-*contact metric manifold.* Moreover, if M is also normal, then, M is called an *indefinite Sasakian manifold* [134, 386]. As in the Riemannian case, we have:

Theorem 7.1.6. *An (ϵ)-almost contact metric manifold M is an indefinite Sasakian manifold if only if*

$$(\nabla_X \phi) Y = g(X, Y)\xi - \epsilon\eta(Y)X, \quad \forall X, Y \in \Gamma(TM). \tag{7.1.7}$$

By replacing Y by ξ in (7.1.7) we get

$$\nabla_X \xi = -\epsilon \phi X, \quad \forall X \in \Gamma(TM) \tag{7.1.8}$$

which implies that $\nabla_\xi \xi = 0$. Thus , we have:

Corollary 7.1.7. *The characteristic vector field ξ on an indefinite Sasakian manifold is a Killing vector field.*

Finally, we present some examples of (ϵ)-Sasakian structures on a semi-Euclidean manifold $\mathbf{R}_q^{2n+1}$ of index q. We need the following notation: $0_{p,k} = p \times k$ null matrix; $I_k = k \times k$ unit matrix. For any non-negative integer $s \le n$ we put

$$\epsilon^a = \begin{cases} -1 & \text{for} & a \in \{1,\ldots,s\}, \\ 1 & \text{for} & a \in \{s+1,\ldots,n\}, \end{cases}$$

in case $s \ne 0$ and $\epsilon^a = 1$ in case $s = 0$. Consider (x^i, y^i, z) as Cartesian coordinates on $\mathbf{R}_q^{2n+1}$ and define with respect to the natural field of frames $\{\partial_{x^i}, \partial_{y^i}, \partial_z\}$ a $(1,1)$ tensor field ϕ by its matrix

$$(\phi) = \begin{pmatrix} 0_{n,\,n} & I_n & 0_{n,\,1} \\ -I_n & 0_{n,\,n} & 0_{n,\,1} \\ 0_{1,\,n} & \epsilon^a y^a & 0 \end{pmatrix}.$$

The 1-form η is defined by

$$\eta = \frac{\epsilon}{2}\left(dz + \sum_{i=1}^{s} y^i dx^i - \sum_{i'=r+1}^{n} y^{i'} dx^{i'} \right), \quad \text{if} \quad s \ne 0$$

and $\eta = \frac{\epsilon}{2}\left(dz - \sum_{1}^{n} y^i\, dx^i \right)$, if $s = o$. The characteristic vector field is $\xi = 2\epsilon\partial_z$. It is easy to check that (ϕ, ξ, η) is an (ϵ)-almost contact structure on $\mathbf{R}_q^{2n+1}$ for each s. Define a semi-Riemannian metric g by the matrix

$$(g) = \frac{\epsilon}{4}\begin{pmatrix} -\delta_{ij} + y^i y^j & -y^i y^{j'} & 0_{s,s} & O_{s,n-s} & y^i \\ -y^i y^{j'} & \delta_{i'j'} + y^{i'} y^{j'} & 0_{n-s,s} & O_{n-s,n-s} & -y^{i'} \\ O_{s,s} & O_{s,n-s} & -I_s & O_{s,n-s} & O_{s,1} \\ 0_{n-s,s} & O_{n-s,n-s} & 0_{n-s,s} & I_{n-s} & O_{n-s,1} \\ y^i & -y^{i'} & O_{1,s} & O_{1,n-s} & I \end{pmatrix}$$

for $s \ne 0$, and for $s = 0$ we get

$$(g) = \frac{\epsilon}{4}\begin{pmatrix} -\delta_{ij} + y^i y^j & 0_{n,n} & y^i \\ O_{n,n} & I_n & O_{n,1} \\ y^i & O_{1,n} & I \end{pmatrix}.$$

An orthonormal field of frames with respect to the above metric is

$$\left\{ \begin{aligned} E_i &= 2\,\partial_{y^i}, & E_{i'} &= 2\,\partial_{y^{i'}}, \\ \phi\,E_i &= 2\,(\partial_{x^i} - y^i\,\partial_z), \\ \phi\,E_{i'} &= 2\,(\partial_{x^i} + y^i\,\partial_z), \xi \end{aligned} \right\}.$$

It is easy to check that the above data provides an (ϵ)-Sasakian structure on $\mathbf{R}_q^{2n+1}$ for any $s \in \{0, \ldots, n\}$. In case $s = 0$ and $\epsilon = 1$ we get the classical Sasakian structure on $\mathbf{R}^{2n+1}$. If $s \neq 0$, then, we either get a spacelike Sasakian structure on $\mathbf{R}_{2s}^{2n+1}$ for $\epsilon = 1$ or a timelike Sasakian structure on $\mathbf{R}_{2(n-s)+1}^{2n+1}$ for $\epsilon = -1$. In particular, for $s = n$ and $\epsilon = -1$ we get a Lorentzian Sasakian structure.

Contact CR-submanifolds. By using the contact metric structure on a Sasakian manifold, one may define invariant and anti-invariant submanifolds, contact CR-submanifolds and slant submanifolds of Sasakian manifolds corresponding to complex submanifolds, totally real submanifolds, CR-submanifolds and slant submanifolds of Kählerian manifolds. In [411], Yano and Kon introduced the notion of a contact CR-submanifold of a Sasakian manifold which is closely similar to the one of a CR-submanifold of a Kählerian manifold defined by Bejancu. Note that contact CR-submanifolds were also defined by Bejancu and Papaghiuc [54], under the name of semi-invariant submanifolds.

Let $(\bar{M}, \bar{g})$ be a $(2m+1)$-dimensional Sasakian manifold with structure tensors $(\bar{\phi}, V, \bar{\eta}, \bar{g})$. Consider a submanifold M of $\bar{M}$ with induced metric tensor field g. First of all, we have:

Proposition 7.1.8. [412] *Let M be an n-dimensional submanifold of a $(2m+1)$-dimensional Sasakian manifold $\bar{M}$. If the structure vector field V is normal to M, then, M is anti-invariant with respect to $\bar{\phi}$, that is, $\bar{\phi}(T_x M) \subset T_x(M)^\perp$ for each point x of M, and $m \geq n$.*

It follows from the above proposition that to study a variety of submanifolds of contact manifolds one must assume that M is an n-dimensional submanifold, tangent to the structure vector field V, of a $(2m+1)$-dimensional Sasakian manifold $\bar{M}$. Then, for any vector field X tangent to M, we put

$$\bar{\phi}X = PX + FX, \qquad (7.1.9)$$

where PX is the tangential part of $\bar{\phi}X$ and FX is the normal part of $\bar{\phi}X$. Similarly, for any vector field N normal to M, we put

$$\bar{\phi}N = tN + fN, \qquad (7.1.10)$$

where tN is the tangential part of $\bar{\phi}N$ and fN is the normal part of $\bar{\phi}N$.

Definition 7.1.9. [411] Let M be a submanifold tangent to the structure vector field V and isometrically immersed in a Sasakian manifold $\bar{M}$. Then M is a *contact CR-submanifold* of $\bar{M}$ if there exists a differentiable distribution $D : x \longrightarrow D_x \subset T_x M$ on M satisfying the following conditions:

1. D is invariant with respect to $\bar{\phi}$, i.e., $\bar{\phi}(D_x) \subset D_x$ for each $x \in M$,

2. the complementary orthogonal distribution $D^\perp : x \longrightarrow D_x^\perp \subset T_x M$ is anti-invariant with respect to $\bar{\phi}$, i.e., $\bar{\phi}(D_x^\perp) \subset T_x M^\perp$ for each $x \in M$.

Set $\dim(\bar{M}) = 2m + 1$, $\dim(M) = 2n + 1$, $\dim(D) = h$, $\dim(D^\perp) = p$ and $codim(M) = q$. If $p = 0$, then, a contact CR-submanifold is an invariant submanifold of $\bar{M}$. If $h = 0$, then, M is an anti-invariant submanifold of $\bar{M}$ tangent to the structure vector field V. If $q = p$, then, a contact CR-submanifold is called a *generic submanifold* of $\bar{M}$. In this case, $\bar{\phi}(T_x M^\perp) \subset T_x M$ for every point x of M. If $h > 0$ and $p > 0$, then a contact CR-submanifold is said to be *proper*. Yano and Kon obtained the following characterization:

Theorem 7.1.10. [411] *In order for a submanifold M, tangent to the structure vector field V, of a Sasakian manifold $\bar{M}$ to be contact CR-submanifold, it is necessary and sufficient that $FP = 0$.*

They also proved the following:

Theorem 7.1.11. [411] *Let M be an $(n + 1)$-dimensional contact CR-submanifold of a $(2m + 1)$-dimensional Sasakian manifold $\bar{M}$. Then:*

1. *The distribution $D^\perp$ is completely integrable and its maximal integral submanifold is a q-dimensional anti-invariant submanifold of $\bar{M}$ normal to V or a $(q + 1)$-dimensional anti-invariant submanifold of $\bar{M}$ tangent to V.*

2. *The distribution D is completely integrable if and only if*

$$h(X, PY) = h(Y, PX), \quad \forall X, Y \in \Gamma(D).$$

Then $\xi \in \Gamma(D)$, where h is the second fundamental form of M. Moreover, the maximal integral submanifold of D is an $(n + 1 - q)$-dimensional invariant submanifold of $\bar{M}$.

It is known that every odd-dimensional sphere S^{2m+1} has Sasakian structure. In this case, we have the following:

Theorem 7.1.12. [411] *Let M be an $(n + 1)$-dimensional contact CR-submanifold of a S^{2m+1} with flat normal connection. If the mean curvature vector of M is parallel and $PA_N = A_N P$ for any vector field N normal to M, then M is S^{n+1} or*

$$S^{m_1}(r_1) \times \ldots \times S^{m_k}(r_k), n + 1 = \sum_i m_i, \sum_i r_i^2, 2 \le k \le n + 1$$

in some S^{n+1+p}, where $m_1, \ldots, m_k$ are odd numbers

Let M be an $(n + 1)$-dimensional contact CR-submanifold of a Sasakian manifold $\bar{M}$. Then, M has *maximal contact CR-dimension* in $\bar{M}$ if $\dim(D_x) = n - 1$. In this case, if there is a non-zero vector U which is orthogonal to V and contained in $D_x^\perp$, then $N = \bar{\phi}U$ must be normal to M. In the case of $\bar{M} = S^{2m+1}(1)$, we have the following:

Theorem 7.1.13. [320] *Let M be an $(n + 1)$-dimensional contact CR-submanifold of $(n - 1)$ contact CR-dimension immersed in a $(2m + 1)$-unit sphere S^{2m+1}.*

If the distinguished normal vector field N is parallel with respect to the normal connection and the following equality $h(X, PY) = -h(PX, Y)$ holds on M, then M is locally isometric to $S^{2n_1+1}(r_1) \times S^{2n_2+1}(r_2)$ $(r_1^2 + r_2^2 = 1)$ for some integers n_1, n_2 with $n_1 + n_2 = \frac{(n-1)}{2}$.

If $n \geq 3$, the above result was improved as follows:

Theorem 7.1.14. [260] *Let M be an $(n + 1)$-dimensional $(n \geq 3)$ contact CR-submanifold of $(n - 1)$-contact CR-dimension immersed in a $(2m + 1)$-unit sphere S^{2m+1}. If the equality $h(X, PY) = -h(PX, Y)$ holds on M, then M is locally isometric to $S^{2n_1+1}(r_1) \times S^{2n_2+1}(r_2)$ $(r_1^2 + r_2^2 = 1)$ for some integers n_1, n_2 with $n_1 + n_2 = \frac{(n-1)}{2}$.*

As in the complex case, a contact CR-submanifold of a Sasakian manifold $\bar{M}$ is called a *contact CR-product*, if it is locally a product of M^T and $M^\perp$, where M^T and $M^\perp$ are the integral submanifolds of $D \oplus \{V\}$ and $D^\perp$.

Theorem 7.1.15. [297] *A contact CR-submanifold of a Sasakian manifold is a contact CR-product if and only if*

$$A_{\bar{\phi}W} X = \bar{\eta}(X)W, \quad \forall X \in \Gamma(D \oplus \{V\}) \quad and \quad W \in \Gamma(D^\perp).$$

If $\bar{M}$ is a Sasakian space form, we have the following non-existence result.

Theorem 7.1.16. [297] *There exists no contact CR-product of a Sasakian space form $\bar{M}(c)$ with $c < -3$.*

On totally umbilical submanifolds we quote the following result:

Proposition 7.1.17. [412] *Let M be a submanifold, tangent to the structure vector field V, of a Sasakian manifold $\bar{M}$. If M is totally umbilical, then M is totally geodesic.*

In fact, the above result tells us that the usual definition of totally umbilical submanifolds does not work for submanifolds of Sasakian manifolds. Therefore, we need the following definition.

Definition 7.1.18. [412] *Let M be a submanifold, tangent to the structure vector field V, of a Sasakian manifold $\bar{M}$. If the second fundamental form of a submanifold is of the form*

$$h(X, Y) = [g(X, Y) - \bar{\eta}(X)\bar{\eta}(Y)]\alpha + \bar{\eta}(X)h(Y, V) + \bar{\eta}(Y)h(X, V) \qquad (7.1.11)$$

for any vector fields X and Y tangent to M, α being a vector field normal to M, then, M is called a totally contact umbilical manifold.

It is easy to show that $\alpha = \frac{1}{n}\operatorname{tr} h = \frac{n+1}{n}H$, where H denotes the mean curvature vector of M. The notion of a totally contact umbilical submanifold of Sasakian manifolds corresponds to that of totally umbilical submanifolds of Kählerian manifolds. If $\alpha = 0$, that is, if h is of the form

$$h(X, Y) = \bar{\eta}(X)h(Y, V) + \bar{\eta}(Y)h(X, V) \tag{7.1.12}$$

then M is said to be *totally contact geodesic*.

Proposition 7.1.19. *Let M be a totally contact umbilical submanifold, tangent to the structure vector field V, of a Sasakian manifold $\bar{M}$. Then M is totally contact geodesic if and only if M is minimal.*

In particular, we have the next result for contact CR-submanifolds.

Theorem 7.1.20. [412] *Let M be a totally contact umbilical proper contact CR-submanifold of a Sasakian manifold. If $\dim(D^{\perp}) > 2$, then, M is totally contact geodesic in $\bar{M}$.*

There are many results on contact CR-submanifolds (also slant submanifolds) of Sasakian manifolds (and other almost contact manifolds). We have quoted some main results for such submanifolds in Riemannian Sasakian manifolds which are needed to compare and understand their counterparts of lightlike geometry. For those interested in knowing more about almost contact manifolds and their non-degenerate contact CR-submanifolds, we recommend the following references: [7, 9, 11, 55, 56, 54, 65, 66, 67, 68, 69, 70, 78, 79, 84, 85, 86, 87, 88, 89, 117, 134, 184, 201, 209, 211, 226, 255, 258, 259, 265, 266, 271, 275, 277, 292, 299, 301, 303, 304, 313, 315, 338, 367, 374, 377, 386, 388, 399, 404, 413].

7.2 Lightlike hypersurfaces

In this section we present up-to-date results on lightlike hypersurfaces of almost contact metric manifolds and indefinite Sasakian manifolds.

Let $(\bar{M}, \bar{\phi}, V, \eta, \bar{g})$ be a $(2m+1)$-dimensional almost contact metric manifold, where $\bar{g}$ is a semi-Riemannian metric of index q, $0 < q < 2m + 1$. Let (M, g) be a lightlike hypersurface of $(\bar{M}, \bar{g})$. Then, for each $\xi_p \in \Gamma(TM^{\perp})$ at $p \in M$, we have $\bar{g}(\xi_p, \xi_p) = 0$. This implies that $\xi_p \in \Gamma(T_pM)$. On the other hand, (7.1.2) implies that $\bar{g}(\xi_p, \bar{\phi}\xi_p) = 0$. Hence $\bar{\phi}\xi_p$ is tangent to M. Thus, $\bar{\phi}(TM^{\perp})$ is a distribution on M of rank 1. Choose a screen distribution $S(TM)$ of M such that it contains $\bar{\phi}\xi$ and V. Using (7.1.2), we have

$$\bar{g}(\bar{\phi}N, \xi) = -\bar{g}(N, \bar{\phi}\xi) = 0, \bar{g}(N, \bar{\phi}N) = 0,$$

where $N \in \Gamma(\text{tr}(TM))$. This implies that $\bar{\phi}N$ is also tangent to M and belongs to $S(TM)$. Moreover, from (7.1.2) we get

$$\bar{g}(\bar{\phi}N, \bar{\phi}\xi) = 1.$$

Thus $\bar{\phi}(TM^{\perp}) \oplus \bar{\phi}(\mathrm{tr}(TM))$ is a non-degenerate vector sub-bundle of $S(TM)$ of rank 2. Then, there exists a non-degenerate distribution $\mathcal{D}_o$ on M such that

$$S(TM) = \{\bar{\phi}(TM^{\perp}) \oplus \bar{\phi}(\mathrm{tr}(TM))\} \perp \mathcal{D}_o, \qquad (7.2.1)$$

where $\mathcal{D}_o$ is an invariant distribution with respect to $\bar{\phi}$. Summing up the above information, we obtain the decompositions

$$TM = \{\bar{\phi}(TM^{\perp}) \oplus \bar{\phi}(\mathrm{tr}(TM))\} \perp \mathcal{D}_o \perp \{V\} \oplus_{\mathrm{orth}} TM^{\perp} \qquad (7.2.2)$$

and

$$T\bar{M} = \{\bar{\phi}(TM^{\perp}) \oplus \bar{\phi}(\mathrm{tr}(TM))\} \perp \mathcal{D}_o \perp \{V\} \perp \{TM^{\perp} \oplus \mathrm{tr}(TM)\}. \qquad (7.2.3)$$

At this point let $(\bar{M}, \bar{\phi}, V, \eta, \bar{g})$ be an indefinite Sasakian manifold. Set

$$\mathcal{D} = TM^{\perp} \oplus_{\mathrm{orth}} \bar{\phi}(TM^{\perp}) \oplus_{\mathrm{orth}} \mathcal{D}_o \qquad (7.2.4)$$

and

$$\mathcal{D}' = \bar{\phi}(\mathrm{tr}(TM)). \qquad (7.2.5)$$

Then, it is obvious that $\mathcal{D}$ is invariant with respect to $\bar{\phi}$. Thus, we have the decomposition:

$$TM = \mathcal{D} \oplus \mathcal{D}'. \qquad (7.2.6)$$

Now, consider local vector fields $U = -\bar{\phi}N$ and $W = -\bar{\phi}\xi$. Then, any vector field on M is expressed as

$$X = SX + QX, \; QX = u(X)U, \qquad (7.2.7)$$

where S and Q are the projection morphisms of TM to $\mathcal{D}$ and $\mathcal{D}'$, respectively. u is a 1-form locally defined on M by

$$u(X) = g(X, W). \qquad (7.2.8)$$

Applying $\bar{\phi}$ to (7.2.7) and using (7.1.2) we get

$$\bar{\phi}X = \phi X + u(X)N, \qquad (7.2.9)$$

where ϕ is a tensor field of type $(1,1)$ defined on M by

$$\phi X = \bar{\phi}SX. \qquad (7.2.10)$$

Then, applying $\bar{\phi}$ to (7.2.9) and using (7.1.2) we obtain

$$\phi^2 X = -X + \eta(X)V + u(X)U. \qquad (7.2.11)$$

Using Sasakian $\bar{M}$, from (7.2.8),(7.2.9), (2.1.14) and (2.1.15) we derive

$$(\nabla_X \phi)Y = \eta(Y)X - g(X,Y)V - B(X,Y)U + u(Y)A_N X, \qquad (7.2.12)$$

$$(\nabla_X u)Y = -B(X, \phi Y) - u(Y)\tau(X), \; \forall X, Y \in \Gamma(TM). \qquad (7.2.13)$$

Lemma 7.2.1. [254] *Let M be a lightlike real hypersurface of an indefinite Sasakian manifold $\bar{M}$.*

1. *If the vector field U is parallel, then, for $X \in \Gamma(TM)$ we have*

$$A_N X = \eta(A_N X)V + u(A_N X)U, \tau(X) = 0. \tag{7.2.14}$$

2. *If the vector field W is parallel, then*

$$A_\xi^* X = \eta(A_\xi^* X)V + u(A_\xi^* X)U, \tau(X) = 0. \tag{7.2.15}$$

Proof. Taking U instead of Y in (7.2.12), we get

$$A_N X = -\bar{\phi}(\nabla_X U) + g(X, U)\xi - B(X, U)U. \tag{7.2.16}$$

Applying ϕ to this equation and using (7.2.11), we have

$$\phi(A_N X) = \nabla_X U - \eta(\nabla_X U)V - u(\nabla_X U)U$$

for every $X \in \Gamma(TM)$. Thus, if U is parallel, we obtain $\phi(A_N X) = 0$. Then from (7.2.9), we get

$$\bar{\phi}(A_N X) = u(A_N X)N.$$

Applying $\bar{\phi}$ and using (7.1.1) we obtain the first equation of (7.2.14). Replacing Y in (7.2.13) by U, we have $(\nabla_X u)U = -u(U)\tau(X)$. Then $u(U) = 1$ implies that $(\nabla_X u)U = -\tau(X)$. Since $(\nabla_X u)U = -u(\nabla_X U) = 0$, we get $\tau(X) = 0$. Now suppose that W is parallel. Replacing Y by ξ in (7.2.12) and using (2.1.16) we obtain $(\nabla_X \phi)\xi = 0$. Hence, we have

$$\nabla_X \phi(\xi) - \phi(\nabla_X \xi) = 0.$$

Then, from (2.1.24) we derive

$$-\nabla_X W + \phi(A_\xi^* X + \tau(X)\xi) = 0.$$

Since W is parallel, we arrive at

$$\phi(A_\xi^* X + \tau(X)\xi) = 0.$$

Hence we derive

$$\phi A_\xi^* X = \tau(X)W.$$

Now, applying ϕ to this equation and using (7.2.11), we obtain

$$-A_\xi^* X + \eta(A_\xi^* X)V + u(A_\xi^* X)U = \tau(X)\phi(W).$$

On the other hand, since $\phi(W) = \bar{\phi}(W) = -\bar{\phi}^2(\xi) = \xi$, we derive

$$-A_\xi^* X + \eta(A_\xi^* X)V + u(A_\xi^* X)U = \tau(X)\xi.$$

Taking the screen and radical parts of this equation, we obtain (7.2.15). $\qquad \square$

From Lemma 7.2.1, we have the following result.

Corollary 7.2.2. [254] *Let M be a lightlike real hypersurface of an indefinite Sasakian manifold $\bar{M}$ such that U and W are parallel vector fields with respect to ∇. Then for any point $p \in M$ the type numbers of M and $S(TM)$ satisfy $t(p) \leq 2$ and $t^*(p) \leq 2$, respectively.*

The next result indicates that the integrability of all distributions involved in the study of lightlike real hypersurfaces is characterized by both second fundamental forms of M and $S(TM)$.

Proposition 7.2.3. *Let M be a lightlike hypersurface of an indefinite Sasakian manifold $\bar{M}$. Then:*

(i) $TM^{\perp} \oplus_{\mathrm{orth}} \bar{\phi}(TM^{\perp})$ *is integrable if and only if*

$$B(X, Y) = 0, \forall X \in \Gamma(\bar{\phi}(TM^{\perp})), Y \in \Gamma(\bar{\phi}(TM^{\perp}) \oplus_{\mathrm{orth}} \mathcal{D}_o). \qquad (7.2.17)$$

(ii) $TM^{\perp} \oplus_{\mathrm{orth}} \bar{\phi}(TM^{\perp}) \oplus_{\mathrm{orth}} \{V\}$ *is integrable if and only if (7.2.17) holds.*

(iii) $\bar{\phi}(\mathrm{tr}(TM)) \oplus \bar{\phi}(TM^{\perp})$ *is integrable if and only if*

$$C(U, W) = C(W, U) \qquad (7.2.18)$$

and
$$B(U, \bar{\phi}X) = C(W, \bar{\phi}X), \forall X \in \Gamma(\mathcal{D}_o). \qquad (7.2.19)$$

(iv) $\mathcal{D}_o$ *is integrable if and only if*

$$C(X, Y) = C(Y, X), \qquad (7.2.20)$$
$$C(X, \bar{\phi}Y) = C(Y, \bar{\phi}X), \qquad (7.2.21)$$
$$B(X, \bar{\phi}Y) = B(Y, \bar{\phi}X), \qquad (7.2.22)$$

for any $X, Y \in \Gamma(\mathcal{D}_o)$.

(v) $TM^{\perp} \oplus_{\mathrm{orth}} \mathcal{D}_o$ *is integrable if and only if (7.2.21) and (7.2.22) hold,*

$$B(X, W) = 0, \forall X \in \Gamma(\mathcal{D}_o) \qquad (7.2.23)$$

and
$$C(\xi, \bar{\phi}X) + B(X, U) = 0. \qquad (7.2.24)$$

(vi) $\bar{\phi}(TM^{\perp}) \oplus_{\mathrm{orth}} \mathcal{D}_o$ *is integrable if and only if (7.2.20) and (7.2.21) hold,*

$$C(U, \bar{\phi}X) = 0 \qquad (7.2.25)$$

and
$$C(X, U) = C(U, X), \forall X \in \Gamma(\mathcal{D}_o). \qquad (7.2.26)$$

(vii) $\bar{\phi}(\mathrm{tr}(TM)) \oplus_{\mathrm{orth}} \mathcal{D}_o$ *is integrable if and only if* (7.2.20) *and* (7.2.21) *hold,*

$$C(U, \bar{\phi}X) = 0 \tag{7.2.27}$$

and

$$C(X, U) = C(U, X), \forall X \in \Gamma(\mathcal{D}_o). \tag{7.2.28}$$

(viii) D *is integrable if and only if* (7.2.22) *and* (7.2.23) *hold and*

$$B(W, W) = 0. \tag{7.2.29}$$

(ix) $\bar{\phi}(\mathrm{tr}(TM)) \oplus_{\mathrm{orth}} TM^{\perp} \oplus_{\mathrm{orth}} \mathcal{D}_o$ *is integrable if and only if* (7.2.21), (7.2.24) *and* (7.2.25) *hold and*

$$B(U, U) = 0. \tag{7.2.30}$$

For totally umbilical lightlike hypersurfaces we first have the following result:

Theorem 7.2.4. [254] *Let M be a totally umbilical lightlike hypersurface of an indefinite Sasakian space form $\bar{M}(c)$. Then $c = 1$ and ρ satisfies the partial differential equations*

$$\xi(\rho) + \rho\tau(\xi) - \rho^2 = 0 \tag{7.2.31}$$

and

$$PX(\rho) + \rho\tau(PX) = 0, \forall X \in \Gamma(TM). \tag{7.2.32}$$

Proof. From (7.1.5), we have

$$4\bar{g}(\bar{R}(X, Y)Z, \xi) = (c - 1)\{\bar{g}(\bar{\phi}Y, Z)\bar{g}(\bar{\phi}X, \xi) + \bar{g}(\bar{\phi}Z, X)\bar{g}(\bar{\phi}Y, \xi) - \bar{g}(\bar{\phi}X, Y)\bar{g}(\bar{\phi}Z, \xi)\}$$

for any $X, Y, Z \in \Gamma(TM)$. Then from (2.4.6) and using (7.2.8), we obtain

$$(\nabla_X B)(Y, Z) - (\nabla_Y B)(X, Z) = \frac{c-1}{4}\{\bar{g}(\bar{\phi}Y, Z)u(X) + \bar{g}(\bar{\phi}Z, X)u(Y) - 2\bar{g}(\bar{\phi}X, Y)u(Z)\} + \tau(Y)B(X, Z) - \tau(X)B(Y, Z). \tag{7.2.33}$$

Replacing X, Y, Z in (7.2.33) by PX, ξ, PZ, respectively, we deduce

$$(\nabla_{PX} B)(\xi, PZ) - (\nabla_\xi B)(PX, PZ) = \frac{c-1}{4}\{\bar{g}(\bar{\phi}\xi, PZ)u(PX) - 2\bar{g}(\bar{\phi}PX, \xi)u(PZ)\} + \tau(\xi)B(PX, PZ) - \tau(PX)B(\xi, PZ).$$

Since M is umbilical, we get

$$-\rho g(\nabla_{PX}\xi, PZ) - \xi(\rho g(PX, PZ)) + \rho g(\nabla_\xi PX, PZ) + \rho g(PX, \nabla_\xi PZ)$$
$$= \frac{c-1}{4}\{-3u(PZ)u(PX)\} + \tau(\xi)\rho g(PX, PZ).$$

Then, by direct calculations, using (2.1.24) and (2.5.1), we get

$$((\rho)^2 - \xi(\rho) - \rho\tau(\xi))g(PX, PZ) = \frac{c-1}{4}\{-3u(PZ)u(PX)\}.$$

Now, taking $PX = PZ = U$, we obtain $c = 1$. On the other hand, substituting $X = \xi$, $Y = PY$ and $Z = PY$ in (7.2.33) with $c = 1$, we derive

$$\{\xi(\rho) - \rho^2 + \rho\tau(PX)\}g(PY, PY) = 0$$

which is (7.2.31) due to $g(PY, PY) \neq 0$ for at least one $Y \in \Gamma(TM)$. Finally, substituting $X = PX$, $Y = PY$ and $Z = PZ$ into (7.2.33) with $c = 1$ and taking into account that $S(TM)$ is non-degenerate, we get

$$\{PX(\rho) + \rho\tau(PX)\}PY = \{PY(\rho) + \rho\tau(PY)\}PX. \qquad (7.2.34)$$

Suppose there exists a vector field X' on some neighborhood of M such that $PX'(\rho) + \rho\tau(PX') \neq 0$ at some point p in the neighborhood. Then, from (7.2.34) it follows that all vectors of the fiber $S(T_pM)$ are collinear with $(PX')_p$. This contradicts $\dim S(T_pM) > 1$. Therefore, (7.2.32) holds which completes the proof. $\square$

From Theorem 7.2.4, we obtain the following result.

Corollary 7.2.5. *There exists no totally umbilical lightlike hypersurface of indefinite Sasakian space forms $\bar{M}(c)$ with $c \neq 1$.*

7.3   Invariant submanifolds

Let $(M, g, S(TM), S(TM^\perp))$ be a lightlike submanifold of an indefinite almost metric manifold $(\bar{M}, \bar{g})$. We put

$$\phi X = PX + FX, \quad \forall X \in \Gamma(TM), \qquad (7.3.1)$$

where PX and FX are the tangential and the transversal parts of ϕX respectively. Moreover, P is skew symmetric on $S(TM)$.

Lemma 7.3.1. *Let M be a lightlike submanifold of an indefinite almost metric manifold $\bar{M}$. If V is tangent to M, then the structure vector field V does not belong to $\text{Rad}\,TM$.*

Proof. Suppose M is a lightlike submanifold and $V \in \Gamma(\text{Rad}\,TM)$. Then, there exists a vector field $N \in \Gamma(\text{ltr}(TM))$ such that $g(N, V) = 1$(or at least $\neq 0$). On the other hand, from (7.1.6) we have

$$\bar{g}(\phi N, \phi V) = \bar{g}(V, N) - \eta(V)\eta(N).$$

Since V is null and $\phi V = 0$ we obtain

$$g(V, N) = 0.$$

This is a contradiction which proves our assertion. $\square$

The above lemma (also see [88]) implies that if M is tangent to the structure vector field V, then, V belongs to $S(TM)$. Using this, we say that M is an *invariant submanifold* of $\bar{M}$ if M is tangent to the structure vector field V and

$$\bar{\phi}(\operatorname{Rad} TM) = \operatorname{Rad} TM, \quad \bar{\phi}(S(TM)) = S(TM). \tag{7.3.2}$$

From (7.1.7), (7.1.8), (7.3.2) and (5.1.15) we get

$$h^l(X, V) = 0, \quad h^s(X, V) = 0, \quad \nabla_X V = PX, \tag{7.3.3}$$
$$h(X, \phi Y) = \phi h(X, Y) = h(\phi X, Y), \ \forall X, Y \in \Gamma(TM). \tag{7.3.4}$$

Proposition 7.3.2. *Let* $(M, g, S(TM), S(TM^\perp))$ *be an invariant lightlike submanifold of an indefinite Sasakian manifold. If the second fundamental forms* h^l *and* h^s *of* M *are parallel then* M *is totally geodesic.*

Proof. Suppose h^l is parallel. Then, we have

$$(\nabla_X h^l)(Y, V) = \nabla_X h^l(Y, V) - h^l(\nabla_X Y, V) - h^l(Y, \nabla_X V) = 0.$$

Thus, using (7.3.3) we have $h^l(Y, PX) = 0$. Similarly, we have $h^s(Y, PX) = 0$, which completes the proof. □

Theorem 7.3.3. *Let* $(M, g, S(TM), S(TM^\perp))$ *be an invariant lightlike submanifold of codimension two of an indefinite Sasakian manifold. Then,* $\operatorname{Rad} TM$ *defines a totally geodesic foliation on* M. *Moreover,* $M = M_1 \times M_2$ *is a lightlike product manifold if and only if* $h^* = 0$, *where* M_1 *is a leaf of the radical distribution and* M_2 *is a semi-Riemannian manifold.*

Proof. Since rank$(\operatorname{Rad} TM) = 2$, $\forall X, Y \in \Gamma(\operatorname{Rad} TM)$ we can write X, Y as linear combinations of ξ and $\phi\xi$, i.e., $X = A_1\xi + B_1\phi\xi$, $Y = A_2\xi + B_2\phi\xi$. Thus by direct calculations, using (5.1.15) we obtain

$$g(\nabla_X Y, \bar{P}Z) = -A_2 A_1 \bar{g}(\xi, h^l(\xi, \bar{P}Z)) - A_2 B_1 \bar{g}(h^l(\phi\xi, \bar{P}Z), \xi)$$
$$- B_2 A_1 \bar{g}(\phi\xi, h^l(\xi, \bar{P}Z)) - B_2 B_1 \bar{g}(h^l(\phi\xi, \bar{P}Z), \phi\xi).$$

Now, by using (5.1.33) and (7.3.4) we derive $g(\nabla_X Y, \bar{P}Z) = 0$. This shows that $\operatorname{Rad} TM$ defines a totally geodesic foliation. Then, the proof of the theorem follows from Theorem 5.1.6. □

Theorem 7.3.4. *Let* $(M, g, S(TM), S(TM^\perp))$ *be an invariant lightlike submanifold of codimension two of an indefinite Sasakian manifold* $\bar{M}$. *Suppose* (M', g') *is a non-degenerate submanifold of* $\bar{M}$ *such that* M' *is a leaf of integrable* $S(TM)$. *Then* M *is totally geodesic, with an induced metric connection if* M' *is so immersed as a submanifold of* $\bar{M}$.

Proof. Since $\dim(\operatorname{Rad} TM) = \dim(\operatorname{ltr}(TM)) = 2$, $h^l(X, Y) = A_1 N + B_1 \phi N$, where A_1 and B_1 are functions on M. Thus $h^l(X, \xi) = 0$ iff $\bar{g}(h^l(X, \xi), \xi) = 0$ and $\bar{g}(h^l(X, \xi), \phi \xi) = 0, \forall X \in \Gamma(TM)$ and $\xi \in \Gamma(\operatorname{Rad} TM)$. From (5.1.33) we have $\bar{g}(h^l(X, \xi), \xi) = 0$. Using (7.3.4) we get $\bar{g}(h^l(X, \xi), \phi \xi) = -\bar{g}(h^l(\phi X, \xi), \xi) = 0$. Similarly, $h^l(X, \phi \xi) = 0$. For M', we write

$$\bar{\nabla}_X Y = \nabla'_X Y + h'(X, Y), \quad \forall X, Y \in \Gamma(TM'),$$

where ∇' is a metric connection on M' and h' is the second fundamental form of M'. Thus, $h'(X, Y) = h^*(X, Y) + h^l(X, Y), \quad \forall X, Y \in \Gamma(TM')$. Also, $g(X, Y) = g'(X, Y), \forall X, Y \in \Gamma(TM)$, which completes the proof. $\qquad \square$

Theorem 7.3.5. *Let $(M, g, S(TM), S(TM^{\perp}))$ be a lightlike submanifold, tangent to the structure vector field V, of an indefinite Sasakian manifold $(\bar{M}, \bar{g})$. If M is totally umbilical, then, M is totally geodesic.*

Proof. Using (7.1.8), (5.1.15), (7.3.1) and the transversal parts we get

$$h^s(X, V) + h^l(X, V) = -FX, \quad \forall X \in \Gamma(TM). \tag{7.3.5}$$

$\phi V = 0$ implies $PV = 0$ and $FV = 0$. Thus from (7.3.5) we have $h^l(V, V) = 0$ and $h^s(V, V) = 0$. Now M totally umbilical, V non-null and (5.3.7), (5.3.8) imply $h^l = 0$ and $h^s = 0$. Thus, M is totally geodesic which completes the proof. $\qquad \square$

Theorem 7.3.6. *Let M be an invariant lightlike submanifold of an indefinite Sasakian manifold $\bar{M}$. Then M is minimal in $\bar{M}$ if and only if $D^l(X, W) = 0$ for $X \in \Gamma(\operatorname{Rad} TM)$ and $W \in \Gamma(S(TM))$.*

Proof. If M is invariant, then $\phi \operatorname{Rad} TM = \operatorname{Rad} TM$ and $\phi S(TM) = S(TM)$, hence $\phi(\operatorname{ltr}(TM)) = \operatorname{ltr} TM$ and $\phi S(TM^{\perp}) = STM^{\perp}$. Then using (7.3.2), (5.1.15) and taking transversal parts, we obtain

$$h(\phi X, Y) = \phi h(X, Y)$$

for $X, Y \in \Gamma(TM)$. Hence we get $h(\phi X, \phi Y) = -h(X, Y)$. Thus

$$\operatorname{tr} h \mid_{S(TM)} = \sum_{i=1}^{m-2r} \varepsilon_i \{ h(e_i, e_i) + h(\phi e_i, \phi e_i) \}$$

$$= \sum_{i=1}^{m-2r} \varepsilon_i \{ h(e_i, e_i) - h(e_i, e_i) \} = 0.$$

On the other hand, from (5.1.28) we get $\bar{g}(h^s(X, Y), W) = \bar{g}(D^l(X, W), Y)$ for $X, Y \in \Gamma(\operatorname{Rad} TM)$ and $W \in \Gamma(S(TM^{\perp}))$. The proof follows from Definition 5.4.1 and Proposition 5.1.3. $\qquad \square$

7.4 Contact CR-lightlike submanifolds

Let $(M, S(TM), S(TM^\perp))$ be a lightlike submanifold of an indefinite Sasakian manifold $\bar{M}$. The next definition is a generalization of lightlike real hypersurfaces of indefinite Sasakian manifolds.

Definition 7.4.1. Let $(M, g, S(TM), S(TM^\perp))$ be a lightlike submanifold, tangent to the structure vector field V, immersed in an indefinite Sasakian manifold $(\bar{M}, \bar{g})$. We say that M is a *contact CR-lightlike submanifold* of $\bar{M}$ if the following conditions are satisfied:

(A) $\operatorname{Rad} TM$ is a distribution on M such that $\operatorname{Rad} TM \cap \phi(\operatorname{Rad} TM) = \{0\}$.

(B) There exist vector bundles D_0 and D' over M such that

$$S(TM) = \{\phi(\operatorname{Rad} TM) \oplus D'\} \perp D_o \perp V,$$
$$\phi D_o = D_o, \ \phi(D') = L_1 \oplus_{\text{orth}} \operatorname{ltr}(TM),$$

where D_0 is non-degenerate and L_1 is a vector sub-bundle of $S(TM^\perp)$.

Thus, we have the decomposition

$$TM = \{D \oplus D'\} \perp V, \quad D = \operatorname{Rad} TM \oplus_{\text{orth}} \phi(\operatorname{Rad} TM) \oplus_{\text{orth}} D_o. \tag{7.4.1}$$

A contact CR-lightlike submanifold is proper if $D_o \neq \{0\}$ and $L_1 \neq \{0\}$. It follows that any contact CR-lightlike 3-dimensional submanifold is 1-lightlike.

Example 1. Let M be a lightlike hypersurface of an indefinite Sasakian manifold $\bar{M}$. For $\xi \in \Gamma(\operatorname{Rad} TM)$, we have $\bar{g}(\phi\xi, \xi) = 0$. Hence, $\phi\xi \in \Gamma(TM)$. Thus, we get a rank 1 distribution $\phi(TM^\perp)$ on M such that $\phi(TM^\perp) \cap TM^\perp = \{0\}$. This enables us to choose a screen $S(TM)$ such that it contains $\phi(TM^\perp)$ as a vector sub-bundle. Consider $N \in \Gamma(\operatorname{ltr}(TM))$ to obtain $\bar{g}(\phi N, \xi) = -\bar{g}(N, \phi\xi) = 0$ and $\bar{g}(\phi N, N) = 0$. Thus, $\phi N \in \Gamma(STM)$. Taking $D' = \phi(\operatorname{tr}(TM))$, we obtain $S(TM) = \{\phi(TM^\perp) \oplus D'\} \perp D_o$, where D_o is a non-degenerate distribution. Moreover $\phi(D') = \operatorname{tr}(TM)$. Hence M is a Contact CR-lightlike hypersurface.

From now on, $(\mathbf{R}_q^{2m+1}, \phi_o, V, \eta, \bar{g})$ will denote the manifold $\mathbf{R}_q^{2m+1}$ with its usual Sasakian structure given by

$$\eta = \frac{1}{2}(dz - \sum_{i=1}^{m} y^i dx^i), V = 2\partial z,$$

$$\bar{g} = \eta \otimes \eta + \frac{1}{4}(-\sum_{i=1}^{\frac{q}{2}} dx^i \otimes dx^i + dy^i \otimes dy^i + \sum_{i=q+1}^{m} dx^i \otimes dx^i + dy^i \otimes dy^i),$$

$$\phi_o(\sum_{i=1}^{m}(X_i \partial x^i + Y_i \partial y^i) + Z \partial z) = \sum_{i=1}^{m}(Y_i \partial x^i - X_i \partial y^i) + \sum_{i=1}^{m} Y_i y^i \partial z,$$

where $(x^i; y^i; z)$ are the Cartesian coordinates. The above construction will help in understanding how the contact structure is recovered in the next three examples.

Example 2. Let $\bar{M} = (\mathbf{R}^9_2, \bar{g})$ be a semi-Euclidean space, where $\bar{g}$ is of signature $(-,+,+,+,-,+,+,+,+)$ with respect to the canonical basis

$$\{\partial x_1, \partial x_2, \partial x_3, \partial x_4, \partial y_1, \partial y_2, \partial y_3, \partial y_4, \partial z\}.$$

Suppose M is a submanifold of $\mathbf{R}^9_2$ defined by

$$x^1 = y^4, \qquad x^2 = \sqrt{1 - (y^2)^2}, y^2 \neq 1.$$

It is easy to see that a local frame of TM is given by

$$Z_1 = 2(\partial x_1 + \partial y_4 + y^1 \partial z), \quad Z_2 = 2(\partial x_4 - \partial y_1 + y^4 \partial z),$$

$$Z_3 = \partial x_3 + y^3 \partial z, \quad Z_4 = \partial y_3, \quad Z_5 = -\frac{y^2}{x^2} \partial x_2 + \partial y_2 - \frac{(y^2)^2}{x^2} \partial z,$$

$$Z_6 = \partial x_4 + \partial y_1 + y^4 \partial z, \quad Z_7 = V = 2\partial z.$$

Hence $\operatorname{Rad} TM = \operatorname{Span}\{Z_1\}$, $\phi_o \operatorname{Rad} TM = \operatorname{Span}\{Z_2\}$ and $\operatorname{Rad} TM \cap \phi_o \operatorname{Rad} TM = \{0\}$. Thus (A) holds. Next, $\phi_o(Z_3) = -Z_4$ implies $D_0 = \{Z_3, Z_4\}$ is invariant with respect to ϕ_o. By direct calculations, we get

$$S(TM^\perp) = \operatorname{Span}\{W = \partial x_2 + \frac{y^2}{x^2}\partial y_2 + y^2 \partial z\} \quad \text{such that} \quad \phi_o(W) = -Z_5$$

and $\operatorname{ltr}(TM)$ is spanned by $N = -\partial x_1 + \partial y_4 - y^1 \partial z$ such that $\phi_o(N) = Z_6$. Hence M is a contact CR-lightlike submanifold.

Let M be isotropic or totally lightlike. Then, $S(TM) = \{0\}$. Hence conditions (A) and (B) of Definition 7.4.1 are not satisfied. Thus it follows that there exist no isotropic or totally lightlike contact CR-lightlike submanifolds.

Proposition 7.4.2. *Let M be a proper contact CR-lightlike submanifold of an indefinite Sasakian manifold $\bar{M}$. Then, D and $D' \oplus D$ are not integrable.*

Proof. Suppose D is integrable. Then $g([X,Y], V) = 0$, for $X, Y \in \Gamma(D)$. On the other hand, from (5.1.23) we derive

$$g([X,Y], V) = \bar{g}(\bar{\nabla}_X Y, V) - \bar{g}(\bar{\nabla}_Y X, V).$$

Then, $\bar{\nabla}$ is a metric connection and using (7.1.8), we have $g([X,Y], V) = g(Y, \phi X) - g(\phi Y, X)$. Hence, $g([X,Y], V) = -2g(\phi Y, X)$. Since M is proper and D_o is non-degenerate we can choose non-null vector fields X ,Y $\in \Gamma(D)$ such that $g(Y, \phi X) \neq 0$, which is a contradiction so D is not integrable. Similarly, $D' \oplus D$ is not integrable which completes the proof.   □

Denote the orthogonal complement sub-bundle to the vector sub-bundle L_1 in $S(TM^\perp)$ by $L_1^\perp$. For a contact CR-lightlike submanifold M we put

$$\phi X = fX + \omega X, \quad \forall X \in \Gamma(TM), \tag{7.4.2}$$

where $fX \in \Gamma(D)$ and $\omega X \in \Gamma(L_1 \oplus_{\mathrm{orth}} \mathrm{ltr}(TM))$. Similarly, we have

$$\phi W = BW + CW, \quad \forall W \in \Gamma(S(TM^\perp) \tag{7.4.3}$$

where $BW \in \Gamma(\phi L_1)$ and $CW \in \Gamma(L_1^\perp)$.

Proposition 7.4.3. *Let M be a contact CR-lightlike submanifold of an indefinite Sasakian manifold $\bar{M}$. Then, $D \oplus_{\mathrm{orth}} V$ is integrable if and only if*

$$h(X, \phi Y) = h(\phi X, Y), \quad \forall X, Y \in \Gamma(D \oplus_{\mathrm{orth}} V).$$

Proof. From (5.1.21), (7.4.2), (7.4.3), (7.1.7) and transversal parts, we get that $\omega(\nabla_X Y) = C h^s(X, Y) + h(X, \phi Y)$, $\forall X, Y \in \Gamma(D \oplus_{\mathrm{orth}} V)$. Thus, $\omega[X, Y] = h(X, \phi Y) - h(\phi X, Y)$, $\forall X, Y \in \Gamma(D \oplus_{\mathrm{orth}} V)$ which completes the proof. $\square$

Proposition 7.4.4. *Let M be a contact CR-lightlike submanifold of an indefinite Sasakian manifold $\bar{M}$. Then $D \oplus_{\mathrm{orth}} V$ is a totally geodesic foliation if and only if*

$$h^l(X, \phi Y) = 0, \quad \text{and} \quad h^s(X, Y) \quad \text{has no components in} \quad L_1. \tag{7.4.4}$$

Proof. By Definition 7.4.1, $D \oplus_{\mathrm{orth}} V$ is a totally geodesic foliation if and only if $g(\nabla_X Y, \phi \xi) = g(\nabla_X Y, W) = 0$ for $X, Y \in \Gamma(D \oplus_{\mathrm{orth}} V)$ and $W \in \Gamma(\phi L_1)$. From (5.1.15) we have $g(\nabla_X Y, \phi \xi) = -\bar{g}(\phi \bar{\nabla}_X Y, \xi)$. Using (7.1.7) and (5.1.15) we get

$$g(\nabla_X Y, \phi \xi) = -\bar{g}(h^l(X, \phi Y), \xi). \tag{7.4.5}$$

In a similar way, we derive

$$g(\nabla_X \phi Y, W) = -g(h^s(X, Y), \phi W). \tag{7.4.6}$$

Thus, from (7.4.5) and (7.4.6) we obtain (7.4.4) which completes the proof. $\square$

Proposition 7.4.5. *Let M be a contact CR-lightlike submanifold of an indefinite Sasakian manifold $\bar{M}$. Then, D' is a totally geodesic foliation if and only if $A_N Z$ has no components in $\phi L_1 \perp \phi(\mathrm{Rad}\,TM)$ and $A_{\phi W} Z$ has no components in $D_o \perp \mathrm{Rad}\,TM$ for $Z, W \in \Gamma(D')$.*

Proof. Note that D' defines a totally geodesic foliation if and only if

$$\bar{g}(\nabla_Z W, N) = g(\nabla_Z W, \phi N) = g(\nabla_Z W, X) = g(\nabla_Z W, V) = 0,$$

$Z, W \in \Gamma(D')$, $N \in \Gamma(\mathrm{ltr}(TM))$ and $X \in \Gamma(D_o)$. From (7.1.8) and (5.1.21) we get

$$g(\nabla_Z W, V) = 0. \tag{7.4.7}$$

On the other hand, $\bar{\nabla}$ is a metric connection and (5.1.21) implies

$$\bar{g}(\nabla_Z W, N) = g(W, A_N Z). \qquad (7.4.8)$$

By using (7.1.8), (7.1.7), (5.1.28) and (7.4.1) we obtain

$$g(\nabla_Z W, \phi N) = g(A_{\phi W} Z, N). \qquad (7.4.9)$$

In a similar way we have

$$g(\nabla_Z W, \phi X) = g(A_{\phi W} Z, X). \qquad (7.4.10)$$

Thus the proof follows from(7.4.8)- (7.4.10). $\qquad\square$

We say that M is a *contact CR-lightlike product* if $D \oplus_{\text{orth}} V$ and D' define totally geodesic foliations in M. This concept is consistent with the classical definition of product manifolds.

Theorem 7.4.6. *Let M be an irrotational contact CR-lightlike submanifold of an indefinite Sasakian manifold $\bar{M}$. Then, M is a contact CR-lightlike product if the following conditions are satisfied:*

(1) $\bar{\nabla}_X U \in \Gamma(S(TM^\perp)), \quad \forall X \in \Gamma(TM)$ *and* $U \in \Gamma(\text{tr}(TM))$.

(2) $A_\xi^* Y$ *has no components in* $D_o \oplus \phi(\text{ltr}(TM)), \quad Y \in \Gamma(D)$.

Proof. If (1) holds, then from (5.1.21) and (5.1.23) we have $A_N X = 0, \quad A_W X = 0$ and $D^l(X, W) = 0$, for $X \in \Gamma(TM)$, $W \in \Gamma(S(TM^\perp))$. These equations imply that D' defines a totally geodesic foliation. From (5.1.28) we get $\bar{g}(h^s(X, Y), W) = -\bar{g}(Y, D^l(X, W)) = 0$. Hence, $h^s(X, Y)$ has no components in L_1. Then, from (5.1.32) and M irrotational we have $\bar{g}(h^l(X, \phi Y), \xi) = -\bar{g}(\phi Y, A_\xi^* X)$ for $X \in \Gamma(TM)$ and $Y \in \Gamma(D)$. Hence, if (2) holds then $h^l(X, \phi Y) = 0$. Thus, using Propositions 7.4.4 and 7.4.5 we conclude that M is a contact CR-lightlike product, which completes the proof. $\qquad\square$

From Theorem 7.3.5, it follows that any totally umbilical lightlike submanifold, tangent to the structure vector field V, of an indefinite Sasakian manifold is always invariant and totally geodesic. This tells that there are no totally umbilical lightlike submanifolds of indefinite Sasakian manifolds other than invariant lightlike submanifolds. Hence, the usual definition of totally umbilical lightlike submanifolds does not work for submanifolds of indefinite Sasakian manifolds. Therefore, as in Section 1, we give the following definition.

Definition 7.4.7. If the second fundamental form h of a submanifold, tangent to the structure vector field V, of an indefinite Sasakian manifold $\bar{M}$ is of the form

$$h(X, Y) = [g(X, Y) - \eta(X)\eta(Y)]\alpha + \eta(X)h(Y, V) + \eta(Y)h(X, V), \qquad (7.4.11)$$

for any $X, Y \in \Gamma(TM)$, where α is a vector field transversal to M, then M is called totally contact umbilical and totally contact geodesic if $\alpha = 0$.

The above definition also holds for a lightlike submanifold M. For a totally contact umbilical M we have

$$h^l(X,Y) = [g(X,Y) - \eta(X)\eta(Y)]\alpha_L$$
$$+ \eta(X)h^l(Y,V) + \eta(Y)h^l(X,V), \tag{7.4.12}$$
$$h^s(X,Y) = [g(X,Y) - \eta(X)\eta(Y)]\alpha_S$$
$$+ \eta(X)h^s(Y,V) + \eta(Y)h^s(X,V), \tag{7.4.13}$$

where $\alpha_S \in \Gamma(S(TM^\perp))$ and $\alpha_L \in \Gamma(\text{ltr}(TM))$.

Lemma 7.4.8. *Let M be a totally contact umbilical proper contact CR-lightlike submanifold of an indefinite Sasakian manifold $\bar{M}$. Then $\alpha_L = 0$.*

Proof. Let M be a totally contact umbilical proper contact CR-lightlike submanifold. Then, by direct calculations, using (5.1.15),(5.1.23), (7.1.7) and taking the tangential parts we have

$$A_{\phi Z}Z + f\nabla_Z Z + \phi h^l(Z,Z) + B h^s(Z,Z) = g(Z,Z)V \tag{7.4.14}$$

for $Z \in \Gamma(\phi L_1)$. Hence, we obtain $\bar{g}(A_{\phi Z}Z, \phi\xi) + \bar{g}(h^l(Z,Z), \xi) = 0$. Using (5.1.28) we get $g(h^s(Z, \phi\xi), \phi Z) + \bar{g}(h^l(Z,Z), \xi) = 0$. Thus, from (7.4.12) and (7.4.13) we derive $-g(Z,Z)\bar{g}(\alpha_L, \xi) = 0$. Since ϕL_1 is non-degenerate we get $\alpha_L = 0$, which completes the proof. $\square$

Theorem 7.4.9. *Let M be a totally contact umbilical proper contact CR-lightlike submanifold of an indefinite Sasakian manifold $\bar{M}$. Then either M is totally contact geodesic or $\dim(\phi L_1) = 1$.*

Proof. Assume M proper totally contact umbilical. (7.1.7), (5.1.15), (7.4.3) and (7.4.4) imply $\omega\nabla_X X + Ch^s(X,X) = 0$ for $X \in \Gamma(D_o)$. Hence,

$$h^s(X,X) \in \Gamma(L_1). \tag{7.4.15}$$

From (7.4.14) and (5.1.28) we have $\bar{g}(h^s(Z,Z), \phi W) = \bar{g}(h^s(Z,W), \phi Z)$ for $Z, W \in \Gamma(\phi L_1)$. Since M is totally contact umbilical we obtain

$$g(Z,Z)\bar{g}(\alpha_S, \phi W) = g(Z,W)\bar{g}(\alpha_S, \phi Z).$$

Interchanging role Z and W and subtracting we derive

$$\bar{g}(\alpha_S, \phi Z) = \frac{g(Z,W)^2}{g(Z,Z)g(W,W)}\bar{g}(\alpha_S, \phi Z). \tag{7.4.16}$$

Considering (7.4.15) the equation (7.4.16) has solutions if either (a) $\dim(L_1) = 1$, or (b) $\alpha_S = 0$. Thus the proof follows from Lemma 7.4.8. $\square$

We know [45] that CR-submanifolds of Kähler manifolds were designed as a generalization of both invariant and totally real submanifolds. Therefore, it is important to know whether contact CR-lightlike submanifolds admit invariant submanifolds (discussed in section 3) and are there any real submanifolds. To investigate this we need the following definition:

Definition 7.4.10. [161] We say that a lightlike submanifold M, of an indefinite Sasakian manifold $\bar{M}$, is screen real submanifold if $\mathrm{Rad}(TM)$ and $S(TM)$ are, respectively, invariant and anti-invariant with respect to ϕ.

Proposition 7.4.11. *Contact CR-lightlike submanifolds are non-trivial.*

Proof. Suppose M is an invariant lightlike submanifold of an indefinite Sasakian manifold. Then we can easily see that radical distribution is invariant. Thus condition (A) of Definition 7.4.1 is not satisfied. Similarly, one can prove the screen real lightlike case is not possible.   □

7.5 Contact SCR-lightlike submanifolds

We know from Proposition 7.4.11 that contact CR-lightlike submanifolds exclude the invariant and the screen real subcases and, therefore, do not serve the central purpose of introducing a CR-structure. To include these two subcases, we introduce a new class, called a *contact screen Cauchy Riemann (SCR)-lightlike submanifold* as follows:

Definition 7.5.1. [161] Let M be a lightlike submanifold, tangent to the structure vector field V, of an indefinite Sasakian manifold $\bar{M}$. We say that M is a contact SCR-lightlike submanifold of $\bar{M}$ if the following holds:

1. There exist real non-null distributions D and $D^{\perp}$ such that

$$S(TM) = D \perp D^{\perp} \perp V, \; \phi(D^{\perp}) \subset (S(TM^{\perp})), \; D \cap D^{\perp} = \{0\}$$

   where $D^{\perp}$ is orthogonal complementary to $D \perp V$ in $S(TM)$.

2. The distributions D and $\mathrm{Rad}(TM)$ are invariant with respect to ϕ.

   It follows that $\mathrm{ltr}(TM)$ is also invariant with respect to ϕ. Hence we have

$$TM = \bar{D} \oplus_{\mathrm{orth}} D^{\perp} \perp V, \quad \bar{D} = D \oplus_{\mathrm{orth}} \mathrm{Rad}(TM). \tag{7.5.1}$$

Denote the orthogonal complement to $\phi(D^{\perp})$ in $S(TM^{\perp})$ by μ. We say that M is a proper contact SCR-lightlike submanifold of $\bar{M}$ if $D \neq \{0\}$ and $D^{\perp} \neq \{0\}$. Note the following features of a contact SCR-lightlike submanifold:

1. Condition 2. implies that $\dim(\mathrm{Rad}\,TM) = r = 2p \geq 2$.

2. For proper M, 2. implies $\dim(D) = 2s \geq 2$, $\dim(D^{\perp}) \geq 1$. Thus, $\dim(M) \geq 5$, $\dim(\bar{M}) \geq 9$.

For any $X \in \Gamma(TM)$ and any $W \in \Gamma(S(TM^\perp))$ we put

$$\phi X = P' X + F' X, \quad \phi W = B' W + C' W, \tag{7.5.2}$$

where $P' X \in \Gamma(\bar{D})$, $F' X \in \Gamma(\phi D^\perp)$, $B' W \in \Gamma(D^\perp)$ and $C' V \in \Gamma(\mu)$.

Example 3. Let M be a submanifold of $\mathbf{R}_2^9$, defined by

$$x^1 = x^2, \, y^1 = y^2, \, x^4 = \sqrt{1 - (y^4)^2}.$$

It is easy to see that a local frame of TM is given by

$$\begin{aligned}
&Z_1 = \partial x_1 + \partial x_2 + (y^1 + y^2)\partial z, \quad Z_2 = \partial y_1 + \partial y_2, \\
&Z_3 = \partial x_3 + y^3 \partial z, \quad Z_4 = \partial y_3, \\
&Z_5 = -y^4 \partial x_4 + x^4 \partial y_4 - (y^4)^2 \partial z, \quad V = 2\partial z.
\end{aligned}$$

Hence $\mathrm{Rad}\, TM = \mathrm{Span}\{Z_1, Z_2\}$ and $\phi_o(Z_1) = -Z_2$. Thus $\mathrm{Rad}\, TM$ is invariant with respect to ϕ_o. Also, $\phi_o(Z_3) = -Z_4$ implies $D = \mathrm{Span}\{Z_3, Z_4\}$. By direct calculations, we get $S(TM^\perp) = \mathrm{Span}\{W = x^4 \partial x_4 + y^4 \partial y_4 + x^4 y^4 \partial z\}$ such that $\phi_o(W) = -Z_5$ and $\mathrm{ltr}(TM)$ is spanned by

$$N_1 = 2(-\partial x_1 + \partial x_2 + (-y^1 + y^2)\partial z), \, N_2 = 2(-\partial y_1 + \partial y_2).$$

It follows that $\phi_o(N_2) = N_1$. Thus $\mathrm{ltr}(TM)$ is also invariant. Hence, M is a contact SCR-lightlike submanifold.

The following results can be easily proved by direct use of Definition 7.5.1.

(1) A contact SCR-lightlike submanifold of $\bar{M}$ is invariant (resp., screen real) if and only if $D^\perp = \{0\}$. (resp. $D = \{0\}$).

(2) Any contact SCR-coisotropic, isotropic and totally lightlike submanifold of $\bar{M}$ is an invariant lightlike submanifold. Consequently, there exist no proper contact SCR or screen real coisotropic or isotropic or totally lightlike submanifolds of $\bar{M}$.

Theorem 7.5.2. *Let M be a contact SCR-lightlike submanifold of an indefinite Sasakian manifold $\bar{M}$. Then the induced connection ∇ is a metric connection if and only if the following hold:*

(1) *$h^s(X, \xi)$ has no components in $\phi(D^\perp)$,*

(2) *$A_\xi^* X$ has no components in D, $\forall X \in \Gamma(TM)$, $\xi \in \Gamma(\mathrm{Rad}\, TM)$.*

Proof. From (7.1.7) we obtain $\bar{\nabla}_X \phi\xi = \phi\bar{\nabla}_X \xi$. Then using (5.1.15), (5.1.27), (7.5.2), we get

$$\nabla_X \phi\xi = B' h^s(X, \xi) + \phi\nabla^{*t}_X \xi - P' A_\xi^* X. \tag{7.5.3}$$

We know that the induced connection is a metric connection if and only if $\mathrm{Rad}\, TM$ is parallel with respect to ∇. Suppose that $\mathrm{Rad}\, TM$ is parallel. Then, from (7.5.3) we have $B'h^s(X, \xi) = 0$ and $P'A_\xi^* X = 0$. Hence $h^s(X, \xi)$ has no components in $\phi(D^\perp)$ and $A_\xi^* X$ has no components in D. Conversely, assume that (1) and (2) are satisfied. Then, from (7.5.3) we get $\nabla_X \phi \xi \in \Gamma(\operatorname{Rad} TM)$. Thus $\operatorname{Rad} TM$ is parallel and ∇ is a metric connection, which completes the proof. $\qquad \square$

From (7.1.7), (5.1.15), (7.5.2) and taking the tangential (respectively, screen transversal, lightlike transversal) parts we have

$$(\nabla_X P')Y = A_{F'Y}X + B'h^s(X, Y) - g(X, Y)V + \eta(Y)X, \qquad (7.5.4)$$
$$(\nabla_X F')Y = C'h^s(X, Y) - h^s(X, P'Y), \qquad (7.5.5)$$
$$\phi h^l(X, Y) = h^l(X, P'Y) + D^l(X, F'Y), \quad \forall X, Y \in \Gamma(TM). \qquad (7.5.6)$$

The following results are similar to those proved in Propositions 7.4.3 and 7.4.4.

Proposition 7.5.3. *Let M be a contact SCR-lightlike submanifold of an indefinite Sasakian manifold $\bar{M}$. Then:*

(1) *The distribution $D^\perp$ is integrable if and only if*

$$A_{\phi X} Y = A_{\phi Y} X, \quad \forall X, Y \in \Gamma(D^\perp).$$

(2) *The distribution $\bar{D} \oplus_{\mathrm{orth}} V$ is integrable if and only if*

$$h^s(X, P'Y) = h^s(P'X, Y), \forall X, Y \in \Gamma(\bar{D}).$$

(3) *The distribution $\bar{D}$ is not integrable.*

Theorem 7.5.4. *Let M be a contact SCR-lightlike submanifold of an indefinite Sasakian manifold. Then, $\bar{D} \oplus_{\mathrm{orth}} V$ defines a totally geodesic foliation in M if and only if $h^s(X, \phi Y)$ has no components in $\phi(D^\perp)$, for $X, Y \in \Gamma(\bar{D} \oplus_{\mathrm{orth}} V)$.*

Proof. From (5.1.15) we have $g(\nabla_X Y, Z) = \bar{g}(\bar{\nabla}_X Y, Z)$ for $X, Y \in \Gamma(\bar{D} \oplus_{\mathrm{orth}} V)$ and $Z \in \Gamma(D^\perp)$. Using (7.1.7) we get $g(\nabla_X Y, Z) = \bar{g}(\bar{\nabla}_X \phi Y, \phi Z)$. Hence we derive $g(\nabla_X Y, Z) = \bar{g}(h^s(X, \phi Y), \phi Z)$ which proves our assertion. $\qquad \square$

Theorem 7.5.5. *Let M be a contact SCR-lightlike submanifold of an indefinite Sasakian manifold $\bar{M}$. Then the following assertions are equivalent:*

(i) *$D^\perp$ defines a totally geodesic foliation on M.*

(ii) *$A_{\phi X} Y$ has no components in $\bar{D}$.*

(iii) *$B'h^s(X, \phi Z) = 0$ and $B'D^s(X, \phi N) = 0$,*

$$\forall X, Y \in \Gamma(D^\perp), \quad Z \in \Gamma(D) \quad and \quad N \in \Gamma(\operatorname{ltr}(TM)).$$

Proof. (i) $\Rightarrow$ (ii): Suppose $D^\perp$ defines a totally geodesic foliation in M. Then, $\nabla_X Y \in \Gamma(D^\perp)$. From (5.1.15) and (7.1.7) we have $g(\nabla_X Y, Z) = \bar{g}(\bar{\nabla}_X \phi Y, \phi Z)$ for $X, Y \in \Gamma(D^\perp)$ and $Z \in \Gamma(D)$. Using (5.1.23) we obtain

$$g(\nabla_X Y, Z) = -g(A_{\phi Y} X, \phi Z). \tag{7.5.7}$$

In a similar way we get

$$g(\nabla_X Y, N) = -g(A_{\phi Y} X, \phi N) \tag{7.5.8}$$

for $N \in \Gamma(\text{ltr}(TM))$ Thus (i)$\Rightarrow$ (ii) follows from (7.5.7) and (7.5.8).

(ii) $\Rightarrow$ (iii) follows from (7.5.7), (7.5.8), (5.1.28) and (5.1.21).

(iii) $\Rightarrow$ (i): By definition of a contact SCR-lightlike submanifold, $D^\perp$ defines a totally geodesic foliation in M if and only if $g(\nabla_X Y, Z) = g(\nabla_X Y, V) = \bar{g}(\nabla_X Y, N) = 0$ for $X, Y \in \Gamma(D^\perp)$, $Z \in \Gamma(\tilde{D})$ and $N \in \Gamma(\text{ltr}(TM))$. From (5.1.15) and (7.1.8) we obtain $g(\nabla_X Y, V) = 0$. Applying a similar method to (7.5.7) and (7.5.8) we get $g(\nabla_X Y, Z) = \bar{g}(h^s(X, \phi Z), \phi Y)$ and $g(\nabla_X Y, N) = -g(D^s(X, \phi N), \phi Y)$. By assumption, $B'h^s(X, \phi Z) = 0$ and $B'D^s(X, \phi N) = 0$. Hence we obtain $g(\nabla_X Y, Z) = 0$ and $g(\nabla_X Y, N) = 0$ which proves the assertion. $\square$

Lemma 7.5.6. *Let M be a contact SCR-lightlike submanifold of an indefinite Sasakian manifold $\bar{M}$. Then we have*

$$h^l(X, V) = 0, \quad \forall X \in \Gamma(TM), \tag{7.5.9}$$

$$\nabla_X V = -\phi X, \quad h^s(X, V) = 0, \quad \forall X \in \Gamma(\bar{D}), \tag{7.5.10}$$

$$\nabla_X V = 0, \quad h^s(X, V) = \phi X, \quad \forall X \in \Gamma(D^\perp). \tag{7.5.11}$$

Proof. Using (7.1.8) and (5.1.15) we get $\nabla_X V + h^l(X, V) + h^s(X, V) = \phi X$ for $X \in \Gamma(TM)$. Then, considering (7.5.1) we get (7.5.9)–(7.5.11). $\square$

Theorem 7.5.7. *Any totally contact umbilical proper contact SCR-lightlike submanifold M of $\bar{M}$ admits a metric connection.*

Proof. From (7.5.6) we obtain $h^l(X, \phi Y) = h^l(\phi Y, X)$, $\forall X, Y \in \Gamma(\tilde{D})$. Using this and (7.4.12) we get $g(X, \phi Y)\alpha_L = g(\phi X, Y)\alpha_L$. Thus, $g(X, \phi Y)\alpha_L = 0$, since D is non-degenerate and $\alpha_L = 0$. Thus, $h^l(X, Y) = \eta(X)h^l(Y, V) + \eta(Y)h^l(X, V)$ for $X, Y \in \Gamma(TM)$. If $X, Y \in \Gamma(D \perp \bar{D})$, from Lemma 7.5.6, then, we obtain $h^l(X, Y) = 0$. If $X \in \Gamma(TM)$ and $Y = V$, then, from (7.5.9) we get $h^l(X, V) = 0$. Thus $h^l = 0$ on M. The assertion follows from (5.1.39). $\square$

Theorem 7.5.8. *Let M be a totally contact umbilical contact SCR-lightlike submanifold of $\bar{M}$. If $\dim(D^\perp) > 1$, then M is contact totally geodesic.*

Proof. The proof is similar to the proof of Theorem 7.4.9. $\square$

Theorem 7.5.9. *There exist no totally contact umbilical proper contact **SCR**-lightlike submanifolds in an indefinite Sasakian form $\bar{M}(c)$ with $c \neq -3$.*

Proof. Suppose M is a totally contact umbilical proper SCR-lightlike submanifold of $\bar{M}$. From Gauss' equation (5.3.11) and (7.1.5) we get

$$\frac{1}{2}(1-c)g(X,X)g(Z,Z) = \bar{g}((\nabla_X h^s)(\phi X, Z), \phi Z)$$
$$-\bar{g}((\nabla_{\phi X} h^s)(X, Z), \phi Z) \qquad (7.5.12)$$

for any $X \in \Gamma(D)$ and $Z \in \Gamma(D^\perp)$, where $(\nabla_X h^s)(\phi X, Z) = \nabla^s_X h^s(\phi X, Z) - h^s(\nabla_X \phi X, Z) - h^s(\phi X, \nabla_X Z)$. Since M is totally contact umbilical we have that $h^s(\phi X, Z) = 0$ and, from (7.5.11) and (7.5.12) we get

$$-h^s(\nabla_X \phi X, Z) = -g(\nabla_X \phi X, Z)\alpha_S - g(\nabla_X \phi X, V)\phi Z.$$

Using (5.1.15) and (7.1.8) we obtain

$$-h^s(\nabla_X \phi X, Z) = -g(\nabla_X \phi X, Z)\alpha_S + g(X,X)\phi Z. \qquad (7.5.13)$$

In a similar way we get

$$-h^s(\phi X, \nabla_X Z) = -g(\phi X, \nabla_X Z)\alpha_S. \qquad (7.5.14)$$

Thus from (7.5.13) and (7.5.14) we have

$$(\nabla_X h^s)(\phi X, Z) = -g(\nabla_X \phi X, Z)\alpha_S + g(X,X)\phi Z - g(\phi X, \nabla_X Z)\alpha_S.$$

On the other hand, since $\bar{g}(\phi X, Z) = 0$, taking the covariant derivative with respect to X, we obtain $g(\nabla_X \phi X, Z) = -g(\phi X, \nabla_X Z)$. Hence we get

$$(\nabla_X h^s)(\phi X, Z) = g(X,X)\phi Z. \qquad (7.5.15)$$

In a similar way we have

$$(\nabla_{\phi X} h^s)(X, Z) = -g(X,X)\phi Z. \qquad (7.5.16)$$

Thus from (7.5.15),(7.5.16) and (7.5.12) we obtain

$$\frac{1}{2}(1-c)g(X,X)g(Z,Z) = 2g(X,X)g(Z,Z).$$

Hence, $(3+c)g(X,X)g(Z,Z) = 0$. Since D and $D^\perp$ are non-degenerate, we can choose non-null X and Z, so $c = -3$ which proves the theorem. $\qquad \square$

Minimal lightlike submanifolds. First of all, Theorem 7.3.5 implies that any totally umbilical lightlike submanifold, with structure vector field tangent to a submanifold, is minimal. Furthermore, from Theorems 7.4.9 and 7.5.8 it follows that a totally contact umbilical contact CR-lightlike submanifold with $(\dim(\phi L_1)) > 1)$ and a totally contact umbilical contact SCR-lightlike submanifolds with $(\dim(D^\perp) > 1)$ are minimal.

Example 4. Let $\bar{M} = (\mathbf{R}_4^{11}, \bar{g})$ be a semi-Euclidean space where $\bar{g}$ is of signature $(-,-,+,+,+,-,-,+,+,+,+)$ with respect to the canonical basis

$$\{\partial x_1, \partial x_2, \partial x_3, \partial x_4, \partial x_5, \partial y_1, \partial y_2, \partial y_3, \partial y_4, \partial y_5, \partial z\}.$$

Suppose M is a submanifold of $\mathbf{R}_4^{11}$ given by

$$x^1 = u^1, \qquad\qquad\qquad y^1 = -u^5$$
$$x^2 = \cosh u^2 \cosh u^3, \qquad y^2 = \cosh u^2 \sinh u^3$$
$$x^3 = \sinh u^2 \cosh u^3, \qquad y^3 = \sinh u^2 \sinh u^3$$
$$x^4 = u^4, \qquad\qquad\qquad y^4 = -u^6$$
$$x^5 = u^1 \cos\theta + u^5 \sin\theta, \qquad y^5 = u^1 \sin\theta - u^5 \cos\theta, \ \ z = u^7.$$

Then it is easy to see that a local frame of TM is given by

$$Z_1 = \partial x_1 + \cos\theta\, \partial x_5 + \sin\theta \partial y_5 + (y^1 + y^5 \cos\theta)\partial z$$
$$Z_2 = \sin\theta \partial x_5 - \partial y_1 - \cos\theta \partial y_5 + y^5 \sin\theta \partial z$$
$$Z_3 = \sinh u^2 \cosh u^3\, \partial x_2 + \cosh u^2 \cosh u^3\, \partial x_3 + \sinh u^2 \sinh u^3\, \partial y_2$$
$$\qquad + \cosh u^2 \sinh u^3\, \partial y_3 + (y^2 \sinh u^2 \cosh u^3 + y^3 \cosh u^2 \cosh u^3)\partial z$$
$$Z_4 = \cosh u^2 \sinh u^3\, \partial x_2 + \sinh u^2 \sinh u^3\, \partial x_3 + \cosh u^2 \cosh u^3\, \partial y_2$$
$$\qquad + \sinh u^2 \cosh u^3\, \partial y_3 + (y^2 \sinh u^3 \cosh u^2 + y^3 \sinh u^2 \sinh u^3)\partial z$$
$$Z_5 = \partial x_4 + y^4 \partial z, \ Z_6 = -\partial y^4, \ Z_7 = 2\partial z.$$

We see that M is a 2-lightlike submanifold with $\operatorname{Rad} TM = \operatorname{Span}\{Z_1, Z_2\}$ and $\phi_o Z_1 = Z_2$. Thus $\operatorname{Rad} TM$ is invariant with respect to ϕ_o. Since $\phi_o(Z_5) = Z_6$, $D = \{Z_5, Z_6\}$ is also invariant. Moreover, since $\phi_o Z_3$ and $\phi_o Z_4$ are perpendicular to TM and they are non-null, we can choose $S(TM^\perp) = \operatorname{Span}\{\phi_o Z_3, \phi_o Z_4\}$. Furthermore, the lightlike transversal bundle $\operatorname{ltr}(TM)$ is spanned by

$$N_1 = 2(-\partial x_1 + \cos\theta\, \partial x_5 + \sin\theta \partial y_5 + (-y^1 + y^5 \cos\theta)\partial z)$$
$$N_2 = 2(\sin\theta \partial x_5 + \partial y_1 - \cos\theta \partial y_5 + y^5 \sin\theta \partial z)$$

which is also invariant. Thus we conclude that M is a contact SCR-lightlike submanifold of $\mathbf{R}_4^{11}$. A quasi orthonormal basis of $\bar{M}$ along M is given by

$$\xi_1 = Z_1, \quad \xi_2 = Z_2, \quad e_1 = \frac{2}{\sqrt{\cosh^2 u^3 + \sinh^2 u^3}} Z_3,$$

$$e_2 = \frac{2}{\sqrt{\cosh^2 u^3 + \sinh^2 u^3}} Z_4, \quad e_3 = 2Z_5, \quad e_4 = 2Z_6, \quad Z,$$

$$W_1 = \frac{2}{\sqrt{\cosh^2 u^3 + \sinh^2 u^3}} \phi_o Z_3, \quad W_2 = \frac{2}{\sqrt{\cosh^2 u^3 + \sinh^2 u^3}} \phi_o Z_4,$$

$$N_1, \quad N_2,$$

where $\varepsilon_1 = g(e_1, e_1) = 1$, $\varepsilon_2 = g(e_2, e_2) = -1$ and g is the degenerate metric on M. By direct calculations and using Gauss' formula (5.1.15) we get

$$h^s(X, \xi_1) = h^s(X, \xi_2) = h^s(X, e_3) = h^s(X, e_4) = 0, h^l = 0, \forall X \in \Gamma(TM)$$

and

$$h^s(e_1, e_1) = \frac{1}{\cosh^2 u^3 + \sinh^2 u^3} W_2, \quad h^s(e_2, e_2) = \frac{1}{\cosh^2 u^3 + \sinh^2 u^3} W_2.$$

Therefore,

$$\operatorname{tr} h_{g|S(TM)} = \varepsilon_1 h^s(e_1, e_1) + \varepsilon_2 h^s(e_2, e_2)$$
$$= h^s(e_1, e_1) - h^s(e_2, e_2)) = 0.$$

Thus M is a minimal contact SCR-lightlike submanifold of $\mathbf{R}_4^{11}$.

Now we prove characterization results for minimal lightlike submanifolds of all the cases discussed in previous sections of this chapter.

Theorem 7.5.10. *A contact SCR-lightlike submanifold M of an indefinite Sasakian manifold $\bar{M}$ is minimal if and only if*

$$\operatorname{tr} A_{W_j}|_{S(TM)} = 0, \ \operatorname{tr}|_{S(TM)} A^*_{\xi_k} = 0 \ on \ D^\perp \ and \ D^l(X, W) = 0$$

for $X \in \Gamma(\operatorname{Rad} TM)$ and $W \in \Gamma(S(TM^\perp))$.

Proof. Since $\bar{\nabla}_V V = -\phi V = 0$, from (5.1.15) we get $h^l(V, V) = h^s(V, V) = 0$. Take an orthonormal frame $\{e_1, \ldots, e_{m-r}\}$ such that $\{e_1, \ldots, e_{2a}\}$ are tangent to D and $\{e_{2a+1}, \ldots, e_{m-2r}\}$ are tangent to $D^\perp$. From Proposition 5.1.3 we know that $h^l = 0$ on $\operatorname{Rad}(TM)$. From (7.5.6), for $Y, Z \in \Gamma(D)$, we have

$$h^l(\phi Y, Z) = \phi h^l(Y, Z).$$

Hence, $h^l(\phi Z, \phi Y) = -h^l(Y, Z)$. Thus $\sum_{i=1}^{2a} h^l(e_i, e_i) = 0$. Since

$$\operatorname{tr} h|_{S(TM)} = \sum_{i=1}^{m-2r} \varepsilon_i(h^l(e_i, e_i) + h^s(e_i, e_i)),$$

M is minimal if and only if

$$\sum_{i=1}^{2a} \varepsilon_i h^s(e_i, e_i) + \sum_{2a+1}^{m-2r} \varepsilon_i(h^l(e_i, e_i) + h^s(e_i, e_i)) = 0.$$

On the other hand, we have

$$
\operatorname{tr} h \mid_{S(TM)} = \sum_{i=1}^{2a} \frac{1}{n-2r} \sum_{j=1}^{n-2r} \varepsilon_i \bar{g}(h^s(e_i, e_i), W_j) W_j
$$

$$
+ \frac{1}{n-2r} \sum_{j=1}^{n-2r} \sum_{i=2a+1}^{m-2r} \varepsilon_i \bar{g}(h^s(e_i, e_i), W_j) W_j
$$

$$
+ \sum_{k=1}^{2r} \frac{1}{2r} \sum_{i=2a+1}^{m-2r} \varepsilon_i \bar{g}(h^l(e_i, e_i), \xi_k) N_k.
$$

Using (5.1.28) and (5.1.32) we get

$$
\operatorname{tr} h \mid_{S(TM)} = \sum_{i=1}^{2a} \frac{1}{n-2r} \sum_{j=1}^{n-2r} \varepsilon_i \bar{g}(A_{W_j} e_i, e_i) W_j
$$

$$
+ \frac{1}{n-2r} \sum_{j=1}^{n-2r} \sum_{i=2a+1}^{m-2r} \varepsilon_i \bar{g}(A_{W_j} e_i, e_i) W_j
$$

$$
+ \sum_{k=1}^{2r} \frac{1}{2r} \sum_{i=2a+1}^{m-2r} \varepsilon_i \bar{g}(A_{\xi_k}^* e_i, e_i) N_k. \tag{7.5.17}
$$

On the other hand, from (5.1.28) we obtain

$$
\bar{g}(h^s(X, Y), W) = \bar{g}(Y, D^l(X, W)), \ \forall X, Y \in \Gamma(\operatorname{Rad} TM) \tag{7.5.18}
$$

and $\forall W \in \Gamma(S(TM^\perp))$. Thus our assertion follows from (7.5.18) and (7.5.19). $\square$

Theorem 7.5.11. *An irrotational screen real lightlike submanifold M of an indefinite Sasakian manifold $\bar{M}$ is minimal if and only if*

$$
\operatorname{tr} A_{W_a} = 0 \quad on \quad S(TM).
$$

Proof. From Proposition 5.1.3 we know that $h^l = 0$. Thus, M is minimal if and only if $h^s = 0$ on $\operatorname{Rad} TM$ and $\operatorname{tr} h^s \mid_{S(TM)} = 0$. Then, the proof follows from Theorem 7.5.10. $\square$

Theorem 7.5.12. *An irrotational contact CR-lightlike submanifold M of an indefinite Sasakian manifold $\bar{M}$ is minimal if and only if:*

(1) $A_\xi^* \phi\xi$ *and* $A_N \phi N$ *has no components in* D'

(2) $D^s(\phi N, N)$ *has no components in* $L_1^\perp$

(3) $\operatorname{tr} A_{W_a} \mid_{D_0 \perp \phi L_1} = 0$, $\operatorname{tr} A_{\xi_k}^* \mid_{D_0 \perp \phi L_1} = 0$

for $N \in \Gamma(\operatorname{ltr}(TM))$ and $\xi \in \Gamma(\operatorname{Rad} TM)$, where $D' = \phi(\operatorname{ltr}(TM)) \perp \phi(L_1)$.

Proof. Let M be an irrotational contact CR-lightlike submanifold of $\bar{M}$. From (5.1.15) and (7.1.7) we have $\bar{g}(h^l(\phi\xi, \phi\xi), \xi_1) = -\bar{g}(\bar{\nabla}_{\phi\xi}\xi, \phi\xi_1)$. Then using (5.1.27) and (5.1.15) we obtain

$$\bar{g}(h^l(\phi\xi, \phi\xi), \xi_1) = -g(A^*_\xi\phi\xi, \phi\xi_1), \quad \forall\xi, \xi_1 \in \Gamma(\operatorname{Rad}TM). \tag{7.5.19}$$

In a similar way, from (5.1.15), (7.1.7), (5.1.27) and (7.5.2) we get

$$\bar{g}(h^s(\phi\xi, \phi\xi), W) = g(A^*_\xi\phi\xi, BW), \quad \forall\xi \in \Gamma(\operatorname{Rad}TM) \tag{7.5.20}$$

and $W \in \Gamma(S(TM^\perp))$. Now, using (5.1.23),(5.1.15) and (7.5.2) we derive

$$h(\phi N, \phi N) = -\omega A_N \phi N + CD^s(\phi N, N), \quad \forall N \in \Gamma(\operatorname{ltr}(TM)). \tag{7.5.21}$$

Then the proof follows from (7.5.19)–(7.5.21) and Theorem 7.5.10. $\qquad\square$

7.6   Generalized contact CR-lightlike submanifolds

In this section we study a new class of lightlike submanifolds, namely, contact generalized Cauchy-Riemann (GCR)lightlike submanifolds of indefinite Sasakian manifolds. We show that this class contains contact CR and SCR-lightlike submanifolds. Therefore, it is an umbrella of all the subcases discussed so far.

Definition 7.6.1. [162] Let $(M, g, S(TM))$ be a real lightlike submanifold of an indefinite Sasakian manifold $(\bar{M}, \bar{g})$ such that V is tangent to M. Then, M is a *contact generalized Cauchy-Riemann (GCR)-lightlike submanifold* if the following conditions are satisfied:

(A) There exist two sub-bundles D_1 and D_2 of $\operatorname{Rad}TM$ such that

$$\operatorname{Rad}TM = D_1 \oplus D_2, \quad \phi(D_1) = D_1, \quad \phi(D_2) \subset S(TM).$$

(B) There exist two sub-bundles D_0 and $\bar{D}$ of $S(TM)$ such that

$$S(TM) = \{\phi(D_2) \oplus \bar{D}\} \perp D_0 \perp \{V\}, \quad \phi(\bar{D}) = \mathcal{L} \perp \mathcal{S} \tag{7.6.1}$$

where D_0 is an invariant non-degenerate distribution on M, $\{V\}$ is the 1-dimensional distribution spanned by V, $\mathcal{L}$ and $\mathcal{S}$ are vector sub-bundles of $\operatorname{ltr}(TM)$ and $S(TM^\perp)$ respectively .

The tangent bundle TM of M is decomposed as

$$TM = \{D \oplus \bar{D}\} \perp \{V\}, \quad D = \operatorname{Rad}TM \oplus_{\operatorname{orth}} D_0 \oplus_{\operatorname{orth}} \phi(D_2). \tag{7.6.2}$$

Proposition 7.6.2. *A contact GCR-lightlike submanifold M, of an indefinite Sasakian manifold $\bar{M}$, is contact CR (respectively, contact SCR-lightlike) if and only if $D_1 = \{0\}$ (respectively, $D_2 = \{0\}$).*

Proof. Let M be a contact CR-lightlike submanifold. Then $\phi \operatorname{Rad} TM$ is a distribution on M such that $\phi \operatorname{Rad} TM \cap \operatorname{Rad} TM = \{0\}$. Therefore, $D_2 = \operatorname{Rad} TM$ and $D_1 = \{0\}$. Hence, $\phi \operatorname{ltr}(TM) \cap \operatorname{ltr}(TM) = \{0\}$. Then it follows that $\phi \operatorname{ltr}(TM) \subset S(TM)$. Conversely, suppose M is a contact GCR-lightlike submanifold such that $D_1 = \{0\}$. Then, $D_2 = \operatorname{Rad} TM$. Hence, $\phi \operatorname{Rad} TM \cap \operatorname{Rad} TM = \{0\}$, that is, $\phi \operatorname{Rad} TM$ is a vector sub-bundle of $S(TM)$. Thus, M is contact CR-lightlike. The other assertion is similar. $\square$

From Example 1 of this chapter, it follows that every lightlike real hypersurface of an indefinite Sasakian manifold $\bar{M}$ is a contact CR-lightlike hypersurface. Also from section 5, it is known that contact SCR-lightlike submanifolds have invariant and screen real lightlike subcases. Thus, we conclude from Proposition 7.6.2 that this class of contact GCR-lightlike submanifolds is an umbrella of real hypersurfaces, invariant, screen real and contact CR-lightlike submanifolds.

Example 5. Consider a semi-Euclidean space $(\mathbf{R}_4^{13}, g)$, where g is of signature $(-,-,+,+,+,+,-,-,+,+,+,+,+)$ with respect to a canonical basis

$$\{\partial x_1, \partial x_2, \partial x_3, \partial x_4, \partial x_5, \partial x_6, \partial y_1, \partial y_2, \partial y_3, \partial y_4, \partial y_5, \partial y_6, \partial z\}.$$

Let M be a 9-dimensional submanifold of $\mathbf{R}_4^{13}$ given by

$$x^4 = x^1 \cos\theta - y^1 \sin\theta, \qquad\qquad y^4 = x^1 \sin\theta + y^1 \cos\theta$$
$$x^2 = y^3, \qquad\qquad x^5 = \sqrt{1 + (y^5)^2}.$$

Then, it is easy to see that $\{\xi_1, \xi_2, \xi_3, e_1, e_2, e_3, Z\}$ given by

$$\xi_1 = \partial x_1 + \cos\theta\, \partial x_4 + \sin\theta\, \partial y_4 + (y^1 + \cos\theta\, y^4)\partial z$$
$$\xi_2 = -\sin\theta\, \partial x_4 + \partial y_1 + \cos\theta\, \partial y_4 - \sin\theta\, y^4 \partial z$$
$$\xi_3 = \partial x_2 + \partial y_3 + y^2 \partial z$$
$$e_1 = \partial x_3 - \partial y_2 + y^3 \partial z, \quad e_2 = 2\partial x_6 + y^6 \partial z$$
$$e_3 = 2\partial y_6,\, e_4 = 2(y^5 \partial x_5 + x^5 \partial y_5 + y^5 \partial z)$$
$$e_5 = 2(\partial x_3 + \partial y_2 + y^3 \partial z), \quad Z = 2\partial z = V$$

is a local frame of TM. Hence M is 3-lightlike with $\operatorname{Rad} TM = \operatorname{Span}\{\xi_1, \xi_2, \xi_3\}$ and $\phi_o \xi_1 = -\xi_2$. Thus, $D_1 = \operatorname{Span}\{\xi_1, \xi_2\}$. On the other hand, $\phi_o \xi_3 = e_1 \in \Gamma(S(TM))$ implies that $D_2 = \operatorname{Span}\{\xi_3\}$. Moreover, $\phi_o e_2 = e_3$. Hence $D_0 = \operatorname{Span}\{e_2, e_3\}$. By direct calculations, we get $S(TM^\perp) = \operatorname{Span}\{W = 2(x^5 \partial x_5 - y^5 \partial y_5 + x^5 y^5 \partial z)\}$. Thus, $\phi_o W = -e_4$. Hence, $\mathcal{S} = S(TM^\perp)$. On the other hand, the lightlike transversal bundle $\operatorname{ltr}(TM)$ is spanned by

$$N_1 = 2(-\partial x_1 + \cos\theta\, \partial x_4 + \sin\theta\, \partial y_4 - y^1 \partial z),$$
$$N_2 = 2(-\sin\theta\, \partial x_4 - \partial y_1 + \cos\theta\, \partial y_4 - y^4 \sin\theta\, \partial z),$$
$$N_3 = 2\,(-\partial x_2 + \partial y_3 + y^2\, \partial z).$$

From this $\mathrm{Span}\{N_1, N_2\}$ is invariant with respect to ϕ_o and $\phi_o N_3 = e_5$. Hence, $\mathcal{L} = \mathrm{Span}\{N_3\}$ and $\bar{D} = \mathrm{Span}\{\phi_o N_3, \phi_o W\}$. Thus, M is a contact GCR-lightlike submanifold of $\mathbf{R}_4^{13}$.

In the following we prove an existence theorem for contact GCR-lightlike submanifolds in an indefinite Sasakian space form $\bar{M}(c)$.

Theorem 7.6.3. *Let $(M, g, S(TM))$ be a lightlike submanifold tangent to the structure vector field V in an indefinite Sasakian $\bar{M}(c)$ with $c \neq 1$ Then, M is a contact GCR-lightlike submanifold of $\bar{M}(c)$ if and only if:*

(a) *The maximal invariant subspaces of $T_p M$, $p \in M$ define a distribution*

$$D = D_1 \oplus_{\mathrm{orth}} D_2 \oplus_{\mathrm{orth}} \phi(D_2) \oplus_{\mathrm{orth}} D_o$$

where $\mathrm{Rad}\, TM = D_1 \oplus_{\mathrm{orth}} D_2$, D_o is a non-degenerate invariant distribution.

(b) *There exists a lightlike transversal vector bundle $\mathrm{ltr}(TM)$ such that*

$$\bar{g}(\bar{R}(X,Y)\xi, N) = 0, \ \forall X, Y \in \Gamma(D_o), \ \xi \in \Gamma(\mathrm{Rad}\, TM), N \in \Gamma(\mathrm{ltr}(TM)).$$

(c) *There exists a vector sub-bundle M_2 on M*

$$\bar{g}(\bar{R}(X,Y)W, W') = 0, \quad \forall W, W' \in \Gamma(M_2)$$

where M_2 is orthogonal to D and $\bar{R}$ is the curvature tensor of $\bar{M}(c)$.

Proof. Suppose that M is a contact GCR-lightlike submanifold of $\bar{M}(c)$, $c \neq 1$. Then $D = D_1 \oplus_{\mathrm{orth}} D_2 \oplus_{\mathrm{orth}} \phi(D_2) \oplus_{\mathrm{orth}} D_o$ is a maximal invariant subspace. Next from (7.1.5), for $X, Y \in \Gamma(D_o)$, $\xi \in \Gamma(D_2)$ and $N \in \Gamma(\mathrm{ltr}(TM))$ we have

$$\bar{g}(\bar{R}(X,Y)\xi, N) = \frac{-c+1}{2}\{g(\phi X, Y)\bar{g}(\phi\xi, N)\}.$$

Since $g(\phi X, Y) \neq 0$ and $\bar{g}(\phi\xi, N) = 0$, $\bar{g}(\bar{R}(X,Y)\xi, N) = 0$. Similarly, from (7.1.5), we get

$$\bar{g}(\bar{R}(X,Y)W, W') = \frac{-c+1}{2}\{g(\phi X, Y)\bar{g}(\phi W, W')\} = 0$$

for $X, Y \in \Gamma(D_o)$, $W, W' \in \Gamma(\phi(\mathcal{S}))$. Conversely, assume that (a), (b) and (c) are satisfied. Then (a) implies that $D = D_1 \oplus_{\mathrm{orth}} D_2 \oplus_{\mathrm{orth}} \phi(D_2) \oplus_{\mathrm{orth}} D_o$ is invariant. From (b) and (7.1.5) we get

$$\bar{g}(\phi\xi, N) = 0 \tag{7.6.3}$$

which implies $\phi\xi \in \Gamma(S(TM))$. Thus, some part of $\mathrm{Rad}\, TM$, say D_2, belongs to $S(TM)$ under the action of ϕ. (7.6.3) also implies $-\bar{g}(\xi, \phi N) = 0$. Hence, a part of $\mathrm{ltr}(TM)$, say $\mathcal{L}$, also belongs to $S(TM)$ under the action of ϕ. On the other hand, (c) and (7.1.5) imply $\bar{g}(\phi W, W') = 0$. Hence we obtain $\phi(M_2) \perp M_2$. Since M_2 is non-degenerate $\bar{g}(\phi W, \phi W') = g(W', W) \neq 0$. Also $\bar{g}(\phi\xi, W) = $

$-\bar{g}(\xi, \phi W) = 0$ implies that $\phi(M_2) \oplus_{\text{orth}} \operatorname{Rad} TM$. This also tells us that ϕM_2 does not belong to $\operatorname{ltr}(TM))$. On the other hand, invariant and non-degenerate D_o implies $g(\phi W, X) = 0$, for $X \in \Gamma(D_o)$. Thus, $M_2 \oplus_{\text{orth}} D_o$ and $\phi(M_2) \oplus_{\text{orth}} D_o$. Moreover, we know that the structure vector field V belongs to $S(TM)$. Then summing up the above results we conclude that

$$S(TM) = \{\phi(D_2) \oplus M_1\} \perp M_2 \perp D_o \perp \{V\}$$

where $\phi(M_1) \subset \operatorname{ltr}(TM)$. Thus, the proof is complete. $\qquad\square$

Let Q, P_0, P_1 and P_2 be the projection morphisms on $\operatorname{Rad} TM$, D_0, $\phi\mathcal{L} = M_1$ and $\phi\mathcal{S} = M_2$, respectively. Then we have

$$X = QX + P_0 X + P_1 X + P_2 X + \eta(X)V$$

for $X \in \Gamma(TM)$. On the other hand, for $X \in \Gamma(TM)$, we write

$$\phi X = TX + \omega X, \tag{7.6.4}$$

where TX and ωX are the tangential and transversal parts of ϕX. Applying ϕ to (7.6.4) we obtain

$$\phi X = TX + \omega P_1 X + \omega P_2 X \tag{7.6.5}$$

where $TX \in \Gamma(D)$, $\omega P_1 X \in \Gamma(\mathcal{L})$ and $\omega P_2 X \in \Gamma(\mathcal{S})$. Similarly,

$$\phi W = BW + CW, \quad W \in \Gamma(\operatorname{tr}(TM)) \tag{7.6.6}$$

where BW and CW are sections of TM and $\operatorname{tr}(TM)$, respectively. Differentiating (7.6.5), and using (5.1.15)–(5.1.23), (7.6.5) and (7.6.6) we get

$$\begin{aligned}(\nabla_X T)Y &= A_{\omega P_1 Y} X + A_{\omega P_2 Y} X + Bh^l(X,Y) + Bh^s(X,Y) \\ &\quad - g(X,Y)V + \eta(Y)X,\end{aligned} \tag{7.6.7}$$

$$\begin{aligned}h^l(X,TY) &= -\nabla_X \omega P_1 Y + \omega P_1 \nabla_X Y - D^l(X, \omega P_2 Y) \\ &\quad + Ch^l(X,Y),\end{aligned} \tag{7.6.8}$$

$$\begin{aligned}h^s(X,TY) &= -\nabla_X \omega P_2 Y + \omega P_2 \nabla_X Y - D^s(X, \omega P_1 Y) \\ &\quad + Ch^s(X,Y), \quad \forall X, Y \in \Gamma(TM).\end{aligned} \tag{7.6.9}$$

Lemma 7.6.4. *Let $(M, g, S(TM))$ be a contact GCR-lightlike submanifold of an indefinite Sasakian manifold $\bar{M}$. Then we have*

$$h^l(X,V) = 0, \quad \nabla_X V = \phi X, h^s(X,V) = 0, \forall X \in \Gamma(D), \tag{7.6.10}$$

$$\nabla_X V = 0, \quad h^s(X,V) = 0, h^l(X,V) = \phi X, \forall X \in \Gamma(M_1),$$

$$\nabla_X V = 0, \quad h^s(X,V) = -\phi X, h^l(X,V) = 0, \forall X \in \Gamma(M_2). \tag{7.6.11}$$

Proof. Using (7.1.8) and (5.1.15) we obtain

$$\nabla_X V + h^l(X, V) + h^s(X, V) = -\phi X$$

for $X \in \Gamma(TM)$. Then, considering (7.6.2) we get (7.6.10)–(7.6.11). $\qquad \square$

Proposition 7.6.5. *Let* $(M, g, S(TM))$ *be a proper contact GCR-lightlike submanifold of an indefinite Sasakian manifold* $\bar{M}$. *Then we have:*

(1) *The distribution* $\bar{D}$ *is integrable if and only if*

$$A_{\phi X} Y = A_{\phi X} Y, \; \forall X, Y \in \Gamma(\bar{D}).$$

(2) *The distribution* $D \oplus_{\mathrm{orth}} \{V\}$ *is integrable if and only if*

$$h(X, TY) = h(TX, Y), \; \forall X, Y \in \Gamma(D \oplus_{\mathrm{orth}} \{V\}).$$

(3) *The distribution* D *is not integrable.*

Proof. For any $X, Y \in \Gamma(\bar{D})$, from (7.6.7) we obtain

$$-T\nabla_X Y = A_{\omega P_1 Y} X + A_{\omega P_2 Y} X + Bh^l(X, Y) + Bh^s(X, Y) - g(X, Y)V.$$

Hence we have $T[X, Y] = A_{\omega P_1 X} Y - A_{\omega P_1 Y} X + A_{\omega P_2 X} Y - A_{\omega P_2 Y} X$, which proves assertion (1). From (7.6.8) and (7.6.9) we get

$$h(X, TY) = \omega P_1 \nabla_X Y + \omega P_2 \nabla_X Y + Ch(X, Y)$$

for $X, Y \in \Gamma(D \oplus_{\mathrm{orth}} \{V\})$. Hence we derive $h(X, TY) - h(Y, TX) = \omega P_1[X, Y] + \omega P_2[X, Y]$ which proves the assertion (2). Suppose that D is integrable. Then, we have $\bar{g}([X, Y], V) = 0$, for $X, Y \in \Gamma(D_0)$. On the other hand, by using metric connection $\bar{\nabla}$ and (7.1.8) we obtain $\bar{g}([X, Y], V) = -2g(\phi Y, X)$. Thus we have $g(\phi Y, X) = 0$. Since D_0 is non-degenerate, this is a contradiction. Hence D is not integrable, which completes the proof. $\qquad \square$

Theorem 7.6.6. *Let* $(M, g, S(TM))$ *be a contact GCR-lightlike submanifold of an indefinite Sasakian manifold* $\bar{M}$. *Then,* $D \oplus_{\mathrm{orth}} \{V\}$ *defines a totally geodesic foliation in* M *if and only if* $h^l(X, \phi Y)$ *has no components in* $\mathcal{L}$ *and* $h^s(X, \phi Y)$ *has no components in* $\mathcal{S}$ *for* $X, Y \in \Gamma(D)$.

Proof. From (7.6.10) we have $\nabla_X V \in \Gamma(D \oplus \{V\})$ for $X \in \Gamma(D)$. Thus $D \oplus_{\mathrm{orth}} \{V\}$ defines a totally geodesic foliation if and only if $g(\nabla_X Y, Z) = g(\nabla_X Y, \phi \xi) = 0$ for $X, Y \in \Gamma(D)$, $Z \in \Gamma(\phi(\mathcal{S}))$ and $\xi \in \Gamma(D_2)$. From (5.1.15) we have $g(\nabla_X Y, Z) = \bar{g}(\bar{\nabla}_X Y, Z)$. Using (7.1.7) we obtain that $g(\nabla_X Y, Z) = \bar{g}(\bar{\nabla}_X \phi Y, \phi Z)$. Then from (5.1.15) we get $g(\nabla_X Y, Z) = \bar{g}(h^s(X, \phi Y), \phi Z)$. This shows that $g(\nabla_X Y, Z) = 0$ if and only if $h^s(X, \phi Y)$ has no components in $\mathcal{S}$. In a similar way, we obtain, $g(\nabla_X Y, \phi \xi) = -\bar{g}(h^l(X, \phi Y), \xi)$ for $X, Y \in \Gamma(D)$ and $\xi \in \Gamma(D_2)$. Hence we see that $g(\nabla_X Y, \phi \xi) = 0$ if and only if $h^l(X, \phi Y)$ has no components in $\mathcal{L}$, which completes the proof. $\qquad \square$

Theorem 7.6.7. *Let* $(M, g, S(TM))$ *be a contact GCR-lightlike submanifold of an indefinite Sasakian manifold. Then,* $\bar{D}$ *defines a totally geodesic foliation in* M *if and only if:*

(i) $A_N Z$ *has no components in* $\phi S \perp \phi(D_2)$.

(ii) $A_{\phi W} Z$ *has no components in* $D_o \perp D_2$, $\forall Z, W \in \Gamma(\bar{D})$.

Proof. Note that $\bar{D}$ defines a totally geodesic foliation if and only if

$$\bar{g}(\nabla_Z W, N) = g(\nabla_Z W, \phi N') = g(\nabla_Z W, X) = g(\nabla_Z W, V) = 0$$

for $Z, W \in \Gamma(\bar{D})$, $N \in \Gamma(\mathrm{ltr}(TM))$, $N' \in \Gamma(\mathcal{L})$ and $X \in \Gamma(D_o)$. First, from (7.1.8) we obtain

$$g(\nabla_Z W, V) = 0. \tag{7.6.12}$$

On the other hand, $\bar{\nabla}$ is a metric connection and (5.1.15) and (5.1.22) imply

$$\bar{g}(\nabla_Z W, N) = g(W, A_N Z). \tag{7.6.13}$$

By using (7.1.8), (5.1.15), (5.1.28) and (7.6.1) we obtain

$$g(\nabla_Z W, \phi N') = g(A_{\phi W} Z, N'). \tag{7.6.14}$$

In a similar way we have

$$g(\nabla_Z W, \phi X) = g(A_{\phi W} Z, X). \tag{7.6.15}$$

Thus the proof follows from (7.6.12)- (7.6.15). $\qquad\square$

We say that M is a *contact GCR-lightlike product* if $D \oplus_{\mathrm{orth}} \{V\}$ and $\bar{D}$ define totally geodesic foliations in M.

Theorem 7.6.8. *A contact GCR-lightlike submanifold* $(M, g, S(TM))$ *of an indefinite Sasakian manifold* $\bar{M}$ *is a product manifold if and only if:*

(a) $Bh(X, Y) = 0$, *for* $X \in \Gamma(TM)$ *and* $Y \in \Gamma(D \oplus \{V\})$.

(b) $A_{\phi Z} W$ *has no components in* D *for* $Z, W \in \Gamma(\bar{D})$.

Proof. From (7.6.7), for $X, Y \in \Gamma(D \oplus \{V\})$ we have

$$\nabla_X TY = T\nabla_X Y + Bh(X, Y) - g(X, Y)V + \eta(Y)X.$$

Thus $\nabla_X TY \in \Gamma(D \oplus \{V\})$ if and only if $Bh(X, Y) = 0$. In a similar way, for $Z, W \in \Gamma(\bar{D})$, from (7.6.7) we get

$$T\nabla_Z W = A_{\phi W} Z + Bh(Z, W).$$

Hence, $\nabla_Z W \in \Gamma(\bar{D})$ if and only if $A_{\phi W} Z$ has no components in D. $\qquad\square$

We say that M is a *proper contact GCR-lightlike submanifold* if $D_2 \neq \{0\}$, $D_1 \neq \{0\}$, $D_0 \neq \{0\}$ and $\mathcal{S} \neq \{0\}$, which has the following features:

(1) The condition (A) implies that $\dim(\operatorname{Rad} TM) \geq 3$.

(2) The condition (B) implies that $\dim(D) \geq 2s \geq 6$ and $\dim(D_2) = \dim(\mathcal{L})$. Thus, $\dim(M) \geq 9$ and $\dim(\bar{M}) \geq 13$.

(3) Any proper 9-dimensional contact GCR-lightlike submanifold is 3-lightlike.

(4) (1) and contact distribution ($\eta = 0$) imply that $\operatorname{ind}(\bar{M}) \geq 4$.

The following result can be easily verified.

Proposition 7.6.9. *There exist no coisotropic, isotropic or totally lightlike proper contact GCR-lightlike submanifold of an indefinite Sasakian manifold.*

Theorem 7.6.10. *There does not exist an induced metric connection of a proper GCR-lightlike submanifold of an indefinite Sasakian manifold $\bar{M}$.*

Proof. Let us suppose that ∇ is a metric connection. Then from Theorem 5.1.4, the radical distribution is parallel with respect to ∇, i.e, $\nabla_X \xi \in \Gamma(\operatorname{Rad} TM)$ for $X \in \Gamma(TM)$ and $\xi \in \Gamma(\operatorname{Rad} TM)$. Now, from (7.1.7) we get

$$\bar{\nabla}_X \phi\xi = \phi\bar{\nabla}_X \xi$$

for $X \in \Gamma(TM)$ and $\xi \in \Gamma(\operatorname{Rad} TM)$. Using contact structure we get

$$\phi\bar{\nabla}_X \phi\xi = -\bar{\nabla}_X \xi - \bar{g}(\xi, \bar{\nabla}_X V)V.$$

Then from (7.1.8) we obtain

$$\phi\bar{\nabla}_X \phi\xi = -\bar{\nabla}_X \xi + \bar{g}(\xi, \phi X)V.$$

Now, choose $X \in \Gamma(\phi\mathcal{L})$ and $\xi \in \Gamma(D_2)$ such that $g(\phi X, \xi) \neq 0$ (since $D_2 \oplus \mathcal{L}$ is a non-degenerate distribution on M, so we can always choose such vector fields). Thus from (5.1.15), (5.1.26), (7.6.5) and (7.6.6) we get

$$-\nabla_X \xi - h(X, \xi) + g(\xi, \phi X)V = \phi\nabla_X^* \phi\xi + \phi h^*(X, \phi\xi) + Bh(X, \phi\xi) + Ch(X, \phi\xi).$$

for $X \in \Gamma(\phi\mathcal{L})$ and $\xi \in \Gamma(D_2)$. The tangential parts from the above equation give

$$T\nabla_X^* \phi\xi + \nabla_X \xi + \phi h^*(X, \phi\xi) + Bh(X, \phi\xi) = g(\xi, \phi X)V.$$

Since $\operatorname{Rad} TM$ is parallel, $\nabla_X \xi \in \Gamma(\operatorname{Rad} TM)$. On the other hand, $T\nabla_X^* \phi\xi + \phi h^*(X, \phi\xi) \in \Gamma(\operatorname{Rad} TM \oplus_{\text{orth}} \phi D_2 \oplus_{\text{orth}} D_0)$ and $Bh(X, \phi\xi) \in \Gamma(\bar{D})$, thus we obtain $\bar{g}(\xi, \phi X)V = 0$. Since $V \neq 0$ and $g(\phi X, \xi) \neq 0$ we have a contradiction so $\operatorname{Rad} TM$ is not parallel and, therefore, ∇ is not a metric connection. $\qquad\square$

Remark 7.6.11. The above theorem shows that the geometry of proper GCR-lightlike submanifolds of indefinite Sasakian manifolds is different from the geometry of contact SCR-lightlike submanifolds and GCR-lightlike submanifolds of indefinite Kähler manifolds, since for this general case the induced connection ∇ on M is never a metric connection. In this respect, these proper GCR-lightlike submanifolds behave similar to the lightlike submanifolds of a semi-Riemannian manifold for which also, in general, there does not exist an induced metric connection.

The proof of the following theorem is similar to the proof of Theorem 7.5.9.

Theorem 7.6.12. *There exists no totally contact umbilical proper contact GCR-lightlike submanifold of an indefinite Sasakian $\bar{M}(c)$ with $c \neq -3$*

Lemma 7.6.13. *Let $(M, g, S(TM))$ be a totally contact umbilical proper contact GCR-lightlike submanifold of an indefinite Sasakian manifold $\bar{M}$. If $D \oplus_{\mathrm{orth}} \{V\}$ is integrable, then, M is totally contact geodesic.*

Proof. Suppose that $D \oplus_{\mathrm{orth}} \{V\}$ is integrable. Then from Theorem 7.6.6 (2) we have $h(X, \phi Y) = h(\phi X, Y)$ for $X, Y \in \Gamma(D_0)$. Thus from (7.4.11) we have $g(X, \phi Y)\alpha = g(\phi X, Y)\alpha$. Hence we have $2g(X, \phi Y)\alpha = 0$. Since D_0 is non-degenerate, we get $\alpha = 0$ which proves the assertion. $\qquad \square$

Lemma 7.6.14. *Let $(M, g, S(TM))$ be a totally contact umbilical proper contact GCR-lightlike submanifold of an indefinite Sasakian manifold $\bar{M}$. Then $\alpha_S \in \Gamma(\mathcal{S})$ and $\alpha_L \in \Gamma(\mathcal{L})$.*

Proof. From (7.6.8) and (7.6.9) we have

$$h^s(X, \phi X) = \omega P_2 \nabla_X X + Ch^s(X, X),$$

$$h^l(X, \phi X) = \omega P_1 \nabla_X X + Ch^l(X, X), \quad \forall X \in \Gamma(D_0).$$

Since M is totally contact umbilical, from (7.6.10), (7.4.12) and (7.4.13) we get

$$\omega P_2 \nabla_X X + g(X, X)C\alpha_S = 0,$$
$$\omega P_1 \nabla_X X + g(X, X)C\alpha_L = 0.$$

Since D_0 is non-degenerate, we derive $C\alpha_S = 0$ and $C\alpha_L = 0$. $\qquad \square$

Theorem 7.6.15. *Let $(M, g, S(TM))$ be a totally contact umbilical proper contact GCR-lightlike submanifold of an indefinite Sasakian manifold $\bar{M}$. Then one of the following holds:*

(i) *The distribution D is totally lightlike,*

(ii) *M is totally contact geodesic,*

(iii) *$\alpha_S = 0$ or $\dim(\mathcal{S}) = 1$ and $D \oplus_{\mathrm{orth}} \{V\}$ is not integrable.*

Proof. If $D_0 = 0$ then D is totally lightlike, which is (i). If $D_0 \neq \{0\}$ and $D \oplus_{\text{orth}}$ $\{V\}$ is integrable, then Lemma 7.6.13 shows that M is totally contact geodesic. If $D \oplus \{V\}$ is not integrable, then from (7.6.7) we have

$$-g(Bh^s(Z, Z), W) = g(A_{\phi Z} Z, W)$$

for $Z, W \in \Gamma(\phi S)$. Thus from (5.1.28) we get

$$g(Bh^s(Z, Z), W) = -\bar{g}(h^s(Z, W), \phi Z).$$

Hence we have

$$\bar{g}(h^s(Z, Z), \phi Z) = \bar{g}(h^s(Z, W), \phi Z).$$

Then using (7.4.13) we get

$$g(Z, W)\bar{g}(\alpha_S, \phi Z) = g(Z, Z)\bar{g}(\alpha_S, \phi W). \tag{7.6.16}$$

Interchanging the roles of Z and W in (7.6.16) we obtain

$$g(Z, W)\bar{g}(\alpha_S, \phi W) = g(W, W)\bar{g}(\alpha_S, \phi Z). \tag{7.6.17}$$

Thus from (7.6.16) and (7.6.17) we derive

$$\bar{g}(\alpha_S, \phi W) = \frac{g(Z, W)^2}{g(W, W)g(Z, Z)}\bar{g}(\alpha_S, \phi W). \tag{7.6.18}$$

Then, since ϕS is non-degenerate, from lemma 7.6.14 and (7.6.18) we conclude that either $\alpha_S = 0$ or Z and W are linearly dependent, which completes the proof. $\qquad \square$

Example 6. Consider a semi-Euclidean space $(\mathbf{R}_4^{15}, g)$, where g is of signature $(-, -, +, +, +, +, +, -, -, +, +, +, +, +, +)$ with respect to a canonical basis

$$\{\partial x_1, \partial x_2, \partial x_3, \partial x_4, \partial x_5, \partial x_6, \partial x_7, \partial y_1, \partial y_2, \partial y_3, \partial y_4, \partial y_5, \partial y_6, \partial y_7, \partial z\}.$$

Let M be a submanifold of $\mathbf{R}_4^{15}$ given by

$$x^1 = u^1 \cosh \alpha, \ x^2 = u^3, \ x^3 = u^1 \sinh \alpha + u^2, \ x^4 = u^3,$$
$$x^5 = \cos u^4 \cosh u^5, \ x^6 = \cos u^6 \cosh u^7, \ x^7 = \sin u^6 \cosh u^7,$$
$$y^1 = -u^2 \cosh \alpha, \ y^2 = u^8, \ y^3 = -u^2 \sinh \alpha + u^1, \ y^4 = u^9,$$
$$y^5 = \sin u^4 \sinh u^5, \ y^6 = \cos u^6 \sinh u^7, \ y^7 = \sin u^6 \sinh u^7, \ z = u^{10}.$$

Then it is easy to see $\{Z_1, \ldots, Z_8, Z\}$ given by

$$Z_1 = \cosh\alpha\,\partial\,x_1 + \sinh\alpha\,\partial\,x_3 + \partial\,y_3 + (y^1\cosh\alpha + y^3\sinh\alpha)\partial\,z,$$

$$Z_2 = \partial\,x_3 - \cosh\alpha\,\partial\,y_1 - \sinh\alpha\,\partial\,y_3 + y^3\,\partial\,z,$$

$$Z_3 = \partial\,x_2 + \partial\,x_4 + (y^2 + y^4)\partial\,z,$$

$$Z_4 = -\sin u^4\cosh u^5\,\partial\,x_5 + \cos u^4\sinh u^5\,\partial\,y_5 + (-y^5\sin u^4\cosh u^5)\partial\,z,$$

$$Z_5 = \cos u^4\sinh u^5\,\partial\,x_5 + \sin u^4\cosh u^5\,\partial\,y_5 + (y^5\cos u^4\sinh u^5)\partial\,z,$$

$$Z_6 = -\sin u^6\cosh u^7\partial\,x_6 + \cos u^6\cosh u^7\,\partial\,x_7 - \sin u^6\sinh u^7\partial\,y_6$$
$$+\cos u^6\sinh u^7\,\partial\,y_7 + (-y^6\sin u^6\cosh u^7 + y^7\cos u^6\cosh u^7)\partial\,z,$$

$$Z_7 = \cos u^6\sinh u^7\partial\,x_6 + \sin u^6\sinh u^7\,\partial\,x_7 + \cos u^6\cosh u^7\partial\,y_6$$
$$+\sin u^6\cosh u^7\,\partial\,y_7 + (y^6\cos u^6\sinh u^7 + y^7\sin u^6\sinh u^7)\partial\,z,$$

$$Z_8 = \partial\,y_2, \ Z_9 = \partial\,y_4, \qquad Z = 2\partial\,z = V$$

is a local frame of TM. Hence we see that M is 3-lightlike with $\mathrm{Rad}\,TM = \mathrm{Span}\{Z_1, Z_2, Z_3\}$ and $\phi_o(Z_1) = Z_2$, thus $D_1 = \mathrm{Span}\{Z_1, Z_2\}$. Moreover, $\phi_o(Z_3) = -Z_8 - Z_9 \in \Gamma(S(TM))$, hence $D_2 = \mathrm{Span}\{Z_3\}$. On the other hand, $\phi_o Z_4 = Z_5$, thus $D_o = \mathrm{Span}\{Z_4, Z_5\}$, which is invariant. It is easy to see $\phi_o Z_6$ and $\phi_o Z_7$ are orthogonal to M and $\{\phi_o Z_6, \phi_o Z_7\}$ is not degenerate. The lightlike transversal bundle $\mathrm{ltr}(TM)$ is spanned by

$$N_1 = 2(-\cosh\alpha\partial\,x_1 - \sinh\alpha\partial\,x_3 + \partial\,y_3 + (y^1\cosh\alpha + y^3\sinh\alpha)\partial\,z),$$

$$N_2 = 2(\partial\,x_3 + \cosh\alpha\partial\,y_1 + \sinh\alpha\partial\,y_3 + y^3\,\partial\,z),$$

$$N_3 = 2(-\partial\,x_2 + \partial\,x_4) + (y^2 - y^4)\partial\,z.$$

Hence it is easy to see that $\phi_o(N_1) = N_2$ and $\phi_o N_3 = Z_8 - Z_9 \in \Gamma(S(TM))$. Thus $\bar{D} = \mathrm{Span}\{\phi_o(Z_6), \phi_o(Z_7), \phi_o(N_3)\}$. Therefore, M is a proper contact GCR-lightlike submanifold of $\mathbf{R}_4^{15}$, with a quasi-orthonormal basis of $\bar{M}$ along M is

$$\xi_1 = \cosh\alpha\partial\,x_1 + \sinh\alpha\partial\,x_3 + \partial\,y_3 + (y^1\cosh\alpha + y^3\sinh\alpha)\partial\,z$$

$$\xi_2 = \partial\,x_3 - \cosh\alpha\partial\,y_1 + \sinh\alpha\partial\,y_3 + y^3\,\partial\,z$$

$$\xi_3 = \partial\,x_2 + \partial\,x_4 + (y^2 + y^4)\partial\,z$$

$$\phi_o(\xi_3) = -\partial\,y_2 - \partial\,y_4, \ Z = V = 2\partial\,z, \ \frac{1}{2}\phi_o N_3 = \partial\,y_2 - \partial\,y_4$$

$$e_1 = \frac{2}{\sqrt{\cosh^2 u^5 - \cos^2 u^4}}\{-\sin u^4\cosh u^5\,\partial\,x_5 + \cos u^4\sinh u^5\,\partial\,y_5$$
$$+ (-y^5\sin u^4\cosh u^5)\partial\,z\}$$

$$e_2 = \frac{2}{\sqrt{\cosh^2 u^5 - \cos^2 u^4}}\{\cos u^4\sinh u^5\,\partial\,x_5 + \sin u^4\cosh u^5\,\partial\,y_5$$

$$+ (y^5 \cos u^4 \sinh u^5) \partial z\}$$

$$X_1 = \frac{2}{\sqrt{\cosh^2 u^7 + \sinh^2 u^7}} \{- \sin u^6 \cosh u^7 \partial x_6 + \cos u^6 \cosh u^7 \partial x_7$$

$$- \sin u^6 \sinh u^7 \partial y_6 + \cos u^6 \sinh u^7 \partial y_7$$

$$+ (-y^6 \sin u^6 \cosh u^7 + y^7 \cos u^6 \cosh u^7) \partial z\}$$

$$X_2 = \frac{2}{\sqrt{\cosh^2 u^7 + \sinh^2 u^7}} \{\cos u^6 \sinh u^7 \partial x_6 + \sin u^6 \sinh u^7 \partial x_7$$

$$+ \cos u^6 \cosh u^7 \partial y_6 + \sin u^6 \cosh u^7 \partial y_7$$

$$+ (y^6 \cos u^6 \sinh u^7 + y^7 \sin u^6 \sinh u^7) \partial z\}$$

$$W_1 = \frac{2}{\sqrt{\cosh^2 u^7 + \sinh^2 u^7}} \{- \sin u^6 \sinh u^7 \partial x_6 + \cos u^6 \sinh u^7 \partial x_7$$

$$+ \sin u^6 \cosh u^7 \partial y_6 - \cos u^6 \cosh u^7 \partial y_7$$

$$(+ - y^6 \sin u^6 \sinh u^7 + y^7 \cos u^6 \sinh u^7) \partial z\}$$

$$W_2 = \frac{2}{\sqrt{\cosh^2 u^7 + \sinh^2 u^7}} \{\cos u^6 \cosh u^7 \partial x_6 + \sin u^6 \cosh u^7 \partial x_7$$

$$- \cos u^6 \sinh u^7 \partial y_6 - \sin u^6 \sinh u^7 \partial y_7$$

$$+ (y^6 \cos u^6 \cosh u^7 + y^7 \sin u^6 \cosh u^7) \partial z\}$$

$$N_1 = 2(- \cosh \alpha \partial x_1 - \sinh \alpha \partial x_3 + \partial y_3 + (y^1 \cosh \alpha + y^3 \sinh \alpha) \partial z)$$

$$N_2 = 2(\partial x_3 + \cosh \alpha \partial y_1 + \sinh \alpha \partial y_3 + y^3 \partial z)$$

$$N_3 = 2(-\partial x_2 + \partial x_4) + (y^2 - y^4) \partial z)$$

where $\epsilon_1 = g(e_1, e_1) = 1$, $\epsilon_2 = g(e_2, e_2) = 1$, $\epsilon_3 = g(X_1, X_1) = 1$ and $\epsilon_4 = g(X_2, X_2) = 1$. Direct calculations and Gauss formula (5.1.15) gives

$$h(\xi_1, \xi_1) = h(\xi_2, \xi_2) = h(\xi_3, \xi_3) = h(e_1, e_1) = h(e_2, e_2) = 0,$$

$$h(\phi_o \xi_3, \phi_o \xi_3) = h(\frac{1}{2} \phi_o N_3, \frac{1}{2} \phi_o N_3) = 0, h^l(X_1, X_1) = h^l(X_2, X_2) = 0,$$

$$h^s(X_1, X_1) = - \frac{1}{\cosh^2 u^7 + \sinh^2 u^7} W_2, h^s(X_2, X_2) = \frac{1}{\cosh^2 u^7 + \sinh^2 u^7} W_2.$$

Therefore,

$$\mathrm{tr}_g \, h_{S(TM)} = h^s(X_1, X_1) + h^s(X_2, X_2) = 0.$$

Thus M is a non-totally geodesic minimal proper contact GCR-lightlike submanifold of $\mathbf{R}_4^{15}$.

Finally, we prove a characterization theorem for minimal contact GCR-lightlike submanifolds. Take a quasi-orthonormal frame

$$\{\xi_1, \ldots, \xi_q, e_1, \ldots, e_m, V, W_1, \ldots, W_n, N_1, \ldots, N_q\}$$

such that $(\xi_1,\ldots,\xi_q,e_1,\ldots,e_m,V)$ belongs to $\Gamma(TM)$. Then take $(\xi_1,\ldots,\xi_q,$ $e_1,\ldots,e_m,V)$ such that $\{\xi_1,\ldots,\xi_{2p}\}$ form a basis of D_1, $\{\xi_{2p+1},\ldots,\xi_q\}$ form a basis of D_2 and $\{e_1,\ldots,e_{2l}\}$ form a basis of D_0. Moreover, we take $\{W_1,\ldots,W_k\}$ a basis of $\mathcal{S}$ and $\{N_{2p+1},\ldots,N_q\}$ a basis of $\mathcal{L}$. Thus we have a quasi-orthonormal basis of M as follows:

$$\{\xi_1,\ldots,\xi_{2p},\xi_{2p+1},\ldots,\xi_q,\phi(\xi_{2p+1}),\ldots,\phi(\xi_q),e_1,\ldots,e_l,\phi e_1,\ldots,\phi e_l,$$
$$\phi W_1,\ldots,\phi W_k,\phi N_{2p+1},\ldots,\phi N_q\}.$$

Theorem 7.6.16. *Let $(M,g,S(TM))$ be a proper contact GCR-lightlike submanifold of an indefinite Sasakian manifold $\bar{M}$. Then M is minimal if and only if*

$$\operatorname{tr} A_{W_i|S(TM)} = 0, \qquad \operatorname{tr} A^*_{\xi_k}|_{S(TM)} = 0$$

and $\bar{g}(Y,D^l(X,W)) = 0$ for $X,Y \in \Gamma(\operatorname{Rad} TM)$ and $W \in \Gamma(S(TM^\perp))$.

Proof. From Proposition 5.1.3, we have $h^l = 0$ on $\operatorname{Rad} TM$. Hence, by Definition 5.4.1 and Definition 7.6.1, it follows that a contact GCR-lightlike submanifold is minimal if and only if

$$\sum_{i=1}^{2l}\varepsilon_i h(e_i,e_i) + \sum_{j=2p+1}^{q} h(\phi\xi_j,\phi\xi_j) + \sum_{j=2p+1}^{q} h(\phi N_j,\phi N_j) + \sum_{l=1}^{k}\varepsilon_l h(\phi W_l,\phi W_l) = 0$$

and $h^s = 0$ on $\operatorname{Rad}(TM)$. From (5.1.28) it follows that $h^s = 0$ on $\operatorname{Rad}(TM)$ if and only if $\bar{g}(Y,D^l(X,W)) = 0$ for $X,Y \in \Gamma(\operatorname{Rad} TM)$ and $W \in \Gamma(S(TM^\perp))$. On the other hand, we have

$$\operatorname{tr} h\,|_{S(TM)} = \frac{1}{q}\sum_{\alpha=1}^{q}\sum_{j=2p+1}^{q} \bar{g}(h^l(\phi\xi_j,\phi\xi_j),\xi_\alpha)N_\alpha + \bar{g}(h^l(\phi N_j,\phi N_j),\xi_\alpha)N_\alpha$$

$$+ \frac{1}{n}\sum_{j=2p+1}^{q}\sum_{a=1}^{n}\varepsilon_a\{\bar{g}(h^s(\phi\xi_j,\phi\xi_j),W_a)W_a + \bar{g}(h^s(\phi N_j,\phi N_j),W_a)W_a\}$$

$$+ \sum_{a=1}^{n}\varepsilon_a\frac{1}{n}\{\sum_{i=1}^{2l}\bar{g}(h^s(e_i,e_i),W_a)W_a + \sum_{l=1}^{k}\bar{g}((h^s(\phi W_l,\phi W_l),W_a)W_a)\}$$

$$+ \sum_{b=1}^{q}\frac{1}{q}\{\sum_{i=1}^{2l}\bar{g}(h^l(e_i,e_i),\xi_b)N_b + \sum_{l=1}^{k}\bar{g}((h^l(\phi W_l,\phi W_l),\xi_b)N_b)\}.$$

Using (5.1.28) and (5.1.39) we obtain

$$
\operatorname{tr} h \mid_{S(TM)} = \frac{1}{q} \sum_{\alpha=1}^{q} \sum_{j=2p+1}^{q} g(A^*_{\xi_\alpha} \phi \xi_j, \phi \xi_j) N_\alpha + g(A^*_{\xi_\alpha} \phi N_j, \phi N_j) N_\alpha
$$

$$
+ \frac{1}{n} \sum_{j=2p+1}^{q} \sum_{a=1}^{n} \varepsilon_a \{ g(A_{W_a} \phi \xi_j, \phi \xi_j) W_a + g(A_{W_a} \phi N_j, \phi N_j) W_a \}
$$

$$
+ \sum_{a=1}^{n} \varepsilon_a \frac{1}{n} \{ \sum_{i=1}^{2l} g(A_{W_a} e_i, e_i) W_a + \sum_{l=1}^{k} g(A_{W_a} \phi W_l, \phi W_l) W_a) \}
$$

$$
+ \sum_{b=1}^{q} \frac{1}{q} \{ \sum_{i=1}^{2l} g(A^*_{\xi_b} e_i, e_i) N_b + \sum_{l=1}^{k} g(A^*_{\xi_b} \phi W_l, \phi W_l) N_b) \},
$$

which proves our assertion. □

Chapter 8

Submanifolds of indefinite quaternion Kähler manifolds

In this chapter, we first recall the structure of indefinite quaternion Kähler manifolds. Then, we give a review of Riemannian submanifolds of quaternion Kähler manifolds. We study the geometry of real lightlike hypersurfaces, the structure of lightlike submanifolds, both, of indefinite quaternion Kähler manifolds and show that a quaternion lightlike submanifold is always totally geodesic. This result implies that the study of lightlike submanifolds, other than quaternion lightlike submanifolds, is interesting. Then, we deal with the geometry of screen real submanifolds in detail. As a generalization of real lightlike hypersurfaces of quaternion Kähler manifolds, we introduce QR-lightlike submanifolds. We show that the class of QR-lightlike submanifolds does not include quaternion lightlike submanifolds and screen real submanifolds. Then, we introduce and study the geometry of screen QR-lightlike and screen CR-lightlike submanifolds as generalizations of quaternion lightlike submanifolds and screen real submanifolds, and provide examples for each class of lightlike submanifolds of indefinite quaternion Kähler manifolds.

8.1 Introduction

Indefinite Quaternion Kähler Manifolds. A quaternionic Kähler manifold is an oriented $4n$-dimensional Riemannian manifold whose restricted holonomy group is contained in the subgroup $Sp(n)Sp(1)$ of $SO(4n)$. These manifolds are of special interest because $Sp(n)Sp(1)$ is included in the list of Berger's work [59] on possible holonomy groups of locally irreducible Riemannian manifolds that are not locally symmetric. It is well known that the twistor theory [329] is closely connected with the existence of canonical quaternionic structures on 4-dimensional oriented semi-Riemannian manifolds. S. Salamon [366] and others extended the theory to $4n$-dimensional quaternionic manifolds. It is also well known that a quaternionic

Kähler manifold M is always Einstein. In this section, we first review indefinite quaternion Kähler manifolds which have been studied by Perez [330] and Perez-Santos [331].

Let $\bar{M}$ be an n-dimensional manifold with a 3-dimensional vector bundle $\mathbb{Q}$ consisting of three tensors J_1, J_2 and J_3 of type $(1,1)$ over $\bar{M}$. Suppose, in any coordinate neighborhood U of $\bar{M}$, there is a local basis $\{J_1, J_2, J_3\}$ of $\mathbb{Q}$ such that

$$J_1^2 = -I, J_2^2 = -I, J_3^2 = -I \tag{8.1.1}$$

and

$$J_2 J_3 = -J_3 J_2 = J_1, J_3 J_1 = -J_1 J_3 = J_2, J_1 J_2 = -J_2 J_1 = J_3. \tag{8.1.2}$$

Such a basis $\{J_1, J_2, J_3\}$ is called a canonical local basis of the bundle $\mathbb{Q}$ in U. We say that the bundle $\mathbb{Q}$ has an *almost quaternion structure* in $\bar{M}$ and $(\bar{M}, \mathbb{Q})$ is called an *almost quaternion manifold* [239] whose dimension is $n = 4m, (m \geq 1)$.

Consider another coordinate neighborhood U' in $(\bar{M}, \mathbb{Q})$ such that $U \cap U' \neq \emptyset$ and $\{J_1', J_2', J_3'\}$ is a canonical basis of $\mathbb{Q}$ in U'. Then, we have

$$J_a' = \sum_{b=1}^{3} S_{ab} J_b, a = 1, 2, 3. \tag{8.1.3}$$

From (8.1.1) and (8.1.2), it follows that (S_{ab}) is an element of the proper orthogonal group $SO(3)$. Thus, every almost quaternion manifold $\bar{M}$ is orientable. Let $\bar{M}$ be an almost quaternion manifold with a canonical local basis of $\mathbb{Q}$ in a coordinate neighborhood U. Assume there exists a system of coordinates (x^h) in each U with respect to J_1, J_2 and J_3 which have components of the form

$$\begin{pmatrix} 0 & -I & 0 & 0 \\ I & 0 & 0 & 0 \\ 0 & 0 & 0 & -I \\ 0 & 0 & I & 0 \end{pmatrix}, \begin{pmatrix} 0 & 0 & -I & 0 \\ 0 & 0 & 0 & I \\ I & 0 & 0 & 0 \\ 0 & -I & 0 & 0 \end{pmatrix} \tag{8.1.4}$$

and

$$\begin{pmatrix} 0 & 0 & 0 & I \\ 0 & 0 & -I & 0 \\ 0 & I & 0 & 0 \\ I & 0 & 0 & 0 \end{pmatrix}, \tag{8.1.5}$$

respectively. Here, I denotes the identity (m, m) matrix. In such a case, the given structure $\mathbb{Q}$ is called an *integrable quaternionic structure*. Suppose g is an indefinite metric on $(\bar{M}, \mathbb{Q})$ such that

$$g(\Phi X, \Phi Y) = g(X, Y), \quad \forall X, Y \in T_p \bar{M}, \quad p \in \bar{M}, \tag{8.1.6}$$

and $\Phi = J_1, J_2, J_3$, with $\{J_1, J_2, J_3\}$ being a basis of $\mathbb{Q}$ at p. Then, $(\bar{M}, g, \mathbb{Q})$ is called an *indefinite almost quaternion manifold* [330]. Just like almost Hermitian manifolds, the condition (8.1.6) puts the following restrictions on the signature of g:

Let E_1 be a non-zero vector field on $(\bar{M}, g, \mathbb{Q})$. If E_1 is timelike, then, $\mathbb{Q} =$ Span$\{E_1, J_1 E_1, J_2 E_1, J_3 E_1\}$ must be 4-dimensional. This condition implies from (8.1.6) that the restriction of g to $\mathbb{Q}(E_1)$ is of signature $(4, 0)$. Similarly, for a spacelike E_1, the restriction g of $\mathbb{Q}(E_1)$ is of type $(0, 4)$. Thus, one can construct a frame field **B** of $\bar{M}$ such that

$$\begin{aligned}
\mathbf{B} = \{&E_1, J_1 E_1, J_2 E_1, J_3 E_1, \ldots, E_{m_1}, J_1 E_{m_1}, J_2 E_{m_1}, J_3 E_{m_1}, \\
&F_1, J_1 F_1, J_2 F_1, J_3 F_1, \ldots, F_{m_2}, J_1 F_{m_2}, J_2 F_{m_2}, J_3 F_{m_2}\}
\end{aligned}$$

where $E_1, \ldots, E_{m_1}$ and $F_1,, \ldots, F_{m_2}$ are spacelike and timelike vector fields, respectively. Consequently, for all three real $J's$, the only possible signature of g is $(4(m_2), 4(m_1))$ with $4m_1 + 4m_2 = 4m$. It follows that the index of g is $s = 4t$, $t \geq 1$. For example, g satisfying (8.1.6) can not be a Lorentzian metric. If the Levi-Civita connection ∇ of $(\bar{M}, g, \mathbb{Q})$ satisfies

$$\begin{aligned}
\nabla_X J_1 &= r(X) J_2 - q(X) J_3, \\
\nabla_X J_2 &= -r(X) J_1 + p(X) J_3, \\
\nabla_X J_3 &= q(X) J_1 - p(X) J_2
\end{aligned} \tag{8.1.7}$$

for any vector field X on $\bar{M}$, then, $\bar{M}$ is called an *indefinite quaternion Kähler manifold*, where p, q and r are certain local 1-forms and $\{J_1, J_2, J_3\}$ is a local canonical basis of $\mathbb{Q}$. The equation (8.1.7) can be written as

$$\nabla_X J_a = \sum_{b=1}^{3} Q_{ab}(X) J_b, a = 1, 2, 3, \tag{8.1.8}$$

for any vector field X on $\bar{M}$, where Q_{ab} are certain 1-forms locally defined on $\bar{M}$ such that $Q_{ab} + Q_{ba} = 0$.

Let $(\bar{M}, g, \mathbb{Q})$ be an indefinite quaternion manifold and $\{J_1, J_2, J_3\}$ be a local basis of $\mathbb{Q}$. Then, the local 2-forms Θ_1, Θ_2 and Θ_3 are defined by

$$\Theta_i(X, Y) = g(X, J_i Y), i = 1, 2, 3, \tag{8.1.9}$$

for any vector fields X and Y on $\bar{M}$. Thus, putting

$$\Omega = \Theta_1 \wedge \Theta_1 + \Theta_2 \wedge \Theta_2 + \Theta_3 \wedge \Theta_3, \tag{8.1.10}$$

we get a 4-form Ω defined globally on $\bar{M}$. Using (8.1.7), (8.1.9) and (8.1.10) we obtain the following theorem.

Theorem 8.1.1. *An indefinite almost quaternion manifold $(\bar{M}, g, \mathbb{Q})$ is an indefinite quaternion Kähler manifold if and only if $\nabla \Omega = 0$.*

In the sequel, we briefly denote an indefinite quaternion Kähler manifold by $\bar{M}$, unless otherwise stated. Let R be the curvature tensor of $\bar{M}$. From (8.1.7), we have

$$
\begin{aligned}
R(X,Y)J_1 Z - J_1 R(X,Y)Z &= \alpha(X,Y)J_2 Z - \beta(X,Y)J_3 Z,\\
R(X,Y)J_2 Z - J_2 R(X,Y)Z &= \gamma(X,Y)J_3 Z - \alpha(X,Y)J_1 Z, \qquad (8.1.11)\\
R(X,Y)J_3 Z - J_3 R(X,Y)Z &= \beta(X,Y)J_1 Z - \gamma(X,Y)J_2 Z,
\end{aligned}
$$

$\forall X, Y, Z \in \Gamma(T\bar{M})$, where α, β and γ are local 2-forms on U given by

$$
\begin{aligned}
\alpha &= dQ_{23} + Q_{31} \wedge Q_{12},\\
\beta &= dQ_{31} + Q_{12} \wedge Q_{23},\\
\gamma &= dQ_{12} + Q_{23} \wedge Q_{31}.
\end{aligned}
$$

Let S be the Ricci tensor of $\bar{M}$. The Bianchi identity and (8.1.11) imply

$$
\begin{aligned}
S(X,Y) &= m\gamma(X, J_1 Y) + \beta(X, J_2 Y) + \alpha(X, J_3 Y),\\
S(X,Y) &= \gamma(X, J_1 Y) + m\beta(X, J_2 Y) + \alpha(X, J_3 Y),\\
S(X,Y) &= \gamma(X, J_1 Y) + \beta(X, J_2 Y) + m\alpha(X, J_3 Y).
\end{aligned}
$$

Hence, if the real dimension of $\bar{M}$ is $4m \geq 8$, we get

$$
\gamma(X,Y) = \frac{S(X, J_1 Y)}{m+2}, \quad \beta(X,Y) = \frac{S(X, J_2 Y)}{m+2},
$$
$$
\alpha(X,Y) = \frac{S(X, J_3 Y)}{m+2}. \qquad (8.1.12)
$$

The proof of the following theorem is commom with the Riemannian case.

Theorem 8.1.2. [331] *Let M be an indefinite quaternion Kähler manifold of real dimension $4m \geq 8$. Then M is Einstein.*

A 4-plane $Q(X)$, spanned by $X, J_1 X, J_2 X$ and $J_3 X$, is called a *quaternionic 4-plane*. A 2-plane in $T_p\bar{M}$, $p \in \bar{M}$ spanned by X, Y is called *half-quaternionic* (respectively, *totally real*) if $\mathcal{K}(X) = \mathcal{K}(Y)$ (respectively, $\mathcal{K}(X) \perp \mathcal{K}(Y)$). If X is a non-null vector field, then $\mathcal{K}(X)$ is a non-degenerate subspace. Recall that, if P is a non-degenerate 2-plane in $T_p M$ spanned by X and Y, its sectional curvature is defined by

$$
\mathcal{K}(X,Y) = \frac{R(X,Y,Y,X)}{g(X,X)g(Y,Y) - g(X,Y)^2},
$$

which, for a half-quaternionic (respectively, totally real) plane is called *quaternionic sectional curvature* (respectively, *totally real sectional curvature*).

Let R_o be the tensor field on $\bar{M}$ given by

$$R_o(X,Y)Z = \frac{1}{4}\{g(Y,Z)X - g(X,Z)Y \tag{8.1.13}$$
$$+ g(J_1Y,Z)J_1X - g(J_1X,Z)J_1Y + 2g(X,J_1Y)J_1Z$$
$$+ g(J_2Y,Z)J_2X - g(J_2X,Z)J_2Y + 2g(X,J_2Y)J_2Z$$
$$+ g(J_3Y,Z)J_3X - g(J_3X,Z)J_3Y + 2g(X,J_3Y)J_3Z\}$$

$\forall X,Y,Z \in \Gamma(T\bar{M})$. Then, R_o does not depend on the basis of $\mathbb{Q}$ chosen on a neighborhood of any point. Moreover, it satisfies the conditions (1.2.9). An indefinite quaternion Kählerian manifold of constant quaternionic sectional curvature is called an *indefinite quaternion space form*. In [331], Perez-Santos proved the following chracterization theorem for an indefinite quaternion space form.

Theorem 8.1.3. [331] *An indefinite quaternion Kähler manifold $\bar{M}$ of real dimension ≥ 8 is an indefinite quaternion space form of quaternionic sectional curvature c if and only if, for any vector fields X,Y and Z on $\bar{M}$,*

$$R(X,Y)Z = \frac{c}{4}\{g(Y,Z)X - g(X,Z)Y \tag{8.1.14}$$
$$+ g(J_1Y,Z)J_1X - g(J_1X,Z)J_1Y + 2g(X,J_1Y)J_1Z$$
$$+ g(J_2Y,Z)J_2X - g(J_2X,Z)J_2Y + 2g(X,J_2Y)J_2Z$$
$$+ g(J_3Y,Z)J_3X - g(J_3X,Z)J_3Y + 2g(X,J_3Y)J_3Z\}.$$

Proof. Let $X \in T_p\bar{M}$, $p \in \bar{M}$, be a non-null vector and c the sectional curvature in $\mathbb{Q}(X)$. Let $R' = R - cR_o$. Then, $R'(Y,Z,Z,Y) = 0$ for any 2-plane $\{Y,Z\}$ in $\mathbb{Q}(X)$. Since $g(X,X)$ is a polynomial function of the coordinates of X in a fixed basis, the zero set of $g(X,X)$ does not contain any open set. Thus, if X is a null vector of $T_p\bar{M}$, there exists a sequence $\{X_n\}$ of non-null vectors such that $\{X_n\} \to X$. Then, $\{J_iX_n\} \to J_iX$. Let Y,Z span a 2-plane of $\mathbb{Q}(X)$. We write

$$Y = a_oX_n + \sum_{i=1}^{3} a_iJ_iX, \ Z = b_oX + \sum_{i=1}^{3} b_iJ_iX.$$

Let $Y_n = a_oX_n + \sum_{i=1}^{3} a_iJ_iX_n$ and $Z_n = b_oX_n + \sum_{i=1}^{3} b_iJ_iX_n$. Then the 2-plane $\{Y_n,Z_n\}$ is contained in $\mathbb{Q}(X_n)$. Hence, $R'(Y_n,Z_n,Z_n,Y_n) = 0$ and

$$R'(Y,Z,Z,Y) = 0. \tag{8.1.15}$$

Thus, (8.1.15) shows that for any $X \in T_p\bar{M}$ and any 2-plane $\{Y,Z\}$ of $\mathbb{Q}(X)$, we have $R'(Y,Z,Z,Y) = 0$. On the other hand, for any $X,Y \in T_p\bar{M}, p \in \bar{M}$, from (8.1.12), we get

$$\gamma'(X,Y) = \gamma(X,Y) - cg(X,J_1Y) = kg(X,J_1Y),$$
$$\beta'(X,Y) = \beta(X,Y) - cg(X,J_2Y) = kg(X,J_2Y), \tag{8.1.16}$$
$$\alpha'(X,Y) = \alpha(X,Y) - cg(X,J_3Y) = kg(X,J_3Y),$$

where $k = (\frac{r}{m+2}) - c$, $4mr$ is the scalar curvature of $\bar{M}$. Let $X \in T_p\bar{M}$ be a non-null vector. From (8.1.11) and (8.1.16), we obtain

$$R'(X, J_1X, J_2X, J_3X) = -R'(X, J_1X, X, J_1X) - kg(X, X)^2 = -kg(X, X)^2,$$
$$R'(X, J_3X, J_1X, J_2X) = -R'(X, J_3X, X, J_3X) - kg(X, X)^2 = -kg(X, X)^2,$$
$$R'(X, J_2X, J_3X, J_1X) = -R'(X, J_2X, X, J_2X) - kg(X, X)^2 = -kg(X, X)^2.$$

Using the above equations and the Bianchi identity, we have $kg(X, X)^2 = 0$. Hence, we derive $k = 0$. Now, (8.1.11) and $k = 0$ imply that

$$R'(X, J_1X, J_2Y, J_3Y) = -R'(X, J_1X, Y, J_1Y). \tag{8.1.17}$$

From the Bianchi identity, we get

$$R'(X, J_1X, J_2Y, J_3Y) = R'(X, J_3Y, X, J_3Y)$$
$$+ R'(X, J_2Y, X, J_2Y), \tag{8.1.18}$$
$$R'(X, J_1X, Y, J_1Y) = R'(X, J_1Y, X, J_1Y)$$
$$- R'(X, Y, Y, X). \tag{8.1.19}$$

Thus, from (8.1.17), (8.1.18) and (8.1.19), we obtain

$$R'(X, Y, Y, X) + R'(X, J_1Y, J_1Y, X) + R'(X, J_2Y, J_2Y, X)$$
$$+ R'(X, J_3Y, J_3Y, X) = 0. \tag{8.1.20}$$

Computing $0 = R'(X + Y, J_1X + J_1Y, J_1X + J_1Y, X + Y)$, $0 = R'(X - Y, J_1X - J_1Y, J_1X - J_1Y, X - Y)$, using the Bianchi identity, (8.1.11), taking into account that $k = 0$ and, then, adding the expressions, we get

$$0 = R'(X, Y, Y, X) + 3R'(X, J_1Y, JY, X). \tag{8.1.21}$$

In the same way, computing $0 = R'(X + J_2Y, J_1X + J_3Y, J_1X + J_3Y, X + J_2Y)$, $0 = R'(X - J_2Y, J_1X - J_3Y, J_1X - J_3Y, X - J_2Y)$ we obtain

$$0 = R'(X, J_2Y, J_2Y, X) + 3R'(X, J_3Y, J_3Y, X). \tag{8.1.22}$$

Also $0 = R'(X + J_3Y, J_1X - J_2Y, J_1X - J_2Y, X + J_3Y)$ and $0 = R'(X - J_3Y, J_1X + J_2Y, J_1X + J_2Y, X - J_3Y)$ imply that

$$0 = R'(X, J_3Y, J_3Y, X) + 3R'(X, J_2Y, J_2Y, X). \tag{8.1.23}$$

Then, from (8.1.22) and (8.1.23) we get

$$R'(X, J_2Y, J_2Y, X) = R'(X, J_3Y, J_3Y, X). \tag{8.1.24}$$

Thus, summing up (8.1.20) and (8.1.21) and (8.1.24) we arrive at $R'(X, Y, Y, X) = 0$. The converse is clear.　　　　　　　　　　　　　　$\square$

Corollary 8.1.4. [331] *Let $\bar{M}$ be an indefinite quaternion Kähler manifold of real dimension $4m \geq 8$. Then $\bar{M}$ is an indefinite quaternion space form of quaternionic sectional curvature $c = \frac{r}{m+2}$ if and only if*

$$R(X,Y)Z = \frac{c}{4}\{g(Y,Z)X - g(X,Z)Y \tag{8.1.25}$$

$$+ \sum_{b=1}^{3} g(J_b Y, Z)J_b X - g(J_b X, Z)J_b Y + 2g(X, J_b Y)J_b Z\}$$

for any vector fields X, Y and Z on $\bar{M}$.

Proposition 8.1.5. [331] *Let $\bar{M}$ be an indefinite quaternion Kähler manifold of real dimension ≥ 8 and X a unit vector of $T_p\bar{M}$, $p \in \bar{M}$. If there exists a local basis of $\mathbb{Q}$, $\{J_1, J_2, J_3\}$ such that*

(i) $\mathcal{K}(X, J_1 X) = \mathcal{K}(X, J_2 X) = \mathcal{K}(X, J_3 X),$

(ii) $R(X, J_1 X, X, J_1 X) = R(X, J_2 X, X, J_2 X) = R(X, J_3 X, X, J_3 X) = 0.$

Then the quaternionic sectional curvature in $Q(X)$ is constant and its value is $r/(m+2)$, $4mr$ is the scalar curvature of $\bar{M}$.

Proof. Using (8.1.11), (8.1.12) and Theorem 8.1.2, we have

$$\mathcal{K}(X, J_1 X) = (\frac{r}{m+2})g(X, X)^2 + R(X, J_1 X, J_2 X, J_3 X),$$

$$\mathcal{K}(X, J_2 X) = (\frac{r}{m+2})g(X, X)^2 + R(X, J_2 X, J_3 X, J_1 X),$$

$$\mathcal{K}(X, J_3 X) = (\frac{r}{m+2})g(X, X)^2 + R(X, J_3 X, J_1 X, J_2 X) \tag{8.1.26}$$

for $X \in \Gamma(T\bar{M})$. Then, the proof follows from (i) and (ii). $\square$

Lemma 8.1.6. [331] *Let $\bar{M}$ be an indefinite quaternion Kähler manifold of real dimension $4m \geq 8$, $p \in \bar{M}$, $X \in T_p\bar{M}$ a unit vector and $\{J_1, J_2, J_3\}$ a local basis of $\mathbb{Q}$ on a coordinate neighborhood of $\bar{M}$ at p. If $\mathcal{K}(Y, \Phi Y)$ is constant for any vector $Y \in Q(X)$ and some $\Phi = J_1, J_2, J_3$, then, the quaternionic sectional curvature on $Q(X)$ is constant.*

Proof. Suppose that $\mathcal{K}(Y, J_1 Y) = a$ is a constant for any vector $Y \in Q(X)$. Let $\mathcal{K}_\theta$ be the sectional curvature of the 2-plane spanned by the vectors $Y = \cos\theta X + \sin\theta J_2 X$, $J_1 Y = \cos\theta J_1 X + \sin\theta J_3 X$. Then

$$\mathcal{K}_\theta = (\cos^2\theta - \sin^2\theta)^2 \mathcal{K}(X, J_1 X) + 4\sin^2\theta\cos^2\theta\mathcal{K}(X, J_3 X)$$

$$+ 4\sin\theta\cos\theta(\cos^2\theta - \sin^2\theta)R(J_3 X, X, X, J_1 X).$$

Hence, we have

$$\mathcal{K}(X, J_1 X) = \mathcal{K}(X, J_3 X) \quad \text{and} \quad R(J_3 X, X, X, J_1 X) = 0. \tag{8.1.27}$$

If we consider the 2-plane spanned by $Y = \cos\theta X + \sin\theta J_3 X$, $J_1 Y = \cos\theta J_1 X - \sin\theta J_3 X$, then we obtain

$$\mathcal{K}(X, J_1 X) = \mathcal{K}(X, J_3 X) \quad \text{and} \quad R(J_1 X, X, X, J_2 X) = 0. \tag{8.1.28}$$

In a similar way, we get
$$R(J_1 X, X, X, J_3 X) = 0. \tag{8.1.29}$$

Then, the result comes from (8.1.27)–(8.1.29) and Proposition 8.1.5. $\qquad\square$

The next result shows how the geometry of an indefinite quaternion Kähler manifold is quite different from the geometry of Riemannian quaternion Kähler manifolds. First, it is obvious that $R(X, \Phi X, \Phi X, X) = 0$, for any null vector $X \in T_p \bar{M}$, for an indefinite quaternion space form.

Theorem 8.1.7. [331] *Let $\bar{M}$ be an indefinite quaternion Kähler manifold of real dimension $4m \geq 8$. If $\forall p \in \bar{M}$ there exists a local basis $\{J_1, J_2, J_3\}$ of $\mathbb{Q}$ at p, such that $R(X, \Phi X, \Phi X, X) = 0$ for any null vector $X \in T_p \bar{M}$ and $\Phi = J_1, J_2, J_3$, then $\bar{M}$ is an indefinite quaternion space form.*

Proof. Let Y, Z be two orthonormal vectors of $T_p \bar{M}$ spanning a totally real 2-plane of signature $(+, -)$. If $\lambda \in \mathcal{R}$, then $\lambda Y + Z$ is a null vector if and only if $\lambda = 1$ or $\lambda = -1$. Hence, we get

$$R(\lambda Y + Z, \lambda\Phi Y + \Phi Z, \lambda\Phi Y + \Phi Z, \lambda Y + Z) = 0. \tag{8.1.30}$$

Then from (8.1.30), we obtain

$$\mathcal{K}(Y, \Phi Y) + \mathcal{K}(Z, \Phi Z) = 8\mathcal{K}(Y, Z) = 8\mathcal{K}(Y, \Phi Z), \tag{8.1.31}$$

$$R(Y, \Phi Z, \Phi Y, Y) + R(Y, \Phi Z, \Phi Y, Y) = 0. \tag{8.1.32}$$

Let β_n and μ_n be two sequences of real numbers such that for any n, $\beta_n > 1$, $\mu_n < -1$, $\lim_n \beta_n = 1$ and $\lim_n \alpha_n = -1$. From (8.1.30) and by continuity we have

$$\lim_n R(\beta_n Y + Z, \beta_n \Phi Y) + R(\Phi Z, \beta_n \Phi Y + \Phi Z, \beta_n Y + Z) = 0, \tag{8.1.33}$$

$$\lim_n R(\mu_n Y + Z, \mu_n \Phi Y) + R(\Phi Z, \mu_n \Phi Y + \Phi Z, \mu_n Y + Z) = 0. \tag{8.1.34}$$

Let $\{\delta_n\}$ be a sequence of positive real numbers such that

$$\lim_n \frac{\delta_n}{(\beta_n^2 - 1)} = 0.$$

From (8.1.33), for any δ_m there exists $n_0(m)$ such that

$$-\delta_m < (\beta_{n(m)}^2 - 1)(\beta_{n(m)}^2 + 1)\mathcal{K}(Y, \Phi Y) - 8(\beta_{n(m)}^2 - 1)\mathcal{K}(Y, \Phi Z)$$
$$+ 4(\beta_{n(m)}^2 - 1))R(Y, \Phi Z, \Phi Y, Y) < \delta_m. \tag{8.1.35}$$

Since $(\beta_{n(m)}^2 - 1)$ is always positive and $n \longrightarrow \infty$ implies $n(m) \longrightarrow \infty$,

$$\lim_m \frac{\delta_m}{\beta_{n(m)}^2} = 0.$$

From (8.1.35) we obtain

$$2\mathcal{K}(Y, \Phi Y) - 8\mathcal{K}(Y, Z) + 4R(Y, \Phi Z, \Phi Y, Y) = 0. \tag{8.1.36}$$

In a similar way, we get

$$2\mathcal{K}(Y, \Phi Y) - 8\mathcal{K}(Y, Z) - 4R(Y, \Phi Z, \Phi Y, Y) = 0. \tag{8.1.37}$$

From (8.1.31), (8.1.36) and (8.1.37) we have $\mathcal{K}(Y, \Phi Y) = \mathcal{K}(Z, \Phi Z) = 4\mathcal{K}(Y, Z) = 4\mathcal{K}(Y, \Phi Z)$ and $R(Y, Z, \Phi Y, Y) = R(\Phi Z, Z, Y, Z) = 0$. Then the proof follows from Lemma 8.1.6. $\qquad\square$

We say that an indefinite quaternion Kähler manifold is *null quaternionically flat*, [199, 273], if $R(U, \Phi U, \Phi U, U) = 0$ for any null vector $U \in T_p\bar{M}$ and $\Phi = J_1, J_2, J_3$. Thus, we can state the above result as:

Corollary 8.1.8. [199, 273] *Let $\bar{M}$ be an indefinite quaternion Kähler manifold of real dimension $4m \geq 8$. Then $\bar{M}$ is an indefinite quaternion space form if and only if it is null quaternionically flat.*

Example 1. Consider $\mathbf{R}^{4n}$ with Cartesian coordinates (x^i, y^i, z^i, w^i), $i = 1, \ldots, n$. Define complex structures $\Phi = J_1, J_2, J_3$ of $\mathbf{R}_4^8$ and a Hermitian metric g as

$$J_1(x_1, y_1, z_1, w_1, \ldots, x_n, y_n, z_n, w_n) = (-y_1, x_1, -w_1, z_1, \ldots, -y_n, x_n, -w_n, z_n),$$
$$J_2(x_1, y_1, z_1, w_1, \ldots, x_n, y_n, z_n, w_n) = (-z_1, w_1, x_1, -y_1, \ldots, -z_n, w_n, x_n, -y_n),$$
$$J_3(x_1, y_1, z_1, w_1, \ldots, x_n, y_n, z_n, w_n) = (-w_1, -z_1, y_1, x_1, \ldots, -w_n, -z_n, y_n, x_n),$$

and

$$g((x_1, y_1, z_1, w_1, \ldots, x_n, y_n, z_n, w_n), (u_1, v_1, t_1, s_1, \ldots, u_n, v_n, t_n, s_n))$$
$$= -\sum_{q=1}^{q}(x_iu_i + y_iv_i + z_it_1 + w_is_i) + \sum_{a=q+1}^{n}(x_au_a + y_av_a + z_at_a + w_as_a).$$

It is easy to see that $(\mathbf{R}^{4n}, \Phi, g)$ defines a flat indefinite quaternion space form and its quaternionic structure is integrable.

For more examples on quaternionic semi-projective spaces and quaternionic hyperbolic space, see:[203].

Submanifolds of Quaternion Kähler Manifolds. Let $(\bar{M}, \bar{J}_a, \bar{g}), a = 1, 2, 3$ be a quaternion Kähler manifold. A submanifold of $\bar{M}$ is defined with respect to the action of complex structures $\bar{J}_a$, $a = 1, 2, 3$ as follows: A submanifold M in a quaternion manifold $\bar{M}$ is called a *quaternion submanifold* (respectively, *totally real submanifold*) if each tangent space of M is carried into itself (respectively, the normal space) by each $\bar{J}_a$. A similar definition can be given for distribution on M. Let M be a Riemannian manifold isometrically immersed in a quaternion manifold $\bar{M}$. A distribution $D : x \longrightarrow D_x \subset T_xM$ is called a *quaternion distribution* if it satisfies the condition $\bar{J}_a(D) \subseteq D$. This means that a distribution D is a quaternion distribution if D is carried into itself by its quaternion structure. It is known that every quaternion submanifold in any quaternion Kähler manifold is always totally geodesic [98]. Therefore, it is more interesting to study a general class of submanifolds than quaternion submanifolds. The first attempt in this direction was made by Barros, Chen and Urbano in [27]. They defined the notion of a quaternion CR-submanifold as follows:

Definition 8.1.9. [27] Let M be a real submanifold of a quaternion Kähler manifold $\bar{M}$. Then M is called a *quaternion CR-submanifold* if it admits a differentiable quaternion distribution D such that its orthogonal complementary distribution $D^{\perp}$ is totally real, i.e., $\bar{J}_a(D_x^{\perp}) \subset T_xM^{\perp}$, a=1, 2, 3 $\forall x \in M$, where $T_xM^{\perp}$ is the normal space of M in $\bar{M}$ at x.

It is easy to see that if $D^{\perp} = 0$ (resp., $D = 0$) then M is a quaternion (resp., totally real) submanifold. A quaternion CR-submanifold is called *proper* if $D \neq 0$ and $D^{\perp} \neq 0$. If M is proper, then Definition 8.1.9 implies the following decomposition:

$$TM = D \perp D^{\perp} \quad \text{and} \quad TM^{\perp} = \bar{J}_1(D^{\perp}) \perp \bar{J}_2(D^{\perp}) \perp \bar{J}_3(D^{\perp}) \perp \nu, \quad (8.1.38)$$

where ν is the invariant complementary sub-bundle to $\bar{J}_1(D^{\perp}) \oplus \bar{J}_2(D^{\perp}) \oplus \bar{J}_3(D^{\perp})$ in $TM^{\perp}$. The following result is similar to Theorem 6.1.12 and it gives a characterization of quaternion CR-submanifolds.

Proposition 8.1.10. [27] *Let M be a submanifold of a quaternion-space-form $\bar{M}(c)$, $c \neq 0$ and $D_x = T_xN \cap \bar{J}_1(T_xM) \cap \bar{J}_2(T_xM) \cap \bar{J}_3(T_xM)$, $x \in M$. Then M is a quaternion CR-submanifold of $\bar{M}$ if and only if either M is totally real or D defines a differentiable distribution of positive dimension such that*

$$\bar{R}(D, D, D^{\perp}, D^{\perp}) = 0$$

where $D^{\perp}$ is the orthogonal complementary distribution of D.

On the integrability of distributions we have [27]:

(i) The totally real distribution $D^{\perp}$ on a quaternion CR-submanifold of a quaternion Kähler manifold is always integrable.

(ii) The quaternion distribution D is integrable if and only if the second fundamental form of M satisfies
$$h(D, D) = 0.$$

For a totally umbilical quaternion CR-submanifold, the following holds:

Proposition 8.1.11. [27] *Every totally umbilical proper quaternion CR-submanifold in a quaternion Kähler manifold $\bar{M}$ is totally geodesic.*

If M is a totally geodesic quaternion CR-submanifold of a quaternion Kähler manifold, then, M is locally the Riemannian product of a totally geodesic quaternion submanifold M^T and totally geodesic totally real submanifold $M^{\perp}$ [27]. A quaternion CR-submanifold is said to be of *minimal codimension* if the sub-bundle ν is trivial, i.e., $TM^{\perp} = \bar{J}_1(D^{\perp}) \oplus \bar{J}_2(D^{\perp}) \oplus \bar{J}_3(D^{\perp})$.

Theorem 8.1.12. [27] *The only quaternion Kähler manifolds which admit totally umbilical proper quaternion CR-submanifolds of minimal codimension are Ricci flat quaternion Kähler manifolds.*

A quaternion CR-submanifold is called a *QR-product* if M is locally the Riemannian product of quaternion submanifold M^T and totally real submanifold $M^{\perp}$ [27]. The next result gives a characterization of QR-products in terms of the curvature tensor $R^{\perp}$ of the normal bundle.

Theorem 8.1.13. [27] *Let M be a quaternion CR-submanifold of a quaternionic space form $\bar{M}(c)$, $c \geq 0$. Then we have*

$$\| R^{\perp} \|^2 = \geq cpq, \tag{8.1.39}$$

where $\dim_Q(D) = p$, $\dim_R(D^{\perp}) = q$. *If the equality of (8.1.39) holds, then M is a QR-product.*

Let M be a quaternion CR-submanifold of a quaternion Kähler manifold $\bar{M}$. For any $X \in \Gamma(TM)$, we put

$$\bar{J}_a X = P_a X + Q_a X, a = 1, 2, 3 \tag{8.1.40}$$

where $P_a X$ and $Q_a X$ are the tangential and the normal parts of $\bar{J}_a X$. Using (8.1.40), following is another characterization of a QR-product:

Theorem 8.1.14. [325] *Let M be a quaternion CR-submanifold of a quaternion Kähler manifold $\bar{M}$. Then M is a QR-product if and only if P_a is parallel, i.e., $(\nabla_X P_a)Y = 0$, for every $X, Y \in \Gamma(TM)$. Here ∇ denotes the induced connection on M.*

Let us now recall a bit of information on real hypersurfaces of quaternion Kähler manifolds. Let $\bar{M}$ be an m-dimensional quaternionic Kähler manifold of constant quaternionic sectional curvature $c \in \mathbf{R} - \{0\}$. Let $\bar{g}$ be the Riemannian metric, $\bar{\nabla}$ the Levi-Civita connection and $\mathbb{Q}$ be the quaternionic Kähler structure of $\bar{M}(c)$ and $\bar{J}_1$, $\bar{J}_2$ and $\bar{J}_3$ be a canonical local basis of $\mathbb{Q}$. Now, let M be an orientable real hypersurface in $\bar{M}$, N a unit normal field on M. We denote the induced Riemannian metric on M by same letter. Let

$$U_a = -\bar{J}_a N, \ a = 1, 2, 3 \tag{8.1.41}$$

be a vector field on an open set O of M. The corresponding 1-form η_a is

$$\eta_a(X) = g(X, U_a). \tag{8.1.42}$$

Let T_a be the skew symmetric tensor field of type $(1,1)$ on O given by

$$\bar{J}_a X = T_a X + \eta_a(X) N \tag{8.1.43}$$

for all vector fields on O. $T_a X$ is the tangential component of $\bar{J}_a X$. From (8.1.41)–(8.1.43), we have

$$T_a U_a = 0, \quad T_a U_{a+1} = U_{a+2}, \quad T_a U_{a+2} = -U_{a+1}, \tag{8.1.44}$$

$$T_a^2 X = -X + \eta_a(X) U_a, \tag{8.1.45}$$

$$T_a T_{a+1} X = T_{a+2} X + \eta_{a+1}(X) U_a, \tag{8.1.46}$$

$$T_a T_{a+2} X = -T_{a+1} X + \eta_{a+2}(X) U_a, \tag{8.1.47}$$

where the index is taken modulo 3. (8.1.44)–(8.1.47) show that (T_a, η_a, U_a), $a = 1, 2, 3$ is an almost contact 3-structure, see [271]. Also observe that

$$\bar{J}_b(U_a) = -\bar{J}_c N = U_c \in \Gamma(TM). \tag{8.1.48}$$

Thus from (8.1.48) we say that quaternion CR-submanifolds do not include real hypersurfaces of quaternion Kähler manifolds contrary to the complex case.

Definition 8.1.15. [46] Let $\bar{M}$ be a quaternion Kähler manifold and M be a real submanifold of $\bar{M}$. Then, M is said to be a QR-submanifold if there exists a vector sub-bundle ν of the normal bundle such that we have

$$\bar{J}_a(\nu_x) = \nu_x \tag{8.1.49}$$

and

$$\bar{J}_a(\nu_x^\perp) \subset T_x M \tag{8.1.50}$$

for $x \in M$ and $a = 1, 2, 3$, where $\nu^\perp$ is the complementary orthogonal bundle to ν in $TM^\perp$.

Let M be a QR-submanifold of $\bar{M}$. Set $D_{ax} = \bar{J}_a(\nu_x^\perp)$. We consider $D_{1x} \oplus D_{2x} \oplus D_{3x} = D_x^\perp$ and 3s-dimensional distribution $D^\perp : x \to D_x^\perp$ globally defined on M, where $s = \dim \nu_x^\perp$. Also we have, for each $x \in M$,

$$\bar{J}_a(D_{ax}) = \nu_x^\perp, \bar{J}_a(D_{bx}) = D_{cx} \tag{8.1.51}$$

where (a, b, c) is a cyclic permutation of $(1, 2, 3)$. Denote by D the complementary orthogonal distribution to $D^\perp$ in TM. Then, D is invariant with respect to the action of $\bar{J}_a$, i.e., we have

$$\bar{J}_a(D_x) = D_x \tag{8.1.52}$$

for any $x \in M$. D is called a quaternion distribution.

It is easy to see that a real hypersurface of a quaternion Kähler manifold $\bar{M}$ is a QR-submanifold with $\nu_x = \{0\}$ and $\nu_x^\perp = \text{Span}\{N\}$.

Recall that a QR-submanifold is called *D-geodesic* if $h(X, Y) = 0$ for $X, Y \in \Gamma(D)$, where h denotes the second fundamental form of M. For the integrability of holomorphic distribution we have the following.

Theorem 8.1.16. [46] *Let M be a QR-submanifold of a quaternion Kähler manifold. Then the following assertions are equivalent:*

1. $h(\bar{J}_a X, Y) = h(X, \bar{J}_a Y)$, $a = 1, 2, 3$ and $X, Y \in \Gamma(D)$.

2. *M is D-geodesic.*

3. *D is integrable.*

Let M be a QR-submanifold of quaternion Kähler manifold $\bar{M}$. Denote by P the projection morphism of TM to the quaternion distribution D and choose a local field of orthonormal frames $\{v_1, \ldots, v_s\}$ on the vector sub-bundle $\nu^\perp$ in $TM^\perp$. Then on the distribution $D^\perp$, we have the local field of orthonormal frames

$$\{E_{11}, \ldots, E_{1s}, E_{21}, \ldots, E_{2s}, E_{31}, \ldots, E_{3s}\} \tag{8.1.53}$$

where $E_{ai} = \bar{J}_a v_i$, $a = 1, 2, 3$ and $i = 1, \ldots, s$. For the integrability of distributions of $D^\perp$, we have the following:

Theorem 8.1.17. [46] *Let M be a QR-submanifold of a quaternion Kähler manifold. Then the following assertions are equivalent:*

(a) *the distribution $D^\perp$ is integrable,*

(b) *$B_{aij}(X) = 0$ for all $a = 1, 2, 3$; $i, j = 1, 2 \ldots, s$, and $X \in \Gamma(D)$,*

(c) *$h(D, D^\perp) \subset \nu$,*

where $B_{aij}(X) = g(\nabla_{E_{ai}} E_{aj}, X)$ and ∇ is the induced connection on M.

As in the complex case, every totally geodesic QR-submanifold is locally a Riemannian product $M^T \times M^\perp$, where M^T and $M^\perp$ are leaves of D and $D^\perp$, respectively[46]. For a totally umbilical QR-submanifold M, we have the following:

Theorem 8.1.18. *Let M be a totally umbilical QR-submanifold of a quaternion Kähler manifold $\bar{M}$.*

(1) *If $\dim(\nu_x^\perp) > 1$ for any $x \in M$, then M is totally geodesic [46].*

(2) *If $\dim(\nu_x^\perp) = 1 \forall x \in M$ and M is not totally geodesic, then it is an extrinsic sphere that inherits a generalized 3-Sasakian structure[50].*

(3) *If $\dim(\nu_x^\perp) = 0$ for any $x \in M$, then M is a totally geodesic quaternion submanifold [50].*

Finally, we have quoted some main results for such submanifolds in Riemannian quaternion Kähler manifolds which are needed to compare and understand their counterparts of lightlike geometry. For those interested in knowing more, we recommend the following references: [27, 46, 50, 98, 108, 115, 199, 232, 234, 239, 264, 283, 280, 279, 300, 319, 321, 330, 331, 352, 364].

8.2   Lightlike hypersurfaces

In this section, we discuss the fundamental properties of real lightlike hypersurfaces of an indefinite quaternion Kähler manifold.

Let $\left(\bar{M}, \bar{J}_a, \bar{g}\right)$, $a = 1, 2, 3$ be a real $4n$-dimensional indefinite quaternion Kähler manifold, where $\bar{g}$ is a semi-Riemannian metric of index $\nu = 4q$, $0 < q < m$. Suppose (M, g) is a lightlike real hypersurface of $\bar{M}$. If ξ is a local section of $TM^{\perp}$, then $g\left(\bar{J}_a\xi, \xi\right) = 0$ and $\bar{J}_a\xi$ are tangent to M. Thus, $\bar{J}_a TM^{\perp}$ is a distribution on M of rank 3 such that $\bar{J}_a TM^{\perp} \cap TM^{\perp} = \{0\}$. Hence we can choose a screen distribution $S(TM)$ such that it contains $\bar{J}_a TM^{\perp}$ as a vector sub-bundle. Consider a local section N of $\mathrm{tr}(TM)$. Since

$$\bar{g}\left(\bar{J}_a N, \xi\right) = \bar{g}\left(N, \bar{J}_a\xi\right) = 0,$$

we deduce that $\bar{J}_a N$ are tangent to M. On the other hand, since $\bar{g}\left(\bar{J}_a N, N\right) = 0$ the components of $\bar{J}_a N$ with respect to ξ vanish. Thus $\bar{J}_a N \in \Gamma\left(S(TM)\right)$. We know that ξ and N are null vector fields satisfying $g(\xi, N) = 1$. Thus, $\bar{J}_a N$ and $\bar{J}_a\xi$ are null vector fields satisfying $\bar{g}\left(\bar{J}_a N, \bar{J}_a\xi\right) = 1$, $a = 1, 2, 3$. Otherwise, we have

$$g(\bar{J}_a N, \bar{J}_b\xi) = 0$$

for $a \neq b$. Hence $\bar{J}_a TM^{\perp} \oplus \bar{J}_a \mathrm{Rad}\, TM$ is a vector sub-bundle of $S(TM)$ of rank 6. Then, there exists a non-degenerate distribution D_0 on M such that

$$S(TM) = \{D_1 \oplus D_2\} \perp D_0$$

where

$$D_1 = \bar{J}_1\xi \oplus_{\mathrm{orth}} \bar{J}_2\xi \oplus_{\mathrm{orth}} \bar{J}_3\xi,$$
$$D_2 = \bar{J}_1 N \oplus_{\mathrm{orth}} \bar{J}_2 N \oplus_{\mathrm{orth}} \bar{J}_3 N.$$

Proposition 8.2.1. *The distribution D_0 is invariant with respect to each $\bar{J}_a$, i.e.,*

$$\bar{J}_a(D_0) = D_0, a = 1, 2, 3.$$

Proof. Since $\bar{J}_a$, $a = 1, 2, 3$, are almost Hermitian structures, we have

$$\bar{g}\left(\bar{J}_a X, Y\right) = -\bar{g}\left(X, \bar{J}_a Y\right)$$

for any $X \in \Gamma\left(D_0\right)$ and $Y \in \Gamma\left(TM\right)$. For $Y = \bar{J}_a\xi$ we get

$$\bar{g}\left(\bar{J}_a X, \bar{J}_a\xi\right) = -\bar{g}\left(X, \bar{J}_a\bar{J}_a\xi\right) = \bar{g}\left(X, \xi\right) = 0$$

and from (8.1.2) we have

$$\bar{g}\left(\bar{J}_a X, \bar{J}_b \xi\right) = -\bar{g}\left(X, \bar{J}_a \bar{J}_b \xi\right) = -\bar{g}\left(X, \bar{J}_c \xi\right) = 0$$

for any $Y = \bar{J}_a \xi \in \Gamma\left(D_1\right)$. Thus, $\bar{J}_a X \perp D_1$. On the other hand, we have

$$\bar{g}\left(\bar{J}_a X, \xi\right) = -\bar{g}\left(X, \bar{J}_a \xi\right) = 0$$

for any $\xi \in \Gamma\left(TM^\perp\right)$. Hence, $\bar{J}_a X \oplus_{\text{orth}} TM^\perp$. From (8.1.2) we obtain

$$\bar{g}\left(\bar{J}_a X, \bar{J}_a N\right) = \bar{g}\left(X, N\right) = 0$$

and

$$\bar{g}\left(\bar{J}_a X, \bar{J}_b N\right) = -\bar{g}\left(X, \bar{J}_a \bar{J}_b N\right) = -\bar{g}\left(X, \bar{J}_c N\right) = 0$$

for any $N \in \Gamma\left(\text{ltr}(TM)\right)$. Thus, $\bar{J}_a X \perp \{\{D_1 \oplus D_2\} \perp TM^\perp\}$. Finally, we derive

$$\bar{g}\left(\bar{J}_a X, N\right) = -\bar{g}\left(X, \bar{J}_a N\right) = 0.$$

Summing up these results we deduce

$$\bar{J}_a X \perp \{\{D_1 \oplus D_2\} \perp (TM^\perp \oplus \text{ltr}(TM))\},$$

that is, $\bar{J}_a(D_0) = D_0, a = 1, 2, 3$, which proves our assertion. □

Thus, the general decompositions (2.1.4) and (2.1.8) become

$$TM = \{(D_1 \oplus D_2) \perp D_0 \oplus_{\text{orth}} TM^\perp\} \tag{8.2.1}$$

and

$$T\bar{M} = \{(D_1 \oplus D_2) \perp D_0 \perp (TM^\perp \oplus \text{ltr}(TM))\}. \tag{8.2.2}$$

Example 2. Consider a hypersurface of $\mathbf{R}_4^8$ given by

$$x_1 = u_1 + u_7 \cos \alpha, \qquad\qquad y_1 = u_2 + b,$$
$$x_2 = u_3 + c, \qquad\qquad y_2 = u_4 + d,$$
$$x_3 = -u_3 \cos \alpha - u_4 \sin \alpha + u_5, \qquad y_3 = u_3 \sin \alpha - u_4 \cos \alpha + u_6,$$
$$x_4 = u_1 \cos \alpha - u_2 \sin \alpha + u_7, \qquad y_4 = u_1 \sin \alpha + u_2 \cos \alpha,$$

where $\alpha \in IR - \{\pi + k\pi, k \in Z\}$. Thus TM is spanned by

$$Z_1 = \partial x_1 + \cos \alpha \, \partial x_4 + \sin \alpha \, \partial y_4,$$
$$Z_2 = \partial y_1 - \sin \alpha \, \partial x_4 + \cos \alpha \, \partial y_4,$$
$$Z_3 = \partial x_2 - \cos \alpha \, \partial x_3 + \sin \alpha \, \partial y_3,$$
$$Z_4 = \partial y_2 - \sin \alpha \, \partial x_3 - \cos \alpha \, \partial y_3,$$
$$Z_5 = \partial x_3, \quad Z_6 = \partial y_3,$$
$$Z_7 = \cos \alpha \, \partial x_1 + \partial x_4.$$

It is easy to check that a hypersurface is lightlike such that $TM^\perp = \text{Span}\{Z_1\}$. Moreover, using the canonical complex structures of $\mathbf{R}_4^8$ we see that $\bar{J}_1 Z_1 = Z_2$, $\bar{J}_2 Z_1 = Z_3$ and $\bar{J}_3 Z_1 = Z_4$. Thus, $\bar{J}_a(TM^\perp), a = 1, 2, 3$ are distributions on M. Now, we consider $V = \cot\alpha\, \partial\, y_1 + \csc\alpha\, \partial\, y_4$. Hence we derive

$$N = -\frac{1}{2}(\partial\, x_1 - 2\cot\alpha\, \partial\, y_1 + \cos\alpha\, \partial\, x_4 - (2\csc\alpha + \sin\alpha)\partial\, y_4).$$

Thus, we have

$$\bar{J}_1 N = -\cot\alpha\, \partial\, x_1 - \frac{1}{2}\partial\, y_1 + (\frac{1}{2}\sin\alpha - \csc\alpha)\partial\, x_4 - \frac{1}{2}\cos\alpha\, \partial\, y_4,$$

$$\bar{J}_2 N = -\frac{1}{2}\partial\, x_2 - \cot\alpha\, \partial\, y_2 + \frac{1}{2}\cos\alpha\partial\, x_3 - (\frac{1}{2}\sin\alpha + \csc\alpha)\partial\, y_3,$$

$$\bar{J}_3 N = \cot\alpha\partial\, x_2 - \frac{1}{2}\partial\, y_2 + (\frac{1}{2}\sin\alpha - \csc\alpha)\,\partial\, x_3 + \frac{1}{2}\cos\alpha\, \partial\, y_3.$$

Hence, M is a real lightlike hypersurface of $\mathbf{R}_4^8$.

Consider the almost quaternion distribution

$$D = \{TM^\perp \oplus_{\text{orth}} D_1\} \oplus_{\text{orth}} D_0. \tag{8.2.3}$$

Denote by S and Q the projection morphisms D and D_2. Then, for $U_a = -\bar{J}_a\xi$ and $V_a = -\bar{J}_a N$, any vector field on M is expressed as

$$X = SX + \sum_{a=1}^{3} f_a(X)V_a, \tag{8.2.4}$$

where $f_a(X)$ are 1-forms locally defined on M by

$$f_a(X) = g(X, U_a). \tag{8.2.5}$$

Apply $\bar{J}_a$ to (8.2.4) and obtain

$$\bar{J}_a X = \phi_a X + f_a(X)N, a = 1, 2, 3 \tag{8.2.6}$$

where $\phi_a X$ are the tangential components of $\bar{J}_a X$. Since $\bar{J}_a^2 = -I$, we have

$$\phi_a^2 X = -X + f_a(X)V_a. \tag{8.2.7}$$

From (8.2.5) we get

$$f_a(V_a) = 1, \ f_a(V_b) = 0, \ a \neq b,$$

$$f_a\phi_a = 0.$$

From (8.1.2) and (8.2.6) we derive

$$\phi_a V_b = \bar{J}_a V_b - f_a(V_b)N = \bar{J}_a V_b = -\bar{J}_a\bar{J}_b N$$
$$= -\bar{J}_c N = V_c.$$

Using (8.2.4) and (8.2.6) we have

$$(f_a \circ \phi_b)X = f_a(\phi_b X) = g(\phi_b X, V_a) = f_c(X).$$

Finally, from (8.1.2), (8.2.4), (8.2.5), (8.2.6) and (8.2.7) we have

$$((\phi_a \circ \phi_b) - V_a \otimes f_b)\, X = \phi_a(\phi_b X) - f_b(X)V_a = \phi_c X.$$

This shows that there exists an almost contact 3-structure on a real lightlike hypersurface of an indefinite quaternion Kähler manifold, (for the definition of almost contact 3-structure, see [271]). Thus, we have the following theorem.

Theorem 8.2.2. *Let M be a lightlike hypersurface of an indefinite quaternion Kähler manifold such that ξ and N are globally defined on M. Then (ϕ_a, V_a, f_a) defines an almost contact 3-structure.*

Remark 8.2.3. From (8.2.4), (8.2.6) and (8.2.7) we see that there exist no almost metric contact 3-structures on lightlike real hypersurfaces of indefinite quaternion Kähler manifolds.

Let $\bar{M}$ be an indefinite quaternion Kähler manifold. Then, by using (8.1.2) and (8.1.8) and Gauss-Weingarten formulae for a lightlike hypersurface, we derive

$$\begin{aligned}
(\nabla_X \phi_a)Y &= f_a(Y)A_N X - B(X,Y)V_a + Q_{ab}(X)\phi_b Y \\
&\quad + Q_{ac}(X)\phi_c Y, \qquad\qquad\qquad\qquad (8.2.8) \\
(\nabla_X f_a)Y &= -B(X, \phi_a Y) - f_a(Y)\tau(X) + Q_{ab}(X)f_b(Y) \\
&\quad + Q_{ac}(X)f_c(Y), \quad \forall X, Y \in \Gamma(TM). \qquad (8.2.9)
\end{aligned}$$

Theorem 8.2.4. *Let $\bar{M}$ be an indefinite quaternion Kähler manifold and M a real lightlike hypersurface of $\bar{M}$. Then, M is totally geodesic if and only if*

$$\begin{aligned}
&C(X, U_a) = 0, \\
&(\nabla_X f_a)Y = 0, \quad \forall X \in \Gamma(TM), Y \in \Gamma(D), \quad and \quad U_a \in \Gamma(D_1).
\end{aligned}$$

Proof. For $Y = V_a \in \Gamma(D_2)$ from (8.2.8) we have

$$(\nabla_X \phi_a)V_a = A_N X - B(X, V_a)V_a + Q_{ab}(X)\phi_b V_a + Q_{ac}(X)\phi_c V_a.$$

Using (8.1.2) we get

$$-\phi_a \nabla_X V_a = A_N X - B(X, V_a)V_a - Q_{ab}(X)V_c + Q_{ac}(X)V_b.$$

For $U_a \in \Gamma(D_1)$, from (2.1.26) and (8.2.6) we derive

$$C(X, U_a) = B(X, V_a).$$

On the other hand, we get

$$(\nabla_X f_a)Y = B(X, Y)$$

for any $Y \in \Gamma(D)$, which completes the proof. $\qquad\square$

Theorem 8.2.5. *Let M be a real lightlike hypersurface of an indefinite quaternion Kähler manifold $\bar{M}$. Then, M is totally geodesic if and only if $A_N X$, $A_\xi^* X$, $\nabla_X^* Z_0 \notin \Gamma(D_2)$ for any $X \in \Gamma(TM)$, $N \in \Gamma(\mathrm{ltr}(TM))$, $\xi \in \Gamma(TM^\perp)$, $Z_0 \in \Gamma(D_0)$.*

Proof. It suffices to show that $h(X,Y) = 0$ for any $X,Y \in \Gamma(TM)$. By the definition of lightlike hypersurfaces, M is totally geodesic if and only if $h(X,V_a) = 0$, $h(X,U_a) = 0$ and $h(X,Z_0) = 0$ for any $X \in \Gamma(TM)$, $V_a \in \Gamma(D_1)$ and $U_a \in \Gamma(D_2)$. From (8.1.2) and (8.2.6) we have

$$B(X,V_a) = -f_a(A_N X).$$

From (8.1.8) we get

$$h(X,U_a) = -\bar{J}_a(A_\xi^* X) + \tau(X)U_a + Q_{ab}(X)V_b + Q_{ab}(X)V_c - \nabla_X U_a$$

and

$$h(X,\bar{J}_a Z_0) = \bar{J}_a(\nabla_X^* Z_0) - C(X,Z_0)U_a - B(X,Z_0)V_a + Q_{ab}(X)\bar{J}_b Z_0 \\ + Q_{ac}(X)\bar{J}_c Z_0 - \nabla_X Z_0.$$

The proof follows from these three equations.   □

Theorem 8.2.6. *Let M be a real lightlike hypersurface of an indefinite quaternion Kähler manifold. Then, D_0 is integrable if and only if*

$$C(X,Y) = C(Y,X), \quad C(X,\bar{J}_a Y) = C(Y,\bar{J}_a X),$$
$$B(X,\bar{J}_a Y) = B(Y,\bar{J}_a X), \quad \forall X,Y \in \Gamma(D_0).$$

Proof. From (8.2.1) we see that D_0 is integrable if and only if , we have

$$\bar{g}([X,Y],V_a) = \bar{g}([X,Y],U_a) = \bar{g}([X,Y],N) = 0$$

for any $V_a \in \Gamma(D_2)$, $U_a \in \Gamma(D_1)$ and $N \in \Gamma(\mathrm{ltr}(TM))$. Now by using (8.1.2) and (8.1.8), we have

$$\begin{aligned}
\bar{g}([X,Y],V_a) &= -\bar{g}([X,Y],\bar{J}_a N) \\
&= -\bar{g}(\bar{\nabla}_X Y, \bar{J}_a N) + \bar{g}(\bar{\nabla}_Y X, \bar{J}_a N) \\
&= \bar{g}(\bar{J}_a \bar{\nabla}_X Y, N) - \bar{g}(\bar{J}_a \bar{\nabla}_Y X, N) \\
&= \bar{g}(\bar{\nabla}_X \bar{J}_a Y - (\bar{\nabla}_X \bar{J}_a)Y, N) - \bar{g}(\bar{\nabla}_Y \bar{J}_a X - (\bar{\nabla}_Y \bar{J}_a)X, N) \\
&= \bar{g}(\bar{\nabla}_X \bar{J}_a Y - Q_{ab}(X)\bar{J}_b Y - Q_{ac}(X)\bar{J}_c Y, N) \\
&\quad - \bar{g}(\bar{\nabla}_Y \bar{J}_a X - Q_{ab}(Y)\bar{J}_b X - Q_{ac}(Y)\bar{J}_c X, N) \\
&= \bar{g}(\bar{\nabla}_X \bar{J}_a Y, N) - \bar{g}(\bar{\nabla}_Y \bar{J}_a X, N).
\end{aligned}$$

From (2.1.23) we get

$$\bar{g}([X,Y],V_a) = \bar{g}\left(h^*(X,\bar{J}_aY),N\right) - \bar{g}\left(h^*(Y,\bar{J}_aX),N\right)$$

or

$$\bar{g}([X,Y],V_a) = C(Y,\bar{J}_aX) - C(X,\bar{J}_aY). \tag{8.2.10}$$

In a similar way, we have

$$\bar{g}([X,Y],U_a) = B(Y,\bar{J}_aX) - B(X,\bar{J}_aY). \tag{8.2.11}$$

Finally, from (2.1.14) we get

$$\bar{g}([X,Y],N) = C(X,Y) - C(Y,X). \tag{8.2.12}$$

Thus, considering (8.2.10), (8.2.11) and (8.2.12) we derive the assertion of the theorem. $\square$

Corollary 8.2.7. *Let M be a real lightlike hypersurface of an indefinite quaternion Kähler manifold. If M is totally geodesic, then, D is parallel.*

Proof. Using (8.2.3) we have, D is parallel if and only if $g(\nabla_X Y, Z) = 0$ for any $X,Y \in \Gamma(D)$, $Z \in \Gamma(D_1)$. From (8.1.2) and (8.1.8) we get

$$\begin{aligned}
g(\nabla_X Y, U_a) &= \bar{g}(\bar{\nabla}_X Y, U_a) = -\bar{g}(\bar{\nabla}_X Y, \bar{J}_a\xi) = \bar{g}(\bar{J}_a\bar{\nabla}_X Y, \xi) \\
&= \bar{g}(\bar{\nabla}_X \bar{J}_a Y - (\bar{\nabla}X\bar{J}_a)Y, \xi) \\
&= \bar{g}(\bar{\nabla}_X \bar{J}_a Y + h(X,Y) - Q_{ab}(X)\bar{J}_b Y - Q_{ac}(X)\bar{J}_c Y, \xi).
\end{aligned}$$

Since M is totally geodesic and $\bar{J}_bY, \bar{J}_cY \in \Gamma(D)$, we have $g(\nabla_X Y, U_a) = 0$. Thus, D is parallel, which completes the proof. $\square$

Lemma 8.2.8. *Let $\bar{M}(c)$ be an indefinite quaternionic space form and M a real lightlike hypersurface of $\bar{M}(c)$. Then, we have*

$$\frac{c}{4}\sum_{a=1}^{3} g(Z,\bar{J}_aY)f_a(X) - g(Z,\bar{J}_aX)f_a(Y) + 2g(X,\bar{J}_aY)f_a(Z)$$

$$= (\nabla_X B)(Y,Z) - (\nabla_Y B)(X,Z) + B(Y,Z)\tau(X) - B(X,Z)\tau(Y), \tag{8.2.13}$$

$$R(X,Y)Z = \frac{c}{4}\{g(Y,Z)X - g(X,Z)Y + \sum_{a=1}^{3} g(Z,\bar{J}_aY)\phi_a(X)$$

$$- g(Z,\bar{J}_aX)\phi_a(Y) + 2g(X,\bar{J}_aY)\phi_a(Z) - B(X,Z)A_N Y + B(Y,Z)A_N X \tag{8.2.14}$$

for any $X,Y,Z \in \Gamma(TM)$.

Proof. Using Gauss-Weingarten formulas and (8.1.14), we obtain the lemma. $\square$

Theorem 8.2.9. *There exists no totally umbilical real hypersurface of $\bar{M}(c)$ with $c \neq 0$.*

Proof. Suppose M is a totally umbilical real hypersurface of $\bar{M}(c)$ with $c \neq 0$. Replacing X, Y, Z by PX, ξ, PZ in (8.2.13) and using (2.5.1) we have

$$\frac{c}{4} \sum_{a=1}^{3} f_a(PZ)f_a(PX) = -B(\nabla_{PX}\xi, PZ) - \nabla_\xi B(PX, PZ)$$

$$+ B(\nabla_\xi PX, PZ) + B(PX, \nabla_\xi PZ)$$
$$+ B(\xi, PZ)\tau(PX) - B(PX, PZ)\tau(\xi).$$

Since M is umbilical, we derive

$$\frac{c}{4} \sum_{a=1}^{3} f_a(PZ)f_a(PX) = -\rho g(\nabla_{PX}\xi, PZ) - \nabla_\xi \left(\rho g\left(PX, PZ\right)\right)$$

$$+ \rho g(\nabla_\xi PX, PZ) + \rho g(PX, \nabla_\xi PZ)$$
$$+ \rho g(\xi, PZ)\tau(PX) - \rho g(PX, PZ)\tau(\xi).$$

Hence, we arrive at

$$\frac{c}{4} \sum_{a=1}^{3} f_a(PZ)f_a(PX) = -\rho g(\nabla_{PX}\xi, PZ) - (\nabla_\xi \rho)\, g\left(PX, PZ\right) - \rho g(PX, PZ)\tau(\xi).$$

Since $\bar{\nabla}$ is a metric connection we have

$$\frac{c}{4} \sum_{a=1}^{3} f_a(PZ)f_a(PX) = \rho g(\xi, \bar{\nabla}_{PX} PZ) - (\nabla_\xi \rho)\, g\left(PX, PZ\right) - \rho g(PX, PZ)\tau(\xi).$$

Then, considering M is umbilical, we obtain

$$\frac{c}{4} \sum_{a=1}^{3} f_a(PZ)f_a(PX) = \left(\rho^2 - (\nabla_\xi \rho) - \rho\tau(\xi) \right) g(PX, PZ).$$

Now, for $PX = PZ = V_1$ we have $c = 0$. This is a contradiction. Thus, M can not be umbilical in $\bar{M}(c)$, $c \neq 0$, which completes the proof.   □

8.3   Screen real submanifolds

As in the indefinite Kähler geometry, lightlike submanifolds of indefinite quaternion Kähler manifolds are defined with respect to the complex structures $\bar{J}_1$, $\bar{J}_2$ and $\bar{J}_3$. A lightlike submanifold M of an indefinite quaternion Kähler manifold

$\bar{M}$ is called a *quaternion lightlike submanifold* if radical distribution and screen distribution of M are invariant with respect to $\bar{J}_a$, $a = 1, 2, 3$, i.e.,

$$\bar{J}_a(\text{Rad}\,TM) = \text{Rad}\,TM \quad \text{and} \quad \bar{J}_a(S(TM)) = S(TM).$$

It is easy to see that lightlike transversal bundle $\text{ltr}(TM)$ is invariant. In a similar way, $\bar{g}(\bar{J}_a X, \xi) = -\bar{g}(X, \bar{J}_a \xi)$ for $X \in \Gamma(S(TM))$ implies that $S(TM)$ is also invariant. One can see that the screen transversal bundle $S(TM^\perp)$ is also invariant. For quaternion lightlike submanifolds of indefinite quaternion Kähler manifolds, we have the following.

Theorem 8.3.1. *A quaternion lightlike submanifold M of an indefinite quaternion Kähler manifold is totally geodesic.*

Proof. From (8.1.8) and (5.1.15) we get

$$\nabla_X \bar{J}_a Y = -h^l(X, \bar{J}_a Y) - h^s(X, \bar{J}_a Y) + Q_{ab}(X)\bar{J}_b Y$$
$$+ Q_{ac}(X)\bar{J}_c Y + \bar{J}_a \nabla_X Y + \bar{J}_a h^l(X, Y)$$
$$+ \bar{J}_a h^s(X, Y), \forall X, Y \in \Gamma(TM).$$

Hence, we obtain

$$h^l(X, \bar{J}_a Y) = \bar{J}_a h^l(X, Y),$$
$$h^s(X, \bar{J}_a Y) = \bar{J}_a h^s(X, Y),$$

which imply that

$$h^l(\bar{J}_a X, \bar{J}_a Y) = -h^l(X, Y), \tag{8.3.1}$$
$$h^s(\bar{J}_a X, \bar{J}_a Y) = -h^s(X, Y). \tag{8.3.2}$$

In a similar way, we derive

$$h^l(\bar{J}_b X, \bar{J}_b Y) = -h^l(X, Y), \qquad h^l(\bar{J}_c X, \bar{J}_c Y) = -h^l(X, Y), \tag{8.3.3}$$
$$h^s(\bar{J}_b X, \bar{J}_b Y) = -h^s(X, Y), \qquad h^s(\bar{J}_c X, \bar{J}_c Y) = -h^s(X, Y). \tag{8.3.4}$$

On the other hand, using (8.1.2), (8.3.1), (8.3.2), (8.3.3) and (8.3.4) we get

$$h^l(\bar{J}_a X, \bar{J}_a Y) = h^l(\bar{J}_b \bar{J}_c X, \bar{J}_b \bar{J}_c Y) = -h^l(\bar{J}_c X, \bar{J}_c Y) = h^l(X, Y) \tag{8.3.5}$$

and

$$h^s(\bar{J}_a X, \bar{J}_a Y) = h^l(X, Y). \tag{8.3.6}$$

Then, (8.3.1), (8.3.2), (8.3.5) and (8.3.6) imply that $h^l(X, Y) = 0$ and $h^s(X, Y) = 0$, i.e., M is totally geodesic, which completes the proof. $\square$

Example 3. Consider in $\mathbf{R}_4^{12}$ a submanifold M given by the equations:

$$x_9 = x_1 \cos\alpha - x_3 \sin\alpha, \qquad x_{10} = x_2 \cos\alpha - x_4 \sin\alpha,$$
$$x_{11} = x_1 \sin\alpha + x_3 \cos\alpha, \qquad x_{12} = x_2 \sin\alpha + x_4 \cos\alpha, \quad \alpha \in (0, \tfrac{\pi}{2}).$$

Then TM is spanned by

$$Z_1 = \partial\, x_1 + \cos\alpha\, \partial\, x_9 + \sin\alpha\, \partial\, x_{11}, \quad Z_2 = \partial\, x_2 + \cos\alpha\, \partial\, x_{10} + \sin\alpha\, \partial\, x_{12},$$
$$Z_3 = \partial\, x_3 - \sin\alpha\, \partial\, x_9 + \cos\alpha\, \partial\, x_{11}, \quad Z_4 = \partial\, x_4 - \sin\alpha\, \partial\, x_{10} + \cos\alpha\, \partial\, x_{12},$$
$$Z_5 = \partial\, x_5, \quad Z_6 = \partial\, x_6, \qquad\qquad Z_7 = \partial\, x_7, \quad Z_8 = \partial\, x_8.$$

Hence M is a lightlike submanifold with $\operatorname{Rad} TM = \operatorname{Span}\{Z_1, Z_2, Z_3, Z_4\}$. Thus, $\operatorname{Rad} TM = TM^\perp \subset TM$, i.e., M is a coisotropic 8-dimensional submanifold of R_4^{12}. Then $S(TM^\perp) = \{0\}$ and the lightlike transversal bundle is spanned by

$$N_1 = \tfrac{1}{2}\{-\partial\, x_1 + \cos\alpha\, \partial\, x_9 + \sin\alpha\, \partial\, x_{11}\},$$
$$N_2 = \tfrac{1}{2}\{-\partial\, x_2 + \cos\alpha\, \partial\, x_{10} + \sin\alpha\, \partial\, x_{12}\},$$
$$N_3 = \tfrac{1}{2}\{-\partial\, x_4 - \sin\alpha\, \partial\, x_9 + \cos\alpha\, \partial\, x_{11}\},$$
$$N_4 = \tfrac{1}{2}\{-\partial\, x_4 - \sin\alpha\, \partial\, x_{10} + \cos\alpha\, \partial\, x_{12}\}.$$

It is easy to see that $\operatorname{Rad} TM$ and $D_0 = \operatorname{Span}\{Z_5, Z_6, Z_7, Z_8\}$ are invariant. Hence $TM = \operatorname{Rad} TM \oplus_{\mathrm{orth}} D_0$ is invariant. Thus M is a quaternion lightlike submanifold.

Theorem 8.3.1 tells that it would be more interesting to study other types of lightlike submanifolds of indefinite quaternion Kähler manifolds. For this aim, we present some results on the geometry of screen real lightlike submanifolds of indefinite quaternion Kähler manifolds. The definition of screen real submanifold is the same as the definition of screen real submanifolds of Kähler manifolds. More precisely, we say that a lightlike submanifold of an indefinite quaternion Kähler manifold is a *screen real lightlike submanifold* if $\operatorname{Rad} TM$ is invariant with respect to $\bar{J}_a$ and $\bar{J}_a(S(TM)) \subseteq S(TM^\perp)$ for all $a = 1, 2, 3$. Let M be a screen real lightlike submanifold of an indefinite quaternion Kähler manifold. Then, for $W \in \Gamma(S(TM))$, we write

$$\bar{J}_a W = B_a W + C_a W, \tag{8.3.7}$$

where $B_a W$ and $C_a W$ are the tangential and screen transversal part of $\bar{J}_a W$.

Example 4. Consider a submanifold M in $\mathbf{R}_4^{12}$ given by

$$\chi(\theta, \varphi, r.s, t) = (\theta, \varphi, r, s, -s, r, -\varphi, \theta, \sin t, 0, \cos t, 0).$$

The tangent bundle of M is spanned by

$$Z_1 = \partial\, x_1 + \partial\, x_8, \quad Z_2 = \partial\, x_2 - \partial\, x_7$$
$$Z_3 = \partial\, x_3 + \partial\, x_6, \quad Z_4 = \partial\, x_4 - \partial\, x_5$$
$$Z_5 = \cos t\, \partial\, x_9 - \sin t\, \partial\, x_{11}.$$

It is easy to see that M is a 4-lightlike submanifold. $\mathrm{Rad}\, TM$ is spanned by Z_1, Z_2, Z_3 and Z_4. Moreover, $\mathrm{Rad}\, TM$ is invariant with respect to the canonical complex structures $\bar{J}_1, \bar{J}_2$ and $\bar{J}_3$. Furthermore, by direct computations, we have

$$\bar{J}_1 Z_5 = \cos t\, \partial x_{10} - \sin t\, \partial x_{12}, \quad \bar{J}_2 Z_5 = \sin t\, \partial x_9 + \cos t\, \partial x_{11},$$

$$\bar{J}_3 Z_5 = \sin t\, \partial x_{10} + \cos t\, \partial x_{12},$$

which span non-degenerate space and this space is orthogonal to TM. Thus, $S(TM^\perp) = \mathrm{Span}\{\bar{J}_1 Z, \bar{J}_2 Z, \bar{J}_3 Z\}$. That is, $S(TM) = \mathrm{Span}\{Z_5\}$ is anti-invariant. Thus, M is a screen real lightlike submanifold.

Let M be a screen real lightlike submanifold of an indefinite quaternion Kähler manifold $\bar{M}$. Using (5.1.15), (5.1.27) and (8.3.7), we have

$$\nabla_X \bar{J}_a \xi = -h^l(X, \bar{J}_a \xi) - h^s(X, \bar{J}_a \xi) + Q_{ab}(X) \bar{J}_b \xi$$
$$+ Q_{ac}(X) \bar{J}_c \xi - \bar{J}_a A_\xi^* X + \bar{J}_a \nabla_X^{*t} \xi$$
$$+ \bar{J}_a h^l(X, \xi) + B_a h^s(X, \xi) + C_a h^s(X, \xi)$$

for $X \in \Gamma(TM)$ and $\xi \in \Gamma(\mathrm{Rad}\, TM)$. Thus, considering (5.1.4), we obtain

$$\nabla_X \bar{J}_a \xi = Q_{ab}(X) \bar{J}_b \xi + Q_{ac}(X) \bar{J}_c \xi + \bar{J}_a \nabla_X^{*t} \xi + B_a h^s(X, \xi), \qquad (8.3.8)$$
$$h^s(X, \bar{J}_a \xi) = -\bar{J}_a A_\xi^* X + C_a h^s(X, \xi), \qquad (8.3.9)$$
$$h^l(X, \bar{J}_a \xi) = \bar{J}_a h^l(X, \xi). \qquad (8.3.10)$$

Then, from (8.3.8) and Theorem 5.1.4, we have the following results:

Theorem 8.3.2. *Let M be a screen real lightlike submanifold of an indefinite quaternion Kähler manifold $\bar{M}$. The induced connection is a metric connection if and only if $h^s(X, \xi)$ has no components in $\bar{J}(S(TM))$ for $X \in \Gamma(TM)$ and $\xi \in \Gamma(\mathrm{Rad}\, TM)$.*

Corollary 8.3.3. *Any irrotational screen real lightlike submanifold of an indefinite quaternion Kähler manifold has the induced metric connection.*

Now, we discuss the integrability of radical and screen distributions of a screen real lightlike submanifold.

Theorem 8.3.4. *Let M be a screen real lightlike submanifold of an indefinite quaternion Kähler manifold $\bar{M}$. Then, the radical distribution is integrable if and only if*

$$B_a h^s(\xi_1, \bar{J}_a \xi_2) = B_a h^s(\bar{J}_a \xi_1, \xi_2), \quad \xi_1, \xi_2 \in \Gamma(\mathrm{Rad}\, TM). \qquad (8.3.11)$$

Proof. From (8.1.8) and (8.1.2) we have

$$\bar{\nabla}_{\xi_1} \xi_2 = -\bar{\nabla}_{\xi_1} \bar{J}_a^2 \xi_2$$
$$= -(\bar{\nabla}_{\xi_1} \bar{J}_a) \bar{J}_a \xi_2 - \bar{J}_a \bar{\nabla}_{\xi_1} \bar{J}_a \xi_2$$
$$= Q_{ab}(\xi_1) \bar{J}_c \xi_2 - Q_{ac}(\xi_1) \bar{J}_b \xi_2 - \bar{J}_a \bar{\nabla}_{\xi_1} \bar{J}_a \xi_2$$

for $\xi_1, \xi_2 \in \Gamma(\mathrm{Rad}\,TM)$. Using (5.1.15), (5.1.27) and taking tangential parts of the obtained equation, we get

$$\nabla_{\xi_1}\xi_2 = Q_{ab}(\xi_1)\bar{J}_c\xi_2 - Q_{ac}(\xi_1)\bar{J}_b\xi_2 - \bar{J}_a\nabla^{*t}_{\xi_1}\xi_2 - B_ah^s(\xi_1, \bar{J}\xi_2). \tag{8.3.12}$$

Changing the roles of ξ_1 and ξ_2 and subtracting, we arrive at

$$\begin{aligned}[\xi_1, \xi_2] &= Q_{ab}(\xi_1)\bar{J}_c\xi_2 - Q_{ac}(\xi_1)\bar{J}_b\xi_2 - \bar{J}_a\nabla^{*t}_{\xi_1}\xi_2 \\ &\quad - B_ah^s(\xi_1, \bar{J}\xi_2) - Q_{ab}(\xi_2)\bar{J}_c\xi_1 + Q_{ac}(\xi_2)\bar{J}_b\xi_1 \\ &\quad + \bar{J}_a\nabla^{*t}_{\xi_2}\xi_1 + B_ah^s(\xi_1, \bar{J}\xi_2). \end{aligned} \tag{8.3.13}$$

Since $\mathrm{Rad}\,TM$ is invariant with respect to $\bar{J}_a$, it is integrable, Then, from (8.3.12) we have (8.3.11). Conversely, if (8.3.11) holds, then, the right side of (8.3.13) belongs to $\mathrm{Rad}\,TM$. Hence, the left side of (8.3.13) also belongs to $\mathrm{Rad}\,TM$. □

Theorem 8.3.5. *Let M be a screen real lightlike submanifold of an indefinite quaternion Kähler manifold $\bar{M}$. If $\mathrm{Rad}\,TM$ is integrable, then, the radical distribution defines a totally geodesic foliation.*

Proof. From (8.3.13), the radical distribution defines a totally geodesic foliation in M if and only if $B_ah^s(\xi_1, \bar{J}_a\xi_2) = 0$. Suppose that $\mathrm{Rad}\,TM$ is integrable. Then, Theorem 8.3.4 implies that $B_ah^s(\xi_1, \bar{J}_a\xi_2) = B_ah^s(\bar{J}_a\xi_1, \xi_2)\ a = 1, 2, 3$. Using this and (8.1.2), we get

$$\begin{aligned} B_ah^s(\xi_1, \bar{J}_a\xi_2) &= B_ah^s(\xi_1, \bar{J}_b\bar{J}_c\xi_2) = B_ah^s(\bar{J}_b\xi_1, \bar{J}_c\xi_2) \\ &= B_ah^s(\bar{J}_c\bar{J}_b\xi_1, \xi_2) = -B_ah^s(\bar{J}_a\xi_1, \xi_2) \end{aligned}$$

which gives $B_ah^s(\xi_1, \bar{J}_a\xi_2) = 0$ that completes the proof. □

Theorem 8.3.6. *Let M be a screen real lightlike submanifold of an indefinite quaternion Kähler manifold $\bar{M}$. Then the screen distribution is integrable if and only if*

$$A_{\bar{J}_aX}Y = A_{\bar{J}_aY}X, \quad \forall X, Y \in \Gamma(S(TM)).$$

Proof. From (8.1.8), (5.1.15), (5.1.23) and (5.1.26) we obtain

$$\begin{aligned} A_{\bar{J}_aY}X &= \nabla^l_X\bar{J}_aY + D^s(X, \bar{J}_aY) - Q_{ab}(X)\bar{J}_bY \\ &\quad - Q_{ac}(X)\bar{J}_cY - \bar{J}_a\nabla^*_XY - \bar{J}_ah^*(X, Y) \\ &\quad - \bar{J}_ah^l(X, Y) - B_ah^s(X, Y) - C_ah^s(X, Y) \end{aligned} \tag{8.3.14}$$

for $X, Y \in \Gamma(S(TM))$. Taking the tangential parts of (8.3.14) we get

$$A_{\bar{J}_aY}X = -\bar{J}_ah^*(X, Y) - B_ah^s(X, Y). \tag{8.3.15}$$

Changing the roles of X and Y in (8.3.16) we derive

$$A_{\bar{J}_aX}Y = -\bar{J}_ah^*(Y, X) - B_ah^s(Y, X). \tag{8.3.16}$$

Then, since h^s is symmetric, from (8.3.15) and (8.3.16) we obtain

$$A_{\bar{J}_a Y} X - A_{\bar{J}_a X} Y = -\bar{J}_a h^*(X, Y) + \bar{J}_a h^*(Y, X). \tag{8.3.17}$$

The proof comes from (8.3.17) and Theorem 5.1.5.   □

Theorem 8.3.7. *Let M be a real lightlike submanifold of an indefinite quaternion Kähler manifold. Then, the screen distribution is parallel in M if and only if $A_{\bar{J}_a X} Y$ has no components in $\operatorname{Rad} TM$ for $X, Y \in \Gamma(S(TM))$.*

Proof. Suppose that $S(TM)$ is parallel. Then, from Theorem 5.1.6, $h^* = 0$. Also, (8.3.16) implies that $A_{\bar{J}_a X} Y = B_a h^s(Y, X)$ which shows that $A_{\bar{J}_a X} Y$ is $S(TM)$-valued. Conversely, assume that $A_{\bar{J}_a X} Y$ has no components in $\operatorname{Rad} TM$. Then, (8.3.16) implies that $\bar{J}_a h^*(X, Y) = 0$. Since the almost complex structures are non-singular, we conclude that $h^* = 0$. Then, Theorem 5.1.6 tells that $S(TM)$ is parallel.   □

Now we study totally umbilical screen real lightlike submanifolds. The proof of the following result comes from (8.3.8) and Theorem 5.1.4.

Theorem 8.3.8. *Let M be a totally umbilical screen real lightlike submanifold of an indefinite quaternion Kähler manifold $\bar{M}$. Then, the induced connection ∇ is a metric connection*

Let M be a screen real lightlike submanifold of an indefinite quaternion Kähler manifold $\bar{M}$. Consider the complementary vector sub-bundle ν to $\bar{J}_a(S(TM))$ in $S(TM^\perp)$. Then, we have

$$S(TM^\perp) = \bar{J}_a(S(TM)) \oplus_{\text{orth}} \nu, a = 1, 2, 3. \tag{8.3.18}$$

Since $\bar{J}_a(S(TM))$ and $S(TM^\perp)$ are non-degenerate, it follows that ν is also non-degenerate. Moreover, it is easy to see that ν is invariant.

Lemma 8.3.9. *Let M be a totally umbilical screen real lightlike submanifold of an indefinite quaternion Kähler manifold. Then, $H^s \in \Gamma(\bar{J}_a(S(TM)))$.*

Proof. Let M be a totally umbilical screen real lightlike submanifold and $X, Y \in \Gamma(S(TM))$. Using (8.1.8), (5.1.4), (5.1.26), (8.3.7) and taking the screen transversal parts of (8.3.18), we have

$$h^s(X, Y) = -Q_{ab}(X)\bar{J}_b Y - Q_{ac}(X)\bar{J}_c Y - \bar{J}_a \nabla_X^* Y - C_a h^s(X, Y).$$

Thus, for $V \in \Gamma(\nu)$, we have

$$\bar{g}(h^s(X, Y), \bar{J}_a V) = -\bar{g}(C_a h^s(X, Y), \bar{J}_a V).$$

Then, a totally umbilical M implies that

$$g(X, Y)\bar{g}(H^s, \bar{J}_a V) = -g(X, Y)\bar{g}(H^s, V). \tag{8.3.19}$$

Since ν is invariant with respect to $\bar{J}_a$ and V is an arbitrary vector field, substituting $\bar{J}_a V$ in (8.3.19) we derive

$$g(X,Y)\bar{g}(H^s,V) = g(X,Y)\bar{g}(H^s,\bar{J}_a V). \tag{8.3.20}$$

Thus, from (8.3.19) and (8.3.20) we obtain

$$g(X,Y)\bar{g}(H^s,V) = 0.$$

$S(TM)$ non-degenerate and $\bar{g}(H^s,V) = 0$ shows that $H^s \in \Gamma(\bar{J}_a(S(TM)))$.   □

Theorem 8.3.10. *Let M be a totally umbilical screen real lightlike submanifold of an indefinite quaternion Kähler manifold. Then, M is either totally geodesic or $S(TM)$ is 1-dimensional.*

Proof. From (8.3.14) and (5.1.28), for $X,Y \in \Gamma(S(TM))$, we have

$$\bar{g}(h^s(X,X),\bar{J}_a Y) = \bar{g}(h^s(X,Y),\bar{J}_a X).$$

Using (5.3.8), we obtain

$$g(X,X)\bar{g}(H^s,\bar{J}_a Y) = g(X,Y)\bar{g}(H^s,\bar{J}_a X). \tag{8.3.21}$$

Changing the role X and Y in (8.3.21), we get

$$g(Y,Y)\bar{g}(H^s,\bar{J}_a X) = g(X,Y)\bar{g}(H^s,\bar{J}_a Y). \tag{8.3.22}$$

Thus, from (8.3.21) and (8.3.22), we have

$$\bar{g}(H^s,\bar{J}_a X) = \frac{g(X,Y)^2}{g(X,X)g(Y,Y)}\bar{g}(H^s,\bar{J}_a X). \tag{8.3.23}$$

Considering Lemma 8.3.9, the solution of the equation (8.3.23) is either $H^s = 0$ or X and Y are linearly dependent, which completes the proof.   □

8.4   QR-lightlike submanifolds

In this section, we introduce and investigate QR-lightlike submanifolds which are a generalization of lightlike real hypersurfaces of indefinite quaternion Kähler manifolds and correspond to Riemannian QR-submanifolds.

Definition 8.4.1. [343] Let M be a lightlike submanifold of an indefinite quaternion Kähler manifold $\bar{M}$. We say that M is a *QR-lightlike submanifold* if the following conditions are fulfilled:

$$J_a \operatorname{Rad} TM \cap \operatorname{Rad} TM = \{0\}, \tag{8.4.1}$$

$$S(TM) = \{J_a \operatorname{Rad} TM \oplus D'\} \perp D_0, \tag{8.4.2}$$

$$J_a(D_0) = D_0, J_a(L_1 \perp L_2) = D' \subset S(TM),$$

where $L_1 = \operatorname{ltr}(TM)$ and L_2 is a vector sub-bundle of $S(TM^\perp)$.

Let M be a QR-lightlike submanifold of an indefinite quaternion Kähler manifold. For $x \in M$, let $D_{ax} = J_a(L_1 \perp L_2)$, $a = 1, 2, 3$. Then, D_{1x}, D_{2x}, D_{3x} are mutually orthogonal vector subspaces of $T_x M$. We consider $D'_x = D_{1x} \oplus D_{2x} \oplus D_{3x}$, then $D' : x \to D'_x$ is globally a $3d$-dimensional distribution defined on M, where $\dim (L_1 \perp L_2)_x = d$. Also we have

$$J_a(D_{ax}) = L_1 \perp L_2 \mid_x, J_a(D_{bx}) = D_{cx},$$

where (a, b, c) is a cycle permutation of $(1, 2, 3)$. Now, we consider

$$D = \{\operatorname{Rad} TM \oplus_{\operatorname{orth}} J_a \operatorname{Rad} TM\} \oplus_{\operatorname{orth}} D_0, \tag{8.4.3}$$

where D_0 is non-degenerate distribution. It is easy to check that D_0 is invariant with respect to each J_a. Thus we have

$$TM = D \oplus D'.$$

From (8.4.3), it follows that D is invariant with respect to the action of J_a and we call D a *quaternion lightlike distribution*. (8.4.3) also implies that D is a $4r$-lightlike almost quaternion distribution on M. Thus we have the following proposition.

Proposition 8.4.2. *Any QR-lightlike 7-dimensional submanifold must be 1-lightlike.*

Example 5. Let $(\bar{M}, J_a, \bar{g})$, $a = 1, 2, 3$ be a real $4n$-dimensional indefinite quaternion Kähler manifold, where $\bar{g}$ is a semi-Riemann metric of index $\nu = 4q$, $0 < q < m$. We suppose that (M, g) is a lightlike real hypersurface of $\bar{M}$, where g is the degenerate induced metric of M. Now, if ξ is a local section of $TM^\perp$, then, $g(J_a\xi, \xi) = 0$ $(a = 1, 2, 3)$ and $J_a\xi$ are tangent to M. Thus $(J_1 TM^\perp \oplus_{\operatorname{orth}} J_2 TM^\perp \oplus_{\operatorname{orth}} J_3 TM^\perp)$ is a distribution on M with rank 3 such that $J_a TM^\perp \cap TM^\perp = \{0\}, a = 1, 2, 3$ that is, the condition (8.4.1) is satisfied. Hence we can choose a screen distribution $S(TM)$ such that it contains $J_a TM^\perp, a = 1, 2, 3$ as a vector sub-bundle. Now we consider the local section N of $\operatorname{tr}(TM)$. Since

$$\bar{g}(J_a N, \xi) = \bar{g}(N, J_a\xi) = 0,$$

we deduce that $J_a N$ are tangent to M. On the other hand, since $\bar{g}(J_a N, N) = 0$ we see that the components of $J_a N$ with respect to ξ vanishes. Thus, $J_a N \in \Gamma(S(TM))$. We know that ξ and N are null vector fields satisfying $g(\xi, N) = 1$, Thus, we deduce that $J_a N$ and $J_a\xi$ are null vector fields satisfying $\bar{g}(J_a N, J_a\xi) = 1$, $a = 1, 2, 3$. Otherwise, we have

$$g(J_a N, J_b\xi) = 0$$

for $a \neq b$. Hence $J_a TM^\perp \oplus J_a(\operatorname{ltr}(TM))$ $(a = 1, 2, 3)$ is a vector sub-bundle of $S(TM)$ of rank 6. Then there exists a non-degenerate distribution D_0 on M such that

$$S(TM) = \{D_1 \oplus D_2\} \perp D_0$$

where

$$D_1 = J_1\xi \oplus_{\text{orth}} J_2\xi \oplus_{\text{orth}} J_3\xi$$

and

$$D_2 = J_1 N \oplus_{\text{orth}} J_2 N \oplus_{\text{orth}} J_3 N.$$

Hence, condition (8.4.2) is also satisfied. Thus, a lightlike hypersurface M of an indefinite quaternion Kähler manifold $(\bar{M}, \bar{g}, \bar{J}_a)$ is a QR-lightlike submanifold.

Example 6. Consider a submanifold M in $\mathbf{R}_4^{12}$ given by the following equations:

$$x_1 = u_1,\ y_1 = u_2,\ x_2 = u_3,\ y_2 = u_4,\ x_3 = u_5,\ y_3 = u_6,\ x_4 = u_7,\ y_4 = u_8$$
$$x_5 = -u_3 \cos\alpha - u_4 \sin\alpha - u_5 u_6 \tan\alpha + u_9$$
$$y_5 = u_3 \sin\alpha - u_4 \cos\alpha + u_5 u_6 + u_{10}$$
$$x_6 = u_1 \cos\alpha - u_2 \sin\alpha - u_7 u_8 \tan\alpha$$
$$y_6 = u_1 \sin\alpha + u_2 \cos\alpha + u_7 u_8$$

where $\alpha \in \mathbf{R} - \left\{\frac{\pi}{2} + k\pi, k \in Z\right\}$. Then the tangent bundle of M is spanned by

$$Z_1 = \partial x_1 + \cos\alpha\, \partial x_6 + \sin\alpha\, \partial y_6$$
$$Z_2 = \partial y_1 - \sin\alpha\, \partial x_6 + \cos\alpha\, \partial y_6$$
$$Z_3 = \partial x_2 - \cos\alpha\, \partial x_5 + \sin\alpha\, \partial y_5$$
$$Z_4 = \partial y_2 - \sin\alpha\, \partial x_5 - \cos\alpha\, \partial y_5$$
$$Z_5 = \partial x_3 - u_6 \tan\alpha\, \partial x_5 + u_6\, \partial y_5$$
$$Z_6 = \partial y_3 - u_5 \tan\alpha\, \partial x_5 + u_5\, \partial y_5$$
$$Z_7 = \partial x_4 - u_8 \tan\alpha\, \partial x_6 + u_8\, \partial y_6$$
$$Z_8 = \partial y_4 - u_7 \tan\alpha\, \partial x_6 + u_7\, \partial y_6$$
$$Z_9 = \partial x_5,\ Z_{10} = \partial y_5.$$

Thus M is a 1-lightlike submanifold with $\operatorname{Rad} TM = \operatorname{Span}\{Z_1\}$. By using canonical complex structures of $\mathbf{R}_4^{12}$ we obtain that $J_1(\operatorname{Rad} TM)$, $J_2(\operatorname{Rad} TM)$ and $J_3(\operatorname{Rad} TM)$ are spanned by Z_2, Z_3 and Z_4, respectively. Hence $J_1(\operatorname{Rad} TM) \cap \operatorname{Rad} TM = \{0\}$, $J_2(\operatorname{Rad} TM) \cap \operatorname{Rad} TM = \{0\}$ and $J_3(\operatorname{Rad} TM) \cap \operatorname{Rad} TM = \{0\}$. Thus condition (8.4.1) is satisfied. Choose $S(TM^\perp)$ spanned by

$$W = a\,\partial x_1 + b\,\partial y_1 - b\,u_8 \sec\alpha\, \partial x_4 - b\,u_8 \sec\alpha\, \partial y_4$$
$$+ (a \cos\alpha - b \sin\alpha)\, \partial x_6 + (a \sin\alpha + b \cos\alpha)\, \partial y_6$$

where a, b are non-zero constants. Take a vector field of $S(TM^\perp)^\perp$ to be

$$V = c\,\partial x_1 + d\,\partial y_1 + A\,\partial x_4 + B\,\partial y_4 + C\,\partial x_6 + D\,\partial y_6$$

where c, d are constants and

$$A = \frac{\frac{a}{b} u_8 \cos \alpha + u_7 \cos \alpha}{(u_7^2 + u_8^2)},$$

$$B = \frac{\frac{a}{b} u_7 \cos \alpha - u_8 \cos \alpha}{(u_7^2 + u_8^2)},$$

$$C = (1 + c) \cos \alpha - d \sin \alpha,$$

$$D = (1 + c) \sin \alpha + d \cos \alpha.$$

It is easy to obtain that $g(Z_1, V) = 1$. Thus, the complementary vector bundle F to $\operatorname{Rad} TM$ in $S(TM^\perp)^\perp$ is spanned by V. Hence, we obtain the lightlike transversal vector bundle

$$\operatorname{ltr}(TM) = \operatorname{Span}\{N = (c - \frac{G}{2}) \partial x_1 + d \partial y_1 + A \partial x_5 + B \partial y_5 + (C - \frac{G}{2}) \partial x_6$$
$$+ (D - \frac{F}{2}) \partial y_6\}$$

where

$$G = \frac{((\frac{a}{b})^2 + 1) \cos^2 \alpha}{u_7^2 + u_8^2} + 1 + 2c.$$

On the other hand by using the canonical complex structures of $\mathbf{R}_4^{12}$ we have

$$J_1 W = -b \partial x_1 + a \partial y_1 + b u_7 \sec \alpha \partial x_4 - b u_8 \sec \alpha \partial y_4 - (a \sin \alpha + b \cos \alpha) \partial x_6$$
$$+ (a \cos \alpha - b \sin \alpha) \partial y_6,$$
$$J_2 W = a \partial x_2 - b \partial y_2 + b u_8 \sec \alpha \partial x_3 - b u_7 \sec \alpha \partial y_3 + (-a \cos \alpha + b \sin \alpha) \partial x_5$$
$$+ (a \sin \alpha + b \cos \alpha) \partial y_5,$$
$$J_3 W = b \partial x_2 + a \partial y_2 + b u_7 \sec \alpha \partial x_3 + b u_8 \sec \alpha \partial y_3 - (a \cos \alpha + b \sin \alpha) \partial x_5$$
$$+ (-a \cos \alpha + b \sin \alpha) \partial y_5,$$

and

$$J_1 N = -d \partial x_1 + (C - \frac{G}{2}) \partial y_1 - B \partial x_4 + A \partial y_4 + (-D + \frac{G}{2}) \sin \alpha \partial x_6$$
$$+ (C - \frac{G}{2}) \cos \alpha \partial y_6,$$
$$J_2 N = (c - \frac{G}{2}) \partial x_2 - d \partial y_2 - A \partial x_3 + B \partial y_3 + (-C + \frac{G}{2}) \cos \alpha \partial x_5$$
$$+ (D - \frac{G}{2}) \sin \alpha \partial y_5,$$
$$J_3 N = d \partial x_2 + (c - \frac{G}{2}) \partial y_2 - B \partial x_3 - A \partial y_3 + (-D + \frac{G}{2}) \sin \alpha \partial x_5$$
$$+ (-C + \frac{G}{2}) \cos \alpha \partial y_5.$$

Since $S(TM^{\perp})$ is spanned by W, $D' = J_1 W \perp J_2 W \perp J_3 W$ is a distribution on M. On the other hand, $\bar{g}(J_1 W, Z_1) = \bar{g}(W, J_1 Z_1) = \bar{g}(W, Z_2) = 0$ and $\bar{g}(J_1 W, W) = 0$ imply that $J_1 W$ is tangent to M. Similarly one can see that $\bar{g}(J_2 W, Z_1) = \bar{g}(J_3 W, Z_1) = 0$. Therefore , we can choose $S(TM)$ to contain $J_1 W, J_2 W$ and $J_3 W$. Finally, since $\bar{g}(J_1 N, Z_1) = \bar{g}(N, J_1 Z_1) = \bar{g}(N, Z_2) = 0$ and $\bar{g}(J_1 N, N) = 0$ we can choose $S(TM)$ to contain $J_1 N, J_2 N$ and $J_3 N$ too. Thus we obtain

$$S(TM) = \{(J_1 Z_1 \oplus_{\text{orth}} J_2 Z_1 \oplus_{\text{orth}} J_3 Z_1) \oplus (J_1 N \oplus_{\text{orth}} J_2 N \oplus_{\text{orth}} J_3 N)\}$$
$$\perp (J_1 W \perp J_2 W \perp J_3 W).$$

Hence, condition (8.4.2) is satisfied, so M is a QR-lightlike submanifold.

Suppose M is an invariant lightlike submanifold of $\bar{M}$. Then, it is easy to check that $J_a(\text{Rad}\,TM) = \text{Rad}\,TM$. Hence, M is not a QR-lightlike submanifold. Next, suppose that M is a screen real lightlike submanifold. Then, one can check that $J_a(\text{Rad}\,TM) = \text{Rad}\,TM$. Hence, a screen real lightlike submanifold also is not a QR-lightlike submanifold. Thus, it follows that the class of QR-lightlike submanifolds contains lightlike real hypersurfaces, but it excludes quaternion lightlike submanifolds as well as screen real lightlike submanifolds. The above example motivates us to consider the following theorem (the proof is common with the proof of Theorem 6.4.3).

Theorem 8.4.3. *Let M be a 1-lightlike submanifold of codimension 2 of a real 4m-dimensional indefinite quaternion Hermitian manifold $(\bar{M}, J_a, \bar{g})$ such that $J_a(\text{Rad}\,TM)$ are distributions on M. Then M is a QR-lightlike submanifold.*

Denote by P the projection morphism of TM to the D and choose a local orthonormal frame $\{v_1, \ldots, v_r, w_1, \ldots, w_k\}$, $k + r = l$ on the vector sub-bundle $L_1 \perp L_2$ in $\text{tr}(TM)$. Then on the distribution D' we have a local field of orthonormal frames

$$\{E_{11}, \ldots, E_{1r}, E_{21}, \ldots, E_{2r}, E_{31}, \ldots, E_{3r}, F_{11}, \ldots, F_{1k}, F_{21}, \ldots, F_{2k}, F_{31}, \ldots, F_{3k}\},$$
$$(8.4.4)$$

where $E_{ai} = J_a(v_i)$, $a = 1, 2, 3$ $i = 1, \ldots, r$ and $F_{aj} = J_a(w_j)$, $j = 1, \ldots, k$. Thus any vector field Y tangent to M can be written locally as

$$Y = PY + \sum_{b=1}^{3} \left\{ \left\{ \sum_{i=1}^{r} \omega_{bi}(Y) E_{bi} \right\} + \left\{ \sum_{j=1}^{k} \omega'_{bj}(Y) F_{bj} \right\} \right\} \qquad (8.4.5)$$

where ω_{bi} and ω'_{bj} are 1-forms locally defined on M by

$$\omega_{bi}(Y) = g(Y, J_b \xi_i) \qquad (8.4.6)$$

and

$$\omega'_{bj}(Y) = g(Y, F_{bj}). \qquad (8.4.7)$$

Applying J_a to (8.4.5), we obtain

$$J_a Y = J_a P Y + \{\sum_{i=1}^{r} \omega_{bi}(Y) E_{ci} - \omega_{ci}(Y) E_{bi} - \omega_{ai}(Y) v_i$$

$$+ \sum_{j=1}^{k} \omega'_{bj}(Y) F_{cj} - \omega'_{cj}(Y) F_{bj} - \omega'_{aj}(Y) w_j\}. \qquad (8.4.8)$$

Let M be a QR-lightlike submanifold of an indefinite quaternion Kähler manifold $\bar{M}$ and $L_2^{\perp}$ be a complementary vector sub-bundle to L_2 in $S(TM^{\perp})$. Then, we have the following decomposition for the transversal bundle of M:

$$\text{tr}(TM) = L_1 \oplus_{\text{orth}} (L_2 \perp L_2^{\perp}). \qquad (8.4.9)$$

Thus, for $V \in \Gamma(S(TM^{\perp}))$, we have

$$J_a V = B_a V + C_a V, \qquad (8.4.10)$$

where $B_a V \in \Gamma(D')$ and $C_a V \in \Gamma(L_2^{\perp})$. Then, using (5.1.14), (5.1.15), (8.4.5), (8.4.8) and (8.4.10) in (8.1.8) and taking the transversal bundle parts we have

$$h(X, J_a P Y) - \omega_{ci}(Y) h(X, E_{bi}) + \omega_{bi}(Y) h(X, E_{ci}) - X(\omega_{ai}(Y)) v_i$$
$$- \omega_{ai}(Y) \nabla_X^l v_i - \omega_{ai}(Y) D^s(X, v_i) + \omega'_{bj}(Y) h(X, F_{cj}) - \omega'_{cj}(Y) h(X, F_{bj})$$
$$- X(\omega'_{aj}(Y)) w_j - \omega'_{aj}(Y) D^l(X, w_j) - \omega'_{aj}(Y) \nabla_X^s w_j = -Q_{ab}(X) \omega_{bi}(Y) v_i$$
$$- Q_{ab}(X) \omega'_{bj}(Y) w_j - Q_{ac}(X) \omega_{ci}(Y) v_i - Q_{ab}(X) \omega'_{cj}(Y) w_j$$
$$- \omega_{ai}(\nabla_X Y) v_i - \omega'_{aj}(\nabla_X Y) w_j + C_a h(X, Y) \qquad (8.4.11)$$

for any $X, Y \in \Gamma(TM)$. Thus for any $X, Y \in \Gamma(D)$, we get

$$h(X, J_a P Y) - C_a h(X, Y) = -\omega_{ai}(\nabla_X Y) v_i - \omega'_{aj}(\nabla_X Y) w_j. \qquad (8.4.12)$$

We say that M is a *D-geodesic QR-lightlike submanifold* of $\bar{M}$ if $h(X, Y) = 0$, for $X, Y \in \Gamma(D)$. Using this definition, we obtain the following result for integrability of the quaternion distribution D.

Theorem 8.4.4. *Let M be a QR-lightlike submanifold of an indefinite quaternion Kähler manifold. Then the following assertions are equivalent:*

(a) $h(J_a X, Y) = h(X, J_a Y)$ $a = 1, 2, 3$ *and* $X, Y \in \Gamma(D)$.

(b) M *is D-geodesic.*

(c) D *is integrable.*

Proof. First, note that by the definition of QR-lightlike submanifold, D is integrable if and only if $g([X, Y], J_a \xi_i) = 0$ and $g([X, Y], F_{aj}) = 0$.

(a)⇒ (b): From (8.1.2), we have

$$h(X, J_a Y) = h(J_a X, Y) = h((J_b o J_c) X, Y)$$
$$= h(J_c X, J_b Y) = h(X, (J_c o J_b) Y)$$
$$= -h(X, J_a Y).$$

Thus we get $h(X, J_a Y) = 0$. For (b) ⇒ (c), from (8.4.12) we obtain

$$\omega_{ai}(\nabla_X Y) v_i + \omega'_{aj}(\nabla_X Y) w_j = 0,$$

$$\omega_{ai}(\nabla_Y X) v_i + \omega'_{aj}(\nabla_Y X) w_j = 0.$$

Hence we have $\omega_{ai}([X, Y]) v_i + \omega'_{aj}([X, Y]) w_j = 0$. Since v_i and w_j are linearly independent we derive $\omega_{ai}([X, Y]) = g([X, Y], J_a \xi_i) = 0$ and $\omega'_{aj}([X, Y]) = g([X, Y], F_{aj}) = 0$ which show that $[X, Y] \in \Gamma(D)$.

(c) ⇒ (a): Suppose that D is integrable. Then from (8.4.12) we have

$$h(X, J_a PY) - C_a h(X, Y) = -\omega_{ai}(\nabla_X Y) v_i - \omega'_{aj}(\nabla_X Y) w_j$$

or

$$h(Y, J_a PX) - C_a h(Y, X) = -\omega_{ai}(\nabla_Y X) v_i - \omega'_{aj}(\nabla_Y X) w_j,$$

which implies that $h(X, J_a Y) = h(J_a X, Y)$.   □

We say that M is *a mixed geodesic QR-lightlike submanifold* if $h(X, Z) = 0$ for any $X \in \Gamma(D)$, $Z \in \Gamma(D')$.

Lemma 8.4.5. *Let M be a QR-lightlike submanifold of an indefinite quaternion Kähler manifold. Then M is mixed geodesic if and only if*

$$A_V X \in \Gamma(D), \tag{8.4.13}$$

$$\nabla^t_X V \in \Gamma(L_1 \perp L_2) \tag{8.4.14}$$

for any $X \in \Gamma(D)$ and $V \in \Gamma(L_1 \perp L_2)$.

Proof. First, since $(L_1 \perp L_2)$ is anti-invariant there exists a non-zero vector field $Z \in \Gamma(D')$ such that $Z = J_a(V)$. By using (5.1.12) we have $h(X, J_a V) = \bar{\nabla}_X J_a V - \nabla_X J_a V$. From (8.1.8) we get

$$h(X, J_a V) = (\bar{\nabla}_X J_a) V + J_a \bar{\nabla}_X V - \nabla_X J_a V$$
$$= Q_{ab}(X) J_b V + Q_{ac}(X) J_c V + J_a \bar{\nabla}_X V - \nabla_X J_a V.$$

From (5.1.14) we derive

$$h(X, J_a V) = Q_{ab}(X) J_b V + Q_{ac}(X) J_c V - J_a A_V X + J_a \nabla^t_X V - \nabla_X J_a V.$$

Using (8.4.8), (8.4.10) and taking the tangent and transversal bundle parts we obtain

$$Q_{ab}(X)J_bV + Q_{ac}(X)J_cV - J_aPA_VX - \sum_{i=1}^{r}\omega_{bi}(A_VX)E_{ci}$$

$$- \omega_{ci}(A_VX)E_{bi} - \sum_{j=1}^{k}\omega'_{bj}(A_VX)F_{cj} - \omega'_{cj}(A_VX)F_{bj}$$

$$+ B_a\nabla_X^tV - \nabla_XJ_aV = 0 \tag{8.4.15}$$

and

$$h(X, J_aV) = \omega_{ai}(A_VX)v_i + \omega'_{aj}(A_VX)w_j + C_a\nabla_X^tV. \tag{8.4.16}$$

Now, if M is mixed geodesic, then from (8.4.16) we have

$$\omega_{ai}(A_VX)v_i + \omega'_{aj}(A_VX)w_j + C_a\nabla_X^tV = 0.$$

Since $(L_1 \perp L_2) \cap L_2^{\perp} = \{0\}$, we get $\omega_{ai}(A_VX)v_i + \omega'_{aj}(A_VX)w_j = 0$ and $C_a\nabla_X^tV = 0$. Thus $\nabla_X^tV \in \Gamma(L_1 \perp L_2)$. Since v_i and w_j are linearly independent we derive $\omega_{ai}(A_VX) = 0, \omega'_{aj}(A_VX)$ that is $g(A_VX, J_a\xi_i) = 0$ and $g(A_VX, F_{aj}) = 0$, hence $A_VX \in \Gamma(D)$. Conversely we suppose that $A_VX \in \Gamma(D)$ and $\nabla_X^tV \in \Gamma(L_1 \perp L_2)$, then from (8.4.16)) we obtain $h(X, Z) = 0$, $X \in \Gamma(D)$ and $Z \in \Gamma(D'.)$ Thus proof of the lemma is complete. □

Theorem 8.4.6. *Let M be a QR-lightlike submanifold of an indefinite quaternion Kähler manifold. Then D' is parallel if and only if A_VY has no components in quaternion distribution D where, $Y \in \Gamma(D')$ and $V \in \Gamma(L_1 \perp L_2)$.*

Proof. Let $\{n_1, \ldots, n_l\}$ $r + k = l$ be a local field of orthonormal frames on the vector sub-bundle $L_1 \perp L_2$. Then we have the local field of orthonormal frames on D',

$$\{K_{11}, \ldots, K_{1l}, K_{21}, \ldots, K_{2l}, K_{31}, \ldots, K_{3l}\}$$

where $K_{as} = J_a(n_s), a = 1, 2, 3$ and $s = 1, \ldots, l$. Thus by using (8.1.8), (5.1.14), (8.4.10) and taking tangent parts we have

$$\nabla_{K_{as}}K_{ap} = Q_{ab}(K_{as})K_{bp} + Q_{ac}(K_{as})K_{cp} - J_aPA_{n_p}K_{as} \tag{8.4.17}$$

$$- \sum_{i=1}^{r}\omega_{bi}(A_{n_p}K_{as})E_{ci} - \omega_{ci}(A_{n_p}K_{as})E_{bi}$$

$$- \sum_{j=1}^{k}\omega'_{bj}(A_{n_p}K_{as})F_{cj} - \omega'_{cj}(A_{n_p}K_{as})F_{bj}$$

$$+ B_a\nabla_{K_{as}}^tn_p.$$

In a similar way we get

$$\nabla_{K_{as}} K_{bp} = Q_{bc}(K_{as})K_{cp} + Q_{ba}(K_{as})K_{ap} - J_b P A_{n_p} K_{as} \tag{8.4.18}$$

$$- \sum_{i=1}^{r} \omega_{ci}(A_{n_p} K_{as})E_{ai} - \omega_{ai}(A_{n_p} K_{as})E_{ci}$$

$$- \sum_{j=1}^{k} \omega'_{cj}(A_{n_p} K_{as})F_{aj} - \omega'_{aj}(A_{n_p} K_{as})F_{cj}$$

$$+ B_b \nabla^t_{K_{as}} n_p,$$

where $p = 1, \ldots, l$. From (8.4.17) and (8.4.18) we have the assertion of the theorem.

$\square$

From Lemma 8.4.5 and (8.4.15) we have the following lemma.

Lemma 8.4.7. *Let M be a mixed geodesic QR-lightlike submanifold of $\bar{M}$. Then we have the following expressions:*

(a) $g(A_{w_j} J_a X + J_a A_{w_j} X, Z) = 0$, $X \in \Gamma(D)$, $Z \in \Gamma(D_0)$, $w_j \in \Gamma(L_2)$.

(b) $\nabla_X F_{11} = -J_1 A_{w_1} X + B_1 \nabla^t_X w_1 + Q_{12}(X)F_{21} + Q_{13}(X)F_{31}$, $X \in \Gamma(D)$.

Let M be a QR-lightlike submanifold of an indefinite quaternion Kähler manifold. Then, we say that M is a *mixed foliate QR-lightlike submanifold* if M is mixed geodesic and D is integrable.

Theorem 8.4.8. *There exists no mixed foliate QR-lightlike submanifold of an indefinite quaternionic space form with positive or negative sectional curvature.*

Proof. Suppose M is a mixed foliate QR-lightlike submanifold of $\bar{M}(c), c \neq 0$. Taking $X, Y \in \Gamma(D_0), Z = F_{11}$ in (5.3.11)and using (8.1.14) we get

$$-\frac{c}{2}g(X, J_1 Y)w_1 = (\nabla_X h)(Y, F_{11}) - (\nabla_Y h)(X, F_{11}).$$

On the other hand, M is mixed foliate and ∇ is torsion free, so we obtain

$$(\nabla_X h)(Y, F_{11}) - (\nabla_Y h)(X, F_{11}) = -h(Y, \nabla_X F_{11}) + h(X, \nabla_Y F_{11}).$$

From Lemma 8.4.7 (b), we derive

$$(\nabla_X h)(Y, F_{11}) - (\nabla_Y h)(X, F_{11}) = -h(Y, -J_1 A_{w_1} X + B_1 \nabla^t_X w_1$$
$$+ Q_{12}(X)F_{21} + Q_{13}(X)F_{31})$$
$$+ h(X, -J_1 A_{w_1} Y + B_1 \nabla^t_Y w_1$$
$$+ Q_{12}(Y)F_{21} + Q_{13}(Y)F_{31}).$$

Since M is mixed geodesic we have

$$(\nabla_X h)(Y, F_{11}) - (\nabla_Y h)(X, F_{11}) = h(Y, J_1 A_{w_1} X) - h(X, J_1 A_{w_1} Y).$$

From Theorem 8.4.4, it follows that M is D-geodesic due to the fact that D is integrable. Thus $(\nabla_X h)(Y, F_{11}) - (\nabla_Y h)(X, F_{11}) = 0$, hence $\frac{c}{2} g(X, J_1 Y) w_1 = 0$. Since D_0 is non-degenerate we have $c = 0$, which completes the proof. $\qquad\square$

In the rest of this section, we study totally umbilical QR-lightlike submanifolds. First, from (8.4.11) we have the following.

Lemma 8.4.9. *Let M be a QR-lightlike submanifold of an indefinite quaternion Kähler manifold. If D_0 is integrable, then*

$$h(X, J_a Y) = h(J_a X, Y), \quad \forall X, Y \in \Gamma(D_0).$$

Theorem 8.4.10. *There exists no totally umbilical QR-lightlike submanifold in an indefinite quaternion Kähler manifold with positive or negative null sectional curvature with respect to ξ and $\bar{\nabla}$.*

Proof. Suppose that M is a totally umbilical QR-lightlike submanifold of $\bar{M}$ with $\bar{K}_\xi \neq 0$. From (8.1.11), we have

$$\bar{g}\left(\bar{R}(X, Y) J_1 Z, J_1 V\right) - \bar{g}\left(\bar{R}(X, Y) Z, V\right) = \alpha(X, Y) \bar{g}(J_1 V, J_2 Z)$$
$$- \beta(X, Y) \bar{g}(J_3 Z, J_1 V)$$

for any $X, Y, Z \in \Gamma(T\bar{M})$ and $V \in \Gamma(L_2)$. Thus, for $X = F_{11}, Y = \xi, Z = \xi$ and $V = F_{11}$ we derive

$$\bar{g}\left(\bar{R}(F_{11}, \xi) J_1 \xi, J_1 F_{11}\right) - \bar{g}\left(\bar{R}(F_{11}, \xi) J_1 \xi, F_{11}\right) = -\alpha(F_{11}, \xi) \bar{g}(w_1, J_2 \xi)$$
$$+ \beta(F_{11}, \xi) \bar{g}(J_3 \xi, w_1).$$

Hence we obtain

$$\bar{g}\left(\bar{R}(F_{11}, \xi) J_1 \xi, w_1\right) + \bar{g}\left(\bar{R}(F_{11}, \xi) \xi, F_{11}\right) = 0. \qquad (8.4.19)$$

On the other hand, from the equation of (5.3.11) we have

$$g\left(\bar{R}(X, Y) Z, W\right) = \bar{g}\left((\nabla_X h^s)(Y, Z), W\right) - \bar{g}\left((\nabla_Y h^s)(X, Z), W\right)$$
$$+ \bar{g}\left(D^s(X, h^\ell(Y, Z)), W\right)$$
$$- \bar{g}\left(D^s(Y, h^\ell(X, Z)), W\right)$$

for any $X, Y, Z \in \Gamma(TM)$ and $W \in \Gamma(S(TM^\perp))$. Since M is a totally umbilical QR-lightlike submanifold, we get

$$\bar{g}\left(\bar{R}(X, Y) Z, W\right) = (\nabla_X g)(Y, Z) \bar{g}(H^s, W) + g(Y, Z) \bar{g}(\nabla_X^t H^s, W)$$
$$- (\nabla_Y g)(X, Z) \bar{g}(H^s, W) - g(X, Z) g(\nabla_Y^t H^s, W)$$
$$+ \bar{g}\left(D^s(X, g(Y, Z) H^\ell), W\right) - \bar{g}\left(D^s(Y, g(X, Z) H^\ell), W\right).$$

Using (5.1.39) we obtain

$$\bar{g}\left(\bar{R}(X,Y)Z,W\right) = \bar{g}\left(h^l(X,Y),Z\right)\bar{g}(H^s,W) + \bar{g}\left(h^l(X,Z),Y\right)\bar{g}(H^s,W)$$
$$+ g(Y,Z)\bar{g}\left(\nabla_X^t H^s,W\right) - \bar{g}\left(h^l(X,Y),Z\right)\bar{g}(H^s,W)$$
$$+ \bar{g}\left(h^l(Y,Z),X\right)\bar{g}(H^s,W) - g(X,Z)g\left(\nabla_Y^t H^s,W\right)$$
$$+ \bar{g}\left(D^s(X,g(Y,Z)H^l),W\right) - \bar{g}\left(D^s(Y,g(X,Z)H^l),W\right).$$

Taking $X = F_{11}, Y = \xi, Z = J_1\xi$ and $W = w_1$ in the above equation, we obtain

$$\bar{g}\left(\bar{R}(F_{11},\xi)J_1\xi,w_1\right) = 0. \tag{8.4.20}$$

Thus, using (8.4.19) and (8.4.20) we derive $K_{\bar{M}}\left(\xi,F_{11}\right) = 0$, which is a contradiction that completes the proof. □

Theorem 8.4.11. *Let M be a totally umbilical QR-lightlike submanifold of an indefinite quaternion Kähler manifold. If D_0 is integrable, then M is totally geodesic.*

Proof. We suppose that D_0 is integrable, then from the previous lemma $h(X,J_aY) = h(J_aX,Y)$ for any $X,Y \in \Gamma(D_0)$. Since M is totally umbilical we get

$$h^s(X,J_aY) = g(X,J_aY)H^s = -g(J_aX,Y)H^s = h^s(J_aX,Y).$$

Thus we have $g(X,J_aY)H^s = 0$. Since D_0 is non-degenerate, we obtain $H^s = 0$. In a similar way, we get $H^l = 0$. Thus M is totally geodesic. □

As a result of this theorem, every totally umbilical QR-lightlike submanifold of an indefinite quaternion Kähler manifold with integrable distribution D_0 has the induced metric connection.

Lemma 8.4.12. *Let M be a totally umbilical QR-lightlike submanifold of an indefinite quaternion Kähler manifold $\bar{M}$. Then we have the following:*

$$g(X,X)H^s = \omega'_{aj}(\nabla_X J_aX)w_j \tag{8.4.21}$$

and

$$g(X,X)H^l = \omega'_{ai}(\nabla_X J_aX)v_i, \forall X \in \Gamma(D_0). \tag{8.4.22}$$

Proof. From (5.1.13) and (8.1.2), we obtain

$$h(X,X) = \bar{\nabla}_X X - \nabla_X X$$
$$= -\bar{\nabla}_X J_a^2 X - \nabla_X X$$
$$= -\left(\bar{\nabla}_X J_a\right)J_aX - J_a\bar{\nabla}_X J_aX - \nabla_X X.$$

Then (5.1.13) and (8.1.8)imply that

$$h(X,X) = -Q_{ab}(X)J_bJ_aX - Q_{ac}(X)J_cJ_aX$$
$$- J_a\left\{\nabla_X J_aX + h(X,J_aX)\right\} - \nabla_X X$$
$$= Q_{ab}(X)J_cX - Q_{ac}(X)J_bX$$
$$- J_a\nabla_X J_aX - \nabla_X X.$$

Using (8.4.8), we derive

$$h(X,X) = Q_{ab}(X)J_cX - Q_{ac}(X)J_bX$$

$$- J_aP\nabla_X J_aX - \sum_{i=1}^{r} \omega_{bi}(\nabla_X J_aX)E_{ci}$$

$$- \omega_{ci}(\nabla_X J_aX)E_{bi} - \omega_{ai}(\nabla_X J_aX)v_i$$

$$+ \sum_{j=1}^{k} \omega'_{bj}(\nabla_X J_aX)F_{cj} - \omega'_{cj}(\nabla_X J_aX)F_{bj}$$

$$- \omega'_{aj}(\nabla_X J_aX)w_j - \nabla_X X.$$

Taking the transversal parts of this equation, we have (8.4.21) and (8.4.22). $\quad\square$

From (8.4.15), we have the following lemma.

Lemma 8.4.13. *Let M be a totally umbilical QR-lightlike submanifold of an indefinite quaternion Kähler manifold $\bar{M}$. If $H^s \in \Gamma(L_2)$ and $H^l \in \Gamma(L_1)$, then*

$$\nabla_X J_a H^s = -J_a PA_{H^s}X + B_a\nabla^t_X H^s + Q_{ab}(X)B_bH^s + Q_{ac}(X)B_cH^s \quad (8.4.23)$$

and

$$\nabla_X J_a H^l = -J_a PA_{H^l}X + B_a\nabla^t_X H^l + Q_{ab}(X)J_bH^l + Q_{ac}(X)J_cH^l \quad (8.4.24)$$

for any $X \in \Gamma(D_0)$.

Theorem 8.4.14. *Let M be a totally umbilical QR-lightlike submanifold of an indefinite quaternion Kähler manifold $\bar{M}$. Then at least one of the following is true;*

(i) *D is totally lightlike,*

(ii) *M is totally geodesic,*

(iii) *D' is not parallel along D_0 with respect to induced connection ∇ (namely, $\nabla_X Y \notin \Gamma(D')$, for $X \in \Gamma(D_0)$, $Y \in \Gamma(D')$) and D_0 is not integrable.*

Proof. $D_0 = 0$ implies D is totally lightlike which is (i). Suppose $0 \neq D_0$ is integrable. Then, from Theorem 8.4.11, M is totally geodesic, which is (ii). Now assume D_0 is not integrable. If $H^l = 0$ and $H^s \in \Gamma(L_2)^\perp$, then, from (8.4.22), M is totally geodesic. Next, for $H^l \neq 0$, $H^s \in \Gamma(L_2)$; so let us suppose that D' is parallel along D_0 with respect to the induced connection ∇. Then, from (8.4.23) and (8.4.24), we have $A_{H^s}X \in \Gamma(D')$, whereas, from (5.1.28), (5.1.32) and (5.1.27) we have, $g(A_{H^s}X, J_a\xi_i) = 0$ and $g(A_{H^s}X, F_{aj}) = 0$. Hence, $A_{H^s}X \in \Gamma(D)$ which is a contradiction. Thus D' is not parallel along D_0 with respect to ∇. $\quad\square$

8.5   Screen QR-lightlike submanifolds

In this section, we introduce a new class, called screen quaternion-real (SQR) lightlike submanifolds of an indefinite quaternion Kähler manifold and investigate the geometry of such submanifolds.

Definition 8.5.1. [354] Let $(M, g, S(TM))$ be a lightlike submanifold of an indefinite quaternion Kähler manifold $(\bar{M}, \bar{g})$. We say that M is a *SQR-lightlike submanifold* of $\bar{M}$ if the following conditions are satisfied:

(i) There exist real non-null vector sub-bundles L and $L^{\perp}$ of $S(TM^{\perp})$ such that

$$S(TM^{\perp}) = L \perp L^{\perp}, \quad \bar{J}_a(L) \subset S(TM), \quad \bar{J}_a(L^{\perp}) = L^{\perp}. \tag{8.5.1}$$

(ii) $\bar{J}_a(\mathrm{Rad}(TM)) = \mathrm{Rad}(TM), a = 1, 2, 3.$

It follows that $\mathrm{ltr}(TM)$ is also invariant with respect to $\bar{J}_a$, $a = 1, 2, 3$, that is

$$\bar{J}_a(\mathrm{ltr}(TM)) = \mathrm{ltr}(TM). \tag{8.5.2}$$

Let M be a screen QR-lightlike submanifold of an indefinite quaternion Kähler manifold. Put $D'_{ap} = \bar{J}_a(L_p)$ and $\dim L_p = s, p \in M$. Then D'_{1p}, D'_{2p} and D'_{3p} are mutually orthogonal vector bundles of M. We consider $D'_p = D'_{1p} \oplus D'_{2p} \oplus D'_{3p}$, a $3s$-dimensional distribution globally defined on M. Also we have

$$\bar{J}_a(D'_{ap}) = L, \quad \bar{J}(D'_{bp}) = D'_{cp}, \quad \text{for each} \quad p \in M,$$

$a = 1, 2, 3$, where (a, b, c) is cyclic permutation of $(1, 2, 3)$. Consider

$$D = \mathrm{Rad}\, TM \oplus_{\mathrm{orth}} D_0$$

which is orthogonal complementary to D' in TM. It is easy to check that D_0 is an invariant non-degenerate distribution. Then, we obtain that D is also invariant with respect to $\bar{J}_a$. We call D and D' the quaternion and anti-quaternion distribution, respectively. Thus, we have

$$TM = D \oplus_{\mathrm{orth}} D'$$

and

$$\mathrm{tr}(TM) = \mathrm{ltr}(TM) \oplus_{\mathrm{orth}} (L \perp L^{\perp}).$$

We say that M is a *proper screen QR-lightlike submanifold* of $\bar{M}$ if $D_0 \neq \{0\}$ and $D' \neq \{0\}$.

Note the following special features:

1. Condition (ii) implies that $\dim(\mathrm{Rad}\, TM) = 4r \geq 4$.

2. For proper M, $\dim(D_0) \geq 4m$ and $\dim(D') \geq 3$.

3. There exists no screen QR-lightlike hypersurface.

Example 7. Consider in $\mathbf{R}_4^{12}$ the submanifold M given by the equations

$$x_1 = x_{11}, \ x_2 = x_{12}, \ x_9 = -x_3, \ x_{10} = -x_4, \ x_8 = \text{constant}.$$

Then the tangent bundle TM is spanned by

$$Z_1 = \partial x_1 + \partial x_{11}, \qquad Z_2 = \partial x_2 + \partial x_{12},$$
$$Z_3 = \partial x_3 - \partial x_9, \qquad Z_4 = \partial x_4 - \partial x_{13},$$
$$Z_5 = \partial x_5, \qquad Z_6 = \partial x_6, \quad Z_7 = \partial x_7.$$

Hence M is a 4-lightlike submanifold with $\mathrm{Rad}\, TM = \mathrm{Span}\{Z_1, Z_2, Z_3, Z_4\}$ and $\mathrm{Rad}\, TM$ is invariant with respect to canonical almost complex structures $\bar{J}_a$ of $\mathbf{R}_4^{12}$. We consider the vector field $W = \partial x_8$ of $S(TM^\perp)$. Then we can obtain that $\bar{J}_1 W = -Z_7, \bar{J}_2 W = -Z_6, \bar{J}_3 W = -Z_5$. Hence D' is spanned by $\{Z_5, Z_6, Z_7\}$. Also the lightlike transversal bundle is spanned by

$$N_1 = \tfrac{1}{2}\{-\partial x_1 + \partial x_{11}\}, \qquad N_2 = \tfrac{1}{2}\{-\partial x_2 + \partial x_{12}\},$$
$$N_3 = \tfrac{1}{2}\{-\partial x_3 - \partial x_9\}, \qquad N_4 = \tfrac{1}{2}\{-\partial x_4 - \partial x_{10}\},$$

which is invariant with respect to $\bar{J}_a, a = 1, 2, 3$. Thus M is a screen QR-lightlike submanifold of $\mathbf{R}_{12}^4$, with $D = \mathrm{Rad}\, TM = \mathrm{Span}\{Z_1, Z_2, Z_3, Z_4\}$ and $D' = \mathrm{Span}\{Z_5, Z_6, Z_7\}$.

Proposition 8.5.2. *A screen QR-lightlike submanifold of an indefinite quaternion Kähler manifold is a quaternion lightlike submanifold if and only if $D' = \{0\}$.*

Proof. Let M be a quaternion lightlike submanifold of an indefinite quaternion Kähler manifold. We can easily check that $\mathrm{Rad}\, TM$ is invariant with respect to $\bar{J}_a$. Therefore, $\mathrm{ltr}(TM)$ is also invariant with respect to $\bar{J}_a$. Hence $\bar{J}_a(S(TM^\perp)) = S(TM^\perp)$, thus $L = \{0\}$. Conversely, let M be a screen QR-lightlike submanifold of $\bar{M}$ such that $D' = \{0\}$. Then $\bar{J}_a(\mathrm{Rad}\, TM) = \mathrm{Rad}\, TM$ and $\bar{J}_a(S(TM)) = S(TM)$. Hence M is a quaternion lightlike submanifold. $\square$

Let M be a screen QR-lightlike submanifold of an indefinite quaternion Kähler manifold. If M is coisotropic, then, $S(TM^\perp) = \{0\}$ implies $L = \{0\}$. Thus we have $TM = D$. Hence M is a quaternion lightlike submanifold. A similar argument shows that a totally lightlike submanifold is also a quaternion lightlike submanifold. On the other hand, considering definitions of screen real and screen QR-lightlike submanifolds, one can conclude that a screen real lightlike submanifold of an indefinite quaternion Kähler manifold is not a screen QR-lightlike submanifold due to $\bar{J}_a(D'_b) = D'_c \subset TM$.

Let M be a screen QR-lightlike submanifold of an indefinite quaternion Kähler manifold. We denote the projection morphism of TM to the quaternion distribution D by S and choose a local field of orthonormal frames $\{v_1, \ldots, v_s\}$ on the vector bundle L in $S(TM^\perp)$. Then we have the local orthonormal frames

$$\{E_{11}, \ldots, E_{1s}, E_{21}, \ldots, E_{2s}, E_{31}, \ldots, E_{3s}\}$$

where $E_{11} = \bar{J}_1(v_1)$. Thus any vector field Y tangent to M can be written as

$$Y = SY + \sum_{b=1}^{3} \omega_{bi}(Y) E_{bi} \tag{8.5.3}$$

where $\omega_{bi}(Y) = g(Y, E_{bi})$. Thus applying $\bar{J}_a$ to (8.5.3) we obtain

$$\bar{J}_a Y = \bar{J}_a SY + \sum_{b=1}^{3} \omega_{bi}(Y) E_{ci} - \omega_{ci}(Y) E_{bi} - \omega_{ai}(Y) v_i. \tag{8.5.4}$$

For any vector field $V \in \Gamma(S(TM^\perp))$ we put

$$\bar{J}_a V = B_a V + C_a V, \ a = 1, 2, 3 \tag{8.5.5}$$

where $B_a V \in \Gamma(D')$ and $C_a V \in \Gamma(L^\perp)$. From the definition of screen QR-lightlike submanifold and using (8.1.8), (5.1.15), (8.5.4) and (8.5.5) we have

$$\nabla_X \bar{J}_a Y = Q_{ab}(X)\bar{J}_b Y + Q_{ac}(X)\bar{J}_c Y + \bar{J}_a S \nabla_X Y$$
$$- \omega_{bi} \nabla_X Y) E_{ci} + \omega_{ci}(\nabla_X Y) E_{bi} + B_a h^s(X, Y), \tag{8.5.6}$$
$$h^l(X, \bar{J}_a Y) = \bar{J}_a h^l(X, Y), \tag{8.5.7}$$
$$h^s(X, \bar{J}_a Y) = \omega_{ai}(\nabla_X Y) v_i + C_a h^s(X, Y), \quad \forall X, Y \in \Gamma(D). \tag{8.5.8}$$

Theorem 8.5.3. *Let M be a screen QR-lightlike submanifold of an indefinite quaternion Kähler manifold. The following conditions are equivalent:*

(1) $h^s(X, \bar{J}_a Y) = h^s(\bar{J}_a X, Y), a \in \{1, 2, 3\}, X, Y \in \Gamma(D)$.

(2) $h^s(X, \bar{J}_a Y) = 0$.

(3) D *is integrable.*

Proof. (1) $\Rightarrow$ (2): Since $\bar{J}_a = \bar{J}_b \circ \bar{J}_c$, we have

$$h^s(X, \bar{J}_a Y) = h^s(X, (\bar{J}_b \circ \bar{J}_c)Y) = h^s(X, \bar{J}_b(\bar{J}_c Y))$$
$$= h^s(\bar{J}_b X, \bar{J}_c Y)$$
$$= h^s(\bar{J}_c \circ \bar{J}_b X, Y)$$
$$= -h^s(\bar{J}_a X, Y).$$

Hence we obtain $h^s(X, \bar{J}_a Y) = 0$. For (2) $\Rightarrow$ (3)) suppose $h^s(X, \bar{J}_a Y) = 0$. From (8.5.8), we obtain $\omega_{ai}(\nabla_X Y)v_i = 0$. Hence, $\omega_{ai}([X, Y]) = 0$, i.e., $[X, Y] \in \Gamma(D)$. (3) $\Rightarrow$ (1): If D is integrable, then, from (8.5.8) we have $\omega_{ai}([X, Y]) = h^s(X, \bar{J}_a Y) - h^s(\bar{J}_a X, Y) = 0$. Hence, $h^s(X, \bar{J}_a Y) = h^s(\bar{J}_a X, Y)$. $\square$

Lemma 8.5.4. *Let M be a screen QR-lightlike submanifold of an indefinite quaternion Kähler manifold. Then we have*

$$\bar{g}(h^s(X, E_{ai}), v_j) = g(A_{v_i}X, E_{aj}) \tag{8.5.9}$$

$$g(A_{v_j}E_{ai}, X) = g(A_{v_i}E_{aj}, X)$$
$$- g(D^l(E_{aj}, v_i), X) + g(D^l(E_{ai}, v_j), X) \tag{8.5.10}$$

for $X \in \Gamma(D)$ and $E_{ai} \in \Gamma(D')$.

Proof. From (5.1.15), (8.1.2) and (8.1.8), we have

$$\bar{g}(h^s(E_{ai}, X), v_j) = -\bar{g}(\bar{\nabla}_X v_i, \bar{J}_a v_j).$$

By using (5.1.23) we derive

$$\bar{g}(h^s(E_{ai}, X), v_j) = g(A_{v_i}X, E_{aj}).$$

On the other hand from (8.5.9) and (5.1.28) we obtain

$$g(A_{v_j}E_{ai}, X) = g(A_{v_i}X, E_{aj}) + g(X, D^l(E_{ai}, v_j)).$$

Using again (8.5.9) we have

$$g(A_{v_j}E_{ai}, X) = \bar{g}(h^s(X, E_{aj}), v_i) + g(X, D^l(E_{ai}, v_j)).$$

Since h^s is symmetric, we derive

$$g(A_{v_j}E_{ai}, X) = \bar{g}(h^s(E_{aj}, X), v_i) + g(X, D^l(E_{ai}, v_j)).$$

Thus taking account of (5.1.28) in this equation, we get (8.5.10). □

Theorem 8.5.5. *Let M be a screen QR-lightlike submanifold of an indefinite quaternion Kähler manifold $\bar{M}$. Then the distribution D' is integrable if and only if*

$$\bar{g}(D^s(E_{ai}, \bar{J}_a N), v_j) = \bar{g}(D^s(E_{aj}, \bar{J}_a N), v_i),$$

$$B_{aj}(X) = 0, \quad D_{aj}(N) = 0$$

for $X \in \Gamma(D_0)$, where $B_{aj}(X) = g(\nabla^*_{E_{ai}}E_{aj}, X)$ and $D_{aj}(N) = g(A_{\bar{J}_c N}E_{ai}, E_{aj})$.

Proof. From (8.1.2), (8.1.8), (5.1.28) and (8.5.10) we have

$$g([E_{ai}, E_{aj}], X) = \bar{g}(\bar{\nabla}_{E_{ai}}E_{aj}, X) - \bar{g}(\bar{\nabla}_{E_{aj}}E_{ai}, X)$$
$$= -\bar{g}(\bar{\nabla}_{E_{ai}}v_j, \bar{J}_a X) + \bar{g}(\bar{\nabla}_{E_{aj}}v_i, \bar{J}_a X)$$
$$= g(A_{v_j}E_{ai}, \bar{J}_a X) - g(A_{v_i}E_{aj}, \bar{J}_a X)$$
$$= 0 \tag{8.5.11}$$

for $X \in \Gamma(D_0)$. In a similar way, we get

$$\bar{g}([E_{ai}, E_{aj}], N) = \bar{g}(D^s(E_{ai}, \bar{J}_a N), v_j) - \bar{g}(D^s(E_{aj}, \bar{J}_a N), v_i). \tag{8.5.12}$$

On the other hand, since $E_{bj} = J_c \bar{J}_a v_j$, from (8.1.2),(5.1.15) and (5.1.26) we obtain

$$g([E_{ai}, E_{bj}], X) = g(\nabla^*_{E_{ai}} E_{aj}, \bar{J}_c X) - g(\nabla^*_{E_{bj}} E_{bi}, \bar{J}_c X) \tag{8.5.13}$$

for $X \in \Gamma(D_0)$. In a similar way,

$$\bar{g}([E_{ai}, E_{bj}], N) = g(A_{\bar{J}_c N} E_{ai}, E_{aj}) - g(A_{\bar{J}_c N} E_{bj}, E_{bi}). \tag{8.5.14}$$

Thus, from (8.5.11), (8.5.12), (8.5.13)and (8.5.14), the proof follows. $\qquad \square$

Theorem 8.5.6. *Let M be a screen QR-lightlike submanifold of an indefinite quaternion Kähler manifold $\bar{M}$. Then D defines a totally geodesic foliation if and only if $h^s(X, \bar{J}_a Y)$ has no components in L for $X, Y \in \Gamma(D)$.*

Proof. From (5.1.15) and (8.1.8) we obtain

$$g(\nabla_X Y, E_{ai}) = \bar{g}(\bar{\nabla}_X Y, E_{ai})$$
$$= -\bar{g}(\bar{J}_a \bar{\nabla}_X Y, v_i)$$

for $X, Y \in \Gamma(D)$. Hence we have

$$g(\nabla_X Y, E_{ai}) = g((\bar{\nabla}_X \bar{J}_a)Y - \bar{\nabla}_X \bar{J}_a Y, v_i).$$

Now, by using (8.1.8) and (5.1.15) we obtain

$$g(\nabla_X Y, E_{ai}) = \bar{g}(h(X, \bar{J}_a Y), v_i)$$

which proves our assertion. $\qquad \square$

Theorem 8.5.7. *Let M be a screen QR-lightlike submanifold of an indefinite quaternion Kähler manifold $\bar{M}$. Then D' defines a totally geodesic foliation if and only if $A_V X$ has no components on D, where $X \in \Gamma(D')$ and $V \in \Gamma(S(TM^\perp))$.*

Proof. Using (8.1.8), (5.1.15), (5.1.28), (8.5.4), (8.5.5) and taking the tangential part we obtain

$$\nabla_{E_{ai}} E_{aj} = Q_{ab}(E_{ai})E_{bj} + Q_{ac}(E_{ai})E_{cj}$$
$$- \bar{J}_a S A_{v_j} E_{ai} - \omega_{bi}(A_{v_j} E_{ai})E_{ci}$$
$$+ \omega_{ci}(A_{v_j} E_{ai})E_{bi} + B_a \nabla^s_{E_{ai}} v_j. \tag{8.5.15}$$

In a similar way we have

$$\nabla_{E_{ai}} E_{bj} = -Q_{cb}(E_{ai})E_{cj} + Q_{ab}(E_{aa})E_{aj}$$
$$- \bar{J}_b S A_{v_j} E_{ai} + \omega_{ci}(A_{v_j} E_{ai})E_{bi}$$
$$+ B_b \nabla^s_{E_{ai}} v_j. \tag{8.5.16}$$

Then, the proof follows from (8.5.15) and (8.5.16). $\qquad \square$

Using Yano-Kon terminology [412] we say that screen QR-lightlike subman-ifold M is a lightlike product if D and D' are its totally geodesic foliations. Thus, from Theorem 8.5.6. and Theorem 8.5.7 we have the following:

Corollary 8.5.8. *Let M be a screen QR-lightlike submanifold of an indefinite qua-ternion Kähler manifold $\bar{M}$. Then, M is a lightlike product if and only if the following conditions are satisfied:*

1. *$A_V X$ has no components in D, $\forall X \in \Gamma(D'), V \in \Gamma(L)$.*

2. *$h^s(X, \bar{J}_a Y)$ has no components in L, $\forall X, Y \in \Gamma(D)$.*

Now we consider totally umbilical screen QR-lightlike submanifolds.

Theorem 8.5.9. *Let M be a totally umbilical screen QR-lightlike submanifold of an indefinite quaternion Kähler manifold $\bar{M}$. Then, the induced connection on M is a metric connection.*

Proof. It is well known that the induced connection ∇ on an r-lightlike submanifold is a metric connection if and only if h^l vanishes identically on M. From (8.1.8), (5.1.15), (8.5.4) and taking the lightlike transversal part we obtain

$$h^l(X, \bar{J}_a Y) = \bar{J}_a h^l(X, Y), \forall X, Y \in \Gamma(D_0).$$

Since M is totally umbilical we get

$$g(X, \bar{J}_a Y)H^\ell = g(X, Y)\bar{J}_a H^\ell.$$

Thus, interchanging the roles of X and Y in this equation and subtracting we have

$$g(X, \bar{J}_a Y)H^\ell = 0.$$

Hence, we derive $H^\ell = 0$ as D_0 is non-degenerate. Thus, we obtain $h^l = 0$. □

Note that the above theorem is not true for any r-lightlike submanifold. Therefore, it is an important property of totally umbilical screen QR-lightlike submanifolds.

Theorem 8.5.10. *Let M be a totally umbilical screen QR-lightlike submanifold of an indefinite quaternion Kähler manifold $\bar{M}$. If $\dim(L) > 1$, then M is totally geodesic.*

Proof. From the previous theorem, we have $H^\ell = 0$. So $H^s = 0$ is enough to show that M is totally geodesic. Using (8.5.8), we have

$$g(X, \bar{J}_a Y)H^s = \omega_{ai}(\nabla_X Y)v_i + C_a h^s(X, Y)$$

for $X, Y \in \Gamma(D_0)$. Hence we obtain

$$2g(X, \bar{J}_a Y)H^s = \omega_{ai}([X, Y])v_i. \tag{8.5.17}$$

For $E_{ai}, E_{aj} \in \Gamma(D')$, in a similar way, we derive

$$g(E_{ai}, E_{aj})H^s = \omega_{ai}(A_{v_j})v_i + C_a \nabla^s_{E_{ai}} v_j. \qquad (8.5.18)$$

Suppose $\dim(L) > 1$. Since L is non-degenerate, it has an orthonormal basis. Thus we can choose vector fields E_{aj}, E_{ai}, $i \neq j$ such that they are orthogonal, then (8.5.18) becomes

$$\omega_{ak}(A_{v_j}E_{ai})v_k + C_a \nabla^s_{E_{ai}} v_k = 0, k \in \{1, \dots, \dim(L)\}.$$

Hence we have

$$\omega_{ak}(A_{v_j}E_{ai}) = 0.$$

Using (5.1.28) we obtain

$$\bar{g}(h^s(E_{ai}, E_{ak}), v_j) = 0.$$

Hence we conclude that

$$H^s \in \Gamma(L^\perp). \qquad (8.5.19)$$

For $X, Y \in \Gamma(D_0)$, if $[X, Y] \in \Gamma(D)$, then from (8.5.17) we derive $H^s = 0$. If $[X, Y] \in \Gamma(D')$ then from (8.5.17) we obtain

$$H^s \in \Gamma(L). \qquad (8.5.20)$$

Then, from (8.5.19) and (8.5.20) we derive $H^s = 0$, i.e., M is totally geodesic. □

Example 8. Consider a submanifold M, in $\mathbf{R}_4^{12}$ with the equations:

$$x_9 = x_1 \sin \alpha - x_2 \cos \alpha, \qquad x_{10} = x_1 \cos \alpha + x_2 \sin \alpha, \qquad x_{11} = x_3 \sin \alpha + x_4 \sin \alpha$$

$$x_{12} = -x_3 \cos \alpha + x_4 \sin \alpha, \qquad x_5 = \sqrt{1 - x_6^2 - x_7^2 - x_8^2}.$$

The tangent bundle of M is spanned by

$$\xi_1 = \partial x_1 + \sin \alpha \, \partial x_9 + \cos \alpha \partial x_{10}, \qquad \xi_2 = \partial x_2 - \cos \alpha \, \partial x_9 + \sin \alpha \partial x_{10},$$
$$\xi_4 = \partial x_3 + \sin \alpha \, \partial x_{11} - \cos \alpha \partial x_{12}, \qquad \xi_4 = \partial x_4 + \sin \alpha \, \partial x_{11} + \sin \alpha \partial x_{12},$$
$$Z_1 = -x_6 \, \partial x_5 + x_5 \, \partial x_6, \qquad Z_2 = -x_7 \, \partial x_5 + x_5 \, \partial x_7,$$
$$Z_3 = -x_8 \, \partial x_5 + x_5 \, \partial x_8.$$

We see that M is a 4-lightlike submanifold and $\operatorname{Rad} TM = \operatorname{Span}\{\xi_1, \xi_2, \xi_3, \xi_4\}$. It is easy to see $\operatorname{Rad} TM$ is invariant with respect to canonical complex structures $\bar{J}_1, \bar{J}_2, \bar{J}_3$. Screen transversal bundle $S(TM^\perp)$ is spanned by

$$W = x_5 \, \partial x_5 + x_6 \, \partial x_6 + x_7 \, \partial x_7 + x_8 \, \partial x_8.$$

By direct calculations, we have

$$U_1 = \bar{J}_1 W = Z_1 - \frac{x_8}{x_5} Z_2 + \frac{x_7}{x_5} Z_3,$$

$$U_2 = \bar{J}_2 W = \frac{x_8}{x_5} Z_1 + Z_2 - \frac{x_6}{x_5} Z_3,$$

$$U_3 = \bar{J}_3 W = \frac{-x_7}{x_5} Z_1 + \frac{x_6}{x_5} Z_2 + Z_3.$$

Hence $D' = \text{Span}\{U_1, U_2, U_3\}$. Thus M is a screen QR-lightlike submanifold. On the other hand, the lightlike transversal bundle is spanned by

$$N_1 = \tfrac{1}{2}\{-\partial x_1 + \sin\alpha\, \partial x_9 + \cos\alpha\partial x_{10}\},$$

$$N_2 = \tfrac{1}{2}\{-\partial x_2 - \cos\alpha\, \partial x_9 + \sin\alpha\partial x_{10}\},$$

$$N_3 = \tfrac{1}{2}\{-\partial x_3 + \sin\alpha\, \partial x_{11} - \cos\alpha\partial x_{12}\},$$

$$N_4 = \tfrac{1}{2}\{-\partial x_4 + \cos\alpha\, \partial x_{11} + \sin\alpha\partial x_{12}\}.$$

Hence lightlike transversal is also invariant. By direct calculations, we have

$$\bar{\nabla}_X \xi_1 = \bar{\nabla}_X \xi_2 = \bar{\nabla}_X \xi_3 = \bar{\nabla}_X \xi_4 = \bar{\nabla}_X N_1 = \bar{\nabla}_X N_2 = \bar{\nabla}_X N_3 = \bar{\nabla}_X N_4 = 0$$

for any $X \in \Gamma(TM)$. On the other hand we have

$$\bar{\nabla}_{U_1} U_1 = \bar{\nabla}_{U_2} U_2 = \bar{\nabla}_{U_3} U_3 = -W,$$

$$\bar{\nabla}_{U_1} U_2 = x_8\partial x_5 + x_7\partial x_6 - x_6\partial x_7 - x_5\partial x_8,$$

$$\bar{\nabla}_{U_1} U_3 = -x_7\partial x_5 + x_8\partial x_6 + x_5\partial x_7 - x_6\partial x_8,$$

$$\bar{\nabla}_{U_2} U_3 = x_6\partial x_5 - x_5\partial x_6 + x_8\partial x_7 - x_7\partial x_8.$$

By using (5.1.15) we obtain

$$h^l = 0, h^s(U_1, U_2) = h^s(U_2, U_1) = h^s(U_1, U_3) = h^s(U_3, U_1) = h^s(X, \xi) = 0,$$

$$h^s(U_1, U_1) = g(U_1, U_1)H^s, \ h^s(U_2, U_2) = g(U_2, U_2)H^s, \ h^s(U_3, U_3) = g(U_3, U_3)H^s,$$

for $X \in \Gamma(TM)$, $Y \in \Gamma(\text{Rad}\,TM)$, where $H^s = -W$. Hence M is a totally umbilical screen QR-lightlike submanifold.

8.6   Screen CR-lightlike submanifolds

In section 5, we have seen that a screen real lightlike submanifold is not a screen QR-lightlike submanifold. In this section we will introduce another class of lightlike submanifolds of an indefinite quaternion Kähler manifold, namely, screen CR-lightlike submanifolds which include screen real lightlike submanifolds as well as quaternion lightlike submanifolds.

Definition 8.6.1. [354] Let $(M, g, S(TM))$ be a lightlike submanifold of an indefinite quaternion Kähler manifold $(\bar{M}, \bar{g}.)$ We say that M is a *screen CR-lightlike submanifold* of $\bar{M}$ if the following conditions are satisfied:

1. There exist real non-null distributions D_0 and D' over $S(TM)$ such that

$$S(TM) = D_0 \perp D', \ \bar{J}_a(D_0) = D_0, \ \bar{J}_a(D') \subset S(TM^\perp), a = 1, 2, 3. \quad (8.6.1)$$

2. $\bar{J}_a(\mathrm{Rad}\, TM) = \mathrm{Rad}\, TM,\ a = 1, 2, 3.$

It follows that $\mathrm{ltr}(TM)$ is also invariant with respect to $\bar{J}_a$, i.e.,

$$\bar{J}_a(\mathrm{ltr}(TM) = \mathrm{ltr}(TM). \quad (8.6.2)$$

Let μ be the orthogonal complementary distribution to $\bar{J}_a D'$ in $S(TM^\perp)$. Note that D_0 and μ are non-degenerate. For a screen CR-lightlike submanifold we have

$$TM = D \oplus_{\mathrm{orth}} D', \quad (8.6.3)$$

where

$$D = \mathrm{Rad}\, TM \oplus_{\mathrm{orth}} D_0. \quad (8.6.4)$$

We say that M is a *proper screen CR-lightlike submanifold* of $\bar{M}$ if $D_0 \neq 0$ and $D' \neq 0$. From (8.6.1), (8.6.2) and (8.6.4), for $X \in \Gamma(TM)$ we can write

$$\bar{J}_a X = \phi_a X + F_a X, \quad (8.6.5)$$

where $\phi X \in \Gamma(D)$ and $F_a X \in \Gamma(\bar{J}_a D')$. For any vector field $V \in \Gamma(S(TM^\perp))$ we put

$$\bar{J}_a V = t_a V + f_a V, \quad (8.6.6)$$

where $t_a V \in \Gamma(D')$ and $f_a V \in \Gamma(\mu)$.

Example 9. Consider in $\mathbf{R}_4^{16}$ the submanifold M given by the equations

$$x_1 = x_{13}, \quad x_2 = x_{14}, \quad x_3 = x_{15}, \quad x_4 = x_{16}$$

$$x_{11} = \sqrt{1 - x_9^2}, \quad x_{10} = \text{constant}, \quad x_{12} = \text{constant}.$$

The tangent bundle of M is spanned by

$$Z_1 = \partial x_1 + \partial x_{13}, \quad Z_2 = \partial x_2 + \partial x_{14}, \quad Z_3 = \partial x_3 + \partial x_{15}$$
$$Z_4 = \partial x_4 + \partial x_{16}, \quad Z_5 = \partial x_5, \quad Z_6 = \partial x_6$$
$$Z_7 = \partial x_7, \quad Z_8 = \partial x_8, \quad Z_9 = \partial x_9 - \frac{x_9}{\sqrt{1 - x_9^2}} \partial x_{11}.$$

Hence M is a 4-lightlike submanifold with $RadTM = \mathrm{Span}\{Z_1, Z_2, Z_3, Z_4\}$ and it is invariant with respect to $\bar{J}_1, \bar{J}_2, \bar{J}_3$. Moreover we can see $D_0 = \{Z_5, Z_6, Z_7, Z_8\}$ is also invariant. It is easy to see that

$$\{Z_1, Z_2, Z_3, Z_4, Z_5, Z_6, Z_7, Z_8, Z_9, W_1 = \bar{J}_1 Z_9, W_2 = \bar{J}_2 Z_9, W_3 = \bar{J}_3 Z_9\}$$

is linearly independent. So

$$\text{Span}\{W_1, W_2, W_3\} = \bar{J}_a(D') = S(TM^\perp).$$

Finally we obtain the lightlike transversal bundle spanned by

$$N_1 = \tfrac{1}{2}\{-\partial x_1 + \partial x_{13}\}, N_2 = \tfrac{1}{2}\{-\partial x_2 + \partial x_{14}\},$$
$$N_3 = \tfrac{1}{2}\{-\partial x_3 + \partial x_{15}\}, N_4 = \tfrac{1}{2}\{-\partial x_4 + \partial x_{16}\}.$$

Thus, we conclude that M is a proper screen CR-lightlike submanifold of $\mathbf{R}_4^{16}$.

Proposition 8.6.2. *A screen CR-lightlike submanifold of an indefinite quaternion Kähler manifold is a screen real lightlike submanifold(resp, quaternion lightlike) if and only if $D_0 = \{0\}$ (resp, $D' = \{0\}$).*

Proof. Let M be a screen real lightlike submanifold of an indefinite quaternion Kähler manifold. Then, the radical distribution is an invariant subspace. Since M is screen real, we have $D_0 = \{0\}$. Conversely, let M be a screen CR-lightlike submanifold such that $D_0 = \{0\}$. Then we have $S(TM) = D'$, since $\bar{J}_a(D') \subset S(TM^\perp)$. We obtain that M is a screen real lightlike submanifold. The other assertion similarly follows. □

Consider a coisotropic submanifold M. Then, $TM = S(TM) \oplus_{\text{orth}} \text{Rad}\,TM$ and $S(TM^\perp) = \{0\}$), so D' is undefined. Therefore, we have $TM = RadTM \oplus_{\text{orth}} D_0$, i.e., it is invariant with respect to $\bar{J}_a, a = 1, 2, 3$. Thus, any coisotropic screen CR-lightlike submanifold of an indefinite quaternion Kähler manifold is a quaternion lightlike submanifold. Similarly, if M is isotropic or totally lightlike, then, M is again quaternion lightlike submanifold.

Theorem 8.6.3. *Let M be a screen CR-lightlike submanifold of an indefinite quaternion Kähler manifold $\bar{M}$. Then:*

1. *D' is integrable if and only if*

$$A_{\bar{J}_a U}V = A_{\bar{J}_a V}U , \forall U, V \in \Gamma(D').$$

2. *D is integrable if and only if*

$$h^s(X, \bar{J}_a Y) = h^s(\bar{J}_a X, Y), \forall X, Y \in \Gamma(D).$$

Proof. From (8.1.8), (5.1.15), (5.1.23), (8.6.5), (8.6.6) and tangential parts we get

$$-A_{\bar{J}_a U}V = \phi_a \nabla_V U + t_a h^s(U, V) \qquad (8.6.7)$$

for $U, V \in \Gamma(D')$. Hence we obtain

$$A_{\bar{J}_a U}V - A_{\bar{J}_a V}U = \phi_a[U, V],$$

thus we get the first assertion. In a similar way, by using (8.1.8), (5.1.15), (5.1.23), (8.6.5), (8.6.6) and taking the screen transversal parts we obtain

$$h^s(X, \bar{J}_a Y) = F_a \nabla_X Y + f_a h^s(X, Y) \tag{8.6.8}$$

for $X, Y \in \Gamma(D)$, hence we obtain the second assertion of the theorem. □

For totally umbilical screen CR-lightlike submanifolds, first we have:

Corollary 8.6.4. *Let M be a totally umbilical screen CR-lightlike submanifold of an indefinite quaternion Kähler manifold $\bar{M}$. Then the induced connection on M is a metric connection.*

The proof is similar to that of section 4 of Chapter 6.

Lemma 8.6.5. *Let M be a totally umbilical proper screen CR-lightlike submanifold of an indefinite quaternion Kähler manifold $\bar{M}$. Then we have*

$$H^s \in \Gamma(\bar{J}_a D'). \tag{8.6.9}$$

Proof. From (8.6.8) and (5.3.8), for $X = Y \in \Gamma(D_0)$ we obtain

$$F_a \nabla_X Y = 0, \ g(X, X) f_a H^s = 0.$$

Since D_0 is non-degenerate we have, at least, a spacelike or timelike vector field, thus $f_a H^s = 0$, which shows us $H^s \in \Gamma(\bar{J}_a D')$. □

Theorem 8.6.6. *Let M be a totally umbilical proper screen CR-lightlike of an indefinite quaternion Kähler manifold. Then*

1. *M is totally geodesic or*

2. *the distribution D' is one-dimensional.*

Proof. From (8.6.7), (5.3.8) and (5.1.28) we obtain

$$g(X, X)\bar{g}(H^s, \bar{J}_a Y) = g(X, Y)\bar{g}(H^s, \bar{J}_a X) \tag{8.6.10}$$

for $X, Y \in \Gamma(D')$. From (8.6.9) we have

$$g(Y, Y)\bar{g}(H^s, \bar{J}_a X) = g(X, Y)\bar{g}(H^s, \bar{J}_a Y). \tag{8.6.11}$$

Thus we have

$$\bar{g}(H^s, \bar{J}_a X) = \frac{g(X, Y)^2}{g(X, X)g(Y, Y)} \bar{g}(H^s, \bar{J}_a X). \tag{8.6.12}$$

Since D' and $S(TM^\perp)$ are non-degenerate , (8.6.10) and (8.6.12) imply $H^s = 0$ or X and Y are linearly dependent. Thus we have proved the theorem. □

Example 10. Let M be the submanifold of $\mathbf{R}_4^{16}$ given in Example 9. Then we have

$$\bar{\nabla}_X Z_1 = \bar{\nabla}_X Z_2 = \bar{\nabla}_X Z_3 = \bar{\nabla}_X Z_4 = 0,$$
$$\bar{\nabla}_X Z_5 = \bar{\nabla}_X Z_6 = \bar{\nabla}_X Z_7 = \bar{\nabla}_X Z_8 = 0$$

for any $X \in \Gamma(TM)$ and

$$\bar{\nabla}_{Z_9} Z_9 = \frac{x_9}{1 - x_9^2} Z_9 - \frac{1}{\sqrt{1 - x_9^2}} W_2.$$

Using Gauss' equation we have

$$h^s(X, Z_1) = h^s(X, Z_2) = h^s(X, Z_3) = h^s(X, Z_4) = 0,$$
$$h^s(X, Z_5) = h^s(X, Z_6) = h^s(X, Z_7) = h^s(X, Z_8) = 0$$

and

$$h^l = 0, \quad h^s(Z_9, Z_9) = g(Z_9, Z_9)H^s$$

where $H^s = -\sqrt{1 - x_9^2}W_2$. Thus M is totally umbilical, with a metric connection.

Theorem 8.6.7. *There exists no proper totally umbilical screen CR-lightlike submanifold in positively or negatively curved indefinite quaternion Kähler manifolds.*

Proof. We suppose that M is a proper totally umbilical screen CR-lightlike submanifold of $\bar{M}$ with $K_{\bar{M}}(X, Y) \neq 0$ for any $X, Y \in \Gamma(TM)$. By direct calculations, using (8.1.11), we have

$$-\bar{g}(\bar{R}(X, Y)X, Y) + \bar{g}(\bar{R}(X, Y)\bar{J}_1 X, \bar{J}_1 Y) = 0 \tag{8.6.13}$$

for $X \in \Gamma(D_0)$, $Z = \bar{J}_1 X \in \Gamma(D_0)$ and $Y \in \Gamma(D')$. By using (5.3.11) and (5.3.13) we have

$$\bar{g}(\bar{R}(X, Y)\bar{J}_1 X, \bar{J}_1 Y) = \bar{g}(H^s, \bar{J}_1 Y)\{-g(\nabla_X Y, \bar{J}_1 X) - g(Y, \nabla_X \bar{J}_1 X) \\ + g(\nabla_Y X, \bar{J}_1 X) + g(X, \nabla_Y \bar{J}_1 X)\}.$$

Then from (5.1.15) we get

$$\bar{g}(\bar{R}(X, Y)\bar{J}_1 X, \bar{J}_1 Y) = \bar{g}(H^s, \bar{J}_1 Y)\{-\bar{g}(\bar{\nabla}_X Y, \bar{J}_1 X) - \bar{g}(Y, \bar{\nabla}_X \bar{J}_1 X) \\ + \bar{g}(\bar{\nabla}_Y X, \bar{J}_1 X) + \bar{g}(X, \bar{\nabla}_Y \bar{J}_1 X)\}.$$

Since $\bar{\nabla}$ is a metric connection we obtain $\bar{g}(\bar{R}(X, Y)\bar{J}_1 X, \bar{J}_1 Y) = 0$. Then, using (8.6.13) we have

$$K_{\bar{M}}(X, Y) = \bar{g}(\bar{R}(X, Y)X, Y) = 0,$$

which is a contradiction. Thus, the proof is complete. □

Remark 8.6.8. Note that among QR-lightlike, screen QR-lightlike and screen CR-lightlike submanifolds there exist no inclusion relations, because a real lightlike hypersurface is a QR-lightlike submanifold, but, it is not a screen QR-lightlike or a screen CR-lightlike submanifold. On the other hand, a screen real lightlike submanifold is a screen CR-lightlike submanifold, of course it is neither screen QR-lightlike nor QR-lightlike. Finally, invariant lightlike submanifolds lie in the intersection of the screen CR-lightlike and screen QR-lightlike submanifold.

Chapter 9

Applications of lightlike geometry

In this chapter we present applications of lightlike geometry in the study of null 2-surfaces in spacetimes, lightlike versions of harmonic maps and morphisms, CR-structures in general relativity and lightlike contact geometry in physics.

9.1 Null 2-surfaces of spacetimes

Let $(\bar{M}, \bar{g})$ be a 4-dimensional spacetime of general relativity. As discussed in Chapter 3, null (lightlike) geodesics are important objects of study in relativity. Here we concentrate on the use of null geodesics in the study of 2-surfaces of spacetimes. First we present some basic information on timelike 2-surfaces which are called photon surfaces explained as follows:

Photon surfaces. Let $C(p)$ be a null curve in $(\bar{M}, \bar{g})$, where $p \in I \subset \mathbf{R}$ is a special parameter, $\{\xi, N, W_1, W_2\}$ is a pseudo-orthonormal Frenet frame along $C(p)$, ξ and N are null vectors such that $\bar{g}(\xi, N) = 1$, $C = \mathrm{Span}\{\xi\}$, and W_1, W_2 are unit spacelike vectors. If N moves along C, then, it generates a ruled surface given by the parameterization $((I \times \mathbf{R}), f)$ where $f : I \times \mathbf{R} \to \bar{M}$ is defined by

$$(p, u) \to f(p, u) = C(p) + uN(p), \quad u \in I_u \subset \mathbf{R}.$$

The above ruled surface is called a *null scroll* which we denote by $\mathcal{S}_c$. It is clear by the above defining equation that the null scroll $\mathcal{S}_c$ is a timelike ruled surface in $\bar{M}$. In particular, if $\mathcal{S}_c$ is ruled by null geodesics, then, there is an important link of the concept of null scrolls [332] with the literature of physics as follows:

In general relativity, null geodesics are interpreted as the world lines of photons and a timelike null scroll is called a *photon surface* if each null geodesic tangent to $\mathcal{S}_c$ remains within $\mathcal{S}_c$ (for some parameter interval). For the construction of photon surfaces, we require that each null scroll will be fully immersed in $\bar{M}$ as opposed to an embedded submanifold, i.e., we allow for self-intersections. This is necessary as our construction methods for photon surfaces may yield ruled surfaces with self-intersections. With this condition, we proceed as follows:

A vector field on $\mathcal{S}_c$ assigns to each point $x \in \mathcal{S}_c$ a vector in the tangent space $T_x\mathcal{S}_c$ whereas a vector field along $\mathcal{S}_c$ assigns to each $x \in \mathcal{S}_c$ a vector field in the tangent space $T_{i(x)}\bar{M}$ where i denotes the immersion $\mathcal{S}_c \rightarrow \bar{M}$. Given a timelike vector field $\mathbf{n}$ on $\mathcal{S}_c$, i.e., $\bar{g}(\mathbf{n}, \mathbf{n}) = -1$, the conditions

$$\bar{g}(\mathbf{n}, \mathbf{v}) = 0 \quad \text{and} \quad g(\mathbf{v}, \mathbf{v}) = 1$$

define a spacelike vector field $\mathbf{v}$ on $\mathcal{S}_c$ uniquely up to sign. Since $\mathbf{n}$ and $\mathbf{v}$ are orthonormal, the relation

$$\xi = \frac{\mathbf{n} + \mathbf{v}}{\sqrt{2}} \quad \text{and} \quad N = \frac{\mathbf{v} - \mathbf{n}}{\sqrt{2}}$$

defines two null vector fields ξ and N on $\mathcal{S}_c$ such that at each point x of $\mathcal{S}_c$, the vectors ξ_x and N_x span the two different null lines tangent to $\mathcal{S}_c$ and

$$g(\xi, N) = 1.$$

Since any choice of $\mathbf{n}$, $\mathbf{v}$ is unique up to sign, so ξ and N are unique up to interchanging. Suppose we replace $\mathbf{n}$ by another vector field $\bar{\mathbf{n}}$ on $\mathcal{S}_c$ such that $\bar{g}(\bar{\mathbf{n}}, \bar{\mathbf{n}}) = -1$ (and take corresponding vector field $\bar{\mathbf{v}}$), then, the following transformation equations will hold:

$$\xi \mapsto f\xi \quad \text{and} \quad N \mapsto f^{-1}N,$$

where f is a nowhere vanishing scalar function on $\mathcal{S}_c$.

Proposition 9.1.1. [191] *A point x in a spacetime manifold $(\bar{M}, \bar{g})$ admits a neighborhood that can be foliated into timelike photon 2-surfaces if and only if on some neighborhood of x there are two linearly independent null geodesic vector fields ξ and N such that the Lie bracket $[\xi, N]$ is a linear combination of ξ and N. In this case, the photon 2-surfaces are the integral manifolds, say $\mathcal{S}_c$, of the 2-surfaces spanned by ξ and N.*

Proof. ξ and N being linearly independent, they generate a timelike 2-surface $\mathcal{S}_c$ at each point x of $\bar{M}$. Then, by the Frobenius theorem these 2-surfaces admit local integral manifolds if and only if the Lie bracket $[\xi, N]$ is a linear combination of ξ and N. Since, by hypothesis, ξ and N are geodesic, it makes sure (by definition) that these integral manifolds, say $\mathcal{S}_c$, are timelike photon 2-surfaces, which proves the if part. To prove the only if part, one just has to verify that the null geodesic vector fields ξ and N, given on each leaf of the respective foliations up to the above transformation, can be chosen such that they make up to two smooth vector fields ξ and N on some neighborhood of x, which completes the proof. $\square$

On the existence of timelike photon surfaces in a spacetime $(\bar{M}, \bar{g})$, one can characterize such surfaces in terms of the second fundamental form $\mathbf{B}$ of a non-null submanifold Σ of $(\bar{M}, \bar{g})$ as follows: We know [317] that $\mathbf{B}$ can be written as

$$\mathbf{B}(u, w) = P^{\perp}(\bar{\nabla}_u w), \quad \forall u, w \in T(\Sigma),$$

and Σ is called totally umbilical if there is a normal vector field Z of Σ such that

$$\mathbf{B}(u, w) = g(u, w)Z, \quad \forall u, w \in \Gamma(T\Sigma).$$

If $Z = 0$, then, Σ is totally geodesic. It is easy to see that the property of being totally umbilical is conformally invariant, whereas the property of being totally geodesic is not. Following is a characterization result for timelike photon 2-surfaces:

Proposition 9.1.2. [191] *A 2-dimensional timelike submanifold Σ of a spacetime manifold is a photon 2-surface if and only if Σ is totally umbilical.*

Proof. Let ξ and N be the two null vector fields on Σ, unique up to the above given transformations and are normalized according as explained before. Suppose Σ is totally umbilical. Then, it is required that $\mathbf{B}(\xi, \xi) = 0$ and $\mathbf{B}(N, N) = 0$. From this it is easy to show that ξ and N are geodesics, so Σ is a photon 2-surface.

Conversely, assume Σ is a photon 2-surface. Let $u = a\xi + bN$ and $w = c\xi + dN$ be any two vector fields on Σ, where a, b, c, d are scalar functions on Σ. Since $\nabla_\xi \xi$ and $\nabla_N N$ are tangents to Σ, the second fundamental form reduces to

$$\mathbf{B}(u, w) = ad \, P^{\perp}(\nabla_\xi N) + bc \, P^{\perp}(\nabla_N \xi).$$

The fact that ξ, N and $[\xi, N]$ must be tangent to Σ implies

$$P^{\perp}(\nabla_\xi N) = P^{\perp}(\nabla_N \xi) =: -Z.$$

On the other hand, $\bar{g}(u, w) = -ad - bc$. Thus, we conclude that the above totally umbilical condition holds for Σ, which completes the proof. □

Remark 9.1.3. For some examples and more details on photon surfaces and their physical use, we refer to [191] and some more referred to therein. The notion of a null geodesic ruled surface was first introduced by Schild [369] in the form of a geodesic null string of a null hypersurface of a 4-dimensional Minkowski or curved spacetime. By null strings we mean 2-dimensional ruled null surfaces on the null cone (with one dimension suppressed) of $\bar{M}$. Since Schild's paper, there has been considerable work done on geodesic and non-geodesic null strings. For example, Ilyenko [235] has presented a twistor description for null 2-surfaces (null strings) in 4D Minkowski spacetime by taking the Lagrangian for a variational principle as a surface-forming null bivector. His proposed formulation is reparametrization invariant and free of any algebraic and differential constraints. An example of a null 2-surface given by the 2-dimensional self-intersecting caustic of a null hypersurface has been studied in this paper (also see many other papers cited therein.) More on null geodesics and photon surfaces is available in [156, Chapter 8].

Half-lightlike surfaces. As discussed in Chapter 4, there is another class of null surfaces, called half-lightlike 2-surfaces, whose physical interpretation has been quite rare. The objective of this section is to present up-to-date results on this class of null surfaces for their possible applications in mathematical physics. As we have seen the importance of totally umbilical surfaces as photon surfaces in relativity, so we concentrate on mathematical results on totally umbilical half-lightlike surfaces in a 4-dimensional spacetime $(\bar{M}, \bar{g})$. Let $(M, g, S(TM))$ be a half-lightlike 2-surface of $(\bar{M}, \bar{g})$. Then, $S(TM)$ is a spacelike screen distribution of rank 1. Suppose $S(TM)$, $\mathrm{Rad}\,TM$, D and $\mathrm{ntr}(TM)$ are locally spanned by U, ξ, u and N respectively. Following is an example:

Example 1. Consider a 4-dimensional spacetime $(\bar{M}, \bar{g})$, with Lorentzian metric $\bar{g}$, of signature $(-, +, +, +)$, which admits a smooth 2-parameter group G, generated by two spacelike Killing vector fields U and V. Suppose $\bar{M}$ also admits a non-Killing null vector field ξ. Then, U and V will span a lightlike surface M defined by $U_{[a}V_{b]}U^a V^b = 0$, $a, b \in \{1, 2, 3, 4\}$. At any x of $T_x M$ we have a unique null vector tangent to $T_x M$ given by

$$\xi = U - \Omega V, \quad \Omega = (V^a V_a)^{-1} U^b V_b, \quad \bar{g}(\xi, U) = 0 = \bar{g}(\xi, V).$$

It is easy to see that M is a half-lightlike surface of $\bar{M}$ and $\mathrm{Rad}\,TM = \mathrm{Span}\{\xi\}$. Since ξ is non-Killing, Theorem 4.1.3 tells us that $D_1 \neq 0$. Thus, M neither admits a metric connection nor is totally geodesic.

Theorem 9.1.4. *Let M be a half-lightlike surface of a 4-dimensional spacetime manifold $(\bar{M}, \bar{g})$. Then we have the following assertions:*

(1) *U and ξ are eigenvector fields for A_ξ^* with respect to the eigenfunction $\lambda_1 = \frac{1}{\|U\|}\bar{g}(h(U, U), \xi)$ and $\lambda_2 = 0$ respectively.*

(2) *U is an eigenvector field for A_{N_1} with respect to the eigenfunction $\alpha = \frac{1}{\|U\|}\bar{g}(h^*(U, U), N)$. Moreover, one of the eigenvalues of A_N is zero.*

Proof. As A_ξ^* is $\Gamma(S(TM))$-valued, it follows that U is an eigenvector field for A_ξ^*. Its eigenfunction follows from (4.1.21). The second part of the assertion (1) follows from (4.1.22). Next, from (4.1.11) we find that A_N is $\Gamma(S(TM))$-valued. Hence U is also an eigenvector field for A_N. Its eigenfunction follows from (4.1.20). Finally, one of the eigenfunctions of A_N is equal to zero since rank $A_N = 1$.   □

Using (4.1.27)–(4.1.33) and (4.1.10) and the field of frames $\{U, \xi, u, N\}$, the structure equations of M can be written as follows:

$$\begin{aligned}
\bar{R}(U, \xi, \xi, U) &= R(U, \xi, \xi, U) + \varepsilon_1(\xi)D_2(U, U) + \epsilon(\varepsilon_1(U))^2 \\
&= (\nabla_\xi D_1)(U, U) - (\nabla_U D_1)(\xi, U) \\
&\quad + \rho_1(\xi)D_1(U, U) + \varepsilon_1(\xi)D_2(U, U) \\
&\quad + \epsilon(\varepsilon_1(U))^2,
\end{aligned} \tag{9.1.1}$$

$$\bar{R}(U,\xi,u,U) = g(\nabla_\xi(A_uU) - \nabla_U(A_u\xi) - A_u[\xi,U],U)$$
$$+ \varepsilon_1(U)E_1(\xi,U) - \varepsilon_1(\xi)E_1(U,U)$$
$$= \epsilon\{(\nabla_\xi D_2)(U,U) - (\nabla_U D_2)(\xi,U)$$
$$+ \rho_2(\xi)D_1(U,U)\}, \tag{9.1.2}$$

$$\bar{R}(U,\xi,N,U) = R(U,\xi,N,U)$$
$$- \rho_2(U)\varepsilon_1(U) - \epsilon\rho_2(\xi)D_2(U,U)$$
$$= g(\nabla_\xi(A_NU) - \nabla_U(A_N\xi)$$
$$- A_N[\xi,U],U),U],U) + \rho_1(U)E_1(\xi,U)$$
$$- \rho_1(\xi)E_1(U,U) - \rho_2(U)\varepsilon_1(U)$$
$$- \epsilon\rho_2(\xi)D_2(U,U), \tag{9.1.3}$$

$$\bar{R}(U,\xi,N,\xi) = R(U,\xi,N,\xi)$$
$$+ \rho_2(\xi)\varepsilon_1(U) - \rho_2(U)\varepsilon_1(\xi)$$
$$= -D_1(U,A_N\xi) + 2d\rho_1(U,\xi)$$
$$+ \rho_2(\xi)\varepsilon_1(U) - \rho_2(U)\varepsilon_1(\xi), \tag{9.1.4}$$

$$\bar{R}(U,\xi,u,\xi) = \epsilon\{(\nabla_\xi D_2)(U,\xi) - (\nabla_U D_2)(\xi,\xi)\}$$
$$= -D_1(U,A_u\xi) + 2d\varepsilon_1(U,\xi)$$
$$\rho_1(U)\varepsilon_1(\xi) - \rho_1(\xi)\varepsilon_1(U), \tag{9.1.5}$$

$$\bar{R}(U,\xi,u,N) = \epsilon\{D_2(U,A_N\xi) - D_2(\xi,A_NU)$$
$$+ 2d\rho_2(U,\xi) + \rho_1(U)\rho_2(\xi)$$
$$- \rho_1(\xi)\rho_2(U)\}$$
$$= \bar{g}(\nabla_\xi(A_uU) - \nabla_U(A_u\xi),N)$$
$$- \epsilon\rho_2([\xi,U]. \tag{9.1.6}$$

By using (4.1.36)–(4.1.39), (4.1.10) and (4.1.22), we obtain the following components of curvature tensor field R of the induced connection ∇ on M:

$$R(U,\xi,\xi,U) = (\nabla_\xi D_1)(U,U) - (\nabla_U D_1)(\xi,U)$$
$$+ \rho_1(\xi)D_1(U,U) \tag{9.1.7}$$
$$= g(\nabla_\xi^*(A_\xi^*U),U) + D_1([U,\xi],U)$$
$$+ \rho_1(\xi)D_1(U,U),$$

$$R(U,\xi,N_1,U) = g(\nabla_\xi(A_{N_1}U) - \nabla_U(A_{N_1}\xi)$$
$$- A_{N_1}[\xi U],U) + \rho_1(U)E_1(\xi,U)$$
$$- \rho_1(\xi)E_1(U,U) \tag{9.1.8}$$
$$= \xi(E_1(U,U)) - U(E_1(\xi,U))$$
$$- E_1([\xi,U],U)$$
$$+ E_1(\xi,\nabla_U^*U) - E_1(U,\nabla_\xi^*U)$$
$$+ \rho_1(U)E_1(\xi,U) - \rho_1(\xi)E_1(U,U),$$

$$R(U,\xi,N,\xi) = -D_1(U, A_{N_1}\xi) + 2d\rho_1(U,\xi)$$
$$= -E_1(\xi, A_\xi^* U) + 2d\rho_1(U,\xi), \tag{9.1.9}$$
$$R(U,\xi,U,U) = -E_1(\xi, U)D_1(U,U). \tag{9.1.10}$$

The components in (9.1.7), (9.1.8), (9.1.9) and (9.1.10) follow from (4.1.38), (4.1.37), (4.1.39) and (4.1.36) respectively. The structure equations of a totally umbilical half-lightlike surface M are

$$\bar{R}(U,\xi,\xi,U) = R(U,\xi,\xi,U)$$
$$= (\nabla_\xi D_1)(U,U) + D_1(\nabla_U \xi, U)$$
$$+ \rho_1(\xi)D_1(U,U),$$
$$\bar{R}(U,\xi,u,U) = g(\nabla_\xi(A_u U) - A_u[\xi, U], U)$$
$$= \epsilon\{(\nabla_\xi D_2)(U,U) + D_2(\nabla_U \xi, U)\}$$
$$\bar{R}(U,\xi,N,U) = R(U,\xi,N,U)$$
$$= g(\nabla_\xi(A_N U) - \nabla_U(A_N \xi) - A_N[\xi, U], U)$$
$$+ \rho_1(U)E_1(\xi, U) - \rho_1(\xi)E_1(U,U),$$
$$\bar{R}(U,\xi,N,\xi) = R(U,\xi,N,\xi)$$
$$= -D_1(U, A_N \xi) + 2d\rho_1(U,\xi),$$
$$\bar{R}(U,\xi,u,\xi) = 0,$$
$$\bar{R}(U,\xi,u,N) = \epsilon\{D_2(U, A_N \xi) + 2d\rho_2(U,\xi) - \rho_1(\xi)\rho_2(U)\}$$
$$= \bar{g}(\nabla_\xi(A_u U), N) - \epsilon\rho_2([\xi, U]).$$

Since the tangent space of a half-lightlike surface M of $\bar{M}$ is a null plane, we define the *null sectional curvature* of M at $x \in M$, with respect to ξ_x, as the real number

$$K_{\xi_x}(M)_x = \frac{R(U_x, \xi_x, \xi_x, U_x)}{g(U_x, U_x)}, \tag{9.1.11}$$

where U_x is an arbitrary non-null vector in $T_x M$. The above definition follows the one given in Beem-Ehrlich [34, page 571] and O'Neill [317, pages 152, 153, 163]. Similarly, define the *null sectional curvature* $\bar{K}_{\xi_x}(T_x M)$ of the null plane $T_x M$ of the Lorentz vector space $T_x \bar{M}$ with respect to ξ_x and ∇ as a real number

$$\bar{K}_{\xi_x}(T_x M) = \frac{\bar{R}(U_x, \xi_x, \xi_x, U_x)}{g(U_x, U_x)}. \tag{9.1.12}$$

On a coordinate neighborhood $\mathcal{U}$ of M denote by $\bar{K}_\xi(TM)_{|\mathcal{U}}$ and $K_\xi(M)_{|\mathcal{U}}$ the *null sectional curvature functions* which associate to each $x \in \mathcal{U}$ the real numbers $\bar{K}_{\xi_x}(T_x M)$ and $K_{\xi_x}(M)_x$, respectively. Since neither of the null sectional curvatures depends on U_x, using (9.1.1) and (9.1.7), we obtain

$$\bar{K}_\xi(T_xM)_{|\mathcal{U}} = K_\xi(M)_{|\mathcal{U}}$$

$$+ \frac{1}{g(U,U)}\{\varepsilon_1(\xi)D_2(U,U) + \epsilon(\varepsilon_1(U))^2\}, \qquad (9.1.13)$$

$$K_\xi(M)_{|\mathcal{U}} = \frac{1}{g(U,U)}\{(\nabla_\xi D_1)(U,U)$$

$$- (\nabla_U D_1)(\xi,U) + \rho_1(\xi)D_1(U,U)\}. \qquad (9.1.14)$$

Hence, in general, the null curvature of M is expressed in terms of the lightlike second fundamental form of M.

If $\bar{M}$ is of constant curvature, by using (9.1.12) and (9.1.13) we obtain the null sectional curvature of M expressed in terms of the screen second fundamental form of M,

$$K_\xi(M)_{|\mathcal{U}} = -\frac{1}{g(U,U)}\{\varepsilon_1(\xi)D_2(U,U) + \epsilon(\varepsilon_1(U))^2\}. \qquad (9.1.15)$$

Comparing (9.1.15) with the classical formula of Gaussian curvature of a non-degenerate surface [149], we may call $K_\xi(M)_{|\mathcal{U}}$, given by (9.1.15), the *null Gaussian curvature* of M on $\mathcal{U}$. Also we note from (9.1.14) and (9.1.15) the lightlike and screen second fundamental forms of a lightlike surface of a spacetime manifold of constant sectional curvature are related by

$$\varepsilon_1(\xi)D_2(U,U) + \rho_1(\xi)D_1(U,U) + \epsilon(\varepsilon_1(U))^2$$

$$= (\nabla_U D_1)(\xi,U) - (\nabla_\xi D_1)(U,U). \qquad (9.1.16)$$

From (9.1.15) and Theorem 4.3.1 (chapter 4), we obtain

Theorem 9.1.5. *The null Gaussian curvature of a totally umbilical half-lightlike surface M of a spacetime manifold vanishes.*

As M is totally umbilical, from (9.1.13) and (9.1.14) we obtain

$$\bar{K}_\xi(T_xM)_{|\mathcal{U}} = K_\xi(M)_{|\mathcal{U}}, \qquad (9.1.17)$$

$$K_\xi(M)_{|\mathcal{U}} = \xi(H_1) - (H_1)^2 + H_1\,\rho_1(\xi). \qquad (9.1.18)$$

Thus we have

Theorem 9.1.6. *Let M be a totally umbilical half-lightlike surface of a spacetime manifold $(\bar{M}, \bar{g})$. Then the null sectional curvature functions $\bar{K}_\xi(TM)$ and $K_\xi(M)$ vanish, if and only if, H_1 is a solution of the partial differential equation*

$$\xi(H_1) - (H_1)^2 + H_1\,\rho_1(\xi) = 0.$$

From (4.1.9) in Theorem 4.3.3 and the above theorem, we obtain

Corollary 9.1.7. *Let M be a totally umbilical half-lightlike surface of a spacetime manifold of constant curvature $(\bar{M}, \bar{g})$. Then the null sectional curvature functions of M and $\bar{M}$ vanish.*

From Theorem 4.1.3 and (4.3.1) and the above theorem, we obtain

Theorem 9.1.8. *Let M be a totally umbilical half-lightlike surface of a spacetime manifold. If the induced connection ∇ on M is a metric connection, then, both the null sectional curvatures $\bar{K}_\xi(TM)$ and $K_\xi(M)$ vanish.*

Null 2-surfaces of a Minkowski space. Suppose (M, g) is a half-lightlike surface of a Minkowski space $(\mathbf{R}_1^4, \bar{g})$, where $\bar{g}$ is the Minkowski metric given by

$$\bar{g}(x, y) = -x^0 y^0 + x^1 y^1 + x^2 y^2 + x^3 y^3.$$

Locally M is given by equations

$$x^A = x^A(u, v), \qquad A \in \{0, 1, 2, 3\}.$$

Then the tangent bundle of M is locally spanned by

$$\left\{ \frac{\partial}{\partial u} = \frac{\partial x^A}{\partial u} \frac{\partial}{\partial x^A} \quad ; \quad \frac{\partial}{\partial v} = \frac{\partial x^A}{\partial v} \frac{\partial}{\partial x^A} \right\}.$$

Considering the vector field

$$\xi = \alpha \frac{\partial}{\partial u} + \beta \frac{\partial}{\partial v},$$

we find that M is half-lightlike, if and only if, the homogeneous linear system with (α, β) as variables,

$$\sum_{a=1}^{3} \left\{ \alpha\left(\left(\frac{\partial x^a}{\partial u}\right)^2 - \left(\frac{\partial x^0}{\partial u}\right)^2 \right) + \beta\left(\frac{\partial x^a}{\partial u} \frac{\partial x^a}{\partial v} - \frac{\partial x^0}{\partial u} \frac{\partial x^0}{\partial v} \right) \right\} = 0,$$

$$\sum_{a=1}^{3} \left\{ \alpha\left(\frac{\partial x^a}{\partial u} \frac{\partial x^a}{\partial v} - \frac{\partial x^0}{\partial u} \frac{\partial x^0}{\partial v} \right) + \beta\left(\left(\frac{\partial x^a}{\partial v}\right)^2 - \left(\frac{\partial x^0}{\partial v}\right)^2 \right) \right\} = 0,$$

has non-trivial solutions. Thus, we can state the following:

Theorem 9.1.9. [149] *The surface M of $\mathbf{R}_1^4$ is half-lightlike, if and only if, on each coordinate neighborhood $\mathcal{U} \subset M$ we have*

$$\sum_{a=1}^{3} (D^{oa})^2 = \sum_{1 \le a < b \le 3} (D^{ab})^2$$

where

$$D^{AB} = \begin{vmatrix} \dfrac{\partial x^A}{\partial u} & \dfrac{\partial x^B}{\partial u} \\ \dfrac{\partial x^A}{\partial v} & \dfrac{\partial x^B}{\partial v} \end{vmatrix}.$$

Next, we choose from the last homogeneous linear system

$$\alpha = \sum_{a=1}^{3} \frac{\partial x^a}{\partial u} \frac{\partial x^a}{\partial v} - \frac{\partial x^0}{\partial u} \frac{\partial x^0}{\partial v},$$

$$\beta = \left(\frac{\partial x^0}{\partial u}\right)^2 - \sum_{a=1}^{3} \left(\frac{\partial x^a}{\partial u}\right)^2, \tag{9.1.19}$$

in case at least one of the quantities from the right-hand side of (9.1.19) is non-zero. Then by direct calculations we obtain that $\mathrm{Rad}\,TM$, $\mathrm{ntr}(TM)$, $S(TM)$ and D are locally spanned on $\mathcal{U}$ by

$$\xi = \xi^A \frac{\partial}{\partial x^A}, \qquad \xi^A = \sum_{B=0}^{3} \epsilon_B D^{AB} \frac{\partial x^B}{\partial u},$$

$$N = \frac{1}{2(\xi^0)^2} \left(-\xi^0 \frac{\partial}{\partial x^0} + \sum_{a=1}^{3} \xi^a \frac{\partial}{\partial x^a} \right), \tag{9.1.20}$$

$$U = \frac{\partial x^0}{\partial v} \frac{\partial}{\partial u} - \frac{\partial x^0}{\partial u} \frac{\partial}{\partial v} = \sum_{a=1}^{3} D^{ao} \frac{\partial}{\partial},$$

$$W = D^{23} \frac{\partial}{\partial x^1} + D^{31} \frac{\partial}{\partial x^2} + D^{12} \frac{\partial}{\partial x^3}$$

respectively, where $\{\epsilon_B\}$ is the signature of the basis $\left\{\frac{\partial}{\partial x^B}\right\}$ with respect to $\bar{g}$. Summing up, we obtain the quasi-orthonormal frames field

$$\{\xi, N, U_o = \frac{1}{\sqrt{\triangle}} U, W_o = \frac{1}{\sqrt{\triangle}} W\}, \qquad \triangle \equiv \sum_{a=1}^{3} (D^{oa})^2.$$

However, in order to simplify the calculations we shall work, from now on, with the frames field $\mathcal{F} = \{\xi, N, U, W\}$.

By direct calculations from the Gauss equation (4.1.7) of M and the Weingarten equation (4.1.8) of $S(TM)$ we derive

$$\rho_1(\xi) = -\frac{1}{2(\xi^0)^2} \sum_{A=0}^{3} P^A \xi^A$$

$$D_1(U, U) = \sum_{a=1}^{3} S^a \, \xi^a,$$

$$D_2(\xi, \xi) = \frac{1}{\triangle} (P^1 D^{23} + P^2 D^{31} + P^3 D^{12})$$

$$= -\varepsilon_1(\xi),$$

$$D_2(U, \xi) = \frac{1}{\triangle} (Q^1 D^{23} + Q^2 D^{31} + Q^3 D^{12})$$

$$= \frac{1}{\triangle}(R^1 D^{23} + R^2 D^{31} + R^3 D^{12})$$

$$= D_2(\xi, U),$$

$$D_2(U,U) = \frac{1}{\triangle}(S^1 D^{23} + S^2 D^{31} + S^3 D^{12}), \tag{9.1.21}$$

$$P^A = (\alpha\frac{\partial\alpha}{\partial u} + \beta\frac{\partial\alpha}{\partial v})\frac{\partial x^A}{\partial u} + (\alpha\frac{\partial\beta}{\partial u} + \beta\frac{\partial\beta}{\partial v})\frac{\partial x^A}{\partial v}$$
$$+ \alpha^2\frac{\partial^2 x^A}{\partial u^2} + 2\alpha\beta\frac{\partial^2 x^A}{\partial u\partial v} + \beta^2\frac{\partial^2 x^A}{\partial v^2}, \tag{9.1.22}$$

$$Q^A = (\frac{\partial\alpha}{\partial u}\frac{\partial x^0}{\partial v} - \frac{\partial\alpha}{\partial v}\frac{\partial x^0}{\partial u})\frac{\partial x^A}{\partial u}$$
$$+ (\frac{\partial\beta}{\partial u}\frac{\partial x^0}{\partial v} - \frac{\partial\beta}{\partial v}\frac{\partial x^0}{\partial u})\frac{\partial x^A}{\partial v} + \alpha\frac{\partial x^0}{\partial v}\frac{\partial^2 x^A}{\partial u^2}$$
$$+ (\beta\frac{\partial x^0}{\partial v} - \alpha\frac{\partial x^0}{\partial u})\frac{\partial^2 x^A}{\partial u\partial v} - \beta\frac{\partial x^0}{\partial u}\frac{\partial^2 x^A}{\partial v^2},$$

$$R^a = (\alpha\frac{\partial^2 x^0}{\partial u\partial v} + \beta\frac{\partial^2 x^0}{\partial^2 v})\frac{\partial x^a}{\partial u}$$
$$- (\alpha\frac{\partial^2 x^0}{\partial u^2} + \beta\frac{\partial^2 x^0}{\partial u\partial v})\frac{\partial x^a}{\partial v} + \alpha\frac{\partial x^0}{\partial v}\frac{\partial^2 x^a}{\partial u^2}$$
$$+ (\beta\frac{\partial x^0}{\partial v} - \alpha\frac{\partial x^0}{\partial u})\frac{\partial^2 x^a}{\partial u\partial v} - \beta\frac{\partial x^0}{\partial u}\frac{\partial^2 x^a}{\partial v^2},$$

$$S^a = (\frac{\partial x^0}{\partial v}\frac{\partial^2 x^0}{\partial u\partial v} - \frac{\partial x^0}{\partial u}\frac{\partial^2 x^0}{\partial^2 v})\frac{\partial x^a}{\partial u}$$
$$- (\frac{\partial x^0}{\partial u}\frac{\partial^2 x^0}{\partial u\partial v} - \frac{\partial x^0}{\partial v}\frac{\partial^2 x^0}{\partial u^2})\frac{\partial x^a}{\partial v}$$
$$+ (\frac{\partial x^0}{\partial v})^2\frac{\partial^2 x^a}{\partial u^2} - 2\frac{\partial x^0}{\partial u}\frac{\partial x^0}{\partial v}\frac{\partial^2 x^a}{\partial u\partial v} + (\frac{\partial x^0}{\partial u})^2\frac{\partial^2 x^a}{\partial v^2},$$

$A \in \{0,1,2,3\}, \quad a \in \{1,2,3\}.$

The following results are straightforward.

Theorem 9.1.10. *A half-lightlike surface M of a Minkowski spacetime $\mathbf{R}_1^4$ is totally umbilical if and only if there exist smooth functions f_1, f_2 on M such that*

$$D_1(U, U) = f_1 g(U, U) = f_1 \triangle,$$
$$D_2(U, U) = f_2 g(U, U) = f_2 \triangle,$$
$$D_2(U, \xi) = D_2(\xi, U) = D_2(\xi, \xi) = 0.$$

Example 2. Consider a surface M in $\mathbf{R}_1^4$ given by

$$x^2 = f(x^1); \qquad x^3 - x^0 = 0,$$

where f is an arbitrary smooth function. By direct calculation we find that M is a half-lightlike submanifold of R_1^4 and we obtain the following frames field on R_1^4

along M:

$$\xi = -(1 + (f'(x^1))^2)\left(\frac{\partial}{\partial x^0} + \frac{\partial}{\partial x^3}\right),$$

$$N_1 = \frac{1}{2(1 + (f'(x^1))^2)}\left(\frac{\partial}{\partial x^0} - \frac{\partial}{\partial x^3}\right),$$

$$U = \frac{\partial}{\partial x^1} + f'(x^1)\frac{\partial}{\partial x^2},$$

$$u = f'(x^1)\frac{\partial}{\partial x^1} - \frac{\partial}{\partial x^2}$$

such that

$$\mathrm{Rad}\,TM = \mathrm{Span}\{\xi\}, \qquad\qquad \mathrm{ntr}(TM) = \mathrm{Span}\{N_1\},$$
$$S(TM) = \mathrm{Span}\{U\}, \qquad\qquad D = \mathrm{Span}\{u\}.$$

By the method of Example 2 in Chapter 4, we obtain

$$D_1 = 0,\, D_2(U,\xi) = D_2(\xi,\xi) = 0,\, D_2(U,U) = -\frac{f''(x^1)}{1 + (f'(x^1))^2}.$$

On the other hand, $\bar{g}\,(U,\,U) = 1 + (f'(x^1))^2$. Therefore

$$D_2\,(U,\,U) = H_2\,\bar{g}\,(U,\,U), \qquad H_2 \equiv \frac{f''(x^1)}{(1 + (f'(x^1))^2)^2}.$$

Thus M is a totally umbilical half-lightlike surface of R_1^4.

9.2 Harmonic maps and morphisms

Introduction. The concept of harmonic maps and morphisms constitutes a very useful tool for both Global Analysis and Differential Geometry. The theory of harmonic maps has been developed in the last fifty years; it is still an active field in differential geometry and it has applications to many different areas of mathematics and physics. The objective of this section is to present latest work on the applications of lightlike geometry in this important topic on harmonicity. For details on Riemannian harmonic maps and morphisms and the terminology used in this section, we refer to the latest book by Baird and Wood [25] and the two bibliographies [80] and [220], respectively. In this section we deal with the case of harmonic maps and morphisms into a class of lightlike manifolds. In the next section we present the latest work on harmonic maps from r-lightlike manifolds to semi-Riemannian manifolds.

Let $(\bar{M}, \bar{g})$ and (M, g) be Riemannian manifolds and $\phi : \bar{M} \to M$ a smooth map from $\bar{M}$ to M. The energy density of ϕ is given by $e(\phi) = \frac{1}{2}|d\phi(x)|^2$, where

$|d\phi(x)|$ is the Hilbert-Schmidt norm of the linear map $d\phi_x : T_x\bar{M} \to T_{\phi(x)}M$. The energy $\epsilon(\phi)$ is defined by

$$\epsilon(\phi) = \int_D e(\phi)\, dv_g \qquad (9.2.1)$$

where dv_g is the volume element of $\bar{M}$ and D is any compact domain of $\bar{M}$. The covariant derivative of the differential $d\phi$, denoted by $\bar{\nabla}\, d\phi$, is called the second fundamental form of ϕ, where $\bar{\nabla}$ is the Levi-Civita connection on $\bar{M}$. The tension field, denoted by $\tau(\phi)$, is defined by the trace of $\bar{\nabla}\, d\phi$. It is known that ϕ is harmonic, that is, ϕ is a critical point of $\epsilon(\phi)$, if and only if $\tau(\phi)$ vanishes. Also, ϕ is a minimal map if $\epsilon(\phi) \leq \epsilon(\phi_i)$ for all ϕ_i in the component of the space of all smooth maps between $\bar{M}$ and M. A subcase of a harmonic map is called a harmonic morphism, first studied by Fuglede [193] and Ishihara [238] and ϕ is a harmonic morphism if it carries germs of harmonic functions to germs of harmonic functions. This means that whenever $\mu : U \to R$ is a harmonic function on M and $\phi^{-1}(U)$ is non-empty, then the composite $\mu \circ \phi : \phi^{-1}(U) \to R$ is also harmonic. Since harmonic functions are solutions of Laplace's equation $\triangle \mu = 0$, any constant map is a harmonic morphism. Also, it is known that a non-constant harmonic morphism only exists if $\dim(\bar{M}) \geq \dim(M)$. The validity of the energy equation (9.2.1) for any semi-Riemannian manifold is not obvious for the following reasons.

(a) The Hopf-Rinow theorem (which guarantees the completeness of all Riemannian metrics for a compact Riemannian manifold) does not hold for an arbitrary semi-Riemannian manifold (see Beem et al. [34]). Therefore, the existence of a compact domain D, of the integral in (9.2.1), for an arbitrary semi-Riemannian manifold $\bar{M}$ is questionable.

(b) The possibility of degenerate(lightlike) fibers. If $\bar{M}$ has boundary $\partial\bar{M}$, then, the null (lightlike) $\partial\bar{M}$ is a physically important case.

Thus, the harmonic maps and morphisms between semi-Riemannian manifolds behave differently and their study is restricted to some classes of semi-Riemannian manifolds. For example, among all Lorentzian manifolds the globally hyperbolic manifolds [34] may be preferred since they do possess compact regions.

Harmonic maps, morphisms into lightlike manifolds. Since for any semi-Riemannian manifold, there is a natural existence of a lightlike subspace (hypersurface or submanifold), whose metric is degenerate, one cannot use the theory of harmonic maps of non-degenerate manifolds for the lightlike case. For example, if the target space M is lightlike, with $\phi : \bar{M} \to M$, then $\phi^{-1}(TM)$ will not exist because the metric of M is degenerate and the tension tensor can not be defined. Similar situation arises if the base space $\bar{M}$ is lightlike. The objective in this section is to explain how this anomaly can be removed if the target space M belongs to a class of lightlike manifolds, called *globally null manifolds* discussed in Section 4 of Chapter 1. The case when the base space $\bar{M}$ is lightlike has been discussed in the next section. We will also explain the reason for the choice of globally null manifolds.

Recall from Chapter 1 that a lightlike manifold (M, g), with $\operatorname{Rad} TM$ of rank one, is called a *globally null manifold* if it admits a global null vector and a complete Riemannian hypersurface. As explained in [134], we choose a class of globally null manifolds which admit a metric connection with respect to the degenerate metric tensor g and $\operatorname{Rad} TM$ is Killing. We need the following result:

Theorem 9.2.1. (Duggal [140] *Let* (M, g) *be an* $(n+1)$*-dimensional lightlike manifold with* $r = 1$. *Then, there exists a quasi-orthonormal frame*

$$F = \{\xi, W_1, \ldots, W_n\}, \quad g(W_a, W_a) = 1, \quad g(\xi, W_a) = 0, \tag{9.2.2}$$

$\forall a \in \{1, \ldots, n\}$, *along a null curve* C, *generated by a null vector field* ξ, *on* M. *Also, there exists a special affine parameter with respect to which* C *is a null geodesic of* M.

Note 1. We refer to the brief Definition 1.4.5 and to [137, 138, 139] for up-to-date information on globally null manifolds with geometric or physical examples.

Definition 9.2.2. Let $(\bar{M}, \bar{g})$ and (M, g) be Riemannian (or semi-Riemannian) and globally null manifolds respectively. Define a harmonic map

$$\phi : \bar{M} \to M' \subset M, \quad \phi(\bar{M}) = M', \tag{9.2.3}$$

where M' denotes a complete Riemannian hypersurface of M.

The above definition is well defined for the energy equation (9.2.1) since the target space M' is Riemannian for which we do have the pullback $\phi^{-1}(TM')$ to define the tension field needed for setting up ϕ as a harmonic map. Observe that the study of semi-Riemannian domain manifolds $\bar{M}$ is subject to restrictions as mentioned in item **(a)** above. The Laplace-Beltrami operator $\triangle_M$ on M is given in local coordinates x^i by

$$\triangle_{\bar{M}} = \frac{1}{\sqrt{|\bar{g}|}} \sum_i^m \partial_{x^i} \left(\sqrt{|\bar{g}|} \sum_i^m \bar{g}^{ij} \partial_{x^j} \right),$$

where $\bar{g} = |det(\bar{g}_{ij})|$. Any harmonic function μ is a C^2-smooth local solution to the Laplace-Beltrami equation $\triangle_{\bar{M}} \mu = 0$. The tension field $\tau(\phi)$ is a vector field along ϕ which assigns each tangent vector $\tau(\phi)(x) = T_{\phi(x)}(M')$ for each point $x \in \bar{M}$. It is the trace of $\nabla d\phi$ where ∇ is the Levi-Civita connection on M. The map ϕ is called a harmonic map if and only if the tension field $\tau(\phi) = 0$. To justify the use of globally null manifolds and the validity of equation (9.2.3), we now discuss the following properties of M to show that the geometry of this class of M essentially reduces to the Riemannian geometry of its hypersurface M'. By definition we assume that (M', g') is a complete spacelike hypersurface of the globally null manifold M, with induced Riemannian metric g'. Also, using Theorem 9.2.1, if we consider a congruence of null geodesics, given by the null vector field $\frac{d}{dp}$, with respect to the distinguished parameter p, the induced metric g' can be expressed by

$$g' = w^1 \otimes w^1 + \ldots + w^{n-1} \otimes w^{n-1} \qquad (9.2.4)$$

where $\{w^1, \ldots, w^{n-1}\}$ is the dual set of an orthonormal basis $\{W_1, \ldots, W_{n-1}\}$ of $\Gamma S(TM)$. Clearly g' being Riemannian metric, its inverse exists and is also Riemannian. Thus, the energy equation is well defined. Moreover, it follows from (9.2.4)that any tensor (including degenerate metric g) on M can be projected onto its screen distribution and all the analysis on M can be done on its integral spacelike hypersurface M', without any loss of the geometry of globally null manifolds.

Recall that the existence problem for harmonic maps concerns the minima and critical points of the energy in homotopy classes of maps defined by (9.2.3). To deal with this problem, we assume that $\bar{M}$ is either compact or has a compact domain D of the integral (9.2.1). Since M' can be a complete integral manifold of $S(TM)$, the existence result for harmonic maps, due to Eells and Sampson [177], must hold for ϕ defined by (9.2.3). Let $\bar{M}$ and M be Riemann surfaces and let M carry a degenerate metric associated to a holomorphic quadratic differential form. Thus, $\dim(M') = 1$. For this case, recall the following existence result:

Theorem 9.2.3. *(Leite [286]) If $\bar{M}$ and M are two Riemann surfaces and M carries a degenerate metric associated to a holomorphic quadratic differential form, then every non-trivial homotopy class contains a Hölder continuous harmonic map.*

Note 2. If $\bar{M}$ and M are two surfaces of the same genus $p \geq 2$, then, Leite's result is a characterization of Teichmüller maps which satisfy Beltrami's equation, whose coefficients involve some holomorphic quadratic differentials. Here, in the above case, the target space of harmonic map ϕ is a 1-dimensional space M' in M.

Note 3. If (B, g_B) is globally hyperbolic, with H_s its Cauchy surface, it is easy to see that (M, g) is a globally null manifold, embedded in B.

Based on Note 3, we consider a harmonic map

$$\phi : \bar{M} \to M' \subset M \xrightarrow{i} B, \quad \phi(\bar{M}) = M' \qquad (9.2.5)$$

where $\bar{M}$ is a Riemannian (or semi-Riemannian) manifold, B and $i(B)$ are globally hyperbolic spacetime and a globally null hypersurface of B respectively. Thus, M will have geometry induced from B. Consequently, the relation (9.2.5) establishes an interplay between harmonic maps and the geometries of globally null manifolds and globally hyperbolic spacetimes. To support this interplay, we present the following physical example.

Example 3. Let $(\mathbf{R}_1^4, \bar{g})$ be the Minkowski spacetime, with metric $ds^2 = -dt^2 + dr^2 + r^2 (d\theta^2 + \sin^2 \theta \, d\phi^2)$, for a spherical coordinate system (t, r, θ, ϕ), which is non-singular if $0 < r < \infty$, $0 < \theta < \pi$, $0 < \phi < 2\pi$. It is well known that $\mathbf{R}_1^4$ is globally hyperbolic. Take two null coordinates $u = t + r$ and $v = t - r$ $(u > v)$. Then, we have

$$ds^2 = -\,du\,dv + \frac{1}{4}\,(u - v)^2\,(d\theta^2 + \sin^2\theta\,d\phi^2\,), \quad -\infty < u,\,v < \infty. \quad (9.2.6)$$

The absence of the terms du^2 and dv^2 in (9.2.6) implies that the two hypersurfaces $\{\,v = \text{constant}\,\}$, $\{\,u = \text{constant}\,\}$ are lightlike. Denote one of these lightlike hypersurfaces by (M, g), where g is the induced degenerate metric of $\bar{g}$. A leaf of the 2-dimensional screen distribution $S(TM)$ is topologically a 2-sphere S^2 with Riemannian metric $d\Omega^2 = r^2\,(d\theta^2 + \sin^2\theta\,d\phi^2\,)$ which is the intersection of two hypersurfaces. Since, by definition, a spacetime admits a global timelike vector field, it follows that both its lightlike hypersurfaces admit a single global null vector field. Thus, there exists a pair of globally null hypersurfaces of $\mathbf{R}_1^4$ and (9.2.5) reduces to

$$\phi : \bar{M} \to S^2 \subset M \xrightarrow{i} \mathbf{R}_1^4, \quad \phi(\bar{M}) = S^2. \quad (9.2.7)$$

Similarly, one can show that there exists a pair of globally null hypersurfaces of Robertson-Walker spacetimes, Reissner-Nordström spacetimes, and Kerr spacetimes, with S^2 a leaf of their respective screen distributions.

Note 4. It is well known that the energy $e(\phi)$ and the equation $\tau(\phi) = 0$ are both preserved by a conformal class of the metric on a surface $\bar{M}$. For this reason, we have set $\dim(\bar{M}) = 2$. Since, in equation (9.2.7), we are dealing with non-degenerate metrics for $\bar{M}$ and S^2 both, the fundamental existence problem of harmonic maps between Riemannian surfaces (see, Eells-Lemaire [168] for the existence problem of harmonic maps from an oriented surface and from a non-oriented surface to S^2) must hold for the harmonic maps defined by (9.2.7).

Harmonic morphisms. Let ϕ be a harmonic map from a smooth Riemannian (or semi-Riemannian) manifold $(\bar{M}, \bar{g})$ into a globally null manifold (M, g), of dimensions m and n respectively, as defined by (9.2.3). If $\bar{M}$ is Riemannian, then, we have the tangent space $T_x\bar{M} = H_x \oplus V_x$, at each point $x \in \bar{M}$, where H_x and $V_x = H_x^{\perp} = \text{Ker}(d\phi_x)$ are its horizontal and vertical spaces, respectively. If $\bar{M}$ is semi-Riemannian, then, $V_x + H_x \neq T_x\bar{M}$ is possible, whenever V_x (or just as well H_x) is degenerate for which ϕ is called a degenerate harmonic map. In this book, we assume that ϕ is non-degenerate.

Definition 9.2.4. A map $\phi : (\bar{M}, \bar{g}) \to (M', g') \subset (M, g)$ is called *horizontally (weakly) conformal* if, for every $x \in \bar{M}$, either $d\phi_x\,|_{H_x}$ is conformal and surjective or $d\phi_x = 0$.

The coefficient of conformality $\lambda = \lambda(x)$ is called the *dilation* of ϕ at x. Clearly, for a non-degenerate ϕ, the possibility $d\phi = 0$ will not arise and, therefore, $\lambda(x) \neq 0$. It is known [193] that harmonic morphisms are characterized as the harmonic maps which are horizontally (weakly) conformal. Such maps pull back harmonic functions to harmonic functions. Since any constant ϕ is obviously a harmonic morphism, for the non-constant case, the existence of harmonic morphisms requires at least the existence of local harmonic functions on M' and $\bar{M}$ both. To deal with this existence problem, we show that there do exist harmonic functions on globally null manifolds. First we deal with the following simple cases.

- Suppose $\dim(M') = 1$. Then, the horizontal weak conformality is trivial and a harmonic morphism is just a harmonic map.

- Suppose $\dim(\bar{M}) = \dim(M') = 2$. Then the harmonic morphisms are exactly (weakly) conformal maps, for which the concept of harmonic morphism into Riemann surfaces is well-defined. Also, see Fuglede [25, 193] for some results on Riemannian and Lorentzian surfaces.

- Suppose $\dim(\bar{M}) = \dim(M') \geq 3$. Then, the harmonic maps are conformal maps with constant dilation [193].

Physical application. Let $(\bar{M}, \bar{g})$ be the Einstein universe which may be described as a 3-sphere S^3 of a fixed radius r, i.e., as the boundary of a 4-dimensional ball, by the equator. Consider a non-constant map

$$\phi : (\bar{M}, \bar{g}) \to S^2 \subset (M, g) \xrightarrow{i} \mathbf{R}^4_1 \qquad (9.2.8)$$

where S^2 is a leaf of a screen $S(T\bar{M})$ of $\bar{M}$. A sub-mapping of (9.2.8), given by

$$\phi|_{S^3} : S^3 \to S^2 \subset \bar{M} \xrightarrow{i} \mathbf{R}^4_1$$

is known as a Hopf map, the most celebrated example of harmonic morphisms, with constant dilation $\lambda = 2$ and minimal fibers. Here we have related the Hopf map with a spacetime and a globally null manifold.

The general theory, with $\dim(\bar{M}) > \dim(M')$, is difficult. For the existence of a harmonic morphism, one must make sure that there is a class of Riemannian (or semi-Riemannian) manifolds $(\bar{M}, \bar{g})$ with local harmonic functions. To justify this, we consider the following special case. Let $\mathbf{R}^{n+2}_1$ be Lorentzian space with a local coordinates system $(x^o, \ldots, x^{n+1})$. Consider a smooth function $\mu : D \to \mathbf{R}$, where D is an open set of $\mathbf{R}^{n+1}$. Then

$$M = \{(x^0, \ldots, x^{n+1} \in \mathbf{R}^{n+2}_1 \ ; \ x^0 = \mu(x^1, \ldots, x^{n+1})\}$$

is a hypersurface of $\mathbf{R}^{n+2}_1$ which is called a Monge hypersurface. Let a natural parameterization on M be given by

$$x^o = \mu(v^0, \ldots, v^n) \ ; \ x^{\alpha+1} = v^\alpha, \ \alpha \in \{0, \ldots, n\}.$$

Hence, the natural frames field on M is globally defined by

$$\partial_{v^\alpha} = \mu'_{x^{\alpha+1}} \partial_{x^0} + \partial_{x^{\alpha+1}}, \ \alpha \in \{0, \ldots, n\}.$$

Then, it is easy to see that $TM^\perp$ is spanned by a global vector

$$\xi = \partial_{x^0} + \sum_{a=1}^{n+1} \mu'_{x^a}\, \partial_{x^a}. \tag{9.2.9}$$

For a hypersurface (see Chapter 2) we know that M is lightlike if and only if $TM^\perp = \mathrm{Rad}\, TM$. This means that ξ, defined by (9.1.9), must be a null vector field if and only if M is lightlike. Now we recall the following results from [149, Section 4].

Proposition 9.2.5. *On a lightlike Monge hypersurface M of $\mathbf{R}_1^{n+2}$, there exists an integrable screen distribution $S(TM)$.*

Proposition 9.2.6. *A lightlike Monge hypersurface M of $\mathbf{R}_1^{n+2}$, is a product manifold $M = L \times M'$, where L is an open subset of a null line and M' is Riemannian.*

Proposition 9.2.7. *A lightlike Monge hypersurface M of $\mathbf{R}_1^4$ is minimal if and only if M admits a harmonic function.*

Note that Proposition 9.2.7 also holds for a general case of $\mathbf{R}_1^{n+2}$, although the proof is very lengthy.

To relate the above three results with the subject matter of this section, we let $\mathbf{R}_1^{n+2}$ admit a globally defined timelike vector field, which means it is a time orientable Minkowski spacetime. This further means that $\mathbf{R}_1^{n+2}$ is a globally hyperbolic spacetime. Since, by Proposition 9.2.6, M' is a Riemannian manifold, the celebrated Hopf–Rinow theorem allows us to assume that M' is a complete Riemannian manifold. Comparing Propositions 9.2.5 and 9.2.6 with the assertions **(a)** and **(b)** of Theorem 1.4.9 and the fact that M admits a single global null vector ξ, M is a globally null Monge hypersurface of a Minkowski spacetime $\mathbf{R}_1^{n+2}$. Consequently, Proposition 9.2.7 implies the following:

Proposition 9.2.8. *A globally null Monge hypersurface M of a Minkowski spacetime is minimal if and only if M admits a harmonic function μ.*

To show the relevance of the above result in the study of harmonic morphisms, we consider a harmonic map

$$\phi : \bar{M} \to M' \subset M \subset \mathbf{R}_1^{n+2}, \quad \phi(\bar{M}) = M', \tag{9.2.10}$$

where $(\bar{M}, \bar{g})$ is a Riemannian (or semi-Riemannian) manifold, M is a globally null Monge hypersurface of $\mathbf{R}_1^{n+2}$ and M' is a Riemannian minimal hypersurface of N. Let $\dim(\bar{M}) = m > n$. We know that the harmonic functions on $\bar{M}$ are those which satisfy the Laplace-Baltrami equation. Thus, the conclusion is that given a harmonic map ϕ, defined by (9.2.10), Proposition 9.2.8 can satisfy the basic requirement (existence of local harmonic functions on $\bar{M}$ and M) for the existence of ϕ as a harmonic morphism. To make sure that there is a class of Riemannian (or semi-Riemannian) manifolds $(\bar{M}, \bar{g})$ with local harmonic functions, there are more than one approaches. We prefer to use the following differential geometric structure, called f-structure on $\bar{M}$, studied by Yano [406]. Also see Fuglede [193] who used a special case of an f-structure, called an almost Hermitian structure, in the study of harmonic morphisms. We first give some brief information on f-structures. In 1963, Yano [406] introduced an f-structure on a C^∞ m-dimensional manifold $\bar{M}$, defined by a tensor field f of type (1,1) which satisfies $f^3 + f = 0$ and has constant rank r which is necessarily even. $T\bar{M}$ splits into two complementary sub-bundles Im f and Ker f and the restriction of f to Im f determines a complex structure on such a sub-bundle. Almost complex and almost contact structures are special cases according as $r = m$ and $r = m - 1$ respectively. Suppose Ker f is parallelizable [152]. Then, there exists a global frame $\{\xi_i\}$ for Ker f with dual 1-forms η^i satisfying:

$$f^2 = -I + \eta^i \otimes \xi_i, \quad \eta^i(\xi_j) = \delta^i_j, \quad 1 \leq i \leq s.$$

It follows that $f\xi_i = 0$, $\eta^i \circ f = 0$. An f-structure on $\bar{M}$ is normal if the tensor field $N_f = [f, f] + 2d\eta^i \otimes \xi_i$ vanishes, where $[f, f]$ is the Nijenhuis tensor of f. Then, there exists a semi-Riemannian metric $\bar{g}$ on $\bar{M}$ such that

$$\bar{g}(X, fY) + \bar{g}(fX, Y) = 0, \quad \forall X, Y \in T(M). \tag{9.2.11}$$

The structure (f, ξ_i, η^i, g) is called a metric f-structure and M is then called a metric framed manifold. $T(\bar{M})$ splits as complementary orthogonal sum of its sub-bundles Im f and Ker f. We denote their respective horizontal and vertical distributions by D and $D^\perp$. In particular, $\bar{g}$ is a Riemannian metric on $\bar{M}$ if both the distributions D and $D^\perp$ are Riemannian. If there is only a single timelike ξ_i and if D is Riemannian, then, $\bar{M}$ is called a Lorentz framed manifold [133]. f extended by $\mathcal{C}$-linearity to the complexified tangent space $T^{\mathcal{C}}(\bar{M})$ has $i, -i, 0$ eigenvalues. Let $H = \{X - ifX | X \in D\} \subset D \otimes \mathcal{C}$, the holomorphic distribution related to the eigenvalue i.

Definition 9.2.9. A metric framed manifold $\bar{M}$ is called a CR-integrable framed manifold if its holomorphic distribution is involutive, that is,

$$[X - ifX, Y - ifY] \in H \text{ for any } X, Y \in D.$$

Proposition 9.2.10. [152] *A metric framed manifold $\bar{M}$ is **CR**-integrable if and only if its horizontal distribution D is normal, that is, the Nijenhuis tensor field $N_f(X, Y) = 0$ for $X, Y \in \mathcal{D}$.*

For a **CR**-integrable framed manifold $(\bar{M}, \bar{g})$, we have

$$\bar{M} = (B \times V, \bar{g}) \ ; \ T\bar{M} = TB \oplus TV, \tag{9.2.12}$$

where (B, g_B) and (V, g_V) are the horizontal and the vertical integral submanifolds (generated by D and $D^\perp$ respectively) of $\bar{M}$. To show the relevance of the above brief note (on f-structures) in the study of harmonic morphisms, we first observe that a **CR**-framed (or integrable framed) manifold $\bar{M}$ has a holomorphic distribution H. This means that $\bar{M}$ admits holomorphic functions whose real parts are locally harmonic functions of its horizontal submanifold (B, g_B) (see equation (9.2.12). Thus, we conclude:

A **CR**-*integrable framed manifold* $(\bar{M}, \bar{g})$ *admits harmonic functions, defined locally on its horizontal submanifold* (B, g_B).

Therefore, we have justified the validity of a basic requirement for the existence of harmonic functions on the manifolds $\bar{M}$ and M' needed for equation (9.2.10). To show that the map ϕ, given by (9.2.10), can be a harmonic morphism, we let $\bar{M} = (B \times V, \bar{g})$ be a **CR**-integrable Lorentz framed manifold and (B, g_B) its Riemannian horizontal submanifold. Also, set

$$\dim(\bar{M}) = m = 2p + s \ (s \geq 1), \ \dim(B) = \dim(M') = n = 2p.$$

Suppose $\mathcal{M}$ denotes the space of all smooth Riemannian metrics on M' and $[h']$ a conformal class of metrics of $\mathcal{M}$. Since $\dim B = \dim M'$ and B is Riemannian, we let $g_B \in \mathcal{M}$. This means that B and M' are conformal to each other. B a horizontal submanifold of $\bar{M}$ implies that $d\phi_x |_{T_x B}$ is conformal for every $x \in B$. Thus ϕ is horizontally (weakly) conformal, which implies that ϕ is a harmonic morphism. If we relax the condition of B conformal to M', ϕ can be a harmonic morphism where we let ϕ pull back harmonic functions of M' to harmonic functions of $\bar{M}$, since their existence has been assured.

In support of the above model, we have the following physical model of a class of 4-dimensional spacetimes which inherit a metric f-structure.

Physical Model. Let $(\bar{M}, \bar{g})$ be a 4-dimensional Einstein Maxwell spacetime of general relativity, with a skew-symmetric (i.e., a 2-form) tensor field $F = (F_{ab})$ on $\bar{M}$ which represents the electromagnetic fields. The complex self-dual electromagnetic tensor field $F^\star$ is defined by

$$F^\star_{ab} = F_{ab} + i \, \tilde{F}_{ab} \quad \tilde{F}_{ab} = \frac{1}{2} \, \epsilon_{abcd} \, F^{cd}, \quad i = \sqrt{-1}.$$

Here, ϵ_{abcd} is the Levi-Civita tensor field. At each point $x \in \bar{M}$ consider the null tetrad $e_a = \{\ell, k, m, \bar{m}\}$ where $\bar{g}(\ell, k) = -1$ and $\bar{g}(m, \bar{m}) = 1$. According to Debney-Zund [120], define the following three complex functions, called Maxwell scalar fields:

$$\phi_0 = 2F_{ab}\ell^a m^b, \quad \phi_1 = F_{ab}(\ell^a k^b + \bar{m}^a m^b), \quad \phi_2 = 2F_{ab}\bar{m}^a m^b.$$

By a Lorentz transformation, one can set $\phi_0 = \phi_2 = 0$ (see Debney-Zund [120]). We are interested in a special case when ϕ_1 is either real or pure imaginary. For this subcase, it is known [120] that its canonical form is given by

$$F_{ab} = -2\text{Re}(\phi_1) \, \ell_{[a}n_{b]} \quad \text{or} \quad 2\,i\,\text{Im}(\phi_1) \, m_{[a}\bar{m}_{b]}, \quad \det(F_{ab}) = 0.$$

Consider a homogeneous spacetime for which ϕ_1 is constant (see Kramer et al [267, page 120]) and set $|\phi_1|^2 = 1$. Define a $(1,1)$-tensor field $f \equiv (f_b^a)$, on the tangent space $T_x(\bar{M})$, at each point $x \in \bar{M}$, such that

$$f_b^a = g^{ac} F_{cb}, \quad i.e., \quad F(X,Y) = \bar{g}(X, fY), \tag{9.2.13}$$

for every $X, Y \in \bar{M}$. Operating f_b^a by itself three times, it is easy to see that f satisfies its own minimum characteristic polynomial equation: $f^3 \pm f = 0$, where the sign $\pm$ depends on the choice of $\mathrm{Im}(\phi)$ or $\mathrm{Re}(\phi)$. We choose $\mathrm{Im}(\phi)$. For a Minkowski space, this condition means that the electric and the magnetic fields are aligned. Thus,

$$f^3 + f = 0, \qquad \mathrm{rank}(f) = 2.$$

F skew-symmetric and (9.2.13) implies f is also skew-symmetric, that is,

$$\bar{g}(X, fY) + \bar{g}(fX, Y) = 0, \qquad \forall X, Y \in \bar{M}.$$

It follows from (9.2.11) and the above two equations that, under some geometric conditions, a 4-dimensional Einstein Maxwell homogeneous spacetime $(\bar{M}, \bar{g}, F)$ inherits a metric f-structure, induced from its electromagnetic tensor field F. Furthermore, using Definition 9.2.9 and Propositions 9.2.6 and 9.2.10, $\bar{M}$ can be endowed with a CR-integrable Lorentz framed structure, denoted by $(f, \bar{g}, H, F)$, such that $\bar{M} = (B \times V, \bar{g})$ is a product manifold, with 2-dimensional horizontal surface (B, g_B). On the existence of such a physical spacetime, we quote the following:

Theorem 9.2.11. [267, page 120] *The only Einstein Maxwell field which is homogeneous and has a homogeneous non-singular Maxwell field is the Bertotti-Robinson solution*

$$ds^2 = A^2(d\theta^2 + \sin^2\theta \, d\phi^2 + dx^2 + \sinh^2 x \, dt^2)$$

for local coordinates (t, x, θ, ϕ) and an arbitrary constant A.

The above solution has two families of orthogonal 2-surfaces having equal and opposite curvatures. Relating these 2-surfaces with D and $D^\perp$, we observe that they are integrable. Thus, Bertotti-Robinson spacetime has an integrable f-structure. Based on the above, we have the following result:

Theorem 9.2.12. *Let $\phi : \bar{M} \rightarrow S^2 \subset (M, h) \subset \mathbf{R}_1^4$ be a map from a homogeneous Bertotti-Robinson spacetime to a Riemann surface S^2 of a globally null hypersurface M of $\mathbf{R}_1^4$. Then, ϕ is a harmonic morphism with respect to a **CR**-integrable Lorentz framed structure $(f, \bar{g}, H, F)$ on $\bar{M}$.*

Note 6. Theorem 9.2.12 will hold if $\mathbf{R}_1^4$ is replaced by those asymptotically flat spacetimes which possess S^2, a leaf of their respective screen distribution of a globally null manifold. Examples are: Schwarzschild, Robertson–Walker, Reissner–Nordström and Kerr spacetimes.

Remark 9.2.13. (a) In the final chapter of a book by Baird and Wood [25] on *Harmonic morphisms between semi-Riemannian manifolds* these authors have discussed the case when the subspaces of the tangent spaces involved may be lightlike (degenerate). For such degenerate cases they have modified the definitions accordingly and proved some characterization theorems of harmonic morphisms as horizontally weakly conformal harmonic maps. They observe that certain harmonic morphisms are simply null solutions of the wave equations.

(b) Pambira [323] has studied *Harmonic morphisms between degenerate semi-Riemannian manifolds* where he used Kupeli's method [273] for the mapping between lightlike manifolds.

(c) The readers may also be interested in the works of Mikhail Gromov and Richard Schoen [219] on *Harmonic maps into singular spaces and p-adic super rigidity for lattices in groups of rank one* and some references in [80] and [220] on harmonic mappings with values in higher-dimensional degenerate submanifolds which appear naturally as solutions of sigma-models with indefinite target spaces and related topics on applications of sigma-models.

9.3 Harmonic maps from lightlike manifolds

In this section, we consider maps from lightlike manifolds. First, we present screen lightlike submersions which can be considered as a lightlike version of Riemannian submersions. Then, as a generalization of lightlike submersions, we define and study screen conformal submersions which can be considered as a lightlike version of horizontally conformal submersions. Finally, we give a definition for harmonic maps from a lightlike manifold to a semi-Riemannian manifold and obtain a characterization of harmonic maps.

Let M and B be Riemannian manifolds. A Riemannian submersion $\phi : M \to B$ is a mapping of M onto B satisfying the following axioms S.1 and S.2:

S.1 ϕ has maximal rank,

S.2 ϕ_* preserves the lengths of horizontal vectors.

For each $b \in B$, $\phi^{-1}(b)$ is a submanifold of M of dimension $\dim M - \dim B$. The submanifolds $\phi^{-1}(b)$ are called fibers and a vector field on M is vertical if it is always tangent to the fibers, horizontal if always orthogonal to the fibers.

The Riemannian submersion was introduced by O'Neill [316] and A. Gray [212]. Since then, it has been an effective tool to describe the structure of a Riemannian manifold. For more information on Riemannian submersions, see: [184] and [413]. Note that the vertical distribution $\mathcal{V}$ of M is defined by $\mathcal{V}_p = \operatorname{Ker} d\phi_p$, $p \in M$. The orthogonal complementary distribution to $\mathcal{V} = \operatorname{Ker} d\phi$ is defined by $\mathcal{H}_p = (\operatorname{Ker} d\phi_p)^\perp$, denoted by $\mathcal{H}$ and called horizontal. Thus it follows that the tangent bundle of M has the decomposition

$$TM = \mathcal{V} \perp \mathcal{H}.$$

Screen lightlike submersions. In case of Riemannian submersion, the fibers are always Riemannian manifolds. However, for the case of semi-Riemannian manifolds, the fibers of ϕ may not be semi-Riemannian manifolds. From the above definition of Riemannian submersions and Section 1.4 of Chapter 1, we have seen that for both Riemannian submersions and lightlike manifolds a natural splitting of the tangent bundle plays a crucial role. For Riemannian submersions, it is the splitting of the tangent bundle of the source manifold into horizontal and vertical part. On the other hand, the tangent bundle of a lightlike manifold is decomposed into radical and screen parts. From this similarity, it is natural to give the following definition of screen lightlike submersion between a lightlike manifold and a semi-Riemannian manifold.

Definition 9.3.1. [357] Let $(M, g_M, S(TM))$ be an r-lightlike manifold and (N, g_N) a semi-Riemannian manifold. We say that a smooth mapping

$$\phi : (M, g_M, S(TM)) \longrightarrow (N, g_N)$$

is a screen lightlike submersion if

(a) at every $p \in M$, $\mathcal{V}_p = \mathrm{Ker}(d\phi)_p = \mathrm{Rad}\,T_pM$, i.e., $d\phi(X) = 0$ for every vector field $X \in \Gamma(\mathrm{Rad}\,TM)$.

(b) At each point $p \in M$, the differential $d\phi_p$ restricts to an isometry of the horizontal space $\mathcal{H}_p = S(TM)_p$ onto $T_{\phi(p)}N$, i.e.,

$$g_{S(TM)}(X_1, X_2) = g_N(d\phi(X_1), d\phi(X_2)), \quad \forall X_1, X_2 \in \Gamma(S(TM)).$$

First of all, from the definition, it follows that $\mathrm{Rad}\,TM$ is integrable. Let Q and P denote the orthogonal projections onto the distributions $\mathrm{Rad}\,TM$ and $S(TM)$, respectively. Obviously, the restriction of the differential $d\phi_p$ to the screen distribution $\mathcal{H}_p = S(TM)_p$ maps that space isomorphically onto $T_{\phi(p)}N$. Then, for any vector $\tilde{X} \in T_{\phi(p)}N$, we say that the vector $X \in S(TM)_p$ is a horizontal lift of $\tilde{X}$, the same as for Riemannian submersions. If $\tilde{X}$ is a vector field on an open subset V of N, then the horizontal lift of $\tilde{X}$ is the vector field $X \in \Gamma(S(TM))$ on $\phi^{-1}(V)$ such that $d\phi(X) = \tilde{X} \circ \phi$. As it is well known, the vector field X is called a basic vector field. Notice that if M is the same as in a semi-Riemannian manifold, then the above definition agrees with the definition of a semi-Riemannian submersion.

Remark 9.3.2. From Definition 9.3.1, it follows that M and N have the same index.

Example 4. Let $R_{2,1,1}^4$ and $R_{0,1,1}^2$ be R^4 and R^2 endowed with the degenerate metric $g_1 = -(dx_3)^2 + (dx_4)^2$ and a Lorentzian metric $g_2 = -(dy_1)^2 + (dy_2)^2$, where x_1, x_2, x_3, x_4 and y_1, y_2 are the canonical coordinates on R^4 and R^2, respectively. Define the map

$$\phi : \qquad R_{2,1,1}^4 \qquad \longrightarrow \qquad R_{0,1,1}^2$$
$$(x_1, x_2, x_3, x_4) \qquad \qquad \left(\tfrac{2x_3+x_4}{\sqrt{3}}, \tfrac{x_3+2x_4}{\sqrt{3}} \right).$$

Then it is easy to see that the kernel of $d\phi$ is

$$\mathrm{Ker}\, d\phi = \mathrm{Rad}\, TM = \mathrm{Span}\{Z_1 = \partial\, x_1, Z_2 = \partial\, x_2\}.$$

By direct computation, we obtain

$$S(TM) = \mathrm{Span}\{Z_3 = \frac{1}{\sqrt{3}}\{\partial\, x_3 + 2\partial\, x_4\}, Z_4 = \frac{1}{\sqrt{3}}\{2\partial\, x_3 + \partial\, x_4\}.$$

On the other hand, we get

$$d\phi(Z_3) = \frac{4}{3}\partial\, y_1 + \frac{5}{3}\partial\, y_2, \ d\phi(Z_4) = \frac{5}{3}\partial\, y_1 + \frac{4}{3}\partial\, y_2.$$

Then, it follows that

$$g_1(Z_3, Z_3) = g_2(d\phi(Z_3), d\phi(Z_3)) = 1$$

and

$$g_1(Z_4, Z_4) = g_2(d\phi(Z_4), d\phi(Z_4)) = -1.$$

Consequently, ϕ is a screen lightlike submersion.

Lemma 9.3.3. *Let $\phi : (M, g_M, S(TM)) \longrightarrow (N, g_N)$ be a screen lightlike submersion and X, Y basic vector fields of M. Then:*

(a) $g_M(X, Y) = g_N(d\phi(X), d\phi(Y)) \circ \phi$.

(b) *The horizontal part $[X, Y]^P$ of $[X, Y]$ is a basic vector field and corresponds to $[\tilde{X}, \tilde{Y}]$.*

(c) *For $\xi \in \Gamma(\mathrm{Rad}\, TM)$, $[X, \xi] \in \Gamma(\mathrm{Rad}\, TM)$.*

Proof. Let X and Y be basic vector fields of M. Then, from Definition 9.3.1 (b), (a) follows. Since $TM = S(TM) \oplus_{\mathrm{orth}} \mathrm{Rad}\, TM$, we write

$$[X, Y] = P[X, Y] + Q[X, Y].$$

Then, the horizontal part $[X, Y]^P$ of $[X, Y]$ is a basic vector field and corresponds to $[\tilde{X}, \tilde{Y}]$, i.e., $d\phi([X, Y]^P) = [d\phi(X), d\phi(Y)]$. On the other hand, since $\mathrm{Rad}\, TM$ is the kernel space of $d\phi$, we have $d\phi([X, \xi]) = [d\phi(X), d\phi(\xi)] = 0$, $\xi \in \Gamma(\mathrm{Rad}\, TM)$, which implies that $[X, \xi]$ belongs to $\mathrm{Rad}\, TM$. □

Theorem 9.3.4. *Let $\phi : (M, g_M, S(TM)) \longrightarrow (N, g_N)$ be a screen lightlike submersion. Then M is a Reinhart (or stationary [273]) lightlike manifold if and only if $\xi(g_M(X, Y)) = 0$ for any $\xi \in \Gamma(\mathrm{Rad}\, TM)$ and X, Y basic vector fields.*

Proof. We first note that the radical distribution is integrable. For $\xi \in \Gamma(\mathrm{Rad}\, TM)$, $X, Y \in \Gamma(TM)$ we have

$$(\mathcal{L}_\xi g_M)(X, Y) = \mathcal{L}_\xi g_M(X, Y) - g_M(\mathcal{L}_\xi X, Y) - g_M(X, \mathcal{L}_\xi Y),$$

where $\mathcal{L}$ denotes the Lie derivative. Then, we obtain

$$(\mathcal{L}_\xi g_M)(X,Y) = \xi\left(g_M(X,Y)\right) - g_M([\xi,X],Y) - g_M(X,[\xi,Y]).$$

Now, if $X \in \Gamma(\operatorname{Rad} TM)$, then $[X,\xi] \in \Gamma(\operatorname{Rad} TM)$ due to $\operatorname{Rad} TM$ is integrable. Thus, if $X,Y \in \Gamma(\operatorname{Rad} TM)$, we obtain

$$(\mathcal{L}_\xi g_M)(X,Y) = 0, \forall \xi, X, Y \in \Gamma(\operatorname{Rad} TM).$$

On the other hand, if $X \in \Gamma(\operatorname{Rad} TM)$ and $Y \in \Gamma(S(TM))$, we get

$$(\mathcal{L}_\xi g_M)(X,Y) = -g_M(X,[\xi,Y]).$$

Hence, we have
$$(\mathcal{L}_\xi g_M)(X,Y) = 0.$$

If $X,Y \in \Gamma(S(TM))$ are basic vector fields, using Lemma 9.3.3 (c), we obtain

$$(\mathcal{L}_\xi g_M)(X,Y) = \xi g_M(X,Y).$$

Thus, from Theorem 1.4.2 and the above equations, we conclude that M is Reinhart if and only if $\xi\left(g_M(X,Y)\right) = 0$, which completes the proof. □

From now on, suppose M is Reinhart unless otherwise stated.

Lemma 9.3.5. *Let $\phi : (M, g_M) \longrightarrow (N, g_N)$ be a screen lightlike submersion. Let also X and Y be basic vector fields on M. Then $(\nabla_X Y)^P$ is the basic vector field corresponding to $\nabla^N_{\tilde{X}} \tilde{Y}$.*

Proof. Since M is Reinhart, it has a torsion free connection ∇ such that g_M is parallel with respect to ∇. Then, from the Kozsul identity, with Z a basic vector field, we have

$$2g_M(\nabla_X Y, Z) = Xg_M(Y,Z) + Yg_M(Z,X) - Zg_M(X,Y)$$
$$- g_M(X,[Y,Z]) + g_M(Y,[Z,X]) + g_M(Z,[X,Y]).$$

Hence, since $d\phi([Z,X]) = [d\phi(Z), d\phi(X)]$, using Lemma 9.3.3, we get

$$2g_M(P\nabla_X Y, Z) = \{\tilde{X}g_N(\tilde{Y},\tilde{Z}) + \tilde{Y}g_N(\tilde{Z},\tilde{X}) - \tilde{Z}g_N(\tilde{X},\tilde{Y})$$
$$- g_N(\tilde{X},[\tilde{Y},\tilde{Z}]) + g_N(\tilde{Y},[\tilde{Z},\tilde{X}]) + g_N(\tilde{Z},[\tilde{X},\tilde{Y}])\} \circ \phi.$$

Then, the right side of the above equation is the Kozsul identity for the Levi-Civita connection ∇^N; thus, we derive

$$g_M(P\nabla_X Y, Z) = g_N(\nabla^N_{\tilde{X}} \tilde{Y}, \tilde{Z}) \circ \phi$$

which proves our assertion. □

Let $\phi : (M, g_M) \longrightarrow (N, g_N)$ be a screen lightlike submersion and $E, F \in \Gamma(TM)$. We now define O'Neill's tensors for ϕ. It will be seen that the situation is very different from a semi-Riemannian submersion. We denote the linear connection on M by ∇. Then since M is Reinhart, we know that $\nabla g_M = 0$. Define a tensor field T as

$$T_E F = P\nabla_{QE} QF + Q\nabla_{QE} PF.$$

It follows that T is a tensor field of type $(1, 2)$.

Lemma 9.3.6. *Let $\phi : (M, g_M) \longrightarrow (N, g_N)$ be a screen lightlike submersion, X, Y be horizontal vector fields and $V, W \in \Gamma(\operatorname{Rad} TM)$. Then we have:*

(a) *T is vertical; $T_E = T_{QE}$, $\forall E \in \Gamma(TM)$.*

(b) *T reverses the radical and screen subspaces.*

(c) *$T_V W = 0$.*

(d) *$T_X Y = 0$.*

(e) *$T_X V = 0$.*

(f) *$T_V X = Q\nabla_V X$.*

Proof. (a), (b), (d), (e) and (f) are clear from the definition of T. So we prove only (c). First, note that $T_V W \in \Gamma(S(TM))$. Then, for $Z \in \Gamma(S(TM))$ we get

$$g_M(T_V W, Z) = g_{S(TM)}(P\nabla_V W, Z) = g_{S(TM)}(\nabla_V W, Z).$$

Taking into account that g_M is parallel with respect to ∇, we get

$$g_M(T_V W, Z) = V g_M(W, Z) - g_{S(TM)}(\nabla_V Z, W) = 0,$$

since $W \in \Gamma(\operatorname{Rad} TM)$. Then non-degenerate $S(TM)$ and $Z, T_V W \in \Gamma(S(TM))$ imply that $T_V W = 0$. $\square$

Thus, from Lemma 9.3.6, we have

$$T_E F = Q\nabla_{QE} PF.$$

Now define another tensor as

$$A_E F = Q\nabla_{PE} PF + P\nabla_{PE} QF$$

which is also a tensor field of type $(1, 2)$.

Lemma 9.3.7. *Let $\phi : (M, g_M) \longrightarrow (N, g_N)$ be a screen lightlike submersion, X, Y horizontal vector fields and $V, W \in \Gamma(\operatorname{Rad} TM)$. Then we have:*

(a) *A is screen; $A_E = A_{PE}$, $\forall E \in \Gamma(TM)$.*

(b) *A reverses the radical and screen subspaces.*

(c) $A_V W = 0$.

(d) $A_X Y = Q\nabla_X Y$.

(e) $A_V X = 0$.

(f) $A_X V = P\nabla_X V = 0$.

Proof. The proof of the lemma is similar to the proof of the previous lemma.   □

Then, from Lemma 9.3.7, we have

$$A_E F = Q\nabla_{PE} PF.$$

Moreover, from Lemma 9.3.6 and Lemma 9.3.7, we obtain the following lemma.

Lemma 9.3.8. *Let* $\phi : (M, g_M) \longrightarrow (N, g_N)$ *be a screen lightlike submersion,* X, Y *horizontal vector fields and* $U, V \in \Gamma(\text{Rad }TM)$. *Then we have:*

1. $\nabla_X Y = P\nabla_X Y + A_X Y$.

2. $\nabla_U V = Q\nabla_U V$.

3. $\nabla_U X = P\nabla_U X + T_U X$.

4. $\nabla_X U = Q\nabla_X U$.

Remark 9.3.9. Notice that Lemma 9.3.8 (2) implies that the fibres are totally geodesic in M. If X is a basic vector field, then $[V, X]$, $V \in \Gamma(\text{Rad }TM)$, is vertical. Hence $P\nabla_V X = P\nabla_X V$. Using Lemma 9.3.8 (4), we obtain $P\nabla_V X = 0$.

Now, we obtain the covariant derivatives ∇T and ∇A. Recall that the covariant derivative of a $(1, 2)$-tensor field B is

$$(\nabla_E B)_G F = \nabla_E(B_G F) - B_{\nabla_E G} F - B_G(\nabla_E F), \forall E, F, G \in \Gamma(TM).$$

Using this, Lemmas 9.3.6 and 9.3.7, we obtain the following result.

Lemma 9.3.10. *Let* $\phi : (M, g_M) \longrightarrow (N, g_N)$ *be a screen lightlike submersion,* X, Y *horizontal vector fields and* $V, W \in \Gamma(\text{Rad }TM)$. *Then we have:*

(a) $(\nabla_V A)_W = 0$.

(b) $(\nabla_X T)_Y = -T_{A_X Y}$.

(c) $(\nabla_X A)_W = 0$.

(d) $(\nabla_V T)_Y = -T_{T_V Y}$.

Now denote $Q\nabla_V W$ by $\hat{\nabla}_V W$, where $V, W \in \Gamma(\text{Rad }TM)$. Also denote the other geometrical objects related to the fibers by $\hat{\ }$. From Lemma 9.3.8 (2), for $U, V, W \in \Gamma(\text{Rad }TM)$, we have

$$\nabla_U \nabla_V W = \hat{\nabla}_U \hat{\nabla}_V W$$

and

$$\nabla_V \nabla_U W = \hat{\nabla}_V \hat{\nabla}_U W.$$

Since $\operatorname{Rad} TM$ is integrable, we have $\nabla_{[U,V]} W = \hat{\nabla}_{[U,V]} W$. Thus we obtain

$$\hat{R}(U,V)W = R(U,V)W.$$

Denote the horizontal lift of the curvature tensor R^N of N by R^*, that is, if X_1, X_2, X_3 and X_4 are basic vector fields of M, we write

$$g_M(R^*(X_1,X_2)X_3, X_4) = g_N(R^N(\tilde{X}_1,\tilde{X}_2)\tilde{X}_3, \tilde{X}_4),$$

where $\tilde{X}_i = d\phi(X_i)$. Also, if X_i and X_j are basic vector fields, we will denote the horizontal lift of $\nabla^N_{\tilde{X}_i}\tilde{X}_j$ by $\nabla^*_{X_i} X_j$.

Lemma 9.3.11. *Let* $\phi : (M, g_M) \longrightarrow (N, g_N)$ *be a screen lightlike submersion,* X, X_1, X_2, X_3 *basic vector fields and* $\xi \in \Gamma(\operatorname{Rad} TM)$. *Then we have*

$$\begin{aligned}
R(X_1, X_2)X_3 = {}& R^*(X_1, X_2)X_3 + A_{X_1}\nabla^*_{X_2}X_3 + Q\nabla_{X_1}A_{X_2}X_3 \\
& - A_{X_2}\nabla^*_{X_1}X_3 - Q\nabla_{X_2}A_{X_1}X_3 - A_{P[X_1,X_2]}X_3 \\
& - P\nabla_{Q[X_1,X_2]}X_3 - T_{Q[X_1,X_2]}X_3
\end{aligned}$$

and

$$R(X, \xi)\xi = Q\nabla_X \hat{\nabla}_\xi \xi - \hat{\nabla}_\xi(Q\nabla_X \xi) - \hat{\nabla}_{[X,\xi]}\xi.$$

Proof. Using Lemma 9.3.8 (1) we have

$$\nabla_{X_1}\nabla_{X_2}X_3 = \nabla_{X_1}\{\nabla^*_{X_2}X_3 + A_{X_2}X_3\} = \nabla_{X_1}\nabla^*_{X_2}X_3 + \nabla_{X_1}A_{X_2}X_3.$$

Since $A_{X_2}X_3 = Q\nabla_{X_2}X_3 \in \Gamma(\operatorname{Rad} TM)$, from Lemma 9.3.7 and Lemma 9.3.8 (1), we get

$$\nabla_{X_1}\nabla_{X_2}X_3 = \nabla^*_{X_1}\nabla^*_{X_2}X_3 + A_{X_1}\nabla^*_{X_2}X_3 + Q\nabla_{X_1}A_{X_2}X_3.$$

In a similar way, we obtain

$$\nabla_{X_2}\nabla_{X_1}X_3 = \nabla^*_{X_2}\nabla^*_{X_1}X_3 + A_{X_2}\nabla^*_{X_1}X_3 + Q\nabla_{X_2}A_{X_1}X_3$$

and

$$\begin{aligned}
\nabla_{[X_1,X_2]}X_3 = {}& \nabla^*_{P[X_1,X_2]}X_3 + A_{P[X_1,X_2]}X_3 + P\nabla_{Q[X_1,X_2]}X_3 \\
& + T_{Q[X_1,X_2]}X_3.
\end{aligned}$$

Thus, from the above equations, we have the first assertion. On the other hand, from Lemma 9.3.8 (2), for $\xi \in \Gamma(\operatorname{Rad} TM)$ and $X \in \Gamma(S(TM))$, we get

$$\nabla_X \nabla_\xi \xi = \nabla_X \hat{\nabla}_\xi \xi.$$

Then, from Lemma 9.3.8 (4), we obtain

$$\nabla_X \nabla_\xi \xi = Q \nabla_X \hat{\nabla}_\xi \xi.$$

In a similar way, we have

$$\nabla_\xi \nabla_X \xi = \hat{\nabla}_\xi (Q \nabla_X \xi).$$

Also, for a basic vector field X, $[X, \xi]$ is vertical. Hence, we get

$$\nabla_{[X,\xi]} \xi = \hat{\nabla}_{[X,\xi]} \xi.$$

Summing up the above equations, we derive the second assertion. □

Theorem 9.3.12. *Let $\phi : (M, g_M) \longrightarrow (N, g_N)$ be a screen lightlike submersion, X_1, X_2 horizontal vector fields spanning 2-planes and $\xi \in \Gamma(\mathrm{Rad}\, TM)$. Then we have*

$$K_\xi(H) = 0$$

and

$$K(X_1, X_2) = K^N(\tilde{X}_1, \tilde{X}_2).$$

Here K and K^N are sectional curvatures of M and N, respectively.

Proof. From (9.1.12) and the second equation of Lemma 9.3.11, it follows immediately that $K_\xi(H) = 0$. To obtain the second formula it suffices to prove

$$g_M(R(X_1, X_2)X_3, X_4) = g_N(R^N(\tilde{X}_1, \tilde{X}_2)\tilde{X}_3, \tilde{X}_4) \circ \phi$$

using basic vector fields since they locally span horizontal vector fields. From Lemma 9.3.11, we obtain

$$g_M(R(X_1, X_2)X_3, X_4) = g_M(R^*(X_1, X_2)X_3, X_4) - g_M(\nabla_{Q[X_1, X_2]} X_3, X_4).$$

Since ∇ is a torsion free connection, we get

$$g_M(\nabla_{Q[X_1, X_2]} X_3, X_4) = g_M([Q[X_1, X_2], X_3], X_4) + g_M(\nabla_{X_3} Q[X_1, X_2], X_4).$$

Since $Q[X_1, X_2] \in \Gamma(\mathrm{Rad}\, TM)$, from Lemma 9.3.3 (c), it follows that

$$[Q[X_1, X_2], X_3] \in \Gamma(\mathrm{Rad}\, TM).$$

Thus, we obtain

$$g_M(\nabla_{Q[X_1, X_2]} X_3, X_4) = g_M(\nabla_{X_3} Q[X_1, X_2], X_4).$$

On the other hand, since g_M is parallel with respect to ∇, we can write

$$g_M(\nabla_{Q[X_1, X_2]} X_3, X_4) = X_3 g_M(Q[X_1, X_2], X_4) - g_M(Q[X_1, X_2], \nabla_{X_3} X_4).$$

Then $Q[X_1, X_2] \in \Gamma(\operatorname{Rad} TM)$ implies that

$$g_M(\nabla_{Q[X_1,X_2]} X_3, X_4) = 0.$$

Thus, using this we obtain

$$g_M(R(X_1, X_2)X_3, X_4) = g_N(R^N(\tilde{X}_1, \tilde{X}_2)\tilde{X}_3, \tilde{X}_4) \circ \phi$$

which completes the proof. □

Remark 9.3.13. Observe that the assertions of Theorem 9.3.12 are not true for a Riemannian submersion as well as for a semi-Riemannian submersion. Indeed, in the Riemannian case, the sectional curvature in the base manifold N increase by the amount $\frac{3\|A_X Y\|^2}{\|X \wedge Y\|^2}$, as presented in [316].

Screen conformal submersions. The notion of Riemannian submersion was generalized to the horizontally conformal submersion. In this case, the second axiom takes the following form: There exists a function $\lambda : M \longrightarrow \mathbf{R}^+$ such that

$$g_B(d\phi(X), d\phi(Y)) = \lambda^2 g(X, Y), \forall X, Y \in \Gamma(\mathcal{H}),$$

where $\mathcal{H}$ is the complementary orthogonal distribution to $\mathcal{V} = \operatorname{Ker} d\phi$ in TM. Thus, by the definition of horizontally conformal submersion, we have the decomposition

$$TM = \mathcal{V} \perp \mathcal{H}.$$

In the rest of this section, first we present a recent work of Sahin [351] who introduced a lightlike version of horizontally conformal submersion. Secondly, as an application we use the theory of lightlike submersion to show the existence of a well-defined concept of harmonic maps from r-lightlike manifolds submersed in a semi-Riemannian manifold.

Let (M, g) and $(\bar{M}, \bar{g})$ be r-lightlike and semi-Riemannian manifolds respectively. Consider a submersion map $\phi : M \longrightarrow \bar{M}$. A vector field X on M is said to be projectable if there exists a vector field $\bar{X}$ on $\bar{M}$ such that $d\phi(X_p) = \bar{X}_{\phi(p)}$, for all $p \in M$. In this case X and $\bar{X}$ are called ϕ-related. We now put $\mathcal{V}_p = \operatorname{Ker} d\phi_p$ and consider a complementary distribution $\mathcal{H}$ to $\mathcal{V}$ in TM. Notice that, since M is a lightlike manifold, $\mathcal{V}^\perp$ is not complementary to $\mathcal{V}$, (in fact, $\mathcal{V}^\perp \subseteq \mathcal{V}$). From discussion on lightlike manifolds we know that complementary distribution $\mathcal{H}$ is non-degenerate.

Definition 9.3.14. Let (M, g) and $(\bar{M}, \bar{g})$ be a lightlike and a semi-Riemannian manifold respectively with a submersion $\phi : M \longrightarrow \bar{M}$. We say that ϕ is a screen conformal lightlike submersion if the following conditions are satisfied:

1. $\operatorname{Ker} d\phi = \operatorname{Rad} TM$.

2. *There exists a function* $\lambda : M \longrightarrow \mathbf{R}^+$ *such that , for* $x \in M$,

$$\bar{g}(d\phi(X), d\phi(Y)) = \lambda^2(x)g(X, Y), \quad \forall X, Y \in \Gamma(S(TM)). \tag{9.3.1}$$

First of all, it follows from the above definition that the function $\lambda^2 : M \longrightarrow$ $\mathbf{R}^+$ is smooth. We say that a screen vector field (i.e., $X \in \Gamma(S(TM))$.) is called basic if it is projectable. Thus, if $\bar{X}$ is a vector field on $\bar{M}$ then there exists a unique basic vector field X on M such that X and $\bar{X}$ are ϕ-related. We say that the vector field X is the screen lift of $\bar{X}$. Also, it is clear from above definition that the index of $S(TM)$ and $\bar{M}$ are same. From now on, we denote the projections onto $\operatorname{Rad} TM$ and $S(TM)$ by Q and P, respectively. Next, we give an example of screen conformal lightlike submersion.

Example 5. Let $R^4_{2,1,1}$ and $\tilde{R}^2_{0,1,1}$ be R^4 and $\tilde{R}^2$ endowed with the degenerate metric $g_1 = -(dx_3)^2 + (dx_4)^2$ and a Lorentzian metric $g_2 = -(dy_1)^2 + (dy_2)^2$, where x_1, x_2, x_3, x_4 and y_1, y_2 are the canonical coordinates on R^4 and $\tilde{R}^2$, respectively. We define the map

$$\phi : \qquad R^4_{2,1,1} \qquad \longrightarrow \qquad \tilde{R}^2_{0,1,1}$$
$$(x_1, x_2, x_3, x_4) \qquad (\sinh x_3 \, \cosh x_4, \cosh x_3 \, \sinh x_4).$$

It is easy to see that the kernel of $d\phi$ is

$$\operatorname{Ker} d\phi = \operatorname{Rad} TM = \operatorname{Span}\{Z_1 = \partial x_1, Z_2 = \partial x_2\}.$$

By direct computation, we obtain

$$S(TM) = \operatorname{Span}\{Z_3 = \partial x_3, Z_4 = \partial x_4\}.$$

On the other hand, we get

$$d\phi(Z_3) = \cosh x_3 \, \cosh x_4 \partial y_1 + \sinh x_3 \, \sinh x_4 \partial y_2,$$
$$d\phi(Z_4) = \sinh x_3 \, \sinh x_4 \partial y_1 + \cosh x_3 \, \cosh x_4 \partial y_2.$$

Then, it is easy to see that

$$g_2(d\phi(Z_3), d\phi(Z_3)) = (\cosh^2 x_3 \, \cosh^2 x_4 - \sinh^2 x_3 \, \sinh^2 x_4) g_1(Z_3, Z_3)$$

and

$$g_2(d\phi(Z_4), d\phi(Z_4)) = (\cosh^2 x_3 \, \cosh^2 x_4 - \sinh^2 x_3 \, \sinh^2 x_4) g_1(Z_4, Z_4).$$

As a result, ϕ is a screen conformal lightlike submersion.

It is known that in the Riemannian case a projection from warped product manifold onto its second factor is a simple example of horizontal conformal submersion. Now, we present similar example for a screen conformal lightlike submersion.

Example 6. Recall the concept of warped product lightlike manifold (see Chapter 1) as follows: let (H, g_H) and (B, g_B) be lightlike and a Riemannian manifold of dimensions n and m, respectively. Here, $\operatorname{Rad} TH$ is of rank r. Let also $f : H \longrightarrow$

$(0, \infty)$ be a function, $\pi : H \times B \longrightarrow H$ and $\sigma : H \times B \longrightarrow B$ the projection maps given by $\pi(x, y) = x$ and $\sigma(x, y) = y$ for every $(x, y) \in H \times B$. Then the product manifold $M = H \times B$ is said to be a lightlike warped product $H \times_f B$, with degenerate metric g defined by

$$g_H(X, Y) = g(d\pi(X), d\pi(Y)) + f(\pi(x, y))g_B(d\sigma(X), d\sigma(Y))$$

for every X, Y of M. It follows that $\operatorname{Rad} TM$ of M still has rank r, but its screen distribution $S(TM)$ is of dimension $m + n - r$. If we consider $\operatorname{rank}(\operatorname{Rad} TH) = n$, then we obtain that M is still a lightlike manifold. But, we see that in this case, $S(TM)$ is of dimension m. We now consider the projection map $\sigma : M = H \times B \longrightarrow B$, then it is obvious that σ is a screen conformal lightlike submersion.

Recall that a distribution $\mathcal{V}$ on a Riemannian manifold is said to be conformal or shear free if, for each $x \in M$,

$$(\pounds_V g)(X, Y) = \nu(V)g(X, Y)$$

for $V \in \Gamma(\mathcal{V})$ and $X, Y \in \Gamma(\mathcal{H})$, where $\mathcal{H}$ is the orthogonal distribution to $\mathcal{V}$ in TM. Notice that, for a non-degenerate manifold, the orthogonal distribution $\mathcal{H}$ to $\mathcal{V}$ is also complementary. For the lightlike case we have the following results:

Lemma 9.3.15. *Let $\phi : (M, g) \longrightarrow (\bar{M}, \bar{g})$ be a screen conformal lightlike submersion and X, Y be basic vector fields of M. Then:*

(a) *The screen part $[X, Y]^P$ of $[X, Y]$ is a basic vector field and corresponds to $[\bar{X}, \bar{Y}]$.*

(b) *For $\xi \in \Gamma(\operatorname{Rad} TM)$, $[X, \xi] \in \Gamma(\operatorname{Rad} TM)$.*

The proof of the above lemma is similar to the proof of Lemma 9.3.3.

Lemma 9.3.16. *Let $\phi : (M, g) \longrightarrow (\bar{M}, \bar{g})$ be a screen conformal lightlike submersion. Then $\operatorname{Rad} TM$ is shear free if*

$$\xi g_N(d\phi(X), d\phi(Y)) = 0, \quad \forall \xi \in \Gamma(\operatorname{Rad} TM) \quad and \quad X, Y \in \Gamma(S(TM)).$$

Proof. We first note that the radical distribution is integrable. Let ∇ be a linear connection on M. For $\xi \in \Gamma(\operatorname{Rad}(TM))$, $X, Y \in \Gamma(TM)$ we have

$$(\pounds_\xi g)(X, Y) = \pounds_\xi g(X, Y) - g(\pounds_\xi X, Y) - g(X, \pounds_\xi Y),$$

where $\pounds$ denotes the Lie derivative. Then, we obtain

$$(\pounds_\xi g)(X, Y) = \xi g(X, Y) - g([\xi, X], Y) - g(X, [\xi, Y]). \tag{9.3.2}$$

Now, if $X \in \Gamma(\operatorname{Rad} TM)$, then $[X, \xi] \in \Gamma(\operatorname{Rad} TM)$ due to $\operatorname{Rad} TM$ is integrable. Thus, if $X, Y \in \Gamma(\operatorname{Rad} TM)$, we obtain

$$(\pounds_\xi g)(X, Y) = 0, \forall \xi, X, Y \in \Gamma(\operatorname{Rad} TM). \tag{9.3.3}$$

On the other hand, if $X \in \Gamma(\operatorname{Rad} TM)$ and $Y \in \Gamma(S(TM))$, from (9.3.2) we get

$$(\pounds_\xi g)(X,Y) = -g(X,[\xi,Y]).$$

But, from Lemma 9.3.15 (b), it follows that $[\xi,Y] \in \Gamma(\operatorname{Rad} TM)$. Hence

$$(\pounds_\xi g)(X,Y) = 0, \forall \xi, X \in \Gamma(\operatorname{Rad} TM) \quad \text{and} \quad Y \in \Gamma(S(TM)). \qquad (9.3.4)$$

If $X,Y \in \Gamma(S(TM))$, using Lemma 9.3.15 (a), from (9.3.2) we have

$$(\pounds_\xi g)(X,Y) = \xi g(X,Y).$$

Then, the screen conformal lightlike submersion ϕ implies that

$$\begin{aligned} (\pounds_\xi g)(X,Y) &= \xi(\lambda^{-2})\bar{g}(d\phi(X), d\phi(Y)) \\ &= \xi(\ln\lambda)g(X,Y) + \xi\bar{g}(d\phi(X), d\phi(Y)) \end{aligned} \qquad (9.3.5)$$

for any $X,Y \in \Gamma(S(TM))$. Then, proof comes from (9.3.3)–(9.3.5). $\qquad\square$

At this point, we give definitions of screen (resp. radical) homothetic maps. These definitions can be seen as a lightlike version of horizontally (resp. vertically) homothetic maps. For the Riemannian case, see [25].

Definition 9.3.17. Let $\phi : M \longrightarrow \bar{M}$ be a screen conformal lightlike submersion. Then, we say that ϕ is screen homothetic if $\operatorname{grad}_{S(TM)}(\lambda^2) = 0, \lambda \neq 0$. Also, we say that ϕ is radical homothetic if $V(\lambda^2) = 0, \lambda \neq 0$.

Remark 9.3.18. The screen conformal lightlike submersion given in Example 5 is radical homothetic. On the other hand, the screen conformal lightlike submersion given in Example 6 is screen homothetic.

From Lemma 9.3.15, Theorem 1.4.2 (with reference to Reinhart manifolds) and the above definition, we have the following:

Theorem 9.3.19. *Let* $\phi : (M,g) \longrightarrow (\bar{M},\bar{g})$ *be a radical homothetic lightlike submersion. Then M is a Reinhart lightlike manifold if $\xi g_N(d\phi(X), d\phi(Y)) = 0$ for $\xi \in \Gamma(\operatorname{Rad} TM)$ and $X,Y \in \Gamma(S(TM))$.*

We now define a tensor field T as

$$T_E F = P\nabla_{QE} QF + Q\nabla_{QE} PF. \qquad (9.3.6)$$

It follows that T is a tensor field of type $(1,2)$. From now on, we suppose that (M,g) is a Reinhart lightlike manifold, unless otherwise stated.

Lemma 9.3.20. *Let* $\phi : (M,g) \longrightarrow (\bar{M},\bar{g})$ *be a screen conformal submersion, X,Y be basic vector fields and $V,W \in \Gamma(\operatorname{Rad} TM)$. Then, we have:*

(a) *T is vertical; $T_E = T_{QE}$.*

(b) T *reverses the radical and screen subspaces.*

(c) $T_V W = 0$.

(d) $T_X Y = 0$.

(e) $T_X V = 0$.

(f) $T_V X = Q \nabla_{QV} P X$.

We now define another tensor as

$$A_E F = Q \nabla_{PE} P F + P \nabla_{PE} Q F \qquad (9.3.7)$$

which is also a tensor field of type $(1,2)$. The tensor field A satisfies the following:

Lemma 9.3.21. *Let* $\phi : (M, g) \longrightarrow (\bar{M}, \bar{g})$ *be a screen conformal submersion,* X, Y *be basic vector fields and* $V, W \in \Gamma(\operatorname{Rad} TM)$. *Then, we have:*

(a) A *is screen;* $A_E = T_{PE}$.

(b) A *reverses the radical and screen subspaces.*

(c) $A_V W = 0$.

(d) $A_X Y = Q \nabla_X Y$

(e) $A_V X = 0$.

(f) $A_X V = P \nabla_X V = 0$.

The proofs of Lemma 9.3.20 and Lemma 9.3.21 are similar to the proof of Lemma 9.3.6, and from them we obtain the following lemma.

Lemma 9.3.22. *Let* $\phi : (M, g) \longrightarrow (\bar{M}, \bar{g})$ *be a screen conformal submersion,* X, Y *be basic vector fields and* $V, W \in \Gamma(\operatorname{Rad} TM)$. *Then, we have:*

1. $\nabla_X Y = P \nabla_X Y + A_X Y$.

2. $\nabla_U V = Q \nabla_U V$.

3. $\nabla_U X = P \nabla_U X + T_U X$.

4. $\nabla_X U = Q \nabla_X U$.

Remark 9.3.23. Lemma 9.3.22(2) implies that fibers are totally geodesic in M. If X is a basic vector field, then $[V, X]$, $V \in \Gamma(\operatorname{Rad} TM)$, is vertical. Hence $P \nabla_V X = P \nabla_X V$. Using Lemma 9.3.22(4), we obtain $P \nabla_V X = 0$.

Now we obtain the screen conformal lightlike version of fundamental equations for horizontally conformal submersion. For the Riemannian case, see:[25]. Denote $Q \nabla_V W$, where $V, W \in \Gamma(\operatorname{Rad} TM)$, by $\tilde{\nabla}_V W$ and the other geometrical objects of fibers, by $\tilde{\ }$. Also denote the screen lift of curvature tensor $\bar{R}$ of $\bar{M}$ by R^*, that is, if X_1, X_2, X_3 and X_4 are screen vector fields of M, we write

$$g(R^*(X_1, X_2)X_3, X_4) = \bar{g}(R^N(\bar{X}_1, \bar{X}_2)\bar{X}_3, \bar{X}_4),$$

where $\bar{X}_i = d\phi(X_i)$. Then, following the proof of Lemma 9.3.11, we have

$$\tilde{R}(U,V)W = R(U,V)W, \tag{9.3.8}$$

$$\begin{aligned}
R(X_1,X_2)X_3 &= R^*(X_1,X_2)X_3 + A_{X_1}\nabla^*_{X_2}X_3 + Q\nabla_{X_1}A_{X_2}X_3 \\
&\quad - A_{X_2}\nabla^*_{X_1}X_3 - Q\nabla_{X_2}A_{X_1}X_3 - A_{P[X_1,X_2]} \\
&\quad - P\nabla_{Q[X_1,X_2]}X_3 - T_{Q[X_1,X_2]}X_3
\end{aligned} \tag{9.3.9}$$

and

$$R(X,\xi)\xi = Q\nabla_X\tilde{\nabla}_\xi\xi - \tilde{\nabla}_\xi(Q\nabla_X\xi) - \tilde{\nabla}_{[X,\xi]}\xi. \tag{9.3.10}$$

Recall that the null sectional curvature $K_\xi(H)$ (see Section 3 of Chapter 6) of a plane H with respect to $\xi \in \Gamma(\mathrm{Rad}\,TM)$ and ∇ is defined by

$$K_\xi(H) = \frac{g_M(R(W,\xi)\xi),W)}{g(W,W)} \tag{9.3.11}$$

where W is an arbitrary non-null vector field in H.

Theorem 9.3.24. *Let $\phi : (M,g) \longrightarrow (\bar{M},\bar{g})$ be a screen conformal submersion, X_1, X_2 be basic vector fields and $\xi \in \Gamma(\mathrm{Rad}\,TM)$. Then we have*

$$K_\xi(H) = 0$$

and

$$K(X_1,X_2) = \frac{1}{\lambda^2}K^*(\bar{X}_1,\bar{X}_2).$$

Here K and K^* are sectional curvatures of M and $\bar{M}$ respectively.

Proof. From (9.3.10) and (9.3.11), it follows that $K_\xi(H) = 0$. On the other hand, from (9.3.9) we obtain

$$g_M(R(X_1,X_2)X_3,X_4) = g_M(R^*(X_1,X_2)X_3,X_4) - g_M(P\nabla_{Q[X_1,X_2]}X_3,X_4)$$

or

$$\begin{aligned}
g_M(R(X_1,X_2)X_3,X_4) &= g_M(R^*(X_1,X_2)X_3,X_4) \\
&\quad - g_M(\nabla_{Q[X_1,X_2]}X_3,X_4).
\end{aligned} \tag{9.3.12}$$

Since M is Reinhart, ∇ is a torsion free metric connection. Hence, we get

$$g(\nabla_{Q[X_1,X_2]}X_3,X_4) = g([Q[X_1,X_2],X_3],X_4) + g(\nabla_{X_3}Q[X_1,X_2],X_4).$$

Since $Q[X_1,X_2] \in \Gamma(\mathrm{Rad}\,TM)$, from Lemma 9.3.2(b), it follows that

$$[Q[X_1,X_2],X_3] \in \Gamma(\mathrm{Rad}\,TM).$$

Thus, we obtain

$$g(\nabla_{Q[X_1,X_2]}X_3, X_4) = g(\nabla_{X_3}Q[X_1, X_2], X_4).$$

On the other hand, since g is parallel with respect to ∇, we can write

$$g(\nabla_{Q[X_1,X_2]}X_3, X_4) = X_3 g(Q[X_1, X_2], X_4) - g(Q[X_1, X_2], \nabla_{X_3}X_4).$$

Then $Q[X_1, X_2] \in \Gamma(\operatorname{Rad} TM)$ implies that

$$g(\nabla_{Q[X_1,X_2]}X_3, X_4) = 0.$$

Thus, using this in (9.3.12) we obtain

$$g(R(X_1, X_2)X_3, X_4) = \frac{1}{\lambda^2}\bar{g}(R^*(\bar{X}_1, \bar{X}_2)\bar{X}_3, \bar{X}_4) \circ \phi$$

which completes the proof. □

Using some of the above results now we discuss the existence of a harmonic map $\phi : M \longrightarrow \bar{M}$. Suppose ∇ and $\bar{\nabla}$ are a linear connection and the Levi-Civita connection on M and $\bar{M}$ respectively. Denote the induced connection by the map ϕ on the bundle $\phi^{-1}(\bar{M})$ by ∇^ϕ. Then the second fundamental form $\nabla d\phi$ of ϕ is defined as

$$\nabla d\phi(X, Y) = \nabla_X^\phi d\phi(Y) - d\phi(\nabla_X Y), \quad \forall X, Y \in \Gamma(TM).$$

In the Riemannian case, a differentiable map f between Riemannian manifolds is called harmonic if $\operatorname{tr}\nabla df = 0$. But in case M is lightlike we know that the trace of the second fundamental form is meaningless on the radical part. To heal this anomaly, recently the second author of this book (Sahin) has done some work in [351] which we now present as follows:

Definition 9.3.25. Let (M, g) be a lightlike manifold and $(\bar{M}, \bar{g})$ a semi-Riemannian manifold. We say that a smooth mapping $\phi : M \longrightarrow \bar{M}$ is harmonic if:

(i) $\nabla d\phi = 0$ on $\operatorname{Rad} TM$.

(ii) $\operatorname{tr}|_{S(TM)} \nabla d\phi = 0$, where the trace is written with respect to g_M restricted to $S(TM)$.

It is known that in the Riemannian as well as semi-Riemannian cases, a minimal isometric immersion is a particular harmonic map [25]. Using the concept of r-lightlike isometric immersion (as discussed in the previous section) we show that the above definition agrees with the non-degenerate cases. Indeed, let (M, g) be a lightlike manifold, $(\bar{M}, \bar{g})$ a semi-Riemannian manifold and $\phi : M \longrightarrow \bar{M}$ an r-lightlike immersion, i.e., $g(X, Y) = \bar{g}(d\phi(X), d\phi(Y))$. Then, from the theory of lightlike submanifolds, we have a decomposition of vector bundles

$$\phi^{-1}(T\bar{M}) = d\phi(TM) \oplus \operatorname{tr}(TM), \quad X = X^{\text{tangent}} + X^{\text{transversal}}$$

into the tangent bundle and transversal bundle. Using $d\phi$ to identify with its image $d\phi(TM)$ in $\phi^{-1}(T\bar{M})$, then for $X, Y \in \Gamma(TM)$ we have $\nabla^\phi_X d\phi(Y) = \bar{\nabla}_X Y$. On the other hand, $d\phi(\nabla_X Y)$ equals the tangential component of $\bar{\nabla}_X Y$. Hence, $\nabla d\phi(X, Y)$ equals the transversal component of $\bar{\nabla}_X Y$. Then it follows that the second fundamental form of an r-lightlike isometric immersion is equal to the second fundamental form of the lightlike submanifold $\phi(M)$ in $\bar{M}$. Thus, considering Definition 5.4.1 and Definition 9.3.25, we conclude that an r-lightlike immersion is harmonic if and only if it is a minimal lightlike immersion.

To investigate the harmonicity of the function ϕ, we let (M, g) be a Reinhart lightlike manifold. This condition is geometrically reasonable, as we know from Chapter 1 that a Reinhart lightlike manifold is equivalent to the existence of a torsion free metric connection ∇ on M such that g is a parallel tensor field with respect to ∇. All these are good properties for discussion on the harmonicity in lightlike geometry. Moreover, we assume that $\phi : M \longrightarrow \bar{M}$ is a screen conformal lightlike submersion where M is Reinhart lightlike. Now we investigate the harmonicity of ϕ.

Lemma 9.3.26. *Let $\phi : M \longrightarrow \bar{M}$ be a screen conformal lightlike submersion. Then,*

$$\nabla d\phi(\xi_1, \xi_2) = 0, \tag{9.3.13}$$

for $\xi_1, \xi_2 \in \Gamma(\mathrm{Rad}\, TM)$ and

$$\nabla d\phi(X, Y) = X(\ln \lambda)d\phi(Y) + Y(\ln \lambda)d\phi(X) - d\phi(\mathrm{grad}\ln \lambda)g_{S(TM)}(X, Y) \tag{9.3.14}$$

for $X, Y \in \Gamma(S(TM))$.

Proof. Let $\xi_1, \xi_2 \in \Gamma(\mathrm{Rad}\, TM)$. Then, $\nabla d\phi(\xi_1, \xi_2) = -d\phi(\nabla_{\xi_1} \xi_2)$. From Lemma 9.3.22 (2), we get $\nabla d\phi(\xi_1, \xi_2) = -d\phi(Q\nabla_{\xi_1} \xi_2)$. Hence, $\nabla d\phi(\xi_1, \xi_2) = 0$, which is (9.3.13). Now let $\{\bar{Z}_i\}$ be an orthonormal frame on an open subset of $\bar{M}$; lift each $\bar{Z}_i$ to a screen vector field Z_i on M, then λZ_i is an orthonormal frame for the screen distribution of M. Let $\bar{X}$ and $\bar{Y}$ be vector fields on an open subset of $\bar{M}$ and X, Y their screen lifts to M. Then we can write

$$P(\nabla_X Y) = \sum_{i=1}^{n} \varepsilon_i g(\nabla_X Y, \lambda Z_i)\lambda Z_i = \lambda^2 \sum_{i=1}^{n} \varepsilon_i g(\nabla_X Y, Z_i)Z_i.$$

Since M is Reinhart, it follows that g is parallel with respect to ∇. As a result of this, we have a Kozsul identity for ∇. Thus we have

$$P(\nabla_X Y) = \frac{\lambda^2}{2} \sum_{i=1}^{n} \varepsilon_i \{ Xg(Y, Z_i) + Yg_M(Z_i, X) - Z_i g_M(X, Y)$$
$$- g(X, [Y, Z_i]) - g(Y, [X, Z_i]) + g(Z_i, [X, Y]) \}.$$

Now, using $g(X, Y) = \frac{1}{\lambda^2} \bar{g}(\bar{X}, \bar{Y})$, we get

$$P(\nabla_X Y) = \frac{\lambda^2}{2} \sum_{i=1}^{n} \varepsilon_i \{ X(\frac{1}{\lambda^2} g(\bar{Y}, \bar{Z}_i)) + Y(\frac{1}{\lambda^2} \bar{g}(\bar{Z}_i, \bar{X})) - Z_i(\frac{1}{\lambda^2} \bar{g}(\bar{X}, \bar{Y})$$
$$- \frac{1}{\lambda^2} \bar{g}(\bar{X}, [\bar{Y}, \bar{Z}_i]) - \frac{1}{\lambda^2} \bar{g}(\bar{Y}, [\bar{X}, \bar{Z}_i]) + \frac{1}{\lambda^2} \bar{g}(\bar{Z}_i, [\bar{X}, \bar{Y}]) \}.$$

Hence, we derive

$$P(\nabla_X Y) = \sum_{i=1}^{n} \varepsilon_i \{ -X(\ln \lambda) \bar{g}(\bar{Y}, \bar{Z}_i) - Y(\ln \lambda) \bar{g}(\bar{Z}_i, \bar{X})$$
$$+ Z_i(\ln \lambda) \bar{g}(\bar{X}, \bar{Y}) + \bar{g}(\bar{\nabla}_{\bar{X}} \bar{Y}, \bar{Z}_i) \} Z_i.$$

Then applying $d\phi$, we get

$$d\phi(P(\nabla_X Y)) = \sum_{i=1}^{n} \varepsilon_i \{ -X(\ln \lambda) \bar{g}(\bar{Y}, \bar{Z}_i) - Y(\ln \lambda) \bar{g}(\bar{Z}_i, \bar{X})$$
$$+ Z_i(\ln \lambda) \bar{g}(\bar{X}, \bar{Y}) + \bar{g}(\bar{\nabla}_{\bar{X}} \bar{Y}, \bar{Z}_i) \} \bar{Z}_i.$$

Hence, we obtain

$$d\phi(P(\nabla_X Y)) = -X(\ln \lambda) \bar{Y} - Y(\ln \lambda) \bar{X}$$
$$+ \sum_{i=1}^{n} \varepsilon_i Z_i(\ln \lambda) \bar{g}(\bar{X}, \bar{Y}) \bar{Z}_i + \bar{\nabla}_{\bar{X}} \bar{Y}.$$

Thus, we arrive at

$$\nabla d\phi(X, Y) = X(\ln \lambda) \bar{Y} + Y(\ln \lambda) \bar{X}$$
$$- \sum_{i=1}^{n} \varepsilon_i (g_{S(TM)}(\operatorname{grad} \lambda, Z_i) \bar{Z}_i)(\lambda^2 g_{S(TM)}(X, Y)).$$

or

$$\nabla d\phi(X, Y) = X(\ln \lambda) \bar{Y} + Y(\ln \lambda) \bar{X}$$
$$- \sum_{i=1}^{n} (\frac{1}{\lambda^2} \varepsilon_i \bar{g}(d\phi(\operatorname{grad} \lambda), \bar{Z}_i) \bar{Z}_i)(\lambda^2 g_{S(TM)}(X, Y)).$$

Hence, we get

$$\nabla d\phi(X, Y) = X(\ln \lambda) \bar{Y} + Y(\ln \lambda) \bar{X}$$
$$- d\phi(\operatorname{grad} \lambda) g_{S(TM)}(X, Y),$$

which proves (9.3.13). $\qquad\square$

Using the above lemma, we get the following theorem.

Theorem 9.3.27. *Let* $\phi : M \longrightarrow \bar{M}$ *be a screen conformal lightlike submersion and* $\dim(S(TM)) = n$. *Then:*

(i) *If* $n = 2$, ϕ *is a lightlike harmonic map.*

(ii) *If* $n \neq 2$, *then* ϕ *is lightlike harmonic* $\Leftrightarrow$ ϕ *is screen homothetic.*

Proof. First of all, from (9.3.13)), we obtain

$$\nabla d\phi = 0 \quad \text{on } \operatorname{Rad} TM. \tag{9.3.15}$$

Let $\{e_1, \ldots, e_n\}$ be a local orthonormal frame for the screen frame for the screen distribution $S(TM)$. Then the trace of the restriction to $S(TM) \times S(TM)$ of the second fundamental form is

$$\operatorname{tr}^P \nabla d\phi = \sum_{i=1}^n \nabla d\phi(e_i, e_i).$$

From (9.3.14), we get

$$\operatorname{tr}^P \nabla d\phi = d\phi \{ \sum_{i=1}^n e_i(\ln \lambda)e_i + e_i(\ln \lambda)e_i - g_M(e_i, e_i)P(\operatorname{grad} \lambda) \}.$$

Thus, we have

$$\operatorname{tr}^P \nabla d\phi = d\phi \{ -(n-2)P(\operatorname{grad} \lambda) = -(n-2)d\phi(\operatorname{grad} \lambda) \}.$$

If $n = 2$, $\operatorname{tr}^P \nabla d\phi = 0$. From this and (9.3.15), it follows that ϕ is harmonic. If $n \neq 2$, then $\operatorname{tr}^P \nabla d\phi = 0$ if and only if $d\phi(\operatorname{grad} \lambda) = 0$.   □

9.4   CR-lightlike geometry in relativity

The use of complex variables in relativity has its roots in a 1906 paper by Poincaré in which he introduced the imaginary coordinate $\sqrt{-1}$, popularly referred to as the imaginary time coordinate of a Minkowski spacetime. However, very soon it became clear that there is no natural place for an imaginary time coordinate for the curved spacetimes of general relativity and its use (even in special relativity) has steadily declined. Nevertheless, it is now well known that the complex techniques in relativity have been very effective tools for understanding spacetime geometry (see Penrose [329, 328], LeBrun [284, 282, 281, 278] Trautman [394] and many more cited therein). Also, complex manifolds have two interesting classes of Kähler manifolds, namely, (i) Calabi-Yau manifolds which have their application in super string theory (see Candelas et al. [90]) and (ii) Teichmuller spaces applicable to relativity (see Tromba [396]). On the other hand, complex numbers have been used in quantum theory as follows: A simple case is that of differential operators such as $\partial/\partial x$ which are needed to represent components of momentum. To make them self-adjoint one has to multiply by $i = \sqrt{-1}$. Also, complex structures appear through Hodge duality in vector and spinor spaces associated with spacetime.

To explore more on this topic, we let (M, g) be a time oriented 4-dimensional spacetime manifold with the Lorentz metric g of signature $(- + ++)$ (although the mathematical results do hold for higher-dimensional Lorentzian manifolds) and expressed in terms of a general coordinate system (x^a), where $(0 \leq a, b, c \ldots \leq 3)$. Suppose $(e_a) = \{e_o, e_1, e_2, e_3\}$ is a local orthonormal real frame field on M. It is known [190] that for each tangent space $T_x M$ at each $x \in M$, there exists a basis $\mathrm{T} = \{l, k, m, \bar{m}\}$ of four null vectors, called the Newman-Penrose (NP) null tetrad such that l, k are real null vectors and $m, \bar{m}$ are complex and its conjugate null vectors, with $g(l, k) = -1$ and $g(m, \bar{m}) = 1$ and all other products are zero. Let $\{\omega^a\} = \{\omega^0, \omega^1, \omega^2, \omega^3\}$ be its dual basis. It is always possible to introduce a NP tetrad locally. Globally, the existence of a NP tetrad is equivalent to the existence of a global orthonormal basis e_a. A null tetrad T is associated with this orthonormal basis as follows:

$$\ell = \frac{1}{\sqrt{2}} (e_o + e_1), \quad k = \frac{1}{\sqrt{2}} (e_o - e_1),$$

$$m = \frac{1}{\sqrt{2}} (e_2 + i\, e_3), \quad \bar{m} = \frac{1}{\sqrt{2}} (e_2 - i\, e_3). \tag{9.4.1}$$

Therefore, the canonical form of the matrix of g is expressed by

$$[g] = \begin{pmatrix} 0 & -1 & 0 & 0 \\ -1 & 0 & 0 & 0 \\ 0 & 0 & 0 & 1 \\ 0 & 0 & 1 & 0 \end{pmatrix}.$$

A Lorentz metric cannot admit an almost Hermitian structure defined by a real endomorphism J such that $J^2 = -I$ and (6.1.2) holds. Indeed, for the real J, satisfying $J^2 = -I$, the eigenvalues of J are i and $-i$ where $i = \sqrt{-1}$, each one of multiplicity 2. As J is real, and J satisfies (6.1.2), the only possible signatures of g are $(2, 2)$. For this reason, as an application to Hermitian and Kählerian structures in relativity, Flaherty [190] modified these structures by using a complex-valued endomorphism J, defined by

$$J = i(\omega^0 \otimes E_0 - \omega^1 \otimes E_1 + \omega^2 \otimes E_2 - \omega^3 \otimes E_3),$$

where $\{E_a\} = \{l, k, m, \bar{m}\}$. It is easy to see that this complex-valued J satisfies the condition (6.1.2) of an almost Hermitian structure with respect to a Lorentz metric g of a spacetime. Flaherty also derived the modified integrability conditions $(\mathbf{N} = 0)$, where for this case $\mathbf{N}$ is the Nijenhuis tensor field with respect to the complex-valued J. Furthermore, he then defined a modified Kählerian structure for the spacetime manifolds M. If the modified integrability condition is satisfied then the Lorentz metric g can be locally expressed as

$$g = A\, dz^0\, d\bar{z}^0 + B\, dz^1\, d\bar{z}^1 + C\, dz^0\, d\bar{z}^1 + D\, dz^1\, d\bar{z}^0, \qquad (9.4.2)$$

for a complex coordinate system $(z^0,\ z^1,\ \bar{z}^0,\ \bar{z}^1)$ and for some functions A, B, C, and D. Well-known physical examples are vacuum spacetimes, which include Schwarzschild and Kerr solutions (see Hawking-Ellis [228, pages 149, 161]. For more details on the use of modified Kählerian structures in relativity (out of the scope of this book) we refer to a book by [190] which also includes an extensive list of related references and many more therein.

As a physical use of CR-lightlike geometry, we first show how to construct a CR-structure for a real $2m$-dimensional Lorentzian manifold $(\bar{M}, \bar{g})$, followed by an induced CR structure on a lightlike hypersurface of a 4-dimensional spacetime of general relativity. According to Penrose [328] $\bar{M}$ has a CR-structure if in the tangent space T_x, at each point $x \in \bar{M}$, a $2n$-real-dimensional subspace H_x of T_x is singled out, called the *holomorphic tangent space*. H_x regarded as n-dimensional complex space and spanned by the vectors $Z_r = X_r + iY_r$ for every: $(1 \leq r \leq n)$ provides a linear operator J, satisfying $J^2 = -1$. Explicitly, $JX_r = -Y_r, JY_r = X_r$ so that $JZ_r = iZ_r$. To complete a basis for the entire T_x, one needs a complementary neighborhood $\mathcal{U}_r$ of x in $\bar{M}$. Newlander and Nirenberg [305] have proved that a real-analytic CR-structure can be realized in the above way if the following integrability relations hold:

$$[Z_r, Z'_r] = \text{ complex linear combination of } Z's$$

where $1 \leq r, r' \leq n$. However, for a C^∞ CR-structure it has been shown in [308] by counterexamples that the above relations are not sufficient. Therefore, in the later case, a *non-realizable CR-structure* may arise. For details on physical spacetimes and non-realizable CR-structure we refer to a paper by Penrose [328]. Here we study only realizable CR-structures. It is known that on a CR manifold there exists a real distribution D, of the subspaces $D_x = Real(H \oplus \bar{H})_x$ such that D is invariant $(JD = D)$.

Related to the focus of this book, we now show how to construct a lightlike CR structure for an oriented $2m$-dimensional Lorentzian manifold $(\bar{M}, \bar{g})$. We need the following result:

Proposition 9.4.1. [317] *For an r-dimensional subspace D_x of a Lorentzian space $T_x\bar{M}$, the following are equivalent:*

(1) D_x *is lightlike.*

(2) D_x *contains a null vector but not any timelike vector.*

(3) $D_x \cap \wedge_x = L_x - 0_x$, *where* $L_x = D_x \cap D_x^\perp$, *the 1-dimensional null space and $\wedge_x$ is the null cone of $T_x\bar{M}$.*

Thus, if the distribution D is lightlike then $D^\perp$ is not the orthogonal complement of D since their sum is not all of $T_x\bar{M}$. Therefore, using the above procedure
one can not realize a CR-structure for $\bar{M}$ so that D is invariant by its structure
tensor J. In view of this, to recover a CR-structure, we proceed as follows:

Since the lightlike D fails to recover a complex structure, its dimension need
not be even. We form the exact sequence

$$0 \to D \to D^\perp \to D^\perp/L \to 0, \quad L = D \cap D^\perp.$$

Fibers of quotient bundle $D^\perp/L$ are $(2m - r - 1)$-dimensional screens S_x.

Physically we construct a lightlike CR-structure for an oriented 4-dimensional
spacetime $\bar{M}$ by setting $m = 2$ for which $r = 1$. Therefore, $D = L$ and 2-
dimensional screen S_x is an oriented plane. Let J act as a rotation in a chosen
plane through 90^0. Then, $J^2 = -1$, and J defines a complex structure on S_x. The
complexified space S_x^c can be represented as a direct sum $S_x^+ \oplus S_x^-$, where

$$S_x^\pm = \{u \in S_x^c : Ju = \pm iu\}.$$

Let $H_x^\pm$ be the subspaces of D_x^c projecting onto $S_x^\pm$ by a canonical map $D_x^c \to S_x^c$.
It is easy to see that

$$H_x^+ \cap H_x^- = L_x^c, \quad H_x^+ + H_x^- = D_x^c.$$

Each $H_x^\pm$ is a (maximal) 2-dimensional lightlike subspace of $T_x^c\bar{M}$. The fact that
a CR structure, with a lightlike distribution D and a maximal holomorphic space,
say H_x^+, can be locally realized on $\bar{M}$, comes from the famous Riemann mapping
theorem: *Any smooth bounded simply connected region in the Argand plane C^1 is
holomorphically identical with a unit disc.*

Note that Poincaré, back in 1907, pointed out that the Riemann mapping theorem for C^1 has no analogue in higher complex dimension. This is why, in general,
non-realizable CR structures may exist [329] for the case of higher-dimensional
complex manifolds.

Relating the above discussion with the theory of CR-lightlike hypersurfaces of
Kähler manifolds, let $\{E_a\} = \{l, k, m, \bar{m}\}$ be the Newman-Penrose (NP) formalism
at each point x of $T_x\bar{M}$. Suppose M is a hypersurface of $\bar{M}$. If the Jacobian
matrix of the map $f : M \to \bar{M}$ is of rank 2 everywhere, then, M is a lightlike
hypersurface of $\bar{M}$. We know from [228] that such a hypersurface can be obtained
by the equation $E_0 = 0$ or $E_1 = 0$ if it is completely integrable. Set $E_0 = 0$.
Examples are lightlike hypersurfaces of asymptotically flat spacetime [228] $\bar{M}$,
where $\bar{M}$ can be endowed with Flaherty's [190] modified Hermitian or Kählerian
structure.

Remark 9.4.2. (a) Since the 1996 Duggal-Bejancu book [149] included very limited
information on the geometry of lightlike CR-submanifolds, we hope that the publication of new and deeper results (as presented in Chapter 6 of this volume) will
stimulate the readers to do more work on applications of lightlike CR-geometry in
physics.

(b) The readers may also be interested in works of Flaherty [190] on applications of modified Hermitian or Kählerian structures in relativity and Penrose
[328] on *Physical spacetime and non-realizable CR-structure* and in particular reference to his work on the *Twistor geometry of light rays* [329], Robinson-Trautman
[335] on Cauchy-Riemann structures in optical geometry and many other related
references therein. Also, see the recent article by Dragomir and Duggal [130] on
indefinite extrinsic spheres.

9.5   Lightlike contact geometry in physics

The theory of contact manifolds has its roots in differential equations, optics and
phase space of a dynamical system (for details see Arnold [9], Maclane [292] and
Nazaikinskii [304] and many more references therein). We refer to basic information
on contact manifolds as presented in Chapter 7. In thermodynamics, there is an
example due to Gibbs which is given by a contact form $du - Tds + pdv$ (u is
the energy, T is the temperature, s is the entropy, p is the pressure and v is the
volume) whose zeros define the laws of thermodynamics. Details may be seen in
Arnold [9]. Also, see a recent paper by Philippe Rukimbira [340] on *Energy, volume
and deformation of contact metrics*. In 1990, Duggal [134] initiated the study of
contact geometry of odd-dimensional spacetime manifolds. We prefer explaining
this concept by means of the following two Models:

Model 1. Consider a $(2n+1)$-dimensional Minkowski spacetime $(\bar{M}, \bar{g})$ with local
coordinates (x^i, y^i, t) and $i = 1, \ldots, n$. $\bar{M}$ being time oriented admits a global
timelike vector field, say ξ. Define a 1-form $\eta = \frac{1}{2}(dt - \sum_1^n y^i dx^i)$ so that $\xi = 2\partial_t$.
With respect to the natural field of frames $\{\partial_{x^i}; \partial_{y^i}, \partial_t\}$, define a tensor field ϕ
of type $(1, 1)$ by its matrix

$$(\phi) = \begin{pmatrix} 0_{n,n} & I_n & 0_{n,1} \\ -I_n & 0_{n,n} & 0_{n,1} \\ 0_{1,n} & -y^i & 0 \end{pmatrix}.$$

Define a Lorentzian metric $\bar{g}$ with line element given by

$$ds^2 = \frac{1}{4}\{\sum_1^n ((dx^i)^2 + (dy^i)^2) - \eta \otimes \eta\}.$$

Then, with respect to an orthonormal basis $\{E_i, E_{n+i}, \xi\}$ such that

$$E_i = 2\partial_i, \quad E_{n+i} = 2\partial_{n+i},$$
$$\phi E_i = 2(\partial_i - y^i \partial_t),$$
$$\phi E_{n+i} = 2(\partial_i + y^i \partial_t),$$

it is easy to verify that the spacetime $(\bar{M}, \bar{g})$ has a Sasakian structure. For the
Riemannian case, we refer to Blair [66, page 99] in which it is shown that R^{2n+1}
is a Sasakian space form with sectional curvature $c = -3$.

Model 2. Let (N, G, J) be a $2n$-dimensional almost Hermitian manifold defined by

$$J^2 = -I, \quad G(J\bar{X}, J\bar{Y}) = G(\bar{X}, \bar{Y}), \qquad \forall \bar{X}, \bar{Y} \in \Gamma(TN),$$

where (N, G) is a compact Riemannian manifold. Construct a globally hyperbolic spacetime $\bar{M} = \{R \otimes N, \bar{g} = -dt^2 + G\}$ (see Section 3 of Chapter 1). Denote a vector field on $\bar{M}$ by $\bar{X} = (\eta(\bar{X})\frac{d}{dt}, X)$ where X is tangent to N, t and η are the time coordinate of R and a 1-form on $\bar{M}$, respectively. Set $\eta = dt$ so that $\xi = (\frac{d}{dt}, 0)$ is a global timelike vector field on $\bar{M}$. Then,

$$\phi(\eta(\bar{X})\frac{d}{dt}, X) = (0, JX),$$

$$\bar{g}\left((\eta(\bar{X})\frac{d}{dt}, X), (\eta(\bar{Y})\frac{d}{dt}, Y)\right) = G(X, Y) - \eta(\bar{X})\eta(\bar{Y}).$$

One can verify that $\bar{M}$ is an almost contact manifold. Thus, an odd-dimensional globally hyperbolic spacetime can carry an almost contact structure. As explained in Model 1, one can show that, with respect to the Lorentzian metric $\bar{g}$, $(\bar{M}, \bar{g})$ can carry a contact structure. Examples: odd-dimensional Minkowski and de Sitter spacetimes, the Lorentz sphere and Robertson-Walker spacetime.

With the above information, one can construct a variety of submanifolds of Lorentzian contact (or Sasakian) manifolds using the theory given in Chapter 7. It is also clear from Model 2 that contact and Hermitian manifolds are interrelated as odd- and even-dimensional manifolds, respectively. Moreover, it is easy to see from the above models that locally any Sasakian manifold is a line bundle over a Kähler manifold. Therefore, Lorentzian Sasakian manifolds are timelike line bundles over indefinite Kähler manifolds.

The traditional way of studying physical objects in 4-dimensional spacetime has recently changed and even- or odd-dimension or higher than 4-dimension of landing space depends on the type of physical problem. Now one needs eleven dimensions in the latest *Theory Of Every Thing* as an attempt to unite all the forces of the universe. Consequently, an odd dimension of contact landing space should work equally well in physical applications.

However, since the mathematical theory of lightlike contact geometry is just being developed (almost all new research results are presented in Chapter 7 of this volume), at present the applications are very rare. There are some isolated examples or papers in which we find physical results on lightlike (or degenerate) objects of contact or odd-dimensional landing manifolds (such as de Sitter spacetime S_1^{2n+1}). Following are the ones we know:

Osserman lightlike hypersurfaces. For this example we refer to the notation and the material discussed in Section 6 of Chapter 3. Consider (M, g) a real lightlike hypersurface of an indefinite almost Hermitian manifold $(\bar{M}, \bar{g}, \bar{J})$, where $\bar{g}$ is a semi-Riemannian metric of constant index. It is easy to check that $\{\xi, N\}$ is being a normalizing pair such that $\operatorname{Rad} TM = \operatorname{Sp}\{\xi\}$ and $\operatorname{ltr}(TM) = \operatorname{Sp}\{N\}$. Moreover, $\bar{J}(TM^\perp) \oplus \bar{J}(\operatorname{tr}(TM))$ is a vector sub-bundle of $S(TM)$. Then,

$$S(TM) = \{\bar{J}(TM^\perp) \oplus \bar{J}(\operatorname{tr}(TM))\} \perp D_0,$$

with D_0 a non-null almost complex distribution with respect to $\bar{J}$. Thus,

$$TM = \{\bar{J}(TM^\perp) \oplus \bar{J}(\operatorname{tr}(TM))\} \perp D_0 \oplus_{\operatorname{orth}} TM^\perp. \tag{9.5.1}$$

Now, consider the almost complex distribution

$$D = \{TM^\perp \oplus_{\operatorname{orth}} \bar{J}(TM^\perp)\} \oplus_{\operatorname{orth}} D_0,$$

and let $\mathcal{S}$ denote the projection morphism of TM on D. Put $U = -\bar{J}N$ and $V = -\bar{J}\xi$. Then, for all $X \in TM$,

$$X = \mathcal{S}X + u(X)U, \tag{9.5.2}$$

with $u = g(\cdot, V)$ a local 1-form on M. It follows that

$$\bar{J}X = FX + u(X)N,$$

with

$$FX = \bar{J}\mathcal{S}X, \quad \forall X \in TM.$$

Clearly, we have

$$F^2 X = -X + u(X)U, \quad u(U) = 1.$$

Thus, provided ξ and N are globally defined on M, (F, u, U) defines an almost contact structure on M.

We construct an algebraic curvature map R^F on M using F as follows:

$$R^F(x, y, z, w) = (\langle Fy, z\rangle + \langle y, Fz\rangle)(\langle Fx, w\rangle + \langle x, Fw\rangle)$$
$$- (\langle Fx, z\rangle + \langle x, Fz\rangle)(\langle Fy, w\rangle + \langle y, Fw\rangle),$$

for all $x, y, z, w \in T_pM$, $p \in M$. It is easy to check that such an R^F is an algebraic curvature map on M. Put $v = \langle \bullet, U\rangle$ and get

$$\langle Fx, y\rangle + \langle x, Fy\rangle = u(x)v(Fy) + u(y)v(Fx). \tag{9.5.3}$$

Now, we compute the pseudo-Jacobi operator $J_{R^F}(x)$ for $x \in \mathcal{S}_p(M)$, $p \in M$. We have for all y in $x^\perp$,

$$J_{R^F}(x)y = R^F(y, x, x, \bullet)^{\sharp_g}$$
$$= [2\langle Fx, x\rangle(\langle Fy, \bullet\rangle + \langle y, F(\bullet)\rangle)$$
$$- (\langle Fy, x\rangle + \langle y, Fx\rangle)(\langle Fx, \bullet\rangle + \langle x, F(\bullet)\rangle)]^{\sharp_g}.$$

Then, using (9.5.3) leads to

$$J_{R^F}(x)y = 2u(x)v(Fx)\left[u(y)(v \circ F)^{\sharp_g} + v(Fy)u^{\sharp_g}\right]$$
$$- (u(x)v(Fy) + u(y)v(Fx))\left[u(x)(v \circ F)^{\sharp_g} + v(Fx)u^{\sharp_g}\right].$$

Using (9.5.2) and the Hermitian structure of $(\bar{M}, \bar{g}, \bar{J})$ we obtain $v \circ F = -\eta$. Then, $(v \circ F)^{\sharp_g} = -\xi$. Also, since $u(\xi) = 0$, we have $u^{\sharp_g} = -\bar{J}\xi$. Thus,

$$J_{R^F}(x)y = u(x)\left[\eta(x)u(y) - \eta(y)u(x)\right]\xi$$
$$+ \eta(x)\left[\eta(x)u(y) - \eta(y)u(x)\right]\bar{J}\xi.$$

This implies

$$J_{R^F}(x) = \left[\eta(x)u(\bullet) - \eta(\bullet)u(x)\right]\left(u(x)\xi + \eta(x)\bar{J}\xi\right).$$

Observe that the pseudo-Jacobi operator $J_{R^F}(x)$, $x \in S_pM$, has values in the holomorphic plane $TM^\perp \oplus_{\text{orth}} \bar{J}(TM^\perp)$.

Proceeding as given in Section 6 of Chapter 3, one can show that (M, g) is a lightlike Osserman hypersurface (with an almost contact structure) of an indefinite almost Hermitian manifold $(\bar{M}, \bar{g}, \bar{J})$.

Lightlike hypersurfaces with conformal structure. Akivis-Goldberg [2] have studied singular points of lightlike hypersurfaces of the de Sitter space S_1^{n+1}. They establish a connection of the geometry of S_1^{n+1} with the geometry of the conformal space C^n as follows: S_1^{n+1} admits a realization on the exterior of an n-dimensional oval hyper-quadratic Q^n of a projective space P^{n+1}. Thus, it is isometric to a pseudo-elliptic space which is exterior to Q^n. Moreover, the interior of Q^n is isometric to the Lobachevski space H^{n+1}. Hence the geometry of Q^n is equivalent to the geometry of the conformal space C^n, which proves the assertion. Using this, the authors proved that the geometry of lightlike hypersurfaces of S_1^{n+1} is directly connected with the geometry of hypersurfaces of C^n. They constructed an invariant normalization and an invariant affine connection of a lightlike hypersurface.

Since S_1^{n+1} can admit a contact (or Sasakian) structure, the above approach of Akivis-Goldberg opens a possible line of research on *lightlike hypersurfaces of a contact (or Sasakian) de Sitter space*. Also, see Fusho and Izumiya [196] who have studied lightlike surfaces of spacelike curves in de Sitter 3-space and investigated the geometric meanings of the singularities of such surfaces.

Bibliography

[1] Abdalla, B. E. and Dillen, F. A Ricci semi-symmetric hypersurface of Euclidean space which is not semi-symmetric. Proc. Amer. Math. Soc. 130, 6(2002),1805–1808.

[2] Akivis, M. A. and Goldberg, V. V. The geometry of lightlike hypersurfaces of the de Sitter space, Acta Appli. Math., 53 (1998), 297–328.

[3] Akivis, M. A. and Goldberg, V. V. Lightlike hypersurfaces on manifolds endowed with a conformal structure of Lorentzian signature, Acta Appl. Math., 57 (1999), 255–285.

[4] Akivis, M. A. and Goldberg, V. V. On some methods of construction of invariant normalizations of lightlike hypersurfaces, Differential Geometry and its Applications, 12 (2000), No. 2, 121–143.

[5] Alexandrov, A. D. Some theorems on partial differential equations of second order, Vestnik Leningrad. Univ. Ser. Mat. Fiz. Him., 9 (1954), No. 8, 3–17.

[6] Anderson, M. T. On stationary vacuum solutions to the Einstein equations, Ann. Henri Poincare, 1, (2000), 977–994.

[7] Al-Aqeel, A. and Bejancu, A., On semi-invariant submanifolds of a Sasakian manifold. Saitama Math. J., 19 (2001), 51–59.

[8] Anderson, M. T., Kronheimer, P. B. and LeBrun, C. Complete Ricci-flat Kähler manifolds of infinite topological type. Comm. Math. Phys. 125, No. 4, (1989), 637–642.

[9] Arnold, V. I. *Contact geometry: The geometrical method of Gibbs's Thermodynamics*, Proc. Gibbs Symposium, Yale Univ., 1989, 163–179.

[10] Arnowitt, R., Deser, S. and Misner, C. W. *The dynamics of general relativity*, in: L. Witten (Ed.), Gravitation: An introduction to current research, Wiley, New York, 1962. Physical Review D, 68 (2003), 14030–10455.

[11] Arslan, K. Ezentas, R., Mihai, I. and Murathan, C. Contact CR-warped product submanifolds in Kenmotsu space forms. J. Korean Math. Soc. 42, No. 5, (2005), 1101–1110.

[12] Ashtekar, A., Beetle, C. and Fairhurst, S. Isolated horizons: a generalization of black hole mechanics, Class. Quantum Grav., 16, (1999), L1–L7.

[13] Ashtekar, A. Beetle, C., Dreyer, O. Krishnan, B., Lewandowski, J. and Wiśniewski, Generic isolated horizons and their applications, Phys. Rev. Lett, 85, (2001), No. 17, 3564–3567.

[14] Ashtekar, A., Beetle, C. and Lewandowski, J. Geometry of generic isolated horizons, Class. Quantum Grav., 19, (2002), 1195–1125.

[15] Ashtekar, A. and Galloway, G. J. Some uniqueness results for dynamical horizons, Adv. Theo. Math. Phys., 9 (2005), No. 6, 1–30.

[16] Ashtekar, A. and Krishnan, B. Dynamical horizons and their properties, Physical Review D, 68, (2003), 14030–10455.

[17] Atindogbe, C. Scalar curvature on lightlike hypersurfaces, Applied Sciences, 11,(2009), 9–18.

[18] Atindogbe, C. and Lionel, B.-B. A note on conformal connections on lightlike hypersurfaces, Conform. Geom. Dyn., 11, (2007), 1–11 (electronic).

[19] Atindogbe, C. and Duggal, K. L. Conformal screen on lightlike hypersurfaces, Int. J. Pure and Applied Math., 11(4), (2004), 421–442.

[20] Atindogbe, C. and Duggal, K. L. Pseudo-Jacobi operators and Osserman lightlike hypersurfaces, Kodai Math. J., 32, No. 1, (2009), 91–108.

[21] Atindogbe, C., Ezin, J.-P. and Tossa, J. Pseudo-inversion of degenerate metrics Int. J. Math. Math. Sci., 55 (2003), 3479–3501.

[22] Atindogbe, C., Ezin, J.-P. and Tossa, J. Lightlike Einstein hypersurfaces in Lorentzian manifolds with constant curvature, Kodai Math. J., 29, No. 1,(2006), 58–71.

[23] Atindogbe, C., Ezin, J.-P. and Tossa, J. Relative nullity foliation of the screen distribution of lightlike Einstein hypersurfaces in Lorentzian spaces. Note Mat., 26, (2006), No. 2, 161-170.

[24] Atiyah, M. F. Hyper-Kähler manifolds, *In Complex Geometry and Analysis*, proceedings, ZPisa 1988, Lext. Notes Math., (ed Villani, V.), 1422, Springer-Verlag Berlin, 1990, 1–13.

[25] Baird, P. and Wood, J. C. *Harmonic Morphisms Between Riemannian Manifolds*, London Mathematical Society Monographs, No. 29, Oxford University Press, 2003.

[26] Barletta, E., Dragomir, S. and Duggal, K. L. *Foliations of Cauchy-Riemann Geometry*, Mathematical Surveys and Monographs of AMS, Volume 140, 2007, ISBN 978-0-8218-4304-8.

[27] Barros, M., Chen, B.-Y. Urbano, F. Quaternion CR-submanifolds of quaternion manifolds. Kodai Math. J. 4 (1981), No. 3, 399–417.

[28] Barros, M. and Romero, A. Indefinite Kähler manifolds, Math. Ann., 261, (1982), 55–62.

[29] Bashir, M. A. On the classification of totally umbilical CR-submanifolds of a Kaehler manifold, Publications De L'institut Math. 51 (65), (1992), 115–120.

[30] Bashir, M. A. Chen's problem on mixed foliate CR-submanifolds, Bull. Austral. Math. Soc., 40, (1989), 157–160.

[31] Beem, J. K. Stability of geodesic structures, Nonlinear Anal. 30(1), (1997), 567–570.

[32] Beem, J. K. Lorentzian geometry in the large, Mathematics of gravitation, Part 1 (Warsaw, 1996), 11-20, Banach Center Publ., 41, Part 1, Polish Acad. Sci., Warsaw, 1997.

[33] Beem, J. K. Causality and Cauchy horizons. Gen. Rel. Grav., 27, (1995), 93-108.

[34] Beem, J. K. and Ehrlich, P. E. *Global Lorentzian Geometry*, Marcel Dekker, Inc. New York, First Edition, 1981. Second Edition (with Easley, K. L.), 1996.

[35] Beem, J. K. and Harris, S. G. The generic condition is generic. Gen. Rel. Grav., 25 (1993), No. 9, 939–962.

[36] Beem, J. K. and Harris, S. G. Non-generic null vectors. Gen. Rel. Grav., 25 (1993), No. 9, 963–973.

[37] Beem, J. K. and Królak, A. Cosmic censorship and pseudo convexity. J. Math. Phys., 33 (1992), No. 6, 2249–2253.

[38] Beem, J. K. and Królak, A. Cauchy horizon end points and differentiability. J. Math. Phys., 39 (1998), No. 11, 6001–6010.

[39] Beem, J. K., Low, R. J. and Parker, P. E. Spaces of geodesics: products, coverings, connectedness. Geom. Dedicata 59 (1996), No. 1, 51–64.

[40] Bejan, C.L. 2-codimensional lightlike submanifolds of almost para-Hermitian manifolds. Differential geometry and applications (Brno, 1995), 7–17, Masaryk Univ., Brno, 1996.

[41] Bejan, C. L. and Duggal, K. L. Global lightlike manifolds and harmonicity, Kodai Math. J., 28, (2005), 131–145.

[42] Bejancu, A. On the geometry of leaves on a CR-submanifold, Ann. St. Univ. AI. Cuza, Iasi, 25, (1979),393–398.

[43] Bejancu, A. Umbilical CR-submanifolds of a Kaehler manifold, Rend. Math. 12, (1980), 439–445.

[44] Bejancu, A. Hypersurfaces of quaternion manifolds. Rev. Roumaine Math. Pures Appl. 28, no. 7, (1983), 567–576.

[45] Bejancu, A. *Geometry of CR-submanifolds*, D. Reidel, 1986.

[46] Bejancu, A. QR-submanifolds of quaternion Kählerian manifolds. Chinese J. Math. 14 (1986), No. 2, 81–94.

[47] Bejancu, A. A canonical screen distribution on a degenerate hypersurface, Scientific Bulletin, Series A, Applied Math. and Physics, 55, (1993), 55–61.

[48] Bejancu, A. Lightlike curves in Lorentz manifolds, Publ. Math. Debrecen, 44 (1994), No. f.1-2, 145–155.

[49] Bejancu, A. Null hypersurfaces of semi-Euclidean spaces, Saitama Math J. 14, (1996), 25–40.

[50] Bejancu, A. and Farran, H. R., On totally umbilical QR-submanifolds of quaternion Kählerian manifolds, Bull. Aust. Math. Soc., 62, (2000), 95–103.

[51] Bejancu, A. and Duggal, K.L. Degenerate hypersurfaces of semi-Riemannian manifolds, Bull. Inst. Politehnie Iasi, 37, (1991), 13–22.

[52] Bejancu, A. and Duggal, K. L. Real hypersurfaces of indefinite Kähler manifolds, Int.. J. Math. and Math. Sci., 16, (1993), 545–556.

[53] Bejancu, A., Ferrández, A. and Lucas P. A new viewpoint on geometry of a lightlike hypersurface in a semi-Euclidean space, Saitama Math. J., (1998), 31–38.

[54] Bejancu, A. and Papaghiuc, N. Semi-invariant submanifolds of a Sasakian manifold. An. Stiint. Univ. Al. I. Cuza Iasi Sect. Mat. (N.S.) 27, (1981), 163–170.

[55] Bejancu, A. and Papaghiuc, N. Normal semi-invariant submanifolds of a Sasakian manifold. Mat. Vesnik 35, No. 4, (1983), 345–355.

[56] Bejancu, A. and Papaghiuc, N. Some results on sectional curvatures of semi-invariant submanifolds in Sasakian space forms. Bull. Math. Soc. Sci. Math. R. S. Roumanie (N.S.) 27(75), No. 2, (1983), 99–110.

[57] Bejancu, A., Yano, K. and Kon, M. CR-submanifolds of complex space form, J. Diff. Geom., 16, (1981), 137–145.

[58] Berger, B.K. Homothetic and conformal motions in spacelike slices of solutions of Einstein's equations, J. Math. Phys., 17 (1976), 1268–1273.

[59] Berger, M. Sur les groupes d'holonomie homogènes de variètèes à connexion affine et des variètèes riemanniennes, Bull. Soc. Math. Fr. 83, (1955),279–330.

[60] Berger, M. *Riemannian Geometry During the Second Half of the Twentieth Century*, Lecture Series, 17, Amer. Math. Soc., 2000.

[61] Berndt, J. Real hypersurfaces in quaternionic space forms. J. Reine Angew. Math. 419, (1991), 9–26.

[62] Besse, A. L. *Einstein Manifolds*, Springer, Berlin, 1987.

[63] Bishop, R. L. and O'Neill, B. Manifolds of negative curvature, Trans. Amer. Math. Soc., 145, (1969), 1–49.

[64] Bishop, R. L. and Goldberg, S. I. On the topology of positively curved Káhler manifolds, Tohoku Math. J., 15, (1963), 359–364.

[65] Blair, D. E. The theory of quasi-Sasakian structures, J. Diff. Geom., I, (1967), 331–345.

[66] Blair, D. E. *Contact Manifolds in Riemannian Geometry*, Lecture Notes in Math., No: 509, Springer Verlag, 1967.

[67] Blair, D. E. Geometry of manifolds with structure group $U(n) \times O(s)$, J. Diff. Geom., 4, (1970), 155–167.

[68] Blair, D. E. *Riemannian Geometry of Contact and Symplectic manifolds*, PM 203, Birkhäuser, 2002.

[69] Blair, D. E. and Chen, B. Y. On CR-submanifolds of Hermitian manifolds, Israel J. Math., 34, (1979), 353–363.

[70] Blair, D. E., Ludden, G.D. and Yano, K. Differential geometric structures on principal toroidal bundles, J. Diff. Geom., 4, (1970), 155–167.

[71] Blažić, N., Bokan, N. and Gilkey, P. A note on Osserman Lorentzian manifolds, Bull. London Math. Soc., 29, (1997), 227–230.

[72] Blažić, N. and Prvanović, M. Almost Hermitian manifolds and Osserman condition, Abh. Math. Sem. Univ. Hamburg, 71, (2001), 35–47.

[73] Bogges, A. *CR-Manifolds and the Tangential Cauchy-Riemann Complex* CRC Press. Boca, Raton, 1991.

[74] Bolós, V. J. Geometric description of lightlike foliations by an observer in general relativity, Math. Proc. Camb. Phil. Soc., 139, (2005), 181–192.

[75] Bonanzinga, V. and Matsumoto, K. Warped product CR-submanifold in locally conformal Kaehler manifold, Periodica Math. Hungar. Mat. 48, (2004), 207–221.

[76] Bonnor, W. B. Null curves in a Minkowski spacetime, Tensor N. S., 20(1969), 229–242.

[77] Bonnor, W. B. Null hypersurfaces in Minkowski spacetime, Tensor N. S., 24 (1972), 329–345.

[78] Boothby, W. M. and Wang, H. C. On contact manifolds, Ann. of Math., 68 (1958), 721–734.

[79] Boyer, Ch., Galicki, K. and Mann, B. The geometry and topology of 3-Sasakian manifolds. J. Reine Angew. Math., 455, (1994), 183–220.

[80] Burstall, L. F. E. and Rawnsley, J. H. *Harmonic maps Bibliography*, http://www.bath.ac.uk/masfeb/harmonic.html.

[81] Cabrerizo, J. L., Carriazo, A., Fernandez, L. M. and Fernandez, M. Riemannian submersions and slant submanifolds, Publicationes Math. Debrecen, 61(3-4) (2002), 523–532.

[82] Cabrerizo, J. L., Carriazo, A., Fernandez,L. M. and Fernandez, M. Semi-slant submanifolds of a Sasakian manifold,Geometriae Dedicata, 78 (1999), 183–1999.

[83] Cabrerizo, J. L. Carriazo, A., Fernandez,L. M. and Fernandez,M. Slant submanifolds in Sasakian manifolds, Glasgow J.Math.,42, (2000), 125–138.

[84] Calin, C. Semi-invariant submanifolds of a normal almost contact metric manifold. Bul. Inst. Politehn. Iasi Sect. I 38(42), No. 1-4, (1992), 23–26.

[85] Calin, C. Contact CR-submanifolds of quasi-Sasakian manifolds. Bull. Math. Soc. Sci. Math. Roumanie (N.S.) 36(84), No. 3-4, (1992), 217–226.

[86] Calin, C. Normal contact CR-submanifolds of a trans-Sasakian manifold. Bul. Inst. Politeh. Ia si. Sect. I. Mat. Mec. Teor. Fiz., 42(46), No, 1-2, (1996), 9–15.

[87] Calin, C. Contact CR-submanifolds of a nearly Sasakian manifold. Demonstratio Math. 29 (2), (1996), 337–348.

[88] Calin, C. *Contributions to Geometry of CR-submanifold*, Thesis, University of Iasi, Iasi, Romania, 1998.

[89] Calin, C. On the existence of degenerate hypersurfaces in Sasakian manifolds, Arab Journal of Mathematical Sciences, vol. 5, No. 1, (1999), 21–27.

[90] Candelas, P., Horowitz, G., Strominger, A. and Witten, E. Vacuum configurations for super strings, Nucl. Phys., B 258, (1985), 46–74.

[91] Carlip, S. Symmetries, horizons, and black hole entropy, Gan. Rel. Grav., 39, (2007), 1519–1523.

[92] Carriazo, A. Fernandez, L. M. and Hans-Uber, M. B. Some slant submanifolds of S-manifolds, Acta Math. Hungar, 107, (4)(2005), 267–285.

[93] Cartan, E. *La Theorie Desk Groupes Finis et Continus et la Geometrie Differentielle*, Gauthier-Villars, Paris, 1937.

[94] Carter, B. Killing horizons and orthogonally transitive groups in spacetimes, J. Math. Phys., 10 (1969), 70–81.

[95] Chandrasekhar, S. *The Mathematical Theory of Black Holes*, Oxford University Press, 1992.

[96] Cheeger, J. and Gromoll, D. The splitting theorem for manifolds of nonnegative Ricci curvature, J. Diff. Geom., 6, (1971), 119–128.

[97] Chen, B. Y. *Geometry of Submanifolds*, Marcel Dekker, 1973.

[98] Chen, B. Y. Totally umbilical submanifolds of quaternion-space-forms. J. Austral. Math. Soc. Ser. A. 26, No. 2, (1978), 154–162.

[99] Chen, B. Y. CR-submanifolds of a Kähler manifold, I-II, J. Diff. Geom., 16(1981), 305-322, 493–509.

[100] Chen, B. Y. Cohomology of CR-submanifolds, Ann. Fac. Sc. Toulouse math. Ser., 3, (1981), 167–172.

[101] Chen, B. Y. Differential geometry of real submanifolds in a Kaehler manifold, Monatsch. für Math.91, (1981), 257–274.

[102] Chen, B. Y. *Geometry of Slant Submanifolds*, Katholieke Universiteit Leuven, 1990.

[103] Chen, B. Y. Riemannian submanifolds, *In: Dillen E, Verstraelen L (eds) Handbook of Differential Geometry*, vol I,(2000), 187–418.

[104] Chen, B. Y. Geometry of warped product CR-submanifolds in Kaehler manifold, Monatsh. Math, 133, (2001), 177–195.

[105] Chen, B. Y. Geometry of warped product CR-submanifolds in Kaehler manifolds II, Monatsh. Mat., 134, (2001), 103–119.

[106] Chen, B. Y. Real hypersurfaces in complex space forms which are warped products, Hokkaido Mathematical Journal, 31, (2002), 363–383.

[107] Chen, B. Y. CR-warped products in complex projective spaces with compact holomorphic factor, Monatsh. Math., 141,(2004), 177–186.

[108] Chen, B. Y. and Houh, C. S., Totally real submanifolds of a quaternion projective space. Ann. Mat. Pura Appl., (4) 120, (1979), 185–199.

[109] Chen, B. Y. and Maeda, S. Real hypersurfaces in nonflat complex space forms are irreducible. Osaka J. Math., 40 (2003), No. 1, 121–138.

[110] Chen, B. Y. and Nagano, T. Harmonic metrics, harmonic tensors and Gauss maps, J. Math. Soc. Japan, Vol. 36, No. 2, (1984), 295–313.

[111] Chen, B. Y. and Ogiue, K. On totally real submanifolds, Trans. Amer. Math. Soc. 193 (1974), 257–266.

[112] Chen, B. Y. and Ogiue, K. Some characterizations of complex space forms in terms of Chern classes, Quart. J. Math., 26, (1975), 459–464.

[113] Chen, B. Y. and Wu, B. Mixed foliate CR-submanifolds in a complex hyperbolic space are non proper, Int. J. Math. and Math. Sci., 11, (1988), 507–515.

[114] Chi, Q. S. A curvature characterization of certain locally rank-one symmetric spaces, J. Diff. Geom., 28, (1988), 187–202.

[115] Chi, Q. S. Quaternionic Kähler manifolds and a curvature characterization of two-point homogeneous spaces, Illinois J. Math., 35, (1991), 408–418.

[116] Chi, Q. S. Curvature characterization and classification of rank-one symmetric spaces, Pacific J. Math., 150, (1991), 31–42.

[117] Chinea, D. and Gonzales, C. A classification of almost contact metric manifolds, Ann. Mat. Appl., IV Ser., **CLVI**, (1990), 15–36.

[118] Coken, A. C. and Ciftci, U. On the Cartan curvatures of a null curve in Minkowski spacetime, Geometry Dedicata, 114, (2005), 71–78.

[119] Coley, A. A. and Tupper, B.O.J. Spacetimes admitting inheriting conformal Killing vector fields, Class. Quantum Grav., 7, (1990), 1961–1981.

[120] Debney, G. C. and Zund, J. D. A note on the classification of electromagnetic fields, Tensor N.S., 22, (1971), 333–340.

[121] Dajczer, M. and Nomizu, K. On the boundedness of the Ricci curvature of an indefinite metric, Bol. Soc. Brasil. Mat., 11, (1980), 267–272.

[122] Defever, F. Descz, R. Senturk, D. Z. Verstraelen, L. Yaprak, S. On problem of P. J. Ryan, Kyungpook Math. J., 37(1997), 371–376.

[123] Defever, F., Descz, R., Senturk, D. Z. Verstraelen, L. and Yaprak, S. P. J. Ryan's problem in semi-Riemannian space forms, Glasgow Math. J., 41(1999), 271–281.

[124] Defever, F. Ricci-semi symmetric hypersurfaces, Balkan J. of Geometry and Its Appl., 5, 1,(2000), 81–91.

[125] Demetrios, C. *The Formation of Black Holes in General Relativity*, EMS Monographs in Mathematics, Volume:4, 2009, 600 pp; ISBN-10: 3-03719-068-X.

[126] Deprez, J. Semi-parallel surfaces in Euclidean space, J. of Geometry, 25 (1985), 192–200.

[127] Deprez, J. Semi-parallel hypersurfaces, Rend. Sem. Mat. Uni. Politec. Torino, 44(2), (1986), 303–316.

[128] Deshmukh, S., Ghazal, T. and Hashem, H. Submersions of CR-submanifolds of an almost Hermitian manifold, Yokohoma Math. J., 40, (1992), 45–57.

[129] Dragomir, S. and Duggal, K. L. Indefinite locally conformal Kähler manifolds, Differential Geometry and Applications, 25, (2007), 8–22.

[130] Dragomir, S. and Duggal, K. L. Indefinite extrinsic spheres, Tsukuba J. Math., 32, No: 2, (2008), 335-348.

[131] Dragomir, S. and Ornea, L. Locally conformal Kähler Geometry, Birkhäuser, Boston, 1998.

[132] Dragomir, S. and Tomassini, G. *Differential Geometry and Analysis on CR manifolds*, Progress in Mathematics, 245, Birkhäuser, Boston, 2006, 487 pp.

[133] Duggal, K. L. Lorentzian geometry of CR-submanifolds, Acta. Appl. Math., 17, (1989), 171–193.

[134] Duggal, K. L. Spacetime manifolds and contact structures, Int. J. Math. & Math. Sci., 13(1990), 545–554.

[135] Duggal, K. L. Lorentzian geometry of globally framed manifolds, Acta. Appl. Math., 19, (1990), 131–148.

[136] Duggal, K. L. Curvature inheritance symmetry in Riemannian spaces with applications to fluid spacetimes, J. Math. Phys., 33(1992), 2989–2997.

[137] Duggal, K. L. Warped product of lightlike manifolds, Nonlinear Analysis, 47(2001), 3061–3072.

[138] Duggal, K. L. Riemannian geometry of half lightlike submanifolds, Math. J. Toyama Univ., 25, (2002), 169–179.

[139] Duggal, K. L. Constant scalar curvature and warped product globally null manifolds, J. Geom. Phys., 43, (2002), 327–340.

[140] Duggal, K. L. Null curves and 2-surfaces of globally null manifolds, Int. J. Pure and Applied Math., 1(4), (2002), 389–415.

[141] Duggal. K. L. Harmonic maps, morphisms and globally null manifolds, Int. J. Pure and Applied Math., 6(4), (2003), 421–436.

[142] Duggal, K. L. On scalar curvature in lightlike geometry, J. Geom. Phys., 57, (2007), 473–481.

[143] Duggal, K. L. A report on canonical null curves and screen distribution for lightlike geometry, Acta. Appl. Math., 95, (2007), 135–149.

[144] Duggal, K. L. On canonical screen for lightlike submanifolds of codimension two, Central European Journal of Mathematics(CEJM), (2007), 710–719.

[145] Duggal, K. L. On existence of canonical screens for coisotropic submanifolds, Int. Electronic J. Geom., 1, (2008), No. 1, 25–32.

[146] Duggal, K. L. Time-dependent black hole horizons on spacetime solutions of Einstein's equations with initial data, M. Plaue and M. Scherfner (Eds.),*Advances in Lorentzian geometry*, Aachen: Shaker Verlag, Berlin, Germany, 2008, 51–61, ISBN 978-3-8322-7786-4.

[147] Duggal, K. L. and Bejancu, A. Lightlike submanifolds of codimension two, Math. J. Toyama Univ., 15(1992), 59–82.

[148] Duggal, K. L. and Bejancu, A. CR-hypersurfaces of indefinite Kaehler manifolds, Acta Appl. Math., 31, (1993), 171–190.

[149] Duggal, K. L. and Bejancu, A. *Lightlike Submanifolds of Semi-Riemannian Manifolds and Applications*, Kluwer Academic, 364, 1996.

[150] Duggal, K. L. and Giménez, A. Lightlike hypersurfaces of Lorentzian manifolds with distinguished screen, J. Geom. Phys., 55, (2005), 107–122.

[151] Duggal, K. L., Ianus, S. and Pastore, A. M. Harmonic maps on f-manifolds with semi-Riemannian metrics, 23rd conference on geometry and topology, Bebes-Bolyai University, Clij-Napoca, (1993), 47–55.

[152] Duggal, K. L., Ianus, S. and Pastore, A. M. Map interchanging f-structures and their harmonicity, Acta Appl. Math., 67, (2001), 91–115.

[153] Duggal, K. L. and Jin, D. H. Geometry of null curves, Math. J. Toyama Univ., 22, (1999), 95–120.

[154] Duggal, K. L. and Jin, D. H. Half lightlike submanifolds of codimension 2, Math. J. Toyama Univ., 22, (1999), 121–161.

[155] Duggal, K. L. and Jin, D. H. Totally umbilical lightlike submanifolds, Kodai Math. J., 26, (2003), 49–68.

[156] Duggal, K. L. and Jin, D. H. *Null Curves and Hypersurfaces of Semi-Riemannian Manifolds*, World Scientific, 2007, 304 pp.

[157] Duggal, K. L. and Jin, D. H. A classification of Einstein lightlike hypersurfaces of Lorentzian space forms, J. Geom. Phys., Submitted, 2009.

[158] Duggal, K. L. and Sahin, B. Screen conformal half-lightlike submanifolds, Int.. J. Math. & Math. Sci., 68, (2004), 3737–3753.

[159] Duggal, K. L. and Sahin, B. Screen Cauchy Riemann lightlike submanifolds, Acta Math. Hungar., 106(1-2) (2005), 137–165.

[160] Duggal, K. L. and Sahin, B. Generalized Cauchy Riemann lightlike submanifolds, Acta Math. Hungar., 112(1-2), (2006), 113–136.

[161] Duggal, K. L. and Sahin, B. Lightlike submanifolds of indefinite Sasakian manifolds, Int. J. Math. Math. Sci., 2007, Art ID 57585, 1–21.

[162] Duggal, K. L. and Sahin, B. Contact generalized CR-lightlike submanifolds of Sasakian submanifolds. Acta Math. Hungar., 122, No. 1-2, (2009), 45–58.

[163] Duggal, K. L. and Sahin, B. On unique existence of screen distribution for r-lightlike submanifolds of semi-Riemannian manifolds, preprint, 2009.

[164] Duggal, K. L. and Sharma, R. *Symmetries of Spacetimes and Riemannian Manifolds*, Kluwer Academic Publishers, 487, 1999.

[165] Duggal, K. L. and Sharma, R. Conformal Killing vector fields on spacetime solutions of Einstein's equations and initial data, Nonlinear Analysis, 63 (2005), e447–e454.

[166] Duggal, K. L. and Sharma, R. Conformal evolution of spacetime solutions of Einstein's equations, Communications in Applied Analysis, 11(2007), No. 1, 15–22.

[167] Eardley, D., Isenberg, J., Marsden, J. and Moncrief, V. Homothetic and conformal symmetries of solutions to Einstein's equations, Comm. Math. Phys., 106, (1986), 137–158.

[168] Eells, J. and Lemaire, L. Another report on harmonic maps, Bull. London Math. Soc., 20, (1988), 385–524.

[169] Ehresmann, C. Sur les varieties presque complexes, Proceedings International Congress of Math., 11, (1950), 412–427.

[170] Eisenhart, L. P. Symmetric tensors of second order whose first covariant derivatives are zero, Tran. Amer. Math. Soc., 25, (1923), 297–306.

[171] Ehrlich, P. E. A personal perspective on global Lorentzian geometry. Analytical and numerical approaches to mathematical relativity, 3-34, *Lecture Notes in Phys.*, 692, Springer, Berlin, 2006.

[172] Ehrlich, P. E., Jung, Y. T., Kim, J. S. and Kim, S. B. Constant curvatures on some warped product manifolds, Tsukuba J. Math., 20(1), (1996), 239-256.

[173] Ehrlich, P. E., Jung, Y. T., Kim, J. S. and Kim, S. B. Volume comparison theorems for Lorentzian manifolds. Geom. Dedicata 73, (1998), No. 1, 39–56.

[174] Ehrlich, P. E., Jung, Y. T., Kim, J. S. and Kim, S. B. Partial differential equations and scalar curvature of warped product manifolds. Nonlinear Anal., 44, (2001), No. 4, Ser. A: Theory Methods, 545–553.

[175] Ehrlich, P. E., Jung, Y. T., Kim, J. S. and Kim, S. B. Jacobians and volume comparison for Lorentzian warped products. *Recent advances in Riemannian and Lorentzian geometries* (Baltimore, MD, 2003), 39-52, Contemporary mathematics, 337, AMS, 2004.

[176] Ehrlich, P. E. and Kim, S. B. Riccati and Index Comparison Methods in Lorentzian and Riemannian Geometry, M. Plaue and M. Scherfner (Eds.), *Advances in Lorentzian geometry*, Aachen: Shaker Verlag, Berlin, Germany, 2008, 63–75, ISBN 978-3-8322-7786-4.

[177] Eells, J and Sampson, J. H. Energie et deformations en geometrie differentiable, Ann. Inst. Fourier (Grenoble), 14, (1964), 61–69.

[178] Eschenburg, J.-H. The splitting theorem for spacetimes with strong energy condition, J. Diff. Geom., 27, (1988), 477–491.

[179] Eschenburg, J.-H. Maximal principle for hypersurfaces, Manuscripta Math., 64, (1989), 55–75.

[180] Erkekoglu, F. Degenerate Hermitian manifolds, Mathematical Physics, Analysis and Geometry, 8, (2005), 361–387.

[181] Etayo, F. On some kinds of submanifolds of an almost Hermitian or para-Hermitian manifold. Libertas Math., 19, (1999), 39–46.

[182] Etayo, F. Fioravanti, M. On the submanifolds of semi-Riemannian almost complex manifolds. Atti Sem. Mat. Fis. Univ. Modena, 46 (1998), No. 2, 535–556.

[183] Etayo, F. Fioravanti, M., Trias, U. R. On the submanifolds of an almost para-Hermitian manifold. Acta Math. Hungar., 85, (1999), No. 4, 277–286.

[184] Falcitelli, M. Pastore, A. M. and Ianus, S. *Riemannian Submersions and Related Topics*, World Scientific, 2004.

[185] Ferrández, A., Giménez, A. and Lucas P. Null helices in Lorentzian space forms, Int. J. Mod. Phys., A16, (2001), 4845–4863.

[186] Ferrández, A., Giménez, A. and Lucas P. Degenerate curves in pseudo-Euclidean spaces of index two, Third International Conference on Geometry, Integrability and Quantization, Coral Press, Sofia, (2001), 209–223.

[187] Ferus, D. Immersions with parallel second fundamental form, Math Z., 140,(1974), 87–93.

[188] Fialkow, A. Hypersurfaces of a space of constant curvature, Ann. of Math. 39 (1938), 762–785.

[189] Fiedler, B. and Gilkey, P. Nilpotent Szabó, Osserman and Ivanov-Petrova pseudo-Riemannian manifolds, AMS Contemporary Math., 337, (2003), 53–63.

[190] Flaherty, E. T. *Hermitian and Kählerian Geometry in Relativity,* Lecture Notes in Physics, 46, Springer-Verlag, Berlin, 1976.

[191] Foertsch, T., Hasse, W. and Perlick, V. Inertial forces and photon surfaces in arbitrary spacetimes, Class. Quantum Grav., 20, 2003, 4635–4651.

[192] Friedman, A. Local isometric imbedding of Riemannian manifolds with indefinite metrics, J. Math. Mech., 10, (1961), 625–649.

[193] Fuglede, B. Harmonic morphisms between Riemannian manifolds, Ann. Inst. Fourier (Grenoble), 28, (1978), 107–144.

[194] Fuglede, B. Harmonic morphisms between semi-Riemannian manifolds, Annales Acad. Scientiarum Fennicae Mathematica, 21, (1996), 31–50.

[195] Fukami, T. and Ishihara, S. Almost Hermitian structure on S^6, Tohoku Math. J., 7, (1955), 151–156.

[196] Fusho, T and Izumiya, S. Lightlike surfaces of spacelike curves in de Sitter 3-space, J. Geom. Phys., 88, No. 1-2, (2008), 19–29.

[197] Galloway, G. J. Maximum principles for null hypersurfaces and null splitting theorem, Ann. Henri Poincaré, 1(2000), No. 3, 543–567.

[198] Galloway, G. J. Lecture notes on Spacetime Geometry, Beijing International Mathematics Research Center, (2007), 1–55.
http://www.math.miami.edu/ galloway/

[199] García-Río, E. and Vazquez-Abal, M. E., On the quaternionic sectional curvature of an indefinite Quaternionic Kähler manifold, Tsukuba J. Math., 2(1995), 273–284.

[200] García-Río, E., Kupeli, D. N. and Váquez-Lorenzo, R. On a problem of Osserman in Lorentzian geometry, Diff. Geom.. Appl., 7, (1997), 85–100.

[201] García-Río, E., Kupeli, D. N. and Váquez-Lorenzo, R. *Osserman Manifolds in Semi-Riemannian Geometry,* Lecture Notes in Mathematics, 1777, Springer-Verlag, Berlin, 2002.

[202] García-Río, E. and Kupeli, D. N. *Semi-Riemannian Maps and Their Applications,* 475, Kluwer Acad. Publ., Dordrecht, 1999.

[203] Gilkey, P. *Geometric Properties of Natural Operators Defined by the Riemann Curvature Tensor,* World Scientific Press, 2001.

[204] Gilkey, P., Ivanova, R. and Stavrov, I. Jordan Szabó algebraic covariant derivative curvature tensors, AMS Contemporary Math., 337, (2003), 65–75.

[205] Gilkey, P., Swann, L., and Vanhecke, L. Isoparametric geodesic spheres and a conjecture of Osserman concerning the Jacobi operator, Quat. J. Math., Oxford, 46, (1995), 299–320.

[206] Goldberg, S. I. A generalization of Kähler geometry, J. Diff. Geom., 6, (1972), 343–355.

[207] Goldberg, S. I. and Kobayashi, S. Holomorphic bisectional curvature. J. Differential Geometry 1, (1967), 225–233

[208] Goldberg, S. I. and Yano, K. On normal globally framed f-manifolds, Tohoku Math. J., 22, (1970), 362–370.

[209] Goldberg, V. V. and Rosca, R. Contact co-isotropic CR submanifolds of a pseudo-Sasakian manifold. Int. J. Math. Math. Sci. 7, No. 2, (1984), 339–350.

[210] Gourgoulhon, E. and Jaramillo, J. L. A $3 + 1$ perspective on null hypersurfaces and isolated horizons, Phys. Rep., 423, (2006), No. 4-5, 159–294.

[211] Gray, A. Some global properties of contact structures, Ann. of Math., II Ser., 69, (1959), 4214–450.

[212] Gray, A. Pseudo-Riemannian almost product manifolds and submersions, J. Math. Mech., 16, (1967), 715–737.

[213] Gray, A. The structure of nearly Kähler manifolds, Math. Ann., 223, (1976), 233-248.

[214] Gray, A. Curvature identities of Hermitian and almost Hermitian manifolds, Tôhoku Math. J., II Ser., 28, (1976), 601–612.

[215] Gray, A. and Hervella, L. M. The sixteen classes of almost Hermitian manifolds, Ann. Mat. Pure Appl., 123, (1980), 35–58.

[216] Greenberg, P. J. Spacetime congruences, J. Math. Anal. Appl., 30, 1970, 128-135.

[217] Greene, R. E. *Isometric Embedding of Riemannian and Pseudo-Riemannian Manifolds*, Memoirs Amer. Soc., 97, 1970.

[218] Greenfield, S. Cauchy Riemann equations in several variables, An. della Scoula Norm. Sup. Pisa, 22, (1968), 275–314.

[219] Gromov, M. and Schoen, R. Harmonic maps into singular spaces and p-adic super rigidity for lattices in groups of rank one, Institite Des Hautes Études Scientifiques, Extrait des Publications Mathématiques, 76, (1992), 165-246.

[220] Gudmundsson, S. *The bibliography of harmonic morphisms*, www.mate matik.lu.su/matematiklu/personal/sigma/harmonic/ bibliography.html

[221] Gunes, R., Sahin, B. and Kilic, E. On lightlike hypersurfaces of a semi-Riemannian space form, Turkish J. Math., 27, (2003), 283–297.

[222] Gunes, R., Sahin, B. and Keles, S. QR-submanifolds and almost contact 3-structure. Turkish J. Math. 24, N. 3, (2000), 239–250.

[223] Haesen, S., Palomo, F. J. and Romero, F. Null congruence spacetimes constructed from 3-dimensional Robertson-Walker spaces, Diff. Geom. and its Appl., 27, (2009), 240-249.

[224] Hall, G. S. *Symmetries and Curvature Structure in General Relativity*, World Scientific, 2004.

[225] Harris, S. G. A characterization of Robertson-Walker spaces by null sectional curvature, Gen. Rel. Grav., 17, (1985), 493–498.

[226] Hasegawa I. and Mihai, I. Contact CR-warped product submanifolds in Sasakian manifolds, Geometriae Dedicata, 102, (2003), 143–150.

[227] Hayward, S. General laws of black-hole dynamics, Phys. Rev., D 49, (1994), 6467-6474.

[228] Hawking, S. W. and Ellis, G.F.R. *The Large Scale Structure of Spacetime*, Cambridge University Press, Cambridge, 1973.

[229] Hernandes, G. On hyper f-structures, Math. Ann., 306, (1996), 205-230.

[230] Honda, K. Some lightlike submanifolds. SUT J. Math. 37 (2001), No. 1, 69–78.

[231] Hopf, H. and Rinow, W. Über den Begriff des vollständigen differential geometrischen Fläche, Comment. Math. Helv., 3, (1931), 209–225.

[232] Houh, C. S. On some submanifolds of quaternion Kaehler manifolds. Soochow J. Math., 4(1978), 107–120.

[233] Ianus, S. Submanifolds of almost Hermitian manifolds, Riv. Mat. Univ. Parma, V Ser., 3, (1994), 123–142.

[234] Ianus, S., Mazzocco, R. and Vilcu, G. E. Real lightlike hypersurfaces of para quaternionic Kähler manifolds, Mediterr. J. Math., 3, (2006), No. 3-4, 581–592.

[235] IIyenko, K. Twistor representation of null 2-surfaces, J. Math. Phys., 10, (2002), 4770–4789.

[236] Ikawa, T. On curves and submanifolds in an indefinite Riemannian manifold, Tsukuba J. Math., 9, (1965), 353–371.

[237] Ishihara, S. Normal structure f satisfying $f^3 + f = 0$, Kōdai Math. Sem. Rep., 18, (1966), 36–47.

[238] Ishihara, T. A. mapping of Riemannian manifolds which preserves harmonic functions, J. Math. Kyoto Univ., 19, (1979), 215–229.

[239] Ishihara, S. Quaternion Kählerian manifolds, J. Diff. Geom., 9, (1974), 483–500.

[240] Ishihara, S. and Yano, K. On the integrability conditions of a structure f satisfying $f^3 + f = 0$, Quart. J. Math., 15, (1964), 217–222.

[241] Jacobson, T. and Kang, G. Conformal invariance of black hole temperature, Class. Quantum Grav., 10, (1993), L201–L206.

[242] Jin, D. H. Null curves in Lorentz manifolds, J. of Dongguk Univ. vol. 18, (1999), 203–212.

[243] Jin, D. H. Fundamental theorem of null curves, J. Korea Soc. Math. Educ. Ser. B: Pure Appl. Math. vol., 7, No. 2, (2000), 115–127.

[244] Jin, D. H. Fundamental theorem for lightlike curves, J. Korea Soc. Math. Educ. Ser. B: Pure Appl. Math. vol., 10, No. 1, (2001), 13–23.

[245] Jin, D. H. Frenet equations of null curves, J. Korea Soc. Math. Educ. Ser. B: Pure Appl. Math. vol.10, No. 2, (2003), 71–102.

[246] Jin, D. H. Natural Frenet equations of null curves, J. Korea Soc. Math. Educ. Ser. B: Pure Appl. Math. vol.12, No. 3, (2005), 71–102.

[247] Jin, D. H. Einstein half lightlike submanifolds with a Killing co-screen distribution, Honam Mathematical J., 30(3) (2008), 487–504.

[248] Jin, D. H. Einstein half lightlike submanifolds of codimension 2, J. Korea Soc. Math. Edu. 16(1) (2009), 31–46.

[249] Jin, D. H. A characterization of screen conformal half lightlike submanifolds, Honam Mathematical J., 31(1) (2009), 17–23.

[250] Kähler, E. Über eine bemerkenswerte Hermitische metric, Abh. Math. Seminar Hamburg, 9 (1933), 173–186.

[251] Kamishima, Y. Uniformization of Kähler manifolds with vanishing Bochner tensor, Acta Math., 172, (1994), 299–308.

[252] Katsuno, K. Null hypersurfaces in Lorentzian manifolds, Math. Proc. Camb. Phil. Soc., 88, (1980), 175–182.

[253] Kang, T. H. Lightlike real hypersurfaces of indefinite quaternionic Kähler manifolds. Indian J. Pure Appl. Math. 33 (2002), No. 12, 1755–1766.

[254] Kang, T. H. Jung, S. D. Kim, B. H. Pak, H. K. Pak, J. S. Lightlike hypersurfaces of indefinite Sasakian manifolds. Indian J. Pure Appl. Math. 34 (2003), No. 9, 1369–1380.

[255] Kenmotsu, K. A class of almost contact Riemannian manifolds, Tohoku Math. J., 24, (1972), 93–103.

[256] Khan, K. A., Khan, V. A. and Husain, S. I. Totally umbilical CR-submanifolds of nearly Kaehler manifolds, Geometriae Dedicata, 50, (1994), 47–51.

[257] Kilic, E, Sahin, B., Karadag, H.B. and Gunes, R. Coisotropic submanifolds of a semi-Riemannian manifold, Turkish J. Math., 28, (2004), 335–352.

[258] Ki, U-H. and Kon, M. Contact CR-submanifolds with parallel mean curvature vector of a Sasakian space form. Colloq. Math. 64 , No. 2, (1993), 173–184.

[259] Kim, H. S. and Pak, J. S. Sectional curvature of contact CR-submanifolds of an odd-dimensional unit sphere. Bull. Korean Math. Soc., 42, (2005), No. 4, 777–787.

[260] Kim, H. S. and Pak, J. S. Certain contact CR-submanifolds of an odd dimensional unit sphere, Bull. Korean Math. Soc. 44 (1), (2007), 109–116

[261] Kobayashi, S. Topology of positively pinched Kähler manifolds, Tôhoku Math. J., 15, (1963), 121–139.

[262] Kobayashi, S. Submersions of CR-submanifolds, Tohoku Math. J., 89, (1987), 95–100.

[263] Kobayashi, S. and Nomizu, K. *Foundations of Differential Geometry*, Vol. 1, Interscience Publish., New York, 1965.

[264] Kobayashi, M. Some results on anti-quaternion submanifolds in Kähler quaternion manifolds. Chinese J. Math. 19, No. 1, (1991), 75–85.

[265] Kobayashi, M. Contact CR products of Sasakian manifolds. Tensor (N.S.), 36, N. 3, (1982), 281–288.

[266] Kobayashi, M. Semi-invariant submanifolds of a certain class of almost contact manifolds. Tensor (N.S.), 43, No. 1, (1986), 28–36.

[267] Kramer, D., Stephani, H., MacCallum, M. and Herlt, E. *Exact Solutions of Einstein's Field Equations*, Cambridge University Press, Cambridge, 1980.

[268] Krishnan, B. *Isolated Horizons in Numerical Relativity*, Ph.D Thesis, Pennsylvania State University, 2002.

[269] Kruskal, M. D. Maximal extension of Schwarzschild metric, Phys. Rev., 119, (1960), 1743–1746.

[270] Kuehnel, W. Conformal transformations between Einstein spaces, In Conformal Geometry, Edited by R. S. Kulkarni and U. Pinkall (vieweg Verlag Braunschweig, Wiesbaden, 1988.

[271] Kuo, Y. Y. On almost contact 3-structure. Tohoku Math. J., 22, (1970), 325–332.

[272] Kupeli, D. N. Notes on totally geodesic Hermitian submanifolds of indefinite Kähler manifolds. Atti. Sem. Mat. Fis. Univ. Modena, XLIII,(1995), 1–7.

[273] Kupeli, D. N. *Singular Semi-Riemannian Geometry*, Kluwer Academic, 366, 1996.

[274] Kwon, J.-H. and Pak, J. S. QR-submanifolds of $(p-1)$ QR-dimension in a quaternionic projective space $QP^{(n+p)/4}$. Acta Math. Hungar., 86(1-2), (2000), 89–116.

[275] Kwon, J. H. and Pak, J. S. On some contact CR-submanifolds of an odd-dimensional unit sphere. Soochow J. Math., 26, (2000), No. 4, 427–439.

[276] Kwon, J.H. and Pak, J. S. QR-submanifolds of $(p-1)$ QR-dimension in a quaternionic projective space $QP^{(n+p)/4}$. Acta Math. Hungar. 86, No. 1-2, (2000), 89–116.

[277] Kwon, J. H. and Pak, J. S. $(n+1)$-dimensional contact CR-submanifolds of $(n-1)$ contact CR-dimension in a Sasakian space form. Commun. Korean Math. Soc., 17, (2002), No. 3, 519–529.

[278] LeBrun, C. Foliated CR manifolds. J. Differential Geom., 22, No. 1, (1985) 81–96.

[279] LeBrun, C. Quaternionic-Kähler manifolds and conformal geometry. Math. Ann., 284, No. 3, (1989), 353–376.

[280] LeBrun, C. On complete quaternionic-Kähler manifolds. Duke Math. J., 63, no. 3, (1991), 723–743.

[281] LeBrun, C. A Kähler structure on the space of string world sheets. Classical Quantum Gravity, 10, No. 9, (1993), L141–L148.

[282] LeBrun, C. Anti-self-dual metrics and Kähler geometry. Proceedings of the International Congress of Mathematicians, Vol. 1, 2 (Zürich, 1994), 498-507, Birkhäuser, Basel, 1995.

[283] LeBrun, C. Fano manifolds, contact structures, and quaternionic geometry. Int. J. Math., 6, No. 3, (1995), 419–437.

[284] LeBrun, C. Mason, L. J. Nonlinear gravitons, null geodesics, and holomorphic disks. Duke Math. J., 136, No. 2, (2007), 205–273.

[285] Leistner, T. Screen bundles of Lorentzian manifolds and some generalizations of pp-waves, J. Geom. Phys., 56(10), (2006), 2117–2134.

[286] Leite, M.-L. Harmonic mappings of surfaces with respect to degenerate metrics, Amer. J. Math., 110, (1988), 399–412.

[287] Levy, H. Symmetric tensors of the second order whose covariant derivatives vanish, Ann. of Maths., 27, (1926), 91–98.

[288] Lewandowski, J. Spacetimes admitting isolated horizons, Class. Quantum Grav., 17, (2000), L53–L59.

[289] Lucas, P. Collection of papers (with his collaborators) on null curves and applications to physics: http://www.um.es/docencia/plucas

[290] Maartens, R. and Maharaj, S. D. Conformal Killing vectors in Robertson-Walker spacetimes, Class. Quant. Grav., 3, (1986), 1005–1011.

[291] Maartens, R., Maharaj, S. D. and Tupper, B. O. J. General solution and classification of conformal motions in static spherical spacetimes. Class. Quant. Grav., 13, (1996), No. 2, 317–318.

[292] MacLane, S. *Geometrical Mechanics* **II**, Lecture Notes, University of Chicago, 1968.

[293] Mangione, V. QR-hypersurfaces of quaternionic Kähler manifolds. Balkan J. Geom. Appl., 8, No. 1, (2003), 63–70.

[294] Marchiafava, S. On the local geometry of a quaternionic Kähler manifold, Boll Un. Mat. Ital., Ser. VII, B5, (1991), 417–447.

[295] Marchiafava, S. Pontecorvo, M. Piccinni, P. (ed) *Proceedings of the 2nd meeting on quaternionic structures in mathematics and physics*, Roma 1999. Dedicated to the memory of A. Lichnerowicz and E. Martinelli, 469, (electronic), 2001.

[296] Martinez, A. and Perez, J. D., Real hypersurfaces in quaternionic projective space, Ann. Mat. Pura Appl., (4), 145, (1986), 355–384.

[297] Matsumoto, K. On contact CR-submanifolds of Sasakian manifolds, Int. J. Math. and Math. Sci, 6(2), (1983), 313–326.

[298] Matsuyama, Y. Complete hypersurfaces with $R.S = 0$ in E^{n+1}, Proc. Amer. Math. Soc., 88, (1983), 119–123.

[299] Mihai, I. Contact CR-warped product submanifolds in Sasakian space forms, Geometriae Dedicata, 109, (2004), 165–173.

[300] Monar, D. Semi-invariant submanifolds of quaternion manifolds. Boll. Un. Mat. Ital. B., 7, No. 4, (1989), 841–855.

[301] Morimoto, A. On normal almost contact structures with a regularity, Tôhoku, Math. J., 16, (1964), 90–104.

[302] Morrow, J. and Kodaira, K. *Complex manifolds*, Holt, Rinehart and Winston, New York, 1971.

[303] Munteanu, M. I. Warped product contact CR-submanifolds of Sasakian space forms, Publ. Math. Debrecen, 66/1-2, (2005), 75–120.

[304] Nazaikinskii, V. E., Shatalov, V. E. and Sternin, B. Y. *Contact Geometry and Linear Differential Equations*, Walter de Gruyter, Berlin, 1992.

[305] Newlander, A. and Nirenberg, L. Complex analytic coordinates in almost complex manifolds, Ann. Math., 65, (1957), 391–404.

[306] Narita, F. CR-submanifolds of locally conformal Kähler manifolds and Riemannian submersions, Colloquium Math. LXX (2), (1996), 165–179.

[307] Niebergall, R. and Ryan, P. J. Real hypersurfaces in complex space forms, Tight and Taut submanifolds, MSRI.

[308] Nirenberg, L. Lectures on linear differential equation, CBMS Regional Conference held at the Texas Technological University, Lubbock, Ser. in Math., 17, A.M.S., 1973.

[309] Nomizu, K. *Fundamentals of Linear Algebra*, McGraw Hill, 1966.

[310] Nomizu, K. On hypersurfaces satisfying a certain condition on the curvature tensor, Tôhoku Math. J., 20, (1986), 46–59.

[311] Nomizu, K. and Ozeki, H. The existence of complete Riemannian metrics, Proc. Amer. Math. Soc., 12, (1961), 889–891.

[312] Nurowski, P. and Robinsom, D. Intrinsic geometry of a null hypersurface, Class. Quantum Grav., 17, (2000), 4065–4084.

[313] Ogiue, K. On almost contact manifolds admitting axiom of planes or axioms of free mobility, Kodai Math. Sem. Rep., 16, (1964), 223–232.

[314] Ogiue, K. Differential geometry of Kähler submanifolds, Advances in Math., 13, (1974), 73–114.

[315] Okumura, M. and Vanhecke, L. A class of normal almost contact CR-submanifolds in C^q. Rend. Sem. Mat. Univ. Politec. Torino, 52, (1994), No. 4, 359–369.

[316] O'Neill, B. The fundamental equations of a submersion, Mich. Math. J., 13, (1966), 458–469.

[317] O'Neill, B. *Semi-Riemannian Geometry with Applications to Relativity.* Academic Press, New York, 1983.

[318] Ornea, L.and Piccinni, P. Locally conformal Kähler structures in quaternionic geometry. Trans. Amer. Math. Soc., 349, No. 2, (1997), 641–655.

[319] Pak, J. S. and Sohn, W. H. Some curvature conditions of n-dimensional QR-submanifolds of $(p-1)$ QR-dimension in a quaternionic projective space $QP^{(n+p)/4}$. Bull. Korean Math. Soc. 40, No. 4, (2003), 613–631.

[320] Pak, J. S., Kwon, J.-H., Kim, H. S. and Kim, Y.-M. Contact CR-submanifolds of an odd-dimensional unit sphere, Geometriae Dedicata, 114, (2005), 1–11.

[321] Pak, J. S. Anti-quaternionic submanifolds of a quaternionic projective space. Kyungpook Math. J., 21, No. 1, (1981), 91–115.

[322] Palomo, F. J. The fibre bundle of degenerate tangent planes of a Lorentzian manifold and the smoothness of the null sectional curvature, Diff. Geom. and Appl., 25, (2007), 667–673.

[323] Pambira, A. Harmonic morphisms between degenerate semi-Riemannian manifolds, Beitrague Zur Algebra und Geometrie, 46 (1), (2005), 261–281.

[324] Papaghiuc, N. Semi-slant submanifolds of a Kaehlerian manifold, An. Stiint. Al.I.Cuza. Univ. Iasi, 40, (1994),55-61.

[325] Papantoniou, B. J. and Shahid, M. H., Quaternion CR-submanifolds of quaternion Kaehler manifolds, Int...Journal of Math and Math.Sci., 27(1), (2001) 27–37.

[326] Patterson, E. M. On symmetric recurrent tensors of the second order, Quart. J. Math., Oxford, 2, (1951), 151–158.

[327] Pedersen, H., Poon, Y. S. and Swann, A. The Einstein-Weyl Equations in Complex and Quaternionic Geometry, Diff.Geo. and Its Appl., (1993), 3309–3321.

[328] Penrose, R. Physical spacetime and non realizable CR-structure, vol. 39, Proc. of symposia in Pure Math., (1983), 401–422.

[329] Penrose, R. The twistor geometry of light rays. Geometry and Physics, Classical Quantum Gravity, 14(1A), (1997), 299–323.

[330] Perez, J. D. Indefinite quaternion space forms. II. An. Stiint. Univ. Al. I. Cuza Iasi Sect. I a Mat., 32, (1986), No. 2, 63–68.

[331] Perez, J. D. Santos, F. G. Indefinite quaternion space forms. Ann. Mat. Pura Appl., (4) 132, (1982), 383–398.

[332] Perlick, V. On totally umbilical submanifolds of semi-Riemannian manifolds, Nonlinear Analysis, 63, 2005, 511–518.

[333] Reinhart, B. *Differential Geometry of Foliations*, Springer-Verlag, Berlin, 1983.

[334] de Rham, G. Sur la réductibilité d'un espace de Riemannian, Comm. Math. Helv. 26, (1952), 328-344.

[335] Robinson, I. and Trautman, A. Cauchy Riemann structure in optical geometry in : *Proc. of the Fourth Marcel Grossmann Meeting on General Relativity*, edited by Ruffini, Elsevier, 1986, 317–324.

[336] Romero, A. and Suh, Y.J. Differential geometry of indefinite complex submanifolds in indefinite complex space forms, Extracta Math., 19(3), (2004), 339-398.

[337] Ronsse, G. S. Generic and skew CR-submanifolds of a Kähler manifold, Bulletin Inst. Math. Acad. Sinica, 18, (1990), 127–141.

[338] Ronsse, G. S. *Submanifolds of Sasakian Manifolds Which are Tangent to the Structure Vector Field*, PH. D. Thesis, Kansas State University, 1984.

[339] Rosca, R. On null hypersurfaces of a Lorentzian manifold, Tensor N.S., 23, (1972), 66–74.

[340] Rukimbira, P. Energy, volume and deformation of contact metrics, Contemporary Mathematics, AMS, Volume 337, (2003), 129–143.

[341] Ryan, P. J. A Class of complex hypersurfaces, Colloquium Math. 26,(1972), 175–182.

[342] Sahin, B. On QR-submanifolds of a quaternionic space form. Turkish J. Math., 25, No. 3, (2001), 413–425.

[343] Sahin, B. QR-lightlike submanifolds of indefinite quaternion Kähler manifold. Indian J. Pure Appl. Math., 33 (2002), No. 11, 1685–1706.

[344] Sahin, B. Warped product lightlike submanifolds, Sarajevo J. Math, 1(14), (2005), No. 2, 157–160.

[345] Sahin, B. Quaternion CR-submanifolds of a locally conformal quaternion Kähler manifold, Ege Uni. Journal of the Faculty of Sci., vol. 28, (2005), 81–90, (arXiv:0708.1160).

[346] Sahin, B. Transversal lightlike submanifolds of indefinite Kaehler manifolds, Analele. Universitatii din Timisoara, XLIV, 1, (2006), 119–145

[347] Sahin, B. Warped product semi-invariant submanifolds of a locally product Riemann manifold, Bull. Mat. Soc. Sci. Math. Roumanie, 49 (97), No. 4 (2006), 383–394.

[348] Sahin, B. Slant submanifolds of an almost product Riemannian manifolds, J. Korean Math. Soc., 43, (2006), No. 4, 717–732.

[349] Sahin, B. Nonexistence of warped product semi-slant submanifolds of Kähler manifolds, Geom. Dedicata, 117, (2006), 195–202.

[350] Sahin, B. Almost locally conformal Kähler product manifolds, Demonstratio Math., 39, (2006), No. 1, 211–218.

[351] Sahin, B. Screen conformal submersions between lightlike manifolds and semi-Riemannian manifolds and their harmonicity, Int. J. Geom. Methods Mod. Phys. 4, No. 6, (2007), 987–1003.

[352] Sahin, B. Slant submanifolds of quaternion Kaehler manifolds. Commun. Korean Math. Soc., 22(1), (2007), 123–135

[353] Sahin, B. Lightlike hypersurfaces of semi-Euclidean spaces satisfying some curvature conditions of semi-symmetric type, Turkish J. Math., 31, (2007), 139–162.

[354] Sahin, B. Lightlike submanifolds of indefinite quaternion Kähler manifolds, Demonstratio Math., 40 (3), (2007), 701–720.

[355] Sahin, B. Slant lightlike submanifolds of indefinite Hermitian manifolds, Balkan Journal of Geometry and Its Appl., 13(1), (2008), 107–119.

[356] Sahin, B. Screen transversal lightlike submanifolds of Kähler manifolds, Chaos, Solitons and Fractals, 38(2008), 1439–1448.

[357] Sahin, B. On a submersion between Reinhart lightlike manifolds and semi-Riemannian manifolds. Mediterr. J. Math., 5, (2008), No. 3, 273–284.

[358] Sahin, B. Every totally umbilical proper slant submanifold of a Kähler manifold is totally geodesic, Results in Mathematics, 54, (2009), 167-172.

[359] Sahin, B. Screen slant lightlike submanifolds, Int. Electronic J. Geom., vol:2, no:1, (2009), 41–54.

[360] Sahin, B. and Gunes, R. Non-existence of real lightlike hypersurfaces of an indefinite complex space form. Balkan J. Geom. Appl. 5 (2000), No. 2, 139–148.

[361] Sahin, B and Gunes, R. Geodesic CR-lightlike submanifolds. Beiträge Algebra Geom., 42 (2001), No. 2, 583–594.

[362] Sahin, B and Gunes, R. Lightlike real hypersurfaces of indefinite quaternion Kaehler manifolds. J. Geom., 75, (2002), No. 1-2, 151–165.

[363] Sahin, B. and Gunes, R. Integrability of distributions in CR-lightlike submanifolds. Tamkang J. Math., 33, (2002), No. 3, 209–221.

[364] Sahin, B. and Gunes, R. QR-submanifolds of a locally conformal quaternion Kaehler manifold. Publ. Math. Debrecen 63, No. 1-2, (2003), 157–174.

[365] Sahin, B., Kilic, E. and Gunes, R. Null helices in $\mathbf{R}_1^3$, Differential Geometry-Dynamical Systems, 3(2), (2001), 31–36.

[366] Salamon, S. Quaternionic Kähler manifolds, Invent. Math., 67, no. 1, (1982), 143–171.

[367] Sasaki, S. *Almost Contact Manifolds*, Lecture notes, Mathematical Institute, Tôhoku University, Volumes I, II, III, 1965–68.

[368] Sato, N. Totally real submanifolds of a complex space form with non-zero parallel mean curvature vector, Yokohoma Math. J. 44, (1997), 1–4.

[369] Schild, A. Classical null strings, Phys. Rev. D, 16(2), (1977), 1722–1726.

[370] Schoen, P. Uniqueness, symmetry and embeddedness of minimal surfaces, J. Diff. Geom., 18, (1983), 791–804.

[371] Schouten, J. A. and Van Kampen, E. R. Zur Einbettungs- und Krümmungs-theorie nichtholonomer Gebilde, Math. Ann., 103, (1930), 752–783.

[372] Shahid, H. M., Shoeb, M. and Sharfuddin, A. Contact CR-product of a trans-Sasakian manifold. Note Mat. 14 (1) (1994), 91-100.

[373] Sharma, R. *CR-submanifolds of Semi-Riemannian Manifolds with Applications to Relativity and Hydrodynamics*, Ph.D. Thesis, University of Windsor, Windsor, Canada, 1986.

[374] Sharma, R. On the curvature of contact metric manifolds, J. Geom., 53, 1995, 179–190.

[375] Sharma, R. Conformal symmetries of Einstein's field equations and initial data, J. Math. Phys., 46(2005), 042502, 1–8.

[376] Sharma, R. and Duggal, K. L. A characterization of affine conformal vector field, C. R. Acad. Sci. Canada., 7, (1985), 201–205.

[377] Singh, G. Contact CR-products of Lorentzian para Sasakian manifold. Bull. Calcutta Math. Soc., 98, (2006), No. 1, 71–78.

[378] Sultana, J. and Dyer, C.C. Conformal Killing horizons, J. Math. Phys., Vol. 45, No. 12, (2004), 4764–4776.

[379] Sultana, J. and Dyer, C. C. Cosmological black holes: A black hole in the Einstein-de Sitter universe, Gen. Rel. Grav., 37(8), (2005), 1349–1370.

[380] Szabo, Z. Structure theorems on Riemannian spaces satisfying $R(X,Y).R = 0$, The local version, J. Differential Geometry, 17(1982), 531–582.

[381] Swift, S. T. Null limit of the Maxwell-Sen-Witten equation, Class. Quantum. Grav., 9, (1992), 1829–1838.

[382] Szekeres, G. On the singularities of a Riemannian manifold. Math. Debreca, 7, (1960), 285–88.

[383] Takagi, R. On homogeneous real hypersurfaces in a complex projective space, Osaka J. math., 10(1973), 495–506.

[384] Takagi, R. Real hypersurfaces in a complex projective space with constant principal curvatures, J. Math. Soc. Japan, 27(1975), 43–53.

[385] Takagi, R. Real hypersurfaces in a complex projective space with constant principal curvatures II, J. Math. Soc. Japan 27, 4, (1975), 507–516.

[386] Takahashi, T. Sasakian manifolds with pseudo-Riemannian metric, Tohoku Math. J., 21, (1969), 271–289.

[387] Tanno, S. Hypersurfaces satisfying a certain condition on the Ricci tensor, Tōhoku Math. J. 21(1969), 297-303.

[388] Tanno, S. Ricci curvatures of contact Riemannian manifolds, Tohoku Math. J., 40(1988), 441–448.

[389] Tashiro, Y. On conformal collineations, Math. J. Okayama Univ., 10, (1960), 75–85.

[390] Tashiro, Y. and Tachibana, S. I. On Fubinian and C-Fubinian manifolds, Kodai Math. Sem. Reports, 15, (1963), 176–178.

[391] Tazawa, Y. Construction of slant submanifolds, Bull. Inst. Math. Acad. Sinica, 22, (1994), 253–266.

[392] Tazawa, Y. Construction of slant submanifolds, Bull. Soc. Math. Belg. 1, (1994), 569–576.

[393] Thomas, T. Y. On closed spaces of constant mean curvature, Amer. J. of Math., 58, (1936), 702–704.

[394] Trautman, A. On complex structures in physics, *On Einstein's path*, New york, 1996, Springer, New york, 1999, 487–501.

[395] Tricerri, F. and Vanhecke, L. Curvature tensors on almost Hermitian manifolds, Trans. Am. Math. Soc., 267, (1981), 365–398.

[396] Tromba, A. J. *Teichmuller Theory in Riemannian Geometry*, Birkhauser-Verlag, Boston, 1992.

[397] Vaisman, I. On locally conformal Kähler manifolds, Israel J. Math., 24, (1976), 338–351.

[398] Vaisman, I. Locally conformal Kähler manifolds with parallel Lee form, Rend. Mat., Roma, V Ser., 12, (1979), 263–284.

[399] Vaisman, I. Conformal changes of almost contact metric structures, In Geometry and Differential Geometry, Proc. Conf. Haifa, (1979), (ed: Artzy, R and Vaisman, I.), Lect. Notes Math., 792, 435–443.

[400] Vaisman, I. Some curvature properties of locally conformal Kähler manifolds, Trans. Am. Math. Soc., 159, (1980), 439–447.

[401] Vaisman, I. On locally and globally conformal Kähler manifolds, Trans. AMS., 2(262), (1980), 533–542.

[402] Van de Woestijne, I. and Verstraelen, L. Semi-symmetric Lorentzian hypersurfaces, Tôhoku Math. J., 39, (1987), 81–88.

[403] Vrănceanu, Gh. Sur quelques points de la théorie des espaces non holonomes, Bull. Fac. Cernăuti, 5, (1931), 177–205.

[404] Wolak, R. A. Contact CR-submanifolds in Sasakian manifolds, a foliated approach. Publ. Math. Debrecen, 56 (1-2), (2000), 7–19.

[405] Wolf, J. A. *Spaces of Constant Curvature*, McGraw-Hill, 1967.

[406] Yano, K. On a structure defined by a tensor field of type $(1, 1)$ satisfying $f^3 + f = 0$, Tensor N. S., 14, (1963), 99–109.

[407] Yano, K. *Geometry of Complex and Almost Complex Manifolds*, Pergamon Press, Oxford, 1965.

[408] Yano, K. *Integral Formulas in Riemannian Geometry*, Marcel Dekker, New York, 1970.

[409] Yano, K. and Ishihara, S. On Integrability conditions of a structure f satisfying $f^3 + f = 0$, Quart. J. Math., 15, (1964), 217–222.

[410] Yano, K. and Kon, M. *Anti-invariant Submanifolds*, Lect. Notes Pure Applied Math., 21, Marcel Dekker, New York, 1976.

[411] Yano, K. and Kon, M. Contact CR-submanifolds, Kodai Math. J., 5, (1982), 238–252.

[412] Yano, K. and Kon, M. *CR-Submanifolds of Kählerian and Sasakian Manifolds*, Birkhauser, 1983.

[413] Yano, K. and Kon, M. *Structures on Manifolds*, World Scientific, 1984.

[414] Yau, S. T. Calabi's conjecture and some new results in algebraic geometry, Proc. Nat. Acad. Sci. USA, 74, (1977), 1798–1799.

Index

Frontiers in Mathematics

This series is designed to be a repository for up-to-date research results which have been prepared for a wider audience. Graduates and post-graduates as well as scientists will benefit from the latest developments at the research frontiers in mathematics and at the "frontiers" between mathematics and other fields like computer science, physics, biology, economics, finance, etc.

■ **Aleman, A. / Feldman, N.S. / Ross, W.T.**, The Hardy Space of a Slit Domain (2009). ISBN 978-3-0346-0097-2

The book begins with an exposition of Hardy spaces of slit domains and then proceeds to several descriptions of the invariant subspaces of the operator multiplication by z. Along the way, we discuss and characterize the nearly invariant subspaces of these Hardy spaces and examine conditions for z-invariant subspaces to be cyclic. This work also makes important connections to model spaces for the standard backward shift operator as well as the de Branges spaces of entire functions.

■ **Avkhadiev, F.G. / Wirths, K.-J.**, Schwarz-Pick Type Inequalities (2009). ISBN 978-3-7643-9999-3

This book discusses in detail the extension of the Schwarz-Pick inequality to higher order derivatives of analytic functions with given images. It is the first systematic account of the main results in this area. Recent results in geometric function theory presented here include the attractive steps on coefficient problems from Bieberbach to de Branges, applications of some hyperbolic characteristics of domains via Beardon-Pommerenke's theorem, a new interpretation of coefficient estimates as certain properties of the Poincaré metric, and a successful combination of the classical ideas of Littlewood, Löwner and Teichmüller with modern approaches.

■ **Duggal, K.L. / Sahin, B.**, Differential Geometry of Lightlike Submanifolds (2010). ISBN 978-3-0346-0250-1

■ **Huber, M.**, Flag-transitive Steiner Designs (2009). ISBN 978-3-0346-0001-9

The monograph provides the first full discussion of flag-transitive Steiner designs. This is a central part of the study of highly symmetric combinatorial configurations at the interface of several mathematical disciplines, like finite or incidence geometry, group theory, combinatorics, coding theory, and cryptography. In a sufficiently self-contained and unified manner the classification of all flag-transitive Steiner designs is presented. This recent result settles interesting and challenging questions that have been object of research for more than 40 years. Its proof combines methods from finite group theory, incidence geometry, combinatorics, and number theory. The book contains a broad introduction to the topic, along with many illustrative examples. Moreover, a census of some of the most general results on highly symmetric Steiner designs is given in a survey chapter.

■ **Kasch, F. / Mader, A.**, Regularity and Substructures of Hom (2009). ISBN 978-3-7643-9989-4

■ **Kravchenko, V.V.**, Applied Pseudoanalytic Function Theory (2009). ISBN 978-3-0346-0003-3

Pseudoanalytic function theory generalizes and preserves many crucial features of complex analytic function theory. The Cauchy-Riemann system is replaced by a much more general first-order system with variable coefficients which turns out to be closely related to important equations of mathematical physics. This relation supplies powerful tools for studying and solving Schrödinger, Dirac, Maxwell, Klein-Gordon and other equations with the aid of complex-analytic methods. The book is dedicated to these recent developments in pseudoanalytic function theory and their applications as well as to multidimensional generalizations.